THE LIBRARY
ST. MARY'S COLLEGE OF MARYLAND
ST. MARY'S CITY, MARYLAND 20686

Photoreception and Vision in Invertebrates

NATO ASI Series

Advanced Science Institutes Series

A series presenting the results of activities sponsored by the NATO Science Committee, which aims at the dissemination of advanced scientific and technological knowledge, with a view to strengthening links between scientific communities.

The series is published by an international board of publishers in conjunction with the NATO Scientific Affairs Division

A	Life Sciences	Plenum Publishing Corporation
B	Physics	New York and London
C	Mathematical and Physical Sciences	D. Reidel Publishing Company Dordrecht, Boston, and Lancaster
D	Behavioral and Social Sciences	Martinus Nijhoff Publishers
E	Engineering and Materials Sciences	The Hague, Boston, and Lancaster
F	Computer and Systems Sciences	Springer-Verlag
G	Ecological Sciences	Berlin, Heidelberg, New York, and Tokyo

Recent Volumes in this Series

Volume 68—Molecular Models of Photoresponsiveness
edited by G. Montagnoli and B. F. Erlanger

Volume 69—Time-Resolved Fluorescence Spectroscopy in Biochemistry and Biology
edited by R. B. Cundall and R. E. Dale

Volume 70—Genetic and Environmental Factors during the Growth Period
edited by C. Susanne

Volume 71—Physical Methods on Biological Membranes and Their Model Systems
edited by F. Conti, W. E. Blumberg, J. de Gier, and F. Pocchiari

Volume 72—Principles and Methods in Receptor Binding
edited by F. Cattabeni and S. Nicosia

Volume 73—Targets for the Design of Antiviral Agents
edited by E. De Clercq and R. T. Walker

Volume 74—Photoreception and Vision in Invertebrates
edited by M. A. Ali

Series A: Life Sciences

Photoreception and Vision in Invertebrates

Edited by
M. A. Ali
University of Montreal
Montreal, Quebec, Canada

Plenum Press
New York and London
Published in cooperation with NATO Scientific Affairs Division

Proceedings of the NATO Advanced Study Institute on
Photoreception and Vision in Invertebrates,
held July 11-24, 1982
at Bishop's University, Lennoxville, Quebec, Canada

Library of Congress Cataloging in Publication Data

NATO Advanced Study Institute on Photoreception and Vision in Invertebrates
(1982: Bishop's University)
Photoreception and vision in invertebrates.

(NATO advanced science institutes series. Series A, Life sciences; v. 74)
"Proceedings of the NATO Advanced Study Institute on Photoreception and Vision in Invertebrates, held July 11-24, 1982, at Bishop's University, Lennoxville, Quebec, Canada"—Verso t.p.
"Published in cooperation with NATO Scientific Affairs Division."
Includes bibliographical references and index.
1. Photoreceptors—Congresses. 2. Vision—Congresses. 3. Invertebrates—Physiology—Congresses. I. Ali, M. A. (Mohamed Ather), 1932-   . II. North Atlantic Treaty Organization. Scientific Affairs Division. III. Title. IV. Series.
QP481.N34 1984   592'.01823   83-27081
ISBN 0-306-41626-3

©1984 Plenum Press, New York
A Division of Plenum Publishing Corporation
233 Spring Street, New York, N.Y. 10013

All rights reserved. No part of this book may be reproduced, stored in a retrieval system, or transmitted in any form or by any means, electronic, mechanical, photocopying, microfilming, recording, or otherwise, without written permission from the Publisher

Printed in the United States of America

PREFACE

> I see a man's life is a tedious one.
>
> Cymbeline, Act III, Sc. 6.

It is well known that the best way to learn a subject is to teach it! Along the same lines one might also say that a pleasant way of learning a subject and at the same time getting to know quite a few of the workers active in it, is to arrange and to attend an Advanced Study Institute (ASI) or a workshop lasting about two weeks. This was and is the wisdom behind the NATO-ASI programme and much as people fear that a fortnight may be too long, before it is over everyone feels that it was too short, especially if the weather had cooperated. Organising this ASI which resulted in this volume has been a very good learning experience. I started my career in research with invertebrates and retained an interest in them over the years due to my teaching a course and working sporadically on various aspects of photoreception in Polychaetes, Crustaceans and Insects. Thus, the thought of organising an ASI on photoreception and vision in invertebrates had been brewing in my mind for the past half a dozen years or so. It was felt that it will be desirable to do a bit of stock taking and discuss possible new approaches to the study of this matter. The ASI was structured along the lines the volume was to be since it would also be a very practical way to examine the subject during the ten working days we had to do it. When the organisation of the ASI was well underway, the Autrum Handbook (Vols. VII/6A, B & C) started coming out but this did not deter my efforts because I was aware that the Handbook was in preparation and indeed many of the persons I had approached to serve as lecturers at the ASI and authors of chapters in the ensuing volume have contributed to the Handbook as well. Further, the main reason was that the nature of this volume was to be different. It was to be the outcome of an ASI and was to be something like a glorified text-book rather than a Handbook. In any case, it turned out to be a pleasant experience and an efficient way of covering the subject in two short weeks. There were many hours of stimulating discussions even outside the lecture room and many friendships were born and joint projects hatched.

As has been mentioned in the volume, of the more than a million species of animals known, over 95% are invertebrates and if one were to take into account the number of individuals, the proportion will be even greater. Of course, if one were to consider biomass the proportion will differ. They occupy all kinds of environments, have every conceivable mode of life, and every imaginable anatomical and physiological adaptation. As such they offer excellent material for studying not only evolutionary processes but also convergent and divergent phenomena.

Invertebrates had not been studied in a concerted manner until lately. Of course, their taxonomy and anatomy are fairly well known. In the case of some economically and medically important groups even some aspects of their life histories and physiology were studied. However, it is only recently that much interest in the study of all aspects of their biology approached in a synthetic manner has accrued. From a fundamental point of view this is because we have realised what a rich source of material and models the invertebrates are for studying any aspect of biology. From an applied point of view also one has understood that they serve as excellent parameters for the study of acute and chronic effects of pollutants and toxic substances that seem to have become a part of our civilisation. Considerable interest has accrued also in the study, particularly of insects in forestry and agriculture. What I have attempted to do in this book is to present a review by various specialists of what is known about the photoreceptive process in different groups along a perspective approach. In a book of this kind a certain amount of duplication is unavoidable and attempts have been made to keep it at a minimum. I asked the authors to be as speculative as they wanted and many indeed have done just that. I had the pleasure of listening to all the lectures. This, along with the editing of the book have considerably increased my knowledge of photoreception and vision and I hope that the reader also will find the book useful.

I wish to explain briefly how the lecturers/authors are selected for the ASI and the ensuing volume. This meeting being a NATO-ASI, lecturers have to be drawn from as many member countries of the alliance as possible. In most cases, travel costs also play an important part. Other significant factors are lecturing and writing ability of the person, his or her ability to deal with a heterogenous group scientifically and socially over a two-week period. Taking all these factors into consideration a list is drawn up and correspondence begins, usually as early as two years before the ASI is to take place and chapters are to be submitted. In spite of such early arrangements one or two persons find out a few months before the event that they cannot attend. They have to be replaced by others who are suitable. The presentations at the ASI are followed by discussions and towards the end a meeting of the authors and the editor takes place. At this meeting matters are discussed openly and suggestions are made to improve the volume or to better balance it. This is generally done by asking one or two of the other participants, generally those whose seminars were found to fit the organisation of the volume, to contribute chapters on subjects along the lines established by consensus. After this, of course, matters are between the editor and individual authors.

PREFACE

The major part of the financial support to hold the ASI came from the Scientific Affairs Division of NATO and I am grateful to Dr. Craig Sinclair for all the help and encouragement he gave. Grants were also given by FCAC of Québec, Natural Sciences and Engineering Council of Canada and the Université de Montréal. I thank the Director of my department for the encouragement and the material support he provided in the organisation of the ASI and the preparation of this volume.

It will be extremely difficult to put in words how much gratitude I have towards my colleague Dr. Mary Ann Klyne for the enormous help she provided so unstintingly not only in the organisation and running of the meeting but also in the editing and production of this book. I also thank Mademoiselle Marielle Chevrefils for typing most of the camera-ready manuscript. My thanks also to Madame Marianne Vèzina-Bélair for proof reading and indexing; Mademoiselle Francine Chatelois for help with the subject index.

It would have been a much more difficult and certainly a more costly job to arrange the meeting without the kind support of Monsieur J.-L. Grégoire, Vice-Principal Administration of Bishop's University and his assistant Mrs. Lillian Garrard. I am grateful to them for everything they did to make my task less arduous and our stay on their campus a very enjoyable one. I should also like to acknowledge the help and encouragement I got from my editor at Plenum Press, Miss Patricia Vann.

Montréal
June 1983

CONTENTS

Prologue M.A. Ali	3
Visual Pigments of Invertebrates D.G. Stavenga and J. Schwemer	11
Natural Polarized Light and Vision Talbot H. Waterman	63
Photoreception in Protozoa, an Overview Pierre Couillard	115
Evolution of Eyes and Photoreceptor Organelles in the Lower Phyla A.H. Burr	131
Photomovement Behavior in Simple Invertebrates A.H. Burr	179
Photoreceptors and Photosensitivity in Platyhelminthes Annie Fournier	217
Photoreceptors and Photoreceptions in Rotifers Pierre Clément and Elizabeth Wurdak	241
Photoréception et Vision chez les Annélides (Photoreception and Vision in Annelids) Martine Verger-Bocquet	289
Photoreceptor Structures and Vision in Arachnids and Myriapods Arturo Muñoz-Cuevas	335
Crustacea M.F. Land	401
The Retinal Mosaic of the Fly Compound Eye Nicolas Franceschini	439

CONTENTS

The Roles of Parallel Channels in Early Visual Processing by the Arthropod Compound Eye Simon Laughlin	457
Functional Neuroanatomy of the Blowfly's Visual System N.J. Strausfeld	483
The Lobula-Complex of the Fly: Structure, Function and Significance in Visual Behaviour Klaus Hausen	523
Behavioural Analysis of Spatial Vision in Insects Erich Buchner	561
Neuroanatomical Mapping of Visually Induced Nervous Activity in Insects by ^{3}H-Deoxyglucose Erich Buchner and Sigrid Buchner	623
The Rules of Synaptic Assembly in the Developing Insect Lamina I.A. Meinertzhagen	635
Morphologie et Développment des Yeux Simples et Composés des Insectes (The Morphology and Development of Simple and Compound Eyes of Insects) Michel Mouze	661
Molluscs M.F. Land	699
Photoreception in Chaetognatha T. Goto and M. Yoshida	727
Photoreception in Echinoderms M. Yoshida, N. Takasu and S. Tamotsu	743
Epilogue M.A. Ali	773
Author Index	789
Subject Index	827
Species Index	851

PROLOGUE

M.A. ALI

Département de Biologie, Université de Montréal

Montréal, P.Q. H3C 3J7 Canada

> You were best to call them generally,
> man by man, according to the scrip.
>
> A Midsummer Night's Dream, Act I, Sc. 2.

The term "invertebrate" is an incongruous one! Normally an object, a plant or an animal, is referred to by a positive characteristic - something it has or possesses, or something it is able to do, or the place it occurs in, its colour etc. Thus, it is strange to refer negatively to the majority of animals as those "without vertebrae"! Had almost all animals possessed vertebrae and a small minority had not, there might then have been a bit of a justification to the name. Also, the Protozoa and the Parazoa (sponges) are lumped with the invertebrates. This situation reflects three facts. First, Anthropocentric or Vertebrocentric attitudes. Quite naturally human being's first pre-occupation was with himself and then with the other mammals and then with the birds and after that the reptiles and so on "down" the line. Thus, we consider the lack in others of a characteristic or characteristics that we possess as a criterion or criteria for recognising them. Examples could be "non-French speaking peoples", "non-Europeans", "non-biologists" etc. This is precisely what we have done with the "in-vertebrates".

Why did we do this? This leads us to the second and third explanations. We know a lot less about invertebrates than we do about vertebrates and, what we do not know about is generally considered less important. Thus, any animal that did not have a backbone was an "invertebrate"! This never got changed because it was (and still is) impossible to find or coin a name that will englobe this immense and heterogenous array of animals. The third point is that it is extremely convenient to retain this term. Purely scientifically speaking, we could do away with it, of course. The Protozoa have been moved to Kingdom

Protista. The others, which belong to the Kingdom Animalia could either be treated as Sub-Kingdoms (Parazoa & Enterozoa) or, even as individual Phyla or even Classes. This will however, go against tradition and will be inconvenient to boot. It is so practical to divide an elementary Animal Biology course into Invertebrate and Vertebrate Biology parts, each taking up one semester. Otherwise we would have to call them Animal Biology I and Animal Biology II and they will not be balanced. As mentioned earlier, we know a lot more about vertebrates and they are a great deal more important to us although they do not even form a full Phylum. Even three Sub-Phyla of the Phylum Chordata are invertebrates (Hemichorda; Urochorda; and Cephalochorda). Therefore, for practical reasons it appears desirable to retain the term, at least for the moment. It is in this sense that it is used in this volume.

According to reliable estimates of the number of known animal species, over 97% are invertebrate ones (Fig. 1). Thus one could see that not only almost all animal species are invertebrate but if the number of individuals could be counted, their importance will be even more staggering. By way of examples, one could mention that there are about 50 000 earthworms per acre of humid and fertile ground; or that the number of free living nematodes is 3 billion in an acre of soil, or that a colony of tropical leaf-cutting ants (Atta) is made up of several million individuals; or yet that the swarms of pasture mosquitoes (Aedes nigromaculus) in an acre of southern California may number 2 millions! Another example of their enormity is that roughly, the insects killed by spiders in one year would weigh as much as the entire human population!

The invertebrates which vary a great deal in their size, as can be seen when one compares a malarial parasite with a giant squid, occupy also all kinds of different niches. They may be found in the driest or the humidest of environments, in the shallowest of puddles or in the deepest of oceanic depths; in the coldest or warmest temperature that an organism (except a virus) can survive in, and in different parts of animals or plants as symbiotes or parasites.

It should be pointed out that with the growing interest in the study of oceans, lakes, rivers and forests for practical purposes, interest in the study of invertebrates has accrued. It has also become important to understand how the light environment varies under different conditions and how animals function in it visually and what any alteration of that light environment, such as increase or reduction in turbidity, does to the normal functioning of the animal.

Invertebrates offer an excellent array of material and models for understanding the processes that have governed the evolution of photoreception and vision. Increase in efficiency and/or complexity, a slow and long term process, has brought about the evolution of an eye spot from generalised sensitivity, then eye spot with optics, then entire cells, then tissues, then organs and then systems specialised for photoreception and vision (Eakin, 1982). The wealth of material that the invertebrates offer permit also innumerable comparative studies which shed light on how adaptive radiation, another but faster evolutionary process, brings about variations in the structure and function of photoreceptors and visual systems. Comparative studies also enable the understanding of conver-

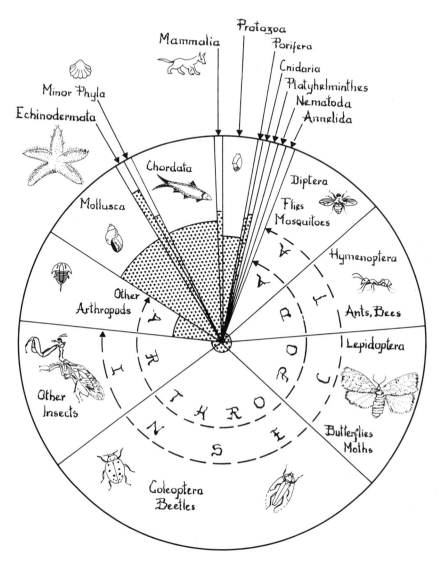

Fig. 1. Diagrammatic presentation of the relative numbers of species in the various animal phyla, some classes and super-orders. The stippled area represents the relative number of known fossils in that sector. In the case of some groups no fossils are available (eg. Annelida). The white area represents living species. One degree equals 3 000 species (after Muller & Campbell, 1954).

gence (e.g. camera eye) and divergence (e.g. photopigments) in the evolution of the photoreceptive process in general.

The primary aim of this volume is to reflect these four aspects i.e.

increase in complexity and/or efficiency, adaptive radiation, convergence and divergence in the evolution of photoreception and vision in invertebrates. A distinction is made between photoreception (e.g. Platyhelminthes, Echinodermata) and vision (e.g. Insecta, Cephalopoda) bearing, however in mind the fact that in many groups it is virtually impossible to differentiate.

To facilitate taxonomic orientation, a mini-classification of the invertebrates is given below.

Kingdom Monera

    Sub-Kingdom Protozoa

        1   Phylum Protozoa

Kingdom Animalia

    Sub-Kingdom Parazoa

        2.   Phylum Porifera   } Metazoa

    Sub-Kingdom Enterozoa

    Acoelomata - radiata

        3.   Phylum Cnidaria   } Coelenterata
        4.   Phylum Ctenophora

    Acoelomata - bilateria

        5.   Phylum Platyhelminthes
        6.   Phylum Rhynchocoela (Nemertinea or Nemertea)

Pseudocoelomata

        7.   Phylum Endoprocta
        8.   Phylum Rotifera (Trochelminthes)
        9.   Phylum Gastrotricha
        10. Phylum Kinorhyncha (Echinodera)
        11   Phylum Acanthocephala
        12. Phylum Nematomorpha
        13. Phylum Nematoda

Coelomata

Super Phylum Annelida (Protostomia)

        14. Phylum Annelida
        15. Phylum Sipuncula
        16. Phylum Priapula
        17. Phylum Echiura
        18. Phylum Onychophora

19. Phylum Arthropoda
   Sub Phylum Chelicerata
      class Merostomata
      class Pycnogonida
      class Arachnida
   Sub Phylum Mandibulata
      class Crustacea
      class Chilopoda
      class Diplopoda
      class Insecta
         Sub class Apterygota
         Sub class Pterygota
            Section Paleoptera
            Section Neoptera
20. Phylum Bryozoa (Ectoprocta)
21. Phylum Phoronida
22 Phylum Brachiopoda
23. Phylum Mollusca

Super Phylum Echinodermata (Deuterostomia)

24. Phylum Chaetognatha
25. Phylum Echinodermata
26. Phylum Chordata

As can be seen from the chapters that follow, the majority of Phyla listed above have not been considered in this book. This is mainly because virtually nothing is known about their photoreception (Ali, Croll & Jaeger, 1978; Ali, Anctil & Cervetto, 1978). In some, such as the endoparasites, that do not have a free living stage, photoreception will probably be of no consequence.

The ensuing chapters (except the first two) have been arranged in the same order as the classification given above. This is because, in a total consideration of all aspects of biology such an arrangement is valid. Of course, if we take photoreception only, the arrangement may have to be different with for example, the echinoderms following the cnidarians, and the Insects coming last. Such an arrangement would also create problems such as putting chaetognaths away from the echinoderms. Thus, to avoid such polemics, the generally accepted sequence in teaching invertebrate biology has been followed.

Since the absorption of light by a photopigment is essential for photoreception, the chapter by Stavenga and Schwemer is placed first. It deals with the general and comparative aspects of photopigments. In the last few years, a number of review chapters and books on photopigments have appeared (Langer, 1972; Dartnall, 1972; Ali & Wagner, 1975; Autrum, 1979; Hamdorf, 1979; Packer, 1982, etc.) and in view of this, this chapter attempts not only to bring us up to date on what has been done but also treats the matter with a tutorial objective in mind. The next chapter by Waterman is also one that covers more than one group and thus is placed in the beginning. A masterly review of polarisation was done by

Waterman a couple of years ago (Waterman, 1981). So he also, in addition to bringing the matter up to date, has taken a fresh perspective of the question in the light of the other presentations and the discussions at the Advanced Study Institute (ASI). At this stage it seems appropriate to make a comment about this book. It is the outcome of a structured tutorial activity which lasted two weeks. The ASI was planned with the book in mind. As such it is not a handbook (in the Handbook of Sensory Physiology sense), nor is it really a review volume rather it is a manual or textbook for workers in this area. In view of this and because of its compactness it is meant to be more easily accessible to a larger number of persons.

The overview of photoreception in Protozoa by Couillard takes into account the appearance of a volume dealing with aneural organisms a few years ago (Lenci, 1980). It also tends to deal with the matter in a general way, and surveys also the literature of the past few years. It blends with the following chapter by Burr who, after dealing principally with the cnidarians and bryozoans analyses the various viewpoints concerning the evolution of photo cells, particularly in the light of the recent debate among Eakin, Vanfleteren and Salvini-Plawen (see Westfall, 1982) and gives his ideas on the matter. In the following chapter he undertakes an in-depth analysis of photobehaviour in invertebrates. This complements to some extent Couillard's contribution and is also a mini- Fraenkel and Gunn (1961) dealing with photoresponses.

Annie Fournier, who has done some very interesting work with trematode larvae reviews what is known about structure and responses in Platyhelminthes. This is followed by the chapter on rotifers by Clément and Wurdak which also deals with structure and function in this interesting group of small animals with which they have done considerable work.

The first group of coelomate Protostomia, the annelids, form the topic of Martine Verger-Bocquet's chapter which, in view of her interests and what is available in the literature has a rather anatomical slant. This reflects, as was pointed out during the ASI by Drach, the paucity of information concerning their photopigments and functioning of photoreceptors.

After this, we start dealing with the arthropods, particularly the insects. As is to be expected, the bulk of the space in the book is given to them, reflecting not only their taxonomic importance (Fig. 1) but also because, among the invertebrates, the insects and the crustaceans are the most studied. The arachnids and myriapods are handled by Muñoz-Cuevas who has attempted to give a synthesis of what is known about the structure and function of photoreceptors in these classes. In his masterly treatment, Land shows for photoreception, particularly the optics of it, in crustaceans, what we try to tell about of their biology. That is, although they are a much smaller class than the insects, the crustaceans are more heterogenous. In physiology (of photoreception) there is of course considerable uniformity and even resemblance within the insects but the types of eye and optics offer an array of magnificent variations. The papers dealing with insects are admittedly more specialised than the others, as can be expected, in view of the wealth of material available. Franceschini treats the functional significance of the retinal mosaic in compound eye using the fly's as a model. A general analysis of the earlier stages of processing in the compound eye with reference to the role played by parallel channels is made by Laughlin. Using some of the most interesting and difficult techniques, Strausfeld has studied the neuro-anatomy of the blowfly's visual system. His chapter is a neat synthesis of the results of his numerous painstaking investigations. Hausen's analysis of what is known about the lobula complex is not only a critical new approach to its structure and function but also a critical view of its significance in visual behaviour of the fly. This is followed by an in-depth analysis by Erich Buchner on the rôle that behaviour plays, and could play, as a criterion for studying spatial vision in insects. Erich and Sigrid Buchner describe how ^{3}H-Deoxyglucose may be used for following nervous activity induced by photoreception. The development of the visual system is obviously of much interest to workers in any aspect of photoreception. With the advent of electron microscopy and sophisticated and reliable marking techniques, the study of the development of synapses has become possible. This forms the object of Meinertzhagen's chapter which handles synaptic formation in the lamina. The structure and development of simple and compound eyes of insects are reviewed by Mouze in the last chapter dealing with insects. The last Protostomian Phylum, Mollusca, is the subject of Land's second contribution in the book. His review emphasises special photoreceptive adaptations and special optical features in this phylum. In invertebrate biology one uses the foot of the molluscs as a model to show adaptive radiation. It seems that one could just as well also use their photoreceptors and eyes for the same purpose with much the same success. Two Deuterostomian Phyla are dealt with in this book. The more interesting (from our point of view) and, simpler Chaetognatha, and the biologically complex Echinodermata which almost entirely abandoned bilateral symmetry and took on a radial one in order to lead a more sedentary life and then "thought better of it" and made attempts to gain back more of that bilateralness (Holothuria). However, in spite of retaining sufficient faculty for photoreception they never developed eyes, in the real sense, or visual systems. These groups are dealt with by Yoshida; chaetognaths with Goto and echinoderms with Takasu and Tamotsu. The last chapter is somewhat of a summary (as we saw it) of the interesting discussion that took place on the last day of the meeting. Six rapporteurs were asked to prepare topics for discussion and there was to be no preoccupation with time. We have attempted to bring out perspectives while summarising the comments made by the rapporteurs and others. It was felt that this will be a more readable form than a transcript.

ACKNOWLEDGEMENTS

I thank my colleagues Mary Ann Klyne and Paul Pirlot for their comments on the first draft of this essay. I alone however, am responsible for any errors.

REFERENCES

Ali, M.A., Anctil, M. & Cervetto, L. (1978) Photoreception. In: Sensory Ecology. Ed. M.A. Ali. New York, Plenum Press, p. 467-502.

Ali, M.A., Croll, R.P. & Jaeger, R. (1978) Phylogenetic survey of sensory functions. In: Sensory Ecology Ed. M.A. Ali. New York, Plenum Press, p. 11-29.

Ali, M A. & Wagner, H J. (1975) Visual pigments: phylogeny and ecology. In: Vision in Fishes, New Approaches in Research. Ed. M A. Ali. New York, Plenum Press, p. 481-516.

Autrum, H (Ed.) (1979) Handbook of Sensory Physiology. Vision in Invertebrates. A: Invertebrate Photoreceptors. Vol. VII/6A. New York, Springer Verlag, 729 pages.

Dartnall, H.J.A. Ed. (1972) Handbook of Sensory Physiology. Photochemistry of vision, Vol. VII/2. New York, Springer Verlag, 810 pages.

Eakin, R.A. (1982) Continuity and diversity in photoreceptors. In: Visual Cells in Evolution. Ed. J.A. Westfall. New York, Raven Press, p. 91-105.

Fraenkel G.S. & Gunn, D.L. (1961) The Orientation of Animals. New York, Dover Publications, 376 pages.

Hamdorf, K. (1979) The physiology of invertebrate visual pigments. In: Handbook of Sensory Physiology. Vol. VII/6A. Ed. H. Autrum. New York, Springer Verlag, p. 145-224.

Langer, H. (1973) Biochemistry and Physiology of Visual Pigments. New York, Springer Verlag, 363 pages.

Lenci, F. (1980) Photoreceptions and sensory transduction in aneural organisms. New York, London, Plenum Press, 422 pages.

Muller, S.W.M. & Campbell, A. (1954) The relative number of living and fossil species of animals. Syst. Zool. 3: 168-170.

Packer, L. (Ed.) (1982) Visual pigments and purple membranes. I. Methods in Enzymology, Biomembranes. Vol. 81. Part H. New York, Academic Press, 902 pages.

Waterman, T.H. (1981) Polarization sensitivity. In: Handbook of Sensory Physiology. Vol. VII/6B. Ed. H. Autrum. New York, Springer Verlag, p. 281-463.
Westfall, J.A. (Ed.) (1982) Visual cells in evolution. New York, Raven Press, 161 pages.

VISUAL PIGMENTS OF INVERTEBRATES

D.G. STAVENGA* and J. SCHWEMER**

*Department of Biophysics, Laboratorium voor Algemene Natuurkunde
Rijksuniversiteit Groningen, The Netherlands
**Institute for Zoophysiology, Ruhr-Universität Bochum
Federal Republic of Germany

> And how exceeding curious and fubtile muft the component parts of the *medium* that conveys light be, when we find the inftrument made for its reception or refraction to be fo exceedingly fmall? we may, I think, from this fpeculation be fufficiently difcouraged from hoping to difcover by any optick or other inftrument the determinate bulk of the parts of the *medium* that conveys the pulfe of light,

Robert Hooke; Micrographia p. 180; 1667.

INTRODUCTION

An understanding of the photopigments is indispensable as, the primary process in photoreception starts with the absorption of light quanta by the photopigment molecules, which in turn triggers the long train of the visual process: molecular transformation, production of transmitter, ionic movements, and often substantial structural changes within the visual cells. Subsequently synaptic transmission of the electric signal to higher order neurons occurs and eventually a behavioural response is elicited, all this being the result of the initial absorption of light quanta.

Since the spectral characteristics of the visual pigments largely determine the spectral sensitivity of the visual cells we will focus our treatise on experimental approaches for determining the spectral qualities of visual pigments. Furthermore, we shall briefly review the rich variety of visual pigments from the view point of sensory ecology; we shall discuss how the spectral properties of invertebrate visual pigments are related to the habitat of the animal they serve. First, however, we shall describe some general characteristics of visual pigments.

GENERAL CHARACTERISTICS OF INVERTEBRATE VISUAL PIGMENTS

1. Rhodopsin and Metarhodopsin

The visual pigments of invertebrates are rhodopsins, the class of chromoproteins which have 11-cis retinaldehyde as chromophore (Fig. 1 a). The chromophore (molecular weight 284 dalton) probably is bound to the protein moiety opsin by a Schiff's base linkage as in vertebrates (Fig. 1 d). Whereas the molecular weight of insect rhodopsin is similar to that of vertebrate rhodopsins (being approximately 38 000 dalton), cephalopod rhodopsins seems to have a somewhat higher molecular weight (Table I).

Free retinal absorbs in the ultraviolet spectral range and opsin at still shorter wavelengths. The main reason for visual pigments absorbing in the longer, so-called visible wavelength range is the protonation of the Schiff's base, whereas further tuning of the absorption spectrum occurs through secondary interactions between retinal and opsin.

As an example, blowfly rhodopsin is presented in Fig. 2. The spectra were obtained in a series of experiments designed to determine the dependence of visual pigment content on the diet of the fly (Paulsen &

TABLE I. Molecular weight MW (dalton) of invertebrate visual pigments.

	MW	Ref.
CEPHALOPODA		
Loligo pealii - squid	49 000	(1)
	45 000	(2)
Todarodes pacificus - squid	51 000	(3)
Watesinia scintillans - squid	49 000	(3)
Sepia officinalis - cuttlefish	45 000	(2)
Octopus vulgaris - octopus	43 000	(4)
Eledone moschata - octopus	43 000	(2)
INSECTA		
Ascalaphus macaronius - owlfly	35 000	(2)
Aedes aegypti - mosquito	39 000	(5)
Calliphora erythrocephala - blowfly	32 500	(6)
Drosophila melanogaster - fruitfly	37 000	(7)

(1) Hagins (1973); (2) Paulsen & Schwemer (1973); (3) Kito et al. (1982); (4) Kropf et al. (1959); (5) Stein et al. (1978); (6) Paulsen & Schwemer (1979); (7) Ostroy (1978).

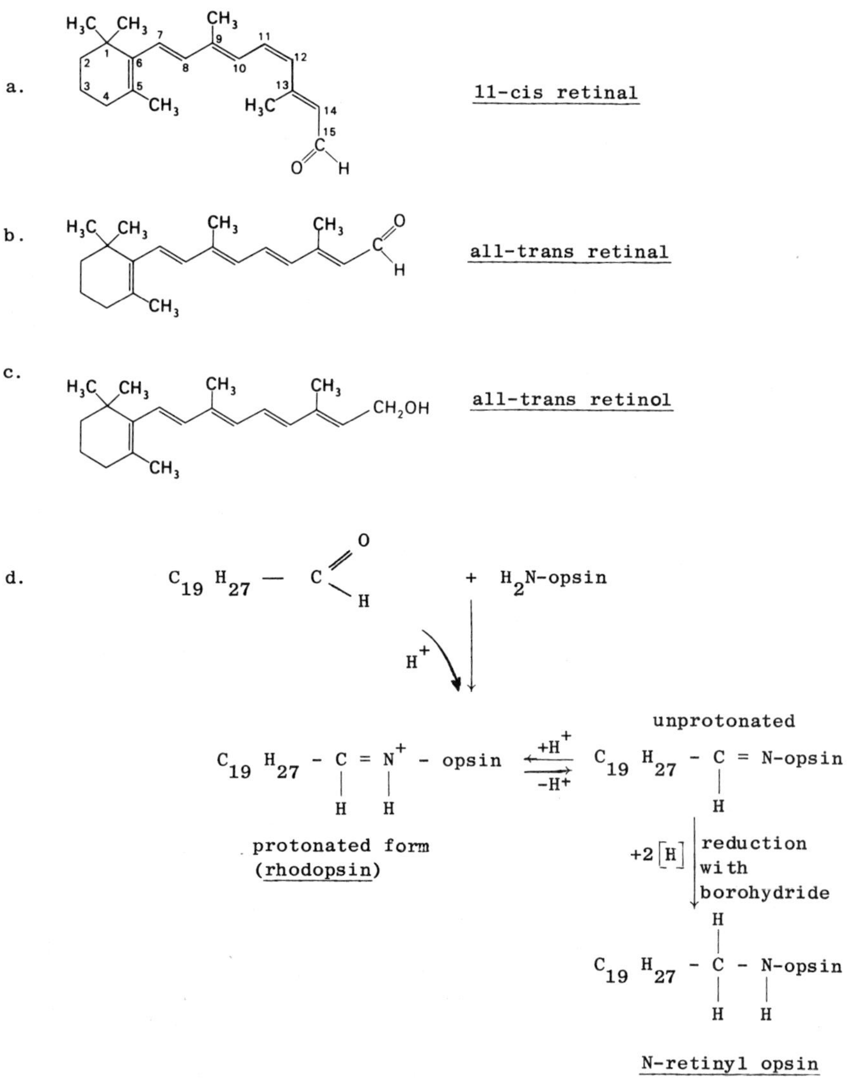

VISUAL PIGMENTS OF INVERTEBRATES 13

a. [structure] 11-cis retinal

b. [structure] all-trans retinal

c. [structure] all-trans retinol

d.
$$C_{19}H_{27}-CHO + H_2N\text{-opsin}$$

$$\xrightarrow{H^+}$$

protonated form (rhodopsin):
$$C_{19}H_{27}-\overset{H}{\underset{H}{C}}=\overset{+}{N}\text{-opsin} \underset{-H^+}{\overset{+H^+}{\rightleftarrows}} C_{19}H_{27}-\overset{H}{C}=N\text{-opsin}\ \text{(unprotonated)}$$

$$\xrightarrow{+2[H]\ \text{reduction with borohydride}}$$

$$C_{19}H_{27}-\overset{H}{\underset{H}{C}}-N\text{-opsin}$$

N-retinyl opsin

Fig. 1. The visual chromophore retinal together with opsin forms the visual pigment. In the native rhodopsin state retinal exists in the 11-cis conformation (a) but in the metarhodopsin state as all-trans retinal (b). Retinal is derived from the alcohol retinol, vitamin A (c) and it is bound to opsin by a Schiff's base, which can be reduced by sodium borohydride.

Schwemer, 1979). It appeared that blowflies reared on bovine liver acquired approximately 20 times more visual pigment than flies reared on heart meat which evidently is related to the fact that liver is rich in

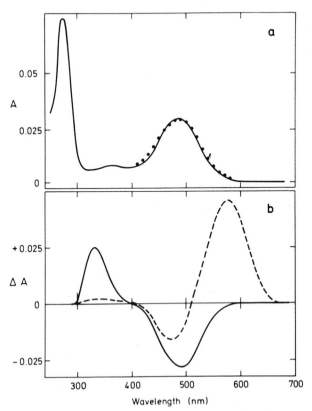

Fig. 2. Absorbance characteristics of digitonin extracts from blowfly photoreceptor membranes. (a) Absorbance difference between rhodopsin-rich and deficient fly retinae. The closed symbols represent a nomogram for a visual pigment with maximum absorbance at 490 nm (Ebrey & Honig, 1977). (b) Absorbance changes recorded after sodium cyanoborohydride was added and the rhodopsin-rich extract had been illuminated with blue light (472 nm):

---------- absorbance change after 2 min of illumination at 8°C, resulting primarily from the formation of metarhodopsin (λ_{max} = 570 nm);

——— absorbance change after 6 min of illumination at 25°C, indicating that N-retinylopsin (λ_{max} = 333 nm) was formed (after Paulsen & Schwemer, 1979).

vitamin A (retinol, Fig. 1 c), whereas the vitamin A content of heart meat is very low. The rhodopsin spectrum (Fig. 2 a) was obtained by extracting retinae of blowflies from both groups, called R^+- and R^-- flies respectively, and measuring the difference in absorbance between these two extracts.

The blowfly spectrum is very similar to that of vertebrate rod rhodopsin and nicely shows the general features of visual pigment spectra: a main, broad band ($\simeq$ 100 nm width at half height) with a maximum in the visible range, called the α-peak, and a subsidiary, smaller band with maximal absorbance in the ultraviolet, called the β-peak. At about 280 nm a pronounced γ-peak is found which is due to the absorption by aromatic amino-acid residues of the protein. Actually, the latter peak indicates that the absence of the chromophore in R^--flies causes not only loss in absorption in the visible, but in the far UV as well, indicating loss in opsin content. A decreased opsin content was directly demonstrated by gel-electrophoresis (Paulsen & Schwemer, 1979).

Rhodopsin molecules, or rather their protein moieties, can be recognized as particles in freeze-fracture preparations of the photoreceptor membrane. Fig. 3 demonstrates that the density of particles in freeze-fractured microvillar membranes of R^+-flies is high, while it is much reduced in R^--flies. The density appears to be 3000 and 600 particles/ m^2, respectively (Miller, Brown, Schwemer, unpublished data). In R^+-flies the average spacing of the visual molecules in the membrane hence is $\sim$20 nm (= 200 Å). The space between the rhodopsin molecules is

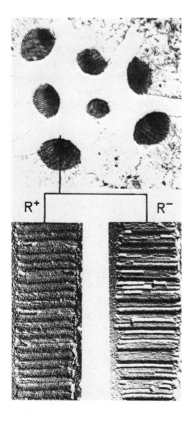

Fig. 3. Cross-section of an ommatidium of the fly Calliphora made by transmission electron microscopy and freeze-fracture preparations of the photoreceptor membrane of flies reared on a vitamin-A rich and vitamin-A deficient diet respectively (Schwemer, 1979).

filled by membrane lipids. The packing of molecules in the visual membrane of invertebrates seems to vary somewhat (for discussion see Hamdorf & Schwemer, 1975, and Hamdorf, 1979).

In the literature visual pigments are often characterized by the molar extinction coefficient ε_{max} (Table II) and the wavelength of maximal absorption λ_{max} (Table III). For example, λ_{max} of blowfly rhodopsin is at about 490 nm and accordingly, the pigment is called R490. Table III shows that peak wavelengths of invertebrate rhodopsins span a wide range, from the ultraviolet up to the orange. As we will see below, the photoproducts of rhodopsin feature the same versatility in absorbing properties.

Light absorption by a rhodopsin molecule causes isomerization of the chromophore retinal from the 11-cis into the all-trans configuration followed by conformational changes in the opsin part. The absorption spectrum of the resulting photoproduct called metarhodopsin depends on pH. The alkaline form, existing at pH $\gtrsim$ 9 invariably absorbs in the ultraviolet ($\lambda_{max} \simeq 380$ nm, close to that of free all-trans retinal, see Table III). Acid metarhodopsin is usually encountered because it is formed at neutral and slightly acid pH. Its absorption spectrum generally peaks in the blue, although a distinct class of yellow-absorbing metarhodopsins exists.

Blowfly metarhodopsin belongs to the latter class: rhodopsin R490 is photoconverted in metarhodopsin with λ_{max} at 570 nm, thus called M570 (Fig. 2), which has a peak absorbance $\sim 1,8$ relative to that of R490. This ratio varies rather for the various invertebrate visual pigment systems (1,3 1,8), but is always distinctly larger than 1. This is most likely due to the stretched (all-trans) chromophore in metarhodopsin, as the absorbance of all-trans retinal is 1,74 times that of the 11-cis form (Table II).

Under physiological conditions metarhodopsins of invertebrates are thermostable at room temperature. This contrasts with vertebrate metarhodopsins which hydrolyse into free all-trans retinal and opsin. Thermostability may seem to be the prerogative of invertebrate metarhodopsins.

2. Photochemistry of Invertebrate Visual Pigments

2.1 Photochromism

Invertebrate visual pigments can be usefully described by the following formalism generally applicable for photochromic pigments (Hamdorf et al., 1968; Schwemer, 1969; Hamdorf & Schwemer, 1975; Stavenga, 1975, 1976; Hochstein et al., 1978; Hamdorf, 1979).

A photochromic pigment has two interconvertible states which can be distinguished by their absorption spectra. The two states of invertebrate visual pigments are rhodopsin (R) and metarhodopsin (M) and the reaction scheme upon illumination is

$$R \underset{k_M}{\overset{k_R}{\rightleftarrows}} M \qquad (1)$$

TABLE II. Molar extinction coefficient ε_{max} (M^{-1} cm^{-1})

	R (11-cis)	M (all-trans)	ref.
retinal	24 900	43 400	(1)
CEPHALOPODA			
Loligo pealii - squid	40 600	59 700	(2)
Todarodes pacificus - squid	35 000	48 000	(3)
Watesinia scintillans - squid	34 000	51 600	(4)
Sepia officinalis - cuttlefish	37 000	47 700	(2)
Octopus vulgaris - octopus	37 000	44 000	(2)
CRUSTACEA			
Orconectes rusticus - crayfish	39 900	54 000	(5)
Procambarus clarkii - crayfish	39 900	54 000	(5)
Libinia emarginata - crab	40 000		(6)
INSECTA			
Calliphora erythrocephala - blowfly	40 000	72 000	(7)
Apis mellifera - honeybee drone	43 000	75 000	(8)

(1) Morton (1972); (2) Brown & Brown (1958); (3) Suzuki et al. (1976); (4) Nashima et al. (1978); (5) Cronin & Goldsmith (1982 b); (6) Hays & Goldsmith (1969); (7) Schwemer (1979); (8) Muri (1979).

TABLE III. Wavelengths of maximum absorbance λ_{max} (nm).

	R(11-cis)	M(all-trans)	method	ref.
retinal	376,5	381	in ethanol	(1)
MOLLUSCA				
CEPHALOPODA				
Sepia officinalis - cuttlefish	492	497	ext	(2)
Sepia esculenta - cuttlefish	486	495	ext	(3)
Sepiella japonica - cuttlefish	500	500	ext	(3)
Loligo pealii - squid	493	500	ext	(2, 4)
Loligo japonica - squid	496	500	ext	(5)
Todarodes pacificus - squid	480	485	ext	(5, 6, 7)
Watasinia scintillans - squid	482	496	ext	(5)
Octopus vulgaris - octopus	475	503	ext	(2, 3)
Octopus ocellatus - octopus	477	512	ext	(3)
Eledone moschata - octopus	470	516	ext	(8)
BIVALVIA				
Pecten irradians - scallop	500	575	ERP	(9)

	R(11-cis)	M(all-trans)	method	ref.
CHELICERATA - XIPHOSURIDA				
Limulus polyphemus - horseshoe crab, median ocellus	360	480	LRP	(10)
Limulus polyphemus - horseshoe crab lateral/ventral eye	520-530	500/515-535	MSP/ERP	(11)/(12)
CRUSTACEAE - CIRRIPEDIA				
Balanus eburneus - barnacle	532	492	LRP/ERP/MSP	(13)/(14)/(15)
Balanus amphitrite - barnacle	532	492	LRP/ERP/MSP	(13)/(14)/(15)
DECAPODA				
Euphausia superba - krill	485	493	ext	(16)
Meganyctiphanes norvegica - krill	488	493	MSP	(16)
Palaemonetes palludosus - fresh water prawn	539	497	ext	(17)
Penaeus duorarum - pink shrimp	516	475	ext	(17)
Homarus americanus - American lobster	515	490	ext	(18)
Panuliris argus - spiny lobster	504	495	ext	(17)
Orconectes rusticus - northern crayfish	535	510	MSP	(19)
Procambarus clarkii - swamp crayfish	535	510	MSP	(19)
Astacus fluviatilis - crayfish	530	500	MSP	(20)
Hemigrapsus edwardsii - mud crab	495	495	ext	(21)
Leptograpsus variegatus - rock crab	513	495	ext	(21)
Libinia emarginata - spider crab	493	498	ext	(21)

(continued)

TABLE III. Wavelengths of maximum absorbance λ_{max} (nm). (Continued)

	R(11-cis)	M(all-trans)	method	ref.
INSECTA - HETEROPTERA				
Gerris lacustris - water strider	350	480	MSP	(23)
	460	520	MSP	(23)
	545	485	MSP	(23)
INSECTA - HYMENOPTERA				
Apis mellifera - honeybee drone	446	505	MSP	(24, 25)
INSECTA - NEUROPTERA				
Ascalaphus macaronius - owlfly	345	475	ext/MSP	(26)/(27)
INSECTA - LEPIDOPTERA				
Deilephila elpenor - sphingid moth	345	480	ext/MSP	(28)/(29)
	440	480	ext/MSP	(28)/(29)
	520	480	ext	(28)
Manduca sexta - tobacco hornworm moth	345	490	ext/MSP	(30)/(31)
	440	490	ext/MSP	(30)/(31)
	520	490	ext/MSP	(30)/(31)
Galleria mellonella - bee moth	510	484	MSP	(32)
Spodoptera exempta - African army-worm moth	350	465/470	MSP/ext	(33)/(34)
	520	485/480	MSP/ext	(33)/(34)
	560	485	MSP	(33)
Apodemia mormo - metalmark butterfly	587	521	MSP	(35)
Vanessa cardui - painted lady	530	490	MSP	(36)

INSECTA - DIPTERA

	R(11-cis)	M(all-trans)	method	ref.
Aedes aegypti - mosquito	515	480	MSP	(37)
Eristalis tenax - dronefly	460	550	MSP	(38)
Drosophila melanogaster - fruitfly	480	580	ext/MSP	(39)/(40)
Sarcophaga bullata - fleshfly	490	575	MSP	(31)
Calliphora erythrocephala - blowfly	490/495	570/580	ext/MSP	(41)/(42)
Musca domestica - housefly	340-345	470	LRP/MSP	(43)/(44)
	430	505	MSP	(45)
	490	580	MSP	(44)/(46)

(1) rev. Morton (1972); (2) Brown & Brown (1958); (3) Takeuchi (1966); (4) Hubbard & St. George (1958); (5) Naito et al. (1981); (6) Yoshizawa & Shichida (1982 a); (7) Hara & Hara (1972); (8) Hamdorf et al. (1968); Schwemer (1969); (9) Cornwall & Gorman (1979); (10) Nolte & Brown (1972); (11) Ostroy (1977); (12) Lisman & Sheline (1976); (13) Hillman et al. (1973); (14) Minke et al. (1973); (15) Minke & Kirschfeld (1978); (16) Denys & Brown (1983); (17) Fernandez (1965); (18) Wald & Hubbard (1957); (19) Cronin & Goldsmith (1982 b); (20) Hamacher & Kohl (1981); (21) Briggs (1961); (22) Bruno et al. (1977); (23) Hamann & Langer (1980); (24) Muri (1978); (25) Bertrand et al. (1979); (26) Gogala et al. (1970); (27) Schwemer et al. (1971); (28) Schwemer & Paulsen (1973); (29) Hamdorf et al. (1972); (30) Schwemer & Struwe, unpubl.; (31) Schwemer & Brown, unpubl.; (32) Goldman et al. (1975); (33) Langer et al. (1979); (34) Schwemer, unpubl.; (35) Bernard, unpubl.; (36) Bernard (1982); (37) Brown & White (1972); (38) Stavenga (1976); (39) Ostroy et al. (1974); (40) Ostroy (1977); (41) Schwemer (1979); (42) Stavenga et al. (1973); (43) Hardie et al. (1979); (44) Kirschfeld (1979); (45) McIntyre & Kirschfeld (1981); (46) Kirschfeld et al. (1977).

with

$$\frac{dC_R}{dt} = -C_R k_R + C_M k_M = -\frac{dC_M}{dt} \qquad (2)$$

Here C_R and C_M are concentrations. The rate constants k_R and k_M equal

$$k_R = \alpha_R \gamma_R I \quad \text{and} \quad k_M = \alpha_M \gamma_M I \qquad (3a)$$

when monochromatic light of quantum flux I is applied; α_R and α_M are the molecular absorbance coefficients and γ_R and γ_M the absolute quantum efficiencies for converting rhodopsin into metarhodopsin and vice versa. The products of α and γ:

$$K_R = \alpha_R \gamma_R \quad \text{and} \quad K_M = \alpha_M \gamma_M \qquad (3b)$$

are called the photosensitivities of the two states.

2.2 Photoequilibrium spectrum

In the closed system of eq. (1) the total concentration of visual pigment $C_{tot} = C_R + C_M$ remains constant so that it is useful to introduce fractions:

$$f_R = C_R/C_{tot} \quad \text{and} \quad f_M = C_M/C_{tot} \qquad (4)$$

When at time $t = 0$ a metarhodopsin fraction f_{M0} exists and a constant flux I is applied, eqs. (2)-(4) solve into

$$f_M(t) = f_{M\infty} + (f_{M0} - f_{M\infty}) e^{-(K_R + K_M) I t} \qquad (5)$$

with

$$f_{M\infty} = \frac{k_R}{k_R + k_M} = \left(1 + \frac{K_M}{K_R}\right)^{-1} \qquad (6)$$

At each adapting wavelength λ_a the photoequilibrium (reached at $t = \infty$) only depends on the ratio of the photosensitivities K_M/K_R and not on the intensity (as can be directly seen from eqs. (2)-(4)).

The isosbestic wavelength λ_{iso} is defined by $\alpha_R(\lambda_{iso}) = \alpha_M(\lambda_{iso})$, and hence

$$f_{M\infty}(\lambda_{iso}) = \left[1 + \frac{\gamma_M(\lambda_{iso})}{\gamma_R(\lambda_{iso})}\right]^{-1} = \left[1 + \phi(\lambda_{iso})\right]^{-1} \qquad (7)$$

where $\phi = \gamma_M/\gamma_R$ is the <u>relative</u> quantum efficiency. Probably ϕ is wavelength independent, or $\phi(\lambda) = \phi(\lambda_{iso})$ (Schwemer, 1969); its value is known for a few visual pigment systems (Table IV).

The wavelength dependency of f_M^∞ (or $f_R^\infty = 1 - f_M^\infty$) is called the saturation spectrum (Hochstein et al., 1978) or photoequilibrium spectrum (Minke & Kirschfeld, 1979).

The metarhodopsin fraction at photoequilibrium produced by illumination at wavelength λ_a normalized to its value at the isosbestic wavelength is

$$Q(\lambda_a) = \frac{f_{M\infty}(\lambda_a)}{f_{M\infty}(\lambda_{iso})} = \left[1 + \phi(\lambda_{iso})\right]\left[1 + \frac{K_M(\lambda_a)}{K_R(\lambda_a)}\right]^{-1} = \frac{1 + \phi(\lambda_{iso})}{1 + \phi(\lambda_a)U(\lambda_a)} \quad (8)$$

with $U = \alpha_M/\alpha_R$ ($U(\lambda_{iso}) = 1$).

2.3 Relaxation spectrum

The photoequilibrium (eq. (6)) is reached with a rate constant (K_R+K_M) (eq. (5)). The wavelength dependency of the sum of the photosensitivities $K_R + K_M$ is called the relaxation spectrum. This spectrum is thus obtained by measuring the rate of photoconversion induced by a given number of quanta at various wavelengths λ_a.

2.4 Difference spectrum

The molecular absorbance coefficients of the visual pigment states can be obtained from transmission measurements. In experimental practice either photostable pigments such as screening pigments, cytochromes etc. contaminate the measurements and/or it is not feasible to calibrate the applied light intensities. A still very useful approach then is to measure the difference in absorbance resulting from a conversion of visual pigment following irradiation. Generally the transmission of a layer of visual pigment with thickness L at wavelength λ (according to the law of Lambert-Beer) is:

$$I_L(\lambda)/I_0(\lambda) = e^{-\left[\alpha_R(\lambda)f_R + \alpha_M(\lambda)f_M\right]C_{tot}L} \quad (9)$$

The difference in absorbance in respect to the case $f_M = 0$ is

$$\Delta E(\lambda) = -^{10}\log\left[I_L/I_L(f_M=0)\right] = 0.43\left[\alpha_M(\lambda) - \alpha_R(\lambda)\right]f_M C_{tot}L \quad (10)$$

From this expression it follows directly that difference spectra obtained with varying f_M have the same shape and are proportional to the

TABLE IV. Relative quantum efficiency $\phi = \gamma_M/\gamma_R$

	ϕ	method	ref.
Todarodes pacificus - squid	0,56	ext	(1)
Eledone moschata - octopus	1,0	ext	(2)
Orconectes rusticus - crayfish	0,71	MSP	(3)
Procambarus clarkii - crayfish	0,71	MSP	(3)
Homarus americanus - lobster	0,76	MSP	(4)
Calliphora erythrocephala - blowfly	0,94/0,93	ext/LRP	(5)/(6)
Drosophila melanogaster - fruitfly	0,71	MSP	(7)
Eristalis tenax - drone fly	0,83	LRP	(8)
Apis mellifera - honeybee drone	0,94	MSP	(9)

(1) Suzuki et al. (1976); (2) Schwemer (1969); (3) Cronin & Goldsmith (1982 b); (4) Goldsmith & Bruno (1973); (5) Schwemer (1969); (6) Hamdorf (1979); (7) Stark & Johnson (1980); (8) Tsukahara & Horridge (1977); (9) Muri (1978).

difference in molecular absorbance coefficients $\alpha_M - \alpha_R$. Furthermore, the absorbance difference at the isosbestic point λ_{iso} always is $\Delta E = 0$.

When absorbance differences are measured at various photoequilibria and at a fixed wavelength λ, then it follows from eqs. (8) and (10) that

$$\frac{\Delta E(\lambda, \lambda_a)}{\Delta E(\lambda, \lambda_{iso})} = \frac{f_M(\lambda_a)}{f_M(\lambda_{iso})} = Q(\lambda_a) \qquad (11)$$

Hence from measurements of absorbance differences one can obtain the photoequilibrium spectrum. By also measuring conversion rates, yielding the relaxation spectrum, it becomes possible to calculate the absorption spectra of rhodopsin and metarhodopsin as well as the relative quantum efficiency ϕ. For practical examples see Schwemer (1969), Stavenga, (1976, 1979), Hamdorf (1979), Stark and Johnson (1980) and Cronin and Goldsmith (1981, 1982 a, b).

EXPERIMENTAL METHODS FOR STUDYING INVERTEBRATE VISUAL PIGMENTS

An exhaustive account of techniques for studying visual pigments was recently published as Vol. 81 of Methods of Enzymology, edited by Packer (1982). We will describe here only those methods which yield the spectral characteristics of visual pigments, i.e. absorption, fluorescence and photosensitivity spectra, as well as quantum efficiencies.

1. Optical Methods

1.1 Spectrophotometry on visual pigment extracts

Spectrophotometry on visual pigment extracts is the classical way for determining the spectral properties of visual pigments. For several reasons, cephalopod eyes are ideal for making extracts with a high yield. Following Hubbard and St. George's lead (1958), a number of Japanese groups have in great detail characterized the photochemical cycle of squid and octopus visual pigments.

We have to remark, however, that caution is necessary for the extraction procedures. Visual pigments can be extracted from purified retinal tissue with the aid of a variety of detergents. It proves that virtually all detergents used for vertebrate rhodopsins affect the molecular properties of invertebrate rhodopsins to a greater or lesser degree. The detergent effect can for instance be noted from the squid literature where λ_{max} of Todarodes rhodopsin is given as 480 or 482 nm (Hara & Hara, 1966; Naito et al., 1981) and that of acid metarhodopsin as 482 or 491 nm (Tokunaga et al., 1975; Naito et al., 1981). (We note that temperature also affects the spectra as λ_{max} of Todarodes rhodopsin at liquid helium temperature is 489 nm, Shichida et al., 1978). It will thus be clear that only very mild agents should be used for extractions; digitonin appears to be the detergent which fulfils that requirement best (see Kito et al., 1982; Schwemer & Langer, 1982).

Only extracts from highly purified retinal receptor membranes contain almost exclusively visual pigment and hence uncontaminated absorption spectra are seldom obtained. Fortunately contaminations are satisfactorily circumvented by taking difference spectra. Fig. 4 presents difference spectra of a digitonin extract from fly retinae (from Schwemer, 1979). Curve 1 is the absorbance change when part of the rhodopsin is photoconverted into the metarhodopsin state. The curve was taken at +5°C and subsequently the extract was warmed to +27°C in the dark. Curve 2 represents the drop in blue absorbance (in respect of the initial state) induced by the warming-up procedure, evidently due to hydrolysis of fly metarhodopsin into free all-trans retinal and opsin. At the same time this curve is the difference spectrum of rhodopsin and retinal. The absorbance spectrum of the bleached rhodopsin can be estimated by a procedure often used in visual pigment studies, namely by approximation with a rhodopsin nomogram (see Fig. 2 a), a method

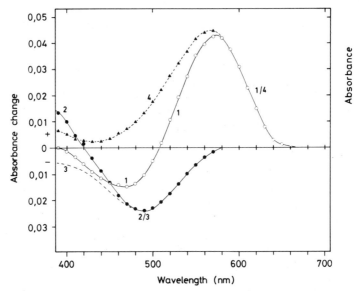

Fig. 4. Absorbance changes in a visual pigment extract (pH = 5,5) after illumination at +5°C and subsequent warming to +27°C.

1: absorbance changes after illumination with 472 nm at +5°C;
2: Absorbance changes in the dark after warming to +27°C due to hydrolysis of metarhodopsin;
3: Nomogram of a rhodopsin with λ_{max} = 490 nm;
4: The metarhodopsin spectrum, obtained by adding nomogram (3) and difference spectrum (1) (Schwemer, 1979; Schwemer & Langer, 1982).

introduced by Dartnall (1953) and improved, among others, by Ebrey and Honig (1977). By adding the nomogram curve 3 to the difference spectrum 1 the metarhodopsin spectrum 4 is obtained. The lability of metarhodopsin in extract at higher temperatures hence can be successfully utilized in obtaining the visual pigment spectra. Some uncertainty remains, however, because digitonin also can substantially affect the molecular characteristics, as is immediately recognized from a comparison with spectrophotometry of the visual pigment in situ. For an extreme example we refer to Goldsmith and Bruno (1973) who found a 20 nm hypsochromic shift induced by digitonin for the visual pigment of the blue crab Callinectes.

The destabilizing effect of extraction is also encountered when the dependency on pH is investigated. In vitro (at a constant temperature of 5°C) the long-wavelength absorbing acid metarhodopsin of blowfly converts into the ultraviolet-absorbing alkaline form when the pH increases above 7; in situ, however, the acid state is found also at pH 7 (Schwemer, 1979). This behaviour appears to be the general rule for invertebrate metarhodopsins (e.g. Hubbard et al., 1965; Schwemer et al., 1971; rev. Goldsmith, 1972, 1975).

1.2 Low-temperature spectrophotometry

Upon light absorption by rhodopsin the molecule undergoes a cascade of thermal transformations. A useful review of the effects of temperature on the photopigment cycle is given by Ali (1975). The cascade can be blocked at various stages by cooling, and in this way it appears that only the first step is allowed at liquid He temperature. Fig. 5 summarizes the results for squid rhodopsin. The nomenclature of the stages is in close analogy with that of the cascade of vertebrate rod rhodopsin (from Yoshizawa & Shichida, 1982 a), but the distinct state mesorhodopsin occurring between lumi- and metarhodopsin has not been found for vertebrates (mesorhodopsin is equivalent to LM- rhodopsin; Ebina et al., 1975; Suzuki et al., 1976).

It is possible to photoregenerate the various states into the rhodopsin state (Shichida et al., 1978). So far, the photoregeneration pathways have been little studied compared to the rhodopsin decay cascade. Nevertheless, a few essential features can be exemplified in the case of the ultraviolet absorbing rhodopsin found in the dorsal eye of the neuropteran Ascalaphus macaronius, the only non-cephalopod rhodopsin studied by low-temperature spectroscopy (Hamdorf et al., 1973). Ascalaphus rhodopsin is photoconverted at room temperature to a blue absorbing acid metarhodopsin. At -50°C, however, after photoconversion the molecule appears locked into an intermediate state, lumirhodopsin. Rhodopsin → lumirhodopsin conversion gives a pronounced increase in absorption (Fig. 6 a, curve 1 → 3) proving cis-trans isomerization takes place. Warming induces the conversion of lumi- to acid metarhodopsin and evokes a large spectral shift in absorption (curve 3 → 2).

The pathway during photoregeneration is given in Fig. 6 b. The first step in photoregeneration is an all-trans → 11-cis transformation of the chromophore (decrease in absorbance, small shift of λ_{max}, curve 2 → 4),

State	λ_{max} (nm)	conversion	time $\tau_{1/e}$
rhodopsin	480	by light	< 19 ps
⤳ hypsorhodopsin	446	thermally	
↓		> $-238°C$	≃ 50 ps
bathorhodopsin	534		
↓		> $-160°C$	≃ 300 ns
lumirhodopsin	515		
↓		> $-65°C$	
mesorhodopsin	486		
↓		> $-20°C$	≃ 10 ms
acid metarhodopsin	482		
$-H^+$ ↓↑ $+H^+$			
alkaline metarhodopsin	367		

Fig. 5. Photochemical cycle of squid (after Yoshizawa & Shichida, 1982 a).

and subsequently a refolding of the molecule into its rhodopsin shape in the dark (Fig. 6 b, curve 4 → 1). Essentially the same sequence has been observed for squid (Suzuki et al., 1972) and octopus visual pigment (Hamdorf et al., 1973), but other paths through various isomers seem possible (see Naito et al., 1981).

Low-temperature spectroscopy has one main methodological advantage: the whole gamut of physico-chemical techniques available nowadays for analysing molecular properties can be applied to the sequential molecular states. Circular dichroism (CD), nuclear magnetic resonance (NMR), Raman and even photo-acoustic spectroscopy on visual pigments are practised in a number of laboratories (see e.g. Yoshizawa & Shichida, 1982 b; Liu & Matsumoto, 1982; Shriver et al., 1982; Callender & Honig, 1977; Boucher & Leblanc, 1981). These approaches are essential for a detailed understanding of the spectral properties of the pigment states as well as of the energetics of the cascadic process.

1.3 Flash photolysis

The decay of rhodopsin through various stages to metarhodopsin can be followed at room temperature with flash photolysis and laser methods, and the decay times given in Fig. 5 were thus obtained. In principle the applied procedure is simple: an exciting flash, inducing rhodopsin conversion, is followed by a test flash after a variable delay. The wavelength of the test flash is set at those wavelengths where a distinct difference in absorption by the various molecular states was found in the low-temperature experiments.

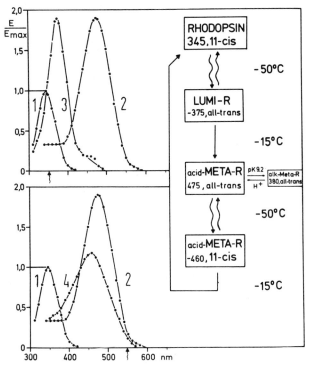

Fig. 6. Photochemical cycle of the owlfly Ascalaphus.

1: Spectral absorbance of rhodopsin R345;
2: acid metarhodopsin;
3: lumirhodopsin;
4: 11-cis metarhodopsin (Hamdorf et al., 1973).

The study of the most rapid conversions clearly requires picosecond laser methods (Fig. 5) and an echelon, a staircase-like mirror, which delivers a train of test flashes after each exciting flash, is often used. Picosecond lasers, however, have their inherent experimental problems and require much expertise; this may be the underlying cause why squid bathorhodopsin with one group has a formation time of ∿ 50 ps (Fig. 5) (Shichida et al., 1978) while according to another group the transformation takes less than 6 ps (see Doukas et al., 1980).

1.4 Partial bleaching of multiple visual pigment systems

An abundant number of animal species have retinae containing more than one photoreceptor type, i.e. more than one type of visual pigment. Digitonin extracts then yield a mixture which can be analysed successfully by applying the knowledge acquired from single visual pigment extracts: selective photoconversion of one of the visual pigments and subsequent photolysing by warming. The experimental results thus obtained for the moth Deilephila elpenor are shown in Fig. 7. There are three visual pigments absorbing in the ultraviolet, blue and green. The absorption spectra of their metarhodopsins, however virtually coincide (see Table III).

2. Spectrophotometry

2.1 Spectrophotometry on visual pigments in situ

In the functioning visual cell the visual pigment molecules are an integral part of the cell membrane. Obviously it is of great interest to study the pigment properties in situ. Attractive objects for in situ spectrophotometry are the retinae of cephalopods, crayfish, and moths where a distinct layer with nicely developed visual membrane exists. That layer can be rather easily isolated and subsequently studied with the procedures applied in analytical spectrophotometry. It thus is found that in situ invertebrate metarhodopsins are thermostable at room temperature (Schwemer et al., 1971; Goldsmith & Bruno, 1973).

For many invertebrate species with small eyes and little visual membrane material, neither extraction nor spectrophotometry of the retina are feasible approaches. The method of choice then is microspectrophotometry (MSP) since it enables absorption measurements on objects of microscopic dimensions.

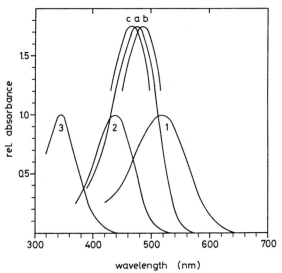

Fig. 7 Scheme of the rhodopsin systems found in the moth Deilephila elpenor. Curves 1, 2 and 3 represent rhodopsins with λ_{max} at 520, 440 and 345 nm, curves a b and c are the corresponding acid metarhodopsins (Schwemer & Paulsen, 1973).

VISUAL PIGMENTS OF INVERTEBRATES

2.2 End-on MSP

The visual pigments of many insect and crustacean eyes are contained in long cylindrical structures, with a diameter of only a few microns, which act as light guides. In end-on MSP, light transmission through the visual photoreceptors is measured. Hamdorf and Langer (1965) were the first to explore this technique in sections of eyes of the blowfly mutant chalky. The light-guiding properties of the rhabdomeres clearly recognizable in eye slices had been revealed previously (De Vries & Kuiper, 1958) and thus, by analogy to knowledge established on vertebrate photoreceptors, pointed to the rhabdomeres as the visual pigment centres. Indeed, Langer and Thorell (1966) measured vertebrate rhodopsin-like absorption curves from single rhabdomeres in blowfly eye slices, so paving the way for the subsequent extensive work on fly visual pigment. Kirschfeld et al. (1977) for instance found in the housefly Musca (white-eyed mutant) with the same technique that the peripheral receptors have a visual pigment R490, M580 essentially identical to the one in blowfly (see also Langer et al., 1982).

Because fly rhabdomeres are optically separate and contain only one visual pigment, these structures are almost ideal for end-on MSP (see Langer et al., 1982). Water striders analysed by Hamann and Langer (1980) offer a similar favourable case. Most invertebrates with compound eyes, however, have rhabdomeres which are fused into a rhabdom and this structure then functions as one optical waveguide. The analysis of individual visual pigment types then is more complicated, but by no means impossible. Yet very little work has been performed along this line. Muri (1978) and Bertrand et al., (1979) performed end-on MSP on the rhabdoms of the dorsal eye parts of honeybee drones and characterised a main blue absorbing rhodopsin R446, M505 together with a small quantity of a UV-rhodopsin.

2.3 Deep pseudopupil

The single rhabdom(ere) measurement can be substantially improved by utilizing the so-called deep pseudopupil phenomenon which enables the sampling of data from hundreds of photoreceptors so that the signal-to-noise ratio is much increased (see Franceschini, 1983 b).

The deep pseudopupil is a superposition of images, created by the facet lenses of the underlying retinulae which is observed near the centre of curvature of the eye. In the eyes of flies, a deep pseudopupil with distinct photoreceptor types is observed, owing to the perfect retinal mosaic (for further details see also Franceschini, 1975; Stavenga, 1979). Hence a large number of homologous photoreceptors can be exquisitely analysed in this way, even in living animals over long periods of time. Utilizing the deep pseudopupil and taking difference spectra, the visual pigments have been characterized for a number of fly species. R495 M580 was determined for the blowfly Calliphora (Stavenga et al., 1973) but R460, M550 appeared to exist in the eyes of the dronefly Eristalis (Stavenga, 1976).

Moreover, the deep pseudopupil facilitates measurement on the intermediate states in the fly visual pigment cycle by flash photolysis. In intact living Drosophila, Kirschfeld et al. (1978) monitored the decay of (presumably) lumirhodopsin to metarhodopsin, having a time constant of $\simeq$ 1 ms (at 5°C $Q_{10} \simeq 2,5$; see Stark et al., 1979). Similar experiments on living blowflies with pulsed dye-lasers revealed the same process plus a much more rapid decay, probably batho → lumirhodopsin, having a time constant of 0,7 µs (room temperature; Kruizinga et al., 1983). Also in the photoregeneration pathway from metarhodopsin to rhodopsin, an intermediate decaying with a time constant of $\sim$ 4 µs was observed (cf. Kirschfeld et al., 1978).

Visually most beautiful pseudopupils are displayed by the eyes of butterflies, as they have interference reflection filters proximal to the rhabdoms. Test light hence can pass the rhabdoms back and forth, and slight changes in the visual pigment composition can be sensitively traced. Using the deep pseudopupil phenomenon, Bernard (1977, 1979, 1982) and Stavenga et al. (1977) demonstrated the existence of several types of visual pigments. Intermediate states living in the order of several tens of seconds were noticed (Stavenga et al., 1977), but they need further characterization.

2.4 Isolated rhabdoms

Applying MSP on isolated rhabdoms of crustaceans, Goldsmith (1978 a, b) and Bruno et al. (1977) analysed the dependence of the photochemical cycle on pH, temperature and chemical agents such as glutaraldehyde and formaldehyde. They found that these substances potentiate photodestruction of metarhodopsin, but pH and temperature affect the visual pigment in an analogous way to that described above for squid and fly. Crabs and lobsters possess visual pigments absorbing maximally around 500 nm in both their rhodopsin and metarhodopsin state but crayfish visual pigments are found at longer wavelengths (Table III).

A yellow absorbing rhodopsin was discovered by Langer et al. (1979) in retinal slices of the moth Spodoptera, following the procedures of Schlecht et al. (1978) on the moth Deilephila who confirmed the work on extracts by Schwemer and Paulsen (1973).

3. Microspectrofluorometry MSF

From microspectrophotometry to microspectrofluorometry is merely a step. MSF has unique advantages and its potentials for visual pigment research have only recently become recognized. Probably the main reason for this late development is that visual pigments are weakly fluorescent so that high excitation intensities are necessary for obtaining detectable signals. Under this condition vertebrate visual pigments are rapidly bleached and reasonably detailed analyses cannot be performed within the experiment time available.

Invertebrate visual pigments do not suffer from this drawback and indeed fluorescence measurements can be most useful in the study of pigment properties. The first reports came from Franceschini (1977; housefly) and Stark et al. (1977; fruitfly) who independently discovered a clearly fluorescing deep pseudopupil in the eyes of flies. This fluorescence was only present in flies reared on a vitamin-A rich diet; it was absent or reduced in vitamin-A deprived flies (Fig. 8).

Subsequent work revealed that in the fly, only metarhodopsin fluoresces whereas rhodopsin fluoresces negligibly (Stavenga & Franceschini, 1981). Hence by measuring fluorescence one can monitor changes in metarhodopsin content induced by the excitation beam, as is demonstrated in Fig. 9. Here the conversion of rhodopsin into metarhodopsin (induced by 456 nm light, Fig. 9 a) as well as metarhodopsin into rhodopsin conversion (by 613 nm light, Fig. 9 b) was monitored both by measuring transmission at 567 nm ($\simeq \lambda_{max}$ of M) and by measuring the emission in the far-red (> 665 nm). Clearly the red emission signal reflects the amount of metarhodopsin present.

Since measurement of fluorescence by MSF can be well performed on completely intact living animals, this method offers the unique possibility of studying visual pigments non-invasively in a wide variety of species. Along this line Kruizinga, Kamman and Stavenga (unpublished) performed photochemical experiments with nanosecond pulsed dye-lasers on blowflies and houseflies (white-eyed mutants) and found that (after rhodopsin conversion) metarhodopsin builds up with a time constant of $\sim$ 200 µs (in confirmation of transmission measurements).

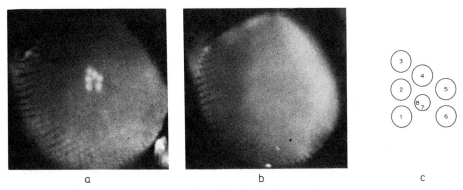

a b c

Fig. 8. UV-induced fluorescence photographed in the deep pseudopupil of the compound eye of completely intact and living white-eyed Drosophila (mutants cinnabar (cn) and brown (bw)), reared on diets high (a) and low (b) in vitamin A. In flies, the rhabdomeres, which are the photopigment containing structures of the photoreceptors, are organised in a characteristic pattern (c). Only the rhabdomeres of cells R1-6 of high vitamin A flies (a) give an observable fluorescence to UV excitation in the deep pseudopupil. The vitamin A-dependent rhabdomere emission is pink. However, when superimposed on the blue autofluorescence of the eye, which is not vitamin A dependent, the fluorescent deep pseudopupil in vitamin A enriched flies has a whitish appearance. No distinct fluorescence is observed from the deep pseudopupil of vitamin A deprived flies (b) (after Stark et al., 1979).

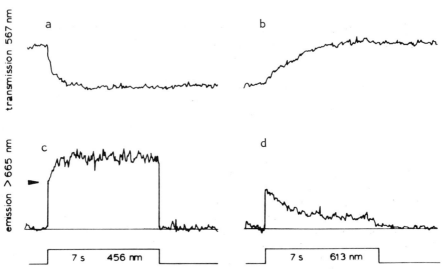

Fig. 9. Transmission and fluorescence measurements performed on the eye of the blowfly Calliphora, mutant chalky. Antidromic transmission was measured at 567 nm, where transmission is high when the rhabdomeres contain little metarhodopsin (Fig. 2 a, start). Creation of metarhodopsin by orthodromic blue (456 nm) light results in a decrease in transmission (a). The orthodromic beam is sufficiently intense to establish a photoequilibrium within a few seconds. The high metarhodopsin fraction is maintained after termination of the 7 s duration orthodromic illumination (a). The metarhodopsin fraction is lowered by red (613 nm) orthodromic light (b). Emission changes (measured above 665 nm) occurring during photoconversion are shown in the lower traces. Increase in emission (c) accompanies metarhodopsin creation and a decrease in emission (d) occurs when metarhodopsin is photoconverted back into rhodopsin. Metarhodopsin emission is seen superimposed upon a background emission due to photostable pigment in the rhabdomeres and surrounding tissue. The base line of the transmission signal coincides with that of the emission. Stray light contributes substantially to the transmission signal (Stavenga et al., 1983).

Apart from metarhodopsin, another fly visual pigment state fluorescing in the red exists (Franceschini et al., 1981; Stavenga & Franceschini, 1981). The excitation and emission spectra of this M' state are presented in Fig. 10 (Stavenga, Franceschini & Kirschfeld, ms). The excitation spectrum of M', peaking at 570 nm, is very similar to the absorbance spectrum of M (Fig. 1). The emission spectrum peaks at

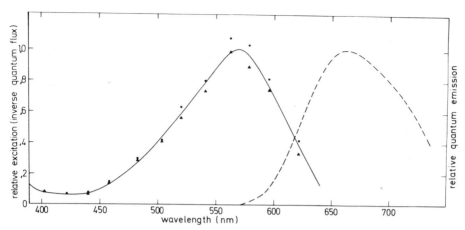

Fig. 10. Excitation (continuous line) and emission (broken line) spectrum for metarhodopsin M' of the housefly Musca (Stavenga et al., 1983).

660 nm. The shape of the emission spectrum of M is essentially identical to that of M', but the emission of M' is $\gtrsim$ 4 times that of M (Stavenga, unpublished).

Within a few years the study of invertebrate visual pigment fluorescence has come of age, as demonstrated especially by the work of Cronin and Goldsmith (1981, 1982 a, b) who performed MSF on isolated rhabdoms of crayfish, partly in combination with absorption measurements by MSP in the same set-up. They determined the excitation and emission spectra of crayfish metarhodopsin as well as the photosensitivity spectra of rhodopsin and metarhodopsin, and the molar extinctions (Tables II, III, IV). Furthermore, the quantum efficiency for converting rhodopsin into metarhodopsin was estimated at $\gamma_R = 0,69$ and for the reverse process $\gamma_M = 0,49$ (Cronin & Goldsmith, 1982 b). A formal method for deriving visual pigment spectra from experimental data is given in the appendix.

4. Optophysiology: Pigment Migration in Visual Cells

Many invertebrate eyes have visual cells with a system of pigment granules which migrate upon a change in illumination (rev. Autrum, 1981). Franceschini (1972) was the first to measure optically the spectral sensitivity of the pigment system in Drosophila and later Bernard and Stavenga (1978, 1979), Bernard (1979) and Bernard and Wehner (1980) studied several insect species. The pigment migration is triggered by rhodopsin conversion, as follows from measurement of the action spectrum (see also Olivo & Chrismer, 1980). Optical measurements on the pigment migration system hence offer an alternative non-invasive means for determining rhodopsin spectra.

5. Electrophysiology

5.1 Late receptor potential

A widely utilized method for inferring the spectral characteristics of visual pigments is to record electrical signals initiated by illumination. Following the absorption of light quanta by rhodopsin molecules, the conductance of the photoreceptor-cell membrane decreases. This conductance decrease leads to a change in transmembrane voltage which can be recorded from many photoreceptors simultaneously (electroretinogram) or from a single photoreceptor by penetrating the cell with a fine glass capillary electrode (single cell recordings). The spectral sensitivity function of a photoreceptor cell $S(\lambda)$, which reflects the absorbance characteristics of the visual pigment $\alpha_R(\lambda)$, is obtained by stimulating the receptor with monochromatic light of different wavelength.

We shall discuss below a number of reasons why the shape of the spectral sensitivity function often deviates from that of the rhodopsin absorption function. First, however, it has to be understood how a spectral sensitivity curve is determined experimentally. An electrophysiological signal is measured, usually a change of the membrane potential, as a response to a monochromatic light stimulus which is delivered to the eye with such a quantum intensity $I_c(\lambda)$ that a signal of a certain amplitude - the criterion response - is evoked. Subsequently this light flux is measured, and its normalized reciprocal value is called the spectral sensitivity.

$$I_c^{-1}(\lambda)/I_c^{-1}(\lambda_{max}) = S(\lambda) \tag{12}$$

It is instructive to consider the most simple case of a photoreceptor with a concentration of rhodopsin molecules C_R which receives a monochromatic light flux $I(\lambda)$. The conversion of rhodopsin molecules into the metarhodopsin state per unit time is then described by (eqs. 2 and 3)

$$\left[\frac{dC_R}{dt}\right]_{R \to M} = K_R(\lambda) I(\lambda) C_R \tag{13}$$

This conversion induces a depolarization of the membrane potential, i.e. the negativity of the cell interior recorded in the dark upon penetration of the receptor (approx. -60 mV) becomes positive. (In a few exceptional cases hyperpolarization occurs, rev. Järvilehto, 1979.) The amplitude of this late receptor potential (LRP) is known to be a function of light intensity and thus a function of the number of converted rhodopsin molecules (Hamdorf, 1979; Hamdorf & Kirschfeld, 1980). The state of adaptation is assumed to be constant during the spectral test procedure. Then

$$LRP = f\left[\left(\frac{dC_R}{dt}\right)_{R \to M}\right] \tag{14}$$

The criterion LRP is thus determined by

$$K_R(\lambda) I_c(\lambda) C_R = K_R(\lambda_{max}) I_c(\lambda_{max}) C_R \qquad (15)$$

and the spectral sensitivity as

$$S(\lambda) = K_R(\lambda)/K_R(\lambda_{max}) \qquad (16)$$

Experimental evidence indicates that the quantum efficiency γ_R generally is independent of wavelength λ. Hence

$$S(\lambda) = \alpha_R(\lambda)/\alpha_R(\lambda_{max}) \qquad (17)$$

and so it may be concluded that the sensitivity spectrum of a visual cell and the rhodopsin absorption spectrum have the same shape.

Fig. 11 presents spectral sensitivity curves obtained by intracellularly recording the membrane potential of so-called peripheral photoreceptor cells of the housefly. The spectral sensitivity in the dark has a peak at $\sim$ 480 nm, close to the rhodopsin peak determined by absorption measurements performed by MSP. An additional high sensitivity in the ultraviolet is at variance with what is known for rhodopsins (see Fig. 2), and this phenomenon demonstrates directly that one must be cautious in interpreting spectral sensitivities purely as rhodopsin absorption.

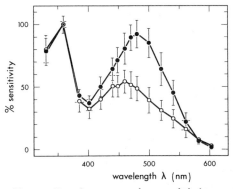

Fig. 11. Normalized spectral sensitivity of peripheral photoreceptor cells of Musca in the dark-adapted (filled circles) and in the light-adapted (open circles) state. The relative decrease in sensitivity in the blue-green and the blue shift is due to the action of the intracellular pupil (Vogt et al., 1982).

5.2 Early receptor potential

At high intensities of illumination the late receptor potential is preceded by the early receptor potential (ERP), a change in membrane potential due to displacement of electrical charges in the visual pigment molecules, also often referred to as fast photovoltage.

The displacement currents in the visual pigment molecules are small but detectable when large numbers of molecules are converted simultaneously and when the molecules are at least partially ordered. The latter can be achieved artificially in model systems, but ordering is already present in the sensory cell membrane Therefore, the ERP is recorded when the visual pigment molecules are massively converted, by a bright flash, within $\sim$ 1 ms. Typically in invertebrate photoreceptors, rhodopsin conversion induces a negative ERP phase and metarhodopsin conversion induces a positive ERP phase, both processes lasting in the order of a few ms. These fast photovoltages can also be measured with extracellular electrodes since a short lived extracellular current is created. Because the potentials are tightly coupled to molecular conversions measurement of their spectral dependence reveals the photosensitivity spectra of the molecular states (rev. Cone & Pak, 1971; Hamdorf, 1979; Järvilehto, 1979).

Minke et al. (1973, 1974) performed a most detailed spectral analysis of the photochemical cycle of the barnacle visual pigment by ERP and Cornwall and Gorman (1979) estimated the spectra of both rhodopsin and metarhodopsin of scallop.

The ERP of flies is enigmatic since only a change in membrane potential correlated with M-conversion exists. This discovery was made by measuring the electroretinogram of fruitflies; together with its lamina-amplified signal, this fly ERP was termed the M-potential (Pak & Lidington, 1974; Minke & Kirschfeld, 1980; Stephenson & Pak, 1980). Why fly rhodopsin is silent in the ERP is a decade-old and yet unsolved question.

The ERP does not require healthy visual cells, it can even be recorded from glutaraldehyde fixed cells (Hagins & McGaughy, 1967). The physiological significance of the ERP in the phototransduction process is unknown.

6. Structural factors affecting the spectral measurements

6.1 Self screening

The most important and often unrealistic assumption implicit to the derivation of eqs. (16) and (17) is that $I_c(\lambda)$, the light flux delivered to the eye, is taken to be the flux actually reaching the visual molecules. In the long cylindrical rhabdom(ere)s of many invertebrates, the flux absorbed by the rhodopsin molecules depends on the x coordinate along the cylinder (Hamdorf et al., 1968). Assuming that only rhodopsin molecules are present:

$$I(x,\lambda) = I(0,\lambda) e^{-\alpha_R(\lambda) C_R x} \tag{18}$$

VISUAL PIGMENTS OF INVERTEBRATES

The number of converted rhodopsin molecules in a slice of thickness dx from a cylinder of cross section A becomes (from eq. (12)):

$$\left[\frac{dC_R}{dt}\right]_{R \to M} A dx = K_R(\lambda) I(x,\lambda) C_R A dx \qquad (19)$$

In a photoreceptor of the length L, the total amount of converted molecules becomes

$$\int_0^L \left[\frac{dC_R}{dt}\right]_{R \to M} A dx = I(0,\lambda) \gamma_R A \left[1 - e^{-\alpha_R(\lambda) C_R L}\right] \qquad (20)$$

The spectral sensitivity accordingly follows (with γ_R = constant):

$$S(\lambda) = \frac{1 - e^{-\alpha_R(\lambda) C_R L}}{1 - e^{-\alpha_R(\lambda_{max}) C_R L}} \qquad (21)$$

Note that this is actually the fraction of absorbed light normalized to the peak value. The implicit assumption here is that rhodopsin conversions throughout the photoreceptors all add equally to the electrical signal (see Hamdorf, 1979).

The consequence of eq. (21) is that with increasing $\alpha_R(\lambda_{max}) C_R L$ the spectral sensitivity becomes more and more flat. For example, let us assume a fly rhabdomere with $\alpha_R(\lambda_{max}) C_R L = 2,3$ (Tsukahara & Horridge, 1977; Stavenga, 1976), i.e. 90% absorption at λ_{max}. Then, at a wavelength where $\alpha_R(\lambda)/\alpha_R(\lambda_{max}) = 0,25$, 44% of the incident light is absorbed, i.e. $S(\lambda) = 0,49$. Hence a doubling of the β-peak (see Fig. 2) with respect to the α-peak results for a visual cell with a long, well-absorbing rhabdomere due to self screening (see also Hamdorf, 1979; Gribakin, 1979; Stowe, 1980).

Obviously, when the concentration of visual pigment molecules is reduced, as is the case for instance with vitamin-A deprivation, eq. (20) eventually becomes equivalent to eq. (6). On the other hand, by careful measurement of the shape of the spectral sensitivity curves together with the absolute sensitivity, it is possible to calculate the total visual pigment density (Stowe, 1980) and also to investigate the extent to which the phototransduction mechanism is independent of the visual pigment concentration in the membrane.

6.2 Filtering by metarhodopsin

Above we indicated the electrophysiological methods for determining the absorption spectrum of rhodopsin. Due to the interconvertible rhodopsin-metarhodopsin system, the spectrum of metarhodopsin can also be estimated by electrophysiology. First, by establishing photoequilibria with monochromatic light of wavelength λ_a the concentration of rhodopsin molecules is reduced to

$$C_R(\lambda_a) = f_R(\lambda_a) C_{tot} \qquad (22)$$

where C_{tot} is the total concentration of visual molecules. By testing the sensitivity with light of a fixed wavelength, where both rhodopsin and metarhodopsin absorb little, the sensitivity as a function of the adapting wavelength is:

$$S(\lambda_a) = f_R(\lambda_a) \qquad (23)$$

The sensitivity is here normalized to the maximal value, i.e. when only rhodopsin exists ($C_R = C_{tot}$). Because at photoequilibrium

$$f_R = \frac{K_M}{K_R + K_M} = U/(1 + \phi U) \qquad (24)$$

and at the isosbestic wavelength

$$f_R(\lambda_{iso}) = \phi/(1 + \phi) \qquad (25)$$

ϕ and subsequently $U = \alpha_M/\alpha_R$ can be calculated from sensitivity measurements. Together with the data of α_R, α_M can then be derived relative to α_R.

Generally, however, the light flux at a test wavelength is modified by the presence of both rhodopsin and metarhodopsin:

$$I(x,\lambda) = I(0,\lambda) e^{-[\alpha_R(\lambda) f_R + \alpha_M(\lambda) f_M] C_{tot} x} \qquad (26)$$

(Again we assume that f_R and C_{tot} are constant throughout the receptor.) Eq. (20) then changes into

$$\int_0^L \left[\frac{dC_R}{dt}\right]_{R \to M} A dx = \int_0^L K_R(\lambda) I(x,\lambda) f_R C_{tot} A dx =$$

$$I(0,\lambda) A \gamma_R \frac{\alpha_R(\lambda) f_R}{\alpha_R(\lambda) f_R + \alpha_M(\lambda) f_M} \{1 - e^{-[\alpha_R(\lambda) f_R + \alpha_M(\lambda) f_M] C_{tot} L}\}$$

$$(27)$$

Clearly sensitivity is then spectrally modulated by metarhodopsin absorption. Nevertheless, one can obviate this effect by measuring sensitivities at the isosbestic wavelength. Then eq. (27) reduces to (20), except for an additional factor f_R, and the sensitivity function $S(\lambda_a) = f_R(\lambda_a)$ (eq. (23)) is obtained again.

The modulation of the spectral sensitivity function at a given $f_R < 1$ was first predicted by Hamdorf and Schwemer (1975), and Tsukahara and Horridge (1977) used this effect for calculating the metarhodopsin spectrum of the dronefly Eristalis. The main advantage of their method over that using $f_R(\lambda_a)$ relies on the fact that the latter method does not provide sufficient accuracy at wavelengths where α_R and α_M strongly differ (see further Tsukahara & Horridge, 1977).

The controllable changes of sensitivity by spectral adaptation, i.e. by changing metarhodopsin content, is not yet generally applied. For practical examples of how to estimate $f_R(\lambda_a)$ from sensitivity measurements we refer to the work on insects by Hamdorf and coworkers: the UV-receptor of Ascalaphus, the blue-green receptor of Calliphora, the green receptor of Deilephila (rev. Hamdorf et al., 1973; Hamdorf, 1979), as well as to the investigations of Bader et al. (1982) on the blue receptor of the honeybee drone. Hertel (1980) could similarly demonstrate that in the branchiopod Artemia salina a blue-green rhodopsin converts into a yellow metarhodopsin.

Less straight-forward but nevertheless convincing evidence for the metarhodopsin spectrum can be obtained through the prolonged depolarizing (or hyperpolarizing) afterpotential which emerges after excessive conversion of rhodopsin and which can be undone by reconversion of metarhodopsin (e.g. Tsukahara et al., 1977; Cornwall & Gorman, 1979; rev. Hamdorf, 1979).

6.3 Filtering by visual pigments in adjacent visual sense cells

The rhabdomeres of compound eyes are usually fused into one optically continuous rhabdom. As before, the proximal photoreceptor layers are spectrally filtered by the more distal layers.

In a fused rhabdom, where visual sense cells with different rhodopsins are stacked on top of each other as in the moth Deilephila (Schlecht et al., 1978), selective depression of sensitivity of the green receptors in the shorter wavelength range can be produced by the distally placed UV and/or blue receptors. Such a layering improves colour discrimination. A different strategy is followed in the long rhabdoms of insects as bees where lateral filtering counteracts the flattening effect of self-screening (Snyder et al., 1973).

6.4 Sensitizing pigments

In eqs. (13) and (14) we have implicitly assumed that the rhodopsin molecules are only converted by light. When the visual cell harbours an additional light absorbing pigment which can transfer absorbed light energy onto the visual pigment, rhodopsin conversion can be enhanced. In fact Kirschfeld et al. (1977; see also Kirschfeld, 1979, and Minke & Kirschfeld, 1979) explain the high UV sensitivity of fly photoreceptors

(Fig. 11) from the presence of an ultraviolet absorbing sensitizing pigment. On the basis of the fine structure in the UV (Gemperlein et al., 1980), Franceschini (1983 a) proposes that the sensitizing effect is due to a retinol-protein complex. In this case $K_A(\lambda)I(\lambda)C_R$ must be added to eq. (13); K_A represents the sensitivity for energy transfer. Since eqs. (15) to (17) do not include sensitizing effects they no longer hold.

6.5 Waveguide modes

An alternative explanation for the high UV-peak was previously put forward by Snyder and Miller (1972) and Snyder and Pask (1973). Their essential point is the recognition that fly rhabdomeres are slender with a cross section of 1-2 μm and have a refractive index larger than that of the surrounding medium. Consequently light travels through the rhabdomeres in guided wave patterns called modes with part of the light propagating outside the boundary of the rhabdomeres. Since the visual molecules are confined within the rhabdomeres, the fraction of light outside is inaccessible to the visual molecules. The fraction outside increases with increasing wavelength and thus light sensitivity in the long wavelength range is depressed with respect to the shorter wavelengths.

The original formulation of the waveguide hypothesis has appeared to be unable to provide satisfactory quantitative predictions (cf. Stavenga & Van Barneveld, 1975). All the same, Pask and Barrell (1980 a, b) showed that waveguide effects in combination with diffraction on the corneal lenses of flies can distinctly modify spectral characteristics and hence such modifications must be considered for invertebrate photoreceptors in general.

6.6 Screening pigments

In one class of central rhabdomeres of flies a strongly blue absorbing carotene exists which makes these rhabdomeres appear yellow (Kirschfeld & Franceschini, 1977). The spectral sensitivity of the proximal receptor cells is thus strongly bathochromically shifted (Hardie et al., 1979).

A much more common filtering occurs by pupil mechanisms. An example is provided by the lower curve of Fig. 11 which represents the spectral sensitivity of housefly peripheral photoreceptor cells in the light-adapted state. In this situation, small granules which contain a highly absorbing photostable pigment, are assembled near the rhabdomere where light is absorbed from the boundary wave. Since the pupil, i.e. the assembly of pigment granules effectively has a blue-green peaking-absorbance (Vogt et al., 1982), the spectral sensitivity in the blue is shifted hypsochromically over several tens of nm (see Hardie, 1979) and is distinctly reduced with respect to the UV-sensitivity (Fig. 11). This demonstrates that screening pigments can substantially alter the spectral sensitivity of the visual cells with respect to the rhodopsin absorption spectrum (see also Stowe, 1980).

A bathochromic shift of 30-35 mm in the spectral sensitivity of crayfish compared to the rhodopsin absorption is caused by the screening pigments of the retina (Goldsmith, 1978 a, b). By modifying the absorption of a single visual pigment in different photoreceptors by different coloured filters colour discrimination would become possible and this is in fact what Leggett (1979) concluded to be the case for crabs, and Kong et al. (1980) for a grasshopper.

6.7 Corneal filters

In contrast with human and many other vertebrates, corneal lenses of invertebrates are usually more transparent in the visible and in the ultraviolet (rev. Miller, 1979). Certain squids have yellow lenses, thus absorbing short wavelength light (rev. Messenger, 1981). Coloured lenses, due to interference can be seen notably in tabanid and dolichopodid flies (rev. Miller, 1979). Whether substantial changes in spectral sensitivity are evoked by the corneal interference filters is not yet established.

6.8 Tapeta

Proximal to the rhabdom in most butterfly species, tracheoles form interference reflection filters (Miller & Bernard, 1968), presumably for improving contrast sensitivity and/or colour vision.

Tapeta consisting of reflecting layers of guanine or of scattering coloured pigments are wide-spread in the animal kingdom, but the spectral consequences are little studied (revs. Miller, 1979; Land, 1981). Also, the reflecting cone envelopes in the superposition eyes of crayfish (Vogt, 1980) are spectrally selective and therefore will influence spectral sensitivity.

REGENERATION OF VISUAL PIGMENT

1. Photoregeneration

Since visual sensitivity is mediated by the rhodopsin molecules it is obviously of vital importance for animals to have control mechanisms for keeping up the rhodopsin concentration. Indeed, great versatility exists in this respect.

A rather elegant and economical method is converting metarhodopsin into rhodopsin by photoregeneration. As metarhodopsin photosensitivity is larger than that of rhodopsin (cf. Table II) a broad band illumination will generally shift a photosteady state towards rhodopsin. Clearly this works well for UV-photoreceptors which look into the blue skies (Hamdorf et al., 1971, 1972). Environmental stray light will specifically convert metarhodopsin when this is bathochromically shifted with respect to its rhodopsin if the stray light is selectively filtered by short wavelength absorbing pigment (Langer, 1975).

Substantial photoconversion occurs at high intensities and then a spectral filter covering rhodopsin will especially be helpful. Not surprisingly then, the blue-green absorbing pupil of flies effectively reduces rhodopsin conversion with respect to metarhodopsin conversion (Stavenga et al., 1973; Stavenga, 1980).

2. Turnover of Photoreceptor Membrane and Visual Pigment

A destruction of the visual membrane and subsequent resynthesis may seem to be an inexpedient way to maintain the functioning of the visual cell, but in fact membrane turnover appears to be a very basic and generally occurring process. Breakdown and resynthesis of the visual membrane of the rhabdomeric microvilli often occurs under control of a light and/or circadian rhythm (revs. Blest, 1980; White et al., 1980; Autrum 1981; Waterman, 1982).

So far, evidence has only been obtained by electron microscopical methods, showing an increase of microvillar volume during the night and decrease during the day (Blest & Day, 1978). Furthermore, light/dark-dependent changes in number and distribution of certain cell organelles have been observed (e.g. White & Sundeen, 1967; White, 1968; Eguchi & Waterman, 1976).

Dramatic changes in rhabdomere volume have been reported for the spider Dinopis (e.g. Blest, 1978; Blest & Day, 1978) and crabs (Nässel & Waterman 1973; Stowe, 1980; Leggett & Stavenga, 1981). In the last case the rhabdom volume was found to be 20 times larger during the night than in the daytime. Similar changes have been reported for Limulus (Barlow et al., 1980) and locust and mantis (Horridge et al., 1981). The main visual function of the change in rhabdom size seems to be to modulate angular sensitivity of the visual cell and thus overall light sensitivity.

From the ultrastructural data it seems obvious that the renewal of photoreceptor membrane also includes the renewal of visual pigment molecules. A demonstration of rhodopsin biosynthesis in an invertebrate is provided by Fig. 12. It should be recalled that flies reared on a vitamin-A deprived diet, R^--flies (Figs. 2, 3), have only a few percent rhodopsin (Schwemer, 1979; see also Razmjoo & Hamdorf, 1976). Injection of 11-cis retinal under dim-red light into the eyes of such deprived animals revealed (in sustained darkness) that rhodopsin concentration increases exponentially with a half time of ~10 hours. On the other hand, after all-trans injection no rhodopsin synthesis is triggered (Fig. 12).

Biosynthesis of rhodopsin was shown to occur not only in R^--flies but in R^+-flies as well. Under normal conditions this biosynthesis keeps pace with a continuous breakdown of visual pigment. The availability of 11-cis retinal is essential for maintaining the visual pigment concentration in the rhabdomeric membrane, and therefore its isomerization from all-trans retinal, which results as a breakdown product, is crucial.

In contrast with the situation in vertebrates, all-trans isomerization into 11-cis is mediated by light in invertebrates. Due to this property the breakdown process can be traced by excluding light of the short wavelength range from the animals.

Since the action spectrum for the isomerization process peaks at about 450 nm it is assumed that all-trans retinal is bound through a Schiff's base to a protein so shifting the absorption maximum of retinal from 380 nm into the blue. This binding is therefore an essential step for isomerization (Schwemer, 1979). Proteins which are favourable candidates for a light-activated isomerizing action on all-trans retinal have been recently isolated from honeybee retinae (Pepe & Cugnoli, 1980; Pepe et al., 1982).

These proteins are probably closely related to the retinochromes, a class of photosensitive pigments found in the retinae of cephalopods and first described in 1965 for squid (rev. Hara & Hara, 1972).

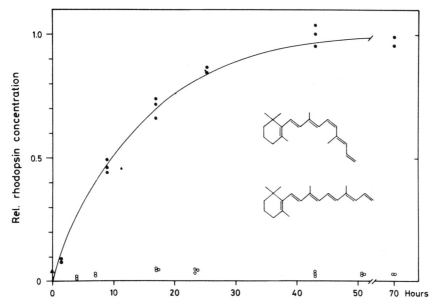

Fig. 12. Rhodopsin concentration in the receptors of vitamin-A deprived flies after injection of 11-cis retinal (closed circles) and all-trans retinal (open circles). The initial rhodopsin concentration is indicated by the triangle at t = 0 (Schwemer, 1979).

All-trans retinal is bound to a protein, which has a molecular weight of $\sim$ 20,000 dalton, through a Schiff's base linkage which is readily attacked by NH_2OH and sodium borohydride in the dark leading to the formation of retinal oxime and probably a retinyl protein, respectively (Hara & Hara, 1972). The molar extinction coefficient of this pigment ε_{max} is approximately 60,800 $M^{-1}cm^{-1}$, ile. $\sim$ 1,5 times that of rhodopsin (Hara & Hara, 1982).

The absorption spectra of cephalopod retinochromes are all very similar, their absorption maxima are slightly bathochromically displaced when compared with those of the rhodopsins (Hara & Hara, 1972). The photochemical cycle of squid retinochrome is shown in Fig. 13. Retinochrome catalyzes the isomerization of all-trans retinal into the 11-cis form, but also other isomers like 13-cis and 9-cis are converted into 11-cis retinal when illuminated in the presence of retinochrome (Hara & Hara, 1982).

ECOLOGY OF VISUAL PIGMENTS

1. Distribution of Visual Pigments Over Invertebrate Orders

As discussed above, the spectral sensitivity of a visual cell is usually very similar to its rhodopsin absorption spectrum. What are the ecological

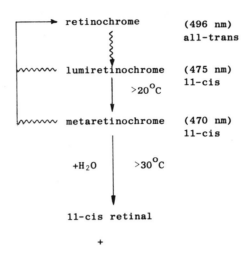

Fig. 13. Photochemical cycle of squid retinochrome (after Hara & Hara, 1982).

forces shaping a photoreceptor's rhodopsin? A definitive answer to this question is not at hand, but sensible speculations can be put forward. In Table III and Fig. 14 we have listed the spectral distribution of visual pigments with known R and M. The spectral sensitivity of several other species has been determined (e.g. Menzel, 1979). From these data we see that cephalopods are remarkably unanimous in tuning the peak of their rhodopsin between 470 and 500 nm, a wavelength range where deep sea and surface waters are brightest. Fishes appear to have evolved visual pigments which capture the greatest possible number of light quanta in any particular light climate under water (rev. Lythgoe, 1971, 1979). Cephalopods appear to have done the same (Messenger, 1981).

Crustaceans have rhodopsins with λ_{max} values scattered between 495 and 540. From table III it is apparent that the marine crabs, lobsters and shrimps absorb at slightly shorter wavelengths than barnacle, fresh water prawn and crayfish. This also may be related to the spectral composition of the environmental light. However, as at least crayfish can rely on more than one visual pigment (Goldsmith, 1972), the benefit of colour discrimination must be considered, but too little is known to provide a reliable explanation.

A quite gratifying insight into the development of spectral sensitivity in relation with the light to be detected is provided by the work by Lall et al. (e.g. 1980, 1982) on fireflies. These insects signal each other by bioluminescence. In all species studied the spectral characteristics of bioluminescence and the electrical signals (ERG) appear to be closely tuned to each other, although the spectra distinctly vary from the green (dark-active species) to the yellow (dusk-active species) wavelength range. The yellow emission (and spectral sensitivity) probably is a means to

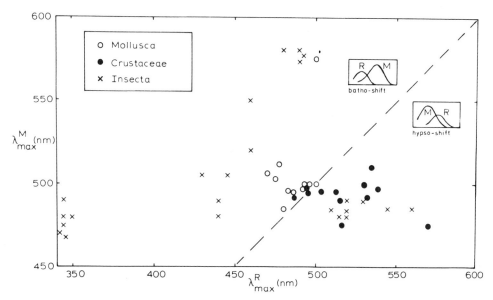

Fig. 14. The peak wavelength λ_{max}^M of metarhodopsin versus λ_{max}^R of rhodopsin for invertebrate visual pigments.

increase the signal-to-noise ratio (contrast) with the low levels of light.

Insects have rhodopsins with maxima spanning the range from ~340 - 590 nm. The spectral composition of the ambient light is obviously not as limited as in the case of deep sea cephalopods. Extensive evidence from behavioural studies shows that insects can rely on colour vision (von Frisch, 1967). Bees for example have trichromatic vision with receptors peaking at 340, 440 and 540 nm (e.g. von Helversen, 1972). Butterflies have extended their vision into the red (e.g. Bernard, 1979).

2. Distribution of Rhodopsins Over the Eye

Quite likely insects also have their receptors tuned to their ecological niches, as may be inferred from the eye regionalization which is noticeable in many insect species. Predatory insects especially frequently have abundant UV and blue sensitive cells in the dorsal part of their eyes. The most exquisite example is the owlfly Ascalaphus (Hamdorf et al., 1971). Presumably, detection of a small spot against the sky is achieved optimally with short wavelength receptors. This understanding of predators (e.g. dragonflies, Laughlin & McGinness, 1978) also holds for males of many insects looking for females (mayfly Atelophlebia, Horridge & McLean, 1978; honeybee drones, Bertrand et al., 1979).

3. Variation of Metarhodopsin Spectra

Table III and Fig. 14 show that the peak wavelengths of virtually all invertebrate metarhodopsins are distributed in a very narrow range of ~ 470-520 nm. An ecological reason for this clustering is obvious for marine animals. For insects the reason could be the abundance of blue light coming from the skies. As noted above, the high absorption of metarhodopsin favours photoregeneration of rhodopsin (Hamdorf et al., 1972). The metarhodopsins of flies and that of the scallop form an exception but, interestingly, flies have screening pigments which act as long wavelength band-pass filters which facilitate photoregeneration.

From the above it will be clear that several factors together determine the optimal functioning of a visual pigment system.

ACKNOWLEDGEMENTS

This study was supported by the Netherlands Organization for the Advancement of Pure Research (Z.W.O.) to D.G.S., and by the Deutsche Forschungsgemeinschaft (SFB 114 and a Heisenberg Fellowship) to J.S. Drs. R.C. Hardie and G.D. Bernard read the manuscript.

APPENDIX: DERIVATION OF VISUAL PIGMENT SPECTRA

We shall exemplify, on the data of Cronin and Goldsmith (1981, 1982 a, b) how optical measurements on invertebrate visual pigments can be evaluated. First, Fig. 15 shows the normalized absorbance spectrum of freshly prepared crayfish rhabdoms with only rhodopsin, together with a spectrum of the photosteady-state mixture produced by illumination at the isosbestic wavelength λ_{iso} = 548 nm. The spectra can be described

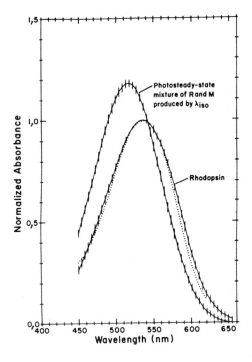

Fig. 15. Normalized absorbance spectra of crayfish rhodopsin (approximated by a nomogram) and of mixture of crayfish rhodopsin and metarhodopsin at a photosteady state produced by irradiation at the isosbestic wavelength (Cronin & Goldsmith, 1982 b).

by

$$A_R(\lambda) = c_1 \alpha_R(\lambda) \quad \text{(A.1a)}$$

and

$$A_{mix}(\lambda) = c_2 \left[f_{R\infty}(\lambda_{iso}) \alpha_R(\lambda) + f_{M\infty}(\lambda_{iso}) \alpha_M(\lambda) \right] \quad \text{(A.1b)}$$

where c_1 and c_2 are proportionality constants. Normalizing the functions at the isosbestic wavelength yields:

$$A_R^*(\lambda) = \alpha_R(\lambda)/\alpha_R(\lambda_{iso}) \quad \text{(A.2a)}$$

and

$$A_{mix}^*(\lambda) = A_R^*(\lambda) \cdot \frac{U(\lambda) + \phi(\lambda_{iso})}{1 + \phi(\lambda_{iso})} \quad \text{(A.2b)}$$

Second, Fig. 16 gives the photoequilibrium spectrum of crayfish metarhodopsin as obtained by fluorescence measurements by Cronin and Goldsmith (1982 a). This spectrum is described by the Q-function (eq. 9):

$$Q(\lambda) = \frac{1 + \phi(\lambda_{iso})}{1 + \phi(\lambda)U(\lambda)} \quad \text{(A.3)}$$

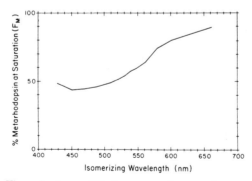

Fig. 16. The photoequilibrium spectrum of crayfish visual pigment. The spectrum was measured in relative units by Cronin & Goldsmith (1982 a) and scaled with $\phi = 0{,}71$ (see Cronin & Goldsmith, 1982

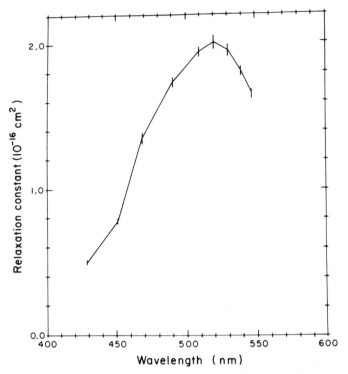

Fig. 17. Relaxation spectrum of crayfish visual pigment as measured by fluorescence (Cronin & Goldsmith, 1982 b).

Third, Fig. 17 presents the relaxation spectrum of crayfish visual pigment determined from photoconversion experiments. The spectrum is described by

$$K_R(\lambda) + K_M(\lambda) = K_R(\lambda)[1 + \phi(\lambda)U(\lambda)] =$$
$$= K_R(\lambda)[1 + \phi(\lambda_{iso})]/Q(\lambda) \tag{A.4}$$

Hence, the photosensitivity spectrum of rhodopsin directly follows from the relaxation spectrum and the photoequilibrium spectrum when the relative quantum efficiency is known:

$$K_R(\lambda) = Q(\lambda)[K_R(\lambda) + K_M(\lambda)]/[1 + \phi(\lambda_{iso})] \tag{A.5}$$

Normalization of the relaxation spectrum at the isosbestic wavelength yields

$$X(\lambda) = \frac{\alpha_R(\lambda)}{\alpha_R(\lambda_{iso})} \cdot \frac{\gamma_R(\lambda)}{\gamma_R(\lambda_{iso})} \cdot \frac{1 + \phi(\lambda)U(\lambda)}{1 + \phi(\lambda_{iso})} \quad (A.6)$$

Combining eq. (A.5) with eqs. (A.2 a), (A.3) and (A.4) yields

$$\gamma_R^*(\lambda) = \frac{\gamma_R(\lambda)}{\gamma_R(\lambda_{iso})} = \frac{X(\lambda)Q(\lambda)}{A_R^*(\lambda)} \quad (A.7)$$

The generally held notion that (absolute and relative) quantum efficiencies are independent of wavelength hence can be directly checked from the spectral data.

Assuming that the relative quantum efficiency is independent of wavelength or $\phi(\lambda) = \phi(\lambda_{iso})$, its value can be derived by combining eqs. (A.2) and (A.3) and solving for $U(\lambda)$. Subsequently then, $U(\lambda)$ follows from eq. (A.3) and hence also $A_M(\lambda) = \alpha_M(\lambda)/\alpha_M(\lambda_{iso})$ follows. The rhodopsin and metarhodopsin spectra calculated by Cronin and Goldsmith (1982 b) with a slightly different procedure as presented here are given in Fig. 18.

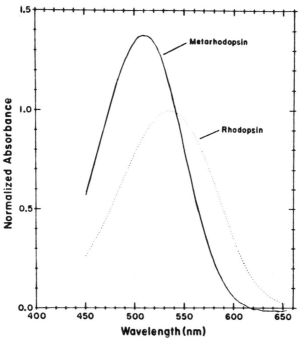

Fig. 18. Rhodopsin and metarhodopsin spectra of crayfish (after Cronin & Goldsmith, 1982 b).

REFERENCES

Ali, M.A. (1975) Temperature and vision. Rev. Can. Biol. 34: 131-186.
Autrum, H. (1981) Light and dark adaptation in invertebrates. In: Handbook of Sensory Physiology, Vol. VII/6C. Ed. H. Autrum. Berlin, Heidelberg, New York, Springer, p. 1-91.
Bader, C.R., Baumann, F., Bertrand, D., Carreras, J. & Fuortes, G. (1982). Diffuse and local effects of light adaptation in photoreceptors of the honey bee drone. Vision Res. 22: 311-317.
Barlow, R.B., Chamberlain, S.C. & Levinson, J.Z. (1980) Limulus brain modulates the structure and function of the lateral eyes. Science 210: 1037-1039.
Bernard, G.D. (1977) Discovery of red-receptors in butterfly retinas. Invest. Ophthalmol. Vis. Sci. 16: Suppl. 61.
Bernard, G.D. (1979) Red-absorbing visual pigment of butterflies. Science 203: 1125-1127.
Bernard, G.D. (1982) Noninvasive optical techniques for probing insect photoreceptors. Meth. Enzymol. 81 (Part H): 752-763.
Bernard, G.D. (1983) Bleaching of photoreceptors in eyes of intact butterflies. Science 219: 69-71.
Bernard, G.D. & Stavenga, D.G. (1978) Spectral sensitivities of retinular cells measured in intact, living bumblebees by an optical method. Naturwissenschaften 65: 442-443.
Bernard, G.D. & Stavenga, D.G. (1979) Spectral sensitivities of retinular cells measured in intact, living flies by an optical method. J. Comp. Physiol. 134: 95-107.
Bernard, G.D. & Wehner, R. (1980) Intracellular optical physiology of the bee's eye. I. Spectral sensitivity. J. Comp. Physiol. 137: 193-203.
Bertrand, D., Fuortes, G. & Muri, R. (1979) Pigment transformations and electrical responses in retinula cells of drone, Apis mellifera ♂. J. Physiol. 296: 431-441.
Blest, A.D. (1978) The rapid synthesis and destruction of photoreceptor membrane by a dinopid spider: a daily cycle. Proc. R. Soc. Lond. 200B: 463-483.
Blest, A.D. (1980) Photoreceptor membrane turnover in arthropods: comparative studies of breakdown processes and their implication. In: The Effects of Constant Light on Visual Processes. Ed. T.P. Williams and B.N. Baker. New York, Plenum Press, p. 217-245.
Blest, A.D. & Day, W.A. (1978) The rhabdomere organization of some nocturnal spiders in light and darkness. Phil. Trans. R. Soc. Lond. 283B: 1-23.
Boucher, F. & Leblanc, R.M. (1981) Photoacoustic spectroscopy of cattle visual pigment at low temperature. Biochem. Biophys. Res. Comm. 100: 385-390.
Briggs, M.H. (1961) Visual pigment of grapsoid crabs. Nature (Lond.) 190: 784-786.
Brown, P.K. & Brown, P.S. (1958) Visual pigments of the octopus and cuttlefish. Nature (Lond.) 182: 1288-1290.
Brown, P.K. & White, R.H. (1972) Rhodopsin of the larval mosquito. J. Gen. Physiol. 59: 401-414.

Bruno, M.S., Barnes, S.N. & Goldsmith, T.H. (1977) The visual pigment and visual cycle of the lobster Homarus. J. Comp. Physiol. 120: 123-142.

Bruno, M.S. & Goldsmith, T.H. (1974) Rhodopsin of the blue crab Callinectes: evidence for absorption differences in vitro and in vivo. Vision Res. 14: 653-658.

Callender, R.H. & Honig, B. (1977) Resonance Raman studies of visual pigments. Ann. Rev. Biophys. Bioeng. 6: 33-55.

Cone, R.A. & Pak, W.L. (1971) The early receptor potential. In: Handbook of Sensory Physiology, Vol. 1. Ed. W.R. Loewenstein. Berlin, Heidelberg, New York, Springer, p. 345-365.

Cornwall, M.C. & Gorman, A.L.F. (1979) Thermally stable photo-interconvertible pigment states in scallop photoreceptor. Invest. Ophthalmol. (Suppl.) 16: 177.

Cronin, T.W. & Goldsmith, T.H. (1981) Fluorescence of crayfish metarhodopsin studied in single rhabdoms. Biophys. J. 35: 653-664.

Cronin, T.W. & Goldsmith, T.H. (1982 a) Photosensitivity spectrum of crayfish rhodopsin measured using fluorescence of metarhodopsin. J. Gen. Physiol. 79: 313-332.

Cronin, T.W. & Goldsmith, T.H. (1982 b) Quantum efficiency and photosensitivity of the rhodopsin $\rightleftarrows$ metarhodopsin conversion in crayfish photoreceptors. Photochem. Photobiol. 36: 447-554.

Dartnall, H.J.A. (1953) The interpretation of spectral sensitivity curves. Br. Med. Bull. 9: 24-30.

Denys, C.J. & Brown, P.K. (1982) Euphausiid visual pigments: the rhodopsins of Euphausia superba and Meganyctiphanes norvegica (Crustacea, Euphausiacea). J. Gen. Physiol. 80: 451-472.

Doukas, A.G., Stefancic, V., Suzuki, T., Callender, R.H. & Alfano, R.R. (1980) Squid bathorhodopsin forms within 10 picoseconds. Photobiochem. Photobiophys. 1: 305-308.

Ebina, Y., Nagasawa, N. & Tsukahara, Y. (1975) An intermediate in the photolytic process of extracted squid rhodopsin. Jap. J. Physiol. 25: 217-226.

Ebrey, T.G. & Honig, B. (1977) New wavelength-dependent visual pigment nomograms. Vision Res. 17: 147-151.

Eguchi, E. & Waterman, T.H. (1976) Freeze-etch and histochemical evidence for cycling in crayfish photoreceptor membranes. Cell. Tissue Res. 169: 419-434.

Fernandez, H.R. (1965) A Survey of the Visual Pigments of Decapod Crustacea of South Florida. Ph.D. Thesis, University of Miami, Coral Gables, Florida.

Franceschini, N. (1972) Sur le Traitement Optique de l'Information Visuelle dans l'Oeil à Facettes de la Drosophile. Thesis, Grenoble.

Franceschini, N. (1975) Sampling of the visual environment by the compound eye of the fly: Fundamentals and applications. In: Photoreceptor Optics. Ed. A.W. Snyder & R. Menzel. Berlin, Heidelberg, New York, Springer, p. 98-125.

Franceschini, N. (1977) In vivo fluorescence of the rhabdomeres in an insect eye. Proc. Int. Union Physiol. Sc. XIII, 237. XXVIIth Int. Congr. Paris.

Franceschini, N. (1983 a) In vivo microspectrofluorimetry of visual pigments. Symp. Soc. Exp. Biol. (In press).
Franceschini, N. (1983 b) The retinal mosaic of the fly compound eye. (This volume).
Franceschini, N., Kirschfeld, K. & Minke, B. (1981) Fluorescence of photoreceptor cells observed in vivo. Science 213: 1264-1267.
Frisch, K. von (1967) The Dance Language and Orientation of Bees. Cambridge, Belknap/Harvard University Press.
Gemperlein, R., Paul, R., Lindauer, E. & Steiner, A. (1980) UV fine structure of the spectral sensitivity of flies visual cells, revealed by FIS (Fourier Interferometric Stimulation). Naturwissenschaften 67: 565-566.
Gogala, M., Hamdorf, K. & Schwemer, J. (1970) UV-Sehfarbstoff bei Insekten. Z. vergl. Physiol. 70: 410-413.
Goldman, L.J., Barnes, S.N. & Goldsmith, T.H. (1975) Microspectrophotometry of rhodopsin and metarhodopsin in the moth Galleria. J. Gen. Physiol. 66: 383-404.
Goldsmith, T.H. (1972) The natural history of invertebrate visual pigments. In: Handbook of Sensory Physiology, Vol. VII/1. Ed. H.J.A. Dartnall. Berlin, Heidelberg, New York, Springer, p. 685-719.
Goldsmith, T.H. (1975) Photoreceptor processes: some problems and perspectives. J. Exp. Zool. 194: 89-102.
Goldsmith, T.H. (1978 a) The spectral absorption of crayfish rhabdoms: Pigment, photoproduct and pH sensitivity. Vision Res. 18: 463-473.
Goldsmith, T.H. (1978 b) The effects of screening pigments on the spectral sensitivity of some crustacea with scotopic (superposition) eyes. Vision Res. 18: 475-482.
Goldsmith, T.H. & Bruno, M.S. (1973) Behaviour of rhodopsin and metarhodopsin in isolated rhabdoms of crabs and lobster. In: Biochemistry and Physiology of Visual Pigments. Ed. H. Langer. Berlin, Heidelberg, New York, Springer, p. 147-153.
Gribakin, F.G. (1979) Cellular mechanisms of insect photoreception. Int. Rev. Cytol. 57: 127-184.
Hagins, F.M. (1973) Purification and partial characterization of the protein component of squid rhodopsin. J. Biol. Chem. 248: 3298-3304.
Hagins, W.A. & McGaughy, R.E. (1967) Molecular and thermal origins of fast photoelectric effects in squid retina. Science 157: 813-816.
Hamacher, K.J. & Kohl, K.D. (1981) Spectroscopical studies of the Astacus visual pigment. Biophys. Struct. Mech. 7: 338.
Hamann, B. & Langer, H. (1980) Sehfarbstoffe im Auge des Wasserlaufer Gerris lacustris. Verh. Dtsch. Zool. G., p. 337.
Hamdorf, K. (1979) The physiology of invertebrate visual pigments. In: Handbook of Sensory physiology, Vol. VII/6A. Ed. H. Autrum. Berlin, Heidelberg, New York, Springer, p. 145-224.
Hamdorf, K., Gogala, M. & Schwemer, J. (1971) Beschleunigung der "Dunkeladaptation" eines UV-Rezeptors durch sichtbare Strahlung. Z. vergl. Physiol. 75: 189-199.
Hamdorf, K. & Kirschfeld, K. (1980) Reversible events in the transduction process of photoreceptors. Nature (Lond.) 238: 859-860.

Hamdorf, K. & Langer, H. (1965) Veränderungen der Lichtabsorption im Fazettenaugen bei Belichtung. Z. vergl. Physiol. 51: 172-184.
Hamdorf, K., Paulsen, R. & Schwemer, J. (1973) Photoregeneration and sensitivity control of photoreceptors of invertebrates. In: Biochemistry and Physiology of Visual Pigments. Ed. H. Langer. Berlin, Heidelberg, New York, Springer, p. 155-166.
Hamdorf, K., Paulsen, R., Schwemer, J. & Täuber, U. (1972) Photoreconversion of invertebrate visual pigments. In: Information Processing in the Visual Systems of Arthropods. Ed. R. Wehner. Berlin, Heidelberg, New York, Springer, p. 97-108.
Hamdorf, K. & Schwemer, J. (1975) Photoregeneration and the adaptation process in insect photoreceptors. In: Photoreceptor Optics. Ed. A.W. Snyder & R. Menzel. Berlin, Heidelberg, New York, Springer, p. 263-289.
Hamdorf, K., Schwemer, J. & Täuber, U. (1968) Der Sehfarbstoff, die Absorption der Rezeptoren und die spektrale Empfindlichkeit der Retina von Eledone moschata. Z. vergl. Physiol. 60: 375-415.
Hara, T. & Hara, R. (1966) Photosensitive pigments found in cephalopod retina. Zool. Mag. Tokyo 75: 264-269.
Hara, T. & Hara, R. (1972) Cephalopod retinochrome. In: Handbook of Sensory Physiology, Vol. VII/1. Ed. H.J.A. Dartnall. Berlin, Heidelberg, New York, Springer, p. 720-746.
Hara, T. & Hara, R. (1973) Biochemical properties of retinochrome. In: Biochemistry and Physiology of Visual Pigments. Ed. H. Langer. Berlin, Heidelberg, New York, Springer, p. 181-191.
Hara, T. & Hara, R. (1982) Cephalopod retinochrome. Meth. Enzymol. 81 (Part H): 827-833.
Hardie, R.C. (1979) Electrophysiological analysis of fly retina. I. Comparative properties of R1-6 and R7 and R8. J. Comp. Physiol. 129: 19-33.
Hardie, R.C., Franceschini, N. & McIntyre, P.D. (1979) Electrophysiological analysis of fly retina. II. Spectral mechanisms in R7 and R8. J. Comp. Physiol. 133: 23-39.
Hays, D. & Goldsmith, T.H. (1969) Microspectrophotometry of the visual pigment of the spider crab Libinia emarginata. Z. vergl. Physiol. 65: 218-232.
Helversen, O. von (1972) Zur spektralen Unterschiedsempfindlichkeit der Honigbiene. J. Comp. Physiol. 80: 439-472.
Hertel, H. (1980) The compound eye of Artemia salina (Crustacea). II. Analysis by electrophysiological methods. Zool. Jb. Physiol. 84: 15-25.
Hillman, P., Dodge, F.A., Hochstein, S., Knight, B.W. & Minke, B. (1973) Rapid dark recovery of the invertebrate early receptor potential. J. Gen. Physiol. 62: 77-86.
Hillman, P., Hochstein, S. & Minke, B. (1972) A visual pigment with two physiologically active stable states. Science 175: 1486-1488.
Hochstein, S., Minke, B., Hillman, P. & Knight, B.W. (1978) The kinetics of visual pigments. I. Mathematical analysis. Biol. Cybern. 30: 23-32.

Horridge, G.A., Duniec, J. & Marçelja, L. (1981) A 24-hour cycle in single locust and mantis photoreceptors. J. Exp. Biol. 91: 307-322.

Horridge, G.A. & McLean, M. (1978) The dorsal eye of the mayfly Atelophlebia (Ephemeroptera). Proc. R. Soc. Lond. 200B: 137-150.

Hubbard, R., Bownds, D. & Yoshizawa, T. (1965) The chemistry of visual photoreception. Cold Spring Harbour Symp. Quant. Biol. 30: 301-315.

Hubbard, R. & St. George, R.C.C. (1958) The rhodopsin system of the squid. J. Gen. Physiol. 41: 501-528.

Järvilehto, M. (1979) Receptor potentials in invertebrate visual cells. In: Handbook of Sensory Physiology, Vol. VII/6A. Ed. H. Autrum. Berlin, Heidelberg, New York, Springer, p. 315-356.

Kirschfeld, K. (1979) The function of photostable pigments in fly photoreceptors. Biophys. Struct. Mech. 5: 117-128.

Kirschfeld, K., Feiler, R. & Minke, B. (1978) The kinetics of formation of metarhodopsin in intact photoreceptors of the fly. Z. Naturforsch. 33c: 1009-1010.

Kirschfeld, K. & Franceschini, N. (1977) Photostable pigments within the membrane of photoreceptors and their possible role. Biophys. Struct. Mech. 3: 191-194.

Kirschfeld, K., Franceschini, N. & Minke, B. (1977) Evidence for a sensitizing pigment in fly photoreceptors. Nature (Lond.) 269: 386-390.

Kito, Y., Naito, T. & Nashima, K. (1982) Purification of squid and octopus rhodopsin. Meth. Enzymol. 81 (Part H): 167-171.

Kong, K.-L., Fung, Y.M. & Wasserman, G.S. (1980) Filter-mediated colour vision with one visual pigment. Science 207: 783-786.

Kropf, A., Brown, P.K. & Hubbard, R. (1959) Lumi- and meta-rhodopsins of squid and octopus. Nature (Lond.) 183: 446-450.

Kruizinga, B., Kamman, R.L. & Stavenga, D.G. (1983) Laser-induced visual pigment conversions in fly photoreceptors measured in vivo. Biophys. Struct. Mech. 9: 299-307.

Lall, A.B., Lord, E.T. & Trouth, C.O. (1982) Vision in the firefly Photuris lucrescens (Coleoptera: Lampyridae): spectral sensitivity and selective adaptation in the compound eye. J. Comp. Physiol. 147: 195-200.

Lall, A.B., Seliger, H.H., Biggley, W.H. & Lloyd, J.E. (1980) Ecology of colours of firefly bioluminescence. Science 210: 560-562.

Land, M.F. (1981) Optics and vision in invertebrates. In: Handbook of Sensory Physiology, Vol. VII/6B. Ed. H. Autrum. Berlin, Heidelberg, New York, Springer, p. 471-592.

Langer, H. (1975) Properties and functions of screening pigments in insect eyes. In: Photoreceptor Optics. Ed. A.W. Snyder & R. Menzel. Berlin, Heidelberg, New York, Springer, p. 429-455.

Langer, H., Hamann, B. & Meinecke, C.C. (1979) Tetrachromatic visual system in the moth Spodoptera exempta. (Insecta: Noctuidae). J. Comp. Physiol. 129: 235-239.

Langer, H., Schlecht, P. & Schwemer, J. (1982) Microspectrophotometric investigation of insect visual pigments. Meth. Enzymol. 81 (Part H): 729-742.

Langer, H. & Thorell, B. (1966) Microspectrophotometry of single rhabdomeres in the insect eye. Exp. Cell Res. 41: 673-677.

Laughlin, S.B. & McGinness, S. (1978) The structures of dorsal and ventral regions of a dragonfly retina. Cell Tissue Res. 188: 427-447.

Leggett, L.M.W. (1979) A retinal substrate for colour discrimination in crabs. J. Comp. Physiol. 133: 159-166.

Leggett, L.M.W. & Stavenga, D.G. (1981) Diurnal changes in angular sensitivity of crab photoreceptors. J. Comp. Physiol. 144: 99-109.

Lisman, J.E. & Sheline, Y. (1976) Analysis of the rhodopsin cycle in Limulus ventral photoreceptors using the early receptor potential. J. Gen. Physiol. 68: 487-501.

Liu, R.S.H. & Matsumoto, H. (1982) Fluorine - labeled retinals and rhodopsins. Meth. Enzymol. 81 (Part H): 694-698.

Lythgoe, J.N. (1972) The adaptation of visual pigment to their photic environment. In: Handbook of Sensory Physiology, Vol. VII/1. Ed. H.J.A. Dartnall. Berlin, Heidelberg, New York, p. 566-624.

Lythgoe, J.N. (1979) The Ecology of Vision. Oxford, Clarendon.

McIntyre, P. & Kirschfeld, K. (1981) Absorption properties of a photostable pigment (P456) in rhabdomere 7 of the fly. J. Comp. Physiol. 143: 3-15.

Menzel, R. (1979) Spectral sensitivity and colour vision in invertebrates. In: Handbook of Sensory Physiology, Vol. VII/6A. Ed. H. Autrum. Berlin, Heidelberg, New York, Springer, p. 503-580.

Messenger, J.B. (1981) Comparative physiology of vision in molluscs. In: Handbook of Sensory Physiology, Vol. VII/6C. Ed. H. Autrum. Berlin, Heidelberg, New York, Springer, p. 93-200.

Miller, W.H. (1979) Ocular optical filtering. In: Handbook of Sensory Physiology, Vol. VII/6A. Ed. H. Autrum. Berlin, Heidelberg, New York, Springer, p. 69-143.

Miller, W.H. & Bernard, G.D. (1968) Butterfly glow. J. Ultrastruct. Res. 24: 286-294.

Minke, B., Hochstein, S. & Hillman, P. (1973) Early receptor potential evidence for the existence of two thermally stable states in the barnacle visual pigment. J. Gen. Physiol. 62: 87-104.

Minke, B., Hochstein, S. & Hillman, P. (1974) Derivation of a quantitative kinetic model for a visual pigment from observations of early receptor potential. Biophys. J. 14: 490-512.

Minke, B. & Kirschfeld, K. (1978) Microspectrophotometric evidence for two photo-interconvertible states of visual pigment in the barnacle lateral eye. J. Gen. Physiol. 71: 37-45.

Minke, B. & Kirschfeld, K. (1979) The contribution of a sensitizing pigment to the photosensitivity spectra of fly rhodopsin and metarhodopsin. J. Gen. Physiol. 73: 517-540.

Minke, B. & Kirschfeld, K. (1980) Fast electrical potentials arising from activation of metarhodopsin in the fly. J. Gen. Physiol. 75: 381-402.

Morton, R.A. (1972) The chemistry of the visual pigments. In: Handbook of Sensory Physiology, Vol. VII/1. Ed. H.J.A. Dartnall. Berlin, Heidelberg, New York, Springer, p. 33-68.

Muri, R.B. (1978) Microspectrophotometry of rhabdomes in the honeybee drone. Neurosci. Lett. Suppl. 1: S410.
Muri, R.B. (1979) Microspectrophotometrie visible et UV des rhabdomes isolés de la rétine du faux-bourdon (Apis mellifera). Ph.D. Thesis, University of Geneva.
Nässel, D.R. & Waterman, T.H. (1979) Massive diurnally modulated photoreceptor membrane turnover in crab light and dark adaptation. J. Comp. Physiol. 131: 205-216.
Naito, T., Nashima-Hayama, K., Ohtsu, K. & Kito, Y. (1981) Photoreactions of cephalopod rhodopsin. Vision Res. 21: 935-941.
Nashima, K., Mitsudo, M. & Kito, Y. (1979) Molecular weight and structural studies on cephalopod rhodopsin. Biochim. Biophys. Acta 579: 155-168.
Nolte, J. & Brown, J.E. (1972) Ultraviolet-induced sensitivity to visible light in ultraviolet receptors of Limulus. J. Gen. Physiol. 59: 186-200.
Nowikoff, M. (1931) Untersuchungen über die Komplexeaugen von Lepidopteren nebst einigen Bemerkungen über die Rhabdome der Arthropoden im Allgemeinen. Z. Wiss. Zool. 138: 1-67.
Olivo, R.F. & Chrismer, K.L. (1980) Spectral sensitivity of screening pigment migration in retinula cells of the crayfish Procambarus. Vision Res. 20: 385-389.
Ostroy, S.E. (1977) Rhodopsin and the visual process. Biochim. Biophys. Acta 463: 91-125.
Ostroy, S.E. (1978) The characteristics of Drosophila rhodopsin in wild type and norp A vision transduction mutant. J. Gen. Physiol. 72: 717-732.
Ostroy, S.E., Wilson, M. & Pak, W.L. (1974) Drosophila rhodopsin: Photochemistry, extraction and differences in the norp A^{P12} phototransduction mutant. Biochem. Biophys. Res. Commun. 59: 960-966.
Packer, L. (Ed.) (1982) Methods in Enzymology, Vol. 81, Biomembranes, Part H. Visual Pigments and Purple Membranes, I. New York, Academic Press.
Pak, W.L. & Lidington, K.J. (1974) Fast electrical potential from a long-lived, long-wavelength photoproduct of fly visual pigment. J. Gen. Physiol. 63: 740-756.
Pask, C. & Barrell, K.F. (1980 a) Photoreceptor optics I: Introduction to formalism and excitation in a lens-photoreceptor system. Biol. Cybern. 36: 1-8.
Pask, C. & Barrell, K.F. (1980 b) Photoreceptor optics II: Application to angular sensitivity and other properties of a lens-photoreceptor system. Biol. Cybern. 36: 9-18.
Paulsen, R. & Schwemer, J. (1972) Studies on the insect visual pigment sensitive to ultraviolet light: retinal as the chromophoric group. Biochim. Biophys. Acta 283: 520-529.
Paulsen, R. & Schwemer, J. (1973) Proteins of invertebrate photoreceptor membranes. Characterization of visual-pigment preparations by gel electrophoresis. Eur. J. Biochem. 40: 577-583.
Paulsen, R. & Schwemer, J. (1979) Vitamin A deficiency reduces the

concentration of visual pigment protein within blowfly photoreceptor membranes. Biochim. Biophys. Acta 557: 385-390.
Pepe, I.M. & Cugnoli, C. (1980) Isolation and characterization of a water-soluble photopigment from honeybee compound eye. Vision Res. 20: 97-102.
Pepe, I.M., Schwemer, J. & Paulsen, R. (1982) Characteristics of retinal-binding proteins from the honeybee retina. Vision Res. 22: 775-781.
Razmjoo, S. & Hamdorf, K. (1976) Visual sensitivity and the variation of total pigment content in the blowfly photoreceptor membrane. J. Comp. Physiol. 105: 279-286.
Schlecht, P., Hamdorf, K. & Langer, H. (1978) The arrangement of colour receptors in a fused rhabdom of an insect. J. Comp. Physiol. 123: 239-243.
Schwemer, J. (1969) Der Sehfarbstoff von Eledone moschata und seine Umsetzungen in der lebenden Netzhaut. Z. vergl. Physiol. 62: 121-152.
Schwemer, J. (1979) Molekulare Grundlagen der Photorezeption bei der Schmeissfliege Calliphora erythrocephala Meig. Habilitationsschrift, Bochum.
Schwemer, J., Gogala, M. & Hamdorf, K. (1971) Der UV-Sehfarbstoff der Insekten: Photochemie in vitro und in vivo. Z. vergl. Physiol. 75: 174-188.
Schwemer, J. & Langer, H. (1982) Insect visual pigments. Meth. Enzymol. 81 (Part H): 182-190.
Schwemer, J. & Paulsen, R. (1973) Three visual pigments in Deilephila elpenor (Lepidoptera, Sphingidae). J. Comp. Physiol. 86: 215-229.
Shichida, Y., Kobayashi, T., Ohtani, H., Yoshizawa, T. & Nagakura, S. (1978) Picosecond laser photolysis of squid rhodopsin at room and low temperatures. Photochem. Photobiol. 27: 335-341.
Shriver, J.W., Mateescu, G.D. & Abrahamson, E.W. (1982) ^{13}C NMR spectroscopy of the chromophore of rhodopsin. Meth. Enzymol. 81 (Part H): 698-703.
Snyder, A.W., Menzel, R. & Laughlin, S.B. (1973) Structure and function of the fused rhabdom. J. Comp. Physiol. 87: 99-135.
Snyder, A.W. & Miller, W.H. (1972) Fly colour vision. Vision Res. 12: 1389-1396.
Snyder, A.W. & Pask, C. (1973) Spectral sensitivity of dipteran retinula cells. J. Comp. Physiol. 84: 59-76.
Stark, W.S., Ivanyshyn, A.M. & Greenberg, R.M. (1977) Sensitivity and photopigments of R1-6, a two-peaked photoreceptor, in Drosophila, Calliphora and Musca. J. Comp. Physiol. 121: 289-305.
Stark, W.S. & Johnson, M.A. (1980) Microspectrophotometry of Drosophila visual pigments: determinations of conversion efficiency in R1-6 receptors. J. Comp. Physiol. 140: 275-286.
Stark, W.S., Stavenga, D.G. & Kruizinga, B. (1979) Fly photoreceptor fluorescence is related to UV sensitivity. Nature (Lond.) 280: 581-583.
Stavenga, D.G. (1975) Derivation of photochrome absorption spectra from absorbance difference measurements. Photochem. Photobiol.

21: 105-110.
Stavenga, D.G. (1976) Fly visual pigments. Difference in visual pigments of blowfly and dronefly peripheral retinula cells. J. Comp. Physiol. 111: 137-152.
Stavenga, D.G. (1979) Pseudopupils of compound eyes. In: Handbook of Sensory Physiology, Vol. VII/6A. Ed. H. Autrum. Berlin, Heidelberg, New York, Springer, p. 357-439.
Stavenga, D.G. (1980) Short wavelength light in invertebrate visual sense cells - Pigments, potentials and problems. In: The Blue Light Syndrome. Ed. H. Senger. Berlin, Heidelberg, New York, Springer, p. 5-24.
Stavenga, D.G. & Barneveld, H.H. van. (1975) On dispersion in visual photoreceptors. Vision Res. 15: 1091-1095.
Stavenga, D.G. & Franceschini, N. (1981) Fly visual pigment states, rhodopsin R490, metarhodopsins M and M', studied by transmission and fluorescence microspectrophotometry in vivo. Invest. Ophth. Vis. Sci. (Suppl.) 20: 111.
Stavenga, D.G., Franceschini, N. & Kirschfeld, K. (1983) Fluorescence of visual pigments studied in the eye of intact flies. (Submitted).
Stavenga, D.G., Numan, J.A.J., Tinbergen, J. & Kuiper, J.W. (1977) Insect pupil mechanisms. II. Pigment migration in retinula cells of butterflies. J. Comp. Physiol. 113: 73-93.
Stavenga, D.G., Zantema, A. & Kuiper, J.W. (1973) Rhodopsin processes and the function of the pupil mechanism in flies. In: Biochemistry and Physiology of Visual Pigments. Ed. H. Langer. Berlin, Heidelberg, New York, Springer, p. 175-180.
Stein, P.J., Brammer, J.D. & Ostroy, S.E. (1978) Renewal of opsin in the photoreceptor cells of the mosquito. J. Gen. Physiol. 74: 565-582.
Stephenson, R.S. & Pak, W.L. (1980) Heterogenic components of a fast electrical potential in Drosophila compound eye and their relation to visual pigment photoconversion. J. Gen. Physiol. 75: 353-379.
Stowe, S. (1980) Spectral sensitivity and retinal pigment movement in the crab Leptograpsus variegatus (Fabricius). J. Exp. Biol. 87: 73-98
Suzuki, T., Sugahara, M. & Kito, Y. (1972) An intermediate in the photoregeneration of squid rhodopsin. Biochim. Biophys. Acta 275: 260-270.
Suzuki, T., Uji, K. & Kito, Y. (1976) Studies on cephalopod rhodopsin: photoisomerization of the chromophore. Biochim. Biophys. Acta 428: 321-338.
Takeuchi, J. (1966) Photosensitive pigments in the cephalopod retina. J. Nara Med. Assoc. 17: 433-448.
Tokunaga, F., Shichida, Y. & Yoshizawa, T. (1975) A new intermediate between lumirhodopsin and metarhodopsin in squid. FEBS Lett. 55: 229-232.
Tsukahara, Y. & Horridge, G.A. (1977) Visual pigment spectra from sensitivity measurements after chromatic adaptation of single fly retinula cells. J. Comp. Physiol. 114: 233-251.
Tsukahara, Y., Horridge, G.A. & Stavenga, D.G. (1977) Afterpotentials in dronefly retinula cells. J. Comp. Physiol. 114: 253-266.

Vogt, K. (1980) Die Spiegeloptik des Flusskrebsauge. J. Comp. Physiol. 135: 1-19.

Vogt, K., Kirschfeld, K. & Stavenga, D.G. (1982) Spectral effects of the pupil in fly photoreceptors. J. Comp. Physiol. 146: 145-152.

Vries, H. de & Kuiper J.W. (1958) Optics of the insect eye. Ann. N. Y. Acad. Sci. 74: 196-203.

Wald, G. & Hubbard, R. (1957) Visual pigment of a decapod crustacean: the lobster. Nature (Lond.) 180: 278-280.

Waterman, T.H. (1982) Fine structure and turnover of photoreceptor membranes. In: Visual Cells in Evolution. Ed. J.A. Westfall. New York, Raven Press, p. 23-41.

White, R.H. (1968) The effect of light and light deprivation upon the ultrastructure of the larval mosquito eye. III. Multivesicular bodies and protein uptake. J. Exp. Zool. 169: 261-278.

White, R.H., Gifford, D. & Michaud, N.A. (1980) Turnover of photoreceptor membrane in the larval mosquite ocellus: rhabdomeric coated vesicles and organelles of the vacuolar system. In: The Effects of Constant Light on Visual Processes. Ed. T.P. Williams and B.N. Baker. New York, Plenum, p. 271-296.

White, R.H. & Sundeen, C.D. (1967) The effect of light and dark deprivation upon the ultrastructure of the larval mosquito eye. I. Polyribosomes and endoplasmatic reticulum. J. Exp. Zool. 164: 461-478.

Yoshizawa, T. & Shichida, Y. (1982 a) Low-temperature spectrophotometry of intermediates of rhodopsin. Meth. Enzymol. 81 (Part H): 333-353

Yoshizawa, T. & Shichida, Y. (1982 b) Low-temperature circular dichroism of intermediates of rhodopsin. Meth. Enzymol. 81 (Part H): 634-642.

NATURAL POLARIZED LIGHT AND VISION

TALBOT H. WATERMAN

Department of Biology
Yale University, Box 6666
New Haven, CT. 06511
U.S.A.

We'll teach you to drink deep ere you depart.

Hamlet, Act I, Sc. 2.

ABSTRACT

Partial linear polarization is a prominent attribute of both scattered and reflected light in nature. Consequently sunlight in the atmosphere and hydrosphere produces substantial polarization patterns in which e-vector directions and (to a more variable extent) degree of polarization are reasonably well predicted by primary Rayleigh scattering. Despite its negligible relevance for normal human vision many organisms have been shown, following von Frisch's pioneer work, to have polarization sensitivity at various levels ranging from dichroism of visual and accessory pigments, to photoreceptor cells, to visual interneurons and ultimately oriented behavior.

By definition polarization sensitivity is the capacity of any biological element to respond differentially to either e-vector direction or degree of polarization of a light stimulus. While this capability shows interesting analogies with spectral sensitivity, the extent to which polarization vision, if indeed it has its own quality, is parallel to color vision remains to be experimentally demonstrated.

Analyzing progress mainly evident in research published during the past two to three years, the present review reports on significant new research in four areas: 1. Measurements of sky polarization in relation to animal vision. 2. Specialization of receptor mechanisms for polarization sensitivity particularly at the retinal and retinular levels. Among other things, work with genetic mutants is beginning to influence this area as it has many others. 3. Information processing (mostly relevant to e-vector direction) remains largely speculative except for some interneuron recordings in crustaceans. 4. Field and laboratory behavioral experiments mainly on bees and ants, but to some extent on flies, continue to define the adaptive applications which animals have evolved for polarization sensitivity.

INTRODUCTION

Recent evidence shows that the partial linear polarization of most natural light is considerably more important biologically than has been generally acknowledged. After its original discovery in the honey bee by von Frisch (1948, 1949), there were early indications that e-vector discrimination is widespread in animals (e.g. Jander and Waterman 1960) and that many if not most insects and their larvae can use sky polarization as a compass (e.g. Wellington et al. 1951).

Current research has reinforced and extended the behavioral as well as the ecological implications of polarized light as a major visual parameter not only in invertebrates (Wolf et al. 1980, Wellington and Fitzpatrick 1981) but also, if less resolutely, in vertebrates (Waterman 1975a, Able 1982, Gould 1982, Phillips and Waldvogel 1982). Because the author has recently published a detailed review and bibliography of polarization sensitivity (Waterman 1981), the present essay attempts primarily to evaluate relevant new research from mid-1980 to early 1983. The three final volumes of the Springer Handbook of Sensory Physiology should be consulted for detailed background information and extensive documentation (Autrum 1979, 1981a,b).

Definitions. Because definitions and symbols may be misunderstood (e.g., Fein and Szuts 1982, pp. 89-92) and various workers in the field tend to use their own, a brief recap of the basic terminology used here may be desirable (Waterman 1981, pp. 283-5). To begin with, polarization sensitivity is a very general term characterizing any biological system which is differentially sensitive to the e-vector orientation (ϕ) or degree of polarization (p) of a polarized light stimulus (as far as current knowledge indicates, only linear polarization needs to be considered in the present context).

Obviously the word differential is critical in this definition since photoreceptor units whether polarization sensitive or not, may absorb photons from both polarized and unpolarized light stimuli. For a given I it is only by having discriminative sensitivity to e-vector orientation that a unit receptor could detect the plane of vibration or alternatively the degree of polarization by gauging the modulation amplitude of $(I_{max}-I_{min})/(I_{max}+I_{min})$ resulting from incident e-vector rotation through 180° relative to the photoreceptor.

Polarization sensitivity as defined, is in some ways analogous to the more familiar <u>spectral sensitivity</u> which again in general defines any biological component or system which responds differentially to the wavelength or purity of a spectral stimulus (Bernard and Wehner 1977). In both cases the first parameters (ϕ,λ) characterize the dominant wavetrains by their plane of oscillation or wavelength respectively, whereas the second parameters specify the fraction of the total intensity contributed by the non-uniform energy distributions of $I(\phi)$ or $I(\lambda)$ in the stimulus. Note that since primary photochemical events are quantal in nature, the relevant intensities should be expressed as photon fluxes (e.g., Dartnall 1975).

While ϕ and p, as well as λ and purity, must for practical use be at least to some degree discriminable from a given intensity (I) of the stimulus, polarization sensitivity has an interesting advantage over spectral sensitivity. This resides in the fact that a single dichroic receptor unit if rotated through 180° can directly discriminate e-vector direction in a stimulus with a stable I. A comparable scan through λ is not possible by known photosensory units.

<u>Comparison with spectral sensitivity</u>. Hence a single univariant photoreceptor is color blind, and two or more spectral receptors are required to determine chromaticity. This is not true in the analogy of rotatable dichroic receptors, but in fact most of the polarization sensitive retinas of several major groups of crustaceans, of cephalopod mollusks and of many, but certainly not all, insects have two orthogonal dichroic input channels (rev., Waterman 1981). Even so, considerable ingenuity has been shown in hypothesizing three channel receptor systems which could account for instant 'polarization vision' (Kirschfeld 1972, Bernard and Wehner 1977, Wunderer and Smola 1982a) where e-vector perception is taken to be analogous to hue recognition in trichromatic color vision (e.g., Bowmaker 1983).

Interrelations between spectral and polarization sensitivities are further complicated by the fact that all polarization sensitive elements have spectral sensitivities; yet all spectrally sensitive components do not necessarily have polarization sensitivity (e.g., typical vertebrate rod and cone outer segments functioning normally in situ). These channeling and information processing mechanisms, which are still poorly understood at least for polarization sensitivity, will be considered further below.

In the general sense defined, polarization sensitivity can refer to the photon absorption function of a single rhodopsin molecule or of some unit of photoreceptor membrane like a rod outer segment or a component disc, a rhabdomere or a component microvillus. Alternatively it could apply to differential receptor potentials, spike responses, or even behavioral outputs like oriented locomotion (polarotaxis) evoked by particular polarization states or their changes. Usually when a quantitative value is given as S_p, it specifies the ratio of maximum to minimum sensitivities obtained by rotating ϕ through 180°, e.g., for absorption it would be the dichroic ratio; for receptor potentials it would be the reciprocal ratio of the numbers of quanta required to evoke equal responses for ϕ_{max} and ϕ_{min}.

The questions of whether or how polarization sensitivity at some primary level is adaptively related to the behavior and ecology of the many organisms known to have it (e.g., Waterman 1973) obviously involve various levels of visual information processing and ultimately motor outputs. The intermediate question of whether the polarization state is perceived as a special quality of an animal's visual 'awareness' is interesting but rather problematical particularly since people themselves with the unaided eye are not usually conscious of (or even able to perceive) either ϕ or p.

Here again the analogy with spectral sensitivity is useful because it is well known that peripheral differential wavelength sensitivity is a necessary, but far from sufficient, component of color vision (e.g., Menzel 1979, Jacobs 1981, Mollon 1982, Zrenner 1983). Indeed, on the basis of behavioral experiments on bees, van der Glas (1978, 1980a) has argued that sky polarization pattern perception is closely linked to color pattern perception (see also Menzel 1979 pp. 556-7). Well oriented worker bee dances, closely similar to those obtained with polarized light, were evoked by overhead unpolarized color patterns comprising smooth transitions between two different short wavelengths including for instance UV and bee purple.

Scope of review. A number of substantial new contributions relevant to polarization sensitivity have in fact recently appeared, ranging from fresh measurements of natural polarization through physiological and theoretical analyses of sensory mechanisms (e.g., Gribakin et al 1979) to field experiments on the resulting behavior. While probably not clearly certifiable as 'breakthroughs', these publications and considerable work in progress demonstrate strong current interest in the visual significance of polarization sensitivity. Nevertheless, a number of important questions remain unanswered and there are several major contradictions in experimental as well as theoretical results.

POLARIZED LIGHT IN NATURE

Natural light almost everywhere is partially linearly polarized (for detailed review and references see Waterman 1981). In the atmosphere and underwater this polarization arises by primary scattering of directional light trains in the medium; at interfaces of air and water (or earth, etc.) polarization is produced by selective reflection. Consequently the orientation (ϕ) and degree (p) of the resultant linear polarization are potentially informative parameters of any cognizant animal's visual world along with intensity (I) and wavelength (λ) of the light.

All of these vary substantially with the bearing (θ) and zenith angle (γ) of the line of sight. Consequently they form characteristic multivariate environmental light patterns whose visual discrimination requires appropriate differential sensory reception and information processing. At any given point ϕ provides axial, but not vectorial, data (Batschelet 1981), while specific values of ϕ (ranging from 0 to 180°) as well as p (ranging from 0 to 1.0) usually occur in more than one sky area.

Sky patterns. Celestial and submarine patterns of ϕ and p are, of course, well known from the point of view of physical optics (e.g. Gehrels 1974, Jerlov 1976), but only recently have detailed sky measurements been made specifically for their relevance to biological problems (Brines 1978). The new quantitative data, particularly germane to the well studied visual physiology and behavior of honey bees and other hymenopterans, have now been published in part (Brines and Gould 1982). Recorded throughout a 5° raster of θ and γ for three bee-relevant wavelengths (350, 500, 650 nm) at various times of day, including different kinds of partial and complete overcast, the resulting sky polarization maps permit several general conclusions to be drawn or reaffirmed.

Although basically determined by Rayleigh scattering of directional sunlight in the atmosphere, sky polarization observed at a given point varies markedly with the sun's position in the sky, with atmospheric turbidity, as well as with the extent and nature of cloud cover, not to mention the potential screening effects of vegetation and landmarks. With regard to the two specific polarization parameters, p in addition to being substantially lower than predicted for ideal primary scattering, often fluctuates so erratically that just a single observation point would not have a high degree of reliability or stability.

In contrast, ϕ deviates less from Rayleigh expectations and is considerably more stable in time. Hence as usually inferred, e-vector orientation or its distribution in space would appear substantially the more reliable reference input in utilizing polarization sensitivity (Brines and Gould 1982). Actually, available evidence indicates that at least for a localized sky area, p is not a critical parameter in orientation provided it is above the threshold for e-vector discrimination (Brines 1978).

Relevance of UV. Honey bees, ants and some other insects (e.g. the scarabeid beetle Lethrus, Gribakin 1981) using sky polarization as a compass must solve the behavioral problem of determining a geographical bearing (for example, the direction of a feeding place or nest) relative to the sun's (or the antisun's) meridian. Because this light compass function is usually believed to be mediated primarily, if not exclusively, by UV receptors in these insects' compound eyes, sky polarization at the UV receptor's λ_{max} (ca. 350 nm) is a matter of particular interest in the new data.

Curiously this near UV range is neither the most strongly polarized spectral band in blue sky light nor is it by a considerable margin close to the maximum of the sky's $I(\lambda)$ photon flux curve. Both of these celestial functions peak at significantly longer wavelengths (e.g., Dartnall 1975) and the quantum spectrum of the sum itself peaks in the orange (610-640 nm, Filippov and Ovchinnikov 1982). Nor is ϕ in the near UV atmospherically the most stable part of the spectrum visible to these hymenopterans. However, note that many terrestrial insects are 2-100x more sensitive to UV than, for example, to green wavelengths (Menzel 1979, p. 554).

Several workers (e.g., Edrich et al. 1979, Brines and Gould 1982, Wehner 1982, pp. 88, 96) have tried to resolve the apparent dilemma these interrelations pose. Their hypotheses depend basically on the idea that a combination of partial linear polarization and a considerable amount of UV rather specifically identify blue sky or more locally scattered directional sun's rays. Presumably with a fully clear daylight sky this might be important mainly for immediate sky-ground discrimination or horizon-fixing.

However, on partly cloudy days with scattered sunlight seen beneath cloud cover or in sunlight illuminated patches under vegetation-cast shade, the UV component might well provide a significantly effective polarized light compass (Brines and Gould 1982). Uncertainty remains, however, whether the p produced by such local Rayleigh scattering will reach the honeybee's threshold of about 10% to discriminate ϕ. Blue sky and earth (e.g. Wehner, 1982, p. 88, Plate 5) or green vegetation (e.g., Silberglied 1979, Burkardt 1982, Prokopy and Owens 1983) differ strongly in their near UV content even though the degree of polarization of light reflected from soil increases as wavelengths shorten while reflectance itself decreases significantly at the same time (Coulson 1968).

UV reflectance patterns are striking in flowers (Daumer 1958) and in insects, particularly Lepidoptera (Eguchi and Meyer-Rochow 1983), but their relevance to polarization sensitivity remains to be tested. In the case of flat water surfaces neither the reflectance nor the polarization by reflection are much affected by wavelengths (Chen and Rao 1968). Furthermore water surfaces do, of course, mirror sky polarization by specular reflection.

High polarization sensitivity ($S_p = 7.0 \pm 1.7$) recorded in UV retinular cells (λ_{max} =350 nm) in ventral ommatidia of the dragonfly Hemicordulia suggest that polarization by Fresnel reflection from the surface of a pond over which it is flying might provide the insect with an artificial horizon (Laughlin 1976). More directly, the waterbug Notonecta, which had earlier been found to have a polarization sensitive ERG for the compound eyes (Lüdtke 1957), has been shown to depend, while flying, on the polarized UV reflection of natural water surfaces for releasing its landing reflex to alight on the pond (Schwind 1983). This appears to depend specifically on a visual stimulus including a horizontally oriented, UV e-vector (Schwind, pers. com.).

Visual contrast. The basic idea of increasing visual contrast under certain conditions by adaptive displacement of the photoreceptor pigment's λ_{max} away from the λ_{max} of the light illuminating an object being discriminated, has been developed and used previously in analyzing underwater vision particularly in fishes (Lythgoe 1966, 1972, 1979, Dartnall 1975, McFarland and Munz 1975, Munz and McFarland 1977). Such offset photopigments can improve the effective S/N for visual discrimination.

A similar explanation has also been effectively applied to bioluminescent communication in fireflies in relation to the daily timing of their periods of flashing activity (Seliger et al. 1982a,b). The analysis correctly predicts the known difference in the bioluminescent λ_{max} for twilight-active and for dusk-active species of these coleopterans. More relevant here is a comparable model which has been developed from the same general principles to predict the photo-receptor λ_{max} best suited for e-vector discrimination both in the sky and underwater (Seliger, Lall and Biggley, in prep.). This model forecasts, among other things, that for the terrestrial case if atmospheric turbidity is high, so that p and hence the S/N for sky polarization are low, an offset visual pigment with λ_{max} = 350n would provide optimal discrimination.

The close approximation of this theoretical λ_{max} to those of known insect UV receptors (rev., Menzel 1979, Table 2a) is striking. Yet presumably it must be in part coincidental since inhibitory interactions between long wavelength stimulation and behavioral e-vector sensitivity have been found in Apis (Brines 1978) and complex coupling between visual input channels may be present in various species even at the retinular level (e.g. Shaw 1981, Horridge et al. 1983). In addition there are many examples (cited above and previously reviewed) of insect (as well as other) photoreceptors with substantial polarization sensitivity in units having their λ_{max} far from the UV. However, the adaptive significance of such receptor characteristics remains to be experimentally demonstrated.

Underwater polarization. On the other hand Seliger et al. (in prep.) predict from their model that in aquatic animals polarization sensitivity should have a λ_{max} near 470nm. This is at least consistent with the absence of significant UV sensitivity in cephalopods (rev., Messenger, 1981) and its rarity in crustaceans (rev., Goldsmith 1972) despite strong e-vector discrimination at longer wavelengths in both groups (rev., Waterman 1981).

The situation in fish is ambiguous in this connection since there is recent evidence for a UV sensitive cone in a cyprinid retina (rev., Bowmaker 1983). Yet goldfish polarization sensitivity measured in the optic tectum seems largely independent of wavelength at least within the spectral range tested, 460-620nm (Waterman and Aoki 1974). This is quite different from rhabdom-bearing eyes where dichroism of visual pigment in microvilli is the basis of their polarization sensitivity (e.g., Waterman et al. 1969).

Overcast sky. In any case to return to the insect sky compass, well oriented bee dances do in fact occur on fully overcast days when any effective sky polarization or other clues related to the sun's specific direction can rather definitely be excluded as mentioned below (Dyer and Gould 1981). Concurrent experiments also show that this seemingly anomalous compass behavior does not involve a magnetic directional sense. Instead they do support a mnemonic mechanism in which correct orientation depends on the bee's remembering the relationship of the sun's bearing at a given time of day to specific landmarks equally visible on both clear and overcast days.

Such piloting is certainly in accord with the relevant meteorological optics, which indicates that extinction coefficients for water droplets in several types of clouds are uniform from 300 to 700 nm (e.g., McCartney 1976). Hence neither the sun's disc nor blue sky polarization would be expected to have selective overcast penetrating capabilities in the near UV. However, photographic evidence that exclusive UV penetration through cloud cover around the sun's direction has been reported (rev., von Frisch 1965, p. 372ff), widely quoted (e.g., van der Glas 1978) but not repeated (e.g., Brines and Gould 1982).

SENSORY MECHANISMS

Considerable new information has been obtained on photoreceptor systems capable of $\underline{e}$-vector discrimination, mainly in arthropods. Analysis of these data on peripheral sensory mechanisms may be conveniently organized under four different subtopics: (1) specialized retinal regions, (2) different cell types within retinulas, (3) rhabdom twisting and (4) photoreceptor membrane turnover. Generally such recent work has proved that almost every relevant phase of the visual mechanism is more complex than it had seemed before.

Specialized retinal regions

Evidence for regional differences in compound eyes occurs at many levels. Eye shape, facet size and orientation, characteristics of the pseudopupil, retinula and rhabdom fine structure, intracellular receptor potentials, spikes in interneurons and visual behavior elicited may all be diagnostic. Such structural and functional differences between ommatidia within compound eyes have long been recognized and are being increasingly evoked as elements in the polarization sensitive mechanisms of arthropods (e.g. Odselius and Nilsson 1983).

Thus retinular cell patterns in the right and left eyes are well known to be mirror images of one another. Yet a given e-vector orientation is perceived by Drosophila as the same for the two eyes (Wolf et al. 1980). Often the retinular cell pattern in the dorsal and ventral halves of the same eye are also distinct in various species, and the gross ocular structure may even differentiate into double eyes (e.g. Schneider et al. 1978, Hardie et al. 1981, Land 1981, Mikkelsen 1981, Zeil 1983). Gradients of facet size and axial separation may be gradual over the eye or, when more acute, give rise to fovea-like regions (Horridge 1980).

In reviewing the celestial orientation problem recently, Wehner (1982) has perceptively recognized that the projection of the sky polarization pattern onto the corresponding retinal mosaic may provide important insights into the underlying visual mechanism. More specifically this approach formulates the important question of what the sky may 'look like' to an animal with photoreceptors having a given operational pattern. In both the honey bee (Apis) and the desert ant (Cataglyphis) there are three retinal regions characterized by different types of ommatidia (Räber 1979, Sommer 1979, Labhart 1980, Wehner 1982).

Selective masking of these areas with opaque paint (or alternatively screening various areas of the sky) shows that under certain conditions the smallest of the three, the dorsal eye margin (containing only a few percent of the total ommatidia), plays a critical role in azimuth direction finding from sky polarization. Here the facet density implies low visual acuity. Experiments with filters demonstrate that near UV wavelengths are both necessary and sufficient for its sky polarization compass function.

The remainder of the dorsal half of the eye in Apis and Cataglyphis contains a second ommatidial type while the ventral half comprises a third. Fine structural and other differences between these three kinds of units will be considered below. Here the effects of selectively painting them over or screening their view of the sky will be summarized. To begin with, monocular individuals orient effectively, even though they may look around more with one eye and follow less linear paths than binocular ones.

Dorsal rim:Hymenoptera. In Cataglyphis the dorsal rim area functioning alone mediates a normal e-vector sky compass; the middle region comprising the rest of the dorsal retina can also do so. In contrast the ventral half (both rim and middle areas painted over) does not seem able to subserve this polarized light compass, even though the ants do turn their heads and bodies so that ventral ommatidia look at the celestial hemisphere. Similarly this ventral eye half apparently does not by itself mediate the sun compass, but the middle area seems to do so.

With the dorsal rim region painted out in Apis azimuth orientation becomes random when restricted sky areas are presented. However, preliminary experiments demonstrate that the second ommatidial type comprising most of the dorsal half of the retina can function by itself in celestial orientation if more extensive sky regions can be seen by the bee. Whether this behavior depends on sky polarization or other visual parameters is not yet certain (Wehner 1982). Curiously, the behavioral restriction of the e-vector sky compass to the dorsal half of the bee retina does not match the distribution of polarization sensitivity indicated by selective screening pigment migration within the retinular cells (Wehner and Bernard 1980).

Measured elegantly by infrared reflectometry this 'pupillary' response to low level light stimulation demonstrated polarization sensitivity values ranging from five to nine for the UV receptors (350nm) whether the ommatidium concerned was in the mid-dorsal or ventral areas of the Apis eye (measurements could not be made in the dorsal rim region, nor in the dorsal frontal retina). At 530nm pupillary responses were quite independent of e-vector direction whereas at 430nm polarization sensitivity was present under some conditions, absent under others. Clearly the retinal distribution of bee UV e-vector detectors seems quite different depending on whether it is judged by sky compass oriented behavior or by pupillary responses.

Hence when speaking of polarization sensitivity, the particular type or level of response in question needs to be specified as previously suggested (Waterman 1981). Other new data reinforce this conclusion. For example, intracellular recording and lucifer yellow marking in the dragonfly Sympetrum show that UV receptors in the ventral eye region, unlike those dorsally located, seem not to be significantly polarization sensitive; in contrast orange receptors (λ_{max}=620nm) do show good e-vector discrimination (Menzel, Meinertzhagen and Kahl, in prep.).

Green receptors, again in this odonate, do not have appreciable polarization sensitivity. Presumably e-vector discrimination by the 620nm units would not be of direct use in a celestial compass because of their downward looking visual field. Yet such long wavelength differential e-vector responses should provide a caveat against frequent general claims that insect polarization sensitivity is restricted solely to the UV and to the dorsal half of the compound eye.

Dorsal rim:other insects. Some evidence suggests that special dorsal rim ommatidia in Diptera (Wada 1974a,b, 1975, Wunderer and Smola 1982a,b), Orthoptera (Burghause 1979, Egelhaff and Dambach 1983) and Lepidoptera (Meinecke 1981) may subserve a similar function to that more fully documented for bees and ants. In

Ascalaphus, a neuropteran, the whole antero-dorsal segment of its double eye has been reported to comprise only UV sensitive elements (Schneider et al. 1978).

In the case of dipterans the apparent conflict discussed below on the cellular basis of polarization sensitivity in Drosophila and in larger flies like Musca might be dependent on regional specialization like the dorsal rim area which has not so far been deliberately tested in research on the two fly types. However, much of the relevant data in various eye mutant flies comes from optomotor experiments which usually involve large laterofrontal but not dorsofrontal visual fields. Hence the possible resolution of the problem just suggested needs to be tested with localized stimuli before the involvement of local retinal specialization can be considered further.

Ventral eye region. The general topic of differential organization of dorsal and ventral eye halves has recently been discussed in the light of marked fine structural specialization in the dorsal part of the eye of the dragonfly Hemicordulia (Laughlin and McGinness 1978). Earlier work had emphasized UV and blue photoreceptor types in the sky viewing part of the odonate retina (Eguchi 1971), but at least in Hemicordulia UV units in the ventral eye half show substantial e-vector sensitivity ($S_p = 7 \pm 1.7$), while blue receptors have somewhat less ($S_p =$ from 3 to 6) discrimination (Laughlin 1976).

It was suggested that this localized strong polarization sensitivity at short wavelengths might function in discriminating sky reflections from natural water surfaces. A role in analyzing reflected polarized light has also been hypothesized for e-vector sensitive units in the ventral sector of Drosophila eyes (Wolf et al. 1980). Demonstration that horizontally polarized UV light like that reflected from a natural water surface induces a landing response in overflying Notonecta has been mentioned above (Schwind 1983).

Current research (Schwind et al., pers. comm.) indicates that this UV, horizontal e-vector evoked landing response is correlated with a localized ventral region of specialized ommatidia in the compound eye. There the two central cells of the open retinula have orthogonal horizontal and vertical microvilli in their respective rhabdomeres; in most of the eye the microvilli in both these rhabdomeres are parallel. In addition microspectrophotometry characterizes the two central cells, specifically within the specialized ventral area, as UV receptors.

Regional differentiation between ommatidia in dipteran eyes is clearly quite complex as evidenced by autofluorescence microscopy (e.g. Franceschini et al. 1981) and by standard transmission electron microscopy (Wunderer and Smola 1982b). In general R_1-R_6 are uniform within retinulas, and over eye regions, but the central rhabdomeres of R_7, R_8 may be quite special. For example, in Calliphora seven different types of R_7 and R_8 occur in four distinct fine structural classes of central rhabdomere. Each has a quite characteristic areal distribution over the eye (Wunderer and Smola 1982b). Most likely related to polarization sensitivity is the special microstructure of ommatidia in the dorsal margin of this compound eye (Wunderer and Smola 1982a). These are characterized by one of the four types of central rhabdomere structure with large non-twisted rhabdomeres apparently adapted specifically for e-vector discrimination.

Specialized retinular cells

Patterns of retinular cells within a single ommatidium vary widely over the many types of arthropod compound eyes not only within the retina of a single individual but between taxa down to the species level and even between the sexes. Probably the liveliest topic in this area is the controversial question of differential function between R_1-R_6 and R_7, R_8 in the neural superposition eye of dipteran insects.

Dipteran open rhabdom. The former six retinular cells have their rhabdomeres peripherally located in a trapezoidal array within the axial intercellular space of the retinula while the two latter have sequential coaxial and centrally located rhabdomeres. Typically the latter are significantly smaller in diameter than peripheral rhabdomeres but in the special dorsal rim ommatidia of some dipterans, as mentioned above, they are substantially larger than usual. At the same time the peripheral rhabdomeres of R_1-R_6 may be smaller or lacking (Wada 1974a).

Two or more channels. As previously reviewed in some detail (Waterman 1981, pp. 392-9) earlier notions that these two sets of photoreceptors in the dipteran retinula comprise a duplex system were already confronted by some apparent anomalies during the late 70's (e.g., Heisenberg and Buchner 1977). Other researchers recognizing the quite different properties of R_7 and R_8 have considered the retinular organization to be a triplex one based primarily on the distribution of three (or more) different photopigments (e.g. Harris et al. 1976, Stark and Carlson 1982). According to one version of the two-channel model (e.g., Kirschfeld and Franceschini 1968) $R_1 - R_6$ comprise a scotopic, low acuity, polarization insensitive system with short axons terminating in the lamina while R_7, R_8 comprise a photopic, high acuity, e-vector discriminating input channel with long axons terminating in the medulla.

Meanwhile the evidence against the validity of this interpretation has increased and has convinced some observers that it has outlived its usefulness at least in Drosophila (Hall, 1982). Indeed in the fruitfly $R_1 - R_6$ now appear almost certainly to be polarization sensitive units even though this may seem anomalous in terms of the special structure of the dorsal rim ommatidia just referred to, as well as the neural superposition principle of visual information processing in flies which could cancel out the e-vector sensitivity of R_1 -R_6 (Kirschfeld 1973, Jährvilehto and Moring 1976). The crucial data come mainly from experiments with visually deficient mutants (Heisenberg and Buchner 1977, Wolfe et al 1980), particularly mutant sevenless (sev) flies which lack R_7 altogether (and would hence also have an abnormally related R_8 whose rhabdomere ordinarily is in series with that of R_7 and has its microvilli perpendicular to those of R_7).

Results with eye mutants. Optokinetic experiments in which the resulting response torque was found to be sinusoidally modulated by rotating a polarizer between a moving pattern of vertical stripes and fixed but walking Drosophila yielded essentially the same responses in sev as in wild type flies (Wolf et al, 1980). This proves that responses to e-vector direction must be mediated at least partially by $R_1 - R_6$. In these data maxima and minima of e-vector influence were at $\pm 45°$ to the flies' horizontal plane. Similar oblique maxima and minima as a function of ϕ were demonstrated by Musca (and Apis) in optomotor experiments (Kirschfeld and Reichardt 1970).

This differs from the landing responses of fixed flying Calliphora presented with a moving periodic grating pattern separated by a rotatable polarizer from the compound eye (Eckert 1983). Measurements with different angular orientations of the stimulus e-vector again modulated the response intensity sinusoidally, but here the maxima and minima were not oriented obliquely but respectively vertically and horizontally relative to the animal's axial coordinates. However, this Calliphora result corresponds to the effect of ϕ direction on the accuracy of visual fixation in flying Musca where minima and maxima again occur respectively at $0°$ and $90°$ relative to the horizontal (Wehrhahn, 1976).

Conflicting evidence. Within this complex array of results Kirschfeld and Reichardt (1970) as well as Wehrhahn (1976) concluded that subsystem R_7, R_8 is responsible for the polarization sensitivity they demonstrated. Eckert (1983) reported that his experiments do not prove whether or not R_7 and R_8 mediate the polarization sensitivity of the landing response, already known in Drosophila and Calliphora to be initiated in unpolarized light by $R_1 - R_6$. Wolf et al. (1980) conclude that R_7 and most likely R_8, are not involved in the modulation they observe of Drosophila's optomotor turning responses to changing ϕ. They do not, however, exclude the participation of R_7, R_8 on a more subtle level or in different kinds of polarization sensitive behavior.

One explanation of these apparently conflicting results might be that Drosophila is different from the 'big' flies like Musca and Calliphora (e.g. Heisenberg and Buchner 1977). A less ad hoc resolution might attribute the ambiguities to a confrontation between genetic behavioral dissection of visual function and more classic optical, neural circuit analysis. For example, the direct approach by intracellular recordings from single identified retinular cells (e.g., Burkhardt 1962, Hardie et al. 1979, Hardie and Kirschfeld 1983) is not only technically difficult, especially for R_7, R_8, but as pointed out above, not necessarily decisive for evaluation at behavioral or adaptive levels.

On the other hand the fine structure, detailed distribution, and physiological properties of the undoubtedly numerous photoreceptor types in mutant Drosophila eyes are probably not yet as well known as they need to be. In the honey bee some beginnings have been made in genetic dissection of the visual system (Gribakin and Chesnokova 1982), but they have not yet been related to polarization sensitivity.

Antenna and screening pigments. The analysis of e-vector discrimination in dipterans is further complicated by the presence of photostable antenna and/or screening pigments which may significantly displace the parent cell's effective spectral sensitivity maximum away from the visual pigment's λ_{max} (McIntyre and Kirschfeld 1981). Polarization sensitivity of the various photoreceptor cells concerned may also interact significantly with the pigments present (Hardie et al. 1979, Kirschfeld 1981 pp. 147-9).

Thus R_1-R_6 in Musca have a visual pigment with its λ_{max} at 500nm supplemented by a higher apparent absorption peak around 350nm contributed by a UV-absorbing photostable accessory pigment. This apparently transfers much of the energy it absorbs to the visual pigment, thus sensitizing it, and its retinular cell to the near UV (Kirschfeld 1979, Vogt and Kirschfeld 1983).

Interestingly enough the UV-sensitizing pigment is not functionally dichroic in R_1-R_6 so that the 350nm peak in the spectral sensitivity curve for Calliphora is not significantly polarization sensitive (Guo 1981a, Vogt and Kirschfeld 1983). Yet some cells of this type have polarization sensitivities between two and three over a broad band of wavelengths from about 400 to 550nm (Guo 1981a). Other R_1-R_6 cells yield polarization sensitivities of less than two over a similar wide range of wavelengths.

Age of the flies and diet both affect e-vector discrimination measured intracellularly in these retinular cells (Guo 1980a,b, 1981b, Vogt and Kirschfeld 1983). Hence if not controlled, these factors may significantly affect the variance of relevant experimental data. Following Jährvilehto and Moring (1976) as well as others, Guo (1981a, p. 282) concludes that this peripheral polarization sensitivity of R_1-R_6 is lost in the neural superposition circuitry of the lamina.

This would seem to be in direct conflict with the behavioral data on sev Drosophila mutants (Wolf et al. 1980) unless the fruit fly is special in the cellular basis of its polarization sensitivity. It also seems to differ importantly from Apis and Cataglyphis because Drosophila's behavioral sensitivity to polarization at wavelengths in the 400-500nm range seems to be about as great as in the near UV even with wild type flies (Wolf et al. 1980).

Complex patterns of R_7 and R_8. Growing knowledge of the relevant properties of R_7 and R_8 reveals an increasingly complex picture of their functional organization. A UV sensitizing pigment like that in R_1-R_6 is apparently present in R_{7y} and R_{8y} of Musca (Hardie and Kirschfeld 1983) and Calliphora (Guo 1981a). Again this does not provide those cells with significant polarization sensitivity in the UV. In addition a blue absorbing accessory pigment (λ_{max} =460nm) does cause an apparent anomaly in R_7 by producing dichroic absorption, as measured by microspectrophotometry, with its maximum perpendicular to the microvillus axes of its rhabdomere (Kirschfeld et al. 1978).

However sensitivity to UV $\underline{e}$-vector direction is present in R_{7p} with its maximal response, as usual for a visual pigment, parallel to the rhabdomere microvilli. This cell type has a λ_{max} near 345nm; its spectral absorption lacks the three peak fine structure characteristic of the accessory pigments and is attributed to a UV peaking rhodopsin.

Still another type of R_7 has been found (Hardie et al. 1981) to occur only in the dorsofrontal region of male houseflies (Musca). This sex-specific photoreceptor (R_{7r}) is remarkably like R_1-R_6; its rhodopsin peaks at about 490nm, but a UV sensitizing pigment contributes a 360nm maximum to its effective spectral sensitivity. Unlike typical R_7's, R_{7r} terminates synaptically not in the medulla but in the lamina, again like R_1-R_6. If its polarization sensitivity parallels its other similarities with the peripheral rhabdomeres, R_{7r} will have considerable $\underline{e}$-vector sensitivity in the blue but little or none in the near UV.

Since R_8 rhabdomeres are optically in tandem with those of overlying R_7's, both their polarization and spectral sensitivities will be subject to filtering effects of the latter (e.g., Hardie et al. 1979, Snyder 1979, Guo 1981a). However, although significant polarization sensitivity has been reported in Calliphora R_8 (Järvilehto and Moring 1976, Hardie et al. 1979, Guo 1981a) the number of cells recorded so far is still small and further work is needed.

At recent count (Hardie et al. 1981) the Musca retinula appears to have five different visual pigments, four of them only in central rhabdomeres, plus two or more accessory pigments. New evidence from Calliphora may prove an exception to the old general rule that all animal photoreceptor pigments are rhodopsins since the photoreceptive chromophore isolated from this compound eye is something other than retinal (Vogt 1983). The implications of this for polarization sensitivity remain to be demonstrated.

Fused (closed) rhabdoms. In the case of fused (closed) rhabdoms there are also new data on specialized retinular cells both in hymenopterans and in decapod crustaceans. For ants and bees there are nine cells within each retinula (e.g., Wehner 1982). In

Cataglyphis these cells are of four types when classified by the axial orientation of their rhabdomeric microvilli (vertical, horizontal and $\pm 45°$ oblique) but apparently fall into only two categories (ultraviolet and green) from the point of view of intracellularly measured spectral sensitivity (Mote and Wehner 1980). Earlier phototactic choice experiments suggested a tetrachromatic system at least for some retinal areas in this ant (Kretz, 1979).

Only two perpendicular directions of microvillus orientation characterize the basic *Apis* retinula, but each one contains blue as well as UV and green receptors. Usually there are eight cells throughout much of the retinula's length, and cells diagonally opposite one another apparently have the same λ_{max} and microvillus orientation. Widely hailed until recently as the e-vector detector, R_9 is short and proximally located throughout all of the ant retina and nearly all of the bee eye.

R_9 types in *Apis*. However, the dorsal rim area of the *Apis* compound eye is a striking exception because its ommatidia have R_9 cells and their rhabdomeres which run the full length of the retinulas. In the rest of the retina R_9 is short and proximal. Its rhabdomere extends over one third or less of the full rhabdom and lies in optical series with the more distal rhabdomere of either R_1 or R_5. Apparently R_9 as well as R_1 and R_5 are the UV receptors crucial for the sky polarization compass. Their synaptic terminations are special, too, being in the medulla rather than in the lamina where the other six terminate.

In the dorsal rim region of *Apis* the microvilli in R_1 and R_5 are parallel to each other whereas those of the long R_9 are perpendicular to their direction. The rhabdomeres and somata of the UV receptors in this small retinal area of nine-long-unit ommatidia are distinctly larger than those of the other six retinular cells. Recent intracellular recordings from *Apis*' dorsal rim have shown that the UV receptors there generally yield polarization sensitivities ranging between S_p=3.8 and >10.0 (av. 6.6) (Labhart 1980). The ϕ_{max}'s for these receptors lie in two orthogonal planes for a given retinal location (presumably corresponding to the microvillus directions in R_1, R_5 vs R_9).

The population of nine-unit ommatidia in the dorsal margin has a characteristic fan-shaped orientation pattern in which the axial plane defined by the microvilli of R_1, R_5 rotates sequentially through about 180° from anterior to posterior within the frontal area (Sommer 1979, Labhart 1980, Wehner 1982). In *Calliphora* too, a fan shaped array of special ommatidia in the dorsal eye margin has been suggested as the special site of dipteran polarization sensitivity (Wunderer and Smola 1982a). These ommatidia are characterized among other things by having significantly larger R_7 and/or R_8 rhabdomeres (Wada 1974b). The problems of relating these facts to polarization sensitivity are referred to above.

In contrast long UV receptors (R_1, R_5) electrically recorded in other parts of the bee retina yield low polarization sensitivities <2.0 (Labhart 1980), although three cells presumed to be basally located, short R_9 elements in this eight-long-unit region, had previously been found (Menzel and Snyder 1974) with polarization sensitivities averaging a high S_p=5.0 (max 9.0). As mentioned above, pupillary responses in the UV have high polarization sensitivities not only in the mid-dorsal area but also in the ventral retina (Wehner and Bernard 1980). Yet only the UV receptors in the nine-long-unit region mediate the sky polarization compass for that retinal area. Does this mean that short R_9 only mediates the pupillary response and thereby provides its e-vector sensitivity?

Interestingly the green receptors of the dorsal rim area (Labhart, 1980) have low polarization sensitivities (S_p=1.8), only somewhat higher than those in the eight long unit region (S_p=1.3). This is consistent with general failure of green receptors in the frontal margin to mediate either the polarized light sky compass or any e-vector sensitivity in the pupillary response (Wehner and Bernard 1980). On the other hand high S_p's might be expected to result from their non-twisted rhabdomeres and must therefore be neutralized by intercellular coupling or some other mechanism. Recordings from blue receptors so far have yielded only low polarization sensitivity <2.0, but the number of cells sampled is too low for firm conclusions (Labhart 1980).

Comparison with Cataglyphis. In some ways Cataglyphis receptor cell specialization shows striking parallels to that in the bee Apis. Thus the dorsal rim area has a special retinula type. In cross section, it has a large dumbbell shaped rhabdom with eight rhabdomeres of which six have microvilli aligned in one axis but the other two are at 90° to that. As in the bee, alignment of the rhabdom's dumbbell axis shifts progressively through 180° from anterior to posterior in the dorsal rim area. However, R_9 in this area, as well as in the rest of the ant's retina, is short and basal. Polarization and spectral sensitivities determined intracellularly in the central retina showed some significant differences from Apis (Mote and Wehner 1980). Other regions have not yet been reported. No specific S_p or λ_{max} data are available for retinular cells in the dorsal margin area nor for R_9 anywhere in Cataglyphis.

For 27 out of the 38 units tested by Mote and Wehner S_p ranged between 2.0 and 4.0 with a total range of 1.0-6.0 (av. S_p between 2.6 and 3.0). This, as just cited, is greater than the intracellular sensitivity of the same eye region in the honey bee (S_p=2.0) but considerably less than that of the UV receptors in the latter's dorsal rim area (av. S_p=6.6). Using this electrophysiological criterion, Cataglyphis polarization sensitivity in the green (av. S_p=2.60) is much higher than the trivial ratios found in Apis. No blue receptors appeared in these ant responses, and only two spectral types were recorded, one in the UV (λ_{max}=347 nm), the other in the green (λ_{max}=506 nm). The latter matches Kretz's (1979) evidence for an R_{505} receptor type, but no units were found of the other three types he postulated: R_{315}, R_{430} and R_{570}. Nearly a third of the cells tabulated by Mote and Wehner (1980) have substantial peaks in the green in addition to a λ_{max} around 347nm. These were judged not to constitute a third retinular cell type. Instead, several lines of evidence suggest that they represent the joint output of electrically coupled adjacent retinular cells. An ionic conductance model predicts that complex changes in ϕ_{max}, S_p and λ_{max} could result from such cell-to-cell interaction (Martin and Mote 1980) but leaves it uncertain whether the implied coupling is normal in the intact eye or is the result of electrode invasion.

Rhabdom twisting

Recently there was an upsurge of interest in the reality and occurence of rhabdom twisting in arthropods. This fascinating phenomenon was first reported in a damsel fly (Ninomiya et al. 1969), then in Apis (Grundler 1974) where it stimulated analyses of its optical consequences, including important implications for polarization sensitivity.

Optical effects. As previously reviewed (e.g. Wehner 1976, Snyder 1979, Waterman 1981), optimum polarization sensitivity at the cellular level in rhabdom bearing eyes depends, among other things, on the precise orientation of rhabdomere microvilli parallel to each other and perpendicular to the rhabdom's optic axis. Consequently, twisting and the resultant dispersal in microvillus axial orientation along each rhabdomere will decrease e-vector selective photon absorption and with it, polarization sensitivity. Basically a smooth 180° twist would completely eliminate a rhabdomere's analyzer properties.

It has been argued that the elaborate twisting patterns actually found in hymenopterans and dipterans are adaptive. According to this view, twisting provides increased sensitivity to unpolarized light as well as eliminates any potentially confusing interactions between stimulus intensity and polarization in the retinular cells involved (Snyder 1973, McIntyre and Snyder 1978). Obviously the demonstrated behavioral polarization sensitivity of many animals (e.g. Waterman 1973), including some with strongly twisted rhabdoms, clearly proves that they have somehow avoided the apparently head-on collision of adaptive objectives implied by such arguments.

Denial and reaffirmation in insects. The conclusion of Ribi (1979, 1980) that reported twists in both Apis and the Calliphora are probably due to a fixation artifact focused attention sharply on this matter. Such a disavowal of twisting naturally was most challenging to those who had used it to develop an understanding of ommatidial optics, particularly with relation to polarization sensitivity (see for examples, Smola and Wunderer 1981a,b; Wehner and Meyer 1981). For both Apis and Calliphora two arguments have been developed to repudiate Ribi's doubts about the normality of rhabdom and rhabdomere twisting.

First the material used by Ribi as evidence against systematic rhabdomere rotation was declared inadequate to support his conclusion (Wehner and Meyer 1981) or actually found to sustain the presence of twist (Smola and Wunderer 1981a,b). Second, careful new quantitative electron microscopy demonstrates beyond a reasonable doubt that twist is real in Apis (Wehner and Meyer 1981), Calliphora, Musca and Drosophila (Smola and Tscharntke 1979; Smola and Wunderer 1981 a,b). Now twisting has also been reported in two nematoceran dipterans (Bibio, Altner and Burckhardt 1981; Ptilogyna, Williams 1981).

Actually the twist pattern may be quite complex both within and between ommatidia. In Apis, for example, two specific patterns of retinular cell arrangement in the main dorsal eye region and in the whole ventral half of the eye are correlated with clockwise and anticlockwise directions of continuous retinular twist in the corresponding ommatidia (Wehner 1976, Wehner and Meyer 1981). In contrast to this extensive eight-long-unit retinal region, the small area on the dorsal eye rim comprises nine-long-unit retinulas with rhabdoms which do not twist (Sommer 1979, Labhart and Meyer 1980, Wehner 1982).

Dipteran rhabdomere twisting. Note that in closed rhabdoms like the bee's, twisting involves helical rope-like rotation of the rhabdom around the ommatidial longitudinal axis. This helix includes the retinular cells as well as their component rhabdomeres. Yet in open rhabdoms like dipteran's, the separate rhabdomeres can rotate directly around their own axes, without the corresponding retinular cell somata doing so (e.g. Smola and Wunderer 1981a,b; Williams 1981). The dipteran twisting pattern can vary between ommatidia and even between retinular cells within one ommatidium. Thus in Calliphora and Drosophila the twisting direction of rhabdomeres R_1-R_3 is opposite that for R_4-R_6 (Smola and Tscharntke 1979) while within rhabdomeres three regions with characteristically different rates of microvillus axis rotation occur.

In Calliphora a number of fine structural features, cited above, suggest that the small dorsal rim region of the eye, like that of Apis, is specialized for polarization sensitivity. With regard to twist, the rim ommatidia have rhabdomeres R_1-R_6 which rotate as in the rest of the retina but have central rhabdomeres of R_7, R_8 which do not rotate, unlike their counterparts elsewhere in the eye (Wunderer and Smola 1982a,b). In Ptilogyna retinulas in the central retina have straight nontwisted R_7 and R_8 rhabdomeres with orthogonal microvilli in sequence. The peripheral rhabdomeres of R_1-R_6 all twist: R_2 and R_5 always in a specific direction while the remainder may rotate either clockwise or counterclockwise (Williams 1981). Still different complex twisting patterns were found in Bibio (Altner and Burkhardt 1981).

No twist in Cataglyphis. A curious anomaly with regard to rhabdomere twisting has been reported in ants. In Cataglyphis, for example, careful serial sections have indicated that twisting is either absent or insignificant throughout their compound eyes (Räber 1979). This, of course, is an animal for which extensive behavioral experiments document (among other things) its highly effective e-vector discrimination, particularly using the blue sky (Wehner 1982).

In contrast the bulldog ant Myrmecia has been reported in detail (Menzel and Blakers 1975) to have rhabdoms with strong clockwise or counterclockwise twisting like Apis (Wehner 1976). We know (references in Waterman 1981) that both formicine and myrmecine ants readily discriminate e-vector directions. In the former, orientation by sky polarization predominates over the sun compass if both are available; whereas in myrmecine ants and Apis, the sun compass is preferred of the two.

No twist in higher crustaceans. There is no evidence for rhabdom twisting in decapod and stomatopod crustaceans, which have been repeatedly studied. While specific quantitative data have not been published, no evidence for either systematic normal or artifactual twisting has ever been found in our extensive light microscopic and electron microscopic studies on many different species of these animals (Figs. 1-4). Recently, in view of the considerable current interest in this topic, we have been collecting our relevant data to document this point (Waterman and Campbell, in prep.)

All decapods and stomatopods (as well as mysids and euphausiids) have closed axial rhabdoms differing from the typical insect case by having interdigitating toothed rhabdomeres (Fig. 4) mainly from seven regular retinular cells (e.g., Waterman 1975b). As originally shown by G.H. Parker (1895), these are neatly fitted together so that bands of parallel photoreceptive microvilli are formed alternately by three and then four of the contributing cells (Fig. 5). These microvillus layers have successive orthogonal orientations of their microvilli which in the central retina are aligned parallel to the horizontal and vertical axes of the animal.

In superposition eyes like crayfish or shrimp the rhabdom is a relatively stout fusiform structure comprising only 25 or so such banded layers. In light micrographs it is relatively easy to locate longitudinal sections of rhabdoms (e.g. Waterman et al 1969, Fig. 1B) in which the midline of alternate rhabdom bands remains centered in the section throughout the whole axial length of the structure. If such rhabdoms were twisting by as little as $0.3°/\mu m$ this midline would appear in the section to be systematically displaced from the center to the edge of the rhabdom between its distal and proximal ends.

NATURAL POLARIZED LIGHT AND VISION 83

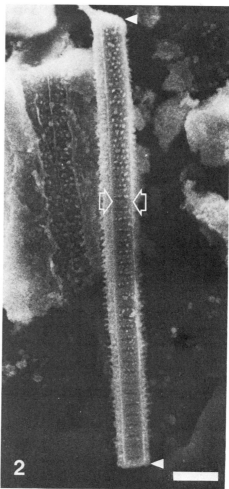

Figures 1,2. Straight, non-twisted crustacean rhabdoms (scanning electron micrographs). Photoreceptor cells should retain inherent e-vector sensivity, decreased or lost by twist. (Waterman and Campbell in preparation; SEM's with Alan Pooley.) Fig. 1 Squilla (stomatopod) rhabdom seen above in square cross-section (asterisk); below, one straight corner extends with no twist for about 60μm (broad arrows). Numerous protoplasmic bridges (sharp arrows) connect rhabdomeres to parent cell cytoplasm. Full length (236μm) Squilla rhabdom with no appreciable twist shown in Waterman 1982. Bar = 5μm. Fig. 2. Ocypode (ghost crab) rhabdom, about 1.5μm wide (rhabdomere of R_1, between open arrows; see Fig. 5B) and 37μm long between pointers. Other non-twisted retinular segments seen up to 186 μm long. Bar = 5μm.

In dissociated retinular cells from the shrimp Penaeus scanning electron micrographs (SEM) clearly show the successive bands of microvilli arising from a single retinular cell in a strictly linear array (Fig. 4). More attenuated rhabdoms also yield the same straight nontwisted geometry (Fig. 2). For instance, in the crab Carcinus the rhabdom is elongated and rod-shaped (ca. 5 μm in diameter and 150 μm long).

In a dissociated retinula of this crab viewed with Nomarski contrast, the corner marking the juncture of alternating microvillus bands can be shown to maintain a nearly constant paraxial location throughout the whole length of the rhabdom. Also in SEM strictly straight retinular cell boundaries are evident (Fig. 3). Even in more elongated rhabdoms, as in the blue crab Callinectes (5x450 μm), a fortunate longitudinal section in TEM will remain parallel to one of the alternating sets through all 250-300 bands of microvilli present (preparations in collaboration with Dr. Y. Toh).

Similarly Nomarski optical review of numerous radial retinal sections in many species of brachyuran crabs, as well as some anomurans, yields no signs of significant twisting either of retinulas or of rhabdoms (Waterman and Campbell, in prep.). In the stomatopod Gonodactylus (Fig. 1) there is also good graphic evidence of remarkably straight, long rhabdoms emphasized in SEM by their sharply square cross-section.

Other depolarizing mechanisms. If the general premise is accepted that rhabdom twisting degrades the potential polarization sensitivity of rhabdomeric photoreceptors, there are, of course, several other mechanisms known which can also reduce or eliminate this capability. A fresh example has recently been reported in the anomuran crab Petrolisthes, (Eguchi et al. 1982). This porcellanid galatheoid has compound eyes with several interesting features: the one of particular relevance here is the presence in the proximal third of the main rhabdom (R_1-R_7) of bidirectional microvillus orientation in each rhabdomere. Thus instead of having alternate bands of orthogonal microvilli contributed respectively by R_1, R_4, R_5 (horizontal microvilli) and R_2, R_3, R_6, R_7 (vertical microvilli), careful examination shows that in Petrolisthes all seven cells (R_1-R_7) contribute parallel microvilli to a single band. Yet the microvillus directions are orthogonal in adjacent bands.

This is functionally similar to the bidirectional microvillus pattern reported earlier for distally located R_8 in the rock crab Grapsus (Eguchi and Waterman 1973), in the spiny lobster Panulirus (Meyer-Rochow 1975), as well as in Petrolisthes (Eguchi et al. 1982). Clearly the effect of having two equal, orthogonal dichroic components in a given rhabdomere should be to cancel the receptor cell's polarization sensitivity but this has not yet been tested.

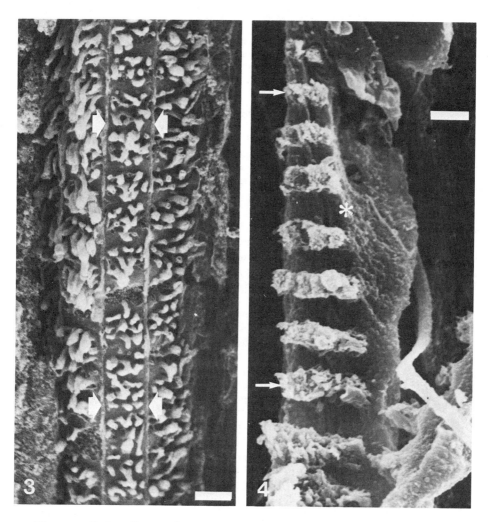

Figures 3,4. Decapod crustacean rhabdoms with non-twisting retinular cell boundaries and rhabdomere geometry (Waterman and Campbell, in preparation; SEM's in collaboration with Alan Pooley). Fig. 3. <u>Carcinus</u> (green crab) rhabdom (ca. 24μm segment) showing two straight vertical paraxial boundaries between neighboring retinular cells. As in Figs. 1,2 numerous protoplasmic bridges project outward towards retinular cell bodies (removed to reveal the central rhabdom). Bar =3μm. Fig. 4. <u>Penaeus</u> (shrimp) toothed rhabdomere shown as spaced-out flattened tufts of microvilli (arrows) emerging from their parent retinular cell (asterisk). No twisting is evident in this 23μm rhabdomere segment (about 50% of its total length). Bar = 2μm.

Obviously a less orderly but equally effective way of 'disarming' a potential polarization analyzer is to have its constituent microvilli oriented about equally in all directions, whether randomly or in a radial array (e.g. Wachmann 1979). Selective adaptation has shown that polarized light can in fact differentially interact with variously oriented microvilli in such a rhabdomere (Meinecke and Langer 1982), but again we do not know how this affects the retinular cell's sensory response.

It has recently been reported that the R_8 rhabdomere of two species of Astacus is made up of microvilli so irregularly arranged as to be probably ineffective as a dichroic analyzer (Krebs and Lietz 1982). Previous results with other crayfish Pacifastacus (Nässel 1976) and Procambarus (Waterman 1977, Fig. 8 in collaboration with Dr. E. Eguchi) had demonstrated that their R_8 rhabdomeres have well aligned horizontal microvilli but no axial banding.

Actually there is already electrophysiological evidence showing that R_8 in Procambarus has substantial polarization sensitivity, $\underline{S}_p = 2.66$ at $\lambda_{max} = 454$nm (Waterman and Fernandez 1970). At that time we did not recognize the violet receptors as being R_8 cells and indeed later obtained selective adaptation evidence that they were among the regular retinular cells R_1-R_7 (Eguchi, et al. 1973). Recent cell marking experiments (Cummins and Goldsmith 1981) and measurements of metarhodopsin fluorescence in isolated main rhabdoms of crayfish (Cronin and Goldsmith 1982) both support the identification of R_8 as the violet cell. In retrospect this is congruent with the horizontal orientation of ϕ_{max} in the majority (5 of 7) of the violet cells previously available for comparison (Waterman and Fernandez 1970).

Also interactions within or between retinulas like optical (e.g. Snyder 1979) or electrical (e.g. Laughlin 1981, Dubs 1982) coupling could by shunting reduce or eliminate differences in primary input channel signals. Membrane particle arrays which could mediate coupling between contiguous retinular cells as well as tight junctions have been observed in the anomuran crab Petrolisthes (Eguchi et al 1982). On the other hand lateral inhibition mediated by extracellular field potentials might sharpen signal differences in neighboring elements. For example, polarization sensitivities as great as 30 recorded in certain retinular cells of the butterfly Papilio have been attributed to such synergistic interactions (Horridge et al. 1983). An earlier model to account for complex changes in ϕ_{max}, $\underline{S}_p$ and λ_{max} due to receptor cell interactions in Cataglyphis was cited above (Martin and Mote 1980).

Predictions from optical theory indicate that a microvillus system might achieve a maximum dichroic ratio of 20 with an optimal alignment of visual pigment absorbing dipoles (Snyder and Laughlin 1975), but dichroic ratios measured microspectrophotometrically are usually less than 6 (e.g. Goldsmith and Wehner 1977). Yet intracellular polarization sensitivities in cases reported before Papilio, have reached 20 (Scylla, Leggett 1978) or 11 or more (Carcinus, Shaw 1969) or 11.9 (Procambarus, Waterman and Fernandez 1970).

On the other hand reports of low or no polarization sensitivity determined intracellularly in a significant fraction of insect retinular cells recorded are common (e.g. Gribakin et al 1979, Gribakin 1981, Meyer-Rochow 1981). Despite an occasional diatribe, our understanding of the relations between polarization sensitivity at various levels and behavioral responses remain unimpressive.

Membrane turnover

Like biological systems in general, photoreceptor membranes are constantly in a state of turnover (e.g. Holtzman and Mercurio 1980, Holtzman 1981, Papermaster and Schneider 1982). While this might not seem particularly relevant to polarization sensitivity, the massive scale of daily rhabdom breakdown and renewal in some arthropods (reviewed in Waterman 1982) indicates that attention to possible interactions is warranted. Substantial changes in the number and length of microvilli, relaxation of their strict regular pattern, breakdown of rhabdomeres, rhabdomes, retinulas or even considerable areas of retina have all been shown to occur on a diurnal schedule in various arthropod eyes (e.g. Blest 1978, Nässel and Waterman 1979, Stowe 1980, Blest et al. 1982, deCouet and Blest 1982, Shaw and Stowe 1982 p. 351, Williams 1982, Piekos and Waterman 1983, Waterman and Piekos, 1983).

Clearly these relate to visual adaptation under periodically changing ambient light conditions and probably to daily endogenous activity cycles in addition to basic renewal turnover. To what extent such membrane changes impair normal visual function, including polarization sensitivity, has not yet been tested. Even if performance level is lowered at certain times of day this may not be operationally important. For example, in vertebrate photoreceptor cell outer segments, breakdown occurs mainly at times just after their daily primary active phase has ended.

Thus cones in vertebrate eyes turn over in the early evening and the rods mainly in the early morning (Young 1978). Clearly the time course of visual discrimination capabilities in rhabdom-bearing eyes needs to be correlated with fine structural changes occuring in their photoreceptor system. Once calibrated, this might provide a direct, but noninvasive, way of measuring the underlying turnover in addition to its primary determination of acuity, thresholds, spectral and polarization sensitivities and other sensory parameters.

INFORMATION PROCESSING

Obviously one of the intriguing aspects of polarization sensitivity is the means whereby the primary visual input (p, ϕ, θ, γ) is processed to provide appropriate motor output and behavior. If this visual capability subserves more than one function, as has been widely surmised, this processing could be of various kinds. A sky compass, for example, should be connected to a locomotor steering function about the yaw (usually dorso-ventral) axis (e.g. Waterman, 1966c); an horizon indicator or visual position stabilizer will have potential outputs around the animal's pitch and roll axes, while a water or wet surface locator or a glare reducing filter might not have any such direct motor consequences.

Relatively little is known about polarization sensitive information transmission at levels above the sensory neurons, although the general consequences of apposition, superposition, and neural superposition types of compound eye have been extensively discussed in terms of input channeling (e.g. Kunze 1979, Hardie et al. 1981, Land 1981a,b, Laughlin 1981, Strausfeld and Nässel 1981, and Wehner 1981). Certainly neuronal circuitry connecting the retina and the optic ganglia to the protocerebrum has been widely studied since Cajal and Hanström made their pioneer contributions. Nevertheless the number of experimentally recorded interneuron responses which can be correlated with changes in polarization parameters is surprisingly small.

Indeed it seems doubtful, despite much research, that there is a single well documented record of a polarization sensitive <u>interneuron</u> in any insect. On the other hand, there are some quite interesting but rather poorly known data for decapod and stomatopod crustaceans (reviewed in Waterman 1981). An important feature of these interneuron responses is that unlike the receptor potentials in retinular cells, differential $\underline{e}$-vector spike responses have not been found for fixed planes of vibration (Waterman and Wiersma 1963, Waterman 1977) but only for rotating $\underline{e}$-vectors (Yamaguchi 1967, Yamaguchi et al. 1976, Leggett 1976, 1978).

<u>Crayfish visual interneurons</u>. While no new publications have appeared in this area recently, there are many new data on <u>Procambarus</u> polarized light responses at the optic nerve level (Yamaguchi, Takahashi and Waterman, in prep.). A brief summary seems appropriate here because so little is known about such neural activity in any invertebrate. Extracellular but effectively single unit responses were recorded in immobilized crayfish stimulated with linear polarized light (Fig. 5). $\underline{E}$-vector rotation rate as well as rotation direction, light intensity and wavelength have all been explored for their effects on visual interneuron responses.

In experiments done with sustaining fibers recorded in the optic nerve a majority of the approximately 300 units studied were

Figure 5. Crayfish (Procambarus) setup for recording polarization sensitivity in single optic nerve interneurons. A. Anterior end of the animal showing three main body axes (L-R, left, right; D-V, dorsal, ventral; A-P anterior, posterior). Linear polarizers (P) in stimulus light-beams, rotated either clockwise (c) or anti-clockwise (ac). Environmental coordinates are H, horizontal; V, vertical. B. Retinular cell patterns in central ommatidia of left (B_1) and right (B_2) eyes are mirror images. Note that although two orthogonal sets of rhabdom microvilli are drawn as if superimposed, the rhabdomeres of R_1, R_4 and R_5 have only horizontal microvilli and those of R_2, R_3, R_5 and R_6 only vertical. (Yamaguchi, Takahashi and Waterman, in preparation).

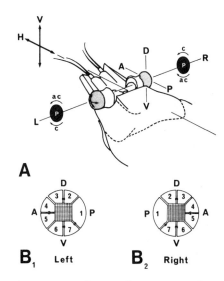

Type 038 neurons which have large receptive fields centered around the posterior dorsal margin of the retina (Wiersma and Yamaguchi 1966). No e-vector discrimination is evident when these are stimulated with a series of flashes with various fixed e-vector directions, but characteristic periodic bursts of spikes result when the e-vector is presented while rotating (Figs. 6-8). In such responses there are always four maxima at $90°$ intervals for each $360°$ polarizer rotation. This is quite different from crayfish receptor cell potentials (e.g. Waterman and Fernandez 1970) and from polarization sensitive spikes recorded in optic ganglia of the mud crab Scylla (Leggett 1978): both of them have their maxima (as well as minima) $180°$ apart.

Opponency not found. Clearly in the crayfish optic nerve the two orthogonal dichroic input channels (Waterman 1966a,c, 1968, Waterman and Horch 1966) have converged on the interneurons recorded; no evidence was found of a single channel 2θ function. This would be expected if only one channel and its constituent three or four specific retinular cells were providing the whole input to the interneuron in question. With careful alignment of an upward vertical reference $0°$ for the polarizer and for the crayfish eye positioned normally in space, the alternate peaks in optic neuron responses can in fact be shown to correspond to successive excitation of the vertical channel (R_2, R_3, R_6, R_7) and the horizontal one (R_1, R_4, R_5).

Hence there is no evidence yet in this crayfish data for an opponent e-vector analysis of the sort hypothesized for λ discrimination in the honey bee by Bernard and Wehner (Wehner 1982) and indeed previously reported to occur in that insect system (reviewed in Menzel 1979). Recently, however, strong differential enhancement of both polarization and spectral sensitivities (measured intracellularly) has been reported for certain retinular cells of the butterfly Papilio (Horridge et al 1983). Such intercellular, or even interommatidial, interactions have not yet been unequivocally documented in crustaceans.

This apparent ambiguity obviously is related to the questions of whether or to what extent polarization sensitivity gives rise to perception in which ϕ and p evoke sensations comparable to hue and saturation in human color vision. Preliminary experiments do imply that responses to simulated intensity patterns are qualitatively different from those to e-vector direction itself, at least for Drosophila (Wolf et al., 1980). In contrast, as cited above bee dance orientations closely paralleling those evoked by polarized light have been demonstrated in response to unpolarized overhead color patterns (van der Glas 1978, 1980a).

Typically in crayfish the responses of the vertical channel (four receptor cells) comprise more numerous and more frequent spikes than those of the horizontal channel (three regular receptor cells) with the same stimulus input (Figs. 6-8). Such channel differences and their time course are strongly dependent on the rate of e-vector rotation ω. At high ω's (e.g. $800°/sec$) the two channels' responses are more closely the same than at low angular velocities (e.g. $16°/sec$.) (Figs. 6-8). As might be expected because of latencies, the phase relations between momentary e-vector direction and the spike burst in response shift extensively (up to $90°$) as ω changed.

However, these phase lags for a given eye are different for clockwise vs. anticlockwise e-vector rotation (Figs. 6,7) and are reversed between right and left eyes. Thus right eye anticlockwise rotations and left eye clockwise rotations show greater phase delays than the opposite rotation directions in the corresponding eyes. Interestingly enough, comparisons show that the dorsal and ventral halves of the same eye yield similar spike patterns in this response.

Wavelength sensitivity. The question of wavelength sensitivity in these responses has been partially investigated because of its relevance to possible interactions between the two polarization channels and the two color channels (violet and yellow-green) known for this eye (e.g. Waterman and Fernandez 1970, Nassel and Waterman 1977, Cummins and Goldsmith 1981). Accordingly, in addition to extensive recordings with white light, comparisons were made of e-vector responses at 430 nm and at 571 nm using narrow band interference filters with these λ_{max}'s. Surprisingly,

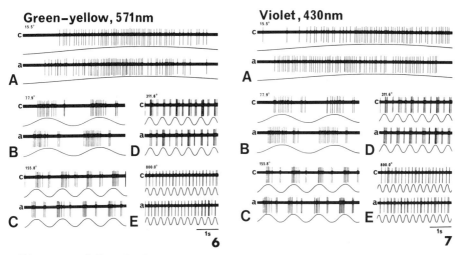

Figures 6,7. Spike responses of <u>Procambarus</u> optic nerve interneurons to compound eye stimulation by rotating e-vector (Fig. 5). Narrow band interference filters were used with λ_{max}'s close to peak sensitivities of the known yellow-green (571nm) and violet (430nm) receptors in <u>Procambarus</u> eyes. Fig. 6. Effects of e-vector angular velocity (given in degrees per sec. for each condition) and direction of rotation (c, clockwise, a, anticlockwise) with λ_{max} at 571nm. Note that both vertical and horizontal polarization channels have converged on the interneuron recorded. Phase relations of responses to e-vector rotations depend both on its direction and angular velocity. Under each spike record a sinusoidal curve monitors e-vector rotation with maxima at 180° intervals. A one-second time scale appears lower right. Fig. 7. Replication of experiment recorded in Fig. 6 except that here the λ_{max} of the stimulus was at 430nm. Closely similar responses were obtained in these two experiments as well as the pair shown in Figs. 8A, 8B. This is surprising in view of the current identification of the retinular cells concerned. (Yamaguchi, Takahashi and Waterman, in preparation).

spike responses to e-vector rotation were closely similar at these two quite different wavelengths (Figs. 6-8).

If R_8, as discussed above, is the violet receptor in each retinula and R_1-R_7 are the yellow-green receptors, short wavelengths should evoke only or mainly horizontal channel responses. However, the expected differential responses have not been found even at low light intensities where saturation would seem unlikely or minimal.

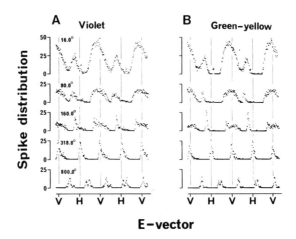

Fig. 8. Experiments similar to those in Figs. 6,7 testing effects of e-vector angular velocity (shown in degrees per sec. for each of four rates) and extreme wavelengths on responses of a single optic nerve axon. Spike frequency distributions measured by counting impulses per sampling time interval spaced over a 360° cycle of e-vector rotation. Thin vertical lines at 90° intervals as well as V's and H's on the abscissa indicate when the e-vector was oriented vertically or horizontally (Fig. 5). As in Figs. 6,7 wavelength shows little differential influence or responses. (Yamaguchi, Takahashi and Waterman, in preparation).

Yet we already know from Leggett's work (1976) on the mud crab Scylla that more than one type of rotating e-vector-sensitive interneuron occurs in crustaceans. Consequently, there remain a number of important unanswered questions for which this promising preparation may well provide answers. A new technique for maintaining viable optic interneurons in the crayfish (Kirk, et al. 1982) should encourage both deeper and more comparative research into the information processing relevant to e-vector discrimination.

BEHAVIORAL RESPONSES

If polarotaxis is defined as a visually directed locomotor or turning response to the e-vector of linearly polarized light (Waterman 1966a,b), such behavior is widespread in both terrestrial and aquatic animals (e.g., Waterman 1973, Part I). In the laboratory and sometimes in the field a stimulating patch of uniformly polarized light is produced with a dichroic filter to provide directional information to the animal concerned. Of course, in the field natural polarized light visible to the animal may range widely from small essentially single e-vector spots seen through apertures or openings between clouds at various bearings and zenith distances to the whole celestial visual field typically including complex patterns of ϕ, p, I and λ as already discussed.

A current example of the last experimental condition is the clear dependence of Cataglyphis on a sky polarization compass when homing on its nest under outdoor conditions where it cannot see the sun's direction but does see extensive blue sky (e.g., Wehner 1982). A wide range of other species including birds (Gould 1982, Able 1982) have been cited for such compass responses. The effective analysis of relatively simple orientation data of this sort should not become mired in the descriptive naivete and terminological hair splitting which sometimes characterize the classic concepts of taxis and kinesis (rev., Schöne 1980). Instead the underlying Loebian concept of directional responses to stimulus input patterns should be brought into line with contemporary biophysics, neurobiology and ethology (see also Burr, this volume).

The basic premise is that the animal is wired and programed so that a functional 'end point' is reached when the animal's sensory coordinates are matched in a particular way to the stimulus pattern coordinates. The effectiveness and interest of an algorithmic or cybernetic approach to simple oriented steering behavior are exemplified by recent stimulating work on visual fixation (Reichardt and Poggio 1980), optical tracking (Collett and Land 1978) and bacterial chemotaxis (Macnab 1978, Koshland 1980).

Laboratory experiments.

While a detailed development of this theme may be out of place a brief report seems appropriate here, on a new quantitative analytical technique applied to spontaneous polarotactic orientation of Daphnia (Wilson and Waterman, unpublished). The original intention was to use this method to explore, among other things, the mechanism of multiple peak orientation ($0°$, $45°$, $90°$, $135°$ relative to the e-vector) in Daphnia and a number of other animals which are strongly polarization sensitive (reviewed in Waterman 1981). While these objectives have not yet been achieved, preliminary studies provided a method by which large batches of quantitative data on movement and orientation can be processed, analyzed, statistically evaluated and graphically presented in a manner not usually practical using less automated methods.

Simply stated, the technique involves recording the relevant behavioral responses on video tape. A 'bugwatcher' system scans this information at appropriate experimental time intervals, recognizes the organism being tracked and, for instance, digitizes its $\underline{x}$, $\underline{y}$ position at these successive moments (Greaves 1975, Wilson 1977). Such time identified loci are fed into a suitable computer program which can then calculate directions and velocities of travel and their rates of change. A considerable variety of data plots can be obtained, ranging from maps of the course covered, the distribution of directions and velocities within a batch of data, to the relationships between two or three of these parameters as well as the effects of ϕ, p, λ, etc. on the recorded behavior.

Examples are given (Figs. 9-12) of such analyses obtained from recordings of Daphnia swimming in a vertical beam of linearly polarized or unpolarized white light. These suggest the scope of this technique which provides an approach to problems which would seem impractical using photographic or manual methods. Its main limitations, in turn, are set first by the need to have a tracking system, like the video camera used here, which can continuously record the animal's progress within an experimental area of appropriate scale. Second, the bugwatcher requires a considerable signal-to-noise ratio to recognize the target reliably from background. Hence in a video system the recorded animal must have substantial contrast against a uniform background. Clearly this technique developed by Davenport (Davenport et al. 1970, Greaves 1975, Wilson 1977) and others has great potential, for studying not only polarotaxis but also other real time orientation and behavior problems.

Field experiments

Ant vector integration. The desert ant Cataglyphis, while foraging, carries out continuous vector integration of its often meandering path away from the nest (Wehner 1982). From any point achieved on this path (which may completely lack landmarks) the ant can home accurately in a straight line, but when displaced becomes completely lost. If it has been carefully picked up and relocated, Cataglyphis runs straight in the right compass direction and for a distance correct for the location from which it was displaced. Of course, this oriented response has become quite inappropriate if the distance and/or angular displacement are substantial. However, under these circumstances the ant initiates an efficient pattern of general search behavior when it fails to encounter the nest in the 'expected' direction and distance (Wehner and Srinivasan 1981).

The reality of the vector integration involved has been demonstrated in Cataglyphis (Wehner 1982), as it had been long ago in the forest ant Formica (Jander 1957), and indeed well before that in bees (von Frisch 1948, review 1965 p. 172ff). As in dead-reckoning generally, such summation must depend on relatively precise information on the direction and length of each leg of the outward path. The mechanism of the ant's distance measurements

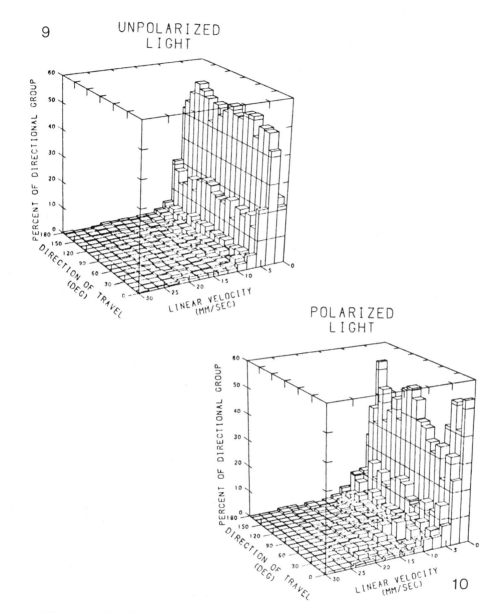

Figures 9,10. Analysis of <u>Daphnia</u> spontaneous azimuth orientation while swimming in a vertical beam of white light from above. Videotaped data were digitized, computer processed and reproduced here as they appeared in three dimensional graphic displays. Percentages of all counts in a given swimming direction are shown as a function of the animal's azimuth heading and linear velocity. (Wilson and Waterman, unpublished.)

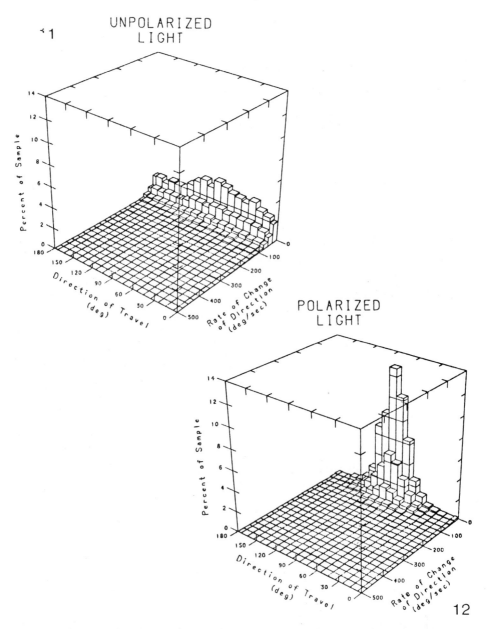

Figures. 11,12. Total sample (%) as a function of <u>Daphnia's</u> azimuth heading and angular velocity. Figs. 9, 11 demonstrate no significant preferential orientation in unpolarized light. Figs. 10,12 show that <u>Daphnia</u> has a strong polaroxis at $90°$ to the downwelling <u>e</u>-vector only within narrow ranges of linear and angular velocities. (Wilson and Waterman, unpublished.)

may be idiothetic (i.e. dependent on kinesthetic interoreceptor data) but remains speculative. In contrast the required compass element has been shown to be celestial and to depend on the visibility of the sun or sky polarization.

For the polarized light compass, near-UV sky light as reviewed above, has been reported to be both necessary and adequate as the visual stimulus. Only the dorsal half of the compound eye can mediate this oriented response. Within the dorsal retina either the dorsal rim region or the rest of the dorsal retina alone appear to be adequate, and such orientation is maintained with only one eye functional (Wehner 1982).

Landmark orientation (rev. Wehner 1981) is quite possible in Cataglyphis, as it is in the honey bee, (e.g. Dyer and Gould 1981, Cartwright and Collett 1982) but does not appear to include much flexible general mapping capability (Collett and Cartwright 1983). Such orientation in the desert ant seems rather to consist of a sequential comparison of observed visual images with remembered patterns experienced in reverse order on the outward journey (Wehner, 1982). It has been clearly proved that both the sun and sky compass responses cited are independent of landmarks in any case. Furthermore, Cataglyphis, unlike most ants, does not make use of pheromonal (or other chemical) cues in its foraging behavior.

Honeybee sky compass. Behavioral work on the honey bee polarized light compass has been substantial during the last few years. Yet many of the basic questions raised by von Frisch's discovery of this phenomenon (1948, review 1965) remain partly or completely unanswered. Recent research has depended mainly on three different workers and their colleagues (e.g. Brines, 1978, van der Glas 1978 and Wehner, 1982) but effective comparisons and evaluation are difficult because major relevant results are only partly published so far. Furthermore, apparently subtantial differences are evident in the orientation behavior observed as well as the conclusions drawn by the different investigators.

Thus Rossel and Wehner (1982) have carried out extensive new experiments on Apis dance orientation to polarized light. These have eliminated some ambiguity in effective stimulus parameters and support a more specific but less complex sky map than the one previously proposed (Rossel et al. 1978). Basically, however, the 1982 analysis can be considered as a confirmation of their earlier data and conclusions. According to this relatively simple model, Apis interprets the solar bearing of any restricted blue sky area (or linearly polarized UV light field) on the basis of an elementary map relating e-vector direction to the angular differences between the bearing of the point observed and that of the sun.

The resulting polarized light compass card makes use of only the sky half away from the sun (see also Edrich et al 1979). Except at sunrise and sunset this half of the clear sky is the more strongly polarized and includes, along the anti-sun's meridian and at $90°$ away from the sun, the point of maximum Rayleigh polarization. This point lies on the equator of the solar spherical coordinate system having the sun and antisun as its poles.

Relative to the observer's earthbound zenith-horizon coordinates, this solar equator marks a band of maximum Rayleigh polarization $90°$ to the sun. At sunrise or sunset at the equinoxes the band follows a north south oriented great circle through the zenith where clear sky polarization is then strongest. As the earth rotates the band of maximum polarization appears to rotate in the opposite direction around the celestial poles at $15°/hr$ (Brines 1980). Accordingly this great circle (at the equinox for an equatorial observer) reaches the western horizon at local noon and rises from the eastern horizon immediately after (Wellington 1974, Waterman 1975a). Obviously the observer's latitude and the season both alter this relatively simple pattern geometry.

Simple sky map. In Rossel and Wehner's model (1982) bees use e-vector orientation along the great circle of maximum polarization to calibrate their sky map. The latter is taken to be invariant both with time of day and with the zenith angle of the sky point observed provided only that this angle does not exceed $60°$. The reference function used by the bees in this mechanism is the relation between e-vector (ϕ) and azimuth (γ) shown by the maximum polarization band e-vector observed at a $30°$ zenith angle. This is a nearly linear (slightly s-shaped) function in which the reference azimuth of the bee's polarized light map is apparently the anti-sun's meridian along which all sky e-vector's are, of course, horizontal.

Like the bee's sun compass, this e-vector compass apparently ignores the zenith angle of the point observed in the celestial hemisphere provided it lies between $0°$ and $60°$. To calibrate the relation between e-vector direction and azimuth, only the horizontal e-vector along the great circle through the zenith and point of maximum polarization is accurately located. For any other azimuth the e-vector direction is taken to be the mean of its value between zenith angles of $0°$ and $60°$.

Vertical e-vectors ($90°$ to those in the reference meridian) are interpreted by the bees as lying $\pm 90°$ from the antisun's azimuth. These would of course be accurate only at sunrise and sunset. Oblique $45°$ e-vector directions are treated as if they lie $\pm 50°$ from the antisun's bearing. This rather simple reduction of the real continuously changing complex e-vector pattern in the sky and its variations (e.g. Brines and Gould 1982) was derived from a systematic study of errors in bee dance directions (Rossel et al. 1978, Rossel and Wehner 1982).

Conflicting error data. Analysis of these orientation errors, which under certain conditions were large, led to the hypothesis for the bee's highly schematic sky map outlined above. The relative simplicity of this navigation mechanism is certainly much simpler than one requiring the solution of celestial spherical triangles or mnemonic storage of detailed e-vector sky patterns and their changes with time of day. Also the skill and care used in designing and executing the experiments certainly lend weight to the argument concerning the compass mechanism. However, as in the monograph on sky navigation in insects (Wehner 1982), neither the protocols nor the quantitative data analysis so far published would seem to do justice to the elegant experiments completed.

Another point which remains unclear is why the systematic errors in polarized light-mediated azimuth orientation studied by Rossel and Wehner have not been previously recognized as important. Actually an earlier study of consistent mistakes in bee dances oriented to the sun compass transposed to the direction of gravity (Lindauer and Martin 1968 Martin and Lindauer 1981) is well known because it was no doubt responsible for the subsequent upsurge of interest in bees' as well as other animals' apparent sensitivity to the earth's magnetic field (Kalmijn 1978, Griffin 1982).

However with regard to polarized light orientation, errors in dance directions observed in their experiments by Brines (1978) and van der Glas (1978) were generally smaller than those underlying the map hypothesis or not detected at all. Also there seems to be considerable variation in the degree to which bimodal or unimodal responses to e-vector have been evoked in various experiments on oriented behavior in the honey bee.

Bimodal responses. Thus Apis, presented with a $10°$ circular field of blue sky (or a similar area illuminated through a linear polarizer), always preferred two dance directions separated by $180°$ (Rossel et al. 1978 p. 9 and Table 1). Except when $\phi = 90°$, one of these dance directions was the major one, but the other at $180°$ to this was favored in a weighted average of 11% of the runs. Note that the $180°$ difference between them only matches the real sky polarization pattern if the sun is on the horizon and the maximum polarization band passes through the zenith.

In contrast during later experiments only occasional, slightly bimodal responses were found to horizontal ($\phi = 0°$) and oblique ($\phi = \pm 45°$) e-vector directions (Rossel and Wehner 1982). Still different bee behavior was observed by van der Glas (in preparation) with bimodal orientation having angles between the two dance directions significantly different from $180°$ except with zenith polarization when they were diametrical.

Brines (1978) generally found that dance directions were mainly unimodal and much more precise than those reported by Rossel et al. (1978) which fitted their model. In other respects, however, the two sets of data are in qualitative agreement (Brines 1978, p.VI-27). Brines concludes that either some sensory or behavioral difference in the strains of bees used or some artifacts in the experimental setups might account for dissimilar orientation preferences observed. Or alternatively the results of Rossel and his colleagues might have been induced by the particular sky area used as a stimulus (Brines 1978, p.VI-13).

As a possible model to account for his data Brines (1978, Chap. VI and App. A) discusses the spherical trigonometry of the scattering triangle established by a polarized point in the sky, the zenith and the sun. This analysis indicates that for an unambiguous determination of the azimuth angle between the sun and the sky point observed four of the six parameters of the triangle need to be known including the sun's zenith distance, the scattering angle ($\underline{a}$) between the direction of the sun's rays and the observer's line of sight, $\emptyset$ and the zenith distance of the point observed.

Obviously, Rossel and Wehner's simplified sky polarization map stands in strong contrast to the apparently rigorous 'cerebral' requirements of this kind of celestial trigonometry. The key question remains, however, why do the Wurzburg and Zurich honeybees make substantial orientation errors (particularly blatant when the sky point observed lies in the solar half of the sky) whereas the Princeton bees apparently do not? On the other hand if the scattering triangle must be solved for sky point polarized light orientation, how could $\underline{a}$ be measured independently?

Actually Brines (1978,p.VI-6) concludes that both physical and behavioral reasons argue against $\underline{a}$ being a feasible parameter for an animal to use in solving this problem. If only the remaining three variables listed above are available, there will generally be two sky points at a given zenith distance which have the same $\emptyset$ values, as already pointed out by von Frisch and others (von Frisch 1965, p.396, Kirschfeld et al. 1975). Brines and Gould (1979) have presented evidence that bees may resolve these ambiguities by applying three particular rules in selecting one of the two possible orientation directions.

1. Sky is discriminated from sun by its size, $\underline{p}$ and % UV. 2. In an ambiguous pair of possible locations the point further from the sun is chosen. 3. In a similar pair of sky points equidistant from the sun, the one to the right is always picked. The zenith seems not resolvable, and both directions are in fact danced. These authors argue in general that some conventional rules are needed by the honeybee since in addition to orientation, communication is involved in their dances. Clearly both the signaller and the responder must understand each other.

<u>Importance of zenith area</u>. None of the recent research on honey bees has referred to Wellington's field demonstration (1974, 1976) that dipterans and many other insects, including some bees and wasps, show a remarkable reduction in straight-course, sky polarization-oriented flight activity when the sun reaches zenith angles of 30° or less. At such times if $\underline{p}$= 0.7, zenith polarization would be about 5% (Waterman 1981, p. 387) well below <u>Apis</u>' threshold (von Frisch 1965).

Like much else in this field von Frisch (1965, p. 248) had previously observed that bee colonies were subject to a mid-day period of relative inactivity extending from 11-12 h to 14-15 h.

Lindauer and others are cited as reporting similar local noon related decreases in the number and vigor of bee dances. These facts were not explicitly correlated with coexisting polarization patterns or sun's zenith angle, as discovered by Wellington, but rather were explained as functions of diurnal (possibly endogenous) activity rhythms.

It would be interesting to examine existing data or carry out new experiments to explain these behavior patterns in more detail. In any case honey bees can be effectively trained to feed specifically at local times around noon (Moore and Rankin 1983), but the particular azimuth cue for this behavior is not known. Presumably the sun itself, if not directly in the zenith, or landmarks could function in direction finding even though sky polarization were not used as a compass during this period of the day.

The implication of Wellington's data is that when p in the sky area used by insects to orient their flight direction falls below a certain percentage rather sharply determined by the sun's zenith angle, a polarization sky compass is no longer used. While it is generally acknowledged (as discussed above) that dorsofrontal visual fields are important in insect polarized sky light menotaxis, the failure of the insects observed by Wellington to use the presumably visually quite adequate ϕ_{max} in the antisum's meridian $90°$ away from the sun seems perplexing in terms of the honeybee models cited above.

For example, when the sun's zenith distance is $30°$, p at this same zenith distance along the antisun's meridian would be 42% if $p_{max} = 0.7$. This should be well above the bee's threshold. Actually Rossel and Wehner (1982) conclude that Apis' internal map of sky polarization is a function of the average e-vector direction in the p_{max} sky band calculated between zenith angles of 0 and $60°$; as plotted this appears to be the same as the ϕ distribution observed around an almucantar with a $30°$ zenith angle.

Polarotaxis. Note that the Rossel and Wehner (1982) celestial polarization map and the proposed mechanism for using it (Wehner 1982) would seem to be rather close to the classic ideas of phototaxis and polarotaxis (Waterman 1966a, Frantsevitch 1980 pp. 210-218). Thus in both the Loebian orientation response and the current hypothesis, a matching of stimulus pattern symmetry with sensory input symmetry is postulated. The former normally is related to environmental coordinate systems while the latter is, of course, set by the individual animal's own coordinate system.

The old notion of menotaxis in which response behavior superimposes (as in a basitaxis) or otherwise sets a compass-like fixed relation between these two coordinate systems would seem quite relevant. Actually the conclusions of Brines (1978), van der Glas, (1978) and Rossel and Wehner (1982) appear basically similar in this regard, however much they may diverge otherwise. In the general case a course dependent fixed angular relation is required between the two coordinate systems, rather than just superposition, because compass orientation will not usually be parallel to the environmental stimulus pattern (e.g. the $\underline{e}$-vector orientation in some part of the sky).

Indeed van der Glas (1978a,b and in prep.) as well as Brines have suggested models where the dynamic relation between the two coordinate systems determines the compass reference direction. These might provide an automatic detector mechanism (Brines 1978) in which spherical trigonometric analysis and perception of polarization patterns are not operationally necessary for the animal to steer its course. On the other hand Wolf et al. (1980 p.190) conclude from their experiments that in _Drosophila_ there is a perceptual representation of ϕ.

Acknowledgements. The author is grateful to a number of colleagues and friends for information on research in progress as well as suggestions on interpreting the current status of polarization sensitivity. He also thanks Ms. Shirley Smith for her skilful help in preparing the typescript.

REFERENCES

Able, K.P. (1982) Skylight polarization patterns at dusk influence migratory orientation in birds. Nature 299: 550-551.

Altner, I., Burkhardt, D. (1981) Fine structure of the ommatidia and the occurrence of rhabdomeric twist in the dorsal eye of male *Bibio marci* (Diptera, Nematocera, Bibionidae). Cell Tissue Res. 215: 607-623.

Autrum, H. (Ed.) (1979) Handbook of Sensory Physiology. Vol. VII/6. Vision in Invertebrates. A: Invertebrate Photoreceptors. Berlin, Springer-Verlag, 707 pp.

Autrum, H. (Ed.) (1981a) Handbook of Sensory Physiology. Vol. VII/6. Vision in Invertebrates. B. Invertebrate Visual Centers and Behavior I. Berlin, Springer-Verlag, 611 pp.

Autrum, H. (Ed.) (1981b) Handbook of Sensory Physiology. Vol. VII/6. Vision in Invertebrates. C: Invertebrate Visual Centers and Behavior II. Berlin, Springer-Verlag, 665 pp.

Batschelet, E. (1981) Circular Statistics in Biology. London, Academic Press, 371 pp.

Bernard, G.D., Wehner, R. (1977) Functional similarities between polarization vision and color vision. Vision Res. 17: 1019-1028.

Blest, A.D. (1978) The rapid synthesis and destruction of photoreceptor membrane by a dinopid spider. A daily cycle. Proc. Roy. Soc. Lond. B 200: 463-483.

Blest, A.D., Stowe, S., Eddey, W., Williams, D.S. (1982) The local deletion of microvillar cytoskeleton from photoreceptors of tipulid flies during membrane turnover. Proc. Roy. Soc. Lond. B 215: 469-479.

Bowmaker, J.K. (1983) Trichromatic color vision: why only three receptor channels? Trends Neurosci. 6: 41-43.

Brines, M.L. (1978) Skylight polarization patterns as cues for honey bee orientation; physical measurements and behavioral experiments. Ph.D. Thesis. Rockefeller University. 393 pp.

Brines, M.L. (1980) Dynamic patterns of skylight polarization as clock and compass. J. Theor. Biol. 86: 507-512.

Brines, M.L., Gould, J.L. (1979) Bees have rules. Science 206: 571-573.

Brines, M.L., Gould, J.L. (1982) Skylight polarization patterns and animal orientation. J. Exp. Biol. 96: 69-91.

Burkhardt, D. (1962) Spectral sensitivity and other response characteristics of single visual cells in the arthropod eye. In: Symposia of the Society for Experimental Biology, 16, Biological Receptor Mechanisms. Ed., J.W.L. Beament, Cambridge, University Press, pp. 86-109.

Burkhardt, D. (1982) Birds, berries and UV. Naturwissenschaften 69: 153-157.

Burghause, F.M.H.R. (1979) Structural specialization in the dorsofrontal region of the cricket compound eye (Orthoptera, Grylloidea). Zool. Jb. Physiol. 83: 502-525.

Cartwright, B.A., Collett, T.S. (1982) How honey bees use landmarks to guide their return to a food source. Nature 296: 560-564.

Chen, H.S., Rao, C.R.N. (1968) Polarization of light on reflection by some natural surfaces. Br. J. Appl. Phys. (J. Phys. D) Ser. 2. 1: 1191-1200.

Collett, T.S., Land, M.F. (1978) How hoverflies compute interception courses. J. Comp. Physiol. 125: 191-204.

Couet, H.G. de, Blest, A.D. (1982) The retinal acid phosphatase of a crab, Leptograpsus: characterization, and relation to the cyclical turnover of photoreceptor membrane. J. Comp. Physiol. 149: 353-362.

Coulson, K.L. (1968) Effect of surface reflection on the angular and spectral distribution of skylight. J. Atmospheric Sci. 25: 759-770.

Cronin, T.W., Goldsmith, T.H. (1982) Photosensitivity spectrum of crayfish rhodopsin measured using fluorescence of metarhodopsin. J. Gen. Physiol. 79: 313-332.

Cummins, D., Goldsmith, T.H. (1981) Cellular identification of the violet receptor in the crayfish eye. J. Comp. Physiol. 142: 199-202.

Dartnall, H.J.A. (1975) Assessing the fitness of visual pigments for their photic environments. In: Vision in fishes. Ed., M.A. Ali, New York, Plenum Press, pp. 543-563.

Daumer, K. (1958) Blumenfarben, wie sie die Bienen sehen. Z. Vergl. Physiol. 41: 49-110.

Davenport, D., Culler, G.J., Greaves, J.O.B., Forward, R.B., Hand, W.G. (1970) The investigation of the behavior of micro organisms by computerized television. IEEE Trans. Biomed. Eng. 17: 230-237.

Dubs, A. (1982) The spatial integration of signals in the retina and lamina of the fly compound eye under different conditions of luminance. J. Comp. Physiol. 146: 321-343.

Dyer, F.C., Gould, J.L. (1981) Honey bee orientation: A backup system for cloudy days. Science 214: 1041-1042.

Eckert, H. (1983) The dependence of the landing response of blowflies, Calliphora, on the e-vector of linearly polarized light. Naturwissenschaften 70: 150-151.

Edrich, W., Neumeyer, C., von Helversen, O. (1979) 'Anti-sun orientation' of bees with regard to a field of ultraviolet light. J. Comp. Physiol. 134: 151-157.

Egelhaaf, A., Dambach, M. (1983) Giant rhabdomes in a specialized region of the compound eye of a cricket: Cycloptiloides canariensis (Insecta, Gryllidae). Zoomorphology 102: 65-77.

Eguchi, E. (1971) Fine structure and spectral sensitivities of retinular cells in the dorsal sector of compound eyes in the dragonfly Aeschna. Z. Vergl. Physiol. 71: 201-218.

Eguchi, E., Meyer-Rochow, V.B. (1983) Ultraviolet photography of forty-three species of Lepidoptera representing ten families. Annot. Zool. Japan. 56:10-18.

Eguchi, E., Waterman, T.H. (1973) Orthogonal microvillus pattern in the eighth rhabdomere of the rock crab Grapsus. Z. Zellforsch. 137: 145-157.

Eguchi, E., Goto, T., Waterman, T.H. (1982) Unorthodox pattern of microvilli and intercellular junctions in regular retinal cells of the porcellanid crab Petrolisthes. Cell Tissue Res. 222: 493-513.

Eguchi, E., Waterman, T.H., Akiyama, J. (1973) Localization of the violet and yellow receptor cells in the crayfish retinula. J. Gen. Physiol. 62: 355-374.

Fein, A., Szuts, E.Z. (1982) Photoreceptors: their role in vision. Cambridge, Cambridge University Press, 212 pp.

Filippov, M.P., Ovchinnikov, A.A. (1982) Quantum optimality of day vision and photocolorimetry. (transl.) Doklady Biophysics 266: 146-149.

Franceschini, N., Hardie, R., Ribi, W., Kirschfeld, K. (1981) Sexual dimorphism in a photoreceptor. Nature 291: 241-244.

Frantsevich, L.I.K. (1980) Visual Analysis of Space in Insects. Kiev, 'Scientific Thought', 288pp. (in Russian)

Frisch, K. von (1948) Gelöste und ungelöste Rätsel der Bienensprache. Naturwissenschaften. 35: 38-43.

Frisch, K. von (1949) Die Polarisation des Himmelslichtes als orientierender Faktor bei den Tänzen der Bienen. Experientia 5: 142-148.

Frisch, K. von (1965) Tanzsprache und Orientierung der Bienen. Berlin, Springer-Verlag, 578 pp.

Gehrels, T. (Ed.) (1974) Planets, Stars and Nebulae. Tucson, University of Arizona Press. 1133 pp.

Glas, H.W. van der (1978) Mechanisms of E-vector orientation in the honeybee. Ph.D. Thesis, Reijksuniversiteit van Utrecht. 186 pp.

Glas, H.W. van der (1980a) Orientation of bees, Apis mellifera, to unpolarized colour patterns, simulating the polarized zenith skylight pattern. J. Comp. Physiol. 139: 225-241.

Glas, H.W. van der (1980b) Models for directional feature extraction in celestial polarization patterns by insects. Proc. Eur. Soc. Comp. Physiol. Biochem. 2: 137-138.

Goldsmith, T.H. (1972) The natural history of invertebrate visual pigments. In: Handbook of Sensory Physiology, VII/1. Ed., H.J.A. Dartnall, Berlin, Springer-Verlag, pp. 685-719.

Goldsmith, T.H., Wehner, R. (1977) Restriction on rotational and translational diffusion of pigment in membranes of a rhabdomeric photoreceptor. J. Gen. Physiol. 70: 453-490.

Gould, J.L. (1982) The map sense of pigeons. Nature 296: 205-211.

Greaves, J.O.B. (1975) The bugsystem: the software structure for the reduction of quantized video data of moving organisms. Proc. IEEE 63: 1415-1425.

Gribakin, F.G. (1981) Automatic spectrosensitometry of photoreceptors in Lethrus (Coleoptera, Scarabaeidae). J. Comp. Physiol. 142: 95-102.

Gribakin, F.G., Chesnokova, E.G. (1982) Changes in functional characteristics of the compound eye of bees caused by mutations disturbing tryptophan metabolism. (transl.) Neurophysiology 14: 57-62.

Gribakin, F.G., Vishnevskaya, T.M., Polyanovskii, A.D. (1979) Polarization and spectral sensitivity of single photoreceptors of the domestic cricket. (transl.) Neurophysiology 11: 483-490.

Griffin, D.R. (1982) Ecology of migration: is magnetic orientation a reality? (book review) Quart. Rev. Biol. 57: 293-295.

Grundler, O.J. (1974) Elektronenmikroskopische Untersuchungen am Auge der Honigbiene (Apis mellifica). I. Untersuchungen zur Morphologie und Anordnung der neun Retinulazellen in Ommatidien verschiedener Augenbereiche und zur Perzeption linear polarisierten Lichtes. Cytobiologie 9: 203-220.

Guo, A. (1980a) Elektrophysiologische Untersuchungen zur Spektral- und Polarisations-empfindlichkeit der Sehzellen von Calliphora erythrocephala. I. Scientia Sinica 23: 1182-1196.

Guo, A. (1980b) Elektrophysiologische Untersuchungen zur Spektral- und Polarisations-empfindlichkeit an den Sehzellen von Calliphora erythrocephala. II. Scientia Sinica 23: 1461-1468.

Guo, A. (1981a) Electrophysiologische Untersuchungen zur Spektral- und Polarisations-empfindlichkeit der Sehzellen von Calliphora erythrocephala. III. Scientia Sinica 24: 272-286.

Guo, A. (1981b) Electrophysiologische Untersuchungen zur Spektral- und Polarisations-empfindlichkeit an den Sehzellen von Calliphora erythrocephala. IV. Scientia Sinica 24: 542-553.

Hall, J.C. (1982) Drosophila neurogenetics. Quart. Rev. Biophysics 15: 223-479.

Hardie, R.C., Kirschfeld, K. (1983) Ultraviolet sensitivity of fly photoreceptors R_7 and R_8: evidence for a sensitising function. Biophys. Struct. Mech. 9: 171-180.

Hardie, R.C., Franceschini, N., McIntyre, P.D. (1979) Electrophysiological analysis of fly retina. II. Spectral and polarization sensitivity in R_7 and R_8. J. Comp. Physiol. 133: 23-39.

Hardie, R.C., Franceschini, N., Ribi, W., Kirschfeld, K. (1981) Distribution and properties of sex-specific photoreceptors in the fly Musca domestica. J. Comp. Physiol. 145: 139-152.

Harris, W.A., Stark, W.S., Walker, J.A. (1976) Genetic dissection of the photoreceptor system in the compound eye of Drosophila melanogaster. J. Physiol. 256: 415-439.

Heisenberg, M., Buchner, E. (1977) The role of retinula cell types in visual behavior of Drosophila melanogaster. J. Comp. Physiol. 117:127-162.

Holtzman, E. (1981) Membrane circulation: an overview. Methods in Cell Biology 23: 379-397.

Holtzman, E., Mercurio, A.M. (1980) Membrane circulation in neurons and photoreceptors: some unresolved issues. Int. Rev. Cytol. 67: 1-67.

Horridge, G.A. (1980) Apposition eyes of large diurnal insects as organs adapted to seeing. Proc. Roy. Soc. Lond. B 207: 287-309.

Horridge, G.A., Marčelja, L., Jahnke, R., Matič, T. (1983) Single electrode studies on the retina of the butterfly Papilio. J. Comp. Physiol. 150: 271-294.

Jacobs, G.H. (1981) Comparative Color Vision, New York, Academic Press, 209 pp.

Jander, R. (1957) Die optische Richtungsorientierung der roten Waldameise (Formica rufa). Z. Vergl. Physiol. 40: 162-238.

Jander, R., Waterman, T.H. (1960) Sensory discrimination between polarized light and light intensity patterns by arthropods. J. Cell. Comp. Physiol. 56: 137-160.

Järvilehto, M., Moring, J. (1976) Spectral and polarization sensitivity of identified retinal cells of the fly. In: Neural Principles in Vision. Eds., F. Zettler, R. Weiler, Berlin, Springer-Verlag, pp. 214-226.

Jerlov, N.G. (1976) Marine Optics. Amsterdam, Elsevier, 231 pp.

Kalmijn, A.J. (1978) Electric and magnetic sensory world of sharks, skates, and rays. In: Sensory Biology of Sharks, Skates, and Rays. Eds., E.S. Hodgson and R.F. Mathewson. Arlington, Office of Naval Research, Dept. of the Navy, pp. 507-528.

Kirk, M.D., Waldrop, B. and Glantz, R.M. (1982) The crayfish sustaining fibers. I. Morphological representation of visual receptive fields in the second optic neuropil. J. Comp. Physiol. 146: 175-179.

Kirschfeld, K. (1972) Die notwendige Anzahl von Rezeptoren zur Bestimmung der Richtung des elektrischen Vektors linear polarisierten Lichtes. Z. Naturforsch. 27b: 578-579.

Kirschfeld, K. (1973) Das neurale Superpositionsauge. Fortschr. Zool. 21: 229-257.

Kirschfeld, K. (1979) The function of photostable pigments in fly photoreceptors. Biophys. Struct. Mechanism 5: 117-128.

Kirschfeld, K. (1981) Bistable and photostable pigments in microvillar photoreceptors. In: Sense Organs. Eds., M.S. Laverack, D.J. Cosens, Glasgow, Blackie and Sons Ltd., pp. 142-162.

Kirschfeld, K., Franceschini, N. (1968) Optische Eigenschaften der Ommatidien im Komplexauge von Musca. Kybernetik 5: 47-52.

Kirschfeld, K., Reichardt, W. (1970) Optomotorische Versuche an Musca mit linear polarisiertem Licht. Z. Naturforsch. 25b: 228.

Kirschfeld, K., Feiler, R., Franceschini, N. (1978) A photostable pigment within the rhabdomere of fly photoreceptors no. 7. J. Comp. Physiol. 125: 275-284.

Kirschfeld, K., Lindauer, M., Martin, H. (1975) Problems of menotactic orientation according to the polarized light of the

sky. Z. Naturforsch. 30c: 88-90.
Koshland, D.E. (1980) Bacterial Chemotaxis as a Model Behavioral System. New York, Raven Press, 193 pp.
Krebs, W., Lietz, R. (1982) Apical region of the crayfish retinula. Cell Tissue Res. 222: 409-415.
Kretz, R. (1979) A behavioral analysis of color vision in the ant Cataglyphis bicolor (Formicidae:Hymenoptera). J. Comp. Physiol. 131: 217-233.
Kunze, P. (1979) Apposition and superposition eyes. In: Handbook of Sensory Physiology. Vol. VII/6A. Ed., H. Autrum. Berlin, Springer-Verlag, pp. 441-503.
Labhart, T. (1980) Specialized photoreceptors at the dorsal rim of the honeybee's compound eye: polarizational and angular sensitivity. J. Comp. Physiol. 141: 19-30.
Labhart, T., Meyer, E.P. (1980) Ultrastructural and electrophysiological studies on a specialized area of the honey bee's eye. Experientia 36: 698.
Land, M.F. (1981a) Optics and vision in invertebrates. In: Handbook of Sensory Physiology. Vol. VII/6B. Ed., H. Autrum, Berlin, Springer-Verlag, pp. 471-595.
Land, M.F. (1981b) Optics of the eyes of Phronima and other deep-sea amphipods. J. Comp. Physiol. 145: 209-226.
Laughlin, S.B. (1976) The sensitivities of dragonfly photoreceptors and the voltage gain of transduction. J. Comp. Physiol. 111: 221-247.
Laughlin, S.B. (1981) Neural principles in the visual system. In: Handbook of Sensory Physiology. Vol. VII/6B. Ed., H. Autrum. Berlin, Springer-Verlag, pp. 133-281.
Laughlin, S., McGinness, S. (1978) The structures of dorsal and ventral regions of a dragonfly retina. Cell Tiss. Res. 188: 427-447.
Leggett, L.M.W. (1976) Polarized light-sensitive interneurons in a swimming crab. Nature 262: 709-711.
Leggett, L.M.W. (1978) Some visual specializations of a crustacean eye. Ph.D. Thesis. Australian National University. 140 pp.
Lindauer, M., Martin, H. (1968) Die Schwereorientierung der Bienen unter dem Einfluss des Erdmagnetfelds. Z. Vergl. Physiol. 60: 219-243.
Lüdtke, H. (1957) Beziehungen des Feinbaues im Rückenschwimmerauge zu seiner Fähigkeit, polarisiertes Licht zu analysieren. Z. Vergl. Physiol. 40: 329-344.
Lythgoe, J.N. (1966) Visual pigments and underwater vision. In: Light as an Ecological Factor. Eds., G.C. Evans, R. Bainbridge, O. Rackham. Oxford, Blackwell Scientific Publications, pp. 375-391.
Lythgoe, J.N. (1972) The adaptation of visual pigments to the photic environment. In: Handbook of Sensory Physiology, Vol. VII/1. Photochemistry of Vision. Ed., H.J.A. Dartnall, Berlin, Springer-Verlag, pp. 566-603.

Lythgoe, J.N. (1979) The Ecology of Vision. Oxford, Clarendon Press, 244 pp.
Macnab, Robert M. (1978) Bacterial motility and chemotaxis: the molecular biology of a behavioral system. CRC Crit. Rev. Biochem. 5: 291-341.
Martin, F.G., Mote, M.I. (1980) An equivalent circuit for the quantitative description of inter-receptor coupling in the retina of the desert ant Cataglyphis bicolor. J. Comp. Physiol. 139: 277-285.
Martin, H., Lindauer, M. (1981) The orientation of bees in the earth's magnetic field. In: Sense Organs. Eds., M.S. Laverack, D.J. Cosens, Glasgow, pp. 328-332.
McCartney, E.J. (1976) Optics of the Atmosphere. New York, Wiley, 408 pp.
McFarland, W.N., Munz, F.W. (1975) The evolution of photopic visual pigments in fishes. III. Vision Res. 15: 1071-1080.
McIntyre, P., Snyder, A.W. (1978) Light propagation in twisted anisotropic media: application to photoreceptors. J. Opt. Soc. Am. 68: 149-157.
McIntyre, P., Kirschfeld, K. (1981) Absorption properties of a photostable pigment (P456) in Rhabdomere 7 of the fly. Comp. Physiol. 143: 3-15.
Meinecke, C.C. (1981) The fine structure of the compound eye of the African armyworm moth, Spodoptera exempta Walk. (Lepidoptera, Noctuidae). Cell Tissue Res. 216: 333-347.
Meinecke, C.C., Langer, H. (1982) Structural reactions to polarized light of microvilli in photoreceptor cells of the moth Spodoptera. Cell Tissue Res 226: 225-229.
Menzel, R. (1979) Spectral sensitivity and color vision in invertebrates. In: Handbook Sensory Physiology VII/6A. Ed., H. Autrum. Berlin, Springer Verlag, pp. 502-580.
Menzel, R., Blakers, M. (1975) Functional organization of an insect ommatidium with fused rhabdom. Cytobiologie 11: 279-298.
Menzel, R., Snyder, A.W. (1974) Polarized light detection in the bee, Apis mellifera. J. Comp. Physiol. 88: 247-270.
Messenger, J.B. (1981) Comparative physiology of vision in molluscs. In: Handbook of Sensory Physiology, VII/6C. Ed., H. Autrum, Berlin, Springer-Verlag, pp. 93-200.
Meyer-Rochow, V. (1975) Larval and adult eye of the western rock lobster, Panulirus longipes. Cell Tissue Res. 162: 439-457.
Meyer-Rochow, V.B. (1981) Electrophysiology and histology of the eye of the bumblebee Bombus hortorum (L.) (Hymenoptera: Apidae). J.R. Soc. New Zealand 11: 123-153.
Mikkelsen, P.M. (1981) Studies on euphausiacean crustaceans from the Indian River region of Florida. I. Systematics of the Stylocheiron longicorne species-group, with emphasis on reproductive morphology. Proc. Biol. Soc. Wash. 94: 1174-1204.
Mollon, J.D. (1982) Color vision. Ann. Rev. Psychol. 33: 41-85.
Moore, D., Rankin, M.A. (1983) Diurnal changes in the accuracy of

the honeybee foraging rhythm. Biol. Bull. 164: 471-482.
Mote, M.I., Wehner, R. (1980) Functional characteristics of photoreceptors in the compound eye and ocellus of the desert ant, Cataglyphis bicolor. J. Comp. Physiol. 137: 63-71.
Munz, F.W., McFarland, W.N. (1977) Evolutionary adaptations of fishes to the photic environment. In: Handbook of Sensory Physiology, Vol. VII/5. The Visual System in Vertebrates. Ed., F. Crescitelli, Berlin, Springer-Verlag, pp. 193-274.
Nässel, D.R. (1976). The retina and retinal projection on lamina ganglionaris of the crayfish Pacifastacus leniusculus (Dana). J. Comp. Neurol. 167: 341-360.
Nässel, D.R., Waterman, T.H. (1977) Golgi EM evidence for visual information channelling in the crayfish lamina ganglionaris. Brain Res. 130: 556-563.
Nässel, D.R., Waterman, T.H. (1979) Massive diurnally modulated photoreceptor membrane turnovers in crab light and dark adaptation. J. Comp. Physiol. 135: 205-216.
Ninomiya, N., Tominaga, T., Kuwabara, M. (1969) The fine structure of the compound eye of a damsel fly. Z. Zellforsch. 98: 17-32.
Odselius, R., Nilsson, D.-E. (1983) Regionally different ommatidial structure in the compound eye of the water-flea Polyphemus (Cladocera, Crustacea). Proc. Roy. Soc. Lond. B 217: 177-189.
Papermaster, D.S., Schneider, B.G. (1982) Biosynthesis and morphogenesis of outer segment membranes in vertebrate photoreceptor cells. In: Cell Biology of the Eye. Ed., D.S. McDevitt, New York, Academic Press, pp. 475-531.
Parker, G.H. (1895) The retina and optic ganglion in decapods especially in Astacus. Mitth. Zool. Sta. Neapel 12: 1-73.
Phillips, J.B., Waldvogel, J.A. (1982) Reflected light cues generate deflector-loft effect. In: Avian Navigation. Eds., F. Papi, H.G. Wallraff. Berlin, Springer Verlag, pp. 190-202.
Piekos, W.B., Waterman, T.H. (1983) Nocturnal rhabdom cycling and retinal hemocyte functions in crayfish (Procambarus) compound eyes. I. Light microscopy. J. Exp. Zool. 225: 209-217.
Prokopy, R.H., Owens, E.D. (1983) Visual detection of plants by herbivorous insects. Ann. Rev. Entomol. 28: 337-364.
Räber, F. (1979) Retinatopographie und Sehfeldtopologie des Komplexauges von Cataglyphis bicolor (Formicidae, Hymenoptera). Diss., Univ. Zürich. 121 pp.
Reichardt, W.E., Poggio, T. (1980) Visual control of flight in flies. In: Theoretical Approaches to Neurobiology. Eds., W. E. Reichardt and T. Poggio. Cambridge, MIT Press, pp. 135-150.
Ribi, W.A. (1979) Do the rhabdomeric structures in bees and flies really twist? J. Comp. Physiol. 134: 109-112.
Ribi, W.A. (1980) New aspects of polarized light detection in the bee in view of non-twisting rhabdomeric structures. J. Comp. Physiol. 137: 281-285.

Rossel, S., Wehner, R. (1982) The bee's map of the e-vector pattern in the sky. Proc. Nat. Acad. Sci. USA. 79: 4451-4455.

Rossel, S., Wehner, R., Lindauer, M. (1978) E-vector orientation in bees. J. Comp. Physiol. 125: 1-12.

Schneider, L., Gogala, M., Draslar, K., Langer, H., Schlecht, P. (1978) Structure of the ommatidia and properties of the screening pigments in the compound eyes of Ascalaphus (Insecta, Neuroptera). Cytobiologie. 16: 274-307.

Schone, H. (1980) Orientierung im Raum. Stuttgart, Wissenschaftliche Verlagsgesellshaft, 377 pp.

Schwind, R. (1983) A polarization-sensitive response of the flying water bug Notonecta glauca to UV light. J. Comp. Physiol. 150: 87-91.

Seliger, H.H., Lall, A.B., Lloyd, J.E., Biggley, W.H. (1982a) The colors of firefly bioluminescence I. Optimization model. Photochem. Photobiol. 36: 673-680.

Seliger, H.H., Lall, A.B., Lloyd, J.E., Biggley, W.H. (1982b) The colors of firefly bioluminescence. II. Experimental evidence for the optimization model. Photochem. Photobiol. 36: 681-688.

Shaw, S.R. (1969) Sense-cell structure and interspecies comparisons of polarized light absorption in arthropod compound eyes. Vision Res. 9: 1031-1040.

Shaw, S.R. (1981) Anatomy and physiology of identified non-spiking cells in the photoreceptor-lamina complex of the compound eye of insects, especially Diptera. In: Neurones Without Impulses. Eds., A. Roberts, B.M.H. Bush, Cambridge, Cambridge University Press, pp. 61-116.

Shaw, S.R., Stowe, S. (1982) Photoreception. In: The Biology of Crustacea, Vol. 3. Ed., D.E. Bliss, New York, Academic Press, pp. 292-367.

Silberglied, R.E. (1979) Communication in the ultraviolet. Ann. Rev. Evol. Syst. 10: 373-398.

Smola, U., Tscharntke, H. (1979) Twisted rhabdomeres in the dipteran eye. J. Comp. Physiol. 133: 291-297.

Smola, U., Wunderer, H. (1981a). Twisting of blowfly (Calliphora erythrocephala Meigen) (Diptera, Calliphoridae) rhabdomeres: an in vivo feature unaffected by preparation of fixation. Int. J. Insect Morphol. and Embryol. 10: 331-344.

Smola, U., Wunderer, H. (1981b) Fly rhabdomeres twist in vivo. J. Comp. Physiol. 142: 43-49.

Snyder, A.W. (1973) Polarization sensitivity of individual retinula cells. J. Comp. Physiol. 83: 331-360.

Snyder, A.W. (1979) The physics of vision in compound eyes. In: Handbook of Sensory Physiology. Vol. VII/6A. Ed., H. Autrum. Berlin, Springer-Verlag, pp. 225-315.

Snyder, A.W., Laughlin, S.B. (1975) Dichroism and absorption by photoreceptors. J. Comp. Physiol. 100: 101-116.

Sommer, E.W. (1979) Untersuchungen zur topografischen Anatomie der Retina und zur Sehfeldtopologie im Auge der Honigbiene,

Apis mellifera (Hymenoptera). Ph.D. Thesis, Univ. Zürich. 180 pp.

Stark, W.S., Carlson, S.D. (1982) Ultrastructural pathology of the compound eye and optic neuropiles of the retinal degeneration mutant (w rdgB^{KS222}) Drosophila melanogaster. Cell Tissue Res. 225: 11-22.

Stowe, S. (1980) Rapid synthesis of photoreceptor membranes and assembly of new microvilli in a crab at dusk. Cell Tissue Res. 211: 419-440.

Strausfeld, N.J., Nässel, D.R. (1981) Neuroarchitecture of brain regions that subserve the compound eyes of Crustacea and insects. In: Handbook of Sensory Physiology. Vol. VII/6B. Ed., H. Autrum. Berlin, Springer-Verlag, pp. 1-132.

Vogt, K. (1983) Is the fly visual pigment a rhodopsin? Z. Naturforsch. 38c: 329-333.

Vogt, K., Kirschfeld, K. (1983) Sensitizing pigment in the fly. Biophys. Struct. Mech. 9: 319-328.

Wachmann, E. (1979) Untersuchungen zur Feinstruktur der Augen von Bockkäfern (Coleoptera, Cerambycidae). Zoomorphologie 92: 19-48.

Wada, S. (1974a) Spezielle randzonale Ommatidien der Fliegen (Diptera: Brachycera): Architektur und Verteilung. Z. Morphol. Tiere. 77: 87-125.

Wada, S. (1974b) Spezielle randzonale Ommatidien von Calliphora erythrocephala Meig. (Diptera, Calliphoridae): Architektur der zentralen Rhabdomeren-Kolumne und Topographie im Komplexauge. Int. J. Insect Morphol. Embryol. 3: 397-424.

Wada, S. (1975) Morphological duality of the retinal pattern in flies. Experientia 31: 921-923.

Waterman, T.H. (1966a) Information channeling in the crustacean retina. In: Proceedings of the symposium on information processing in sight sensory systems. Ed., P.W. Nye. Pasadena, California Institute of Technology, pp. 48-56.

Waterman, T.H. (1966b) Specific effects of polarized light on organisms. In: Environmental Biology, Eds., P.L. Altman and D.S. Dittmer, Bethesda, Fed. Am. Soc. Exp. Biol., pp. 155-165.

Waterman, T.H. (1966c) Systems analysis and the visual orientation of animals. Am. Sci. 54: 15-45.

Waterman, T.H. (1968) Systems theory and biology--view of a biologist. In: Systems Theory and Biology. Ed., M.D. Mesarovic. New York, Springer-Verlag, pp. 1-37.

Waterman, T.H. (1973) Responses to polarized light: Animals. In: Biology Data Book (Edition 2), vol. 2. Eds., P.L. Altman, D.S. Dittmer. Bethesda, Fed. Amer. Soc. Exp. Biol., pp. 1272-1289.

Waterman, T.H. (1975a) Natural polarized light and e-vector discrimination by vertebrates. In: Light as an Ecological Factor: II. Eds., G.C. Evans, R. Bainbridge, O. Rackham, Oxford, Blackwell, pp. 305-335.

Waterman, T.H. (1975b) The optics of polarization sensitivity.

In: Photoreceptor Optics. Eds., A.W. Snyder and R. Menzel. Berlin, Springer-Verlag, pp. 339-371.

Waterman, T.H. (1977) The bridge between visual input and central programming in crustaceans. In: Identified Neurons and Behavior in Arthropods. Ed., G. Hoyle. New York, Plenum, pp. 371-386.

Waterman, T.H. (1981) Polarization sensitivity. In: Handbook of Sensory Physiology. Vol. VII/6B. Ed., H. Autrum. Berlin, Springer-Verlag, pp. 281-471.

Waterman, T.H. (1982) Fine structure and turnover of photoreceptor membranes. In: Visual Cells in Evolution. Ed., J. Westfall. New York, Raven, pp. 23-41.

Waterman, T.H., Aoki, K. (1974) $\underline{E}$-vector sensitivity patterns in the goldfish optic tectum. J. Comp. Physiol. 95: 13-27.

Waterman, T.H., Fernandez, H.R. (1970) $\underline{E}$-vector and wavelength discrimination by retinular cells of the crayfish Procambarus. Z. Vergl. Physiol 68: 154-174.

Waterman, T.H., Horch, K.W. (1966) Mechanism of polarized light perception. Science. 154: 467-475.

Waterman, T.H., Piekos, W.B. (1983) Nocturnal rhabdom cycling and retinal hemocyte functions in crayfish (Procambarus) compound eyes. II. Transmission electron microscopy and acid phosphatase localization. J. Exp. Zool. 225: 219-231.

Waterman, T.H., Wiersma, C.A.G. (1963) Electrical responses in decapod crustacean visual systems. J. Cell. Comp. Physiol. 61: 1-16.

Waterman, T.H., Fernandez, H.R., Goldsmith, T.H. (1969) Dichroism of photosensitive pigment in rhabdoms of the crayfish Orconectes. J. Gen. Physiol. 54: 415-432.

Wehner, R. (1976) Structure and function of the peripheral pathway in hymenopterans. In: Neuronal Principles in Vision. Eds., F. Zettler, R. Weiler. Berlin, Springer-Verlag, pp. 280-333.

Wehner, R. (1981) Spatial vision in arthropods. In: Handbook of Sensory Physiology. Vol. VII/6C. Ed., H. Autrum. Berlin, Springer-Verlag, p. 287-617.

Wehner, R. (1982) Himmelsnavigation bei Insekten. Neurophysiologie und Verhalten. Neujahrsbl. Naturf. Ges. Zürich 184: 1-132.

Wehner, R., Bernard, G.D. (1980) Intracellular optical physiology of the bee's eye. II. Polarizational sensitivity. J. Comp. Physiol. 137: 205-214.

Wehner, R., Meyer, E. (1981) Rhabdomeric twist in bees - artefact or in vivo structure? J. Comp. Physiol. 142: 1-17.

Wehner, R., Srinivasan, M.V. (1981) Searching behavior of desert ants, genus Cataglyphis (Formicidae, Hymenoptera). J. Comp. Physiol. 142: 315-338.

Wehrhahn, C. (1976) Evidence for the role of receptors R7/8 in the orientation behaviour of the fly. Biol. Cybern. 21: 213-220.

Wellington, W.G. (1974) A special light to steer by. Nat. Hist.

83: 46-53.

Wellington, W.G., Fitzpatrick, S.M. (1981) Territoriality in the drone fly, Eristalis tenax (Diptera: Syrphidae). Can. Entomologist. 113: 695-704.

Wellington, W.G., Sullivan, C.R., Henson, W.R. (1951) Polarized light and body temperature level as orientation factors in the light reactions of some hymenopterous and lepidopterous larvae. Can. J. Zool. 29: 339-351.

Wiersma, C.A.G., Yamaguchi, T. (1966) The neuronal components of the optic nerve of the crayfish as studied by single unit analysis. J. Comp. Neurol. 128: 333-358.

Williams, D.S. (1981) Twisted rhabodmeres in the compound eye of a tipulid fly (Diptera). Cell Tissue Res. 217: 625-632.

Williams, D.S. (1982) Ommatidial structure in relation to turnover of photoreceptor membrane in the locust. Cell Tissue Res. 225: 595-617.

Wilson, R.S. (1977) Light elicited behavior of the marine dinoflagellate, Ceratium dens. Ph.D. Thesis, University of California, Santa Barbara. 160 pp.

Wolf, R., Gebhardt, B., Gademann, R., Heisenberg, M. (1980) Polarization sensitivity of course control in Drosophila melanogaster. J. Comp. Physiol. 139: 177-191.

Wunderer, H., Smola, U. (1982a) Fine structure of ommatidia at the dorsal eye margin of Calliphora erythrocephala Meigen (Diptera:Calliphoridae): an eye region specialized for the detection of polarized light. Int. J. Insect Morphol. Embryol. 11: 25-38.

Wunderer, H., Smola, U. (1982b) Morphological differentiation of the central visual cells R7/8 in various regions of the blowfly eye. Tissue and Cell 14: 341-358.

Yamaguchi, T. (1967) Mechanism of polarized light perception and its neural processes through the optic nerves in crayfish. Zool. Mag. 76: 443 (Abstr.)

Yamaguchi, T., Katagiri, Y., Ochi, K. (1976) Polarized light responses from retinula cells and sustaining fibers in the mantis shrimp. Biol. J. Okayama Univ. 17: 61-66.

Young, R.W. (1978) Visual cells, daily rhythms, and vision research. Vision Res. 18: 573-578.

Zeil, J. (1983) Sexual dimorphism in the visual system of flies: the divided brain of male Bibionidae (Diptera). Cell Tissue Res.: 229: 591-610.

Zrenner, E. (1983) Neurophysiological Aspects of Color Vision in Primates. Berlin, Springer-Verlag, 218pp.

PHOTORECEPTION IN PROTOZOA, AN OVERVIEW

PIERRE COUILLARD

Département de Sciences Biologiques
Université de Montréal
C.P. 6128, Succursale "A"
Montréal, Québec, Canada H3C 3J7

> Cease to persuade, my loving Proteus.
> The Two Gentlemen of Verona, Act I, Sc. 1.

I remember Lynn Margulis showing in a seminar, that the lowly Prokaryotes, Bacteria and blue-green Algae, had invented most everything in Biochemistry, well before we, the Eukaryotes, came onto the scene. Photosynthesis, glycolysis, protein synthesis, dicarboxylic acid cycle, oxidative phosphorylation, you name it and some microbe has it!

In the same vein, it is my purpose here to demonstrate that the Protozoa, the earliest Prokaryotes, have also done some innovating of their own in the field of photoreceptor design. Of course, this is not unexpected. Protozoa are the result of one to two billion years of unceasing evolution without having bothered to complicate matters by going multicellular (Rogers, 1976). This also accounts for their extraordinary diversity. A recent classification scheme (Levine et al., 1980) recognizes no less than seven phyla within the group! So, there is probably as much difference, phylogenetically speaking, between an amoeba and a paramecium as between a fish and a slug! Thus, analogous structures in different groups will not necessarily imply close evolutionary relationships; as in the higher groups, it could just as well be a case of convergent evolution.

The subject of protozoan photoreception is not new in NATO-ASIs, it was part of my presentation at an earlier Lennoxville NATO-ASI (Couillard, 1978). Another NATO-ASI held in Versilia Italy in September 1979 on Photoreception and sensory transduction in aneural organisms, gives a very thorough coverage of the subject (Lenci & Colombetti, 1980). For a review of the earlier literature see Halldal (1964).

Phototactic responses

Short of electrophysiologically monitoring membrane events, as was done, for example, with the photoresponse of Stentor (Wood, 1976), the simplest way of recognizing photoreception in Protozoa is by its effects on cell movement. Light responses can assume a variety of modalities as summarized in Table I. The net result is positive or negative phototaxis. The animal either seeks the light source and accumulates in lighted areas or disperses away from it. These evolutionary acquired responses have a high survival value; photosynthetic organisms can optimize their anabolism, predatory species can find higher prey concentrations and photosensitive forms can avoid radiation damage

The organisms we shall be concerned with live on the photic zones of fresh or ocean waters, penetrated by parallel rays from an infinitely distant source (the Sun) with no inherent gradient-forming properties, compared, for example, to a close point source, except in shade situations. Background noise may be very high: reflections and refractions from an often wavy surface, diffraction by suspended particulates, local convection which alter the orientation of organisms. Except for the very smallest, protozoa are large enough to be unaffected by Brownian motion.

TABLE I

TERMINOLOGY OF BEHAVIOURAL RESPONSES

RESPONSE: Any stimulus induced alteration in activity
(Step up (on) as step down (off))

PHOBIC RESPONSE: A change in stimulus intensity elicits transient alteration in activity. Organism eventually resumes original behavior. (adaptation).
(shock reaction)
(stop response)
(motor response)

N.B. If very long adaptation time, indistinguishable from kinesis.

KINESIS: Effect on steady state of activity.
No adaptation. (Positive or negative)

ORTHOKINESIS: Stimulus alters linear velocity of organism.

KLINOKINESIS: Stimulus alters frequency of directional changes.

Carlile (1980), Diehn et al. (1977), Doughty & Diehn (1980)

Photoreception in Phytoflagellates

We shall deal first with phototaxis in green Flagellates, a group claimed both by algologists because of their photosynthetic ability and by protozoologists for their motility. For these flagellated plant-like unicells, phototactic adaptations present a very high adaptive value. It should not come as a surprise, therefore to find in this group of Protozoa the most advanced photosensory organelles.

Simple photoreceptor models

It was originally assumed that all that was needed for phototaxis was a simple pigment spot or stigma, superficially located in the front of the cell and somehow transmitting a quantitative signal to the locomotor apparatus. It does not take much to show that such a device is quite inefficient. As seen in Fig. 1-a, as long as the stigma lies within the illuminated hemisphere, there is little change in signal intensity with direction of movement. It is only when the direction of the cell relative to the light brings the stigma in the dark hemisphere that self shading brings about an attenuation of the signal (Fig. 1-b). Variations with time of stigma illumination are small unless the organism undergoes a drastic change of direction with respect to the light.

The situation improves somewhat if the organism rotates on its directional axis. Such rotation is frequent in motile Protozoa, either to compensate for body asymmetry, as in Paramecium, or to counteract an asymmetric orientation of the locomotor apparatus as in Euglena (Fig. 4) or in Dinoflagellates (Fig. 7). The net result is that the cell moves in an helicoidal pathway around its general axis of direction.

As shown in Fig. 2, this significantly improves the efficiency of our simple photodetector especially if it is situated close to the equator of the cell relative to the polar axis of rotation and direction. Indeed, a subequatorial detector undergoes cyclic changes of illumination if the organism deviates in any extent from a true phototactic heading. Information-wise, this oscillating signal is much more effective. As we

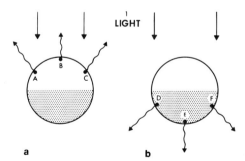

Fig. 1. A simple, non-directional photoreceptor in a non-rotating flagellate has low phototactic discrimination. Identical steady signals will result from path directions (A, B or C) or (D, E or F).

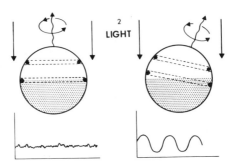

Fig. 2. In a rotating cell, a non-directional photoreceptor will be more efficient if close enough to cell equator. Slight deviations from true light path will induce signal fluctuations via self-shading.

said before, the rotating stigma must be close to the cell's equator for small deviations from true path to be detected.

Further improvement is gained if our rotating cell acquires a directional photodetector (Fig. 3). In this case, self-shading loses much of its importance and high-amplitude cyclic signals are obtained from anywhere in the front part of the cell unless the organism points directly towards the light source, in which case, the signal becomes steady.

In a very provocative paper on algal phototaxis, Foster and Smyth (1980) give a convincing demonstration that phytoflagellate unicells (motile algae and gametes of some multicellular algae), have developed a remarkable variety of complex photodetectors. These can be assimilated to scanning (rotating) directional light antennae. Further information on the structure and functions of phytoflagellate photodetectors can also be found in Greuet (1982). A useful classification of algal eyespots with respect to their ultrastructure and degree of association with chloroplasts has been proposed by Dodge (1969).

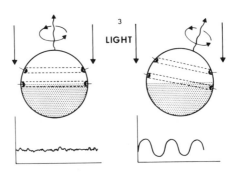

Fig. 3. A directional photoreceptor in a rotating cell will be efficient in all positions. Signal will be stable only under true light path orientation (left).

According to Foster and Smyth (1980), directional photoreceptors (light antennae) of phototactic algae correspond to four basic types:

I. <u>Simple screen eyespots:</u>

In Euglenophyceae the stigma is a 3 - 7 u m disc of randomly agglutinated carotenoid globules, independent of the chloroplast (Heelis et al., 1980). This eyespot, though it may contain traces of flavins, appears to be inactive photochemically (Heelis et al., 1981). It lies lateral to the wall of the reservoir, a deep invagination of the front tip of the cell from which emerge the short, vestigial, and the much longer propulsive flagella (Fig. 4). At the level of the eyespot, the active flagellum swells into the paraflagellar body, a crystalline array of flavoproteins (Benedetti & Checcucci, 1975; Checcucci et al., 1976; Benedetti & Lenci, 1977; Doughty & Dienh, 1980). The paraflagellar body can be considered as a dichroic crystal photodetector (Foster & Smyth, 1980), a characteristic which is probably responsible for the reported sensitivity of <u>Euglena</u> to polarized light (Bound & Tollin, 1967).

The functioning of this directional photodetector is well-known: (Wolken & Shih, 1958; Jahn & Bovee, 1968): <u>Euglena</u> rotates about its directional axis (1 - 2 Hz) as it progresses. If it diverges from the light direction, the paraflagellar body becomes shaded by the stigma for part of every turn. This causes the flagellum to deflect the cell laterally towards the light, each turn contributing a small correction to the trajectory of the cell. When the euglena finally points towards the light, eyespot shading of the paraflagellar body no longer occurs and no directional corrections intervene.

Similar light antennae have been described in <u>Phacus, Trachelomonas</u> and <u>Eutreptiella</u>.

In Chrysophyceae (golden algae), Xanthophyceae (yellow-green algae) and flagellated gametes of Phaeophyceae (multicellular marine brown algae) (Fig. 5). The stigma again acts as a simple shading device but it is situated within the chloroplast. The photoreceptor proper is still the paraflagellar body but without the crystalline structure found in

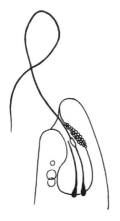

Fig. 4. The simple shading eyespot of <u>Euglena</u>.

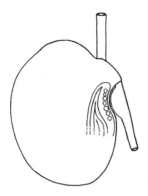

Fig. 5. The simple screen eyespot of Xanthophyceae zoospores. (Modified from Greuet, 1982).

Euglenophyceae. Light antenna orientation is normal to cell path. According to Greuet (1982), the region of apposition between the flagellar photoreceptor and the plasma membrane overlying the stigma is "synapse-like" and represents a case of coaptation. Whether or not this has functional implications is not said. The presence of a functional synapse between neighboring areas of the same cell's membrane would represent a unique situation in neurophysiology.

In Eustigmatophyceae, the paraflagellar body has evolved into a distinct extension of the flagellum, the paraflagellar button (Fig. 6), closely apposed to the plasma membrane at the level of the screening eyespot.

In all these cases, shading of the flagellar photoreceptor by the stigma in the course of cell rotation, brings about a step-wise reorientation of the cell path by the flagellar apparatus.

II. <u>Stigma acting as quarterwave interference reflector.</u>

In Chlorophyceae (green algae like <u>Volvox</u> and <u>Chlamydomonas</u>), the stigma is part of the chloroplast and contains from 2 to 9 layers of pigment globules separated by layers of unpigmented ground plasm. Taking into account the spacing of these layers and their respective

Fig. 6. The paraflagellar button of Eustigmatophyceae. <u>(Polyedriella)</u> (Modified from Greuet 1982).

indices of refraction, Foster and Smyth (1980) have calculated that they should act as quarterwave interference reflectors. Indeed, these stigmata do reflect light as shown (ibidem) by a photograph of Volvox under epiillumination.

The photoreceptor proper is believed to be a specialized, slightly thicker patch of plasma membrane just overlying the reflector stigma (Walne & Arnott, 1967). The identity of the photoreceptor pigment is not known. Such a design makes for a very efficient, directional light antenna. Light from the front is reflected back upon the sensitive membrane area thus doubling the exposure, while light from the back, already attenuated by passage through the cytoplasm, is kept from impinging upon the detector by reflection and by absorption. The reduced phototactic ability of stigma-less mutants of Chlamydomonas is attributed to shading of the membrane photoreceptors by the chloroplast (Morel-Laurens & Feinlab, 1983). Similar quarterwave reflecting eyespots are found in some Dinoflagellates like Peridinium (Fig. 7). In all cases, the cell rotates along its axis of motion and the light antenna is equatorial or subequatorial, being directed normal to the cell path. The effect of periodic illumination of the light antenna in Chlamydomonas at least, involves calcium coupled modulation of flagellar movement (Schmidt & Eckert, 1976; Hyams & Borisy, 1978).

III. Slab waveguide photoreceptors shaded by stigma.

In Cryptophiceae (Cryptomonas, Chroomonas), the entire light antenna lies deep within the cell, and is contained in an extension of the chloroplast (Fig. 8). The photoreceptor proper consists in a stack of thylakoid discs which contain the photopigment, in this case, phycoerythrin (Watanabe & Furuya, 1974). The discs are perpendicular to the cell's long axis. By themselves, these alternate layers of refractile discs,

Fig. 7. The quarter-wave interference reflector eyespot of Chlorophyceae and Dinophyceae. Here, the Dinoflagellate Peridinium. (modified from Foster & Smyth, 1980).

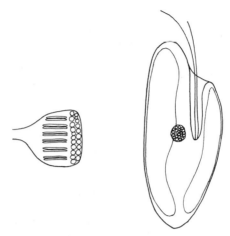

Fig. 8. The slab waveguide photoreceptor of Cryptophyceae (Cryptomonas). (Modified from Foster & Smyth, 1980).

separated by less refractile ground plasm, can act as a slab waveguide, transmitting preferentially light that comes in parallel to the plane of the discs. The addition, at one end of the stack, of a light-screening stigma, absorbing light coming from the opposite direction gives directionality to the antenna.

IV. The ocelloid of Warnowiidae.

This family of planktonic marine Dinoflagellates is exceptional in many respects: Lacking photosynthetic pigments, they feed by phagocytosis. They are rather rare (Francis, (1967) reports 1 - 40 cell/1 S.W. in shore waters off La Jolla, Calif.) and very fragile, which explains the rarity of behavioral studies. All are characterized by the possession of an ocelloid which appears, at the light microscope level, as a pigment spot (melanosome) overlaid by a hyaline hemisphere (hyalosome) which may or may not protrude from the cell surface (Fig. 9). The ocelloid is always

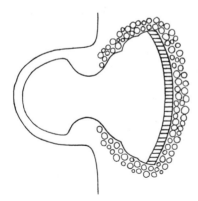

Fig. 9. The ocelloid of Warnowiaceae. (highly simplified from Foster & Smyth, 1980).

lateral to the long (locomotory) axis of the body and its position may vary from postequatorial to subapical according to species. Likewise the orientation of the ocelloid with respect to the swimming path of the cell, may also vary from species to species.

This extraordinary structure, thoroughly described by Greuet in many publications (Greuet, 1968 a, b, 1978, 1982), represents a unique case of evolutionary convergence between an intracellular photoreceptor and the eyes of vertebrates and cephalopods. It comes complete with cornea, lens, vitreous body, crystalline retinoid, backed by a pigment cup or quarterwave reflecting layer (Fig. 9). The lens is clearly a converging system, with a refractive index of 1,52, a field of view of 30° and an alleged capacity to form images in the plane of the retinoid (Francis, 1967). Indeed, in a seldom-observed species, Leucopsis cylindrica, there is a composite hyalosome, consisting of some 15 "lenses" converging upon a single retinoid, apparently devoid of pigment cup (Greuet, 1968 a).

At the least, this ocelloid can clearly function as a directional light antenna, in the Foster and Smyth (1980) sense. Thus it accounts for the strong positive phototaxis of Warnowiidae (Greuet, 1982). For the moment, one fails to understand which selective pressures might have directed the evolution of such a sophisticated organelle, having seen that phototaxis can effectively be achieved with far simpler structures. Warnowiidae are non-photosynthetic, auxautrophs which feed by phagocytosis; their ocelloid may enable them to reach high concentrations of phototactic prey, or even to detect individual prey organisms at short range (when they shade the photoreceptor). The highly ordered, palissadic structure of the retinoid may render it sensitive to polarized light (Greuet, 1982). If an image is really formed on the retinoid, we fail to see, in the present state of our knowledge, how an integrative computer could exist to analyze it within the cell.

Colonial phytoflagellates: Volvox.

A Volvox aureus colony may contain as many as 10 000 cells (zoids) interconnected by protoplasmic strands, on the surface of a gelatinous sphere. The colony rotates along its antero-posterior axis and is positively phototactic (Schletz, 1976).

We have already seen that the Volvox zooid has a directional, reflection-type photodetector at the base of its two locomotor flagella. Under normal circumstances, when the colony's polar axis is oriented parallel to the light direction, all zooids are evenly illuminated throughout the rotation cycle of the colony and movement proceeds unaffected towards the light. If the colony deviates from light direction, forward hemisphere cells undergo cyclic changes in direction and intensity of illumination with each rotation of the colony. Seen at the level of the individual zooid, this brings about a photophobic response, the flagella of the anterior hemisphere zooids stop beating when rotation brings them into the illuminated side, as the illumination level increases. For the colony as a whole, the resulting unbalance of motile forces deflects the motion towards the light direction (Gerisch, 1959). Lowering the temperature inverts the photophobic response (Sakaguchi & Tawada, 1977).

Photoreception in Ciliates

Viewed in the context of the Phytoflagellates, photoreception processes in Ciliates seem quite primitive despite the high degree of evolution of the Ciliates in other respects. Most ciliates lack visible pigmentation and will only react to extreme light intensities. Best known of pigmented ciliates are members of the spirotrich genera Stentor and Blepharisma.

Stentor. (For a general treatment of the genus, see Tartar, 1961).

This genus of trumpet-shaped spirotrich ciliates contains species of many shades and degrees of pigmentation including one, S. polymorphus, which often harbours symbiotic Zoochlorellae. Only two naturally-pigmented species, S. coeruleus and S. niger have been investigated for their phototaxis.

S. coeruleus, often said to be one of the most spectacular of Protozoa, measures up to 2 mm when extended. It owes its blue color to cortical stripes of pigment granules (0,3 - 0,7 μm dia) alternating with clear stripes centered on ciliary rows (kineties). It can attach by its pointed end to the substratum and feed in an extended configuration or swim about, assuming a pear-shaped form while rotating (0,5 - 1; Hz) in reaction to the metachronal beat of its peristomial row of membranelles. Holt and Lee (1901) were first to report that S. coeruleus accumulates in shaded areas. The reaction was studied in detail by Mast (1906). S. coeruleus reacts to a strongly-increasing light gradient by a typical photophobic response involving a sudden ciliary reversal, an inversion of the beat polarity of the cilia. This reaction, which has been well studied in Paramecium (Naitoh & Kaneko, 1972; Naitoh, 1974), is due to a depolarization of the ciliary membrane accompanied by calcium influx.

Wood (1976) has studied the electrophysiological events accompanying photophobic responses in S. coeruleus. Working with animals impaled by intravacuolar microelectrodes, he found that light can elicit three distinct responses:

1 - Graded photic receptor potentials reaching occasionally 10-20 mV,
2 - In extended animals, threshold receptor potential elicit all-or-none action potentials associated with body contraction by subcortical fibrillar myonemes.
3 - In pear-shaped stentors (swimming form), regenerative responses, varying from a few mV to 60 mV, which are associated with ciliary reversal in peristomial membranelles and somatic cilia.

The action spectrum of the ciliary response matches the absorption spectrum of the cortical stripe pigment, stentorin, a mesonaphtodianthrone If stentors are bleached in 0,3 M caffeine, the light stimulation threshold rises by a factor of 20. According to Wood (1976), this makes stentorin a true photoreceptor pigment, not merely a photosensitizer. Repeated light challenges induces habituation in Stentor (Wood, 1973).

Song (1981) and Song et al. (1981) have elucidated the molecular events leading to the measured electrophysiological and ciliary responses of S. coeruleus to a positive light gradient. These involve transient proton flux from pigment granules to cytoplasm leading to a transient receptor potential which in turn elicits a regenerative action potential accompanied by Ca^{++} influx. As in Paramecium, this Ca^{++} influx triggers ciliary reversal.

Blepharisma. (For a general coverage of the genus, see Giese, 1973).

When cultivated in the dark, this spirotrich ciliate accumulates a pink pigment, blepharismin, in subpellicular granules. Blepharismin is closely related, chemically, to stentorin. Blepharisma does not exhibit a clear photophobic response (Song, 1981), although we have some evidence of phototactic behavior; in cultures, the ciliates are bottom-dwellers in the daytime while they are randomly distributed in the fluid space during the night.

Blepharismin is clearly a photosensitizing pigment. Well-pigmented cells are killed by strong light in the presence of oxygen. Purified blepharismin can also photosensitize unpigmented protozoa such as Paramecium

It is of interest that chemical analogues of stentorin and blepharismin also occur in plants, e.g. weeds of the genus Hypericum and Buckwheat (Fagopyrum). Hypericum is dreaded by farmers as it causes an often fatal hypersensitivity to light in cattle and sheep which consume it.

Ciliates with symbiotic algae, Paramecium bursaria.

Members of the genus Paramecium are not naturally pigmented and are indifferent to light of normal intensities. P. bursaria, which, in nature, always contains symbiotic zoochlorellae, is positively phototactic (Gelber, 1956). The association is mutually advantageous; within their homeostatic shelter, algal symbionts find a source of raw materials (including host respiratory CO_2) for photosynthesis and are transported to favourable light environments. The host cell, in turn, obtains photosynthetic end products, mostly in the form of maltose, and oxygen (Brown & Nielsen, 1974). It seems also that noxious algal metabolites protect the paramecium from predation (Berger, 1980).

The mechanism of the phototactic response is simple. A resting P. bursaria is not affected by sudden increases in illumination. If however the light level is suddenly decreased, a phobic response is initiated in the form of a ciliary reversal and the paramecium backs up. This same ciliary reversal in decreasing light explains why paramecia show more frequent direction changes while swimming down a light gradient than when moving towards the light source (klinokinesis). This lead to accumulation of P. bursaria in illuminated areas (Saji & Oosawa, 1974). The phototactic response is associated with the presence of algae in the cell. Preliminary action spectra from photoaccumulation of P. bursaria were established by Pado (1972). Riboflavins and/or carotenoids (max. 450 nm) seem to be involved for short term light exposures (2 min). Over longer periods (3,5 h) a photosynthetic-like spectrum is obtained. Chlorella, a non-motile alga being devoid of specific photoreceptor (eyespot), chloroplast pigments must be involved

In the paramecium, each alga lies within a separate vacuole (Karakashian et al., 1968). We are still a long way from understanding how photoreception by algal chloroplasts can lead to membrane electrophysiological phenomena associated with ciliary reversal in the host. The rather long delay in response, 1,5 - 2 s. (Pado, 1972), is probably accounted for by diffusion processes in linking algal receptors and ciliate membrane effectors. The Paramecium - Chlorella system lends itself to a variety of experimental approaches. Controlling ciliate nutrition and light regime allow the number of algae per paramecium to be controlled down to nil. Algae can be grown in vitro and axenic paramecia can be reinfected at will. Thus light response could be correlated with the number of algal symbionts per host cell.

Photoresponses in an unpigmented protozoon, Amoeba proteus.

This giant, uninucleate amoeba is found in shallow, shady, tranquil fresh waters. Diffuse light causes an increase in rate of locomotion, up to 15 000 lux (Mast & Stahler, 1937). Under lateral illumination, the cells orient parallel to the light (Mast, 1911, 1931), a position in which self shading is maximized. This is attributed to the non specific effects of longer wavelengths on cell temperature. More precisely, if an advancing pseudopod is challenged with a light spot, it stops and diverts laterally. The light shock is most effective if applied to the region of granular cytoplasm just posterior to the hyaline cap, in front of an active pseudopod (Mast, 1930, 1932, 1964). The shorter wavelengths are especially effective (Mast, 1911). Earlier studies by Harrington and Leaming (1900) found a maximum effect on cytoplasmic streaming at 450 - 490 nm. This was later corrected to 515 nm (in the green) by Hitchcock (1961) though on the basis of very unconvincing action spectra.

These Amoeba photoresponses were attributed by Mast (1932) to a gelating effect of light on cortical endoplasm. Similar conclusions were reached, under controlled conditions, by Grebecki (1981) and Grebecka (1982).

As to why visible light, in its shorter wavelengths, should induce protoplasmic contraction in an apparently colorless protozoon, we must remember that all mitochondria contain essential respiratory pigments, the flavins, which absorb maximally at 450 nm and the cytochromes, with their Soret band at 420 nm. Flavins are quite generally involved as photoreceptors not only in Phytoflagellates, as we have seen, but also in plants and invertebrates (Ninnemann, 1980). As for cytochromes, Kato et al. (1981) have shown conclusively that light at 420 nm activates respiration of mouse liver mitochondria. Mitochondria can accumulate calcium and are believed to regulate protoplasmic concentration of this important cation. Calcium itself probably controls protoplasmic contractility (Durham, 1974), particularly in Amoeba (Taylor et al., 1973). It seems logical to suggest that light absorption by flavins and cytochromes could induce calcium release by mitochondria and consequent local activation of the contractile system.

REFERENCES

Ali, M.A. (Ed.) (1978) Sensory Ecology: Review and Perspectives. New York, Plenum Press. 597 pages.

Benedetti, P.A. & Checcucci, A. (1975) Paraflagellar body (PFB) pigments studied by fluorescence microscopy in Euglena gracilis. Plant Sci. Lett. 4: 47-51.

Benedetti, A. & Lenci, F. (1977) In vivo microspectrofluorometry of photoreceptor pigment in Euglena gracilis. Photochem. Photobiol. 26: 315-318

Berger, J. (1980) Feeding behaviour of Didinium nasutum on Paramecium bursaria with normal or apochlorotic zoochlorellae. J. Gen. Microbiol. 118: 379-404.

Borle, A.B. (1973) Calcium metabolism at the cellular level. Fed. Prod. 32: 1944-1950.

Bound K.E. & Tollin, G. (1967) Phototactic response of Euglena gracilis to polarized light. Nature (Lond.) 216: 1042-1044.

Brown, J.A. & Nielsen, P.J. (1974) Transfer of photosynthetically produced carbohydrate from endosymbiotic chlorellae to Paramecium bursaria. J. Protozool. 21: 469-470.

Carlile, M.J. (1980) Sensory transduction in aneural organisms. In: Photoreception and SensoryTransduction in Aneural Organisms. Ed. Lenci, F. & Colombetti, G. New York, Plenum Press, p. 1-22.

Checcucci, A., Colombetti, G., Ferrari, R. & Lenci, F. (1976) Action spectra for photoaccumulation of green and colorless Euglena. Evidence for identification of photoreceptor pigments. Photochem. Photobiol. 23: 51-54.

Couillard, P. (1978) Taxes in unicells. Especially protozoa. In: Sensory Ecology. Ed. Ali, M.A. New York, Plenum Press, p. 31-54.

Diehn, B., Feinleib, M., Haupt, W., Hildebrand, E. & Lenci, F. (1977) Terminology of behavioural response of motile microorganisms. Photochem. Photobiol. 26: 559-560.

Dodge, J.D. (1969) A review of the fine structure of algal eyespots. Br. Phycol. J. 4: 199-210.

Doughty, M.D. & Diehn, B. (1980) Flavins as photoreceptor pigments for behavioral responses. In: Structural Bonding, n°41, Sensory Physiology. Ed. Dunitz et al. New York, Springer Verlag.

Durham, A.C.H. (1974) A unified theory of the control of actin and myosin in non-muscle movements. Cell 2: 123-136.

Forget, J. & Couillard, P. (1983) La cinétique de la vacuole contractile chez Amoeba proteus: Effets de la lumière panchromatique. Can. J. Zool. 61: 518-523.

Foster, K.W. & Smyth, R.D. (1980) Light antennae in photosynthetic algae. Microbiol. Rev. 44: 572-630.

Francis, D. (1967) On the eyespot of the Dinoflagellate Nematodinium. J. Exp. Biol. 47: 495-501.

Gelber, B. (1956) Investigations of the behavior of Paramecium aurelia. III. The effect of the presence and absence of light on the occurrence of a response. J. Genet. Psychol. 88: 31-36.

Gerisch, G. (1959) Die Zelldifferenzierung bei Pleodorina californica und die organization der Phytomonadenkolonien. Arch. Ptotistenk. 104: 292-358.

Giese, A.C. (1973) Blepharisma. Stanford, Stanford University Press, 366 pages.

Grebecka, L. (1982) Local contraction and the new front formation site in Amoeba proteus. Protistologica 18: 397-402.

Grebecki, A. (1981) Effects of localized photic stimulations in amoeboid movement and their theoretical implications. Eur. J. Cell Biol. 24: 163-175.

Grenet, C. (1968) Leucopsis cylindrica nov. gen. nov. sp. Peridimia Warnowiidae Lindemann: Considérations phylogénétiques sur les Warnowiidae. Protistologica 4: 419-422.

Greuet, C. (1968) Organisation ultrastructurale de l'ocelle de deux Péridiniens Warnowiidae, Erythropsis pavillardi, Kofoid et Sweeney et Warnowia pulchra Schiller. Protistologica 4: 209-230.

Greuet, C. (1978) Organisation ultrastructurale de l'ocelloïde de Nematodinium. Aspect phylogénétique de l'évolution du photo-récepteur des Péridiniens Warnowiidae Lindemann. Cytobiologie 17: 114-136.

Greuet, C. (1982) Photorécepteurs et phototaxie des Flagellés et des stades unicellulaires d'organismes inférieurs. Ann. Biol. 21: 98-141.

Halldal, P. (1964) Phototaxis in Protozoa. In: Biochemistry and Physiology of Protozoa. III. Ed. S.H. Hunter. New York, Academic Press. p. 277-296.

Harrington, N.R. & Leaming, E. (1900) The reactions of Amoeba to light of different colors. Am. J. Physiol. 3: 9-66

Heelis, D.V., Heelis, P.F., Bradshaw, F., & Phillips, G.O. (1981) Does the stigma of Euglena gracilis play an active role in the photoreception processes of this organism? A photochemical investigation of isolated stigmata. Photobiochem. Photobiophys. 3: 77-81.

Heelis, D.V., Heelis, P.F., Kernick, N.A. & Phillips, G O. (1980) The stigma of Euglena gracilis strain 2: An investigation into the possible occurrence of carotenoids and nucleic acids. Cytobios. 29: 135-143.

Hitchcock, L. (1961) Color sensitivity of the amoeba revisited. J. Protozool. 8: 322-324.

Holt, E.B. & Lee, F.S. (1901) The theory of phototactic response. Amer. J. Physiol. 4: 460-468.

Hyams, J.S. & Borisy, G.G. (1978) Isolated flagellar apparatus of Chlamydomonas: characterization of forward swimming and alteration of waveform and reversal of motion by calcium ions in vitro. J. Cell. Sci. 33: 235-253

Jahn, T.L. & Bovee, E.C. (1968) Locomotion and motile responses of Euglena. In: The Biology of Euglena. Vol. I. Ed. D.E. Buetow. New York, Academic Press, p. 45-108.

Karakashian, S.D., Karakashian, M.W. & Rudzinska, M.A. (1968) Electron microscopic observations on the symbiosis of Paramecium bursaria and its intracellular algae. J. Protozool. 15: 113-128.

Kato, M., Shinzawa, K. & Yoshikawa, S. (1981) Cytochrome oxidase is a possible photoreceptor in mitochondria. Photobiochem. Photobiophys. 2: 263-269.

Lenci, F. & Colombetti, G. (1980) Photoreception and Sensory Transduction in Aneural Organisms. (NATO-ASI). New York, Plenum Press, 422 pages.

Levine, N.D., Corliss, J.O., Cox, F.E.G., Deroux, G., Grain, J., Honigberg, B.M., Ceedale, G.F., Loeblich, A.R., Lom, J. Lynn D., Merinfeld, D., Page, E.G., Poljansky, G., Sprague, V., Vaura, J. & Wallace, F.G. (1980) A newly revised classification of the Protozoa. J. Protozool. 27: 37-58.

Mast, S.O. (1906) Light reactions in lower organisms. I. Stentor coeruleus. J. Exp. Zool. 3: 359-399.

Mast, S.O. (1911) Light and the Behaviour of Organisms. London, New York, John Wiley & Sons.

Mast, S.O. (1930) Response of Amoeba to localized photic stimulation. Anat. Rec. 47: 283-284.

Mast, S.O. (1931) The nature of response to light in Amoeba proteus. Z. vergl. Physiol. 15: 137-147.

Mast, S.O. (1932) Localized stimulation, transmission of impulses and the nature of response in Amoeba. Physiol. Zool. 5: 1-15.

Mast, S.O. (1964) Motor responses in unicellular animals. V. In: Protozoa in Biological Research. Ed. G.N. Calkins & F.M. Summers. New York, Hafner Publishing Co., p. 271-351.

Mast, S.O. & Stahler, N. (1937) The relation between luminous intensity, adaptation to light and rate of locomotion in Amoeba proteus. Biol. Bull. 63: 126-133.

Morel-Laurens, N.M.L. & Feinlab, M.E. (1983) Photomovement in an "eyeless" mutant of Chlamydomonas. Photochem. Photobiol. 37: 189-194.

Naitoh, Y. (1974) Bioelectric basis of behavior in Protozoa. Am. Zool. 14: 883-893.

Naitoh, Y. & Kaneko, H. (1972) Reactivated Triton-extracted models of Paramecium: Modification of ciliary movement by calcium ions. Science 176: 523-524.

Nichols, K.M. & Rikmenspoel, R. (1980) Flagellar waveform reversal in Euglena. Expt. Cell Res. 129: 337-381.

Ninnemann, H. (1980) Blue photoreceptors. BioSci. 30: 166-170.

Pado, R. (1972) Spectral activity of light and phototaxis in Paramecium bursaria. Acta Protozool. 11: 387-393.

Rogers, J. (1976) When did evolution take its biggest step? New Scientist 71: 333.

Saji, M. & Oosawa, F. (1974) Mechanism of photoaccumulation in Paramecium bursaria. J. Protozool. 21: 556-561.

Sakaguchi, H. & Tawada, K. (1977) Temperature effect on the photoaccumulation and phobic response of Volvox aureus. J. Protozool. 24: 284-288.

Schletz, K. (1976) Phototaxis bei Volvox pigment system der lichtrichtungsperzeption. Z. Pflanzenphysiol. 77: 189-211.

Schmidt, J.A. & Eckert, R. (1976) Calcium couples flagellar reversal to photostimulation in Chlamydonomas reinhardii. Nature (Lond.) 262: 713-715.

Song, P.S. (1981) Photosensory transduction in Stentor coeruleus and related organisms. Biochem. Biophys. Acta 639: 1-29.

Song, P.S., Walker, E.B. & Yoon, M.J. (1980) Molecular aspects of photoreceptor functions in Stentor coeruleus. In: Photoreception and Sensory Transduction in Aneural Organisms. (NATO-ASI). Ed. F. Lenci & G. Colombetti. New York, Plenum Press. p. 241-252.

Song, P.S., Walker, B., Auerbach, R.A. & Robinson, J. (1981) Proton release from Stentor photoreceptors in the excited states. Biophy. J. 35: 551-555.

Tartar, V. (1961) The Biology of Stentor. New York, Pergamon Press. 413 pages..

Taylor, D.L., Condeelis, J.S., Moore, P.L. & Allen, R.D. The contractile basis of amoeboid movement. I. Chemical control of motility in isolated cytoplasm. J. Cell. Biol. 59: 378-394.

Tuffrau, M. (1957) Les facteurs essentiels du phototactisme chez le Cilié hétérotriche Stentor niger. Bull. Soc. Zool. Fr. 82: 354-356.

Walne, P.L. & Arnott, H.J. (1967) The comparative ultrastructure and possible function of eyespots: Euglena granulata and Chlamydomonas. Planta 77: 325-353.

Watanabe, M. & Furuya, M. (1974) Action spectrum of phototaxis in a cryptomonad alga, Cryptomonas sp. Plant Cell Physiol. 15: 413-420.

Wolken, J.J. & Shih, E. (1958) Photomotion in Euglena gracilis. I. Photokinesis and phototaxis. J. Protozool. 5: 39-46.

Wood, D.C. (1973) Stimulus specific habituation in a Protozoan. Physiol. Behav. 11: 345-354.

Wood, D.C. (1976) Action spectrum and electrophysiological responses correlated with the photophobic response of Stentor coeruleus. Photochem. Photobiol. 24: 261-266.

EVOLUTION OF EYES AND PHOTORECEPTOR ORGANELLES IN THE LOWER PHYLA

A. H. BURR

Department of Biological Sciences
Simon Fraser University
Burnaby, B.C.
Canada, V5A 1S6

> The eye altering alters all.
>
> William Blake, The Mental Traveller, 1800.

INTRODUCTION

How could such a complex organ as the vertebrate eye have evolved by natural selection of numerous, successive, slight modifications? Charles Darwin posed this question but could not answer it satisfactorily because of the rather limited knowledge of invertebrate eyes in his day.

However he indicated how it should be answered:

"If numerous gradations from a perfect and complex eye to one very imperfect and simple, each grade being useful to its possessor, can be shown to exist; if further, the eye does vary ever so slightly and the variations be inherited, which is certainly the case; and if any variation or modification in the organ be ever useful to an animal under changing conditions of life, then the difficulty of believing that a perfect and complex eye could be formed by natural selection, though insuperable by our imagination, can hardly be considered real" (Darwin, 1859).

For the vertebrate eye, though genetically inheritable variations are known, the question cannot be answered even today because of the lack of examples to fill the huge gap between the relatively primitive pigment-cup eyes of chordate ancestors and the fully-developed lens eye of the simplest vertebrates. Fortunately, as we now know, the lens eye has evolved independently several other times, and stepwise evolution is suggested by the existence of intermediate grades along those distinct lines.

Examples of well-developed lens eyes of alciopid annelids, certain gastropods, cephalopods, crustacea and arachnids are described by Land (1981; 1983a,b). In this chapter I intend to show how eyes could have evolved using examples chosen from the lower invertebrate phyla. The topic splits conveniently into two: evolution of the eye as an organ, and evolution of photoreceptor organelles.

The gradation in complexity of known light-receiving organs is so fine that it is difficult to draw a line indicating what is an eye and what is commonly referred to as an ocellus. I will avoid the problem by referring to, as an eye, any organ more complex than a single, unpigmented photoreceptor cell. The term 'photoreceptor' is commonly used to refer to either a photoreceptive organ, cell or organelle. Since it is the organelle that is specialized to detect photons, I will call it the photoreceptor and refer to the other structures by 'eye' and 'photoreceptor cell'.

EVOLUTION OF EYES

The most common eye type, the pigment-cup eye, apparently has evolved in several different ways. Examples from Bryozoa, Nematoda and Cnidaria will be used to illustrate several possible sequences. Examples from Cnidaria will also indicate one way a lens eye might have been formed from a pigment-cup eye.

Bryozoan line

Though the adult bryozoan is sessile, the larvae have a brief swimming phase during which positive and/or negative phototaxis is observed. Pigment spots are observed in the epidermis, the number and structure depending on species. In the 7 species studied by Woollacott and coworkers, a single ciliated photoreceptor cell is found in each pigment spot. This is associated with other epithelial cells in what Hughes and Woollacott identify as 3 levels of topological complexity (Hughes & Woollacott, 1980; Zimmer & Woollacott, 1977; Woollacott & Zimmer, 1972). A few more levels of morphological complexity are discernable and I have taken the liberty to arrange them in an order that suggests one possible sequence by which a pigment-cup eye could have evolved.

The simplest example is found in Bugula simplex and B. pacifica. Their larvae have 3 and 2 pairs of eyes, respectively. The sensory cell lies flush with the surface of the epithelium and because it is not invaginated, the tufts of 50-100 cilia that project from it lie above the surface (Fig. 1). The clump of cilia, thought to be the photoreceptor, would be shaded by the pigment found in the putative photoreceptor cell and adjacent corona cells. It would appear that the cilia of one of the corona cells (ciliated locomotory cells of the larvae) have become modified for photoreception and the cytoplasm of the cells of the immediate region has become pigmented to provide directional sensitivity.

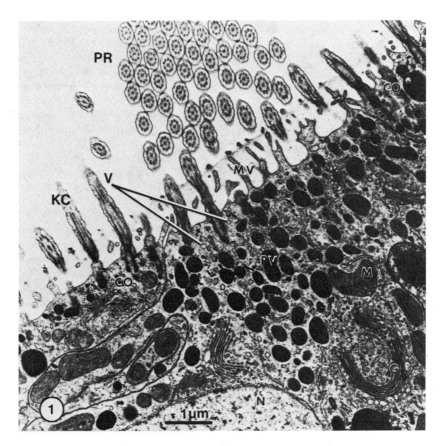

Fig. 1. Eye of Bugula simplex larvae (Bryozoa). A tuft of probably photoreceptive cilia (PR) projects from a single cell which connects to the nervous system. Pigment granules (PV) are present in both this cell and neighboring corona cells (CO). The latter bears kinocilia (KC). Branching microvilli (MV) project from both cell types. G, golgi network; M, mitochondrion; V, membrane vesicles. After Hughes and Woollacott (1980).

An earlier step in the evolution might have been a photoreceptor cell without shading pigment. An example may yet be found in bryozoa because larvae of certain species are known to be photosensitive though they lack pigment spots (Hughes & Woollacott, 1980).

A possible later step is illustrated by the eyes of B. stolonifera in which the apical surface of the photoreceptor cell is indented to form a shallow cup. Otherwise the eye structure is similar to that of B. simplex and B. pacifica. A deeply-indented sensory cell is found in B. turrita. The pair of posteriolateral eyes are separately located, but the anteromedial ones (Fig. 2) are nearly adjacent. In this species each eye appears to be composed of a single cell in which are incorporated both the shading and photosensory functions. The advantage of sinking the photoreceptor into a cup, besides a protective one, is that the directionality of photoreception is improved. The 2 anteromedial eyes appear to collect light entering through different but overlapping sectors of space (Fig. 2).

If the photoreceptor lay deeper in the pigment cup the directionality would be further improved. This occurs in the eyes of B. neritina and Tricellaria occidentalis. In these eyes, the sensory cell is at the base of an epidermal invagination, the third topological grade of Hughes & Woollacott (1980).

Among the eyes of these 2 species, 3 grades of structural complexity are seen. The posterolateral eyes of T. occidentalis are composed of a single sensory cell at the base of a depression. Modified parts of two corona cells form the sides and rim. In the cytoplasm of all 3 cells, the pigment granules are concentrated towards the depression. The two corona cells are further modified from their normal form by the absence of cilia in the pigmented zone.

The anteromedian eye of T. occidentalis is constructed of 2 modified corona cells and 2 sensory cells. Of the bryozoan larvae described to date, this is the only example of a multireceptor-cell pigment-cup eye. Since the photoreceptor cilia interdigitate, the 2 cells would receive light from the same directions. The advantage of such an increase in complexity therefore, is not clear in this case.

In the eyes of B. neritina we find a significant increase in complexity. The surface of the pigment cup is made up of a single receptor cell at the base and walls of modified corona cells as before. The pigment, however, is located in special pigment cells as is common in most pigment-cup eyes. In B. neritina these are subepidermal in location (Woollacott & Zimmer, 1972).

From these bryozoan examples one can reconstruct a plausible stepwise sequence for the evolution of pigment-cup eyes. Of the species included, only T. occidentalis is not in the same genus, and that example could be discarded without affecting the conclusion. Nevertheless, inferring an evolutionary sequence from modern examples is rather a risky venture, and the following sequence should be regarded as only one possibility which could have occurred: 1) formation of an unpigmented photoreceptor cell from a ciliated motor cell, 2) addition of shading pigment to the photoreceptor cell and neighbouring epithelial cells, 3) indentation of the receptor cell, 4) invagination of the epithelium to form a deeper cup, and in several steps, 5) increased specialization of photoreceptor cells, epithelial cells and pigmented cells of the eye. Each of these successive slight increases in complexity can be seen to be useful to the organism.

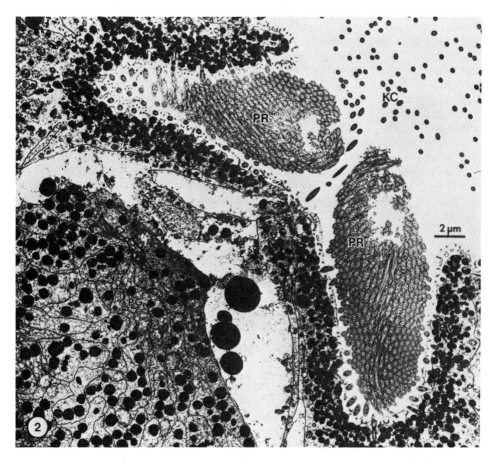

Fig. 2. Anteromedial eyes of Bugula turrita larvae (Bryozoa). The two pigment-cup eyes are separated by projections of underlying cells. KC, PR, PV as in Fig. 1. PG, pyroform gland. After Hughes and Woollacott (1980).

Further, individual variation of the structures is observed (Hughes and Woollacott, 1980). Therefore Darwin's criteria for "believing" that a pigment-cup eye can evolve by natural selection are met in Bryozoa.

Nematode lines

A small number of marine and aquatic nematodes have a noticeable pair of pigment spots in the anterior. A variety of structures has been noted and the morphology in 5 species has now been described at the electron microscope level (Siddiqui & Viglierchio, 1970; Croll et al., 1972; 1975; Burr & Burr, 1975).

Negative phototaxis has been observed in 3 of these species and photoklinokinesis has been observed in at least one species that lacks pigment spots (Burr, 1983).

Each eye contains a single photoreceptor and a shading pigment which is usually located within an anterior cell of the pharynx. The three examples in Fig. 3 indicate the range of morphologies described so far. The following evolutionary sequence could have occurred: 1) formation of the photoreceptor, 2) pigmentation of a nearby pharyngeal cell, 3) evagination of the pigmented cell, 4) formation of a pigment-cup from the pigmented cell. Steps 2, 3 and 4 are represented by the modern examples shown in Fig. 3.

A number of differences are noticeable between the nematode and bryozoan lines though the morphological result is the same: a cup-shaped shading structure containing a single photoreceptor. In the bryozoan larvae (excepting B. neritina), the pigmentation occurs in the photoreceptor cell or another epithelial cell, both of which are very similar to the ciliated motor cell (corona cell) of the epidermis and probably evolved from it. In the nematodes, however, the pigment and photoreceptor cells have originated from totally different cells which were already highly differentiated for other purposes. The photoreceptor cell probably evolved from a neuron whereas the pigment is located in the pharynx, the cells of which have a very different embryological lineage, at least in Caenorhabditis elegans (J.E. Sulston, E. Schierenberg, J.G. White, J.N. Thompson and G. von Eherenstein, personal communication). The pharynx is enclosed by a basal lamina which prevents any contact with the photoreceptor cell, whereas the 2 cell types of Bryozoa are in direct contact (Woollacott & Zimmer, 1972).

The existence of unique eye structures suggests that evolution of nematode eyes may have occurred independently at least 4 times. In Deontostoma californicum (Siddiqui & Viglierchio, 1970), Chromadorina bioculata (Croll et al., 1972), Chromadorina sp. (Croll et al., 1975) and Enoplus anisospiculus (Bollerup & Burr, 1979; Burr, unpublished), the shading pigment is concentrated in a region of a marginal cell of the pharynx, which continues to also maintain its normal function to support the cuticular lining of the lumen. The photoreceptor organelle is lamellar and a ciliary axoneme is absent. In Oncholaimus vesicarius, on the other hand, the pigment is located in the anterior of a pharyngeal muscle cell, the bulk of which is filled with myofilaments for its role in swallowing (Burr & Burr, 1975). The photoreceptor is a cluster of modified cilia. A third possible line has culminated in Araeolaimus elegans, in which the shading pigment is located outside the pharynx and is different ultrastructurally and spectrally from the pharyngeal pigment (Croll et al., 1975). The photoreceptor is lamellar. A fourth possible line is represented by Mermis nigrescens in which the presumptive shading pigment is located in large expansions of the anterior hypodermis (Ellenby & Smith, 1966; Croll et al., 1975; Burr, unpublished) and is an oxyhemoglobin (Ellenby, 1964; Burr et al., 1975), whereas the

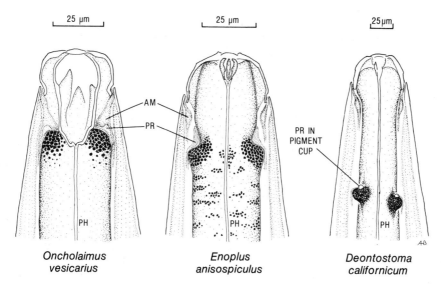

Fig. 3. Eye structure in 3 nematode species. A. In Oncholaimus vesicarius a multiciliary photoreceptor (PR) occurs in an olfactory organ anterior to a pigmented region in a pharyngeal muscle cell. Original, based on live worms and electron micrographs (Burr & Webster, 1971; Burr & Burr, 1975). B. In Enoplus anisospiculus a multilamellar, epigenic photoreceptor (PR) lies anterior to a pigmented evagination of a pharyngeal marginal cell. Original, based on live worms and unpublished electron micrographs. C. In Deontostoma californicum the pigmented region of a marginal cell is cup-shaped. Original, based on Siddiqui and Viglierchio (1970) and Hope (1967). PH, pharynx; AM, amphid, an olfactory organ.

pigment of the pharyngeal shading structures is probably a melanin (Bollerup & Burr, 1979). A photoreceptor has not yet been identified in Mermis, but the nature of the positive phototaxis (Cobb, 1926; 1929; Croll, 1966; Burr, to be published) suggests that a photoreceptor is shaded by the cylindrical pigmented region.

Cnidarian lines

The Cnidaria include the Hydrozoa, Scyphozoa (true jellyfish) and Anthozoa (sea anenomes). Their life cycles include 2 forms: polyps which are photosensitive but probably lack eyes, and medusae. They are thought to be phylogenetically more primitive than other metazoans, having only ectodermal and endodermal tissue and a limited degree of organ development. Surprisingly, one of the few evolutionary lines that lead to a lens eye occurs in medusae of certain Scyphozoa.

Eyes ranging in complexity from the flush epithelial grade to multireceptor pigment-cup eyes occur in the free-swimming medusae of Hydrozoa. Two examples are illustrated in Figs. 4 and 5. In Leuckartiara octona the ciliary photoreceptor cells are flush with the epidermis, as in Bugula simplex, but have only one cilium per cell (Fig. 4). The shading pigment is located in other epithelial cells and many photoreceptor cells occur in each eye (Singla, 1974).

In Bougainvillia principis (Fig. 5) the epidermis is invaginated to form a pigment cup. Each photoreceptor cell dendrite passes through the pigmented layer and projects 1-3 cilia into the cup. Microvilli, formed from the ciliary membrane, lie in clumps at the base of the pigment cup and probably constitute the photoreceptors. Branched projections of the pigment cells form a transparent mass that fills the center of the cup. Whereas the bryozoan pigment cup was open to the exterior, that of Bougainvillia is closed by a layer of epidermal cells (Singla 1974).

Eye complexity intermediate between Leukartiara and Bougainvillia is found in Polyorchis penicillatus and Sarsia tubulosa (Eakin & Westfall, 1962; Singla & Weber, 1982a,b). A pigment cup is formed of pigment and receptor cells. Each receptor cell projects a single cilium into the cup. Microvilli form from the ciliary membrane. In both species, microvilli also evaginate from the pigment cell membrane, from the apical surface and from a projection ("distal process"). These microvilli intermingle with the receptor cell microvilli in the cup. Distal to the presumed photosensory region of ciliary microvilli, the photoreceptor cilia of Sarsia swell into large vesicles and are covered by a mucous layer. In Polyorchis, on the other hand, the pigment cells form a similar layer of swollen tips. In both examples the pigment-cup eye is otherwise open; it is not enclosed by an epidermal layer as in Bougainvillia.

Bougainvillia eyes may have evolved from undifferentiated epithelium in the following sequence, a modification of one suggested by Singla (1974): 1) formation of photoreceptor cells in epithelium (There are several light-sensitive examples that lack pigment spots). 2) pigmentation of neighbouring cells to provide directional sensitivity (Leukartiara), 3) invagination of the epithelium (Polyorchis, Sarsia), 4) formation of distal processes of the pigment cells and microvillar elaborations of ciliary membrane (Polyorchis, Sarsia), 5) modification of the tip of distal processes into a large vesicle (Polyorchis) or branched projections (Bougainvillia) to form a transparent mass, 6) enclosure of the pigment cup by projection (Cladonema radiatum, Weber, 1981) or migration (Bougainvillia) of adjacent epithelial cells. All of the examples are Anthomedusae (Hydrozoa. Cnidaria).

The pigment-cup eyes of Polyorchis, Sarsia, Bougainvillia and Cladonema are examples of the everted type: one in which the photoreceptor cells enter the eye by penetrating through the layer of pigment cells. Inverted pigment-cup eyes occur in another

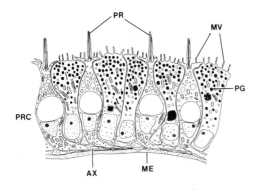

Fig. 4. Eye of <u>Leukartiara octona</u> medusae (Hydrozoa, Cnidaria). A single cilium (PR) and several microvilli (MV) project from each of many photoreceptor cells (PRC). Pigment granules (PG) are found only in the pigment cells. AX, axon; ME, mesoglea. After Singla (1974).

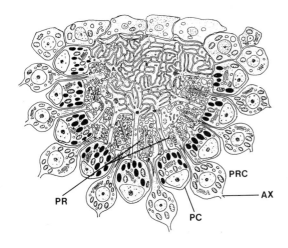

Fig. 5. Pigment-cup eye of <u>Bougainvillia principis</u> medusae (Hydrozoa, Cnidaria). A dendrite of the photoreceptor cell (PRC) passes between pigment cells (PC) and projects 1-3 cilia from its apical surface. Microvilli (PR) are formed from the ciliary membrane. Tubular projections of pigment cells (PC) branch to fill the center of the cup with a transparent füllmasse. AX, axon. After Singla (1974).

hydrozoan medusa, <u>Tiaropsis multicirrata</u> (Singla, 1974) and in the smaller eye of two types in a scyphozoan medusa, <u>Aurelia aurita</u> (Yamasu & Yoshida, 1973). In this type, only pigment cells form the cup and the numerous photoreceptor dendrites enter via the cup opening. Unlike in the other, larger eye of <u>Aurelia</u> and in other medusan eyes, the pigment cells of these 2 eyes are of endodermal origin and are separated from the photoreceptors by mesoglea. They could represent a second line of evolution in medusae.

Evolution of a lens

The transparent mass of the <u>Bougainvillia</u> eye, or füllmasse, may help maintain its shape by filling-up space. The swollen vesicles in <u>Polyorchis</u> and <u>Sarsia</u> may serve the same function. This type of structure is common in invertebrate eyes and is regarded as being a possible forerunner of a lens. In another hydromedusan, Cladonema radiatum, the füllmasse of a similar eye becomes filled with spherical granules during development. These later change into compact crystalline bodies (Weber, 1981). A later stage in evolution of the lens may be represented by the distal eye of the cubomedusan, Tamoya bursaria (Cnidaria, Scyphozoa) described by Yamasu and Yoshida (1976). This eye (Fig. 6 A,B) has many structural similarities to that of Bougainvillia and Cladonema. It is everse, the putative photoreceptors are villi derived from ciliary membrane, the pigment cells project distal processes through the region of ciliary microvilli, and the eye is closed by a layer of epithelial cells. The lens of Tamoya replaces the füllmasse and it appears to have partly a space-filling function. The cells of the lens are separated from the retina by a capsule of extracellular material.

The eye of Tamoya is obviously considerably more complex than the other examples, with a shaped lens and a retina composed of hundreds of photoreceptors. To evolve from an ancestor like Bougainvillia, or Cladonema however, no new tissues or structures need be added, only further proliferation of photoreceptor and pigment cells and a modification of the füllmasse into a lens. By natural selection, the optimum shape should evolve easily. The cells that make up the lens of Tamoya, however, have an uncertain origin. Because they may have arisen by further specialization of some of the pigment cells that form the füllmasse of Bougainvillia or Cladonema, it would be interesting if it could be shown in Tamoya that the lens cells have an ontogeny common with that of pigment cells.

In comparison with the eyes of fish or squid, Tamoya's eye is still quite primitive. Still lacking are a well-developed iris diaphragm, a focussing mechanism, muscles of ocular movement, and geometry for optimum resolution of the image. For possible sequences in the evolution of lenses of some higher invertebrates see Salvini-Plawen and Mayr (1977) and Land (1981, 1983a,b).

The question of whether or not the distal eye of Tamoya can actually form an image is an interesting one. The formation of a retina containing hundreds of photoreceptor cells would appear to be a waste of energy unless information could be derived from at least a low resolution image. The proximal edge of the lens, however, is adjacent to the retina and this usually indicates that an image cannot be formed on the retina (Land, 1981). However, the curvature of the lens is greater for the region of distal surface not covered by pigment cells (Fig. 6B), and the ratio f/r (Land 1981, p. 514) estimated from the drawing ranges from 2 to 3,8 if f and r are measured from the center of curvature to the receptor organelle layer and the distal surface, respectively. With a Matthiessen's ratio as high as 3,8 an image could be formed within the layer by a lens material of reasonable refractive index. The range of f/r reflects the thickness of the receptor organelle layer. Because of this thickness the image could never be in focus

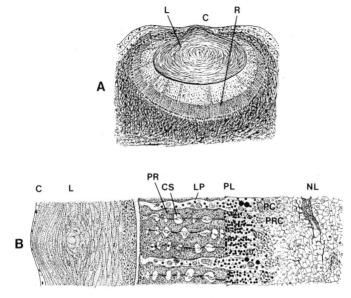

Fig. 6. Distal eye of <u>Tamoya bursaria</u> (Scyphozoa, Cnidaria). A, distal eye; B, details of structure. The eye is composed of a cornea (C), lens (L) and a retina which includes a receptor layer, pigmented layer (PL) and nerve layer (NL). Pigmented photoreceptor cells (PRC) project a cilium which swells at intervals (CS) and evaginates numerous microvilli (PR). Pigment cells (PC) send a long process (LP) through the layer of microvilli. After Yamasu and Yoshida (1976).

everywhere in the layer, and consequently the information collected by the retinal cells would have a low resolution.

The lens is structurally inhomogeneous with rounded cells in the center and flattened cells peripherally (Fig. 6B). If a lens is optically inhomogeneous as well, with a higher refractive index in the center, as in the fish, the ideal lens without spherical aberration is approached (Land, 1981). Demonstration of the presence of an advanced lens design in such a phylogenetically primitive organism would be of considerable interest.

Compound eyes and other eye types

Compound eyes appear only in the higher Protostoma: annelids, molluscs and arthropods, and they will be treated by others in this volume. Clusters of small pigment-cup eyes, seen in some Platyhelminthes, may be precursors of this type.

In terrestrial animals the cornea can act as the primary refracting surface. This eye type is common in terrestrial vertebrates and arachnids and occurs also in land snails and insect larvae (Land, 1981). The compound eye predominates in adult insects. Eyes that incorporate image-forming reflectors are found only in the scallop, Pecten, and the giant deep-sea ostracod Gigantocypris (Land 1981, 1983a,b).

Conclusions

The examples chosen from 3 unrelated phyla illustrate not only that stepwise evolution of eyes is a reasonable possibility, but equally important, that more than one independent line leading to pigment-cup eyes must have occurred within two of the phyla. In a review of all of the phyla, Salvini-Plawen and Mayr (1977) enumerated numerous examples of this among the "40 or 65" phyletic lines they identified.

In 2 of the lines I described above, both photoreceptor and pigment cells appear to have originated from relatively undifferentiated ciliated epithelial cells, but the everse pigment-cup eyes that resulted are different in many details. In the nematode examples, the eyes appear to have evolved from cells that were previously in an advanced state of differentiation for other purposes. In these and other known examples there is much fodder for the theorist.

The scyphozoans (Cnidaria) provide the only example of a lens eye in the lower Metazoa. Why are they lacking in such important groups as the platyhelminthes, aschelminthes, echinoderms and lower chordates even though pigment-cup eyes are common? (See Fig. 7 for possible relationships). Why did lens eyes arise in only a few isolated lines of the higher phyla, e.g. fishes, cephalopod molluscs, prosobranchs (gastropod molluscs), and alciopids (polychaete annelids) (Land, 1981; 1983a)? The occurrence in Cnidaria and absence in many phylogenetically more advanced phyla indicates that phylogenetic or tissue grade is unimportant; neither mesodermal tissue nor a coelom is required. Indeed, the occurrence of a lens eye in the predatory dinoflagellate, Erythropsis (Protista), shows that even a multicellular grade is unnecessary (Couillard, 1983).

A common feature of all the organisms with lens eyes is a fast-moving, free-swimming habit and moderate to large size. All but the cubomedusan are active predators. By contrast there are no examples of lens eyes in the phyla that are predominantly sessile, slow-moving, benthic, interstitial or parasitic. There may be selection towards medium resolution eyes of other types in these organisms. An example is the scallop, which has image-forming mirror optics in lieu of a lens (Land, 1981; Land, 1983a). Other examples are given by Salvini-Plawen and Mayr (1977).

The very small size of rotifers may have prevented their eyes from developing past the pigment-cup stage even though they are active swimmers. Their eyes, at about 5 μm in diameter, are 1/10 the size of the lens eye of Erythropsis. Since resolution is limited by eye size (Land, 1981), selection pressures for small body size may oppose pressures for increasing eye resolution and formation of lens eyes.

EVOLUTION OF PHOTORECEPTORS

Darwin's criteria can also be applied to the question of how photoreceptors evolved. Richard Eakin, over the past 20 years, has made a noble attempt to demonstrate a stepwise sequence in two lines of evolution, a ciliary line following along the deuterostomate phyla to the vertebrates, and a rhabdomeric line following along the protostomate phyla to the arthropods and molluscs (Eakin, 1963, 1979). His observations, ideas and review articles have been a major catalyst to research in the area. However his original idea is now contradicted by many exceptions. Ciliary and rhabdomeric organelle types appear to have newly arisen many times, even both types within the same phylum. With the increasing number of examples being described in the lower and minor phyla, other relationships are becoming more evident. New theories, proposed by Vanfleteren and Coomans (1976), Salvini-Plawen and Mayr (1977) and Clément (1980) nicely explain many observations. Nevertheless Eakin's is the only proposal so far that attempts to explain the noticeable trend towards rhabdomeric receptors in the Annelida, Mollusca, and Arthropoda.

The diversity of photoreceptor structure has been reviewed several times (Eakin, 1972; Salvini-Plawen & Mayr, 1977; Coomans, 1981; Vanfleteren, 1982). In this chapter I will present current ideas that appear to me to be most significant to the understanding of photoreceptor evolution and then will add a conceptual model which may be increasingly useful as the problem unfolds. Some of the ideas are traceable to Eakin (1979; 1982), Vanfleteren (1982) and Salvini-Plawen (1982). However the emphasis, the model and other ideas of this chapter are new to the field. Examples will be chosen primarily from lower phyla that are not reviewed in detail in other chapters of this volume.

Ciliary and rhabdomeric photoreceptors

All known photoreceptors have originated in cells of epidermal origin. The most common and best known are the ciliary and rhabdomeric types, distinguishable by whether or not the organelle is formed from the membrane of a cilium or of microvilli, respectively. Here I depart from Eakin's definition of rhabdomeric photoreceptors: all types in which the organelle is derived from the cell membrane proper. By restricting rhabdomeric photoreceptors to those formed of microvilli, I omit only one rare type included by Eakin's definition, to which I will give another name (see section on epigenous photoreceptors).

Eakin's definition assumes that photoreceptors composed of microvilli originally arose de novo by evagination of the cell membrane. It is much more likely that microvilli developed first for other purposes and acquired photosensitivity secondarily. This is what Eakin proposes for photoreceptors derived of cilia. As for cilia, the cytoskeleton of photoreceptor microvilli resembles other types. The central cytoskeleton of squid rhabdomeric microvilli consists of a core of actin filaments connected to the membrane by side-arms. Thus it resembles the cytoskeleton of the microvilli of the intestinal brush border, except for its much smaller diameter (Saibil, 1982). The core filament bundle of leech rhabdomeres is larger. The apparent absence of a bundle in some photoreceptor microvilli may be due to inadequate fixation (Saibil, 1982). It would be hard to explain how the cylindrical shape can be maintained without such a cytoskeleton, and it is likely that the core is a characteristic of all microvilli, however modified. It is only in the membrane envelope where qualitative differences between microvilli are seen.

The distinction between rhabdomeric and ciliary photoreceptors is not always clear. In many medusae (Cnidaria), the organelle is composed of a clump of microvilli that project from the membrane of a cilium. Though a few microvilli also originate from the apical cell membrane in some medusae (Singla, 1974; Singla & Weber, 1982a) the bulk of them project from the membrane of the cilium. In Tamoya the ciliary microvilli sometimes coalesce into compact parallel-hexagonal arrays as do the rhabdomeric microvilli in the rhabdomeres of higher protostomes (Yamasu & Yoshida, 1976). In many arthropods, the rhabdomeric organelles commonly appear adjacent to a cilium and the two arise simultaneously during development (Vanfleteren, 1982). Critical to classification in these confusing cases is the location of the boundary between the photosensitive membrane and the rest of the plasma membrane. This will be the subject of a later section.

Epidermal cells of all metazoa are similar in being able to produce microvilli and cilia on their external surface. Therefore, the genome of all metazoa must contain the genetic information necessary for their manufacture and maintainance. Cilia and microvilli are specialized domains of the external cell membrane and they contain the cytoplasmic and membrane components that maintain their special shape and function.

The evolution of photoreceptors probably occurred in the following stages: 1) evolution of a photosensitive cell membrane, 2) localization of photosensitivity in a particular domain of the cell membrane such as the ciliary or microvillar membrane 3) further increase in surface area of the membrane and 4) increase in organization of the organelle. Only stages 3 and 4 can be inferred from metazoan examples; the other stages must have occurred earlier in evolution.

Evolution of photosensitivity

Photosensitive behavior is clearly present in bacteria and Protista of many types. A rhodopsin-like, retinal-protein photopigment in Halobacterium is known to be involved in photobehavior but rhodopsin has not yet been positively identified in Protista (Hildebrand, 1978). Behavioral action spectra implicate a carotenoprotein, perhaps rhodopsin, as the photopigment of some phytoflagellates such as Volvox, Chlamydomonas and Platymonas (evidence discussed by Nultsch & Häder, 1979; Foster & Smyth, 1980). At least 2 additional classes of photopigment are implicated for other unicell species.

Much less work has been done on lower Metazoa. Since rhodopsin is the photopigment of vertebrates, arthropods and molluscs, it can be inferred that rhodopsin arose in a common ancestor. It is reassuring that ocellar potentials of a planarian (Turbellaria, Platyhelminthes, Brown et al., 1968) and of two hydromedusans (Weber 1982a,b) have action spectra resembling the absorbance spectrum of rhodopsin. From schema such as that of Fig. 7 one can infer that rhodopsin arose in Protista or a primitive metazoan.

Some components of sensory transduction now being discovered in unicells resemble those of vertebrates and arthropods. Paramecium has been the most extensively studied. Mechanical stimulation to the anterior opens a Ca^{++} channel and depolarizes the cell. Touching the posterior opens a K^+ channel and hyperpolarizes the cell. A depolarizing receptor potential triggers a regenerative depolarization by opening Ca^{++} channels located in the ciliary membrane. The resulting increase in intraciliary Ca^{++} concentrations inactivates the Ca^{++} channels and reverses ciliary beating (Eckert & Brehm, 1979). A similar, but light-induced, electrical response coupled to ciliary reversal has been recorded in another ciliate, Stentor (Wood, 1976). Receptor potentials have not yet been recorded intracellularly in flagellates, however several lines of evidence indicate that photopotentials and Ca^{++} are involved in light-induced reversals (Nultsch & Häder, 1979; Foster & Smyth 1980).

In Metazoa, intracellular recordings have been obtained of chordate, arthropod and mollusc photoreceptor potentials but none in the other phyla. Of those studied, most ciliary photoreceptors hyperpolarize due to light-activated closing of Na^+ channels in the ciliary membrane, and most rhabdomeric photoreceptors depolarize due to light-activated opening of Na^+ channels. However the hyperpolarization of the ciliary photoreceptor of the scallop (Pecten, Mollusca) is due to the opening of K^+ channels (McReynolds & Gorman, 1974; Gorman & McReynolds, 1978). This is probably the case also in a rhabdomeric photoreceptor in Salpa democratica (Tunicata, Chordata. Gorman et al., 1971; McReynolds & Gorman, 1975) and in the ciliary photoreceptor of the pineal organ of the trout (Vertebrata. Tabata et al., 1975).

We can conclude that, in Metazoa and some Protista, receptor potentials are generated by the modulation of ion permeabilities in a variety of ways. Until similar studies are done in other phyla,

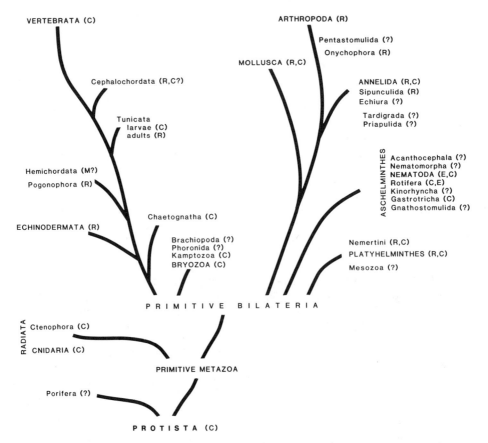

Fig. 7. Distribution of photoreceptor types among the phyla. C, ciliary; R, rhabdomeric; E, epigenous; M, mixed; ?, none yet identified or identification debatable. After Vanfleteren (1982).

however, it will be impossible to infer how each type of transduction mechanism has evolved.

Localization of photosensitivity

The localization of photosensitivity in special domains of the cell membrane occurs even in phytoflagellates. A primitive example is shown in Fig. 8, an electron micrograph of a freeze-fracture preparation of Chlamydomonas reinhardii. A cluster of 8-12 nm intramembrane particles (IMP's) at high density ($6500/\mu m^2$) are apparent in the protoplasmic face (p-face) of the split membrane. These are restricted to the region of cell membrane overlying the shading pigment or eyespot (Melkonian & Robenek (1980a). IMP's 16-20 nm in diameter are excluded from this zone. Newly released

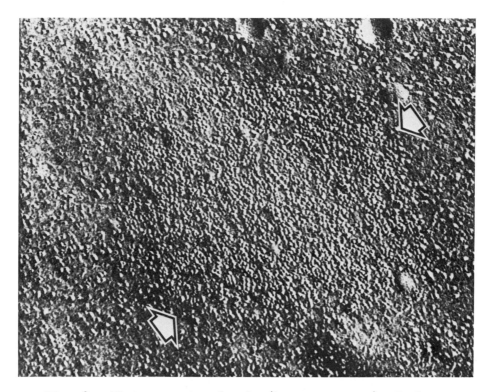

Fig. 8. Photoreceptive domain (between arrows) of plasma membrane of <u>Chlamydomonas reinhardii</u> (phytoflagellate, Protista). In the p-face leaflet, a high concentration of 8-12 nm particles is seen in the region overlying the stigma. The size and density of particles outside this region differs. After Melkonian and Robenek (1980a).

zoospores of the green alga <u>Chlorosarcinopsis gelatinosa</u> have a similar specialized zone which, unlike the shading pigment, disappears at the time the zoospore settles and becomes no longer phototactic (Melkonian & Robenek, 1980b). As yet there is no proof that the IMP's of these patches in the cell membrane are composed of photopigment, however similar p-face IMP's in vertebrate rod outer segment and crayfish microvillar membranes are probably aggregates of rhodopsin (see below).

There is a selective advantage to localization of the photopigment in a zone where it can be shaded by a pigment spot. It provides a simple mechanism for detecting the direction of the incident light. Without localization there would always be unshaded regions of the cell membrane that are light sensitive, and directional sensitivity would be poor. In multireceptor pigment-cup eyes, the greater the localization, the better the resolution would be.

For localization to exist, the intramembrane particles must somehow be prevented from diffusing over the surface of the membrane. They could be tied together somehow or they may be held by some kind of molecular fence. Various components of the membrane or the underlying cytoskeleton have been proposed for this role (Nicholson, 1979; Cherry, 1979; Satir, 1980; Saibil, 1982).

Ciliary domain. Another restricted domain of the plasma membrane is the envelope of the cilium (or flagellum). Sooner or later, advantage would be taken of this pre-existing structure for localization of photosensitivity. This may have occurred in an ancestor of the phytoflagellate Chromulina psammobia (Fig. 9), in which an entire modified flagellum is shaded by a trough-shaped pigment spot (Fauré-Fremiet & Rouiller, 1957). Localization to the ciliary membrane must have occurred early in metazoan evolution, if not in Protista, since it is evident in photoreceptor cells of medusae (Cnidaria). The density of 8-10 nm IMP's is much higher in p-face leaflets of the ciliary shaft and ciliary microvilli than in those of the photoreceptor cell body or of the microvilli and cell body of pigment cells (Takasu & Yoshida, 1982).

Localization of rhodopsin to ciliary membranes is evident in rod and cone cells. IMP's observed in the p-face are characterized by a high density (4400-4700 per μm^2), relatively random distribution and homogeneous particle size in the 8-12 nm range. There is a variety of evidence that these IMP's must consist of rhodopsin (for references see Besharse & Pfenninger, 1980). Rhodopsin is an intrinsic component of both discs and outer segment plasma membrane and is known to traverse the membrane. The p-face 8-12 nm IMP's are thought to be aggregates of 4-5 rhodopsin molecules. Individual molecules protrude from the intra-discal surface (Roof & Heuser, 1982). Several investigators have shown similar-appearing 8-12 nm p-face IMP's in membranes reconstituted from purified rhodopsin and lipid.

IMP's with the same size and density are present in the p-faces of the membrane discs, outer segment plasma membrane, ciliary stalk membrane and the periciliary region of the inner segment plasma membrane. In contrast, the p-face IMP's of the plasma membrane outside the periciliar region, including that of the calycal processes, are less homogeneous in size, smaller on the average and 2/3 the density (Besharse & Pfenninger, 1980). The border of the ciliary domain is probably located just inside the ring of calycal processes (Fig. 10). This segregation of IMP's is the same in both rods and cones. Because rhodopsin is known to diffuse laterally in the membranes of rod and cone outer segments (Poo & Cone, 1974) one must assume that some kind of molecular fence prevents it from escaping from the ciliary domain. It is clear that the fence is not associated with the ciliary necklace or

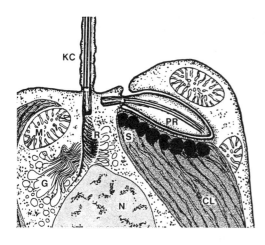

Fig. 9. Eye of <u>Chromulina psammobia</u> (phytoflagellate, Protista). Anterior to the stigma (s) lies an entire modified cilium (PR), the putative photoreceptor. CL, chloroplast; G, golgi network; KC, kinocilium; M, mitochondrion; N, nucleus; R, ciliary rootlet. After Fauré-Fremiet (1961).

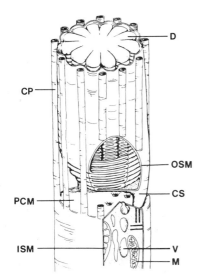

Fig. 10. Junction between inner and outer segments of an amphibian rod photoreceptor. Membrane vesicles (V) appear to fuse with the periciliary membrane (PCM). From there, photoreceptor membrane flows up the ciliary stalk (CS) to the outer segment membrane (OSM), from which the membrane discs (D) are formed at the base of the outer segment. The calycal processes (CP) lie outside the domain of the ciliary membrane. M, mitochondrion. After Besharse and Pfenninger (1980).

bracelet particles which are universally present in all cilia: they are in the wrong location - on the ciliary stalk (Röhlich, 1975).

There are at least 3 advantages of the ciliary membrane for localization of photosensitivity. 1) It is a restricted membrane domain where integral membrane proteins such as rhodopsin can be localized, 2) the genetic information necessary for its construction and maintainance is already available in the genome of metazoan epidermal cells, and 3) its surface area is large, therefore the amount of photopigment that can be shaded is greater than for a planar patch such as in <u>Chlamydomonas</u>. A possible fourth advantage, certain ion transport and regulatory proteins, useful in phototransduction, may already have been segregated to these domains. For example, calcium-ion channels and pumps are localized to the membrane of kinocilia (Eckert & Brehm, 1979; Dentler, 1981). Another advantage of localizing photosensitivity in a cilium is suggested by Fig. 9. The stiffness provided by the axoneme probably facilitates indentation of the cell by the cilium. It may provide needed support to ciliary organelles of metazoan sensory cells.

Microvillar domain. The microvillar membrane has the same advantages. Though microvilli occur on the apical surface of epithelial cells of Hydromedusae (Cnidaria), the lowest phyla in which rhabdomeric photoreceptors have been found are Platyhelminthes and Echinodermata. Thus the colonization of the microvillar domain by photopigment could have first occurred in a common primitive bilateral ancestor (see Fig. 7).

In the crayfish, Procambarus (Fernández & Nickel 1976; Eguchi & Waterman, 1976), the p-face of the microvillar membrane contains 8-10 nm particles in high density. Digitonin, which extracts rhodopsin from membranes, gradually removed the IMP's. In the perimicrovillar plasma membrane of the dendrite, the IMP's have a much lower density even right up to the base of each microvillus (Fernández & Nickel, 1976). Either some kind of molecular fence must be located at the base of the microvillus or the particles are not mobile. The latter is suggested by other studies (Goldsmith & Wehner, 1977; Saibil, 1982). The plasma membrane remote from the rhabdom contains p-face particles in a broader range of sizes and with a smaller mean diameter (Eguchi & Waterman, 1976). In Drosophila, on the other hand, the p-face IMP density is about the same in the microvillar and plasma membranes of the retinula cells. Vitamin-A deprivation and 3 mutations decreased the partical density of both membrane domains as well as rhodopsin concentration (Schinz et al., 1982). Vitamin A-deficient blow flies have substantially fewer than normal IMP's in the rhabdomal membranes, and the rhodopsin content is lower by the same proportion (Boschek & Hamdorf, 1976). In the land snail, Helix aspersa, 7 nm p-face IMP's are more abundant in microvillar membranes of dark-adapted rhabdomeres than after prolonged treatment with light (Brandenberger et al., 1976). In the ant, however, 8,5 nm p-face IMP's of green- and UV-sensitive rhabdomes are not affected by various amounts of light-adaptation (Nickel & Menzel, 1976). Differences in the light treatments of the two studies may explain the divergent results.

Thus the p-face IMP's are correlated with the presence of rhodopsin in rhabdomeres of several higher invertebrates. In the crayfish they appear to be confined to the microvillar domain, whereas in Drosophila they are not. There have been no freeze-fracture studies of rhabdomeric photoreceptors of lower invertebrates. However there is reason to believe that one would find evidence of localization. First, localization of photosensitivity is present in Protista, as already described. Second, in pigment-cup eyes, so common among lower invertebrate phyla, there is a clear advantage to localization of photoreceptive membranes to the interior of the cup. A photoreceptor cell would lose its directional sensitivity if the plasma membrane of the dendrite, soma and axon were photosensitive. Third, if a means of localizing visual pigment to microvilli were not developed early, the septate junctions could provide a fence. Septate junctions, equivalent to the vertebrate tight junction, are present in all invertebrate phyla except Chordata (where vertebrate-like tight junctions are found) and form a gasket-like seal around the apical circumference of cells lining an external or internal surface (Green & Bergquist, 1982). One might expect these junctions to be as effective a fence as the tight junctions of vertebrate olfactory cells cited in the next section. At what point in evolution photosensitivity became limited to the microvillar membrane is an interesting question for future study.

Chemosensory membranes. The advantages provided by the ciliary or microvillar membrane domains should also be useful for localization of chemosensitivity or mechanosensitivity. In metazoa, the apical end of neurons specialized for these functions usually project one or both of these structures. Surprisingly, however, the fence of vertebrate olfactory dendrites appears to be located at the tight junctions formed with the adjacent supporting cells rather than in the more restricted domain of the cilia or microvilli. P-face IMP's 8-12 nm in diameter are distributed over the entire apical surface of the dendrite (Kerjaschki & Hörandner, 1976; Menco et al., 1976; Usukura & Yamada, 1978). The density depends on species. Large, rod-shaped IMP's predominate in the apical membrane of adjacent supporting cells and IMP's were rare in motile cilia of "respiratory" cells of the same olfactory epithelium. Thus the 8-12 nm p-face IMP's probably have an olfactory function and, by analogy with the rhodopsin IMP's, may be chemoreceptive proteins.

Where both microvilli and cilia are present on olfactory dendrites, the particles are found in the membranes of both structures. These particles are rare on the lateral surfaces beyond the tight junctions. Apparently, in olfactory cells, it is sufficient to limit chemosensitivity to the entire exposed surface, and the microvilli and cilia both serve to increase the surface area.

In the newt the olfactory cells are of two types: one with several cilia and short microvilli and the other with only long microvilli. The apical surfaces and processes of both types contain the same 11 nm, randomly-distributed, p-face IMP's. However they occur at a higher density in the nonciliary type (Usukura & Yamada, 1978). Non-ciliated microvillar olfactory cells have been found in several other vertebrates (Bannister, 1965; 1968) and are analogous to rhabdomeric photoreceptors. The microvilli never become as densely packed or tightly organized as they do in rhabdomeres. This reflects the difference in selection pressure. For efficient chemoreception the microvillar surfaces must be accessible to diffusing odorant molecules. For photoreception this is not a requirement and microvilli can be packed together to maximize the likelihood of capturing a photon.

Mixed-type photoreceptor cells. In view of the probable occurrence of chemosensory ciliary and microvillar projections in the same olfactory receptor cell, could photoreceptor cells exist in which both structures are photosensitive? This question calls to mind the examples mentioned earlier that are hard to classify. Where is the fence located in the photoreceptor cells of the hydromedusae Leukartiara and Polyorchis? Both microvilli and a cilium arise from the apical membrane of these species and in the latter species these microvilli intermingle with microvilli formed of the ciliary membrane (Singla & Weber, 1982a). The freeze-etch study of Takasu and Yoshida (1982) stops short of distinguishing between these two domains. In asteroid eyes, the microvilli are now known to arise from the apical membrane, not the ciliary membrane (Eakin & Brandenburger, 1979), but could the membrane of the adjacent cilium also be photosensitive? Eakin and Brandenburger argue that only the microvilli become disrupted by strong light, but could not the axoneme prevent disruption in cilia? A better indication comes from a freeze-fracture study: 8-12 nm p-face IMP's are present in high density in the microvillar and not the ciliary membrane (Yoshida, personal communication). The asteroid photoreceptor, therefore, would appear to be rhabdomeric, not a mixed type. It has been suggested (Brandenburger et al., 1973) that the photoreceptor cell of the tornaria larvae of a hemichordate is a mixed-type.

Synthesis and turnover of photoreceptive membranes

In the preceding section the occurrence of localized domains for photoreception was documented, and the need for a fence to segregate the domains was established. Also needed is a mechanism for inserting newly synthesized sensory proteins and other membrane proteins, specifically, within these domains. Along with this is the need to remove excess or worn out material from the domains. These mechanisms are now being studied intensively in both vertebrate and arthropod eyes. I will only summarize the prevailing view and leave discussion of arguments and evidence to the many reviews. I will then review the current evidence for synthesis and turnover in eyes of turbellarians and other lower invertebrates.

Vertebrates. Even though membrane turnover in rods and cones has received much attention in the last decade, the current understanding of some steps is still somewhat speculative (reviews: Whittle, 1976; Holtzman & Mercurio, 1980; Olive, 1980; Papermaster

& Schneider, 1982). Opsin, evidently, is synthesized on granular endoplasmic reticulum (GER) of the inner segment, and from then on, remains integrally associated with membrane. It is glycosylated initially on the ribosomes and terminally in the Golgi apparatus. GER and agranular endoplasmic reticulum (ER) are both involved with lipid metabolism. Labelled opsin or lipids appear in the ciliary stalk plasma membrane and flow to the outer segment where the membrane forms into the membrane discs. The discs were formerly thought to originate by invagination of the outer segment membrane but new evidence indicates that disc surfaces develop by evagination, and the opposing paired membranes are zippered by a separate mechanism (Steinberg et al., 1980). Incorporation of the 11-cis retinal, synthesized in the pigment epithelium, occurs within the outer segments (Bok et al.,1977). Membrane discs of rods slowly migrate up the outer segment and are ultimately shed and then phagocytized by the pigment epithelium. Since cone disc membranes retain their connection to the outer segment plasma membrane in most vertebrates, new opsin and lipid can diffuse throughout the outer segment. Cone outer segment tips are also shed and phagocytized. In a diurnal cycle, rods shed their tips in the morning after illumination begins, and cones shed theirs in the evening soon after it ends.

The mechanisms of transport of opsin and membrane from GER to golgi and from golgi to the ciliary stalk are still being debated. There is much evidence suggesting that membrane vesicles are the carriers (Papermaster & Schneider, 1982). These are seen near the distal end of the inner segment, near the Golgi apparatus and in the intervening ellipsoid region. It is thought that these structures insert membrane by fusion, as in exocytosis, with the periciliary membrane. Since this region lies within the fence around the ciliary membrane domain, it is only necessary for the added membrane to flow from there along the plasma membrane via the ciliary stalk to the outer segment. IMP's of the appropriate size distribution but at a lower density have been observed in these vesicles (Besharse & Pfenninger, 1980). Immunohistochemically demonstrable opsin is present (Papermaster et al, 1979), and labeled protein appears in vesicles near the cilium at the appropriate time. Pinocytotic coated vesicles that are commonly seen near the periciliary membrane may selectively remove material from the membrane, such as excess lipid deposited by the vesicles.

An essential step in bulk transport by exocytotic fusion of vesicles is the recognition of the target membrane (Palade, 1975). Presumably rhodopsin-containing vesicles fuse with the periciliary membrane, and Na^+/K^+ ATPase-containing vesicles fuse with the lateral membrane of the inner segment. Membrane recognition systems have been proposed for other cells and it is thought that the Golgi apparatus adds this component to the vesicle (Palade, 1975; Whaley & Dauwalder, 1979).

Arthropods and Molluscs. Membrane turnover in rhabdomeric photoreceptors is less well understood (for a recent review see Waterman, 1981). The cytoplasm near the rhabdomere typically contains GER, elements of ER, vesicles and tubules of ER, coated vesicles, multivesicular bodies and lysosomes. The role of each of these actors in the synthesis, transport and degradation of the rhabdomere is still uncertain. The amount of each component as well as the size of the rhabdomere varies with time of development, time of day and diurnal or long-term exposure to light or darkness, and it is by the control of these variables that the mechanism is being explored. The current view is that new rhabdomeric membrane may reach the base of microvilli via vesicles in some species and directly via flow along a tubular or ellipsoidal reticular system in others. Golgi complexes are comparatively rare in arthropod photoreceptor dendrites and are little affected by light or dark. "Photic vesicles" originate from Golgi saccules in photoreceptors of the snail (Brandenburger & Eakin, 1970) where, unlike in vertebrates, vitamin A is incorporated.

There is good support for the hypothesis that in arthropods membrane removal occurs by pinocytosis at the base of the microvilli, forming coated vesicles, and degradation occurs in the photoreceptor cell via a sequence of multivesicular bodies, lysosomes and residual bodies. Extracellular shedding followed by phagocytosis by the photoreceptor cell, hemocytes or other cells is an alternative route. Further discussion of the complexity and great variety of turnover mechanisms in arthropod and mollusc photoreceptors should be sought in Waterman (1981).

Lower invertebrates. Evidence of turnover in lower invertebrate phyla comes chiefly from studies of the effect of light and darkness on turbellarian (Platyhelminthes) eyes. Many of these studies were done long before the recent explosion of interest in photoreceptor membrane turnover and concerned the effects of long-term darkness on various Tricladida (Röhlich & Török, 1962; Röhlich & Tar, 1968; Carpenter et al., 1974b). A recent investigation of changes during normal diurnal light/dark periods was done on a rhabdocoel (Bedini et al., 1977). The microvilli degenerate in the dark and regenerate in the light. As they degenerate the microvilli become shorter and their tips disorderly. At their base, tubules, vesicles, multivesicular bodies and broken membranes are observed. Vesicles, large vacuoles and multilamellar whorls are distributed in the cytoplasm. Bristle-coated vesicles, normally seen pinching off from the microvillar base, are absent. When 12-hour darkness-treated eyes are exposed to light the microvilli are reconstructed with most of the change occurring between the 5th and 10th minute! The rapidity of restoration is even greater than observed in a dinopid spider, in which it occurs at dusk (Blest, 1978). The authors suggest that preassembled membranes are stored in the cytoplasm for the rapid regeneration at dawn. Membrane appears to be transported to the base of the microvilli via vesicles.

Membrane degradation in the normal light-adapted eye of turbellarians can be inferred from the presence of coated vesicles (apparently originating by pinocytosis of the proximal end of the microvillar membrane), multivesicular bodies and lysozomes. Transport of new membranes in the fully light-adapted eye may utilize the same mechanisms as observed in regeneration; smooth-surfaced ER vesicles are present. In the light-adapted eye of D. dorotocephala, Carpenter et al. (1974a) observed ER in the form of long columns, parallel to the axis of the dendrite, which appear to merge with the microvilli. This has also been observed in other triclads (Stewart, 1966).

We can surmise that the mechanism of turnover in rhabdomeric photoreceptors is essentially the same in turbellarians and arthropods: degradation via a sequence of pinocytotic vesicles, MVB's and lysosomes, and bulk transport of new membrane via vesicular or tubular elements of the endoplasmic reticulum. Certainly the necessary equipment is there. The dynamics are still to be shown conclusively.

Development or regeneration of ciliary receptors has not been studied in lower invertebrates. The only hint of the mechanism of synthesis, transport and degradation of ciliary membrane comes from the structure of the apical cytoplasm. The nematode olfactory dendrite appears to have a particulary high rate of membrane turnover, as inferred from an extensive tubular and vesicular reticulum leading up to the periciliary membrane and evidence of shedding at the distal end of the cilium exposed to the sea water (Burr & Burr, 1975). Because of the cuticle, phagocytotic recovery of the shed membranes would be impossible.

Since the Cnidaria are regarded to be an early offshoot in the evolution of the Bilateria from phytoflagellates, it is interesting that their cilia contain the same structures involved in turnover of the ciliary membrane of vertebrates. This should be no surprise, since the same structures are well developed in phytoflagellates too. In Tamoya, the possibility that shed ciliary membranes are phagocytized by the pigment cell is suggested by the presence of many large multivesicular bodies and myelinated bodies (lysosomes?) in the long extension which extends though the network of microvilli derived from the ciliary membranes (Yamasu & Yoshida, 1976). Thus these pigment cells of the lowly Cnidaria may be analogous in this function to the pigment epithelial cells of vertebrates!

Increase in surface area

Because of the advantages of increased photosensitivity, natural selection has favored an increase in the surface area of the photosensitive membrane. Ciliary or microvillar membranes can be increased in area either by increasing the number of organelles per cell, by enlargement, or by pleating of the organelle membrane in various ways. In order to preserve the shading properties of simple eyes, or the resolution of multi-receptor eyes, the larger membrane area would have to occupy a small volume. Selection is likely, therefore, to lead to a high membrane density in photoreceptors.

Multiple cilia have been observed in Bryozoa, Kamptozoa and in a nematode, however in these examples the organelles are likely to have evolved from cells which already contained multiple cilia. In bryozoan larvae the photoreceptive cilia have axonemes that are morphologically identical to those of the motile cilia of the neighbouring corona cells from which they probably evolved (Hughes & Woollacott, 1980). In Bugula simplex, both cell types have numerous cilia with the same spacing (Fig. 1). In the nematode example, Oncholaimus vesicarius, a clump of 10 cilia arise from a single dendrite in an olfactory organ but are enclosed and lie adjacent to a pigment spot (Fig. 3) rather than (like the similar chemoreceptors) passing through a cuticular opening to the exterior (Burr & Burr, 1975). Their location, and the nature of the phototaxis of this organism (Burr, 1979), implicate the cilia as photoreceptors. Their structural similarity to the chemoreceptive cilia and location in the olfactory organ suggest that they may have evolved by modification of one of the chemoreceptor cells (Burr & Burr, 1975). If so, they probably inherited the multiple cilia since the chemoreceptor cells are multiply ciliated. The multiple cilia of the chemoreceptors are undoubtedly a result of selection pressures for increased chemosensitivity.

Enlargement of a cilium is described in a kamptozoan (syn. entoproct). The unpleated multiple cilia lack arms on the a-microtubles, suggesting that the photoreceptor has been modified more than those of the closely related phylum, Bryozoa. The cilia also expand in diameter 1 µm above the basal body, giving the ciliary membrane a modest increase in surface area (Woollacott & Eakin, 1973). "Ampulla-shaped cilia" are found in rotifers (Clément, 1980) and the phytoflagellate, Chromulina (Fig. 9). Flattened cilia or "sacs" (Eakin, 1972) found in certain polychaete worms also represent a simple enlargement of the ciliary membrane area.

A much higher membrane area per unit volume can be obtained by pleating the ciliary membrane. The microvillar projections from cilia of cnidarian medusae have already been noted and illustrated (Figs. 5-6). It is interesting that a possible prior evolutionary step is illustrated by Leukartiara which has a single, unpleated cilium as a putative photoreceptor (Fig. 4). Pleated ciliary photoreceptors are found in many animal phyla. The various forms have been called lamellae, paddles, digits, tubules and discs (Eakin, 1972; Salvini-Plawen & Mayr, 1977). Lamellar expansions of the ciliary membrane that project into a cavity invaginated into the photoreceptor cell are found in Rotifera and Platyhelminthes (Clément, 1980; Clément & Wurdak, 1983; Fournier, 1983). These resemble similar structures, called phaosomes, of rhabdomeric photoreceptor cells of Annelida and Pogonophora (Eakin, 1972; Vanfleteren, 1982).

In the case of rhabdomeric organelles, the surface area of the microvilli has been enlarged by increasing their number, by elongation and by branching (Eakin, 1972). Pleating of microvillar membranes is not observed. A high density is possible without pleating because of the already small size of microvilli.

Evolution of organelle complexity

In many phyletic lines there is evidence that photoreceptor membranes, after an increase in surface area, have been modified further towards greater organization and specialization. First, photoreceptor cells of the most advanced phyla have exquisitely structured organelles such as the rhabdomeres of crustacea (Fig. 11), insects and cephalopods, and the rods and cones of vertebrates. Secondly, in many lines, one can arrange examples in order of increasing complexity. My example is chosen from Cnidaria. From Leukartiara (Fig. 4) to Polyorchis the ciliary membrane changes from the unpleated form to one projecting loose microvilli which have varying diameters (Eakin, 1972). From Polyorchis to Sarsia the ciliary microvilli increase in density, and from Sarsia to Bougainvillia (Fig. 5) to Cladonema one can discern an increase in order towards a columnar arrangement of the organelle. In the scyphomedusan Tamoya the column is lengthened further (Fig. 6B). Intermittent swellings of the ciliary membrane, accompanied by a periodic fraying of the axoneme, provide a surface from which microvilli project. The microvilli have a more even diameter and sometimes are grouped into parallel - hexagonal clumps reminiscent of the arrangement in arthropod and mollusc rhabdomeres. It is striking that along the same stepwise sequence a comcomitant increase in specialization, complexity and organization occurs in the pigment cells and the eye as a whole. A comparison of the central nervous systems of these examples would be interesting.

The various forms that photoreceptor organelles have taken are catalogued by Eakin (1972). The rationale for each form is not

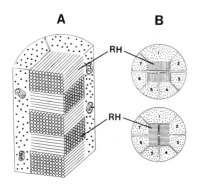

Fig. 11. Rhabdom of compound eye of the crayfish Procambarus clarkii (Crustacae, Arthropoda). A. Stereo diagram showing perpendicular orientation of the microvilli projecting from two photoreceptor cells. B. Transverse sections at two levels showing the organization of the rhabdomeres (RH) of the 7 photoreceptor cells of one ommatidium of the compound eye. After Eguchi (1965).

always obvious, except for the ones in which parallel microvilli are oriented perpendicular to the direction of propagation. Such a structure is likely to be sensitive to the direction of polarization of light. Two such rhabdomeres, oriented perpendicular to each other and located in different cells, are common in Crustacea and Insecta. Behavioral and physiological studies have demonstrated their role in the detection of polarization angle of light (Waterman, 1981). Parallel microvilli are also found in the interesting eyes of the Müller's larvae of Pseudoceros canadensis (Turbellaria, Platyhelminthes). As seen in Fig. 12, the right eye contains 3 rhabdomeres, oriented at different angles. Theoretically, it could detect polarization angle. The left eye, for some reason, contains a ciliary organelle as well (Eakin & Brandenburger, 1981).

Many photoreceptor forms appear to minimize polarization sensitivity. Though the disc membranes of rods and cones are perpendicular to the direction of propagation, rotational diffusion

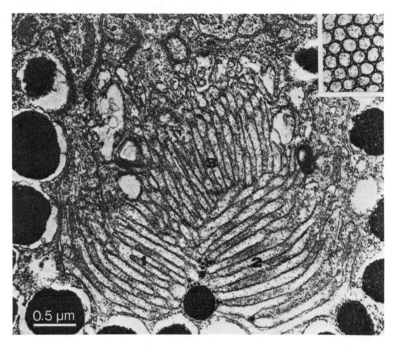

Fig. 12. Right-hand pigment-cup eye of Müller's larvae of Pseudoceros canadensis (Turbellaria, Platyhelminthes). Ordered microvilli of the 3 rhabdomeric receptors are shown. The left-hand eye has, in addition, another receptor composed of a clump of cilia (not shown). INSET, transverse section of microvilli. After Eakin and Brandenburger (1980).

randomizes the absorption vector of rhodopsin in the disc plane. The pigment molecules of Dugesia lugubris (Turbellaria, Platyhelminthes) are unlikely to be sensitive to the polarization vector because the microvillar membranes are parallel to the direction of propagation (Röhlich & Török, 1961). Disordered microvilli, such as those of the pigment-cup eye of the sipunculid worm (Fig. 13) may actually be "organized" in such a way to minimize sensitivity to polarization.

To make an ordered cellular structure, the cytoskeleton and extracellular material must be precisely designed and located. The ability to do this is evident even in the lowest phyla, cnidaria and platyhelminthes. Versatility and precision, however, appear to be greatest in the higher phyla.

Epigenous photoreceptor organelles

In addition to rhabdomeric and ciliary organelles, there are a few examples of photoreceptive organelles, in Nematoda and Rotifera, that are not composed of ciliary or microvillar membranes, being lamellar extensions of the apical membrane of a dendrite. Salvini-Plawen and Mayr (1977) propose that these are independently evolved from deep-lying, cerebral cells already differentiated for a neural function, and call them "ganglionic diverticular" organelles. In this category they also include other acilious examples that have microvillar membrane elaborations. This is puzzling because these microvilli appear the same, within variation, as rhabdomeric microvilli. Eakin (1972, 1979) includes both types of ganglionic organelle in his definition of a rhabdomeric receptor: any which are formed of non-ciliary plasma membrane. Salvini-Plawen (1982) admits of the difficulty of deciding whether ganglionic organelles arose from an acilious cerebral cell or by modification of a rhabdomeric cell. Also, Clément (1980) disagreed with Salvini-Plawen's new category, pointing to the historical difficulty of identifying cerebral neurons of Rotifera and to the presence of "cerebral" neurons bearing ciliary organelles. I propose that only clear morphological characters be used in the classification of photoreceptors, not presumed origins. Thus 'rhabdomeric' should refer to those organelles that are modified microvilli and 'ciliary' to those that are modified cilia. Accordingly, the microvillar type of "ganglionic" photoreceptor should be classified as a rhabdomeric photoreceptor. The lamellar type of acilious photoreceptor, then, is left unnamed. I propose the term 'epigenous' [growing on the surface] for any type of photoreceptor organelle that is formed of the cell membrane proper, not of a microvillar or ciliary membrane. This would include the special lamellar type identified above or any other form of projection of the cell membrane such as the cylindrical ones identified in the rotifer, Rhinoglena frontalis (Clément, 1980; Clément & Wurdak, 1983). A membrane patch like that of Chlamydomonas (Fig. 8) may be a precursor of epigenous photoreceptors. On the other hand,

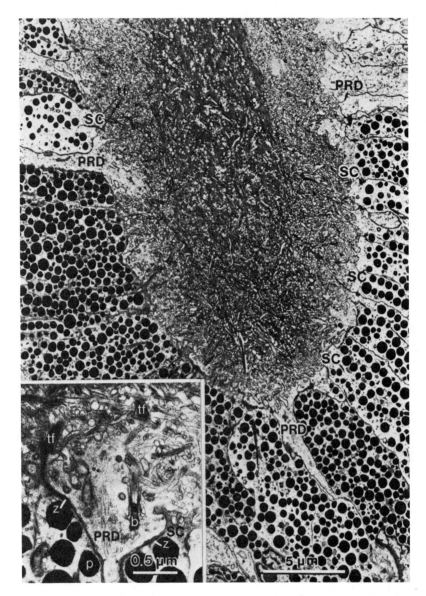

Fig. 13. Everse pigment-cup eye of <u>Phascolosoma agassizii</u> (Sipunculida), a marine worm. Photoreceptor dendrites (PRD) penetrate through the layer of pigmented supporting cells (SC). INSET. Detail of apical part of photoreceptor cell showing randomly arrayed microvilli (PR) and adventitious cilium (AC) that project from it. b, basal body; p, pigment granule; tf, tonofilaments; z, "zonula adhaerentes" (probably a septate junction). After Hermans and Eakin (1969).

epigenous organelles could be collapsed cilia which have had their axonemes suppressed during development.

Distribution among the phyla and trends

Four basic types of photoreceptor have been identified: ciliary, rhabdomeric, mixed and epigenous. In which phyla are they found? Figure 7 presents our current knowledge as summarized recently by Vanfleteren (1982). It is readily seen that the ciliary type has been found in all but five of the many phyla examined. There is a chance that ciliary examples may yet be found in some of these five because in only one of them, Arthropoda, have more than a few eyes been investigated. The rhabdomeric type has been found in all but 9 of the metazoan phyla investigated, and of the exceptions, only Vertebrata has been extensively studied.

Not only are both rhabdomeric and ciliary photoreceptors widespread, but each type appears to have arisen many times in metazoan phylogeny. Either ciliary or rhabdomeric organelles can occur in apparently cenogenic (newly arisen) eyes, as well as in well-established lines. Salvini-Plawen and Mayr (1977) argue that this disproves Eakin's hypothesis that there are 2 main lines of photoreceptor evolution.

There are some clear trends, however. Vertebrates have only ciliary, and Arthropods only rhabdomeric, photoreceptors in spite of the common occurrence of the opposite type of chemoreceptors and mechanoreceptors in these taxa. In the Platyhelminthes and Annelida, rhabdomeric photoreceptors predominate. In Annelida and Mollusca only rhabdomeric photoreceptors occur in cerebral eyes (Eakin, 1982). The ciliary photoreceptors of these phyla are found in caudal, branchial, epidermal, tegmental or mantel eyes. These are more likely to be cenogenic since they evolved in tissues that were not orginally photoreceptive (Salvini-Plawen & Mayr, 1977; Rosen et al., 1978; Eakin 1979, 1982).

Thus there is excellent evidence for a major trend towards rhabdomeric organelles in the protostomate line leading from ancestors of Platyhelminthes to Annelida, Mollusca and Arthropoda (Fig. 7). Minor trends are identifiable in two cases, each involving only two related phyla: 1) the clumped cilia of Bryozoa and Kamptozoa, and 2) the ciliary discs of ascidian tadpole larvae and vertebrates.

Though ciliary examples may yet be found, the apparent lack of ciliary photoreceptors in Echinodermata (Yamamoto & Yoshida, 1978; Eakin & Brandenburger, 1979; Eakin, 1982) has severely shortened the "ciliary line of evolution" originally proposed by Eakin (1963). The presence of ciliary photoreceptors in cephalochordates remains to be proven (Vanfleteren, 1982). What is left is the minor trend mentioned above.

The two other basic types of photoreceptor, epigenous and mixed, have only a few examples and appear to occur rarely (Fig. 7). Epigenous lamellar organelles have been discovered in Nematoda and Rotifera, and a cylindrical organelle has been found in

Rotifera. These appear to be no more nor less cenogenic than other types in the same phyla.

The mixed type has been proposed in only two phyla, Cnidaria (Singla, 1974) and Hemichordata (Brandenburger et al, 1973). Still needed is some kind of proof that the cilium and the (non-ciliary) microvilli are both photoreceptive. There are many rhabdomeric photoreceptors which have adventitious cilia adjacent to the microvillous organelle. In most cases it is unknown whether the ciliary membrane is also photosensitive. Freeze-fracture electron microscopy appears to be a promising approach to this problem; it has shown that olfactory neurons are usually of the mixed type (see earlier section on chemosensory membranes).

Conclusions and questions

Evolution of ciliary and rhabdomeric organelles may have occurred in four stages: 1) evolution of photosensitivity (in Protista or early Metazoa), 2) localization to restricted membrane domains such as the membrane of cilia (in Protista) or microvilli (in primitive Bilateria), 3) increase of surface area of microvillar or ciliary membranes by increase in number, enlargement, or in the case of ciliary types, pleating of the membrane, and 4) increase in complexity by greater organization and more precise form. Following Darwin, step-wise sequences increasing in membrane area and organelle complexity can be arranged for photoreceptors, provided closely related examples within phyla are chosen. Steps 3) and 4) are paralleled by increases in specialization and complexity of pigment cells and the eye as a whole.

As is the case for eyes, photoreceptors appear to have arisen many times in metazoa: many independent lines can be distinguished. Despite this, two major trends can be distinguished in the distribution of organelle type and complexity among phyla. One is towards greater versatility and precision of organelle structure as one proceeds toward higher phyla. The other is the selection, predominantly, of rhabdomeric organelles in cerebral eyes of the Platyhelminthes, Annelida, Arthropoda and Mollusca. Two minor trends toward modified cilia are also identifiable.

How can ciliary or rhabdomeric photoreceptors have newly arisen so many times? It must have been vastly easier for these organelles to evolve than, for example, mitochondria or endoplasmic reticulum. Why is versatility and precision of organelle structure so much greater in the higher phyla? Why are photoreceptors exclusively ciliary in Vertebrata and rhabdomeric in Arthropoda? Comparative anatomy, by itself, appears not to be able to answer these questions. In the next section I propose a mechanism of evolution that may help.

COMPONENT-SELECTION MODEL FOR DIFFERENTIATION AND EVOLUTION OF RECEPTOR ORGANELLES

What components are essential for a receptor organelle in its role as a transducer? In the preceding sections I have introduced four types: a localized membrane domain for the receptor protein with a system for membrane renewal, the modifications that determine the special shape of the microvillar, ciliary, or epigenic membrane, the receptor protein itself, and the auxiliary components of transduction. These four classes are listed in Table 1 with examples. Though undoubtedly oversimplified, the list provides a useful basis for discussion.

The expression of each of these components in a particular receptor cell depends on the activation of structural genes that transcribe messenger RNA's for synthesis of the required proteins. Though cells of a given organism probably possess identical genomes (Laskey, 1979), they may have widely divergent morphology and function. In addition, form or function may change during the life of the cell. These facts are explained by assuming that genes in different cells are activated in different combinations and at different times. The mechanism of gene regulation in eukaryotes is still a matter of speculation, however several models that assume a hierarchical regulatory system form a useful framework for explaining many aspects of cell differentiation and evolution.

The component selection model is proposed here with the hope of providing some insight and provoke some thought as to how photoreceptors may have evolved. At the least it appears to tie together a variety of observations. Though it was inspired by the well-known Britten-Davidson model and is here described in similar terms, it is not dependent on the particular molecular mechanisms proposed by Britten and Davidson (1969, 1971). Indeed these have been modified recently (Davidson & Britten, 1979) without affecting the hierarchial form. The evidence for gene control mechanisms and models has been reviewed by Lewin (1980) and Darnell (1982).

The component-selection model assumes that there is a hierarchy of regulatory genes which ultimately determines the particular combination of structural genes that are expressed in a given cell at a given time. Intermediate in the hierarchy would be a 'master gene' that, when activated, causes the designated component in each class of Table 1 to be expressed. This regulation could come about through any of a variety of molecular processes. Because of its simplicity, the original mechanism of the Britten-Davidson model was adapted for illustration in Table 2. Many copies of an 'activator RNA' are transcribed from the master gene, diffuse throughout the genome, and bind to 'receptor genes' (Table 2). The binding could activate an adjacent 'structural gene', in which case messenger RNA for a protein is transcribed. Alternatively, binding to a receptor gene adjacent to an 'integrator gene' would stimulate the transcription of many copies of another activator RNA. The latter, in turn, would bind to other

Table 1: Essential components of receptor organelles.

Class	Examples
1. membrane domain, its associated cytoskeleton and systems for renewal	ciliary, microvillar, apical, or epigenic domains.
2. modifications to organelle structure	components that determine the special shape of ciliary, rhabdomeric or epigenic photoreceptors, mechano- or chemoreceptors of each species.
3. receptor protein	photoreceptor, chemoreceptor or mechanoreceptor proteins.
4. components of transduction	cGMP and Ca^{++} control systems, controllable ion channels, etc.

Table 2: One possible mechanism of component selection.

Genome Segment	Segments active in Blue-sensitive cone cell	Segments active in rod cell
Master 1	C MC BCO PCC ... (Master gene)	
Master 2	C MC BCO PCC (activator RNA being transcribed)	C MR RO PCR ...
	activator RNA bound to receptor gene	
1*	C MC BCO PCC ... 1 2 3 4 5 ...	
2a	11 12 ... 21 22 ...	
2b		11 12 ... 31 32 ...
3a	gene for blue cone opsin	
3b		gene for rod opsin
4a	41 42 ... 51 52 ...	
4b		41 42 ... 61 62 ...
	receptor genes / structural or integrator genes	

* The numbers correspond to the component classes of Table 1.

dispersed receptor genes. Activation of the master gene ultimately results in the activation of a unique battery of structural genes, which in turn produce the mRNA for the needed protein components.

The specificity of an activator RNA is determined by segments of its base-sequence that are complementary to segments of one or more receptor genes (Table 2). The role proposed for activator RNA could equally be carried out by diffusible regulatory proteins coded by activator RNA (Britten & Davidson, 1969). Alternatively the activator RNA (now called integrating transcript) could complex with a mRNA precursor in the nucleus and affect its post-transcriptional processing (Davidson & Britten, 1979). In either case the hierarchial structure of the model is unaffected.

In the component-selection model, one master gene controls the expression of all of the essential components (Table 1) of one type of receptor cell. Other types of receptor cell would result from activation of a different master gene. Production of components common to every cell (Golgi apparatus, ER, ribosomes, etc.) would be controlled by other master genes. Master genes that program special features of other cell types (e.g. muscle) would not be activated in a photoreceptor cell. At a higher level in the hierarchy, inducing agents, morphogens, hormones or other regulatory genes might select the appropriate master genes to be activated in a particular cell and at the appropriate time during development. The selection of a specific master gene may occur in a multistep process during development and may involve several different inducing agents.

Control of differentiation of photoreceptor cells

According to the component-selection model, for example in a cell destined to become a blue-sensitive cone cell, a master gene (Master 1) turns on: A) an integrator gene (genome segment 1, Table 2) that controls all aspects of the production and renewal of a normal cilium: B) an integrator gene (segment 2a) that controls all modifications to the cilium, including disc formation, suppression of the central pair of microtubules in the axoneme, etc.: C) a structural gene (segment 3a) that transcribes messenger RNA for the opsin of blue-sensitive cones, and: D) an integrator gene (segment 4a) that controls production of all physiological components specific to phototransduction in the blue cone. Also synthesis of the chromophore of rhodopsin, retinal₁ (reviewed for vertebrates and invertebrates by Fein and Szuts, 1982) in the adjacent pigment epithelium cell would somehow have to be induced.

In a rod cell of the same organism, a different master gene (Master 2) is activated (Table 2). The same integrator gene for differentiation of a cilium (on genome segment 1) would be activated. Some of the modifications to the cilium would be the same and some would differ (compare segments 2a and 2b). A different receptor protein would be selected (segment 3b). Also, some of the transducing components may differ (segments 4a and 4b), though blue cone responses may be more similar to those of rods than other cone types (Gouras & Zrenner, 1979; Zrennar & Gouras, 1979).

The essential feature of the component-selection model is that the nucleotide sequence of a master regulatory gene determines which component in each class of Table 1 the cell is destined to produce, by turning on the appropriate regulatory and structural genes. The timing could be controlled by additional signals. The master gene is like a restaurant patron ordering from an a la carte menu. What is eventually put on the table depends on what he selects for a soup, salad, meat, vegetable and a dessert. Another patron could order a different combination.

During development of an organ, the particular structure that develops often depends on positional information. This is so also in the retina. Cone cells predominate in the fovea, becoming increasingly intermixed with rod cells with distance from the center. The positional information in this 2 dimensional field may be provided by the concentration gradient of a radially diffusing morphogen, which directly or indirectly induces cone master genes and represses rod master genes.

Thus the component selection model can provide explanations for the specificity of genome expression, as illustrated for rod and cone cells. It should be understood that, in the absence of any solid evidence, these explanations are rather soft. Other observations discussed earlier in this chapter can also be "explained":

1) Both ciliary and rhabdomeric photoreceptors are sometimes found in the same organism. Explanation: The genetic information for both cilia and microvilli is coded in the genome of all animals. The master gene selected in one photoreceptor cell can be different from that selected in another. Two master genes of the same genome could differ only in the membrane domain that they select. (Class 1 of Table 1). If the domains are modified, they could differ also in the modifications that they select (Class 2).

2) In O. vesicarius, cilia having the same structural modifications and located in the olfactory organ are either chemoreceptoral or photoreceptoral. Explanation: The master gene which selects components for the chemoreceptor cell could be identical to that of the photoreceptor cell except for the segment which selects receptor protein (Class 3). The transduction mechanism in chemoreceptors and photoreceptors could (theoretically) be otherwise identical.

3) In bryozoan larvae, photoreceptor cilia and kinocilia of neighbouring cells have identical axonemes and the relatively unmodified ciliary membranes are nearly identical (Hughes & Woollacott, 1980). Explanation: The 2 master genes could be the same except for the addition of a segment that selects photoreceptor pigment (Class 3) and perhaps a difference in the segment selecting components of Class 4. The slightly different appearance of the ciliary membranes in the two cases

could be due simply to the addition of photoreceptor protein to the membrane.

Component selection in evolution of photoreceptors

Hierarchial models of gene regulation can be used to explain evolution-related phenomena as well as differentiation (Britten & Davidson, 1969, 1971; Valentine & Campbell, 1975; Galau et al., 1976). The composition of the battery of structural genes activated by an integrator gene can be changed by mutation in either the integrator gene or the receptor genes. Mutations affecting regulatory genes or other processes higher in the regulatory hierarchy can affect whole batteries of structural genes, causing them to be expressed in situations where they were not previously.

Mutations in the regulatory system can occur in 2 ways, by base substitution ("point" mutations) or by chromosomal rearrangements. Base substitution could affect the binding of a segment of the copied activator RNA (or a translated regulatory protein) to some of its former receptor genes, or cause it to bind to new sites. By chromosomal rearrangements, genes or portions of regulatory genes could be transposed to new relationships. The resulting changes could create novel patterns involving many structural genes. Some potentially disadvantageous new patterns might not be expressed immediately and could be preserved in the genome until further changes or new selection pressures allow them to be expressed successfully. Evidence for unexpressed "pseudogenes" has been reviewed recently (Little, 1982). There is evidence for interchange of sequence blocks between these and active genes of the same family (Dover, 1982; Jeffreys, 1982).

What do these ideas mean in terms of the component-selection model for photoreceptors? By simple modification of the master gene or receptor genes, a component can be deleted, replaced, or a new one added to a cell. For example, a suitable point mutation in the BCO segment of Master 1 (Table 2) could cause the transcribed activator RNA (or a translated regulatory protein) to bind to the receptor for the rod opsin gene instead of that for blue-cone opsin. Thus the spectral sensitivity of a photoreceptor could be modified in a simple way, bypassing the much slower process of new mutational experiments with the structural gene. Other putative changes in the components selected can explain some of the puzzling observations described earlier in this chapter:

1) Photoreceptors appear to have newly arisen many times.
Explanation: The genes for photoreception are present in the genome of most animals, since, with few exceptions they are expressed in at least one stage of their life cycle. A variety of simple mutations in the regulatory system could lead to photosensitivity appearing in a new cell or at a different stage. Such mutations can explain rapid evolutionary modification (Valentine & Campbell, 1975).

Examples: a) A single-point mutation in a master gene could result in the chance selection of a different receptor protein (Class 3, Table 1). A chemoreceptor organelle could thereby be converted into a photoreceptor, as may have happened in O. vesicarius (see example 2 in preceding section on differentiation). Either a ciliary or microvillar chemoreceptor cell could be converted in this way. If this had selective advantage, further mutation and natural selection could add refinements. b) One mutation could introduce a photoreceptor protein into the membrane domain of a kinocilium. If this is not deleterious to the organism, a second mutation could later add the physiological components of phototransduction. This may have happened in corona cells of bryozoan larvae (see example 3 of preceding section). Bryozoan photoreceptor cells could also have evolved as in the following example. c) The apical membrane of a neuron could be made photosensitive by addition of photoreceptor components of classes 3 and 4 (Table 1). Later, surface area may have been increased by addition of i) cilia as in bryozoan larvae and Leukartiara, or ii) microvilli as in rhabdomeric receptors. The cilia or microvilli would resemble those of other cells in the same organism until modifications are evolved.

2) Most photoreceptors are either modified cilia or modified microvilli. Epigenous organelles are rare. Explanation: The structural components of epigenous organelles may have to be created de novo. The addition of microvilli or cilia to the apical surface of a dendrite, according to the component selection model, can occur simply by one suitable mutation in a regulatory gene.

3) The photoreceptors and eyes that evolved in the higher phyla are much more precisely structured than those of lower phyla, in spite of the dearth of stepwise sequences that cross the boundaries between phyla. Explanations: a) A general increase in the number and variety of protein components may have occurred. A well-documented example is the increase in hemoglobin varieties in vertebrates. This is thought to have occurred by structural gene duplications and subsequent modification by point mutations. From primitive fish to man the number of active hemoglobin genes per genome increased from one to 9 (Jeffreys 1982). Genes for the ϵ- and ζ-hemoglobin subunits are selected in embryonic red cells, α- and γ-chains in fetal red cells and α- and β-chains in adult red cells. Most of these genes are paired. Each hemoglobin type has slightly different physiological properties. In addition to these active genes 3 pseudogenes for globin are present in the human genome which are not expressed, having been silenced by numerous mutations. Many mutants of the active genes, some deleterious like that for sickle cell

hemoglobin, are also present in the gene pool of man. Similar gene duplication and subsequent modification may also have provided the 3-4 rhodopsins with different absorption spectra that are found in arthropods and vertebrates having colour vision. The process may also have produced a variety of more refined building-blocks of cell structure. The structural gene for the protein strands that tie the rims of rod outer-segment discs to each other and to the ciliary membrane (Roof & Heuser, 1982) may not be present in the genome of ascidians (Tunicata), in which larval photoreceptors consist of discs which are not so precisely organized as in vertebrates (Eakin, 1972). The membrane junctions between microvilli of squid rhabdomeres (Saibil, 1982) appear not to occur in lower phyla (see Fig. 13). b) Increases in the size of the regulatory hierarchy probably have occurred (Britten & Davidson, 1969). With the potential increase in control of differentiation that this would provide, there can be a greater number of possible combinations of structural genes in batteries coded by the genome. It may enable more precise control of the location of the components that modify cell structure and more precise timing and coordination of their synthesis.

The increase in the variety of the components of cell structure and in the size of the regulatory hierarchy can have occurred independently of the evolution of eyes or photoreceptors. Herein may lie the explanation of the trend towards greater complexity in photoreceptors and eyes which occurs in spite of the lack of stepwise sequences across phyla.

4) Though photoreceptor cells have frequently arisen de novo in Arthropoda and Vertebrata, no ciliary photoreceptors have been found in the first phylum nor rhabdomeric photoreceptors in the second. Explanation: This cannot have been due to a loss of the ability to combine the components of transduction with the other type of membrane domain, because ciliary chemo- and mechanoreceptors are common in Arthropoda, and microvillar chemo- and mechanoreceptors exist in Vertebrata. In the well-developed main eyes of these 2 groups a change of photoreceptor type probably would be so disruptive as to be strongly selected against. However extraocular photoreceptors exist in both groups which are not a part of any complex organ (Yoshida, 1979), e.g. the (rhabdomeric) ventral photoreceptors of Limulus (Clark et al., 1969) and the (ciliary) carotid photoreceptors of vertebrates (Kikuchi & Aoki, 1982). Why has not the opposite type of photoreceptor appeared in a few of these examples? Could there be something about the organization of the genome of vertebrates and arthropods that makes modification of a well-established gene battery unlikely?

Predictions

Some predictions of the model are worth mentioning. First, since one or only a few mutations affecting component selection may imbue a ciliary or microvillar membrane with photosensitivity, and since such mutations are unlikely to prevent reproduction, one might expect the modified master genes to exist in the gene pool of a species. Thus photosensitive mechano- or chemoreceptor cells may be found in some individuals. Second, experimentally induced mutations should affect the regulation of photoreceptor cell structure. Some observations of chemotaxis-defective mutant strains of Caenorhabditis elegans (Lewis & Hodgkin, 1977; Ward, 1978) are interesting in this regard. Most of the mutant strains had morphological defects in sensory cells or their supporting cells. In two mutants, the ciliary axonemes distal to the basal bodies were lacking in all the ciliated sensory neurons that were investigated. In two other mutants, among other effects, the microvillar or "finger-like" projections of a putative sensory cell were greatly reduced in length and number. In another, ciliary rootlets were restored in some sensory cells that normally lack them, but there was also widespread distortion of a variety of other structures of the body. It would appear that at least some of these mutations are located in the regulatory hierarchy but at the present stage of investigation it is difficult to infer anything as to how the hierarchy is structured.

Conclusions

Britten and Davidson (1969) speculated that evolution can occur in three ways: 1) modification of structural genes, 2) changes to the regulatory hierarchy that produce new combinations of structural genes and 3) enlargement of the regulatory hierarchy. These ideas in the form of a novel, component-selection model provide plausible explanations for many questions arising from a comparison of metazoan photoreceptors. In particular, the puzzling occurrence of rhabdomeric and ciliary types in closely related organisms, or in the same organism, can now be explained.

These explanations, however, are of the "soft" or hermeneutic type (Stent, 1977) which is almost inevitable with questions concerning highly complex systems. Further, the component-selection model, as presented, is highly simplified. There is no assurance that a "menu" resulting from evolution of regulatory systems is nearly so concise or orderly as that of Table 1. The value of the model is that it provides a simple and logical framework which ties together otherwise unrelated or puzzling observations. Future modification will be inevitable as other observations are considered and the predictions are tested. The model, as applied to the evolution of photoreceptors, makes very clear that our understanding of the evolution of photoreceptors will improve considerably as the mechanisms of gene regulation in eukaryotes are elucidated.

ACKNOWLEDGEMENTS

The author is grateful to Richard M. Eakin, Merrill B. Hille and Michael J. Smith for reading the manuscript and offering suggestions. The author also appreciates the careful work of Wendy Yee in preparing the manuscript camera-ready for publication.

REFERENCES

Bannister, L.H. (1965) The fine structure of the olfactory surface of teleostean fishes. Quart. J. Microsc. Sci. 106: 333-342.

Bannister, L.H. (1968) Fine structure of the sensory endings in the vomero-nasal organ of the slow-worm Anguis fragilis. Nature (Lond.) 217: 275-276.

Bedini, C., Ferrero E. & Lanfranchi, A. (1977) Fine structural changes induced by circadian light-dark changes in photoreceptors of Dalyelliidae (Turbellaria: Rhabdocoela). J. Ultrastruct. Res. 58: 66-77.

Besharse, J.C. & Pfenninger, K.H. (1980) Membrane assembly in retinal photoreceptors. 1. Freeze-fracture analysis of cytoplasmic vesicles in relationship to disc assembly. J. Cell Biol. 87: 451-463.

Blest, A.D. (1978) The rapid synthesis and destruction of photoreceptor membrane by a dinopid spider: a daily cycle. Proc. Roy. Soc. Lond. B 200: 463-483.

Bok, D., Hall, M.O. & O'Brien, P. (1977) The biosynthesis of rhodopsin as studied by membrane renewal in rod outer segments. In: International Cell Biology 1976-1977. Ed. B.R. Brinkley & K.R. Porter. New York, Rockefeller Univ. Press, p. 608-617.

Bollerup, G., & Burr, A.H. (1979) Eyespot and other pigments in nematode esophageal muscle cells. Can. J. Zool. 57: 1057-1069.

Boschek, C.B. & Hamdorf, K. (1976). Rhodopsin particles in the photoreceptor membrane of an insect. Z. Naturforsch. 31c: 763 and facing plate.

Brandenburger, J.L. & Eakin, R.M. (1970) Pathway of incorporation of vitamin A $^3H_2$ into photoreceptors of a snail Helix aspersa. Vision Res. 10: 639-653.

Brandenburger, J.L., Eakin, R.M. & Reed, C.T. (1976) Effects of light- and dark-adaptation on the photic microvilli and photic vesicles of the pulmonate snail Helix aspersa. Vision Res. 16: 1205-1210.

Brandenburger, J.L., Woollacott, R.M. & Eakin, R.M. (1973) Fine structure of eyespots in tornarian larvae (Phylum: Hemichordata). Z. Zellforsch. Mikrosk. Anat. 142: 89-102.

Britten, R.J. & Davidson, E.H. (1969) Gene regulation for higher cells: a theory. Science 165: 349-357.

Britten, R.J. & Davidson, E.H. (1971) Repetitive and non-repetitive DNA sequences and a speculation on the origins of evolutionary novelty. Quart. Rev. Biol. 46: 111-138.

Brown, H.M., Ito, H. & Ogden, T.E. (1968) Spectral sensitivity of the planarian ocellus. J. Gen. Physiol. 51: 255-260.

Burr, A.H. (1979) Analysis of phototaxis in nematodes using directional statistics. J. Compar. Physiol. 134: 85-93.

Burr, A.H. (1983) Photomovement behavior in simple invertebrates (This volume).

Burr, A.H. & Burr, C. (1975) The amphid of the nematode Oncholaimus vesicarius: Ultrastructural evidence for a dual function as a chemoreceptor and photoreceptor. J. Ultrastruct. Res. 51: 1-15.

Burr, A.H., Schiefke, R. & Bollerup, G. (1975) Properties of a hemoglobin from the chromatrope of the nematode Mermis nigrescens. Biochim. Biophys. Acta 405: 404-411.

Burr, A.H. & Webster, J.M. (1971) Morphology of the eyespot and description of two pigment granules in the esophageal muscle of a marine nematode, Oncholaimus vesicarius. J. Ultrastruct. Res. 36: 621-632.

Carpenter, K.S., Morita, M. & Best, J.B. (1974a) Ultrastructure of the photoreceptor of the planarian Dugesia dorotocephala. I. Normal eye. Cell Tissue Res. 148: 143-158.

Carpenter, K.S., Morita, M. & Best, J.B. (1974b) Ultrastructure of the photoreceptor of the planarian Dugesia dorotocephala. II. Changes induced by darkness and light. Cytobiol. 8: 320-338.

Cherry, R.J. (1979) Rotational and lateral diffusion of membrane proteins. Biochim. Biophys. Acta 559: 289-327.

Clark, A.W., Millecchia, R. & Mauro, A. (1969) The ventral photoreceptor cells of Limulus. I. The microanatomy. J. Gen. Physiol. 54: 289-309.

Clément, P. (1980) Phylogenetic relationships of rotifers, as derived from photoreceptor morphology and ultrastructural analyses. Hydrobiol. 73: 93-117.

Clément, P. & Wurdak, E. (1983) Photoreceptors and photoreception in rotifers (This volume).

Cobb, N.A. (1926) The species of Mermis. J. Parasitol. 13: 66-72 and Plate II.

Cobb, N.A. (1929) The chromatropism of Mermis subnigrescens, a nemic parasite of grasshoppers. J. Wash. Acad. Sci. 19:159.

Coomans, A. (1981) Phylogenetic implications of the photoreceptor structure. In: Origine dei Grandi Phyla dei Metazoi. Rome, Accademia Nazionale dei Lincei, p. 23-68.

Couillard, P. (1983) Photoreception in Protozoa, an overview (This volume).

Croll, N.A. (1966) A contribution to the light sensitivity of the "chromatrope" of Mermis subnigrescens. J. Helminthol. 40: 33-38.

Croll, N.A., Evans, A.A.F. & Smith, J.M. (1975) Comparative nematode photoreceptors. Comp. Biochem. Physiol. 51A: 139-143.
Croll, N.A., Riding, I.L. & Smith, J.M. (1972) A nematode photoreceptor. Comp. Biochem. Physiol. 42A: 999-1009.
Darnell, J.E. (1982) Variety in the level of gene control in eukaryotic cells. Nature (Lond.) 297: 365-371.
Darwin, C. (1859) On the Origin of Species. London, John Murray (Facsimile of First Edition: Cambridge, Harvard Univ. Press, 1966), p. 186-187.
Davidson, E.H. & Britten, R.J. (1979) Regulation of gene expression: possible role of repetitive sequences. Science 204: 1052-1059.
Davidson, E.H., Galau, G.A., Angerer, R.C. & Britten, R.J. (1975) Comparative aspects of DNA organization in Metazoa. Chromosoma (Berl.) 51: 253-259.
Davidson, E.H., Klein, W.H. & Britten, R.J. (1977) Sequence organization in animal DNA and a speculation on hnRNA as a coordinate regulatory transcript. Devel. Biol. 55: 69-84.
Dentler, W.L. (1981) Microtubule-membrane interactions in cilia and flagella. Int. Rev. Cytol. 72: 1-47.
Dover, G. (1982) Molecular drive: a cohesive mode of species evolution. Nature 299: 111-117.
Eakin, R.M. (1963) Lines of evolution of photoreceptors. In: General Physiology of Cell Specialization. Ed. D. Mazia & A. Tyler. New York, McGraw-Hill.
Eakin, R.M. (1972) Structure of invertebrate photoreceptors. In: Handbook of Sensory Physiology, Vol. VII/1. Ed. H.J.A. Dartnall. Berlin, Springer-Verlag, p. 625-684.
Eakin, R.M. (1979) Evolutionary significance of photoreceptors: In retrospect. Amer. Zool. 19: 647-653.
Eakin, R.M. (1982) Continuity and diversity in photoreceptors In: Visual Cells in Evolution. Ed. J.A. Westfall. New York, Raven Press, p. 91-105.
Eakin, R.M. & Brandenberger, J.L. (1979) Effects of light on ocelli of seastars. Zoomorphol. 92: 191-200.
Eakin, R.M. & Brandenburger, J.L. (1981) Unique eye of probable evolutionary significance. Science 211: 1189-1190.
Eakin, R.M. & Westfall, J.A. (1962) Fine structure of the photoreceptors in the hydromedusan Polyorchis penicillatus. Proc. Nat. Acad. Sci. U.S. 48: 826-833.
Eckert, R. & Brehm, P. (1979) Ionic mechanisms of excitation in Paramecium. Ann. Rev. Biophys. Bioeng. 8: 353-383.
Eguchi, E. (1965) Rhabdom structure and receptor potentials in single crayfish retinular cells. J. Cell Comp. Physiol. 66: 411-430.
Eguchi, E. & Waterman, T.H. (1976) Freeze-etch and histochemical evidence for cycling in crayfish photoreceptor membranes. Cell Tissue Res. 169: 419-434.

Ellenby, C. (1964) Hemoglobin in the "chromatrope" of an insect parasite nematode. Nature (Lond.) 202: 615-616.
Ellenby, C. & Smith, L. (1966) Hemoglobin in a marine nematode. Nature (Lond.) 210: 1372.
Fauré-Fremiet, E. (1961) Cils vibratiles et flagelles. Biol. Rev. 36: 464-536.
Fauré-Fremiet, E. & Rouiller, C. (1957) Le flagelle interne d'une Chrysomonadale: Chromulina psammobia. Comptes Rend. Acad. Sci. Fr. 244: 2655-2657.
Fein, A. & Szuts, E.Z. (1982) Photoreceptors: Their Role in Vision. Cambridge, Cambridge Univ. Press, p. 104-109.
Fernández, H.R. & Nickel, E.E. (1976) Ultrastructural and molecular characteristics of crayfish photoreceptor membranes. J. Cell Biol. 69: 721-732.
Foster, K.W. & Smyth, R.D. (1980) Light antennas in phototactic algae. Microbiol. Rev. 44: 572-630.
Fournier, A. (1983) Photoreceptors and photosensitivity in Platyhelminthes (This volume).
Galau, G.A., Chamberlin, M.E., Hough, B.R., Britten, R.J. & Davidson, E.H. (1976) Evolution of repetitive and nonrepetitive DNA. In: Molecular Evolution. Ed. F.J. Ayala. Sunderland, Mass., Sinauer Assoc., Inc, p. 200-224.
Goldsmith, T.H. & Wehner, R. (1977) Restrictions on rotational and translational diffusion of pigment in the membranes of a rhabdomeric photoreceptor. J. Gen. Physiol. 70: 453-490.
Gorman, A.L.F. & McReynolds, J.S. (1978) Ionic effects on the membrane potential of hyperpolarizing photoreceptors in scallop retina. J. Physiol. 275: 345-355.
Gorman, A.L.F., McReynolds, J.S. & Barnes, S.N. (1971) Photoreceptors in primitive chordates: fine structure, hyperpolarizing receptor potentials, and evolution. Science 172: 1052-1054.
Gouras, P. & Zrenner, E. (1979) The blue sensitive cone system. Exerpta Medica Int. Cong. Ser. 450: 379-384.
Green, C.R. & Bergquist, P.R. (1982) Phylogenetic relationships within the invertebrata in relation to the structure of septate junctions and the development of "occluding" junctional types. J. Cell Sci. 53: 279-305.
Hermans, C.O. & Eakin, R.M. (1969) Fine structure of the cerebral ocelli of a sipunculid, Phascolosoma agassizii. Z. Zellforsch. 100: 325-339.
Hildebrand, E. (1978) Bacterial phototaxis. In: Taxis and Behavior. Ed. G.L. Hazelbauer. London, Chapman and Hall, p. 35-73.
Holtzman, E. & Mercurio, A.M. (1980) Membrane circulation in neurons and photoreceptors: some unresolved issues. Int. Rev. Cytol. 67: 1-67.

Hope, W.D. (1967) Free-living marine nematodes of the genera Pseudocella Filipjev, 1927, Thorocostoma Marion, 1980, and Deontostoma Filipjev, 1916 (Nematoda: Leptosomatidae) from the west coast of North America. Trans. Amer. Microsc. Soc. 86: 307-334.

Hughes, R.L. & Woollacott, R.M. (1980) Photoreceptors of bryozoan larvae (Cheilostomata, Cellularoidea). Zool. Scripta 9: 129-138.

Jeffreys, A.J. (1982) Evolution of globin genes. In: Genome Evolution. Ed. G.A. Dover & R.B. Flavell. London, Academic Press, p. 157-176.

Kerjaschki, D. & Hörandner, H. (1976) The development of mouse olfactory vesicles and their cell contacts: a freeze-etching study. J. Ultrastruct. Res. 54: 420-444.

Kikuchi, M. & Aoki, K. (1982) The photoreceptor cell in the pineal organ of the Japanese common newt. Experentia 38: 1450-1451.

Land, M.F. (1981) Optics and vision in invertebrates. In: Handbook of Sensory Physiology, Vol. VII/6B. Ed. H. Autrum. Berlin, Springer-Verlag, p. 471-592.

Land, M.F. (1983a) Mollusca. (This volume).

Land, M.F. (1983b) Crustacea. (This volume).

Laskey, R.A. (1979) Biochemical processes in early development. In: Companion to Biochemistry, Vol 2. Ed. A.T. Bull, J.R. Lagnado & J.O. Thomas. London, Longman Group Ltd., p. 137-160.

Lewin, B. (1980) Gene Expression, Vol. 2, 2nd Edition. New York, Wiley & Sons, p. 949-957.

Lewis, J.A. & Hodgkin, J.A. (1977) Specific neuroanatomical changes in chemosensory mutants of the nematode Caenorhabditis elegans. J. Comp. Neurol. 172: 489-510.

Little, P.F.R. (1982) Globin pseudogenes. Cell 28: 683-684.

McReynolds, J.S. & Gorman, A.L.F. (1974) Ionic basis of hyperpolarizing receptor potential in scallop eye: increase in permeability to potassium ions. Science 183: 658-659.

McReynolds, J.S. & Gorman, A.L.F. (1975) Hyperpolarizing photoreceptors in the eye of a primitive chordate Salpa democratica. Vision Res. 15: 1181-1186.

Melkonian, M. & Robenek, H. (1980a) Eyespot membranes of Chlamydomonas reinhardii: a freeze-fracture study. J. Ultrastruct. Res. 72: 90-102.

Melkonian, M. & Robenek, H. (1980b) Eyespot membranes in newly released zoospores of the green alga Chlorosarcinopsis gelatinosa (Chlorosarcinales) and their fate during zoospore settlement. Protoplasma 104: 129-140.

Menco, B.P.M., Dodd, G.H. & Davey, M. (1976) Presence of membrane particles in freeze-etched bovine olfactory cilia. Nature (Lond.) 263: 597-599.

Nickel, E. & Menzel, R. (1976) Insect UV- and green-photoreceptor membranes studied by freeze-fracture technique. Cell. Tissue Res. 175: 357-368.

Nicolson, G.L. (1979) Topographic display of cell surface components and their role in transmembrane signaling. Current Topics in Devel. Biol. 13: 305-338.

Nultsch, W. & Häder, D.-P. (1979) Photomovement of motile microorganisms. Photochem. Photobiol. 29: 423-437.

Olive, J. (1980) The structural organization of mammalian retinal disc membranes. Int. Rev. Cytol. 64: 107-169.

Palade, G. (1975) Intracellular aspects of the process of protein synthesis. Science 189: 347-358.

Papermaster, D.S. & Schneider, B.G. (1982) Biosynthesis and morphogenesis of outer segment membranes in vertebrate photoreceptor cells. In: Cell Biology of the Eye. Ed. D.S. McDevitt. New York. Academic Press, p. 475-531.

Papermaster, D.S., Schneider, B.G. & Besharse, J.C. (1979) Assembly of rod photoreceptor membranes. Immunocytochemical and autoradiographic localization of opsin in smooth vesicles of the inner segment. J. Cell Biol. 83: 275a.

Poo, M.P. & Cone, R.A. (1974) Lateral diffusion of rhodopsin in the photoreceptor membrane. Nature (Lond.) 247: 438-441.

Röhlich, P. (1975) The sensory cilium of retinal rods is analogous to the transitional zone of motile cilia. Cell Tissue Res. 161: 421-430.

Röhlich, P. & Tar, E. (1968) The effect of prolonged light deprivation on the fine structure of planarian photoreceptors. Z. Zellforsch. 90: 507-518.

Röhlich, P. & Török, L.J. (1961) Elektronenmikroskopische untersuchungen des auges von planarien. Z. Zellforsch. 54: 362-381.

Röhlich, P. & Török, L.J. (1962) The effect of light and darkness on the fine structure of the retinal clubs of Dendrocoelum lacteum. Quart. J. Microsc. Sci. 103: 543-548.

Roof, D.J. & Heuser, J.E. (1982) Surfaces of rod photoreceptor disc membranes: integral membrane components. J. Cell Biol. 95: 487-500.

Rosen, M.D., Stasek, C.R. & Hermans, C.O. (1978) The ultrastructure and evolutionary significance of the cerebral ocelli of Mytilus edulis, the bay mussel. The Veliger 21: 10-18.

Saibil, H. (1982) An ordered membrane-cytoskeleton network in squid photoreceptor microvilli. J. Mol. Biol. 158: 435-456.

Salvini-Plawen, L.V. (1982) On the polyphyletic origin of photoreceptors. In: Visual Cells in Evolution. Ed. J.A. Westfall. New York, Raven Press, p. 137-154.

Salvini-Plawen, L.V. & Mayr, E. (1977) On the evolution of photoreceptors and eyes. Evolutionary Biol. 10: 207-263.

Satir, B.H. (1980) The role of local design in membranes. In: Membrane-Membrane Interactions. Ed. N.B. Gilula. New York, Raven Press, p. 45-58.

Schinz, R.H., Lo, M-V.C., Larrivee, D.C. & Pak, W.L. (1982) Freeze-fracture study of the Drosophila photoreceptor membrane: mutations affecting membrane particle density. J. Cell Biol. 93: 961-969.

Siddiqui, I.A. & Viglierchio, D.R. (1970) Ultrastructure of photoreceptors in the marine nematode Deontostoma californicum. J. Ultrastruct. Res. 32: 558-571.

Singla, C.L. (1974) Ocelli of hydromedusae. Cell Tissue Res. 149: 413-429.

Singla, C.L. & Weber, C. (1982a) Fine structure studies of the ocelli of Polyorchis penicillatus (Hydrozoa, Anthomedusae) and their connection with the nerve ring. Zoomorphol. 99: 117-129.

Singla, C.L. & Weber, C. (1982b) Fine structure of the ocellus of Sarsia tubulosa (Hydrozoa, Anthomedusae). Zoomorphol. 100: 11-22.

Steinberg, R.H., Fisher, S.K. & Anderson, D.H. (1980) Disk morphogenesis in vertebrate photoreceptors. J. Comp. Neurol. 190: 501-518.

Stent, G.S. (1977) Introduction: cerebral hermeneutics. In: Function and Formation of Neural Systems. Ed. G.S. Stent. Berlin, Dahlem Konferenzen, p. 13-20.

Stewart, A. (1966) Ultrastructure of the eyes in three freshwater triclad worms. Amer. Zool. 6: 615-616.

Tabata, M., Tamura, T., & Niwa, H. (1975) Origin of the slow potential in the pineal organ of the rainbow trout. Vision Res. 15: 737-740.

Takasu, N. & Yoshida, M. (1982) Freeze-fracture study of medusan ocelli. Jap. J. Ophthalmol. 26: 90-91.

Usukura, J. & Yamada, E. (1978) Observations on the cytolemma of the olfactory receptor cell in the newt. Cell Tissue Res. 188: 83-98.

Valentine, J.W. & Campbell, C.A. (1975) Genetic regulation and the fossil record. Amer. Sci. 63: 673-680.

Vanfleteren, J.R. (1982) A monophyletic line of evolution? Ciliary induced photoreceptor membranes. In: Visual Cells in Evolution. Ed. J.A. Westfall. New York, Raven Press, p. 107-136.

Vanfleteren, J.R. & Coomans, A. (1976) Photoreceptor evolution and phylogeny. Z. Zool. Syst. Evolut.-forsch. 14: 157-169.

Ward, S. (1978) Nematode chemotaxis and chemoreceptors. In: Taxis and Behavior. Ed. G.L. Hazelbauer. London, Chapman and Hall, p. 141-203.

Waterman, T.H. (1981) Polarization sensitivity. In: Handbook of Sensory Physiology, Vol. VII/6B. Ed. H. Autrum. Berlin, Springer-Verlag, p. 281-469.

Weber, C. (1981) Structure, histochemistry, ontogenetic development, and regeneration of the ocellus of *Cladonema radiatum* Dujardin (Cnidaria, Hydrozoa, Anthomedusae). J. Morphol. 167: 313-331.

Weber, C. (1982a) Electrical activity in response to light of the ocellus of the hydromedusan, *Sarsia tubulosa*. Biol. Bull. 162: 413-422.

Weber, C. (1982b) Electrical activities of a type of electroretinogram recorded from the ocellus of a jellyfish *Polyorchis penicillatus* (Hydromedusae). J. Exp. Zool. 223: 231-243.

Whaley, W.G. & Dauwalder, M. (1979) The golgi apparatus, the plasma membrane and functional integration. Int. Rev. Cytol. 58: 199-245.

Whittle, A.C. (1976) Reticular specializations in photoreceptors: a review. Zool. Scripta 5: 191-206.

Wood, D.C. (1976) Action spectrum and electrophysiological responses correlated with the photophobic response of *Stentor coeruleus*. Photochem. Photobiol. 24: 261-266.

Woollacott, R.M. & Eakin, R.M. (1973) Ultrastructure of a potential photoreceptor organ in the larva of an entoproct. J. Ultrastruct. Res. 43: 412-425.

Woollacott, R.M. & Zimmer, R.L. (1972) Fine structure of a potential photoreceptor organ in the larva of *Bugula neritina* (Bryozoa). Z. Zellforsch. 123: 458-469.

Yamamoto, M. & Yoshida, M. (1978) Fine structure of the ocelli of a synaptid holothurian, *Opheodesoma spectabilis*, and the effects of light and darkness. Zoomorphol. 90: 1-17.

Yamasu, T. & Yoshida, M. (1973) Electron microscopy on the photoreceptors of an anthomedusa and a scyphomedusa. Publ. Seto Mar. Lab. 20: 757-778.

Yamasu, T. & Yoshida, M. (1976) Fine structure of complex ocelli of a cubomedusan, *Tamoya bursaria* Haeckel. Cell. Tissue Res. 1970: 325-339.

Yoshida, M. (1979) Extraocular photoreception. In: Handbook of Sensory Physiology, Vol. VII/6A. Ed. H. Autrum. Berlin, Springer-Verlag, p. 581-640.

Zimmer, R.L. & Woollacott, R.M. (1977) Structure and classification of gymnolaemate larvae. In: Biology of Bryozoans. Ed. R.M. Woollacott & R.L. Zimmer. New York, Academic Press, p. 57-90.

Zrenner, E. & Gouras, P. (1979) Blue-sensitive cones of the cat produce a rodlike electroretinogram. Invest. Ophthalmol. 18: 1076-1081.

PHOTOMOVEMENT BEHAVIOR IN SIMPLE INVERTEBRATES

A. H. BURR

Department of Biological Sciences
Simon Fraser University
Burnaby, B.C.
Canada, V5A 1S6

> It is easier to be wise on behalf of others
> Than to be so for ourselves.
>
> La Rochefoucauld, Maxims, 1665.

INTRODUCTION

In the last decade we have seen a considerable increase in research on sensory systems in simple organisms. On the one hand are the intensive studies of physiological mechanisms of behavior in bacteria (Berg, 1975; Goy & Springer, 1978; Hildebrand, 1978; Koshland, 1980) and single-celled eukaryotes (Nelson & Kung, 1978; Darmon & Brachet, 1978; Diehn, 1979; Nultsch & Häder, 1979). On the other hand are neurobiological studies of opisthobranch molluscs (Kandel, 1979; Alkon, 1980), daphnia (Macagno et al., 1973; Young, 1981) and leeches (Nicholls et al., 1977; Fernández & Stent, 1980). Somewhere in between is the behavioral and neuroanatomical work begun on nematodes (Ward, 1978; Wright, 1980; Dusenbery, 1980a, Burr, 1979), and rotifers (Clément & Wurdak, 1983). A variety of studies of marine invertebrates are reviewed by Creutzberg (1975).

The concomitant interest in sensory behavior has brought to light some problems with the existing nomenclature. One difficulty is that several sets of terms describing movement behavior are in current use. Another is that the definitions of terms have been either unclear or divergent. A good example is the term 'kinesis'. 'Photokinesis' was introduced 100 years ago by Englemann to refer to the dependence of the velocity of photosynthetic bacteria on the intensity of the stimulus. Until very recently, microbiologists have continued to use 'kinesis' with this meaning (e.g. Nultsch, 1970). Observations on animals led Gunn et al. (1937) to expand the term to include all undirected movement responses, including those that adapt to the stimulus. They introduced 'orthokinesis' to refer to the effect of a stimulus on speed and 'klinokinesis' to effects on the rate of change of direction. These meanings were promoted by Fraenkel and Gunn (1940, 1961) and are understood by most zoologists today. However for a period 'klinokinesis' was confused, even by Gunn himself, with the migration that can occur as a consequence of a kinesis (Gunn, 1975). In some publications it is sometimes not clear which meaning was intended; both usages have appeared even in the same paper. Recently, microbiologists have adopted 'orthokinesis' and 'klinokinesis' but still include only non-adapting responses in their definitions (Diehn et al., 1977).

Thus there are 4 meanings of 'kinesis' in the literature, 2 of which are in current use. The distinguishing feature for microbiologists is that the responses are non-adapting; directed responses to a stimulus gradient are included. In contrast, the distinguishing feature for the followers of Fraenkel and Gunn is that the responses are undirected; adapting responses are included. Between the 2 schools there are also conflicting definitions of 'taxis' and there exist terms that are not in common.

In addition to problems with the nomenclature used to classify movement behavior, the klinokinetic mechanism for migration over a gradient of stimulus intensity, proposed by Ullyott (1936a,b), has never been satisfactorily demonstrated for metazoans (Stasko & Sullivan, 1971; Gunn, 1975). Only for bacteria migrating on chemical gradients (Berg & Brown, 1972) has anyone demonstrated all the criteria, including that individual changes of direction be undirected in the presence of the gradient (following Fraenkel & Gunn's meaning).

On light gradients there has been no satisfactory demonstration of klinokinetic migration with any organism. One reason for this is a poor understanding of what is needed to demonstrate this mechanism. Another is the near impossibility of eliminating all horizontal scattered light (Stasko & Sullivan, 1971), together with the choice of animals that respond sensitively to the direction of the scattered light. In many cases the information-gathering capability of the eye under investigation was not fully appreciated in the design and interpretation of experiments.

In this chapter, I hope to [a] refine the terminology to make it more serviceable and acceptable to all biologists, [b] clarify what kinds of information can be gathered by each type of eye structure, [c] illustrate [a] and [b] with examples from studies of nematode behavior and [d] show how the distinction between a klinokinetic migration and a taxis can be made using track data.

SIMPLE MOVEMENT BEHAVIOR

The experimental study of behavior is a starting point for learning how information gathered by sensory organs is processed by nervous systems. The outstanding example of this is the influence of human psychophysical studies on later neuroanatomical and neurophysiological research. To study the effect of stimulus parameters in other animals, one must either train the animal to respond in a measurable way or rely on innate reactions. In both cases, usually, simple responses are preferred even in organisms with complex behavior. In the remainder of this chapter I will focus on simple movement responses.

Though there is great diversity in the mechanisms of movement, there are a limited number of ways an organism can move, as pointed out by Jander (1970). Either rotational motion changes the orientation of one or more body axes relative to the surrounds, or translational motion changes the locus of the body in space. A translation may occur in 3 dimensional space or be constrained to one or two dimensions. Translation can be considered to have distance and directional components. A movement response is an alteration in the translation or orientation of an organism. These can be changed only in direction, amount, speed, frequency or timing.

Movement response terminology

Because the total number of possible movement responses is small, it is not surprising that simple movement responses of bacteria, flagellates, nematodes, insects and mammals naturally fall in the same categories. Noticing this, several biologists over the last century have attempted to classify such responses. The classification schemes have not been without their problems, however, as described in several reviews (Fraenkel & Gunn, 1940, 1961; Hand & Davenport, 1970; Stasko & Sullivan, 1971; Gunn, 1975; Nultsch, 1975).

At least 2 schemes are currently in use. One, originating with Kühn (1919. See Nultsch, 1975 for a summary) is used by microbiologists. The other, a modification of Kühn's system by Fraenkel & Gunn (1940, 1961) is commonly used by zoologists. Recently an ad hoc Committee on Behavioral Terminology, convened at a conference of microbiologists, proposed a new set of terms which combines some of the stronger features of the two systems (Diehn, et al., 1977; Hildebrand, 1978; Diehn, 1979). In short, Fraenkel and Gunn's terms: 'klinokinesis', 'orthokinesis' and 'taxis' were adopted, Kühn's 'phobotaxis' was changed to 'phobic response', and Kühn's definition of 'kinesis' was adopted.

Though intended for microbiologists, the committee expressed the hope that their new terminology would be adopted universally. However in its present form it is inadequate for metazoan behavior for several reasons. First, it retains Kühn's definition of kinesis restricting the term to non-adapting movement responses to stimuli. It is understandable that by doing so, the microbiologists are able to separate the non-adapting photobehavior energized by photosynthesis from behavior dependent on other physiological processes. However this does not work for metazoans because non-adapting kineses exist that have nothing to do with photosynthesis, for example the humidity reactions of Peripatopsis (Onychophora, Arthropoda. Bursell & Ewer, 1950).

Secondly, all adapting alterations in movement are called phobic responses by the committee (Diehn et al., 1977), replacing Kühn's term 'phobotaxis'. Thus, transient changes in speed and ordinary turning responses are inexplicably lumped together with what they call a typical phobic response: a change in direction which follows a "change in linear velocity, normally a stop response". The primary distinction between their kinesis and their phobic response is whether or not the response adapts. This creates a major problem for responses with a moderate to long adaptation time. The committee suggests that the choice of terms, in this case, be "left up to the individual investigator." In the modified terminology I am proposing, all response terms include both adapting and non-adapting behavior, a premise which avoids this problem and is more realistic physiologically. The distinguishing feature of a kinesis becomes that it is undirected behavior, as defined by Gunn et al. (1937). With this meaning, 'kinesis' has been useful in animal studies.

A problem with both the committee's terminology and that of Fraenkel and Gunn is that there is no term that clearly distinguishes the "typical phobic response" from other turns. These are innate sequences of behavior that begin with a stop or sudden decrease in speed and typically result in a major change in direction. They come as close to being a biological universal as any phenomenon: the pattern is found in bacteria, ciliates, flagellates, sponge larvae and metazoans, including insects and vertebrates. Even in lower invertebrates and ciliates the change of direction can be a directed one (see nematode example later in this chapter). Therefore the term 'klinokinesis' is inappropriate, contrary to the claim of Gunn et al. (1937) and Fraenkel and Gunn (1940, 1961). 'Phobic reactions' as defined by Diehn et al. (1977) is too general and 'phobic' has undesirable connotations in English. The terms, trial and error reaction, shock reaction, titubant reaction, avoiding reaction, reversal bout, tumbling and twiddles are not suitable because they describe a particular form of the reaction rather than its general form and some are objectionable as they have anthropomorphic connotations in English (See Note 7 of Appendix).

I therefore am proposing a new term to designate this behavior which refers to its most important operational feature - a significant change in direction of movement. The terms, ecclisis and ecclitic response come from the Greek noun εκκλισις, a turning out of one's course. It is intended to apply to all response patterns in which a change of direction is preceded by a stop of movement, whether they are spontaneous or elicited by external stimuli, adapting or non-adapting, and whether the change of direction is directed or undirected.

In reviewing the literature on movement behavior I was impressed by the amount of disagreement or confusion as to the meaning of movement terms. In most cases this can be traced to imprecise definitions. In the list that follows I have attempted to improve the definitions, making them as operational and phenomenological as possible. Any further inadequacies I hope will be brought out by future discussion.

In the case of stimuli (such as light) that have a directional property the terminology does not provide for the distinction between a taxis due to the direction of the stimulus and one due to an intensity gradient. Taxis as defined by Diehn et al. (1977) inexplicably excludes responses to gradients, which would mean that 'chemotaxis' and 'thermotaxis' would cease to exist. I propose that 'taxis' include both types of directed migration (following Fraenkel & Gunn, 1940, 1961) and suggest that 'directional taxis' and 'gradient taxis' be used whenever the distinction needs to be made.

Proposed movement terminology

In summary, I have started with the terminology of Diehn et al. (1977) and Fraenkel and Gunn (1940, 1961) and ended up somewhere in between. I have dispensed with 'phobic response' and proposed the new term 'ecclitic response' with a more definite meaning. I have adopted a version of Fraenkel and Gunn's definition of kinesis and have refined other definitions in the following list. Later in this chapter, some nematode examples are described, and some statistical methods for distinguishing between klinokinetic migration and taxes are suggested. An appendix of notes is included at the end. My hope is that this terminology will not only be acceptable to both micro- and metazoan biologists, but also be clearer and more serviceable than existing ones.

STIMULUS. An agent (Note 1) capable of eliciting a behavioral response.

The physical nature of the stimulus can be indicated by a prefix to the term describing the response, for example, photo, thermo, electro, geo, mechano, magneto, or chemo.

A stimulus usually has more than one property that can be varied independently (Note 2). The modifiers "step-up" or "step-down" can be added where an increase or decrease in stimulus intensity elicits the response. Thus an ecclitic response to a sudden increase in light intensity can be called a step-up photoecclitic response.

STIMULUS DIRECTION. The direction of propagation (of sound or electromagnetic stimuli) or direction of force (due to other directional stimuli. See Note 2).

STIMULUS GRADIENT. At any point in the spatial distribution of stimulus intensity, the direction and/or magnitude of maximum rate of increase in intensity.

STIMULUS FIELD. The region where stimulus parameters are controlled. In a <u>uniform field</u> the parameters are constant. In a <u>gradient field</u> the gradient of one parameter (usually intensity) varies with location in a regular, controlled manner.

RESPONSE. Any alteration in the translational or rotational motion of an organism (Note 3).

A response may be spontaneous or elicited by a stimulus, adapting or non-adapting, directed or undirected. A response may be considered to be comprised of a sequence of unitary responses.

UNITARY RESPONSE. One of a set of simple responses which form the basic movement repertoire of an organism.

Common examples are: <u>start</u> (of forward motion), <u>turn</u>, <u>stop</u>, <u>reversal</u> (where the locomotory machinery is operated in reverse), <u>reversal turn</u> (a 180° turn).

DIRECTED RESPONSE. Any response that consistently results in a decrease in the current deviation from the mean direction of migration. The mean direction is usually related to the stimulus direction or gradient.

For example the dorsal turn made by <u>Euglena</u> whenever the cell is oriented so that the stigma shades the photoreceptor (Diehn, 1979) is a directed turn, since the immediate result is a trajectory more closely oriented towards the direction of positive phototaxis.

KINESIS. Behavior comprised of undirected responses that are dependent on the intensity or temporal change in intensity of a stimulus (Note 4).

Though they may occur in a gradient and may indirectly cause migration or aggregation in a gradient, kineses are usually tested by varying the intensity of a uniform field of stimulus. Where a stimulus such as light has a direction, it is directed perpendicular to the plane of allowable movement (Note 5). The term "non-directional illumination" should not be used for this. A kinesis may adapt or may not adapt to the stimulus (Note 6). In a <u>positive</u> kinesis the rate is higher in the presence of the stimulus, in a <u>negative</u> kinesis it is lower (Diehn <u>et al</u>. 1977).

ORTHOKINESIS. A kinesis in which translational motion is affected.

Commonly measured are average speed or proportion of individuals in a population that are active at one time.

KLINOKINESIS. A kinesis in which rotational motion is affected.

Commonly measured are frequency of turns or amount of turning per unit time or per unit track length ("rate of change of direction"). Turning angle should also be measured (Jones, 1971).

ECCLITIC RESPONSE (Note 7). A change in motion characterized by a stop of locomotion, a change of direction and a resumption of on-going motion.

The response may be spontaneous or elicited by a change in stimulus intensity. The frequency of ecclitic responses may adapt or may not adapt to a constant stimulus level. In detail, the pattern is typical of the species and usually is the same (within variation) regardless of the nature of the stimulus. A pause and reversal may occur before the change of direction. The new direction may be the result of rotational diffusion during the pause (in the case of bacteria), or due to a turn. The turn may be directed or undirected. Since the net response is a change of direction, undirected ecclises are a type of klinokinesis, and directed ecclises can result in a taxis.

ECCLISIS. Behavior characterized by ecclitic responses.

MIGRATION. The trend of movement of an individual or population in the mean direction.

The mean direction usually has a definite relationship to a stimulus gradient or direction; however the relation can be changeable. Migration can be a taxis or can be due to a kinesis (Note 4).

TAXIS. Migration oriented with respect to the stimulus direction or gradient (Note 8) which is established and maintained by directed turns.

A taxis is positive if net movement is towards the source or up-gradient, negative if in the opposite direction. Transverse and oblique taxes and light compass reactions (mnenotaxes) are included (Note 9).

KLINOTAXIS. A taxis which results from directed responses to sequential samples of stimulus intensity or direction.

Because of body motion and/or morphology, the samples are obtained with the eye or receptor organ located at successive spots in the stimulus field or pointing toward different regions of space. Klinotaxis is distinguished by sampling movements, such as lateral movements of the head or antennae.

TROPOTAXIS. A taxis which results from directed responses to two (or in 3 dimensions, three) simultaneous samples of stimulus spatial distribution or directional distribution.

Sampling movements are not apparent. For example, in positive directional phototropotaxis on a plane, turning continues in one direction until two shaded photoreceptors are equally illuminated. This orientation is maintained with only slight deviations. In the presence of 2 light sources, movement is directed between them. If one photoreceptor is desensitized, turning away from the blinded side ("circus movement") occurs in a beam of light (Note 10).

TELOTAXIS. A taxis due to directed responses to information gathered by a raster of many receptors. Directed turning occurs until the "fixation area" of the raster is exposed to the directional stimulus.

Sampling motions are not apparent, the direction of one of two lights is selected (provided they can be resolved), and unilateral blinding does not interfere with the taxis (Note 10).

ORIENTATION BEHAVIOR. Orientation of body axes with respect to stimulus direction, which is established and maintained by directed rotations.

Similar to taxes but translation need not occur. Includes the dorsal light reaction and gravity orientation described by Fraenkel and Gunn (1940, 1961).

ACCUMULATION BEHAVIOR. Accumulation or dispersal from a region of higher or lower stimulus intensity.

Accumulation can be a consequence of any of several types of response. For example, accumulation in a region of higher light intensity can be due to a step-up negative orthokinesis (which results in more time being spent in the zone), step-down photoecclitic responses (which lower the likelihood of leaving the zone) or a positive phototaxis to light scattered from the region.

INFORMATION-GATHERING CAPABILITY AND EYE STRUCTURE

The relationships between eye structure, information-gathering capability and movement behavior are indicated in Table 1. Note how quickly the information capability and photomovement possibilities increase as eye structure becomes more complex. With only 2 pigment-spot eyes having one photoreceptor each, much ecologically-important information can be obtained. All photomovement except telotaxis is possible, provided the organism moves only in 2 dimensions. Tropotaxis in 3 dimensions would require 3 such eyes (with 2 eyes, all directions in the plane of symmetry between the 2 eyes would be confused). This may explain why pelagic larvae often have 3 or more pigment spot eyes.

Also note (Table 1) that only a single unshaded photoreceptor is needed for migration on a gradient of light intensity, or for accumulation or dispersal from a region of uniform intensity. The best-studied examples, the photobacteria (Hildebrand, 1978), gather the necessary information by sequential sampling at points along their path. The temporal changes in light intensity experienced as they move govern the orthokinesis or klinokinesis. Larger organisms which can make sequential samples laterally, are theoretically capable of klinotaxis on a gradient with only one unshaded photoreceptor. A possible nematode example (Heterodera spp.) that utilizes head-swinging in making directed turns on a gradient will be described later. Presumably proprioceptive information about head position is utilized with the temporal

Table 1. Minimum eye structure, information-gathering capability and possible movement responses.[a]

Minimum eye structure needed		Information-gathering capability	Possible photomovement responses						
			Migration by kinesis		Klinotaxis		Tropotaxis		Telotaxis
No. PR	No. shading-pigment structures		Ortho-	Klino-	Grad.	Dir.	Grad.	Dir.	(Dir.)
1	0	Intensity change in Int. timing of change gradient direction (sequentially)	x	x	x				
2 separated	0	gradient direction	x	x	x	x	x		
2 dichroic	0	polarization	x	x	x	x			
2 dichroic + 1	0	polarization	x	x	x	x		x	
2-3 spectral	0	colour	x	x	x				
1	1	dir. dist. of int. (sequentially)	x	x	x	x			
2 dichroic + 1	1	dir. dist. of pol.	x	x	x	x	x		
2-3 spectral	1	dir. dist. of colour	x	x	x	x			
2	2	dir. dist. of int., change	x	x	x	x	2D ⎡ If sepa-		
3	3	dir. dist. of int., change	x	x	x	x	3D ⎣ rated		
1 raster		borders, form, change	x	x	x	x	3D		x
2 rasters		stereoscopic vision	x	x	x	x	3D		x

[a] photomovement terms used are defined in the text. dir.= directional or direction; dist.= distribution; grad.= gradient; Int.= intensity; pol.= polarization; PR= photoreceptors (in separate cells in the case of multicellular organisms).

information about light intensity. Two, sufficiently separated, photoreceptors are required before simultaneous sampling can reveal information about a light gradient. I know of no examples where this behavior has been demonstrated.

Two nearby, identical, unshaded photoreceptors cannot gather any more information than one. Two with different spectral sensitivities can provide information about spectral distribution (Table 1); many (but not all) colours could be discriminated by simultaneous comparison of the 2 responses. At least 3 are needed for full colour discrimination as in man, and more would provide further colour information (Gouras & Zrenner, 1981). Two receptor units with polarization sensitivity and differing angles of maximum polarization sensitivity are sufficient for a polarotaxis by sequential sampling. Simultaneous unambiguous identification of the plane of polarization requires, in addition, an independent receptor without (or with) polarization sensitivity (Waterman, 1981, p. 374-5). A tropotaxis relative to the polarization angle would be possible (Table 1). Without shading pigment, the colour or polarization of incident light from all directions would be analyzed.

With the simple addition of a pigment, dense enough and close enough to cast a shadow on the photoreceptor, information about the direction of a single light source can be obtained. In the natural environment, where light comes from many directions, this device would provide information about the directional distribution (intensity vs. direction) of light incident on the organism (Table 1). In the simplest case, a change in body or head orientation results in a change in intensity of the light falling on the organelle, due to the shadow. This, together with information about head or body orientation, is sufficient for making a directed turn. By sequential sampling and turning operations, the organism can adjust its body orientation towards the brightest or dimmest region of space. A well-studied example of this behavior is the photoklinotaxis of Euglena (Diehn, 1979). In effect, its movement is oriented with respect to the directional distribution of intensity by a process of sequential sampling. Note the distinction between this photoklinotaxis, in which directional distribution of intensity is sampled, and a photoklinotaxis in response to a gradient of intensity, in which spatial distribution is sampled sequentially.

With one shading pigment spot but 2-3 spectral organelles, or 2 dichroic and 1 other photoreceptor, the directional distribution of colour or polarization of light can also be sampled sequentially and a directional klinotaxis to this kind of information could result (Table 1). Orientation by sequential sampling is slow, therefore there is an advantage to having two photoreceptor - shading pigment combinations that can simultaneously sample two different regions in the directional distribution of light. This information makes possible orthotaxis, which is more direct than klinotaxis. Efficient polarotaxis can be achieved if each eye contains 2 dichroic photoreceptors, oriented perpendicular to each other, plus 1 other photoreceptor. That this may occur in the Müller's larvae of Pseudoceros canadensis (Turbellaria, Platyhelminthes) is suggested by the morphology (Eakin & Brandenburger, 1981; Fig. 13 in Burr, 1983).

Two photoreceptor - pigment spot combinations, sampling different regions of the directional distribution, are also the minimum structure needed in order for a motionless organism to detect a change in the directional distribution of light (Table 1), such as might be caused by a moving predator or prey organism. The more such eyes, the larger the visual field that can be monitored at a useful level of resolution. Rows of such simple eyes are observed in certain Turbellaria. An array of 10 000 ocelli scattered over the curved hypertrophied siphon of the giant clam, Tridacna, is thought to enable detection of an approaching predator at a distance sufficient for a timely closing of the valves (Fankboner, 1981). Such an array is called a raster (Schöne, 1975). An array of photoreceptors lining the inner surface of a pigment cup, as in planarian eyes (Fig. 1a), is another type of raster. The relative position of each photoreceptor in the raster is an index of the relative direction of the sample of light that it receives. The position and excitation of every organelle constitute the information that is utilized in the detection of borders and forms. Changes in excitation of adjacent organelles indicate the movement of borders. Understandably, the analysis of all this information, and the discrimination of movement of the "image" of a predator from movement due to the animal's own locomotion, requires a relatively sophisticated centralized nervous system. Would this be limiting in Tridacna or Planaria? This has never been tested.

One limitation on the information available from any photoreceptor - shading pigment combination is its resolving power. This can be measured by the reciprocal of the angle between 2 point sources of light that can just be distinguished by the animal. For a pigment-cup eye, whether one or many photoreceptors line the cup, the maximum resolving power in reciprocal radians is approximately "f"/A where A is the diameter of the pupil and "f" is the axial length of the eye from the center of the pupil to the photoreceptor (Land, 1981). For Planaria maculata eyes (Fig. 1a), with a 25 μm depth and 30 μm aperture, the angle between 2 just resolvable point sources is predicted to be about 69°. The spacing between photoreceptors is 10 μm, about the same as the predicted minimum useful spacing (Land, 1981). Making the aperture smaller or the axial length greater will increase resolving power until the diffraction limit is reached. However the intensity received is decreased according to A^2. Lowering the intensity may also limit resolution, as discussed more fully by Land (1981). The resolution of an organism's response to 2 point sources will also depend on the precision of its movements and the resolution of the proprioceptive information.

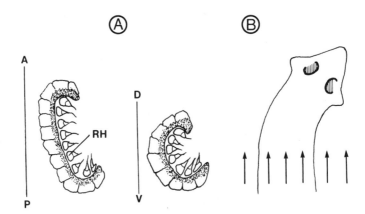

Fig. 1. Pigment-cup eye of <u>Planaria maculata</u> (Turbellaria, Platyhelminthes). A. Frontal and transverse sections. B. Positioning and illumination of eyes for 45° turn from direction away from a light source. A-P, longitudinal axis of worm; D-V, dorsoventral line; RH, rhabdomere. After Taliaferro (1920).

Though one can predict, from eye structure, much of what information may be available to the nervous system of a simple organism, one cannot predict whether the information will be used or how it will be used in modifying behavior. Furthermore, whether or not one can observe use of the information depends on the design of the experiment. This is illustrated by experiments with <u>Planaria maculata</u> by Taliaferro (1920). When a horizontal beam of light was directed perpendicular to the body axis, the animal turned directly away from the source without sampling movements. With continued illumination from this direction, it moved away the source. However sampling movements were necessary, in the form of 45° turns or a stop-and-twist movement. This behavior would be classified as a negative klinotaxis to the light direction. The information necessary, as indicated by Table 1, could be gathered by a single photoreceptor shaded by a pigment spot, facing forward with an acceptance angle of 270°. In fact, planaria have 2 pigment-cup eyes with a raster of perhaps 200 receptors per cup (Taliaferro, 1920) each of which has approximately a 69° acceptance angle (Land, 1981). The cups face laterally and slightly forward, such that the animal receives no orienting stimulus for negative phototaxis unless the head is turned at least 45° from the direction of locomotion away from the source (Fig. 1b). This explains why little of the information capability of the planarian eye is used in the negative phototaxis. Whether the capability <u>can</u> be used, and <u>how</u> it may be used, is not brought out by the experiment.

The visual fields of the eyes overlap anteriorly (Fig. 1b), and Taliaferro (1920) found that illumination of the contralaterally directed region of one eye elicits the same turn response as illumination of the ipsilateral eye from the same direction. The resulting turn away from the source has a minimum size regardless of initial orientation. The visual system appears to be sophisticated enough for a telotaxis. No one has thought to try, as a stimulus, a dark spot in an otherwise uniform directional distribution of light. A positive scoto<u>telo</u>taxis may be possible in an animal that has, to date, been shown to be capable of only a negative photo<u>klino</u>taxis. The limitation appears to be in the design of the experiment, not the eye!

I conclude that investigation of the photomovement of an organism can be usefully guided by knowledge of the structure of the organism's eyes and predictions of their information-gathering capability. Experimental verification of the predictions should be an important aspect of any thorough study.

NEMATODE EXAMPLES

Recently, several laboratories have turned to nematodes for the study of neural systems because of their simplicity. The total number of neurons is thought to be about 300, including the cerebral ganglion, whether the species is 1 mm long (<u>Caenorhabditis elegans</u>) or 100 mm long (<u>Ascaris</u>). These include typical sensory, motor, and interneurons. <u>C. elegans</u> has been particularly intensively studied because of the ease of culture, short life cycle (3 days), hermaphroditic reproduction (genetically identical strains are easily maintained) and small size (Edgar & Wood 1977; Riddle, 1978). <u>C. elegans</u> has become one of the principal model systems for the study of neurogenesis (Anderson <u>et al</u>., 1980). The lineage, location and morphology of every one of the approximately 800 somatic cells has now been described. Some of the neuroanatomy has been published (Ward <u>et al</u>., 1975; Ware <u>et al</u>., 1975; White <u>et al</u>., 1976) and the remainder is being prepared by J. White and coworkers. Ward (1978) and Dusenbery (1980a) have reviewed the behavioral studies. The description of cell lineages is being prepared by J. Sulston and coworkers.

In all, about 80 receptor neurons have been identified in <u>C. elegans</u>. Of these, 58 are located at the anterior tip (Ward <u>et al</u>., 1975; Ware <u>et al</u>., 1975). Most are thought to be chemoreceptors because of their connection to the exterior. Receptor organelles embedded in the cuticle are presumed to be mechanoreceptors. Other receptors have no obvious function. Temperature, light and oxygen receptors, indicated by behavioral studies, have not yet been identified.

Other nematode species have been studied for their special features. Certain marine nematodes with pigment-spot eyes have been chosen for studies of phototaxis and photoreceptors (Burr & Burr, 1975; Burr, 1979; Croll, 1966b). The sensory anatomy and behavior of many plant and animal parasitic nematodes (reviewed by Croll & Sukhdeo, 1981; Coomans & De Grisse, 1981) are interesting for comparative analysis.

Information capacity of sensory system

The eyes are comparatively simple. Most nematodes lack any structures that could be identified as an eye at the light microscope level. The few that do, have 2 lateral spots of dense pigment in the anterior, sometimes cup-shaped, each associated with a single photoreceptor organelle (Burr & Burr, 1975; Croll et al., 1972; Siddiqui & Viglierchio, 1970). For an illustration see Fig. 3 of another chapter of this volume (Burr, 1983).

Some nematodes appear to be photosensitive although they lack pigment spots, e.g. larvae of the hookworms Ancylostoma tubaeforme (Croll & Al-Hadthi, 1972) and Trichonema sp. (Croll 1965, 1966a, 1970, 1971). There have been no attempts to identify the photoreceptors in these species. I have recently found Caenorhabditis elegans to be photosensitive (Burr & Boyle, to be published). Though the sensory neurons have all been described it will be a challenge, without an adjacent pigment spot as a label, to prove which ones are photosensitive.

Without the pigment spots, it is unlikely that a nematode could obtain information about the directional distribution of light, since its semi-transparent body would permit the photoreceptor to receive light from any point. Therefore telotaxis and tropotaxis would be unlikely responses and klinotaxis if it occurs must be based on sequential samples at different points in a gradient.

With a pair of pigment-spot eyes, the directional distribution of light could be sampled. As usually studied, however, tropotaxis would not be observable. This is because nematodes locomote by propagating body waves in the dorso-ventral plane, therefore when confined to a planar surface the nematodes move while lying on their sides. The laterally located eyes, consequently cannot provide simultaneous samples that would be useful in any taxis in the plane of allowable movement. Phototropotaxis, if it is possible in nematodes, would only be observable in a 3 dimensional medium. The use of klinotaxis in the dorsal ventral plane and tropotaxis in the lateral plane would be an interesting possibility. However both of the pigment-spot eyes of Oncholaimus vesicarius (see Fig. 3 of Burr, 1983) have approximately the same anterior view.

The most prominent chemoreceptive organs are the anteriorly-located amphids, which have openings to the exterior through the lateral cuticle. A tropotaxis based on information gathered by these organs would not normally be observable for the reason given above for the phototropotaxis. Also, Ward estimates that these organs and the inner labial papillae are too close together to be useful in a tropotaxis (Ward, 1978). A few chemoreceptors located along the sides and tail (Coomans & De

Grisse, 1981; Wright, 1980), including the phasmids present in the tail of some nematodes, could be involved in anterior-posterior comparisons, but this has not yet been shown.

From what is known about the morphology of nematode receptors, one can predict that the information capacity of the system appears to be sufficient for kineses and klinotaxes. A tropotaxis would be a possibility only under the special circumstances mentioned.

Motion responses

How an animal responds to the information depends ultimately on its repertoire of unitary responses. That of C. elegans has been described by Croll (1975). The subset involved in movement on a substrate such as agar is representative of most nematodes. Forward locomotion is by posteriorly propagating body waves. This normal movement is interrupted occasionally by spontaneous ecclitic responses which typically consist of a stop, pause, reversal and turn. During a pause, the body waveform is held until locomotion resumes. During a reversal, anteriorly propagating waves drive the animal backwards. With both types of wave, the entire body moves in the dorsoventral plane, including the anterior end where most of the receptor organs are located. Therefore during normal movement the receptors are swung to one side then the other of the path of movement. Additional smaller turns of just the head have lower amplitude and higher frequency. Using these 2 types of lateral motion, sequential sampling could yield information about the spatial distribution of stimulus intensity across the path. Also the pigment-spot eyes, when present in the anterior, would be aimed alternatively to one side then the other of the forward direction and sequential sampling could provide information about the directional distribution of light intensity as well as the spatial distribution in a gradient.

The direction of movement of nematodes is changed by turns of varying magnitude that occur during forward movement or after a reversal. The larger turns ('omega waves', Croll, 1975) nearly always follow reversals as a part of ecclitic responses. Ecclitic responses are elicited by changes in chemical concentration, temperature or light intensities or upon collision with an object. Spontaneous ecclitic responses are probably triggered by internal stimuli rather than uncontrolled external stimuli in the substrate, since the interval between them is the same on agar as in water (Croll, 1975).

When moving over a surface where the stimuli are uniform, C. elegans and other nematode species often move in large loops or spirals. Looping behavior could provide an additional mechanism of sequential sampling which would be useful in very shallow gradients because of its larger range.

Photokinesis

Photokinesis has been demonstrated in uniform fields of

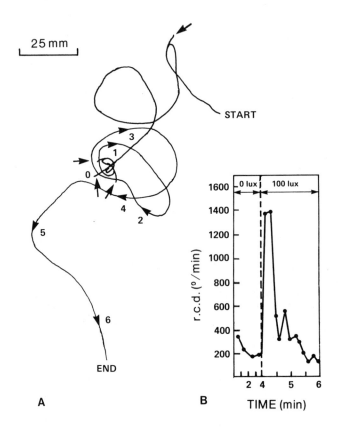

Fig. 2. Klinokinesis of the planarian Dendrocoelum dorotocephala (Turbellaria, Platyhelminthes) in a uniform field of light. After Stasko and Sullivan (1971). A. Track in 4 min of darkness then 6 min of 100 lux starting at 0. One minute intervals in the light indicated by numerals at arrowheads. Arrows, probable ecclitic responses. B. Mean rate of change of direction (●) for 5 animals. Sudden illumination occurred at 4 min as in A. Note the scale change.

perpendicular light for a wide variety of microorganisms and lower invertebrates. Klinokinesis with adaptation is shown for planaria in Fig. 2a,b. Despite this evidence of capability, as tested in uniform fields, there are few well-documented cases of accumulation in a light gradient due to klinokinesis, despite many attempts. The problem is a technical one; it is nearly impossible to create a gradient of perpendicular illumination that is devoid of horizontal light due to scatter, and most organisms that have been tested preferentially respond phototactically to the horizontal light. The efforts of several investigators with planaria are well documented by Stasko and Sullivan (1971) and commented on by Gunn (1975). On the other hand the thorough study of phototaxis by Taliaferro (1920) showed that these organisms respond in a relatively complex way to the light direction, using multireceptor (raster) pigment-cup eyes.

The simplicity of nematode eyes that lack shading pigment makes these organisms promising candidates for the study of simple photobehavior without the complication of a sensitivity to light direction. Orthokinesis (Croll, 1966a; 1970), klinokinesis (Croll, 1965) and photoecclisis (Croll, 1971) have been studied in hookworm infective larvae by the usual tests in a uniform field of perpendicular illumination. C. elegans also exhibits photoecclisis (Burr & Boyle, to be published). Lacking shading pigment, it failed to respond phototactically to obliquely incident uniform illumination which did produce a strong phototaxis in O. vesicarius (Burr, to be published).

As yet, no study of klinokinetic migration in a gradient field has been done with nematodes, however 4 tracks were published by Croll (1971) of hookworms responding to what was probably a shallow, radial gradient-field of perpendicular light. Upon a step increase in intensity, the first reaction was an ecclitic response. Further ecclitic responses followed at intervals but the main contributors to the rate of change of direction are other turns and tight loops as is also true of the planarian (Fig. 2a). After a period of 1-4 min during which little or no migration occurred, a consistent trend towards the perimeter of the light spot is observed. In a region on one track a series of tight loops are observed that have a longer leg where the worm was moving down the gradient. This is similar to the pattern postulated by Fraenkel and Gunn (1940, 1961) for migration in a gradient field by a klinokinesis with adaptation, and nearly identical to the track predicted by a computer model of this behavior (Green, 1977). Further investigation is warranted, as this may provide the first clear example in a metazoan of migration by photoklinokinesis.

Photoecclitic responses

A step-up in light intensity sometimes elicits a response that closely resembles ecclitic responses to other stimuli: a stop usually followed by a reversal, pause and turn. With continued illumination ecclitic responses occur for a short time at a higher frequency than in the dark. Croll (1971) observed in infective larvae of a hookworm that the reversal frequency adapted back to the dark level (0,2 per minute) within the time resolution of his experiment (10 min). Similar ecclitic responses were observed by Burr and Boyle (to be published) in experiments where one could rule out the possibility that the response was elicited by a light-induced temperature change in the media. The probability of a reversal during a 10 s sample period was measured in the dark and light. Light increased the probability to 0,22 from a dark level of about 0,11. While this was not a large effect, it was significant. For O. vesicarius, light at the same intensity (1,7 mW cm^{-2}) increased the probability from 0,16 to 0,35. This intensity is one tenth that which was used to elicit phototaxis from O. vesicarius in an earlier study (Burr, 1979).

Ecclitic responses to other stimuli

Ecclitic responses of nematodes to stimuli other than light have the same basic pattern as those to light, but in detail they may depend on the nature of the stimulus. Croll (1975, 1976) observed that, for ecclitic responses to collision with a bead, the resulting change of direction was less on the average than for spontaneous ecclisis. The turns did not seem to be directed: the number of turns away from the side of contact was approximately equal to the number towards that side (Croll, 1976).

Dusenbery (1980b) has developed a method of studying reversal responses to step changes in the concentration of chemical stimuli. The worm is tethered by holding the tail in a suction pipet and the position of its shadow is measured by photocells. Normal forward movements, pauses, reversals and large turns can be discerned on a multichannel recording. The effect of chemical stimuli were studied by changing concentrations in solutions pumped passed the tethered worm. A step reduction in the concentration of NaCl from 50 mM caused a large increase in the probability of a reversal. This behavior adapted back to a basal level in about one minute. A step increase in concentration caused a decrease in the probability, and this response adapted more slowly. A transient increase in the frequency of reversals also occurs to a decrease in oxygen concentration (Dusenbery, 1980c).

Phototaxis

Phototaxis has been observed in several nematode species having pigment spots (Burr, 1979; Croll, 1966b; Chitwood & Murphy, 1964). In the statistical study by Burr (1979) it was shown that O. vesicarius preferentially moves in the direction away from the source of obliquely-incident uniform illumination. In a similar experiment with C. elegans and O. vesicarius under identical conditions, O. vesicarius behaved as before while C. elegans showed no directional preference, as did both species in the absence of light (Burr, to be published). These observations add strong support to the proposal that the lateral pigment spots function in the detection of the direction of light. Other evidence is the presence of a receptor organelle adjacent and anterior to each pigment spot, a position that would be predicted from the behavior (Burr & Burr, 1975). A similar behavioral/morphological correlation exists for Chromadorina bioculata (Croll 1966b; Croll et al., 1972) and Enoplus anisospiculus (Burr, 1979; Burr to be published). Because of the very small size of these organisms, evidence from electrical recordings will be difficult to obtain.

Thermotaxis and chemotaxis

Insight into phototactic behavior can be provided by responses to other stimuli. For example, since it is possible for C. elegans to direct its movements according to a temperature or chemical gradient, one might expect to find a similar tactic orientation to a gradient of light intensity.

Both linear and radial temperature gradient fields were tested by Hedgecock and Russell (1975). Whether C. elegans was placed at a higher or lower temperature in the gradient field, it moved to the temperature to which it had been acclimated. The published tracks were more or less parallel to the gradient with apparently normal wave motion and minor (< 30°) corrections. Once the acclimation temperature (Note 11) was reached the worms accurately moved along isothermal lines, a behavior which may be different from orientation along the gradient (Hedgecock & Russell, 1975). A decrease in the frequency of ecclitic responses is observed during isothermal tracking (Hedgecock & Russell, 1975; Rutherford & Croll, 1979).

On radial gradient fields of chemical attractants, C. elegans tracks are oriented up the gradient (Ward, 1973). In the published tracks, minor correcting turns of < 30° are observed and occasional large turns occur as ecclitic responses. This chemotaxis replaces the looping behavior observed in the absence of a gradient. Mutants with blisters covering the phasmids, but not the amphids, oriented normally to chemical gradients, but mutants with amphid-covering blisters did not. This appears to rule out a tropotaxis with simultaneous comparison of head and tail receptors, at least for the chemical stimuli tested. A mutant that moved slowly with normal head-swings oriented as well as the wild-type, therefore Ward concluded that sequential comparisons at points separated by forward movement is an unlikely mechanism (Ward, 1976). Mutants with dorsal or ventral bends at the anterior tip moved diagonally in the gradient field in such a way that the anterior tip was directed up-gradient. The spiral track was composed of frequent turns and tight loops which apparently were made to correct for the rudder-like effect of the bent head. These mutant studies show clearly that orientation of the head, not the tail, with respect to the gradient is an important component of the taxis. To conclude that the behavior is a klinotaxis, Ward must assume that tropotaxis is unlikely because of the close proximity of the anterior chemoreceptors and the equivalent positioning of the amphids during locomotion on one side.

Since a mutant unable to make the fine lateral head movements can still migrate up-gradient, these movements appear not to be essential to taxis. However the track (Ward, 1976) indicates that this mutant may have had to use a different sampling motion. The details of head movement during taxis of wild type and mutants needs to be studied.

Once the center of a radial gradient is reached by C. elegans its position at the maximum concentration is maintained by frequent turning and ecclitic responses. It is difficult to tell from the published tracks whether these responses are directed (klinotaxis) or undirected (klinokinesis).

In tracks of male Heterodera spp. orienting to a gradient of sex attractant (Green, 1966), a variety of sampling behaviors can be distinguished. In Fig. 3, male A initially moved diagonally up the gradient by normal locomotory movement. At point a, lateral probings appear as points at the extremes of lateral swings. When the translation was down-gradient, at b, ecclitic responses resulted in larger turns that reoriented the worm up-gradient. This type of correction can be seen also in tracks of C. elegans. Male B (Fig. 3), after an initial series of loops and turns, became oriented up-gradient. This direction was maintained until the worm neared the female. Interestingly, lateral probings appear at about the same distance from the female as for track A. Male C approached a female by a spiral path punctuated by ecclitic responses that resulted in large turns. These 3 tracks indicate that 2 types of turns are used to correct the direction of movement during a taxis: smaller turns synchronized with the locomotory undulations and the larger turns of ecclitic responses. Both types appear to be directed with respect to the gradient, decreasing the angle between the body axis and the direction to the female in most cases. For this reason it is difficult to agree with Green (1966) that the observed ecclitic responses are a klinokinesis. More tracks, analysed quantitatively, would be helpful.

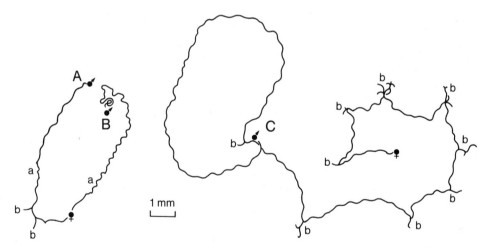

Fig. 3. Chemotaxis of 3 Heterodera spp. males on a radial gradient field of attractant secreted by a female. Males started at ♂, female motionless at ♀. a, lateral probing; b, ecclitic responses. After Green (1966).

The track of male C, like some others published by Green (1966), suggest that the worm had a consistent slew to one side, which was corrected only by ecclitic responses. Green (1977) used a computer model to simulate this behavior. With an inherent, constant slew the model worm moved forward so long as the observed concentration (actual concentration multiplied by an adaptation factor) equalled or exceeded a constant threshold ($C_{obs} \geq T$). An ecclitic response occurred when $C_{obs} < T$. Locomotion stopped and the anterior was swung from side to side until $C_{obs} \geq T$ whereupon the model worm turned at a fixed angle towards the side where this was detected. Such a change of direction is a directed one; it would be undirected if the direction of turn relative to the body axis was based instead on the toss of a coin. C_{obs} is continually adjusted to account for adaptation, forcing the condition $C_{obs} = T$ to occur at progressively higher real concentrations. By adjusting parameters, the model nematode can be made to mimic the track of male C (Fig. 3). As Green states, the model does not prove that it is the only possible mechanism or the one actually operating in the organism being studied, but it shows that such a mechanism can work and under what conditions.

Adaptation and independence of sensory input

The question of whether adaptation of behavior is due to adaptation of the receptor or some other neuron is open to question. If due to the receptor, then the effect of stimulation of another receptor that elicits the same response should be independent of adaptation to the first stimulus. Croll (1966a) observed that the orthokinetic response of Trichonema infective larva could be elicited by mechanical stimulation even after the light stimulus was no longer effective. Further study along these lines would be fruitful.

Drives and habituation

Various factors affect the predisposition of C. elegans to emit a given response. Magnitude of the attraction of C. elegans to 10 mM NaCl was greatest when tested at the temperature of acclimation (Dusenbery et al., 1978). One could postulate that at other temperatures the worms were less disposed toward responding to chemical stimuli. In another study (Dusenbery, 1980c), Dusenbery observed that the ecclitic response to a step reduction in oxygen concentration is stronger after the first 20 minutes in his apparatus. He showed that this was due to time in the apparatus rather than to effect of stimulation with low oxygen and speculated that this might be due to food deprivation.

A major change in thermotaxis was observed after C. elegans adults had been starved for 4-6 hours (Hedgecock & Russell, 1975). Instead of accumulating at the acclimation temperature, the worms actually dispersed from this temperature. It is evident from these examples that both responsiveness and the nature of a response can be affected by an internal state of the organism not induced by the stimulus.

Other observations indicate that responsiveness may be affected by stimulus repetition. Ward observed that after C. elegans has remained in the center of a radial gradient field for a "brief" period, it ceases to respond to decreasing stimulus concentration and migrates to the periphery (Ward, 1973). The lower the attractant concentration the less time the worm spends in the center. After a "brief" time at the periphery it again responds to the gradient to repeat the cycle. Ward regards this as resembling a classical habituation to repeated stimuli. We now know from Dusenbery's study of chemoecclitic responses (Dusenbery, 1980b) that adaptation to chemical stimuli occurs in about a minute. From the tracks published by Ward it appears that at least 15 min is spent in the center before dispersal. Thus adaptation can probably be ruled out.

Reacclimation of C. elegans to a new temperature takes 4-6 hours (Hedgecock & Russell, 1975). Could this be due to habituation? Hedgecock and Russell postulate from their mutant studies that migration to an acclimation temperature may involve two opposing drives, one directing positive thermotaxis when the temperature is lower and another directing negative thermotaxis when the temperature is higher. Since responses of thermophilic but not cryophilic mutants were affected by acclimation temperature, they suggest that only the positive drive is reset during reacclimation.

In the nematode studies, several new and powerful techniques have been introduced to the investigation of movement behavior and underlying mechanisms, but several major questions have yet to be answered conclusively. With regard to the classification of movement behavior, 2 questions may be resolved by further nematode studies: [1] In metazoa, can klinokinesis cause accumulation in a gradient field of perpendicular illumination? and [2] Can phototaxis occur in a gradient field of perpendicular illumination? Answers will depend on the development of objective criteria to distinguish between klinokinetic migration and taxes.

DISTINGUISHING BETWEEN KLINOKINETIC MIGRATION AND A TAXIS

A variety of mechanisms can be postulated for either a taxis or migration by a klinokinesis. In both cases either the frequency or the magnitude of turning could be controlled. In both cases the turns could be either coordinated with forward movement, or occur as part of an ecclitic response. In both cases, a net movement along a gradient could result. To add to the confusion, both types of migration could be operating in the same experiment, and the proportion of each could depend on the steepness of the gradient.

The distinction between the 2 behaviors was indicated in the original definitions (Gunn et al., 1937; Fraenkel & Gunn, 1940, 1961, Table 4). In a taxis, turning behavior is directed, on the average, with respect to the stimulus direction or gradient and for a klinokinesis it is undirected. The relationship of the turn to the body axes or prior direction of movement may have a bearing on the mechanism but it is not the distinguishing feature. This can be illustrated by 2 hypothetical cases.

Let the mean direction of migration be the same as the direction of the stimulus or its gradient, and let θ be the angle between this direction and the direction of movement of an individual just prior to the turn. Thus θ is the current deviation from the environmental direction.

Case 1.   Turn direction is unbiased; turn magnitude is controlled. If the magnitude is greater whenever the turn is in the direction of decreasing θ, then turning is directed. Let μ_+ and μ_- be the mean turn magnitudes in the direction of increase and decrease, respectively. One must use an inference test to determine if μ_- is significantly greater than μ_+. Several two-sample tests are available for directional data (Mardia, 1972; Batschelet, 1981).

Case 2.   Turn magnitude is unbiased; turn direction is controlled. If a turn is more likely to occur in the direction of decreasing θ, then turning is directed. Let N_+ and N_- be the number of turns in the direction of increasing or decreasing θ, respectively. Significance can be inferred using the binomial test or its normal approximation (Zar, 1974). If $N_-/(N_+ + N_-)$ is significantly greater than 0,5 (i.e. $N_- > N_+$) then one can conclude that turning is probably directed.

These 2 cases apply for responses both to a gradient field and to a directional stimulus. If either turn magnitude or turn direction proves to be biased, the resulting migration is at least partly a taxis. Male C of Fig. 3 appears to behave in a gradient field as in Case 2. An ecclitic response occurs at each of the corners of the spiral polygonal track. Since all 9 turns are towards decreasing θ, $N_- = 9$ and $N_+ = 0$, and directedness is highly significant at P = 0,002. The migration is at least partly a taxis.

One could find that neither turn magnitude nor turn direction is significantly biased towards the direction of migration, i.e. that the null hypotheses $H_0: \mu_- = \mu_+$ and $H_0: N_- = N_+$ can be accepted. But only under certain conditions can one then conclude that turning is undirected. The problem is that the probability of accepting the null hypothesis when one of the alternative hypotheses, $H_1: \mu_- > \mu_+$ or $H_1: N_- > N_+$ is in fact true is unknown and could be very large. This "Type II" error becomes smaller with increasing sample size. It can be calculated exactly only if the actual alternative distribution is known, which is seldom the case.

Some calculations based on binomial probabilities make this

clearer (For background see Sokal and Rohlf [1969]). If the probability of a turn towards decreasing θ was actually $p_- = 0,75$ the binomial test would miss detecting that $p_- > 0,50$ in 70% of samples of size 9 and in 27% of samples of size 25. In either case, the outcome of accepting the null hypothesis $H_0: p_- = 0,5$ would have an unsatisfactorily high chance of occurring when in fact $p_- = 0,75$. All one can conclude is that the sample <u>could</u> have come from a binomial distribution with $p_- = 0,5$.

How large a sample size is needed before one can be reasonably sure that turning direction is unbiased. This can be estimated once an arbitrary criterion of directedness is chosen: a value of p_- above which a turn is considered to be directed. For $p_- \geqslant 0,60$ a sample size $n \geqslant 265$ is needed in order to ensure that the test misses detecting $H_1: p_- > 0,5$ in less than 5% of samples. If $H_0: p_- = 0,5$ can be accepted for a sample of at least this size one would be in error less than 5% of the time by concluding that turning direction is not biased. For $p_- \geqslant 0,55$, $n = 1077$ and for $p_- \geqslant 0,51$, $n = 27\,055$. Thus $p_- \geqslant 0,60$ would appear to be the more practical criterion for the majority of experiments. Minimum sample size for other criterion values, p, can calculated from

$$\sqrt{n} = 1,645\,[0,5 + \sqrt{(p)(1-p)}\,]/(p - 0,5).$$

To show that turning magnitude is unbiased, one must similarly choose an arbitrary criterion of directedness and estimate a minimum sample size. The estimation is easy if one has first shown that turning direction is probably not biased, by the above method. Then, $N_+ \approx N_- \approx 0.5\,n$. Since a large n is involved, a normal distribution can be assumed to be a good approximation for the distribution of angles about μ_- or μ_+. The variances s_-^2 and s_+^2 are calculated in the conventional way as if the measured angles were numbers on a line. For a given criterion difference in μ_- and μ_+, d, and a common sample variance $s^2 = s_-^2 + s_+^2$, a sample size of at least $n = 43,30\,s^2/d^2$ is needed in order to ensure that the test misses detecting $H_1: \mu_- - \mu_+ > 0$ in less than 5% of samples. Therefore to show that turning magnitude is unbiased, a sample size of, say, $n = 100$ is taken, and if $H_0: \mu_- = \mu_+$ is accepted at $P > 0,05$ then s^2 is calculated from the data and n estimated from the formula. If n turns out to be greater than the initial sample size then more samples are taken, H_0 tested, s^2 calculated, n estimated, and so on until the estimated minimum sample size is less than the sample size taken. If H_0 can still be accepted then one can reasonably conclude that turning magnitude is unbiased. As can be seen from the formula, with $d = 10$ and $s = 30$, $n = 390$.

Summarizing, if one can reject the null hypothesis with a small sample size for <u>either</u> turning magnitude or direction, then the conclusion: turning is significantly directed, is arrived at relatively easily. Such was the case with the above example, male C. To demonstrate that turning is undirected, however, tests must indicate that <u>both</u> turning direction and turning magnitude are unbiased. The two H_0 must be accepted for a sample of sufficiently large size, given reasonable criteria of directedness.

If turning can be considered to be undirected, the migration behavior is not a taxis, and must be due to a kinesis. A migration due to klinokinesis could result if the turning magnitude ($\mu_+ + \mu_-$) or the probability of turning depends on the change in intensity experienced along the track. For example, if the probability of (undirected) turning is greater whenever stimulus intensity decreases, a migration towards higher stimulus intensity would result. A migration due to orthokinesis would occur if speed of translation depends on intensity.

That migration in a gradient field can, under appropriate conditions, be partly a taxis and partly due to klinokinesis, will be made clear with a further hypothetical example inspired by Green's 1966 data. For males A and C (Fig. 3), the occurrence of the ecclitic responses, which were shown above to include a directed turn, appears to depend on the variation in the concentration encountered by an animal as it moves about. The ecclitic responses appear to be initiated only when movement is down-gradient. If this still occurs on a shallower gradient but the turns become undirected, a klinokinesis would result. The following model illustrates how this might occur.

Movement in a gradient field of stimulus intensity is assumed. As in the ecclitic responses of Heterodera spp., the turn follows a period of head waving which Green (1977) postulates is sampling behavior. Whether a directed or undirected turn follows may depend on whether or not the change in stimulus intensity, ΔI, experienced during the head-waving produces a suprathreshold signal. In the model ΔI produces a receptor potential, ΔV, which must be greater than ΔV_t for a directed turn to be possible. The magnitude of ΔI depends on the angle between the gradient direction and the body orientation and on G, the magnitude of the gradient. For the sake of simplicity the angle is approximated by ω, the angle between the gradient direction and the direction of movement prior to the turn. This is illustrated in Fig. 4A. The receptor potential is equal to ΔI times dV/dI, the receptor sensitivity, which can be changed by adaptation to the ambient stimulus intensity. Thus the dependence of ΔV on ω and s, the span of the head waving movement, is given by $\Delta V = (dV/dI) G s \sin \omega$. A plot of the absolute value of ΔV vs ω is given in Fig. 4B. Also plotted is the threshold potential ΔV_t which is constant. The inner shaded region indicates the range of body orientations over which ΔV is above the threshold needed to elicit a directed turn. If the gradient magnitude G is decreased, ΔV is decreased and this range is made smaller. The gradient can be so shallow that ΔV is always below the threshold regardless of body orientation (Fig. 4C).

As is the case for Heterodera in a chemical gradient (Fig. 3), the ecclitic response occurs whenever the model worm moves in the direction of decreasing stimulus intensity, i.e. for $90° < \omega$

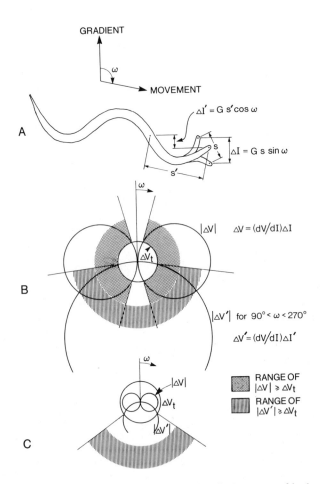

Fig. 4. Hypothetical model of limiting conditions for sampling on a gradient of magnitude G. Predictions for sampling at 2 points during forward movement and during head-waving are both illustrated. A. Geometry of sampling. s and s' are the spans for sampling during head-waving and during forward movement, respectively. ΔI and $\Delta I'$ are the differences in stimulus intensity experienced. ω is the direction of movement with respect to the gradient. B. Change in receptor potential, ΔV and $\Delta V'$, plotted as a function of ω. The receptor potential change corresponding to a behavioral threshold, ΔV_t, is constant. C. As in B but for a lower gradient magnitude. ΔV is below threshold for all orientations while $\Delta V'$ is above threshold for some.

> 270°. Samples taken at successive points along the path may have a greater span, s', than for head waving. The change in stimulus intensity encountered, $\Delta I'$, is proportional to s' and cos ω (Fig. 4A). The receptor potential that elicits an ecclitic response, therefore, is $\Delta V' = (dV/dI) G s' \cos ω$. This is plotted in Figs. 4B and C. Assuming the same threshold level as for directed turns, ΔV_t, the range of body orientations over which an ecclitic response is elicited is indicated by the outer shaded range of Figs. 4B and C.

A number of predictions are suggested by this model:

[1] Within the range where ecclitic turns are elicited there will always be some body orientations for which the gradient cannot be detected by head waving. These occur when the nematode is moving almost directly down-gradient (Fig. 4B). The range increases as the steepness of the gradient decreases. Any turns from these body orientations will be undirected.

[2] Since ecclitic responses are initiated only when 90° < ω < 270°, those ecclitic responses that result in undirected turns will contribute a klinokinetic component to migration in a gradient field. The relative amount of this component will depend on the steepness of the gradient.

[3] Those ecclitic responses that produce a directed turn will contribute a klinotaxic component to the migration. These will occur for values of ω where the shaded areas overlap (Fig. 4B). Note that the initiation of directed turns depends on the variation in intensity in exactly the same way as for undirected turns. Both a taxis and migration due to klinokinesis in a gradient field can utilize a similar intensity-dependent mechanism for detecting when the course strays significantly from the preferred direction. What distinguishes the 2 types is whether or not the turn that it triggers is directed.

[4] The initiation of ecclitic responses may drop below threshold for very shallow gradients. But this would occur at smaller gradients than for the directed turning response so long as the span between comparison points for initiation, s', is larger than that for directed turning, s. For a certain range of gradient magnitudes, there will be certain course directions for which ecclitic responses can be initiated but turns cannot be directed (Fig. 4C). Then, migration will be due purely to klinokinesis.

[5] Since taxes are more efficient than migration due to klinokinesis, one might expect to find alternative strategies to occur when ΔV due to head-waving falls below threshold in a shallow gradient:
   a) Head waving is continued until adaptation increases ΔV to above threshold (by raising the sensitivity,

dV/dI). This option was incorporated into Green's model (Green, 1977). However, only limited adaptation should occur in this case because adaptation depends on the ambient intensity, which hardly varies in a shallow gradient, not on ΔI.

b) Head waving is supplemented by forward and reverse movements to increase the sampling span and hence ΔV. That this may have occurred during some of the ecclitic responses of male C (Fig. 3) is suggested by the tracks.

c) Variation in track pattern as an organism moves in a changing gradient field is a common observation. This may be due to changes in sampling strategy. Juveniles and mutants of C. elegans with shortened heads (and hence a smaller head-swinging span), have a higher frequency of ecclitic responses than adults tested in the same gradient field (Ward, 1978). Ward also reports that orientation of adults is not as direct at lower gradient magnitudes as in steeper gradients.

[6] For gradients so shallow that ecclitic responses are not initiated, other modes of behavior may predominate, such as looping and spontaneous ecclitic responses.

This simple model is intended only for illustration. In the face of good data some of the assumptions may need to be reexamined. Variation between species can be expected. Directed turns during travel more directly up-gradient, as in Fig. 3A and B, or during orientation to the direction of light may have very different explanations. It is hoped that this example will encourage others to test for directed turns in the tracks they obtain and to suggest suitable models that incorporate both the sampling and turning behavior observed.

Recommendations

After one has demonstrated migration with respect to a stimulus gradient, e.g. by use of directional statistics, one should examine the tracks to determine if: [a] speed of translation is dependent on the stimulus gradient (time as well as distance would have to be recorded); [b] initiation of turning depends on the prior direction of movement with respect to the stimulus gradient and [c] turning is directed with respect to stimulus gradient by a bias in turning frequency or magnitude. Without this minimum analysis one cannot conclude whether the migration is a taxis or due to orthokinesis or klinokinesis.

Further information can be obtained by comparison of behavior in gradients with behavior in uniform stimulus fields for which directed turning cannot occur, and with responses to fields of differing gradient magnitude. The outcome will be a clearer understanding of the mechanism of migration, whether it is a taxis or due to klinokinesis. Sampling behavior can be investigated by the detailed study of head movements during migration. Proprioceptive signals indicating head position are probably important. Comparisons between different species and between wild type and mutants may help to select or eliminate various possibilities.

SUMMARY AND OUTLOOK

A revised set of movement terms has been presented in the hope that it will be more serviceable than existing terminology and will be acceptable to both microbiologists and metazoan biologists. Simple statistical tests have been suggested as an aid in classification of motion responses. Applied to the identification of taxes and klinokinetic migration, much of the uncertainty and confusion surrounding these terms can be removed.

A description of nematode studies has served to illustrate some simple movement behavior, and to review work in which some new tools are being applied. Particularly promising are the studies of behavioral mutants (e.g. Ward, 1976) and video recording (Croll, 1975). Also promising are methods developed for the study of microorganisms, particularly automated tracking techniques (e.g. Berg & Brown, 1972) and computer techniques for analyzing video recordings (Greaves, 1975, 1977; Lang et al., 1979). It is only recently that any of these methods have been applied to quantitative analysis of metazoan behavior (Lang et al., 1979). Wider application of such techniques is expected; a similar system is being developed for the study of nematode behavior (Dusenbery, personal communication). The ease with which detailed quantitative data can be obtained will, among other things, encourage the application of statistical methods to movement behavior (e.g. see Berg & Brown, 1972).

Analysis of video recordings will also lead the investigation in another direction: toward a more detailed description of movement behavior. Though the existing terminology will continue to be useful in classification of movement behavior, they are inadequate for describing more than the general features. Even the simple behavior of a nematode is far richer than indicated by, for example, "step-up positive klinokinesis." One can expect in the near future the introduction of new formats for the description of invertebrate behavior. Particularly promising is the Eshkol-Wachmann movement notation now being evaluated for mammalian behavior (Golani, 1976). Originally designed for choreography, the semi-quantitative notation can convey movement details simultaneously in several coordinate systems. Applied to the behavior of an organism orienting to the gradient or direction of light, one might use the notation to describe a sequence of movements in relation to the animal's body axes, in relation to the arena, and in relation to the movement of the gradient or direction vector relative to the animal's eyes. Taken together, these 3 descriptions would facilitate the interpretation of the behavior in terms of the management of input. Also such a description could provide a solid base for future neural explanations of the behavior, since all of the outward manifestations of motor activity could be described in a concise notation.

With the wider application of these new methods we are beginning to see a renaissance of movement studies. Descriptive and quantitative analysis of response sequences should lead to a better understanding of the mechanisms of migration in stimulus fields. Investigation of the effects of different stimuli or different internal states on a given response should reveal more of how sensory information is processed in simple organisms. In combination with neurophysiological approaches, movement studies on simple invertebrates may reveal some fundamental principles of how nervous systems control behavior.

APPENDIX: NOTES ON TERMINOLOGY

Note 1. Some authors have been more restrictive in the definition of stimulus. The ad hoc Committee on Behavioral Terminology (Diehn et al., 1977) defined it as a "quantity of energy or matter". Stasko and Sullivan (1971) regarded stimulus, like stimulation, as implying "an interaction between stimulating agent and sensory receptor". However, such definitions appear to be counter to common usage and create major problems when dealing with more complex stimuli, such as a rotating striped pattern. In operational definitions of behavioral terminology, one should avoid terms that depend on presumed physiological properties or inputs. At this level, it is the organism that is stimulated by the stimulus, not its parts.

Note 2. A stimulus has properties which may be varied independently. The number of properties depends on the nature of the stimulus. Temperature can be varied in intensity, timing, and spatial distribution. A chemical stimulus, can in addition, be varied in a qualitative way by changing its chemical properties. Light, gravitational, electrical, magnetic and mechanical stimuli all have a direction. Light can be varied in intensity, timing, spatial distribution, wavelength or spectral distribution, direction or directional distribution and in degree, direction, and spatial distribution of polarization. On the other hand, gravity is effectively constant in direction and magnitude. The prefix, "polaro" is used to indicate a response to the polarization of light, and "scoto" to indicate a response to a dark region in the directional distribution of light.

Note 3. Thus the definition is here restricted to motion responses.

Note 4. Some authors have incorrectly used kinesis to refer to the migration or accumulation resulting from a kinesis (Gunn, 1975). Where that meaning is intended, I suggest using 'orthokinetic' or 'klinokinetic' as modifiers of 'migration' or 'accumulation'.

Note 5. This is done in the hope that a response to the stimulus direction is made impossible, but in some cases it may merely be made non-observable.

Note 6. Thus this definition of a kinesis is more general than that understood by microbiologists (Diehn et al., 1977; Nultsch, 1975), but equivalent to that of Fraenkel and Gunn (1940, 1961).

Note 7. Ecclitic response and ecclisis come from the Greek noun εκκλισις whose primary meaning is a turning out of one's course. They have no previous meaning in English. Terms now used are unsatisfactory for various reasons. 'Avoidance response', 'shock response' and 'phobic response', have undesirable connotations that do not fit a response that can be spontaneous. 'Phobic response' includes any transient alteration in activity, i.e. other types of turning behavior (Diehn et al., 1977). 'Reversal bout' (Croll 1975), 'tumbling' and 'twiddles' (Berg, 1975) are not universal enough. 'Klinokinesis' is inadequate because it includes other types of undirected turns, and because a term is needed that includes both the undirected and directed responses that otherwise are so similar. 'Trial and error response' is also restricted to undirected responses. 'Titubant reaction', introduced by Ewer and Bursell (1950) has unfortunate meanings in English ("staggering, reeling or stumbling as if tipsy; marked by wavering, vacillating", Webster's Third New International Dictionary, 1971) and the authors gave it an ambiguous definition. One interpretation is that it refers only to directed responses (Fraenkel & Gunn, 1940, 1961, note 7). Another is that it refers only to undirected responses (Stasko & Sullivan, 1971, p. 110). In any event, a term that includes both response types is needed.

The adoption of 'ecclitic response' solves many problems: [1] a name is given to this very characteristic and widespread behavior which does not conflict with any existing terms. [2] It avoids the above-mentioned problems caused by using "klinokinesis" or "phobic response" for this behavior. [3] Biologists readily recognize an ecclitic response, but few have tested to see if it is a directed or undirected response. The new term includes both possibilities, and can be applied without implicating a kinesis or taxis. Where the distinction needs to be made, 'directed ecclitic response' or 'undirected ecclitic response' can be used.

Note 8. Both stimulus direction and stimulus gradient were included in the definition of Fraenkel & Gunn (1940, 1961) but for some reason Diehn et al. (1977) have omitted gradients in their definition. Apparently they have overlooked the fact that 'chemotaxis' and 'thermotaxis' will then cease to exist. 'Directional taxis' or 'gradient taxis' should be used where the distinction needs to be made.

Note 9. This generalization follows the definition of taxes introduced by Diehn et al. (1977). Fraenkel and Gunn (1940, 1961) included only positive and negative taxes, putting light compass reactions in a separate category. The latter fall naturally under klino-, tropo- or telotaxis if the definition is more general. A mutation affecting body form can convert a positive chemoklinotaxis to an oblique one (see section: Nematode Examples, Thermotaxis and chemotaxis).

Note 10. The two-light and unilateral-blinding tests are commonly cited as distinguishing between tropo- and telotaxis. However they are not conclusive. For example in telotaxis, the movement would be directed between 2 light sources if they cannot be resolved by the eye. This behavior is replaced by selection of the direction of one source as the organism approaches the lights. Also, even in a klinotaxis one of 2 lights could be selected if the 2 are well resolved by the combination of the eye and sampling movements. Unilateral blinding seems rather crude and indirect. Better tests are needed. If telotaxes cannot be reliably distinguished by operational criteria, perhaps this term should be dropped from movement terminology. It has been pointed out that tropotaxis and telotaxis are not in principle different (Schöne, 1975; Land, 1981, p. 584).

Note 11. The term 'eccritic temperature' is often used for the acclimation temperature, despite the fact that 'eccritic' has a completely different meaning in English: an agent that promotes excretion. I recommend that it no longer be used.

ACKNOWLEDGEMENTS

I appreciate the advice of several statisticians of the Mathematics Department, particular Richard Routledge, Barry Spurr and Michael Stephens. Also, E. Wyn Roberts and Marta Maftei helped by suggesting possible greek roots from which 'ecclisis' was selected. Larry M. Dill and Pierre Clément provided stimulating discussions. My research is supported by grants from The Natural Sciences and Engineering Research Council of Canada. Dr. M.A. Ali provided helpful editorial comments. I am grateful to Wendy Yee for patiently preparing the camera-ready manuscript.

REFERENCES

Alkon, D.L. (1980) Cellular analysis of a gastropod (Hermissenda crassicornis) model of associative learning. Biol. Bull. 159: 505-560.

Anderson, H., Edwards, J.S. & Palka, J. (1980) Developmental neurobiology of invertebrates. Ann. Rev. Neurosci. 3: 97-139.

Batschelet, E. (1981) Circular Statistics in Biology. London, Academic Press, p.93-128.

Berg, H.C. (1975) Bacterial behavior. Nature (Lond.) 254: 389-392.

Berg, H.C. & Brown, D.A. (1972) Chemotaxis in Escherichia coli analyzed by three-dimensional tracking. Nature (Lond.) 239: 500-504.

Burr, A.H. (1979) Analysis of phototaxis in nematodes using directional statistics. J. Comp. Physiol. 134: 85-93.

Burr, A.H. (1983) Evolution of eyes and photoreceptor organelles in the lower phyla. (This volume).

Burr, A.H. & Burr, C. (1975) The amphid of the nematode Oncholaimus vesicarius: Ultrastructural evidence for a dual function as a chemoreceptor and photoreceptor. J. Ultrastruct. Res. 51: 1-15.

Bursell, E., & Ewer, D.W. (1950) On the reaction to humidity of Peripatopsis moseleyi (Wood-Mason). J. Exp. Biol. 26: 335-353.

Chitwood, B.G. & Murphy, D.G. (1964) Observations on two marine monhysterids - their classification, cultivation and behavior. Am. Microscop. Soc. Trans. 83: 311-329.

Clément, P. & Wurdak, E. (1983) Photoreceptors and photoreception in rotifers. (This volume).

Coomans, A. & De Grisse, A. (1981) Sensory structures. In: Plant Parasitic Nematodes, Vol III. Ed. B.M. Zuckerman & R.A. Rhode. New York, Academic Press, p. 127-174.

Creutzberg, F. (1975) Orientation in space: animals. Invertebrates. In: Marine Ecology, Vol II, Part 2. Ed. O. Kinne. London, Wiley, p. 555-655.

Croll, N.A. (1965) The klinokinetic behavior of infective Trichonema larvae in light. Parasitol. 55: 579-582.

Croll, N.A. (1966a) Activity and the orthokinetic response of larval Trichonema to light. Parasitol. 56: 307-312.

Croll, N.A. (1966b) The phototactic response and spectral sensitivity of Chromadorina viridis (Nematoda, Chromadorida) with a note on the nature of the paired pigment spots. Nematol. 12: 610-614.

Croll, N.A. (1970) Sensory basis of activation in nematodes. Exp. Parasitol. 27: 350-356.

Croll, N.A. (1971) Movement patterns and photosensitivity of Trichonema spp. infective larvae in non-directional light. Parasitol. 62: 467-478.

Croll, N.A. (1975) Components and patterns in the behavior of the nematode Caenorhabditis elegans. J. Zool. (Lond.) 176: 159-176.
Croll, N.A. (1976) When Caenorhabditis elegans (Nematoda: Rhabditidae) bumps into a bead. Can. J. Zool. 54: 566-570.
Croll, N.A. & Al-Hadthi, I. (1972). Sensory basis of activity in Ancylostoma tubaeforme infective larvae. Parasitol. 64: 279-291.
Croll, N.A., Riding, I.L. & Smith, J.M. (1972) A nematode photoreceptor. Comp. Biochem. Physiol. 42A: 999-1009.
Croll, N.A. & Sukhdeo, M.V.K. (1981) Hierarchies in nematode behavior. In: Plant Parasitic Nematodes, Vol 3. Ed. B.M. Zuckerman & R.A. Rohde. New York, Academic Press, p. 227-251.
Darmon, M. & Brachet, P. (1978) Chemotaxis and differentiation during the aggregation of Dictyostelium discoidium amoebae. In: Taxis and Behavior. Ed. G.L. Hazelbauer. London, Chapman and Hall, p. 101-139.
Diehn, B. (1979) Photic responses and sensory transduction in protists. In: Handbook of Sensory Physiology, Vol. VII/6A. Ed. H. Autrum. Berlin, Springer-Verlag, p. 23-68.
Diehn, B., Feinleib, M., Haupt, W., Hildebrand, E., Lenci, F., & Nultsch, W. (1977) Terminology of behavioral responses of motile microorganisms. Photochem. Photobiol. 26: 559-560.
Dusenbery, D.B. (1980a) Behavior of free-living nematodes. In: Nematodes as Biological Models, Vol. 1. Ed. B.M. Zuckerman. New York, Academic Press, p. 127-158.
Dusenbery, D.B. (1980b) Responses of the nematode Caenorhabditis elegans to controlled chemical stimulation. J. Comp. Physiol. 136: 327-331.
Dusenbery, D.B. (1980c) Appetitive response of the nematode Caenorhabditis elegans to oxygen. J. Comp. Physiol. 136: 333-336.
Dusenbery, D.B., Anderson, G.L. & Anderson E.A. (1978) Thermal acclimation more extensive for behavioral parameters than for oxygen consumption in the nematode Caenorhabditis elegans. J. Exp. Zool. 206: 191-197.
Eakin, R.M. & Brandenburger, J.L. (1981) Unique eye of probable evolutionary significance. Science 211: 1189-1190.
Edgar, R.S. & Wood, W.B. (1977). The nematode Caenorhabditis elegans: A new organism for intensive biological study. Science 198: 1285-1286.
Ewer, D.W. & Bursell, E. (1950) A note on the classification of elementary behavior patterns. Behav. 3: 40-47.
Fankboner, P.V. (1981) Siphonal eyes of giant clams (Bivalva: Tridacnidae) and their relationship to adjacent zooxanthellae. The Veliger 23: 245-249.
Fernandez, J. & Stent, G.S. (1980) Embryonic development of the glossiphoniid leech Theromyzon rude: Structure and development of the germinal bands. Devel. Biol. 78: 407-434.

Fraenkel, G.S. & Gunn, D.L. (1940, 1961) The Orientation of Animals. Oxford, Clarendon Press; New York, Dover.
Golani, I. (1976) Homeostatic motor processes in mammalian interactions: a choreography of display. In: Perspectives in Ethology, Vol. 2. Ed. P.P.G. Bateson & P.H. Klopfer. New York, Plenum, p. 69-134.
Gouras, P. & Zrenner, E. (1981) Colour vision: a review from a neurophysiological perspective. In: Progress in Sensory Physiology, Vol. 1. Ed. H. Autrum, D. Ottoson, E.R. Perl & R.F. Schmidt. Berlin, Springer-Verlag, p. 139-179.
Goy, M.F. & Springer, M.S. (1978) In search of the linkage between receptor and response: the role of a protein methylation reaction in bacterial chemotaxis. In: Taxis and Behavior. Ed. G.L. Hazelbauer. London, Chapman and Hall, p. 1-34.
Greaves, J.O.B. (1975) The bugsystem: the software structure for the reduction of quantized video data of moving organisms. Proc. Inst. Elect. Electron. Engr. 63: 1415-1425.
Greaves, J.O.B. (1977) Television, computers and the behavior of sick bugs. In: Proc. Fifth New England Bioengineering Conference. Ed. M.R. Cannon. New York, Pergamon, p. 171-173.
Green, C.D. (1966) Orientation of male Heterodera rostochiensis Woll. and H. schachtii Schm. to their females. Ann. App. Biol. 58: 327-339.
Green, C.D. (1977) Simulation of nematode attraction to a point in a flat field. Behav. 61: 130-146.
Gunn, D.L. (1975) The meaning of the term 'klinokinesis'. Anim. Behav. 23: 409-412.
Gunn, D.L., Kennedy, J.S. & Pielou, D.P. (1937) Classification of taxes and kineses. Nature (Lond.) 140: 1064.
Hand, W.G. & Davenport, D. (1970) The experimental analysis of phototaxis and photokinesis in flagellates. In: Photobiology of Microorganisms. Ed. P. Halldal, London, Wiley-Interscience, p. 253-282.
Hedgecock, E.M. & Russell, R.L. (1975) Normal and mutant thermotaxis in the nematode Caenorhabditis elegans. Proc. Nat. Acad. Sci. U.S.A. 72: 4061-4065.
Hildebrand, E. (1978) Bacterial phototaxis. In: Taxis and Behavior. Ed. G.L. Hazelbauer. London, Chapman and Hall, p. 35-73.
Jander, R. (1970) Ein Ansatz zur modernen Elementarbeschreibung der Orientierungshandlung. Z. Tierpsychol. 27, 771-778.
Jones, F.R.H. (1971) The response of the planarian Dendrocoelum lacteum to an increase in light intensity. Anim. Behav. 19: 269-276.
Kandel, E.R. (1979) Cellular insights into behavior and learning. Harvey Lect. 73: 19-92.
Koshland, D.E. Jr. (1980) Bacterial chemotaxis in relation to neurobiology. Ann. Rev. Neuroscience 3: 43-75.

Kühn, A. (1919) Die Orientierung der Tiere in Raum. Jena, Gustav Fischer.
Land, M.F. (1981) Optics and vision in invertebrates. In: Handbook of Sensory Physiology Vol. VII/6B. Ed. H. Autrum. Berlin, Springer-Verlag, p. 471-592.
Lang, W.H., Forward, R.B. Jr. & Miller, D.C. (1979) Behavioral responses of Balanus improvisus nauplii to light intensity and spectrum. Biol. Bull. 157: 166-181.
Macagno, E.R., Lopresti, V. & Levinthal, C. (1973) Structure and development of neuronal connections in isogenic organisms: variations and similarities in the optic system of Daphnia magna. Proc. Nat. Acad. Sci. U.S.A. 70: 57-61.
Mardia, K.V. (1972) Statistics of Directional Data. London, Academic Press, p. 152-162, 196-206.
Nelson, D.L. & Kung, C. (1978) Behavior of paramecium: chemical, physiological and genetic studies. In: Taxis and Behavior. Ed. G.L. Hazelbauer. London, Chapman and Hall, p. 75-100.
Nicholls, J.G., Wallace, B. & Adal, M. (1977) Regeneration of individual neurons in the nervous system of the leech. In: Synapses. Ed. G.A. Cottrell & P.N.R. Usherwood. New York, Academic Press, p. 249-263.
Nultsch, W. (1970) Photomotion of microorganisms and its interaction with photosynthesis. In: Photobiology of Microorganisms. Ed. P. Halldal. London, Wiley, p. 213-245.
Nultsch, W. (1975) Phototaxis and photokinesis. In: Primitive Sensory and Communication Systems: The Taxes and Tropisms of Micro-Organisms and Cells. Ed. M.J. Carlile. London, Academic Press, p. 29-90.
Nultsch, W. & Häder, D.P. (1979) Photomovement of motile microorganisms. Photochem. Photobiol. 29: 423-437.
Riddle, D.L. (1978) The genetics of development and behavior in Caenorhabditis elegans. J. Nematol. 10: 1-16.
Rutherford, T.A. & Croll, N.A. (1979) Wave forms of Caenorhabditis elegans in a chemical attractant and repellant and in thermal gradients. J. Nematol. 11: 232-240.
Schöne, H. (1975) Orientation in space: animals. General introduction. In: Marine Ecology, Vol. II, Part 2. Ed. O. Kinne. London, Wiley, p. 499-553.
Siddiqui, I.A. & Viglierchio, D.R. (1970) Ultrastructure of photoreceptors in the marine nematode Deontostoma californicum. J. Ultrastruct. Res. 32: 558-571.
Sokal, R.R. & Rohlf, F.J. (1969) Biometry. San Francisco, W.H. Freeman, p. 155-166.
Stasko, A.B. & Sullivan, C.M. (1971) Responses of planarians to light: an examination of klinokinesis. Anim. Behav. Monogr. 4: 45-124.
Taliaferro, W.H. (1920) Reactions to light in Planaria maculata with special reference to the function and structure of the eyes. J. Exp. Zool. 31: 59-116.

Ullyott, P. (1936a) The behavior of Dendrocoelum lacteum I. Responses at light-dark boundaries. J. Exp. Biol. 13: 253-264.

Ullyott, P. (1936b) The behavior of Dendrocoelum lacteum II. Responses in non-directional gradients. J. Exp. Biol. 13: 265-278.

Ward, S. (1973) Chemotaxis by the nematode Caenorhabditis elegans: identification of attractants and analysis of the response by use of mutants. Proc. Nat. Acad. Sci. U.S.A. 70: 817-821.

Ward, S. (1976) The use of mutants to analyze the sensory nervous system of Caenorhabditis elegans. In: The Organization of Nematodes. Ed. N.A. Croll. London, Academic Press, p. 365-382.

Ward, S. (1978) Nematode chemotaxis and chemoreceptors In: Taxis and Behavior. Ed. G.L. Hazelbauer. London, Chapman and Hall, p. 141-203.

Ward, S., Thomson, N., White, J.G., & Brenner, S. (1975) Electron 313-338.

Ware, R.W., Clark, D., Crossland, K., & Russell, R.L. (1975) The nerve ring of the nematode Caenorhabditis elegans: Sensory input and motor output. J. Comp. Neurol. 162: 71-110.

Waterman, T.H. (1981) Polarization sensitivity. In: Handbook of Sensory Physiol., Vol. VII/6B. Ed. H. Autrum. Berlin, Springer-Verlag, p. 281-469.

White, J.G., Southgate, E., Thomson, J.N. & Brenner, S. (1976) The structure of the ventral nerve cord of Caenorhabditis elegans. Phil. Trans. Roy. Soc., Ser. B., 275: 327-348.

Wright, K.A. (1980) Nematode sense organs. In: Nematodes as Biological Models, Vol. 2. Ed. B.M. Zuckerman. New York, Academic Press, p. 237-295.

Young, S. (1981) Behavioral correlates of photoreception in Daphnia. In: Sense Organs. Eds. M.S. Laverack & D.J. Cosens. Glasgow, Blackie, p. 49-63.

Zar, J.H. (1974) Biostatistical Analysis. Englewood Cliffs, N.J. Prentice-Hall, p. 287-290.

PHOTORECEPTORS AND PHOTOSENSITIVITY IN PLATYHELMINTHES

ANNIE FOURNIER

Departement de Biologie Animale
Centre Universitaire
Avenue de Villeneuve
66025 Perpignan, Cedex, France

ABSTRACT

Many kinds of photoreceptors are known in Platyhelminthes. The ultrastructural relationships between the photoreceptors of adult turbellarians and free-living larvae of parasitic platyhelminthes explain their phylogenetic relationships. The discovery of a photoreceptor which is clearly ciliary and rhabdomeric, in the pelagic larvae of turbellarians, is important on account of the situation of the phyla among the Metazoa.

The studies on the different kinds of photosensitivities are considered differently according to the phylogenetic classes:-
1. At the experimental level in Turbellaria: the planarian is an object of experimentation for the study of the classical photoresponses (e.g. photo-klino-kinesis);
2. At the ecological level in the free-living larvae of Monogenea and Trematoda: the study of photosensitivity leads to the study of larval behaviour, closely related to the host behaviour. Light is one of the most important factor in the processes favouring the host location (spatially and temporally). It will be subsequently associated with other parameters which will then allow the host-contact and host-penetration.

INTRODUCTION

Platyhelminthes, comprise a primitive and highly varied class, the Turbellaria, which live in all media; and five classes of parasites: Temnocephala, Monogenea, Trematoda, Cestodaria and Cestoda. Numerous theories have been proposed to explain the evolutionary relationships among the different groups of Platyhelminthes. It is currently admitted that at least the Monogeneans and Trematodes are derived from a rhabdocoelian type of Turbellarian ancestor.

Photoreceptive structures can be interpreted in words of phylogenetic relations. The study of photosensitivity of parasitic platyhelminthes is particularly important for knowing the mechanisms favouring the transmission to the host. Among the Turbellarians light plays a preponderant role in the distribution of species. Considerable work has been carried out on the photosensitivity of species belonging to the main orders: Polyclada, Triclada, Proseriata and Rhabdocoela.

Among platyhelminth parasites, the adult stages are partially (ectoparasites) or totally (endoparasites) excluded from the action of light. A phase of free or carried parasitic dispersion assures the perpetuation of the cycle, that is to say the change of host-medium. Light acts directly at the level of the free stage in an aquatic medium. In this review I shall only consider Monogenea and Trematoda, eliminating Cestoda, where no photoreceptor structure has yet been demonstrated. Fig. 1 schematically shows the biological cycle of a monogenean and a trematode.

In the simple monogenean cycle, with only one host, there is only one free stage, the ciliated swimming larva or oncomiracidium. In the more complex trematode cycle, where at least two hosts are obligatory,

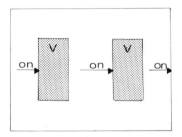

Monogenea

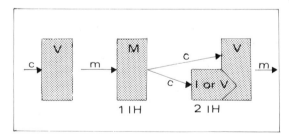

Trematoda

Fig. 1: The free-swimming larvae in the life-cycle of Monogenea (oncomiracidia) and Trematoda (miracidia and cercariae).

c: cercaria; I: Invertebrate; IH: intermediate host; M: Mollusc; m: miracidium; on: oncomiracidium; V; Vertebrate.

there are two free stages: the miracidium (ciliated swimming larva) which ensures the passage from the Vertebrate to the Mollusc; and the cercaria (swimming larva with a muscular tail), produced by the Mollusc, which ensures the passage to the definitive host in the case of two-host cycles or to the second intermediate host in the case of cycles with more than two hosts. Oncomiracidia, miracidia and cercariae have very short life spans; beyond 24 hours, if they survive, their activity is greatly reduced.

PLATYHELMINTH PHOTORECEPTORS

Ultrastructural data on photoreceptors are available only for a relatively small number of species, but these are varied in their systematic position. Five types of photoreceptors are known in Platyhelminthes. We distinguish the eyespots visible on the living animal, which correspond to the pigmented photoreceptors, from the non-pigmented sensorial structures, which are presumed to be photoreceptors by analogy with comparable structures known in other branches.

1. Eyespots or Pigmented Photoreceptor

a. Rhabdomeric pigmented photoreceptors

At present, all known pigmented photoreceptors in adult Turbellaria and free Platyhelminth larvae are rhabdomeric.

i. Adult Turbellaria

In Polyclads and Triclads, photoreceptors are often numerous, small and dispersed (nuchal, tentacular and cerebral eyes). In certain polyclads (e.g. Dugesia) they tend to be clustered: both eyes are then of considerable size. In Rhabdocoela, there are usually two eyes, situated near the nervous ganglia.

The eyespots of Polyclads (MacRae, 1966), Triclads (Press, 1959; MacRae, 1964; Röhlich & Török, 1961; Kishida, 1965, 1967a, b; Carpenter et al., 1974a; Durand & Gourbault, 1975, 1977) and Rhabdocoels (Bedini et al., 1973) have been described at an ultrastructural level. They are of the inverted type and always composed of two characteristic elements: pigmented cells and sensory cells.

The pigmented cells always contain large homogeneous, spherical granules which are highly osmophilic and surrounded by a membrane. Their chemical nature is unknown. In Triclads of the genus Dugesia, which have only one pair of large eyespots (110 X 75 μm), several dozen grouped cells from a wide and deep pigmentary cup. A similar number of sensory cells are associated with this structure (Fig. 2). In Polyclads and Triclads, whose eyes are small and often numerous, each eyespot consists of only one pigmented cell, associated with one or two sensory cells.

Sensory cells are bipolar: the pericaryon is outside the ocellus and the axon is directed towards the nerve centres. Inside the pigmentary cup the distal extremity forms a regularly ordered group of microvilli which constitutes the rhabdomere. The cytoplasmic zone below the rhabdomere contains numerous pinocytotic vesicles, mitochondria with clear matrix and neurotubules oriented parallel to the axis of the cell. The most

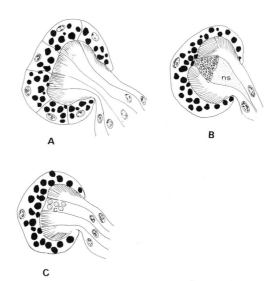

Fig. 2: Different kinds of pigmented rhabdomeric ocelli in Turbellaria. **A**: Polyclada, after Durand and Gourbault (1977). **B**: Dalyellidae, after Bedini et al. (1973). **C**: Muller's larva of Polyclada (Pseudoceros canadensis), after Eakin and Brandenburger (1981).
c: cilia; **ns**: neurosecretory cell.

proximal zone is characterised by a considerable density of multivesicular bodies, lyosome-like bodies, saccules of smooth endoplasmic reticulum and Golgi bodies. There is little glycogen.

Septate junctions join the plasma membranes of the sensorial and pigment cells.

The two eyespots of two species of Dalyellidae (Turbellaria, Rhabdocoela) show certain originalities in comparision with the ocelli of Polyclads and Triclads (Bedini et al., 1973): (1) three lamellar cytoplasmic projections of the pigment cell in the ocellar cavity delimit an anterior and posterior ocellar chamber; (2) two sensory cells penetrate into the cavity of each pigment cell: their distal end is organised into regular microvilli which constitute two orthogonal rhabdomeric bundles in the anterior cell and only one rhabdomere in the posterior cell; (3) the extension of a 4th, probably neurosecretory, cell is inserted between the dendrites of the two sensory cells and the extremity of the pigment wall. This kind of formation, which has never been described for other eyes, appears enigmatic (Fig. 2). These differences are very important from a phyletic point of view as they indicate considerable similarities between this kind of eye and that of larval trematodes, as we shall see below.

ii. Free larvae of platyhelminth parasites

Eyespots are characteristic of most free stages: monogenean

oncomiracidia and trematode miracidia and cercariae. They are always situated below the tegument in the dorsal region of the cerebral mass.

Most oncomiracidia of monogeneans have ocelli. Bychowsky (1957), Baer and Euzet (1961) and Llewellyn (1972) have described the kind, number and distribution of eyes. In Monopisthocotylea, there are usually two pairs of ocelli with "crystalline lenses". Their orientation is characteristic: the anterior pair is directed postero-laterally and the posterior pair antero-laterally. Euzetrema knoepffleri is the only representative of the sub-group currently known not to have lenses. Most Polyopisthocotylea have a pair of eyes without a crystalline lens. In Diclidophoridae, Llewellyn has described lipid droplets associated with each eyespot which serve as temporary lenses. In a few cases, there are two pairs of lens-less eyes (all Polystomatidae).

In Trematodes, miracidia and cercariae often have ocelli. However sometimes in the same species, only one of the two free stages possesses pigmented photoreceptors. Miracidia with ocelli have only one pair of closely-spaced eyes which form an X-shaped spot, sometimes with formations of sensory or pigment origin, playing the role of the lens. In certain species the left and right spots are asymmetric, in which case the left one is always larger. Cercariae most often have two symmetric eyespots. A few notocotyliidean and opistorchioidean cercariae have three (one lateral pair and a lone median eyespot). Lenticular formations have been observed.

Pigmentary spots of free stages of a dozen monogenean and tremadoda species have been described by electron microscopy (Kearn & Baker, 1973; Fournier, 1975; Kearn, 1978; Fournier & Combes, 1978; Isseroff, 1964; Pond & Cable, 1966; Isseroff & Cable, 1968; Rees, 1975).

In eyespots of all free-living stages of parasitic platyhelminthes (apart from those of monogenean polystomatidean larvae) the pigment cell granules are in all respects comparable with those described in turbellarians, and their chemical nature equally unkown. Pigment cells of monogenean Polystomatidae are different. They contain a pigment localised on platelets situated in about fifteen concentric rings whose nature ia unknown. In the polarising microscope these cells appear birefringent. This disposition is comparable in all respects with that described by Land (1966) in the mollusc Pecten maximus: it is characteristic of the "mirror eyes" occasionally found in other phyla (Rotifers, Archiannelids, Molluscs and Arthropods). The originality of this structure in monogeneans is that it appears characteristic of all species of the family Polystomatidae (Plate II A).

The number and arrangement of each of the constitutive elements (pigmentary and sensory) of the eyespot are variable (Fig. 3; Plate I A, B).

In the case of distinct eyespots (oncomiracidia of monogenean, Monopisthocotylea and Polystomatidae) each is composed of one pigment and one sensory cells: the latter forms two orthogonal rhabdomeric bundles in the cupule of the posterior eyespots of Monopisthocotylea.

In the case of the median eyespot (oncomiracidia of monogenean Polyopisthocotylea, miracidia and cercariae of some trematodes) several arrangements exist: (1) The convex sides of two pigment cells lie adjacent, one (or two) sensory cells filling the concavity with its rhabdomeric

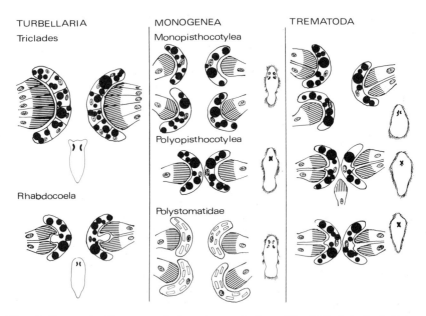

Fig. 3: Synoptic diagram of the pigmented ocelli of Platyhelminthes.

extensions; (2) The two adjacent pigment cells extend dorsoventral protusions which divide the concavity into anterior and posterior chambers, thus isolating the rhabdomeres derived from the two sensory cells; (3) A disymmetry appears with a rhabdomere, derived from a fifth sensory cell lodged in a hollow and which is no longer lateral but dorsomedian to the left-hand pigment cell. An orthogonal arrangement of the rhabdomeres derived from the anterior and posterior sensory cells is very often found; (4) The disymmetry is even greater when the left-hand pigment eye is formed from two pigment cells.

The cytoplasm of the sensory cell outside the pigmentary cup is characterised by a considerable accumulation of dense mitochondria, in addition to the vesicles and neurotubules. In the region of the pericaryon, the Golgi bodies and saccules of endoplasmic reticulum are well developed. Glycogen is much more abundant than in Turbellaria.

Lenticular formations attached to these photoreceptors have been described for several cases:
- low density bulbous cytoplasmic portions of the sensory cell at the opening of the pigmentary cup in miracidia (e.g. Philophthalmus);
- a single cell with an excentric nucleus, with apparently empty

Plate I:

A. Median photoreceptor of Echinostoma togoensis miracidium (Trematoda) (X 9 000).

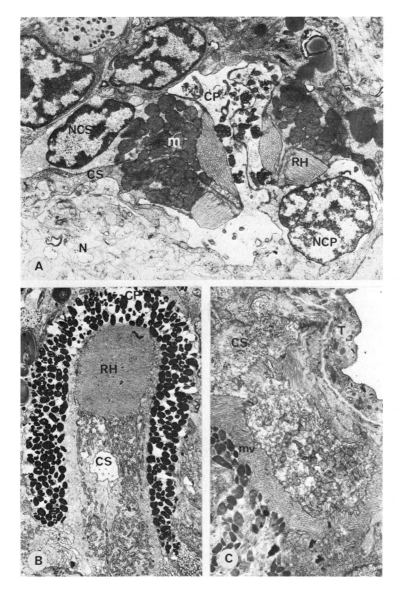

CP: pigmented cell; CS: sensory cell; m: mitochondria; mv: microvilli; NCP: nucleus of the pigmented cell; NCS: nucleus of the sensory cell; N: neuropilum; RH: rhadomere; T: tegument.

B. Lateral pigmented eyespot of a trematode cercaria (Notocotylidae) (X 3 000).
C. Rhabdomeric sensory receptor without pigment shield of Echinostoma togoensis cercaria (X 6 000).

cytoplasm, situated below the sensory cell in the case of a median eyespot of the tri-ocellate cercaria of Macrovestibulum;
- crystalline lenses of monogenean Monopisthocotylea oncomiracidia, described by Kearn and Baker (1973). These formations, 6-10 μm in diameter, are not of a cellular nature, but are composed of a rigid material with crystalline structure. The presence of such a lens is unique among platyhelminthes.

b. Mixed and ciliary pigmented photoreceptors

Certain species of polyclad Turbellaria have ocellate pelagic larval forms: Müller and Götte larvae. Two kinds of ocelli, cerebral and epidermal, have been described in these species by Ruppert (1978) and Eakin and Brandenburger (1981).

The cerebal ocelli of the Müller's larva of the polyclad, Pseudoceros canadensis, present at the originality of a marked structural disymmetry; in effect, while the right eyespot is of an exclusively rhabdomeric type (one pigment cell and three sensory cells), the left is mixed: in addition to the pigment cell and the three rhabdomeric sensory cells, a fourth sensory cell possesses many cilia. The presence of both ciliary and rhabdomeric cells in a cerebral eyespot is completely unique in a platyhelminth.

The single epidermal ocellus is composed of two cells: one contains large spherical pigment granules as well as modified cilia forming flattened lamellae oriented perpendicularly to the direction of the light; the other covers this first one; its function is debatable, a lens for Ruppert (1978); or a support for the lamellae of the pigment cell for Eakin and Brandenburger (1981).

2. Non-Pigmented Structures, Presumed Photoreceptive

a. Rhabdomeric receptors (Plate I C)

Bedini and Lanfranchi (1974) described "retinal clubs" lodged in the cerebal capsule of two otoplanid species (Turbellaria, Proseriata). These structures represent the discoidal processes of a bipolar cell connected to the neuropilum. The tubular whorls of the plasma membrane lying perpendicular to the disk surface form a rhabdomere.

In platyhelminth parasites, comparable receptors have been found in the monogenean Diplozoon paradoxum by Kearn (1978) and the cercariae of Echinostoma togoensis by Fournier (unpublished results) (Plate I C). Kearn (1978) considers that this structure may intervene in the control of photoperiodic processes, as "monitors of day (or night) length". A comparable structure exists in the polychaete annelid Armandia brevis (Herman & Cloney, 1966).

b. Ciliary receptors (Plate II B)

Such receptors have been described in both free and parasitic Turbellaria and Trematoda. In two species of Turbellaria Proseriata it consists of a group of about a hundred cilia (with a "9+0" structure) with no root, inside a cellular cavity (Ehlers & Ehlers, 1977). In platyhelminth parasites, these receptors have less cilia, equally lodged in the intracellular cavity of a neurone. Discovered for the first time, by Lyons (1972), in

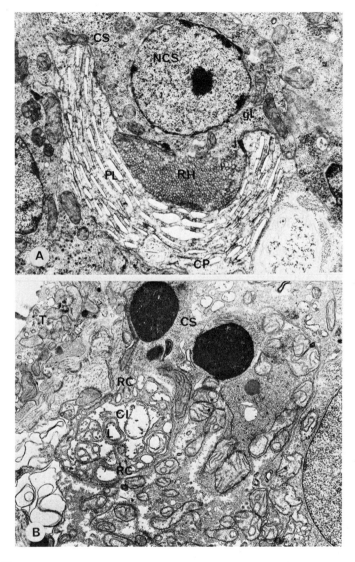

Plate II:

A. Mirror eye of Polystoma integerrimum (Monogenean, Polystomatidae) (X 9 000).
B. Sensory receptor, ciliary in nature, of Echinostoma togoensis (Trematoda) (X 12 000).

   CI: intracellular cavity; CP: pigmented cell; CS: sensory cell; gl: glycogen; j: septate junction; L: lamellar processes; NCS: nucleus of the sensory cell; Pl: platelets; RC: ciliary root; RH: rhabdomere.

the oncomiracidium of the monogenean Entobdella soleae they have since been found in numerous other species: miracidia and cercariae of Schistosoma mansoni (Short & Gagne, 1975), miracidia of Diplostomum spathaceum (Brooker, 1972), of Fasciola hepatica (Wilson, 1970). Very similar in structure to the "phaosomes" of polychaete annelids, they are presumed by analogy to be photoreceptors. They belong to the same phyletic line as the phaosomes which are also known in Rotifers (Clément, 1980).

3. What Do We Know About the Function of these Photoreceptors?

The study of changes taking place at the level of sensory and pigment cells when the photoreceptor is subjected to variations in light stimulation has demonstrated the functional role of the eyespots.

Experiments on different Triclad species (Rühlich & Török, 1962; MacRae, 1966; Stewart, 1966; Rühlich & Tar, 1968; Carpenter et al., 1974b; Durand & Gourbault, 1977) during a period of long adaptation to darkness has shown: random pigment migration throughout the cell, disorganisation of microvilli, swellings of the vacuoles of smooth endoplasmic reticulum (SER), shortening of sub-rhabdomeric mitochondria and accumulation of glycogen. All these changes are very rapidly reversible. Based on these observations, Rühlich and Török (1962), and Carpenter et al. (1974b) suggest a direct coupling of the photoreceptive processes of the rhabdomere and the metabolic processes of the mitochondria in the sub-rhabdomeric region (the increased incidence of glycogen in the dark could result from a decrease in oxidative phosphorylation reflected by the shrunken appearance of the mitochondria). In addition, the obvious relationship between the SER lamellae, the microvilli and the vesicles, as well as the swelling of the SER following light deprivation, are interpreted by Rühlich and Török (1962) as an argument which implicates the SER in photoreception (the reticulum vacuoles contain a photosensitive substance which is discharged into the microvilli during exposure to light). The mechanisms of migration of the pigment and by which the pigment cell receives information about the light are unknown.

During a normal lighting cycle (LD 12 : 12) the same modifications (disorganisation of the microvilli, vacuolarisation of the SER, accumulation of glycogen, shrinking of the mitochondria) have been described in Dalyella viridis (Turbellaria, Rhabdocoela). A retinomotor response (enlargement of the pigmentary cup and movement of the sensory cell in relation to the cup in the dark) has been demonstrated. For Bedini et al. (1977) these changes imply a metabolic process of disorganisation, storage and turnover of microvilli membranes, as well as a retinomotor response, and have physiological significance in the life of the animal.

We may ask whether comparable phenomena occur in the free-living stages of platyhelminth parasites given their very short lifetime (24 h or less). No data are available on this subject. The only modifications which have been observed occur during the life-cycle: when the larva encounters the host, it encounters conditions of obscurity which are considerable for ectoparasites and total for endoparasites. In Trochopus pini and Entobdella soleae (Monogenea, Monoposthocotylea) the crystalline lens disappears as as soon as the larva attaches itself to the skin of the fish (Kearn, 1971). In Euzetrema knoepffleri, another monogenean, although the pigment cell persists, the rhabdomeric microvilli are completely disorganised as soon as the larva begins its renal migration (Fournier, 1975). No photoreceptor structure has been described for adult monogeneans or trematodes, apart from a few remains of pigment granules in the parenchyma of certain monogeneans.

4. Phyletic Significance of Photoreceptor Structures

Five kinds of cerebal photoreceptors are currently known in platyhelminthes. The most common are of the rhabdomeric type: the rhabdomeric line is merely dominant, as ciliary photoreceptors (phaosomes and mixed photoreceptors) are also found in this group.

The mixed type of photoreceptor in pelagic larvae of polyclad Turbellaria are particularly worth consideration, given the phyletic position of the paltyhelminthes among the Metazoa. For Eakin and Brandenburger (1981) the rhabdomeric line is an innovation which appeared in the Turbellaria, thus beginning the separation between the ciliary and rhabdomeric lines as Eakin defined them in 1968. The theory of ciliary induction of the rhabdomere (Vanfleteren & Coomans, 1978; Vanfleteren, 1982) currently has no supporting arguments in this phylum.

Several kinds of photoreceptors exist in the same parasite species, either at different levels of the cycle (miracidium and cercaria of trematodes, which have the same genome, most often have different photoreceptors) or at the same stage. With the exception of the "mirror eyes" of the monogenean Polystomatidae, all adult turbellarian photoreceptors have also been found in free-living larvae of monogeneans and trematodes.

The similarity of photoreceptor structure between adult Turbellaria and free-larvae of platyhelminth parasites indicates their phylogenetic relationship. This is not surprising as the photoreceptor of monogeneans and trematodes are structures of the free-living stage which disappear as soon as the larva acquires a parasitic life-style. All these kinds of photoreceptors are also found in other phyla: Rotifers, Annelids, Molluscs or Arthropods. We in fact have several phylogenetic lines, as Clément (1980) has shown for Rotifers.

Finally, the similarity of eyespots in Dalyellidae (Turbellaria, Rhabdocoela) and those of certain monogeneans and trematodes larvae is an ultrastructural argument for the hypothesis of common rhabdocoel-type ancestor. We may ask, within the phylum: What was the ancestral disposition of eyespots? Which of the schemas of Dalyellidae or monogeneans with four eyespots represents the primitive situation? (Fig. 4).

THE DIFFERENT PHOTOSENSITIVITES OF PLATYHELMINTHES

There are two kinds of photosensitivity: one, implicated in displacement and orientation, which determines a spatial distribution; the other, responsible for hatching or emission rhythms of larval stages, which determines a temporal distribution.

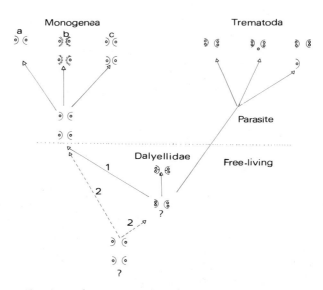

Fig. 4: Two hypotheses for the phylogenetic relationships of the pigmented rhabdomeric ocelli, from Turbellaria to parasitic Platyhelminthes.

1. Photosensitivity and Spatial Location

a. Turbellaria

It is impossible to cite all the work concerning motor orientation responses to light in the planarians. Viaud (1950, 1951) achieved a synthesis, enriched by his own results. Marine planarians are characterised by positive phototropism, due to the presence of symbionts (Zoochorelles). Freshwater planarians, which have been most intensely studied, are mostly photo-negative. When exposed to light, they first undergo a "dynamogenic action" independent of the definitive direction of the reaction. When the intensity or duration of the exposure increases, the movements become progressively slower. The planarians thus move faster in the dark, which indicates an inhibitory action of light. The orientation is determined by the eyes in ocullate planarians. When blinded, the same planarians show the same reactions, although much slower. Their orientation is then presumed to be due to dermatoptical sensitivity, or sensitivity of the whole body surface to light.

The study of the speed of locomotion in monochromatic light shows: a reduction in speed in the central region of the spectrum in ocullate species, a reduction in speed under violet light in the same species when blinded and an increase in photo-kinesis from red to violet in the absence of motor inhibition.

Viaud (1948) thus confirms the presence of two photosensitivities previously defined in Daphnia and Rotifers. One, visual, of maximum efficiency in green light, which determines orientation and inhibition in ocullate planarians; the other, corresponding to the dermatoptical sense, sensitive to short wavelengths, which determines the orientation of blind or blinded planarians and photo-kinesis in all planarians.

While this work constitutes a considerable mass of experimental data, it must be admitted that the obsolete terminology which is used covers very ill-defined notions derived from experiments in which the quality and quantity of stimuli were not adequately controlled.

Another group of important investigations on the response of Turbelleria to light concerns the study of photo-klino-kinesis. Historically, the work of Ullyot (1936a, b) was the basis of a vast debate on this subject. Studying the action of sudden illumination of vertical light on the locomotion of the planarian Dendrocoelum lacteum adapted to darkness, Ullyot proposed the hypothesis that the selection mechanism of a non-directional stimulation gradient is due to a photo-klino-kinesis with adaptation. Based on these results, Fraenkel and Günn (1940) defined photo-klino-kinesis as the frequency or rate of turning per unit time, dependent on intensity of stimulation. It was intended to cover a description of how "an organism could by random undirected movements, unaided by ortho-kinesis or any taxis or higher behaviour progress along a smooth gradient to a zone in which it then tended, still moving, to remain".

Stasko and Sullivan (1971), after a very critical review of all studies of klino-kinesis in Turbellaria, re-examined this concept under better defined experimental conditions. The response to a sudden increase in uniform vertical light intensity, described by Ullyot for Dendrocoelum lacteum, was found by these authors for another triclad, Dugesia dorotocephla. The more an animal is adapted to low intensities the more the response is rapid and greater. However, they showed that this photo-klinetic response with adaptation cannot explain the selection of a stimulation gradient. In effect, when submitted to two successive increases in intensity (10^{-4} to 1 LX, then 1 LX to 100 LX) Dugesia do not respond to the second stimulus. The orientation of the planarian is determined by the rays of lateral or directorial light, reflected by the walls or particles suspended in the medium, allowing them to move away from the zone most exposed to the light, even in the absence of vertical light.

Harden Jones (1971) using more elaborate experimental apparatus (infra-red recording), showed that, in fact, the response of D. lacteum to increasing light intensities is composed of both a klino-kinetic component (increase in frequence and angle of the change of direction) and an ortho-kinetic component (increase in speed). He also demonstrated that the decrease in klino-kinesis at very low intensities corresponded to a decrease in the change of direction angle and not of its frequence.

Verheijen and Gerbsen Schoorl (1975) recommenced the experiments on Dendrocoela lacteum, using a more precise approach to the different parameters used to quantify photo-klino-kinesis. They note that it is very difficult to produce ideal stimulation in order to study this phenomenon (e.g. non-directional light gradient). Despite the development of an apparatus in which the animal moves in a horizontally isotropic light field and the use of a light level which depends on the changes in the radial position of the animal, these authors consider that their results are insufficiently detailed (distance moved, angles, frequency) and propose standardisation with a mathematical model.

Photo-klino-kinesis appears as a very difficult behaviour to study experimentally. Gunn (1975), after all these controversies, even writes that the definition of klino-kinesis cannot be applied to Metazoans!

b. Parasitic platyhelminthes

i. Characteristics of swimming of free-living stages

In parasitic platyhelminthes numerous data exist on the photosensitivity of free-living larvae, but the experimental conditions are not always well defined. Several publications are much more precise. These results do not give rise to theoretical notions, as in Turbellaria, but rather to a knowledge of larval behaviour, allowing a better understanding of the ethological selection mechanisms of the host.

The very short free stage of parasitic platyhelminthes may be divided into two successive steps:-
- a dispersion step, immediately after hatching;
- a longer step, consisting of a search for the host, several hours after hatching (Saladin, 1979).

These two steps are essential in the process of facilitating a contact with the host (Combes, 1980).

The means of displacement of free-living stages varies between species. Some swim by rotating about their longitudinal axis, following a straight (corkscrew swimming of Schistosoma mansoni miracidia) or spiral (Diplozoon paradoxum oncomiracidia, Fasciola hepatica miracidia) trajectory. Others do not rotate about their axis, having a straight trajectory (Euzetrema knoepffleri oncomiracidia), or following a more or less loose helix (Polystoma pelobatis oncomiracidia). Depending on the species, some cercariae swim continuously, whereas the majority swim intermittently, alternating phases of activity and passivity (Schistosoma mansoni).

Oncomiracidia and miracidia move by means of cilia, while cercarial locomotion is assured by muscular contractions of the tail. No link with nerve cells has ever been found for ciliated cells. These have voluminous mitochondria and, in most cases, considerable reserves of glycogen. The diminution of these reserves has been demonstrated in aging miracidia. The cilia is thus an autonomous system. Thus, for a given species, swimming behaviour varies with age. Schistosoma mansoni miracidia have a rapid linear speed during the first hour (2,27 mm/s) which, after a slight reduction, remains steady for 5 hours (2,00 mm/s) and after 8 hours it is only 1,52 mm/s. While linear speed decreases, angular speed increases, even in the absence of the mollusc (Mason & Fripp, 1976).

Different physiochemical characteristics characterise the aquatic medium on hatching. Light appears to be the most important stimulus in determining the spatial distribution of larvae.

ii. Orientation with respect to light

In most known cases light determines the orientation of free stages. Numerous positive or negative photo-tactic responses have been described. Thus, Chapman (1974) showed that Cryptocotyle lingua cercariae are capable of detecting precisely the orientation of a light source and move towards it by sinusoidal swimming movements which makes them deviate very little from the light source. This orientation is characteristic of larvae having pigmented photoreceptors. In effect, Donges (1964) showed that the cercariae of Posthodiplostomum cuticola, which have ocelli with pigmented cups, are sensitive to the direction of the light, while cercariae of Apatemon sp. which have rhabdomeric ocelli lacking pigmentary cups, are photosensitive, but not to the direction of the rays. Photo-taxis has nevertheless been described for miracidia and cercariae of Schistosoma mansoni, which lack such pigmentary photoreceptors, by Croll (1973), Wright (1974a) and Mason and Fripp (1977). In most of their experiments the conditions are such that cercariae detect the relative intensities of two light sources in addition to their orientation, thus it is hard to separate photo-taxis in the strict sense of the term from the ortho-kinetic component of the movement.

These photo-tactic reactions in the same species change with age. Kearn (1980) observed among oncomiracidia of Entobdella soleae, which were actively attracted by light at hatching the appearance of individuals showing a photo-negative response. This alternating behaviour ensures readjustments in the vertical positioning of the larvae; as they become older, negative photo-taxis becomes the dominant behaviour.

iii. Speed of movement and light intensity

Numerous authors have demonstrated an obvious relationship between light intensity and larval swimming activity. An increase in intensity can provoke either an acceleration of linear speed (Haas, 1969; Mason & Fripp, 1976), or a decrease in this speed (Donges, 1963; Chapman, 1974; Wright, 1974b). The work of Saladin (1980) on photo-kinetic responses of Schistosoma mansoni is the most complete of these studies. He showed that S. mansoni cercariae are even active in the dark, swimming without any effect of intensity up to a certain threshold (10 $\mu W/cm^2$). Saladin suggests the existence of a pacemaker functioning in the absence of light in order to initiate sporadic swimming and suppress motor responses during the passive phases independently of any stimulation. This pacemaker is accelerated by intensities between the threshold and 4 000 $\mu W/cm^2$, the increasing activity of the cercariae being due to a shortening of the passive phase (DP) compared with the active phase (DA). Although the light intensity decreases both the DP and DA, it has a greater influence on the DP. Above 4 000 $\mu W/cm^2$ the increase in intensity leads to the suppression of photo-kinetic behaviour. At this moment the shortening of the DA is probably due to muscular fatigue (Fig. 5).

A pacemaker has equally been envisaged in cercariae of Cryptocotyle lingua by Chapman (1974) and Proterotrema macrostoma by Uglem and Prior (1978).

In many cases it has been shown that the temperature plays a role in modifying the photoresponse. Thus, at 15° C, the ortho-kinetic response of S. mansoni miracidia is inhibited (Mason & Fripp, 1977). As soon as the

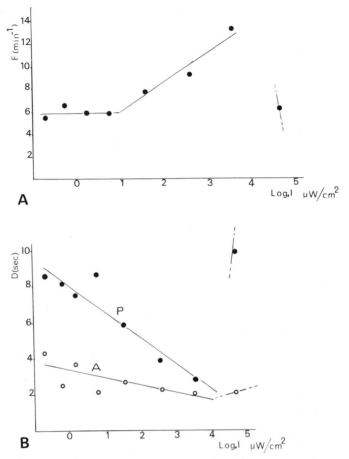

Fig. 5: <u>Schistosoma mansoni</u> cercariae: **A**; swimming frequence (**F**) and irradiance; **B**: A-phase and P-phase duration and irradiance, after Saladin (1980).

temperature increases the photo-negative behaviour of <u>S. haematobium</u> miracidia is reversed (Shiff, 1974).

iv. "Off" response

Immediate responses to sudden changes in light intensities have been observed in many parasitic platyhelminth larvae. The "off" kind of response seems to be dominant (Miller & MacCoy, 1930; Donges, 1963; Chapman, 1974). Saladin (1980) showed that, whereas <u>Schistosoma mansoni</u> cercariae react to both abrupt increases or decreases in intensity the "off" response is twofold greater. He presumes that the support for this behaviour is the phaosome (presumed photoreceptive ciliary structure).

The study of the relative sensitivity of trematode larvae to monochromatic light has shown that the greatest photoresponses are situated between λ = 500-550 nm (Yasuraoka, 1954; Wright et al., 1972; Engel & Davenport, 1973; Wright, 1974a, b; Macgeshin, 1977; Kennedy, 1979). However, it is difficult to determine from these results whether it is the taxic or kinetic component which is important. An additional peak at 650 nm is seen for miracidia of S. mansoni by Wright (1974a) and Bunodera mediovitellata by Kennedy (1979). On the other hand the relationship between these action spectra and the sensitive pigment have been little studied. For Saladin (1980) the action spectrum of the "off" response of S. mansoni cercariae, which shows a double peak, is due to two pigments affecting swimming behaviour: one at 498 nm, corresponding to the action spectrum of the carotenoid pigment of the rhodopsin family, the other at 576 nm, corresponding to an absorption of rays by a non-carotenoid pigment which transforms the energy into muscular contraction.

v. Function of responses to light

Many authors (Bevelander, 1933; Donges, 1963; Haas, 1969; Kennedy, 1979; Kearn, 1980) have shown that the different photosensitivities have an essential effect at hatching on the mechanisms favouring contact with the host by influencing larval dispersion and selection of a medium corresponding to an ecological preferendum of the host. Kennedy (1979), in the case of the cycle of the trematode Bunodera mediovitellata, showed clearly the equivalence existing between the behaviour of the two free-living stages (the miracidia have negative photo-taxis and positive geo-taxis, and the cercariae positive horizontal photo-taxis) and the micro-habitat of their natural hosts (molluscs living on dark bottoms for the miracidia; phrygane larvae found in shallow waters which are well illuminated for the cercariae).

The "off" response is equally interpreted as a facilitating behaviour: the shadow of a passing fish makes Posthodiplostomum cuticola cercariae very active: Donges (1964) believes that the rapid adaptation noted after the "off" response minimises fruitless responses to regular shadows caused by moving vegetation.

2. Photosensitivity and Temporal Location of Parasitic Platyhelminth Larvae

This type of photosensitivity exists among Turbellaria as well as in monogeneans and trematodes. I shall only consider it for these two classes of parasites.

The hatching of monogenean and trematode eggs may be delayed for long periods by storing eggs which are ready to hatch in the dark. The participation of a photoreceptive structure does not seem to be necessary.

In platyhelminth parasites biological rhythms exists at various levels of the cycle in which the photoperiod, and often the thermoperiod, intervene as synchronisers of hatching and emergence rhythms.

In the monogenean, Entobdella soleae, Kearn (1973) showed that egg hatching occurs during the first hours following the illumination period, whatever the duration of the latter (LD 12 : 12; 16 : 8; 8 : 16). When incubated in total darkness the larvae hatch without rhythmicity. If they are placed in continuous light or darkness at the beginning of hatching the emission rhythm persists, implying the existence of an endogenous component. Comparable results have been obtained for other species (MacDonald & Coombs, 1978; Bundy, 1981). Studying host behaviour Kearn (1973) suggested that such rhythmicity of hatching favours maximal contact between the oncomiracidia and Solea solea, which remains inactive on the sandy bottom during the day.

More recently it has been shown for many species that the maximum number of larvae released corresponds temporally to a particular behaviour of the host. Thus, in Entobdella hippoglossi (Kearn, 1974) and Diplozoon homoion gracile (MacDonald & Jones, 1978) the emission maxima are respectively at night or at dusk, the targets being fish which have their resting period at night. In Polystoma integerrimum (MacDonald & Combes, 1978) hatching in the middle of the day assembles larvae at a moment when the host tadpoles are concentrated at the edges of the pool.

In trematodes, cercarial emergence rhythms appear to be mainly synchronised by the photoperiod and thermoperiod (see Theron, 1982 for a review). In their absence the rhythmicity is completely desynchronised. The inversion of the photoperiodic rhythms induces an inversion of the acrophases. In the case of Ribeiroia marini cercariae, Theron (1975) showed that the synchronisation of the cercarial nocturnal emission of rhythm results exclusively from the inhibiting action of light. In the absence of light, cercariae which leave the sporocyst accumulate within the mollusc. The biological rhythm of cercarial emission depends on means which are closely related to the activity of the host which is to be infected. In the case of cercariae released during the day (e.g. S. mansoni) the diurnal acrophase corresponds to the hours when human activity in the aquatic medium of endemic zones is at its highest (washing, irrigation, bathing). For nocturnal cercariae (e.g. Ribeiroia marini) the nocturnal acrophase increases the probability of contact and infestation of the host fish during its nocturnal resting period. (Combes & Theron, 1977).

The mechanisms implicated in these chronobiological processes are completely unknown, as well as the location and nature of the photopigments on receptors.

REFERENCES

Baer, J.G. & Euzet, L. (1961) Classe des Monogènes. In: Traité de Zoologie. Ed. P.P. Grassé. Vol. 4. Paris, Masson. p. 243-325.

Bedini, C., Ferrero, E. & Lanfranchi, A. (1973) Fine structure of the eyes in two species of Dalyellidae (Turbellaria : Rhabdocoela). Monitore Zool. Ital. 7: 51-70.

Bedini, C., Ferrero, E. & Lanfranchi, A. (1977) Fine structural changes induced by circadian light-dark cycles in photoreceptors of Dalyellidae (Turbellaria : Rhabdocela). J. Ultrastruct. Res. 58: 66-77.

Bedini, C. & Lanfranchi, A. (1974) The fine structure of photoreceptor in two otoplanid species (Turbellaria: Proseriata). Z. Morph. Tiere 77: 175-186.

Bevelander, G. (1933) Response to light in the cercariae of Bucephalus

elegans. Physiol. Zool. 6: 289-305.
Brooker, B.E. (1972) The sense organs of Trematode miracidia. J. Linn. Soc. Lond. 51 (suppl. 1): 171-180.
Bundy, D.A.P. (1981) Periodicity in the hatching of digenean eggs: a possible circadian rhythm in the life-cycle of Transversotrema patialense. Parasitol. 83: 13-22.
Bychowsky, B.E. (1957) Monogenetic Trematodes. Their Systematics and Phylogeny. Akad. Nauk. SSSR 1-509 (in Russian). English translation by W.J. Hargis & P.C. Oustinoff (1961). Washington, American Institute of Biological Sciences.
Carpenter, K.S., Morita, M. & Best, B. (1974a) Ultrastructure of the photoreceptor of the planarian Dugesia dorotocephala. I. Normal eye. Cell Tissue Res. 148: 143-158.
Carpenter, K.S., Morita, M. & Best, B. (1974b) Ultrastructure of the photoreceptor of the planarian Dugesia dorotocephala. II. Changes induced by darkness and light. Cytobiol. 8: 320-338.
Chapman, H.D. (1974) The behaviour of the cercaria of Cryptocotyle lingua. Z. Parasitenk. 44: 211-226.
Clément, P. (1980) Phylogenetic relationships of Rotifers, as derived from photoreceptor morphology and other ultrastructural analyses. Hydrobiol. 73: 93-117.
Combes, C. (1980) Les mécanismes de recrutement chez les Métazoaires parasites et leur interprétation en termes de stratégies démographiques. Vie et Milieu 30: 55-63.
Combes, C. & Theron, A. (1977) Rhythmes d'émergence des cercaires de Trématodes et leur intérêt dans l'infestation de l'homme et des animaux. Inst. Biol. Publicaciones especiales 4, Vol. in memorian Prof. Caballero y Caballero. p. 141-150.
Croll, N.A. (1973) Parasitism and Other Associations. London, U.K., Pitman Medical.
Donges, J. (1963) Reizphysiologische Untersuchungen an der cercarie von Posthodiplostomum cuticola (V. Nordmann 1832) Dubois 1836, dem Erreger des Diplostomatidenmelanoms der Fische. Verh. deustsch. Zoologische Gesellschaft. 27: 216-248.
Donges, J. (1964) Des Lebenszyklus von Posthodiplostomum cuticola (V. Nordmann 1832) Dubois 1836 (Trematoda, Diplastomatidea). Z. Parasitenk. 24: 169-248.
Durand, J.P. & Gourbault, N. (1975) Etude cytologique des organes photorécepteurs de Dendrocoelopsis (Amyadenium) chattoni (Triclade Paludicole Hypogé). Ann. Speleol. 30: 129-135.
Durand, J.P. & Gourbault, N. (1977) Etude cytologique des organes photorécepteurs de la Planaire australienne Cura pinguis. Can. J. Zool. 55: 381-390.
Eakin, R.M. (1968) Evolution in Photoreceptors. In: Evolutionary Biology, II. Amsterdam, North Holland Publ. Co. p. 194-242.
Eakin, R.M. & Brandenburger, J.L. (1981) Fine structure of the eyes of Pseudoceros canadensis (Turbellaria, Polycladida). Zoomorphol. 98: 1-16.
Ehlers, B. & Ehlers, U. (1977) Ultrastruktur pericetbraler cilienaggregate bei Dicoelandropora atriopapillata Ax und Notocaryoplanella

glandulosa Ax (Turbellaria, Proseriata). Zoomorphol. 88: 163-174.
Engel, R.E. & Davenport, D. (1973) The investigation of light responses in larval flatworms. Am. Zool. 13: 1337.
Fournier, A. (1975) Les taches pigmentaires de l'oncomiracidium d'Euzetrema knoepffleri (Monogenea): Ultrastructure et évolution au cours du cycle biologique. Z. Parasitenk. 46: 203-209.
Fournier, A. & Combes, C. (1978) Structure of photoreceptors of Polystoma integerrinum (Platyhelminthes, Monogenea). Zoomorphol. 91: 147-155.
Fraenkel, G.S. & Gunn, D.L. (1940) The Orientation of Animals: Kinesis, Taxes and Compass Reactions. Clarendon Press. 352 pages.
Gunn, D.L. (1975) The meaning of the term "klinokinesis". Anim. Behav 23: 409-412.
Haas, W. (1969) Reizphysiologische Untersuchuntgen an cercarien von Diplostonum spathaceum. Z. vergl. Physiol. 64: 254-287.
Harden Jones, F.R. (1971) The response of the planarian Dendrocoelum lacteum to an increase in light intensity. Anim. Behav. 19: 269-276.
Herman, C.O. & Cloney, R.A. (1966) The fine structure of the prostomial eyes of Armandia brevis (Polychaeta Opheliidae). Z. Zellforsch. mikrosk. Anat. 72: 583-596.
Isseroff, H. (1964) Fine structure of the eyespot in the miracidium of Philophthnalmus megalurus (Cort, 1914). J. Parasitol. 50: 549-554.
Isseroff, H. & Cable, R.M. (1968) Fine structure of photoreceptors in larval trematodes. A comparative study. Z. Zellforsch. mikrosk. Anat. 86: 511-534.
Kearn, G.C. (1971) The attachment site, invasion route, and larval development of Trochopus pini, a monogenean from the gills of Trigla hirundo. Parasitol. 63: 513-525.
Kearn, G.C. (1973) An endogenous circadian hatching rhythm in the monogenean skin parasite Entobdella soleae and its relationships to the activity rhythm of the host. Parasitol. 66: 101-122.
Kearn, G.C. (1974) The effects of fish mucus on hatching in the monogenean parasite Entobdella soleae from the skin of the common sole, Solea solea. Parasitol. 68: 173-188.
Kearn, G.C. (1978) Eyes with, and without, pigment shields in the oncomiracidium of the monogenean parasite Diplozoon paradoxum. Z. Parasitenk. 157: 35-47.
Kearn, G.C. (1980) Light and gravity responses of the oncomiracidium of Entobdella soleae and their role in host location. Parasitol. 81: 71-89.
Kearn, G.C. & Baker, N.O. (1973) Ultrastructure and histochemical observations on the pigmented eyes of the oncomiracidium of Entobdella soleae, a monogenean skin parasite of the common sole, Solea solea. Z. Parasitenk. 41: 239-254.
Kennedy, M.J. (1979) The responses of miracidia and cercariae of Bunodera mediovitellata (Trematoda : Allocreadiidae) to light and gravity. Can. J. Zool. 57: 603-609.
Kishida, Y. (1965) The ultrastructure of the eyes in Dugesia japonica. I. The distal portion of the visual cell. Zool. Mag. Tokyo 74: 149-155.
Kishida, Y. (1967a) Electron microscopic studies on the planarian eye. I.

Fine structure of normal eye. Sci. Rep. Kanazawa Univ. 12: 75-110.
Kishida, Y. (1967b) Electron microscopic studies on the planarian eye. II. Fine structure of the regenerating eye. Sci. Rep. Kanazawa Univ. 12: 111-142.
Land, M. (1966) A multilayer interference reflector in the eye of the scallop Pecten maximus. J. Exp. Biol. 45: 433-447.
Llewellyn, J. (1972) Behaviour of monogeneans. J. Linn. Soc. Lond. 51 (suppl. 1): 19-30.
Lyons, K.M. (1972) Sense organs of monogeneans. J. Linn. Soc. Lond. 51 (suppl. 1): 181-199.
MacDonald, S. & Combes, C. (1978) The hatching rhythm of Polystoma integerrimum, a monogenean from the frog Rana temporaria. Chronobiol. 5: 277-285.
MacDonald, S. & Jones, A. (1978) Egg-laying and hatching rhythms in the monogenean Diplozoon homoion gracile from the southern barbel (Barbus meridionalis). J. Helminthol. 52: 23-28.
Macgeachin, W.T. (1977) Responses of miracidia of the liberian strain of Schistosoma mansoni to monochromatic light. 29th Annual Meeting, Annual Midwest Conference of Parasitologists, Manhattan, Kansas. Abst.
MacRae, E.K. (1964) Observations on the fine structure of photoreceptor cells in the planarian Dugesia tigrina. J. Ultrastruct. Res. 10: 334-349.
MacRae, E.K. (1966) The fine structure of photoreceptors in a marine flatworm. Z. Zellforsch. mikrosk. Anat. 75: 469-484.
Mason, P.R. & Fripp, P.J. (1976) Analysis of the movements of Schistosoma mansoni miracidia by dark ground photography. J. Parasitol. 62: 721-727.
Mason, P.R. & Fripp, P.J. (1977) The reactions of Schistosoma mansoni miracidia to light. J. Parasitol. 63: 240-244.
Miller, H.M. Jr. & MacCoy, O.R. (1930) An experimental study of the behaviour of Cercaria floridensis in relation to its fish intermediate host. J. Parasitol. 16: 185-197.
Pond, G.C. & Cable, R.M. (1966) Fine structure of photoreceptors in three types of ocellate cercariae. J. Parasitol. 52: 483-493.
Press, N. (1959) Electron microscope studies of the distal portion of a planarian retinular cell. Biol. Bull. Mar. Biol. Lab., Woods Hole 117: 511-517.
Rees, G.R. (1974) Studies on the pigmented and unpigmented photoreceptors of the cercaria of Cryptocotyle lingua (Creplin) from Littorina littorea (L.). Proc. R. Soc. Lond. 188: 121-138.
Rühlich, P. (1966) Sensitivity of regeneration and degenerating planarian photoreceptors to osmium fixation. Z. Zellforsch. 73: 165-173.
Rühlich, P. & Tar, E. (1968) The effect of prolonged light-deprivation on the fine structure of planarian photoreceptors. Z. Zellforsch. mikrosk. Anat. 90: 507-518.
Rühlich, P. & Török, J. (1961) Elektronenmikroskopische untersuchungen des Anges von planarian. Z. Zellforsch. mikrosk. Anat. 54: 362-381.
Rühlich, P. & Török, J. (1962) The effect of light and darkness on the fine structure of the retinal clubs of Dendrocoelum lacteum. Quart. J.

Microsc. Sci. 104: 543-548.
Ruppert, E.E. (1978) A review of metamorphosis of Turbellarian larvae. In: Settlement and Metamorphosis of Marine Invertebrate Larvae. Ed. F.S. Chia & R.E. Rice. New York, Elsevier/North Holland Biomedical Press, p. 65-81.
Saladin, K.S. (1979) Behavioural parasitology and perspectives on miracidial host-finding. Z. Parasitenk. 60: 197-210.
Saladin, K.S. (1980) Behavioural Manifestations of Photosensibility in Cercariae of Schistosoma mansoni (Digenea : Schistosomatidae). Ph. D. Thesis. Florida State University. 202 pages.
Shiff, C.J. (1974) Seasonal factors influencing the location of Bulinus (Physopsis) globosus by miracidia of Schistosoma haematobium. J. Parasitol. 60: 578-583.
Short, R.B. & Gagne, H.T. (1975) Fine structure of a possible photoreceptor in cercariae of Schistosoma mansoni. J. Parasitol. 61: 69-74.
Stasko, A.B. & Sullivan, C.M. (1971) Responses of planarian to light: an examination of klinokinesis. Anim. Behav. Monogr. 4: 47-124.
Stewart, A. (1966) Ultrastructure of the eyes in three freshwater Triclad worms. Am. Zool. 6: 615.
Theron, A. (1975) Chronobiologie des cercaires de Ribeiroia marini (Faust et Hoffman, 1934) parasite de Biomphalaria glabrata: action de la photopériode sur le rythme d'émission. Acta Trop. 32: 309-316.
Theron, A. (1982) Le Compartiment Cercaire dans le Cycle de Schistosoma mansoni Sambon, 1907. Ecologie de la Transmission Bilharzienne en Guadeloupe. Thèse Doctorat d'Etat. Univ. Perpignan. 506 pages.
Uglem, G.L. & Prior, D.J. (1978) Electrophysiology and swimming behaviour of Proterometra macrostoma cercariae. 53rd Annual Meeting, American Society of Parasitologists, Chicago, Illinois. Abstr.
Ullyott, P. (1936a) The behaviour of Dendrocoelum lacteum. I Responses at light and dark boundaries. J. Exp. Biol. 13: 253-264.
Ullyott, P. (1936b) The behaviour of Dendrocoelum lacteum. II. Responses in non-directional gradients. J. Exp. Biol. 13: 265-278.
Vanfleteren, J.R. (1982) A monophyletic line of evolution? Ciliary induced photoreceptor membranes. In: Visual Cells in Evolution. Ed. J. Westfall. New York, Raven Press, p. 107-136.
Vanfleteren, J.R. & Coomans, A. (1976) Photoreceptor evolution and phylogeny. Z. Zool. Systemat. Evol. Forsch. 14: 157-169.
Verheijen, F.T. & Gerssen Schoorl, K.H.J. (1975) Photoklinokinesis in planarian? Nether. J. Zool. 25: 433-453.
Viaud, G. (1948) Le phototropisme et les deux modes de la photoréception. Experientia 3: 81-88.
Viaud, G. (1950) Recherches expérimentales sur le phototropisme des planaires. Behav. 2: 163-216.
Viaud, G. (1951) Le phototropisme chez les Cladocères, les Rotifères et les Planaires. Ann. Biol. 55: 355-378.
Wilson, R.A. (1970) Fine structure of the nervous system and specialized nerve endings in the miracidium of Fasciola hepatica. Parasitol. 60: 399-410.

Wright, D.G.S. (1974a) Responses of miracidia of Schistosoma mansoni to an equal energy spectrum of monochromatic light. Can. J. Zool. 52: 857-888.

Wright, D.G.S. (1974b) Responses of cercariae of Trichobilharzia ocellata to white light, monochromatic light and irradiance reduction. Can. J. Zool. 52: 575-579.

Wright, D.G.S., Lavigne, D.M. & Ronald, K. (1972) Responses of miracidia of Schistosoma douthitti (Cort, 1914) to monochromatic light. Can. J. Zool. 50: 197-200.

Yasuraoka, K. (1954) Ecology of the miracidium of Fasciola hepatica. Jap. J. Med. Sci. Biol. 7: 181-192.

PHOTORECEPTORS AND PHOTORECEPTIONS IN ROTIFERS

PIERRE CLÉMENT and ELIZABETH WURDAK
Laboratoire Histologie et Biologie Tissulaire
LA CNRS 244, RCP CNRS 657 and CMEABG
Université Lyon I. 69622 Villeurbanne Cedex
France

> Yes, forsooth; I wish you joy o'the worm.
>
> Anthony and Cleopatra, Act V, Sc. 2.

INTRODUCTION

1. <u>What use are their eyes to them?</u>

Every rotifer has eyes and ocelli. Most frequently, there is one eye on the brain and two anterior ocelli in the rotatory apparatus.

However, rotifers do not "see"; they perceive neither form nor movement. It is only through chance encounters that these small aquatic animals contact their food or their sexual partner which they then recognize as a result of tactile or chemical stimulation (Clément <u>et al.</u>, 1983).

So what is the use of these eyes or ocelli which are so striking since they are the only colored spots in these transparent animals? The question remains unanswered for many species. However, in certain cases several different responses to light have been shown to exist. Light has a direct influence on locomotor behavior (taxes and kineses) and it modulates the reproductive cycle by inducing the development of bisexually reproducing animals which produce dormant resting eggs. In the absence of such stimulation, these rotifers reproduce by parthenogenesis. Parthenogenesis leads to the rapid multiplication of rotifers in ponds and lakes where they often constitute the most abundant element of the plankton.

Photoreception probably plays an important role in the horizontal and vertical distribution of planktonic rotifers and in the colonization of specific biotopes. (Rotifers may be pelagic or periphytic in their habitat or restricted to the water layer adhering to mosses, lichens and soils). Preissler (1977, 1980) has shown, for example, that two species of planktonic rotifers avoid the zone of shadow next to the shore in laboratory and field situations.

Positive phototactic responses have been known to occur in rotifers since the beginning of the century (Jennings, 1901) and their detailed observation continues to the present day (cf. section IV below). They probably serve to bring the animals into a zone where they are most likely to encounter the food item. A case in point is Notommata copeus. This periphytic species is phototactic only when it is hungry (Clément, 1977 a, c).

2. What use are their eyes to us?

Rotifers are a favorite object of laboratory and field study for a variety of reasons, they are aesthetically pleasing in their form and their movements and they represent a biological model which is without parallel in the animal kingdom.

All descendants of a female are isogenetic because parthenogenesis is mitotic in these animals (King, 1977). Thus, they form a clone, which can easily be maintained under laboratory conditions. Culture methods are well established for many species of rotifers (Pourriot, 1965, 1977, 1980). In addition, the parameters leading to the appearance of males are also known in several cases (Gilbert, 1977; Clément et al., 1977, 1981; Pourriot & Clément, 1981). Consequently, these animals may be used to study the interaction between genome and environment at the neurobiological and behavioral level.

Rotifers are extremely small. They measure 0,1 to 1 mm in length. They consist of approximately one thousand highly specialized cells. They are perfectly eutelic: there is no cellular division from birth to death. Consequently, they may be regarded as a small scale model of a complex organism which is of particular interest to neurobiologists. The rotifer brain is made up of about 200 cells including the sensory neurons which terminate in the epidermis. The sensory organs consist of very few cells (Clément et al., 1983). Each eye or ocellus, for example, contains one or, at the most, two photoreceptor neurons (cf. sections I and II below).

The ultrastructural study of the rotifers appears to be very promising. The technique of serial, thin, sectioning, applied to an analysis of the brain of Asplanchna brightwelli by Ware (1971) and Seldon (1972) in the laboratory of C. Levinthal (Levinthal & Ware, 1972) has shown a remarkable constancy in the position of the neurons and synapses.

The study of photoreceptors and photoreceptions of rotifers is only at the beginning. The results obtained to date deal, on the one hand, with the ultrastructure of the photoreceptor or presumed photoreceptor organs (photoreceptive function has not been submitted to electrophysiological testing yet). These results are summarized in the three sections which follow. On the other hand, there are observations concerning the different responses of rotifers to light: behavioral (locomotor) responses (section IV) and the effect of light on the reproductive cycle (section V).

The chemical nature of the sensory and accessory pigments in rotifers is unknown. Nonetheless some information has been obtained through the analysis of light-dependent locomotor responses under monochromatic illumination in the visible and the near ultra-violet range, and through the application of the technique of microspectrophotometry to the accessory pigments. These observations are summarized in section IV.

I - ULTRASTRUCTURE OF THE CEREBRAL EYES

1. Light microscope observations

Cerebral eyes have been known to exist in rotifers since the nineteenth century. They are described and localized with precision in articles dating from the end of the 19-th and the beginning of the 20-th century. Their red color facilitates detection because the rest of the animal is transparent. In his synthesis, Remane (1929-1933) makes the following observations. Cerebral eyes are totally absent in certain groups of rotifers: Seisonidae, Flosculariacea and Collothecacea. When present, the cerebral eye is unpaired as a general rule. However, in certain bdelloïds and in the monogonont Epiphanes senta it is paired. The situation then approaches that found in Synchaeta or Brachionus where the X-shaped pigment cup of the single eye suggests the fusion of two eyes. The red eye spot is always located at the periphery of the brain. In the Notommatidae, the Trichoceridae and the Synchaetidae it is linked to the retrocerebral apparatus as well. Certain authors (De Beauchamp, 1907, 1909) have supposed that the retrocerebral apparatus plays an active role in photoreception, but it was shown by electron microscopy that this apparatus consists exclusively of classical mucous glands (Clément, 1977 a, b, 1980).

A characteristic feature of the early observations is that their authors equate the red pigment spot with the photoreceptive structure. The vocabulary utilized illustrates the confusion fully. Remane (1929-1933) followed the terminology of his day (that of De Beauchamp, 1909, 1965, for example) in calling the red pigmented spots existing, in addition to the cerebral eye, in the brain or in the anterior ciliated cells "supernumerary eyes". Viaud (1940, 1943 a, b) spoke of "blind" Asplanchna whenever the red pigment of the cerebral eye disappeared. This disparition is the consequence of a lack of carotenoids in the diet over several generations (Birky, 1964).

Plate (1925) (cited in Remane, 1929-1933) was the only worker to describe the structure of the cerebral eye in rotifers correctly; it consists of two cells, one pigmented and the other sensory. However, in his synthesis Remane (1929-1933) refutes this interpretation. He emphasizes that "all investigations speak against such a structure and show that all parts of the eye are located in a single cell". This cell was considered to be similar to certain ganglionic cells of the brain. The rotifer brain is now known to possess, in addition to the neurons, epithelial cells at its periphery which could correspond to this type (Clément, 1977 a, b).

The same authors described a lens situated in front of the pigmented spot. This "strongly refractive spherical corpuscule" undoubtedly represents what we interpret at present as the photoreceptive parts of the eye on the basis of its ultrastructure.

2. The cerebral eye of Asplanchna brightwelli

This eye was the first rotifer photoreceptor to be studied with the aid of the electron microscope (Eakin & Westfall, 1965). Afterwards, Eakin (1968, 1972, 1982) always considered it as being representative of rotifers in general, although this structure has not been found in other rotifers observed by electron microscopy since (Clément, 1980; Cornillac, 1982 and unpublished observations to be presented later - Fig. 11).

More recently, animals deficient and non-deficient in carotenoids have been investigated (Cornillac, 1982; Wurdak, unpublished). When the diet contains adequate carotenoids (the animals are fed other rotifers which ingest unicellular green algae in their turn), the pigment cup is red. On a deficient diet (the animals are fed paramecia), it is nearly colorless or slightly brown.

Ware (1971) and Seldon (1972) have carried out a three-dimensional reconstruction of adult and embryonic brain cells, respectively, of Asplanchna brightwelli females from serial ultrathin sections. It is a pity that they repeated the mistake of the earlier investigators of the century and confused the eye with its pigmented cell.

The cerebral eye of Asplanchna brightwelli is composed of two cells. The pigmented cell (Fig. 1) contains the red, semi-spherical cup whose opening, measuring 5 μm across, is directed towards the front of the brain. It is an epithelial cell; its parts are evenly disposed around the axis of symmetry of the eye. It contains either a single nucleus segregated into four lobes (Ware, 1971) or two nuclei each of which is divided into two lobes (Fig. 1). The pigments are located within platelets which are arranged in concentric semi-circular layers. Microfilaments, measuring 7 nm in diameter, are regularly disposed midway between two successive layers. A preliminary comparison of pigmented and non-pigmented eyes shows certain differences in platelet organization. In red eyes there are nine to ten layers of platelets; the layers are separated from each other by a distance of 70-100 nm (Fig. 2, 3). In eyes lacking pigment the number of layers is inferior (there are 5-7) and they are more distinct (125-150 nm) from each other (Fig. 1, 4).

The presence of a large number of vesicles at the periphery of the layers of platelets in deficient animals indicates a disturbance in platelet synthesis which might contribute to the observed reduction in the number of pigment layers.

The configuration of the pigmented platelets in Asplanchna calls to mind the multilayered mirrors described by Land (1965, 1966, 1979) in Pecten and by Fournier and Combes (1978) in a platyhelminth. Multilayered mirrors occur in other zoological groups as well (cf. in this volume). If the refractive index of the cytoplasm is taken to be 1,3-1,5, maximal reflection is obtained at an incident illumination whose wavelength is between 300 and 600 nm for non-carotenoid-deficient and between 650 and 900 nm for carotenoid-deficient animals.

The morphology of the pigmented platelets of Asplanchna is

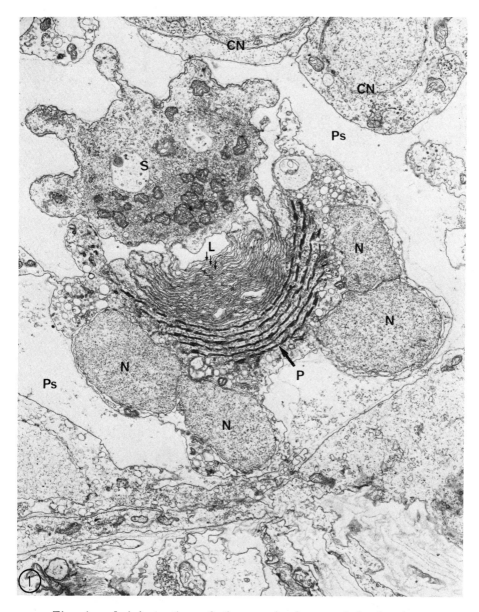

Fig. 1. Axial section of the cerebral eye of Asplanchna brightwelli (female deficient in carotenoids). (x 9 000).

N: nucleus of the pigmented cell; P: accessory pigment platelets, arranged in six concentric layers; S: sensory neuron; L: stacked photoreceptive lamellae originating from the sensory neuron; CN: cerebral neuron; Ps: pseudocoel (i.e. general body cavity in rotifers).

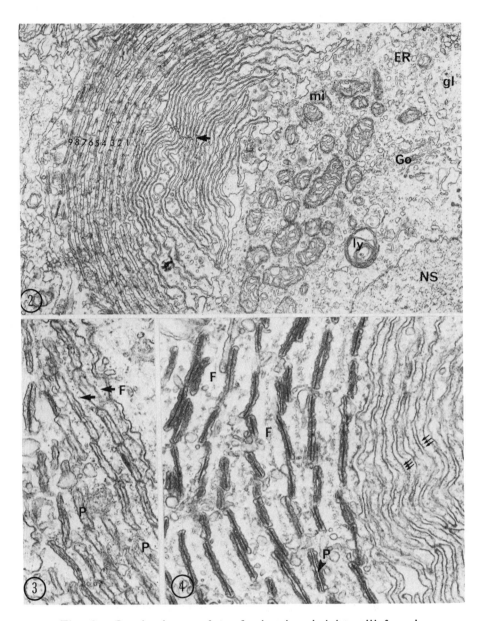

Fig. 2. Cerebral eye of an <u>Asplanchna brightwelli</u> female, not deficient in carotenoids. The accessory pigments are disposed in nine to ten concentric layers (1 to 9). The stacked photoreceptive lamellae come from the two extrem-

somewhat different from that of the analogous components of the multilayered mirrors cited above. The platelet is surrounded by a membrane and it contains a membrane-enclosed, flattened, bag-like structure in the middle whose contents are often virtual (Fig. 3, 4). The two walls of the sack are closely apposed, if not actually fused, together. The sack is surrounded by a matrix which is electron-dense in deficient (Fig. 4) and electron-lucent in non-deficient (Fig. 3) animals.

The sensory neuron projects beyond the space delimited by the pigment cup (Fig. 1). It bears lamellar expansions which are piled up like onion scales in this extracellular space. Lamellae coming from opposite sides of the neuron abut each other near the axis of the symmetry of the eye (Fig. 2). The cytoplasm of the sensory neuron contains many mitochondria, glycogen-like granules and dense-cored vesicles which probably enclose neurotransmitters. The sensory cell is joined to the pigmented cell by desmosomes (Wurdak, unpublished). It sends two axons (Ware, 1971) to the neuropile of the brain (Fig. 11). No ciliary structure is observable in the sensory neuron in the course of its development (Seldon, 1972).

3. The cerebral eye of Trichocerca rattus

Described by Clément (1975, 1977 a, 1980), the eye of T. rattus differs from that of A. brightwelli in the following respects (Fig. 5).

- It caps the retrocerebral gland, while remaining at the periphery of the brain at the same time.

- The pigmented cell is not symmetrical; it contains a single bilobed nucleus.

- The accesory pigment cup has the same size as that of A. brightwelli (the diameter of the opening is about 6 μm), but it is formed of a single layer of closely juxtaposed pigment granules which enlarge and coalesce as the animal ages.

---

ities of the sensory neuron and meet head on in the axial plane of the eye (arrow). (x 18 000).

gl: glycogen like granules; Go: dictyosome of the Golgi apparatus; ly: secondary lysosome; mi: mitochondrion; NS: nucleus of the sensory cell; ER: endoplasmic reticulum.

Fig. 3. Accessory pigments in a non carotenoid - deficient A. brightwelli female. (x 45 000).

Fig. 4. Accessory pigments in a carotenoid-deficient female. Microfilaments (F) present in the cytoplasm between the pigment platelets (P). The platelet matrix is electron-opaque in carotenoid-deficient animals. The small arrows in Fig. 4 indicate fibrillar material attached to the membrane of the stacked photoreceptive lamellae. (x 45 000).

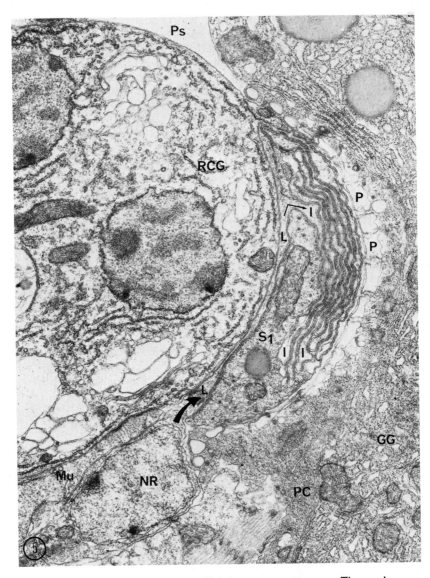

Fig. 5. Cerebral eye of <u>Trichocerca rattus</u>. The relay neuron whose nucleus (NR) is seen at the lower left, has a poorly developed perikaryon. It continues (thick arrow) as a dendritic blade (L). The latter, in turn, gives rise to dendritic lamellae (l) which are piled up in the cytoplasm of the sensory neuron (S_1). The accessory pigment granules (P) form a single layer in the cytoplasm of the pigmented cell (PC). (x 11 500).

- The sensory neuron is smaller. It is lodged entirely within the cavity formed by the accessory pigment cup. It contains glycogen-like granules, mitochondria, and neuro-transmitter vesicles. It has no axon. It is penetrated by 8-9 piled dendritic lamellae (Fig. 5). The lamellae interdigitate with narrow layers of sensory neuron cytoplasm.

- They originate from a dendritic blade which follows the contours of the apex of the retrocerebral gland, on one side, and of the concave part of the sensory neuron, on the other side. This blade comes from a small neuron which Clément did not describe in his original observations (1975). The axon of this neuron joins the cerebral neuropile. The perikaryon is poorly developed (Fig. 5, 11). Precisely aligned microtubules and filamentous structures underlying the plasma membrane are present within the dendritic lamellae (Clément, 1975).

4. The cerebral eye of Brachionus calyciflorus

The eye of B. calyciflorus is more complex and symmetrical in its organization. It consists of four cells (Figs. 6, 7, 8), (Cornillac, 1982).

1- A symmetrical pigment cell contains a pigmented mass hollowed out by two cavities which are separated from each other by a pigmented wall. The opening of each cavity is 5 to 7 µm wide and the cavity on the left is deeper than the one on the right. The spherical pigment granules measure 0,1 to 0,6 µm in diameter. They are present either dispersed in the cytoplasm or arranged in a single layer lining the two cavities under the plasma membrane

2 and 3- A sensory neuron that has no axon lies within each of the two cavities. Its cytoplasm includes mitochondria, glycogen-like granules, dense-cored vesicles and characteristic formations of the endoplasmic reticulum which are likely to be the photoreceptive organelles (Fig. 6, 7). The latter appear as cylindrical bodies consisting of closely apposed tubules of the endoplasmic reticulum surrounded by an external membrane which is also a part of the endoplasmic reticulum (Figs. 7, 8). In addition, there are cisternae of the smooth endoplasmic reticulum located directly underneath the plasma membrane. The cytoplasm at these sites is reduced to a very narrow (13 nm) layer. This arrangement probably serves to couple the plasma membrane and the reticulum as it has been proposed for neurons in other organisms (cf. Henkart et al., 1976).

4- Lastly, an unpaired medial neuron sends a dendritic lamella into each of the sensory neurons. This lamella follows the contours of the bottom of each of the cavities of the pigment cup from which it is separated by a narrow (20-30 nm) layer of sensory neuron cytoplasm. The elongated nucleus is perpendicular to the axis of the eye. It occupies a large part of the cell body. The cytoplasm is less rich in inclusions than that of the sensory neurons. The medial neuron gives off two symmetrical

---

GG: gastric gland; Ps: pseudocoel; RCG: retrocerebral (mucous) gland with several nuclei; Mu: muscle around the retrocerebral gland.

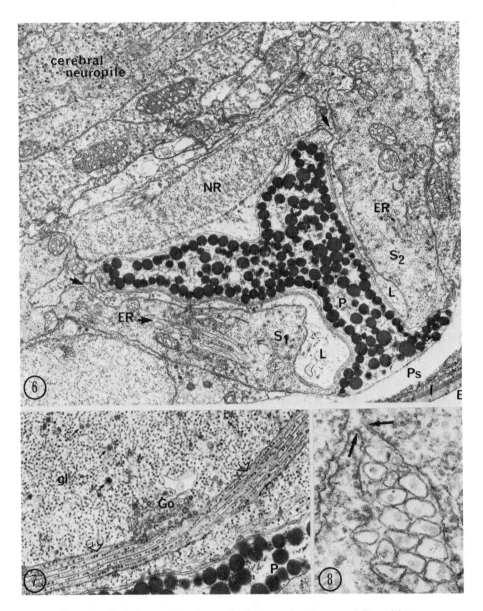

Fig. 6. Axial section through the cerebral eye of <u>Brachionus calyciflorus</u>. The accessory pigment granules form a mass hollowed out by two cavities each of which contains a sensory neuron (S_1 and S_2). The relay neuron has an elongated nucleus (NR) lying perpendicular to the axis of the eye; the cytoplasm of this neuron continues (arrows) as two

5. The cerebral eye of Brachionus plicatilis (Figs. 9, 10)

The cerebral eye of B. plicatilis closely resembles that of B. calyciflorus in its structure (Amsellem, Luciani, Clément unpublished). The accessory pigment cell bears two cavities which are occupied by two sensory neurons having a dense cytoplasm. Glycogen-like granules are particularly abundant, but the cylindrical formations of smooth endoplasmic reticulum are absent. There is no isolated, medial neuron serving as a relay between the sensory neurons and the brain. A neuron of this type is completely enclosed by the cytoplasm of the sensory neuron. It bears a dendritic blade whose shape and location are identical to those of B. calyciflorus. This neuron has been seen on only one side of the animal (Fig. 9). We have not been able to confirm the presence of a second relay neuron, on the other side, due to gaps in the series of sections; however, it appears highly probable that such a neuron exists since no continuities have been observed between the first relay neuron and the dendritic blade which is symmetrical to it. The last two structures are always separated by an extracellular space of about 20nm. Therefore, we conclude that there are two symmetrical relay neurons, each of which gives off an axon to the cerebral neuropile.

A particular feature of this eye is the stack of membranes lying along its axis between the medial pigment wall and the brain. These membranes represent extensions of the sensory neurons. They are the cytoplasmic layers which separate the dendritic blade of the relay neuron from the pigment cup. The cytoplasmic layers coming from the two sensory neurons are rolled up into a single structure giving rise to four

---

dendritic blades (L) located in the cytoplasm of each of the two sensory neurons. The latter contain unusual configurations of the endoplasmic reticulum (ER) which are better seen at high magnification in Figs. 7 and 8. (x 14 500).

Ps: pseudocoel; I: integument; E: external medium.

Fig. 7. Axial section of a cylindrical specialization of the endoplasmic reticulum (large arrows) in a sensory neuron of the eye of B. calyciflorus. (x 27 000).

P: accessory pigments; Go: dictyosome; gl: glycogen-like granules; small arrows: dense-cored vesicles (neurotransmitters). (x 27 000).

Fig. 8. Transverse section of the same type of cylindrical specialization of the endoplasmic reticulum in the sensory neuron of the eye of B. calyciflorus. The arrows indicate the continuity between the nuclear membrane and the membrane which encircles the juxtaposed cylinders of endoplasmic reticulum. (x 113 400).

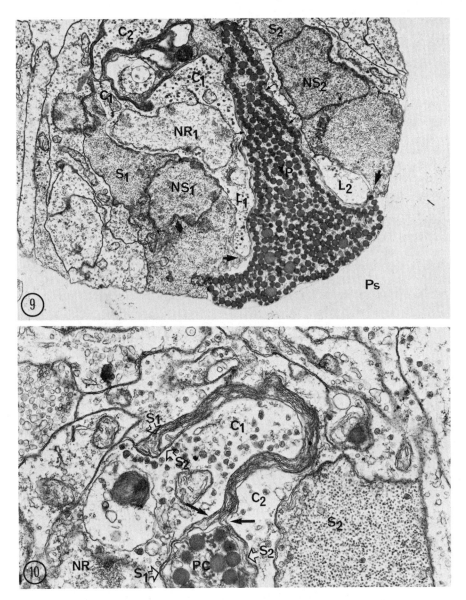

Fig. 9. Cerebral eye of <u>Brachionus plicatilis</u>. (x 9 000).

Fig. 10. Cerebral eye of <u>B. plicatilis</u>. The accessory pigments (P) form two symmetrical cavities each of which holds a sensory neuron (their nuclei are indicated by NS_1 and NS_2; their cytoplasm by S_1 and S_2). A relay neuron is lodged in each sensory neuron. NR_1: nucleus of relay neuron. C_1 and C_2 cytoplasm of the two relay neurons which form in each case, a cytoplasmic blade L_1 and L_2.

plasma membranes lying side by side. However, the actual number of membranes counted is eight. This is due to the fact that each cytoplasmic extension contains a flattened cisterna of smooth endoplasmic reticulum (Fig. 10).

6. Comparison of the cerebral eyes of monogonont rotifers (Fig. 11)

The purpose of making this comparison is to point out possible homologies among the structures described above. A certain unity of organization underlies these four types of eyes which appear very dissimilar at first.

The most primitive arrangement is paired. In B. plicatilis both sensory and relay neurons are paired. In B. calyciflorus only the sensory neurons are paired; the relay neuron is single. It is located in an axial position and it has a symmetrical structure (two axons).

The cerebral eye of T. rattus is diagrammatically one half that of B. calyciflorus. Pigment cup and pigmented cell are non-symmetrical. There is only one sensory neuron and one relay neuron which are likewise non-symmetrical. The number of dendritic lamellae is increased to 8-9 instead of there being just one per cavity as in Brachionus. The lamellae are lodged within the cytoplasm of the sensory neuron.

The cerebral eye of A. brightwelli represents another direction from the brachionid prototype. The sensory neurons are no longer present; only the medial neuron remains and it takes on a sensory function. Both the symmetry of its dendritic lamellae and the symmetry of the pigmented cell bear out the observation that the Asplanchna sensory neuron is situated in the axis of the eye (Figs. 1, 11). The pigment platelet assembly corresponds to an enlargement of the axial pigmented wall seen in Brachionus. In Trichocerca only the pigments lining one cavity have been conserved.

With respect to the structures assumed to be photoreceptive, the endoplasmic reticulum takes a preponderant role in the most primitive eyes: reticular formations arranged as cylindrical bodies in B. calyciflorus, flattened cisternae underlying the membrane of the rolled cytoplasmic extensions in B. plicatilis. Stacked cytoplasmic membranes also intervene in photoreception in T. rattus. In this case the membranes of the sensory neuron are separated by dendritic lamellae from the relay

---

The cytoplasm of the sensory neuron (S_1 and S_2) continues (Fig. 9: large arrows) between these blades (L_1 and L_2) and the pigmented cell (Fig. 9: fine arrows). The cytoplasmic extensions of the two sensory cells S_1 and S_2 meet on the axis of symmetry of the eye over the pigmented cell (PC) (Fig. 10). At the level of this juxtaposition (black arrows: Fig. 10) each of these extensions contains a cisterna of smooth endoplasmic reticulum giving rise to a total of 8 membranes lying side by side (Fig. 10). (x 23 000).

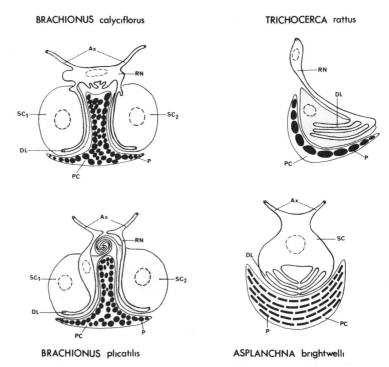

Fig. 11. Diagram of the cerebral eye of monogonont rotifers currently under investigation with the aid of the electron microscope. - An epithelial cell contains the pigment-cup (Pc) composed of pigment granules or platelets (P). - The sensory neurons lacking axons are stippled; there are two (SC_1 and SC_2) in Brachionus, one in Trichocerca and none in Asplanchna. - The relay neuron (RN) whose axon (Ax) or axons lead to the neuropile of the brain is shown in white. It bears dendritic lamellae (DL) lodged in the sensory neuron in Brachionus and Trichocerca or piled up in the extracellular space in Asplanchna.

neuron (Cornillac, 1982). The cerebral eye of A. brightwelli is simpler and more similar to classical rhabdomeric photoreceptors found in other invertebrates. The stacked dendritic lamellae are presumed to be the photoreceptive structures.

7. The cerebral eyes of a bdelloid rotifer, Philodina roseola (Figs. 12, 13, 14)

The bdelloids constitute a very homogeneous group. They differ from other rotifers in several ultrastructural and behavioral aspects (Clément, 1980; Amsellem & Ricci, 1982). They live in mosses, soils and in bodies of fresh water. Unlike other rotifers, they can move over surfaces by striding and they can dry up along with the mosses and lichens which they inhabit and remain for years in a state of anhydrobiosis. They reproduce exclusively by parthenogenesis, no bdelloid male has ever been observed.

P. roseola has two flat, red pigment spots on the sides of its brain. The ultrastructure of these eye spots shows that the accessory pigment granules are less precisely disposed than in the cerebral eyes of monogonont rotifers (Figs. 13, 14). An extension of a sensory neuron projects two short cilia into the space between the pigmented epithelial cell and the rest of the brain. The cilia are ampulla-shaped and an electron-dense matrix is present around the axoneme on the side nearest the brain. The axoneme is of the classical 9 x 2 + 2 type, but these are not kinocilia. These cilia are very similar to the parabasal apparatus of certain unicellular green algae such as Euglena (cf. Osafune & Schiff, 1980 for a recent ultrastructural description; Foster & Smyth, 1980; and Couillard, 1983, for a review) and to the short cilium described by Fauré-Fremiet in a phytoflagellate (Fauré-Fremiet & Rouiller, 1957; Fauré-Fremiet, 1961. Fig. 13). In the flagellates the parabasal apparatus is juxtaposed to the stigma which is composed of carotenoid pigments. This indicates clearly that the matrix of this structure has a photoreceptive function. The same argument may be applied to Philodina. As in the flagellates, the electron-dense substance around the axoneme is thought to be the morphological basis of photoreception.

II - ULTRASTRUCTURE OF THE ANTERIOR OCELLI

1. Light microscope observations

Paired anterior ocelli are present in many rotifers. As in the case of the cerebral eyes, there is a red pigment spot associated with each ocellus. Remane in his synthesis of 1929-1933 describes "apical eyes" situated at the center of the anterior ciliated crowns and "lateral eyes" likewise found on the apical part of the animals but closer to the external ciliated crown called the cingulum. In species of small size it is difficult to distinguish between these two categories. Anterior eyes are sometimes present in rotifers which do not have a cerebral eye: in the Flosculariacea and the Collothecaecea (the genus Filinia, for example), in Rhinoglena (a genus related to Brachionus). However, they can also coexist along with the cerebral eye: in certain species of Asplanchna and Trichocerca, for example. Finally, several rotifer genera were considered to have no eyes at all, since they bear no pigment spot. Hence the photoreceptors could not be detected with the light microscope. This group includes certain bdelloids and monogononts, certain adult sessile females of the order Gnesiotroque and the Seisonidae.

The anterior eyes are nearly always paired, exceptions to this rule are extremely rare. Remane cites four or five species, out of 2500, in which the anterior eye is unpaired or in which supernumerary anterior eyes are present. All the early authors thought that these eyes were derived from an anterior epithelial cell of the rotatory apparatus by intracellular differentiation (Remane 1929-1933). So widespread is this misconception that Salvini-Plawen and Mayr (1977) considered them as a

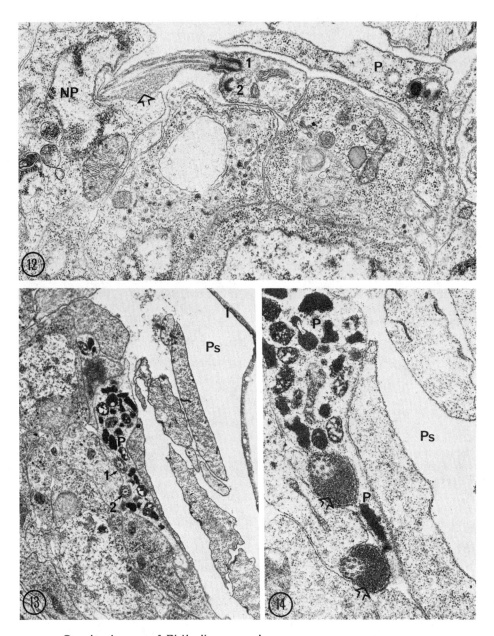

Cerebral eyes of Philodina roseola:

Fig. 12. (x 21 900).

Fig. 13. (x 10 900).

Fig. 14. At the periphery of the brain, a sensory nerve fiber

PHOTORECEPTORS IN ROTIFERS

separate line of epithelial photoreceptors. The true nature of these eyes was revealed by electron microscopy.

2. The anterior ocelli of Rhinoglena frontalis (Figs. 15, 16)

The ocelli of Rhinoglena frontalis were classed as apical eyes by Stossberg (1932) and Remane (1929-1933). They are part of a paired anterior sensory complex which is situated lateral to the apical receptors, next to the outlets of the retrocerebral apparatus in front of the rostrum of the animals (Clément, 1980; Clément et al., 1983). The pigment cup consists of a few granules lying side by side in a single layer. It is situated within the cytoplasm of an epithelial cell of the rotatory apparatus. The opening measures 2,5 - 3 µm in diameter and it is directed toward the front of the animal. The dendrite bearing the photoreceptive structure penetrates the same epithelial cell along with other sensory endings. Together they form a receptor complex of which the ocellus is just one element. The sensory neurons are located in the brain. The photoreceptive structure is rhabdomeric. Each tubular rabdomere is like a leaf inserted on a branch which represents the dendritic ending (Fig. 16). The rhabdom is housed in a closed spherical cavity of the epithelial cell. Septate junctions anchor the nerve ending to this epithelial cell on the side opposite the pigment cup.

3. The anterior ocelli of Asplanchna brightwelli (Figs. 17, 18, 19, 20)

No pigmented structure is observable in the rotatory apparatus of A. brightwelli with the light microscope.

Under the electron microscope it is seen that the apical portion of the lateral sensory receptors ("horns") bears ampulla-shaped cilia (Wurdak, unpublished). These cilia resemble the ampulla-shaped cilia of the cerebral eye of Philodina roseola (Fig. 14). Therefore, we consider that they are photoreceptive by reason of analogy. The lateral sensory organ is a complex structure: a supporting epithelial cell shelters several sensory nerve endings which terminate either in short, more or less modified, cilia or in microvilli. The most important of these sensory dendrites, running along the axis of the horn, bears at its apex about thirty ampulla-shaped cilia. The cilia are separated from the external medium by the thin, permeable cuticle which is secreted by the anterior ciliated cells of the rotary apparatus. Under ordinary circumstances only the cuticle is visible under the scanning electron microscope (Fig. 18). However, in rare

---

bears 2 short cilia (1 and 2) which are not connected to ciliary rootlets. Near their base these cilia have a classical structure, Figs. 12 and 13; distally they enlarge to form ampullae which contain an electron dense material around the axoneme (arrows). The pigmented cell separates the cilia from the pseudocoel. It has a nucleus and accessory pigment granules (P) of irregular shape. I = integument. (x 24 700).

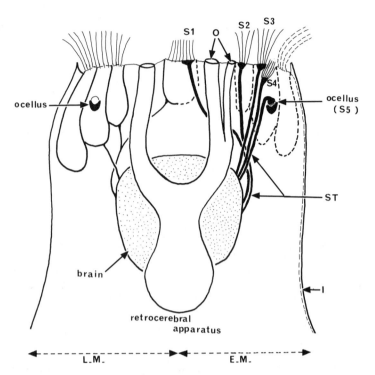

Fig. 15. Rhinoglema frontalis. Diagrammed at left are the observations possible under the light microscope (L.M.) in vivo (after Stossberg, 1932, Remane, 1933). At right the observations carried out with the aid of the electron microscope (E.M.) (Clément et al., 1983). S_1 - S_4 represent the tips of the sensory neurons at the anterior end of the animal. The ocellus (S_5) is also a sensory nerve ending, only the accessory pigment cup is epithelial

E: anterior epithelial cell; I: integument; S.T.: sensory nerve fibers (the corresponding nerve cell bodies are located in the brain). The retrocerebral apparatus opens to the exterior via two pairs of canal (o).

instances it is removed during sample preparation and the underlying structures are disclosed (Fig. 19). The cilia are very short; they measure 1,5 μm in length. They are dilated to form an ampulla having a diameter of 1,0 μm, which contains an electron-dense matrix. The axoneme does not contain central tubules and the peripheral doublets diverge and

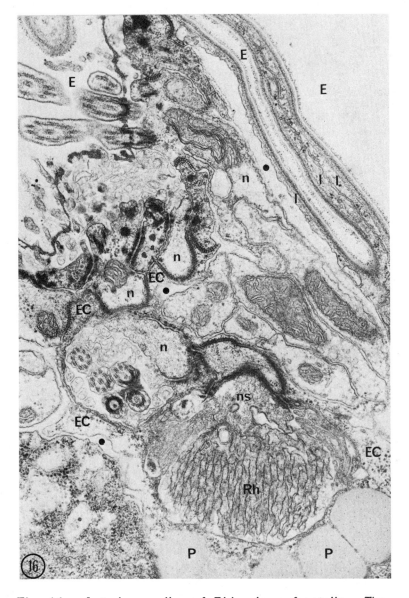

Fig. 16. Anterior ocellus of Rhinoglema frontalis. The accessory pigment granules (P) are located within an anterior epithelial cell (EC) which also encloses nerve fibers (n and ns). These sensory nerve endings extend to the external medium (E) or terminate, as in the case of the rhabdomeres (Rh) of the ocellus originating from the sensory nerve fiber (ns), in a closed cavity formed by the epithelial cell (EC). (x 25 000).   I: integument;   ● : pseudocoel.

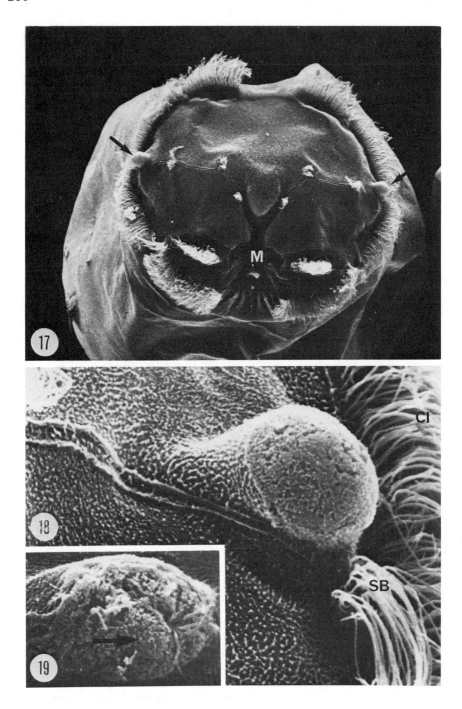

Fig. 17. Anterior view of an Asplanchna brightwelli female. The mouth (M) is ventral. The head of the animal is encircled by a crown a locomotor cilia, the cingulum, and bears many sensory receptors. The arrows point to one of them, the lateral sensory complex. (x 360).

Fig. 18 & Fig. 19. The lateral sensory complex is in the shape of a horn juxtaposed to the cilia of the cingulum (CI) and a sensory bristle (SB). In Fig. 19 the thin cuticular covering over the horn is missing. The underlying sensory cilia are therefore visible, in particular the thirty odd ampulla-shaped cilia (arrow) representing the photoreceptive part of the sensory complex. (Fig. 18, x 4 500; Fig. 19, x 4 500).

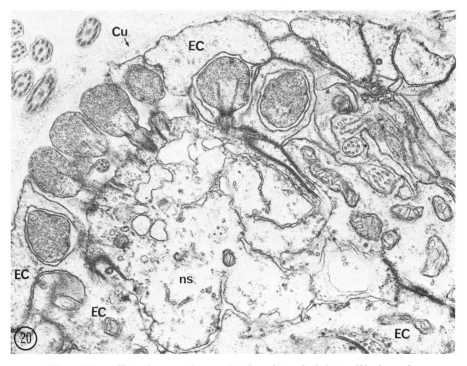

Fig. 20. The horn of an Asplanchna brightwelli female sectioned obliquely. A nerve ending (ns) bears ampulla-shaped cilia containing electron dense material. The cilia are separated from the external medium by a thin cuticle (Cu). The nerve ending and the ampulla-shaped cilia are surrounded by a supporting epithelial cell (EC). (x 23 000).

penetrate into the matrix in going from the base toward the tip of the cilium (Fig. 20).

4. The anterior ocelli of Trichocerca rattus (Figs. 21, 22)

There are two sensory complexes located on either side of the outlets of the retrocerebral apparatus in T. rattus. They are not visible under the light microscope and they are not associated with a pigment spot. Their fine structure is similar to that of the lateral sensory receptors of A. brightwelli even though the whole complex does not have the shape of a horn. An epithelial cell supports several sensory nerve endings which bear short cilia or microvilli (Clément, 1977 a, b). The biggest dendrite in each sensory organ terminates in ampulla-shaped cilia which resemble those present in the anterior ocelli of A. brightwelli (Fig. 20) in all respects. However, there are fewer ampulla-shaped cilia in T. rattus. By analogy with the cilia described in the ocellus of phytoflagellates and in the cerebral eye of Philodina roseola, these cilia are considered to be photoreceptive (Clément, 1980).

5. The phaosome of Philodina roseola (Fig. 23)

The bdelloid rotifer P. roseola has, at the tip of its rostrum, a number of sensory receptors including a complex organ lying directly beside the outlets of the retrocerebral apparatus (Clément et al., 1983). At the base of this organ, there is a nerve ending linked to an epithelial supporting-cell by desmosomes. The nerve ending is hollow. It encloses a cavity having the form of an elongated sphere. Its diameter is between 2 and 3 um. This cavity has no opening to the exterior. It contains piled, compressed membraneous sacks. Each sack is, in effect, a cilium flattened out in the shape of a fan. The peripheral tubules become the ribs of the fan. The base of the cilium has a normal diameter, but the central tubules are lacking. It is of the $9 \times 2 + 0$ type (Fig. 23). Occasionally, the ciliary membrane gives off lateral extensions beyond the fan-like structure. These membraneous sheets have no microtubular framework. A piling up of membranes, of ciliary or non-ciliary origin, within the closed cavity of a sensory nerve ending represents a typical phaosome. We consider the phaosome of P. roseola to be a photoreceptor by reason of analogy with the phaosomes present in other zoological groups (Platyhelminthes: Fournier, 1983; Annelids: Verger-Bocquet, 1983; Pogonophores: Nørrevang, 1974).

Fig. 21. Axial section of the anterior ocellus of Trichocerca rattus. (x 38 000).

Fig. 22. Transverse section of the anterior ocellus of T. rattus. An epithelial cell (EC) harbors several sensory nerve endings (n and ns). One of these endings (ns) terminates in ampulla-shaped cilia filled with an electron dense matrix. The axoneme of the cilia does not have any central tubules.

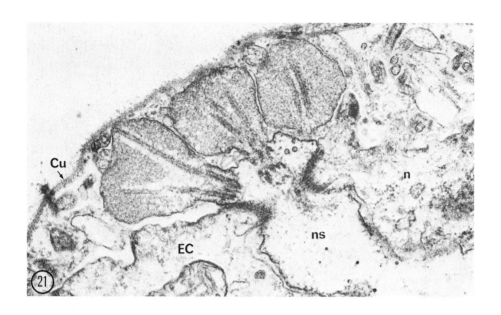

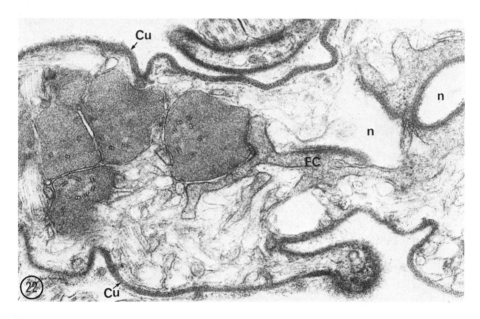

The peripheral doublets are farther apart from each other in the apical part of the cilium. (x 38 000).

Cu: thin cuticle separating the sensory structures from the external medium

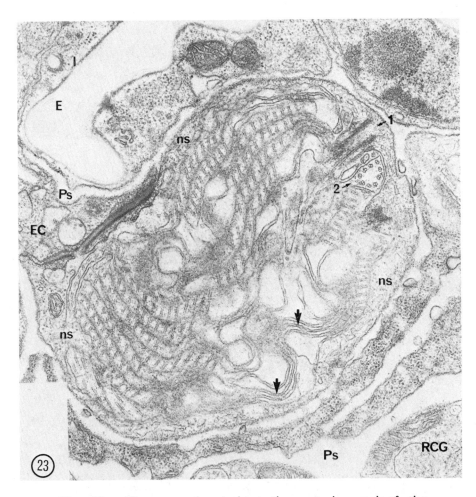

Fig. 23. Phaosome located at the anterior end of the bdelloid rotifer <u>Philodina roseola</u>. The sensory nerve ending (ns) linked to an anterior epithelial cell (EC) by desmosomes, contains a closed spherical cavity in which flattened cilia are piled up. The base of these cilia, seen in longitudinal (1) and transverse (2) section, does not have any central tubules. The flattened part of these cilia contains singlet tubules lying side by side which are continuous with those of the axoneme. The ciliary membrane extends further to form lamellae devoid of tubules (arrows). The phaosome is located in the pseudocoel (Ps) near the outlets of the retrocerebral glands (RCG). (x 33 000).

I: integument; E: external medium.

III - PHYLOGENESIS OF PHOTORECEPTORS

Several conflicting theories dealing with the position of photoreceptors in evolution are being considered currently. Eakin (1963, 1968, 1982) proposed two major lines in photoreceptor evolution: a ciliary line found in present-day radiates and deuterostomes and a rhabdomeric line represented by flatworms, aschelminths and protostomes. Vanfleteren and Coomans (1975) and Vanfleteren (1982) maintain, in opposition to Eakin, that photoreceptor differentiation is always dependent on ciliary induction at least during embryogenesis. Salvini-Plawen and Mayr (1977) have proposed a third theory, often termed "aphyletic", which is contradictory to the two preceeding ones. It is based on the description of 40 to 65 types of photoreceptive structures which were selected independently during evolution. Salvini-Plawen (1982) indicates that this "polyphyletic" elaboration of photoreceptors results in four different structural types:
   a) unmodified cells with cilia and microvilli
   b) cells with modified cilia
   c) microvillar cells with vestigial cilia (rhabdomeric type)
   d) cells that have become aciliate before elaboration (ganglion diverticular types)

Clément (1980) expressed his objections to the hypotheses of Eakin, Vanfleteren and Coomans and Salvini-Plawen and Mayr. He advanced a polyphyletic theory based on his own observations of rotifer photoreceptor morphology. The existence of both ciliary and rhabdomeric photoreceptors in rotifers (occasionally within the same animal), as well as in many other invertebrates, contradicts the diphyletic theory of Eakin. The absence of ciliary induction in the cerebral eyes of monogonont rotifers is confirmed by an ultrastructural study of the embryos of Asplanchna (Seldon, 1972). This observation refutes the hypothesis of Vanfleteren; it is more in agreement with the fourth type of photoreceptive structure proposed by Salvini-Plawen (1982). However, the distinction made by Salvini-Plawen between photoreceptors of epidermal origin (structures b and c listed above) and those of nervous origin, lacking cilia, (d-ganglion diverticular type) is not valid for rotifers since certain cerebral receptors are ciliated in this group (Philodina). Moreover, the perikarya of the anterior, "epidermal" photoreceptors are localized in the brain. Finally, the polyphyletism which we advocate is opposed to the view that photoreceptors are always newly-acquired structures without phylogenetic significance. Rotifer photoreceptors may be assigned to three phyletic lines:
1) Ampulla-shaped cilia having an electron-dense matrix. Such cilia have been described in a phytoflagellate (Fig. 24, Fauré-Fremiet & Rouiller, 1957; Fauré-Fremiet, 1961). They are similar to the parabasal apparatus of the locomotory flagella of other phytoflagellates. The structural similarities among these modified cilia and flagella permit us to suppose that they are related and they constitute a phyletic line.
2) The phaosome of Philodina is a structure consisting of a pile of membranes of ciliary origin within the closed cavity of a nerve

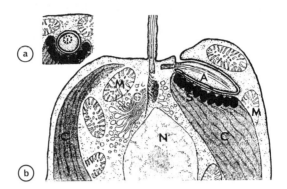

Fig. 24. Anterior portion of the phytoflagellate Chromulina psammobia (after Fauré-Fremiet & Rouiller, 1957). a: transverse section of the ampulla-shaped cilium. It contains an electron-dense matrix around the axoneme. It is lodged in an invagination of the plasma membrane. b: axial section. The ampulla-shaped cilium (A) is apposed to the accessory pigment granules which make up the stigma(s).

C: chloroplast; M: mitochondrion; N: nucleus.

ending. Such structures also exist in the platyhelminthes, the pogonophores and in the annelids. There is a gradual regression of the cilia. In the most evolved forms they may be represented only by ciliary rootlets. Once again we have a phyletic line which may be expressed in several zoological groups through the retention of common genetic determinants.

3) Pigment-cup eyes of the monogonont rotifers: The four cerebral eyes which have been described to date are very different from each other, particularly with respect to the structures which are assumed to be photoreceptive. In the most primitive forms (Brachionus) the endoplasmic reticulum seems to play an important role in photoreception. This role is taken over by the plasma membrane in the more evolved forms (Trichocerca, Asplanchna). Figure 11 illustrates possible homologies among these four types of cerebral eyes. They may all be considered to belong to a phyletic line characteristic of rotifers.

In summary, rotifers constitute a zoological group which is as primitive as the platyhelminthes (Clément, 1980). There is great structural variation in their photoreceptors; no single solution has been adopted as in the arthropods or the vertebrates. Rotifers represent an evolutionary blind alley. Their body organs, consisting of only a few cells, are highly specialized and eutelic. However, they attest to an initial diversity of possible solutions to the problem of photoreception in the animal kingdom.

IV - BEHAVIORAL RESPONSES TO LIGHT STIMULATION

1. Different types of behavioral responses to light

Rotifers do not "see" in the most common sense of the word: even in a state of hunger, they can approach a food item during swimming without deviating from their path. They have to contact the food directly before feeding behavior can be initiated (Gilbert, 1976 a, b, 1977, 1980; Clément, 1977 a, b, c; Clément et al., 1983). The same holds for their reproductive behavior: a male has to enter into contact with a female before he will make an attempt at fecundation. Furthermore, no "off" response has been observed in rotifers: the animals do not retract following a sudden change in illumination.

The only light-dependent behavioral responses which have been described in rotifers to date, are those concerned with the characteristics of their motion. Since the 19-th century, the occurrence of positive phototaxis has been noted in several planktonic rotifers, in particular by Jennings (1901). These rotifers are attracted by a light source. It is this response which has received the most attention until now (Viaud, 1940, 1943 a; Menzel & Roth, 1972; Clément, 1977 a, b; Hertel, 1979; Cornillac, 1982; Cornillac et al., 1983). Certain species or certain clones present a negative phototaxis under the same conditions of illumination (Viaud, 1940; de Beauchamp, 1965; Clément, 1977 a). Other species are indifferent to light. Clément (1977 a, c) has shown that one of these species, Notommata copeus, usually considered to be non-phototactic, is attracted by a light source when the females are subjected to fasting and are not creeping along an algal filament which they normally feed upon.

Clément (1977 a, c) has also shown that the percentage of females that are swimming or creeping varies with light quality in the same species under the same nutritional conditions. In a study using monochromatic, isoquantic light he demonstrated that the number of swimming females increases in going from blue to red. Variations in swimming speed have been measured during positive phototaxis by Viaud (1940, 1943), Clément (1977 a, c), Hertel (1979) and Cornillac (1982). The speed varies with light intensity and with wavelength (cf. below). Finally, Preissler (1977, 1980) has shown that two species of planktonic rotifers avoid the zone of shadow on the lakeshore. This field observation was tested experimentally in the laboratory. No hypothesis was given concerning the mechanism underlying this reaction

2. Variations in locomotor responses as a function of light intensity

a) Phototaxis

Viaud (1940) has shown in B. calyciflorus that a beam of white light attracts the animals to an extent proportional to its intensity relative to a control beam placed at right angles to it. He did not measure the light intensities he used in his experiment. Menzel and Roth (1972), Hertel (1979) and Cornillac (1982) made similar observations using relative light intensities.

Cornillac (1982) measured the minimum light intensity capable of provoking positive phototaxis in B. calyciflorus and in A. brightwelli. The light source utilized was a mercury lamp whose maximal emission is between 500 and 600 nm. The minimal threshold is between 2 and 4 x 10^{-10} W.m^{-2} for A. brightwelli. For B. calyciflorus it was found to be inferior to 0,04 x 10^{-10} W.m^{-2}. This result is surprising in view of the fact that A. brightwelli has a multilayered mirror-like eye-cup, whereas B. calyciflorus does not. However, the threshold values may be explained by supposing that in B. calyciflorus the single relay neuron sums the influx of the two sensory neurons before transmitting it to the brain (Fig. 11). In A. brightwelli there is only one sensory neuron in the cerebral eye (Fig. 11).

b) Photokinesis

Viaud (1940) indicates that linear speed increases with light intensity in B. calyciflorus females subjected to monochromatic illumination. Due to the fact that he used narrow wavelength-range filters he could obtain only very weak light intensities. Clément (1977 a, c) demonstrated that speed decreased when the intensity was changed from 160 to 420 pE in the same species. He used monochromatic light of 475, 500 and 543 nm. At these wavelengths 160 pE equals 0,4, 0,39, 0,37 W.m^{-2}; 420 pE equals 1,06, 1,00 and 0,92 W.m^{-2}, respectively. Therefore, it is possible that as light intensity is steadily increased there is, first, an increase in linear speed and then, a decrease.

A. brightwelli, on the other hand, responds to strong light intensities by accelerating (Cornillac, 1982). Its linear speed increases as the light intensity is varied from 21 to 200 W.m^{-2}. As shown by the minimal threshold values above, the eyes of B. calyciflorus and A. brightwelli differ in their sensitivity to weak illumination. A difference is also observed at high intensities.

3. Wavelength-dependent variations in locomotor response

a) Phototaxis

The experimental setup utilized for the study of phototaxis in rotifers varies from one worker to another. Cornillac (1982; Cornillac et al., 1983) uses monochromatic, isoquantic light. She counts the number of animals drawn toward the light after a period of time which is sufficiently long to allow even the slowest of them to reach the illuminated sector of her experimental vessel. The experimental vessel is a rectangular chamber. It is divided into three compartments by means of movable partitions. The animals are placed into the middle compartment at the beginning of the experiment. Menzel and Roth (1972) and Hertel (1979) employ a Y-shaped chamber and relative light intensitiies. For each wavelength, they establish a curve response as a function of light intensity. They assume that this relationship is linear and they obtain by extrapolation the intensity value needed to attract 50 per cent of the animals. Viaud (1940, 1943 a) made use of two isoenergetic light beams, one white and the other monochromatic, directed at right angles to each other. Under these conditions the rotifers orient themselves according to Loeb's rule: in a direction intermediate to the direction of the two beams. The effectiveness of the colored light is expressed by the angle of deviation (tg θ) defined as the difference between the angles formed by the axis of swimming of the animal and each of the two light beams.

The results (Figs. 25 and 26) vary according to the technique and the species employed. Viaud's technique has two drawbacks: it is not adapted for testing in the ultraviolet range, and it is not very sensitive. It works only in the most effective wavelengths. B. calyciflorus females show a less clear-cut orientation at wavelengths superior to 560 nm than they do in the 500 to 560 nm range, and consequently escaped Viaud's detection. Nonetheless, wavelengths tested in the 580 and 660 nm range provoke

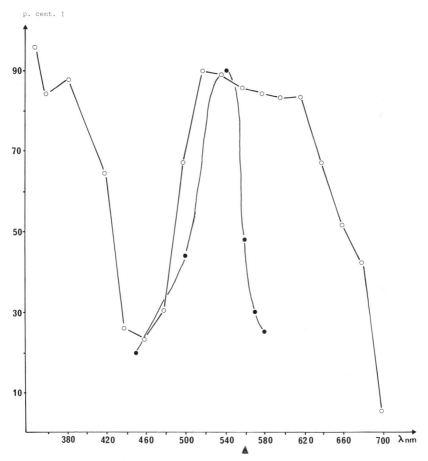

Fig. 25. Variations in positive phototaxis as a function of wavelength in Brachionus calyciflorus. The y-axis indicates, for our graph (o——o) the percentage of animals in compartment A of the experimental vessel. For the graph of Viaud (●——●) it gives the values for the tangent of θ, θ being the orientation angle of the swimming axis of the animal.

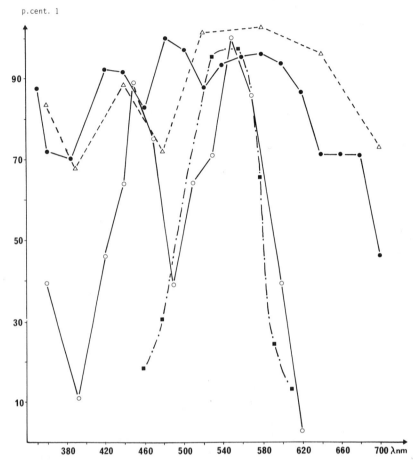

Fig. 26. Variations in positive phototaxis as a function of wavelength in Asplanchna brightwelli (●———●), Asplanchna priodonta (△- - -△ and ○———○), Asplanchna girodi (■—·—■). The y-axis indicates, for our graph (●———●), the percentage of animals in compartment A; in the illuminated branch for the graph of Menzel and Roth (△- - -△) and Hertel (○———○). For the graph of Viaud (■—·—■) it gives the values for the tangent of θ (θ is the angle of orientation of the swimming axis of the animals).

positive phototaxis under the experimental conditions used by Cornillac (1982). The same holds for Asplanchna in the 600 to 700 nm range. Hertel's response curves (Fig. 26) are similar to those of Viaud even though he used a different species of Asplanchna. The results of Cornillac (1982) are fairly close to those of Menzel and Roth (1972) (Fig. 26).

In summary, three types of wavelength-dependent variations were

observed in rotifers (Figs. 25, 26):
- The Asplanchna brightwelli type (Cornillac, 1982): All wavelengths in the 350 to 700 nm range are effective.
- The Brachionus calyciflorus type (Cornillac, 1982): Only wavelengths in the 350 to 420 nm range and the 520 and 660 nm range are effective
- The Filinia longiseta type (Menzel & Roth, 1972): The effective wavelengths lie in the 350 to 580 nm range. In Filinia the pigment-cup eyes are not cerebral; they are anterior ocelli.

b) The respective roles of the accessory and the sensory pigments in phototaxis.

The absorption spectrum of the accessory pigments of the cerebral eyes of B. calyciflorus (Cornillac et al., 1983) and A. brightwelli was determined with the technique of microspectrophotometry. In B. calyciflorus the absorption is maximal between 350 and 580 nm and remains worthy of note up till 640 nm (Fig. 27). Therefore, the accessory pigments are responsible for the drop in phototaxis at wavelengths beyond 660 nm and for the diminution in phototaxis between 580 and 640 nm. However, the absorption spectrum of the accessory pigment does not explain the lack of the response between 420 and 500 nm. The hiatus is probably due to the characteristics of the sensory pigment (s). They are sensitive only to wavelengths in the 350 to 420 nm range and to those in the 500 to 660 nm range and possibly to those beyond 660 nm. In A. brightwelli (Cornillac, 1982; Fig. 28) the absence of phototaxis at wavelengths superior to 700 nm is due, in part, to the accessory pigment(s). The sensory pigment is sensitive to all wavelengths in the 350 to 700 nm range and possibly even above. The response observed in the 350 to 700 nm range is mediated by the joint intervention of the sensory and accessory pigments.

It is possible to abolish the red color in the cerebral eye of A. brightwelli by culturing the animals on a diet deficient in carotenoids (cf. above). Under these conditions, the phototactic response disappears at all wavelengths except at 360 and 480 nm (Fig. 29). The peaks observed at 600 nm and at 680 nm do not differ significantly from the control values. The remaining reaction is due to the interaction of the non carotenoid accessory pigment with the sensory pigment.

c) Photokinesis

Variations in swimming speed were measured by Viaud (1940, 1943) in three species of clearly phototactic planktonic rotifers. He observed a slight decrease in linear speed in going from the blue to the red end of the spectrum (Fig. 30). However, his experimental setup was inadequate because the monochromatic light he used was isoenergetic rather than isoquantic, and the far red light he employed as background illumination may have influenced swimming speed. Clément (1977 a, c) repeated Viaud's experiments on B. calyciflorus using isoquantic monochromatic light without any background illumination (Fig. 30). He noted a steady increase in speed in going from the blue to the red end of the visible spectrum. Cornillac (1982) employed a highly sophisticated technique in

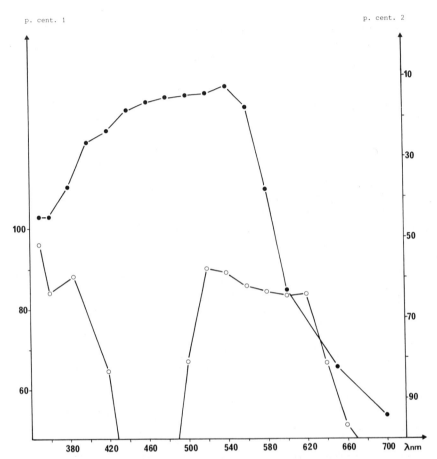

Fig. 27. Wavelength-dependent variations in positive phototaxis (o———o) and the transmission of the pigment-cup of the cerebral eye (•———•) of Brachionus calyciflorus. p. cent. 1: percentage of animals in compartment A. p. cent. 2: percent transmission of the pigment-cup of the cerebral eye.

which the rotifer's swimming path is automatically tracked and recorded (Coulon et al., 1983). A. brightwelli females deficient in carotenoids were used for these studies in order to minimize the interaction between photokinesis and phototaxis. The swimming speed over non-rectilinear paths was accurately measured and it was found to rise as the light was changed from blue to red using three different 100 nm-wavelength-range filters. There is no significant variation in r.c.d. (rate of change of direction) under the same experimental conditions. This result needs to be confirmed on a larger sample since the values obtained were highly dispersed.

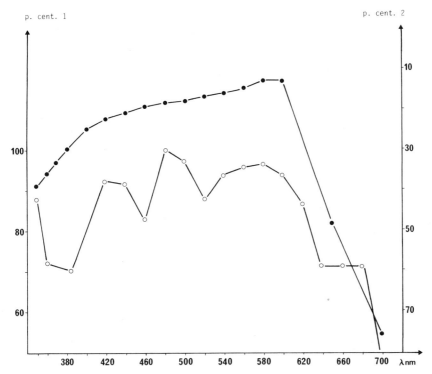

Fig. 28. Wavelength-dependent variations in positive phototaxis (o———o) and the transmission of the pigment-cup of the cerebral eye (●———●) of Asplanchna brightwelli. p. cent. 1: percentage of animals in compartment A. p. cent. 2: percent transmission of the pigment-cup of the cerebral eye.

4. The concepts of phototaxis, orthokinesis and klinokinesis

The terminology comes from the classical synthesis of Fraenkel and Gunn (1940, reedited in 1961) which is based on the work of Kühn (1919, 1929), Loeb (1918) and Viaud (1938). The same terms are currently used to describe the swimming motion of unicellular organisms (Diehn et al., 1977). In their critical analysis, Stasko and Sullivan (1971) contend that the definitions of Fraenkel and Gunn do not take into account the mechanisms underlying these responses. Moreover, they question the results of Ullyot (1936) on which Fraenkel and Gunn relied for their definition of the term klinokinesis. Later Gunn (1975) redefined this term in such a manner that it became applicable only to the results obtained in bacteria. For Gunn, the existence of klinokinesis in multicellular organisms has not been convincingly demonstrated up until now. In order that a stimulus dependent variation in r.c.d. may be qualified as klinokinesis it should not be accompanied by a variation in speed nor by oriented movements (Gunn, 1975).

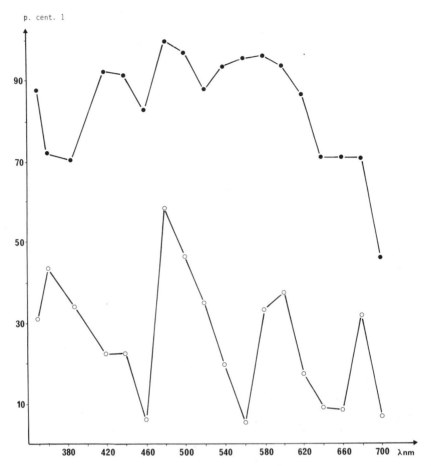

Fig. 29. Variations in positive phototaxis as a function of wavelength in Asplanchna brightwelli females having a pigmented (●——●) and a non-pigmented (○——○) eye-cup. p. cent. 1: percentage of animals in compartment A of the experimental vessel.

In monogonont rotifers swimming is accomplished through the action of the locomotor cilia of the cingulum. Their infraciliature is in direct relation with the muscles (Clément, 1977 a, b; Luciani, 1982; Cornillac, 1982). These muscles are richly innervated (Clément, 1977 a). The speed of swimming and the changes in direction may be regulated through muscular activity since the ciliated cells themselves are not innervated (Clément, 1977 a; Luciani, 1982; Cornillac, 1982). The observed correlation between linear speed and angular speed in Asplanchna and in Brachionus is understandable since only one effector system for swimming is present in these animals. The more a female turns, the more she slows

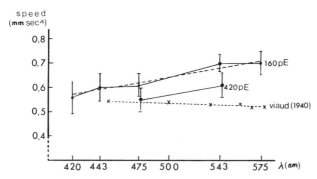

Fig. 30. Real speed of <u>Brachionus</u> <u>calyciflorus</u> females placed in different monochromatic radiations (isoquantic spectra of 160 and 420 pE). The unknown intensities used by Viaud probably have a lower energy. For the three curves, the real speed is calculated after taking into account the different parameters of the helicoidal swimming path of the females.

down (Luciani, 1982; Cornillac, 1982). However, other muscles, inserted on the cells of the rotatory apparatus or on the integument of the trunk may intervene to carry out movements of orientation and changes in direction.

Among all the sensory receptors present in rotifers, the eyes or ocelli possessing an accessory pigment-cup are the ones responsible for phototaxis. The problem is to determine whether it is the same or some other photoreceptors that mediate photokinesis (linear and angular speed). The results obtained regarding, on the one hand, the spectral characteristic of the sensory pigments involved in the phototaxis of <u>A</u>. <u>brightwelli</u> and <u>B</u>. <u>calyciflorus</u>, and the variations in linear speed, as a function of wavelength in the same species, on the other hand, (cf. above and Figs. 25 and 27) suggest that different pigments participate in the two responses. In <u>A</u>. <u>brightwelli</u> there is ultrastructural evidence for two types of photoreceptors: the cerebral eye and the two anterior ocelli. Each one of these receptors contains a single sensory neuron. It is, therefore, conceivable that the two sensory pigments related to phototaxis and photokinesis are located in two different receptors. Our results are too preliminary in nature to determine whether two different sensory pigments, located in separate receptors perhaps, underlie the observed changes in linear and angular speed (Fig. 31).

In conclusion we feel than an exact definition of the concepts utilized to characterize the variations in swimming speed and direction as a function of light, and the interactions among these concepts should also take account of the underlying mechanisms. Which muscles, other than those connected to the infraciliature of the cingulum are involved in phototaxis and in r.c.d.? Which nervous pathways control these muscles? Are these nervous pathways informed by two or by three different

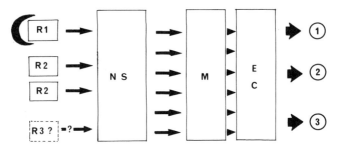

Fig. 31. From photoreceptors to locomotor responses in rotifers.

NS: nervous system; M: muscles; E: epidermis: anterior epithelial cells and syncytial integument; C: cilia of cingulum and ciliary rootlets.

The photoreceptor R_1 provided with an accessory pigment-cup is involved in response 1: orientation of the body of the animal or phototaxis. Responses 2 (speed of movement or orthokinesis) and 3 (angular speed or klinokinesis) are probably controlled by a photoreceptor other than R_1, either the ocelli R_2, or R_2 and R_3.

photoreceptors? Rotifers appear to be material of choice for the study of these problems.

V - OTHER RESPONSES OF ROTIFERS TO LIGHT

1. The influence of photoperiod on the appearance of males

Pourriot (1963) has shown that in three periphytic, phytophagous species, (Notommata copeus, Notommata codonella and Trichocerca rattus) males appeared only when days were long. In the region around Paris from which the data were collected parthenogenetic reproduction is accompanied by sexual reproduction in spring and summer. Mictic females, morphologically similar to the amictic females responsible for female parthenogenesis, lay haploid male eggs. The males are reduced in size and lack a functional digestive tract. Upon being fertilized by these males, the mictic females produce resting eggs provided with a protective shell which allows them to resist freezing and dessication. The resting eggs enter into diapause. They assure the survival of the species.

The influence of different parameters of the photoperiod on the appearance of mictic females has been studied in detail in Notommata copeus (Pourriot & Clément, 1973 a, b; Clément & Pourriot, 1976 a, b). The rate of mictic female production varies when monochromatic isoquantic light of different wavelengths is employed (Fig. 32). The pigment sensitive to photoperiod differs from the pigments involved in phototaxis and photokinesis: wavelengths in the 500 to 600 nm range are

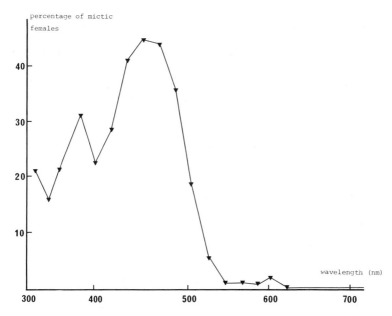

Fig. 32. Variation in the percentage of mictic female production in Notommata copeus over the 300 to 700 nm range.

nearly ineffective and long wavelengths (600 nm) are totally ineffective.

The minimal threshold of intensity, measured in both white and monochromatic light, (Clément & Pourriot, 1974) is comparable to the threshold of positive phototaxis in Asplanchna brightwelli. An intensity equivalent to that of the full moon cannot provoke the appearance of mictic females in N. copeus. It is possible, therefore, that the photoreceptor sensitive to the photoperiod consists of a single sensory neuron.

The influence of light intensity was also tested in white and monochromatic light (Pourriot & Clément, 1973 a, b; Clément & Pourriot, 1976 a, b; Clément, 1977 a). Under a 16:8 light-dark cycle, the rate of mictic female production rises very rapidly in the lower intensity range; it then stabilizes in the medium intensity range and drops off gradually at high intensities.

However, the effect of photoperiod on N. copeus is modulated by several exogenous and endogenous factors:
- grouping and population density (Clément & Pourriot, 1973 a, b. 1975; Clément, 1977 a).
- food quality (Clément, 1977 a; Pourriot et al., 1982),
- age of the parents (Clément & Pourriot, 1975, 1976 b),
- age of the grandparents (Clément & Pourriot, 1979),
- culture conditions for the grandparents and other predecessors (Clément & Pourriot, 1980),

- clone (Clément & Pourriot, 1976 a, 1979)

Photoperiod exercises a direct influence not over ovogenesis itself, but over a photoreceptor connected to the nervous system which participates in the control of ovogenesis (Clément & Pourriot, 1972). To understand the mechanism underlying this influence the most important parameter to take into consideration is the length of illumination. This variable has been tested using white light (a neon "daylight" tube of constant intensity).

In a 24-hour cycle, the rate of mictic female production rises linearly in going from 14h30 to 17h of illumination per day (Pourriot & Clément, 1975; Fig. 33). Below 14h30 reproduction is totally parthenogenetic. Between 17h (corresponding to the maximum amount of daylight in France) and 24h, the percentage of mictic females stays high, but it shows a slight dip at 22h (Fig. 33).

The results obtained using shorter periods of illumination and obscurity (1h, 2h, etc...) have given rise to a series of hypotheses regarding the mechanisms involved in the photoperiod effect (Pourriot et al., 1982).

- Contrary to the situation observed in arthropods (Lees, 1968; Saunders, 1976) a photoperiod skeleton is ineffective in N. copeus (Fig. 34).
- An illumination of 6 times 2 hours, with an hour of darkness between each 2-hour period gives results comparable to 15h of continuous daily illumination, not 12. In the same fashion 7 times 2 hours corresponds to 17, and not 14, hours of illumination (Fig. 34).

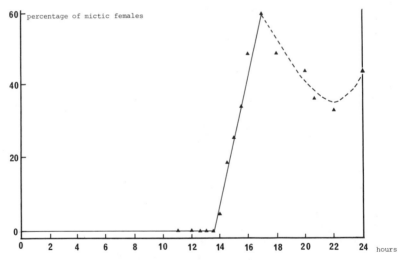

Fig. 33. Variation in the percentage of mictic females as a function of the duration of daily illumination expressed in hours (Pourriot & Clément, 1975).

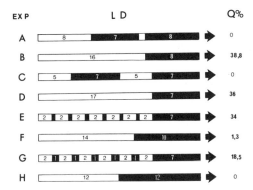

Fig. 34. The influence of continuous and discontinuous illumination over several 24-hour cycles on the appearance of mictic females in the F_1 generation (Q %) in the rotifer Notommata copeus. The parental females are placed under a definite set of light conditions at the time of birth and maintained there for the following 10 days, at which point they no longer lay eggs and are considered to be old. A certain proportion (Q %) of their F_1 descendants are mictic females which lay male eggs (example B, D, E, F, G).

In conclusion, Pourriot et al. (1981) suggest that photoperiod exerts an hourglass-type of influence on N. copeus. A period of illumination provokes the flow of the sand granules in the hourglass, but this flow is not brought to an immediate halt when obscurity follows illumination: the period of inertia is comparable to a lengthening of the illumination by 30 minutes.

2. The hatching of subitaneous eggs

The rate of reproduction (Ro) varies with different exogenous factors, but light has not been shown to influence it (Pourriot & Clément, 1981; Clément et al., 1981). In certain sessile and colonial rotifers light can provoke a degree of synchronization in the hatching of subitaneous eggs (Champ, 1976; Champ & Pourriot, 1977 a, b; Wallace, 1980).

Field and laboratory observations have prompted Champ and Pourriot to emit the following hypothesis on the mechanism underlying this synchronization. An inhibitory substance synthetized in the egg is destroyed by light. Hatching occurs when the concentration of this substance reaches a threshold value. The hypothesis rests on the observation that groups of eggs maintained under photoperiods with progressively longer dark intervals (i.e. L-D of 16:8, 14:10, 12:12) require longer periods of light before hatching occurs.

3. The hatching of resting eggs

Most rotifer resting eggs have to undergo a period of obligatory latency (2 to 3 weeks in Brachionus rubens: Pourriot et al., 1981) before any hatching takes place. This period is lengthened in the absence of

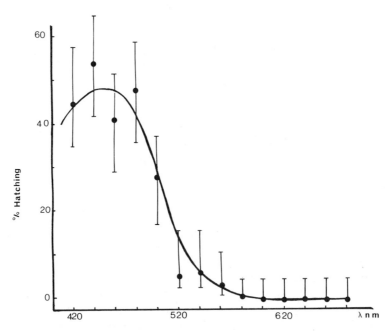

Fig. 35. Wavelength-dependent variation in the percentage of hatching in the 400 to 700 nm range, among resting eggs of Brachionus rubens after 2 months of storage in the dark at 5°C (Blanchot & Pourriot, 1982 a).

factors which can lift dormancy. These factors are numerous and vary according to the species and clone (Pourriot & Snell, 1983).

In Brachionus rubens light is indispensable for the hatching of resting eggs. The absence of light as well as, extreme temperatures (10°C and 22°C) block hatching. Nonetheless, after a 3 to 4 month period of inactivity, resting eggs begin to hatch even in darkness and cold (Pourriot & Clément, 1981; Blanchot & Pourriot, 1982 a). The effective wavelengths fall into the blue region (Blanchot & Pourriot, 1982 a). Maximum hatching is attained in the 400 to 480 nm range (Fig. 35). A much weaker reaction (6% hatching) is noted in the green between 500 and 520 nm. Red light is completely ineffective.

Pourriot and coworkers propose that a hatching enzyme exists whose synthesis is accelerated by light, but slowed down by extreme temperatures (Blanchot & Pourriot, 1982 b; Pourriot & Clément, 1981).

CONCLUSIONS

The study of rotifer photoreceptors and photoreceptions opens two avenues of research for the future.

1. Simple models for neurobiological experimentation

A rotifer is a natural system based on a small number of neurons. The 200 neurons in the brain fulfill the sensory-motor coordination of the whole animal. The small size of these neurons has discouraged investigators from implanting microelectrodes in them. However, it may be possible to make electrophysiological recordings from the largest of these neurons such as that of the cerebral eye of Asplanchna which is easily localized because it is adjacent to the accessory pigment cup.

The small size of the animal on the other hand, is advantageous for carrying out the three-dimensional reconstruction of its entire nerve net on the basis of ultrathin serial sections. The nerve tracts leading from the photoreceptive neurons in the brain (Fig. 36) to the mono or bicellular muscles, inserted on the integument or on the ciliary rootlets of the rotatory apparatus, may be traced in this manner. This information can be applied to interpret the behavioral responses of the animals to light.

Finally, the genetic uniformity of animals belonging to the same clone would allow us to follow the development of nerve connections and of the nerve network as a function of the age and life history of the individual.

2. Phylogenesis and evolution

In rotifers, it is possible to study the variability of sensory structures, nerve pathways and behavioral responses among animals having the same or different genotype. Such a study, in conjunction with investigations in ecology and population genetics, may resolve the problem of analyzing the role of behavior in the mechanisms of speciation and evolution.

Two dominant traits which we have underlined in this chapter are related to the very primitive position of the rotifers in the animal kingdom:
- a) the weak performance of their photoreceptive system, due in large measure to the fact that each eye or ocellus is composed of only one or two sensory neurons (Fig. 36),
- and b) the diversity of the structures specialized for photoreception (ampulla-shaped cilia, lamellae, formations of the endoplasmic reticulum, rhabdoms, phaosomes), probably linked to different sensory pigments, to different photosensitivities and to different responses to light.

Everything seems to indicate that rotifers are the living reminders of a vast, primitive testing stand for different photoreceptive structures among multicellular animals possessing bilateral symmetry. In the more advanced zoological groups such as the arthropods or the vertebrates only certain photosensitive pigments and photoreceptive structures were preserved. The major innovation that took place between the rotifers and these groups was the multiplication of sensory neurons in the photoreceptive organ and of the relay neurons leading to the brain. The Platyhelminthes illustrate the beginning of neuron multiplication which reaches full manifestation in the Annelida. In these groups there is still a considerable diversity among photoreceptive structures (see Fournier, 1983; Verger-Bocquet, 1983).

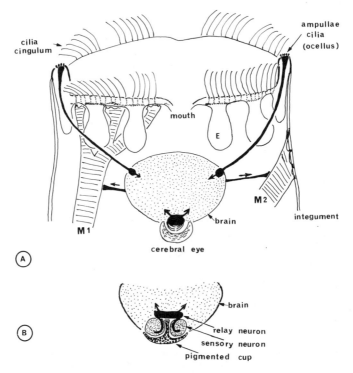

Fig. 36. A - <u>Asplanchna brightwelli</u>. The three neurons responsible for the transfer of information relevant to light to the rest of the brain are situated at its periphery. There is one in the cerebral eye. The two others are connected directly to the ampulla-shaped cilia of the anterior ocelli. The brain innervates two categories of muscles: those (type M_1) inserted on the infraciliature of the cilia of the cingulum which are borne by the anterior epithelial cells (E) and those (type M_2) inserted on the integument.

B - <u>Brachionus calyciflorus</u>. As in <u>Asplanchna</u>, two axons lead from the cerebral eye toward the neuropile of the brain, but they take their origin from a relay neuron, itself informed by two photoreceptive neurons.

ACKNOWLEDGEMENTS

We thank A. Cornillac, A. Luciani and especially J. Amsellem for making available to us some of their results which we have incorporated in this review and for having assisted us in the preparation of the manuscript

This work was made possible by grants from the University Lyon I and by the CNRS (RCP 657, GIS Physiologie sensorielle, LA 244).

Electron microscopical observations were carried out in the Center of Electron Microscopy Applied to Biology and Geology (C.M.E.A.B.G.) in the University Lyon I, France.

REFERENCES

Amsellen, J. & Clément, P. (1977) Correlation between ultrastructural features and contraction rates in rotiferan muscles. I. - Preliminary observations on longitudinal retractor muscles in Trichocerca rattus. Cell. Tissue Res. 181: 81-90.

Amsellem, J. & Clément, P. (1980) A simplified method for the preparation of rotifers for transmission and scanning electron microscopy. Hydrobiol. 73: 119-122.

Amsellem, J. & Ricci, C. (1982) Fine structure of the female genital apparatus of Philodina (Rotifera, Bdelloidea). Zoomorphol. 100: 89-105.

Beauchamp, P. de (1907) Morphologie et variations de l'appareil rotateur des Rotifères. Arch. Zool. Exp. Gén. 6: 1-29.

Beauchamp, P. de (1909) Recherches sur les Rotifères: les formations tégumentaires et l'appareil digestif. Arch. Zool. Exp. Gén. 10: 1-410.

Beauchamp, P. de (1965) Classe des Rotifères. In: Traité de Zoologie, Anatomie, Systématique, Biologie. Ed. P.P. Grassé. Paris, Masson et Cie. Tome IV, Fasc. III, p. 1225-1379.

Birky, C.W. Jr. (1964) Studies on the physiology and genetics of the rotifer Asplanchna. - I - Methods and physiology - J. Exp. Zool. 155: 273-292.

Blanchot, J. & Pourriot, R. (1982 a) Influence de trois facteurs de l'environnement, température, lumière et salinité sur l'éclosion des oeufs de durée d'un clone de Brachionus plicatilis (O F M) (Rotifère). C.R. Acad. Sci. Fr. 295: 243-246.

Blanchot, J. & Pourriot, R. (1982 b) Effets de l'intensité d'éclairement et de la longueur d'onde sur l'éclosion des oeufs de durée de Brachionus rubens Ehr. (Rotifère). C.R. Acad. Sci. Fr. 295: 123-126.

Champ, P. (1976) Etude des populations d'un Rotifère épiphyte dans la Loire. Thèse Doct. Spécialité, Université Paris VI, 81 pages.

Champ, P. &Pourriot, R. (1977 a) Reproductive cycle in Sinantherina socialis. Arch. Hydrobiol. Beih. 8: 184-186.

Champ, P. & Pourriot, R. (1977 b) Particularités biologiques et écologiques du Rotifère Sinantherina socialis (Linné). Hydrobiol. 55: 55-64.

Clément, P. (1975) Ultrastructure de l'oeil cérébral d'un Rotifère, Trichocerca rattus. J. Microsc. Biol. Cell. 22: 69-86.

Clément, P. (1977 a) Introduction à la Photobiologie des Rotifères dont le Cycle Reproducteur est Contrôlé par la Photopériode. Approches Ultrastructurale et Comportementale. Thèse doctorat d'Etat, Univ. Lyon I, 7716, 262 p.

Clément, P. (1977 b) Ultrastructural research on rotifers. Arch. Hydrobiol. Beih. 8: 270-297.

Clément, P. (1977 c) Phototaxis in rotifers (action spectra.). Arch. Hydrobiol. Beih. 8: 67-70.

Clément, P. (1980) Phylogenetic relationships of rotifers, as derived from photoreceptor morphology and other ultrastructural analysis. Hydrobiol. 73: 93-117.

Clément, P., Luciani, A. & Pourriot, R. (1981) Influences exogènes sur le cycle reproducteur des Rotifères. Bull. Soc. Zool. Fr. 106: 255-262.

Clément, P. & Pourriot, R. (1972) Photopériodisme et cycle hétérogonique chez certains Rotifères Monogonontes. I. Observations préliminaires chez Notommata copeus. Arch. Zool. Exp. Gén. 113: 41-50.

Clément, P. & Pourriot, R. (1973 a) Influence de la densité de population sur la production de femelles mictiques induite par la photopériode chez Notommata copeus (Rotifère). C.R. Acad. Sci. Fr. 276: 3151-3154.

Clément, P. & Pourriot, R. (1973 b) Mise en évidence d'un effet de masse et d'un effet de groupe dans l'apparition de phases de reproduction sexuée chez Notommata copeus (Rotifère). C.R. Acad. Sci. Fr. 227: 2533-3154.

Clément, P. & Pourriot, R. (1974) Photopériodisme et cycle hétérogonique chez Notommata copeus (Rotifère). III - Recherche du seuil minimal d'éclairement. Arch. Zool. Exp. Gén. 115: 641-650.

Clément, P. & Pourriot, R. (1975) Influence du groupement et de la densité de population sur le cycle de reproduction de Notommata copeus (Rotifère). I. Mise en évidence et essai d'interprétation. Arch. Zool. Exp. Gén. 116: 375-422.

Clément, P. & Pourriot, R. (1976 a) Influence du groupement et de la densité de population sur le cycle de reproduction des Rotifères. II. Comparaison de deux souches de Notommata copeus Ehr. Arch. Zool. Exp. Gén. 117: 5-14.

Clément, P. & Pourriot, R. (1976 b) Photopériodisme et cycle hétérogonique chez le Rotifère Notommata copeus. IV. Influence de l'intensité d'éclairement en lumière monochromatique. Variations du pourcentage de femelles mictiques au cours de la ponte des femelles parentales. Arch. Zool. Exp. Gén. 117: 205-224.

Clément, P. & Pourriot, R. (1979) Influence de l'âge des grands-parents sur l'apparition des mâles chez le Rotifère Notommata copeus. Internat. J. Invert. Repr. 1: 89-98.

Clément, P. & Pourriot, R. (1980) About a transmissible influence through several generations in a clone of the Rotifer Notommata copeus Ehr. Hydrobiol. 73: 27-31.

Clément, P., Rougier, C. & Pourriot, R. (1977) Les facteurs exogènes et endogènes qui contrôlent l'apparition des mâles chez les Rotifères. Bull. Soc. Zool. Fr. 101 (suppl. 4): 86-95.

Clément, P., Wurdak, E. & Amsellem, J. (1983) Behaviour and sensory receptors in Rotifers. Hydrobiol. (In press)

Cornillac, A. (1982) Yeux Cérébraux et Réponses Motrices à la Lumière chez Brachionus calyciflorus et Asplanchna brightwelli (Rotifères). Thèse Doct. Spécialité, Univ. Lyon I., 1212, 86 pages.

Cornillac, A., Wurdak, E. & Clément, P. (1983) Phototaxis in monochromatic light and microspectrophotometry of the cerebral eye of the Rotifer Brachionus calyciflorus. Hydrobiol. (In press)
Couillard, P. (1983) Photoreception in Protozoa: An overview. (This volume).
Coulon, J., Charras, J.P., Chasse, J.L., Clément, P., Cornillac, A., Luciani, A. & Wurdak, E. (1983) An experimental system for the automatic tracking and analysis of rotifer swimming behaviour. Hydrobiol. (In press).
Diehn, B., Feinleib, M., Haupt, W., Hildebrand, E., Lenci, F. & Nulisch, W. (1977) Terminology of behavioral responses of motile microorganisms. Photochem. Photobiol. 26: 559-560.
Eakin, R.M. (1963) Lines of evolution of photoreceptors. In: General Physiology of Cell Specialization. Ed. D. Mazia & A. Tyler. New York, McGraw-Hill, p. 393-425.
Eakin, R.M. (1968) Evolution of photoreceptors. In: Evolutionary Biology. Ed. T. Dobzhansky, M.K. Hecht & W.C. Steere. New York, Appleton-Century-Crofts, p. 194-242.
Eakin, R.M. (1972) Structure of Invertebrate Photoreceptors. In: Handbook of Sensory Physiology Vol. VII. Ed. H.J.A. Dartnall, Berlin, Heidelberg, New York, Springer Verlag. p. 625-684.
Eakin, R.M. (1982) Continuity and diversity in Photoreceptors. In: Visual Cells in Evolution. Ed. J.A. Westfall. New York, Raven Press, p 91-105.
Eakin, R.M. & Westfall, J.A. (1965) Ultrastructure of the eye of the Rotifer Asplanchna brightwelli. J. Ultrastruct. Res. 12: 46-62.
Fauré-Fremiet, E. (1961) Cils vibratiles et flagelles. Biol. Rev. 36: 464-536.
Fauré-Fremiet, E. & Rouiller, C. (1957) La flagelle interne d'un Chrysomonadale: Chromulina psammiobia. C.R. Acad. Sci. Fr. 244: 2655-2657.
Foster, K.W. & Smyth, R.D. (1980) Light antennas in phototactic Algae. Microbiol. Rev. 44: 572-630.
Fournier, A. (1979) Les photorécepteurs des Plathelminthes parasites. Colloque "La vision chez les Invertébrés", C.N.R.S., Paris 1979. (In press)
Fournier, A. (1983) Photoreceptors and photosensitivity in Platyhelminthes. (This volume)
Fournier, A. & Combes, F. (1978) Structure of photoreceptors of Polystoma integerrimum (Platyhelminthes, Monogena). Zoomorphol. 91: 147-155.
Fraenkel, G.S. & Gunn, D.L. (1961) The Orientation of Animals. A reprint of the Oxford Edition, 1940, with extra notes. New York, Dover Press.
Gilbert, J.J. (1976 a) Selective cannibalism in the rotifer Asplanchna sieboldi: Contact recognition of morhotype and clone. Proc. Nat. Acad. Sci. USA 73: 3233-3237.
Gilbert, J.J. (1976 b) Sex specific cannibalism in the rotifer Asplanchna sieboldi. Science 194: 730-732.
Gilbert, J.J. (1977) Mictic-female production in Monogonont Rotifers.

Arch. Hydrobiol. Beih. 8: 142-155.
Gilbert, J.J. (1980) Feeding in the rotifer Asplanchna: Behavior, cannibalism, selectivity, prey defenses and impact on rotifer communities. In: Evolution and Ecology of Zooplankton Communities. Ed. W.C. Kerfoot. Hanover, New Hampshire and London, England, University Press of New England, p. 158-172.
Gunn, D.C. (1975) The meaning of the term "Klionokinesis". Anim. Behav. 23: 409-412.
Henkart, M., Landis, D.M.D. & Reese, T.S. (1976) Similarity of junctions between plasma membranes and endoplasmic reticulum in muscle and neurons. J. Cell Biol. 70: 338-347
Hertel, H. (1979) Prototaktische Reaktion von Asplanchna priodonta bei monochromatischen Reizlight. Zeitsch. Natur. 34: 148-152.
Hyman, L.H. (1951) Class Rotifera. In: The Invertebrates. New York, McGraw-Hill Book Co. Vol. 3, p. 59-151.
Jennings, H.S. (1901) On the significance of spiral swimming in organisms. Am. Nature 35: 369-378.
King, C.E. (1977) Genetics of reproduction, variation and adaptation in rotifers. Arch. Hydrobiol. Beih. 8: 187-201.
Koste, W. (1978) Rotatoria. Berlin, Brontraeger. Vol. 2, 673 p., 234 pl.
Kuhn, A. (1919) Die Orientierung der Tiere im Raum. Iena, 71 p.
Kuhn, A. (1929) Phototropismus und Phototaxis der Tiere. Bethes Handb. norm. path. Physiol. 12: 17-35.
Land, M.F. (1965) Image formation by a concave reflector in the eye of the Scallop, Pecten maximus. J. Physiol. 179: 138-153.
Land, M.F. (1966) A multilayer interference reflector in the eye of the Scallop Pecten maximus. J. Exp. Biol. 45: 433-447.
Land, M.F. (1979) Des animaux dotés d'yeux à mirroir. Pour la Science, 16: 28-37
Lees, A.D. (1968) Photoperiodism in insects. In: Photophysiology Current Topics. Ed. A.C. Giese. New York, London, Academic Press 4: 47-138.
Levinthal, C. & Ware, R. (1972) Three dimensional reconstruction from serial sections. Nature (Lond.) 236: 207-210.
Loeb, J. (1918) Forced Movements, Tropisms and Animal Conduct. Philadelphia and London, J.B. Lippincott Co., 209 pages.
Luciani, A. (1982) Contribution à l'Etude du Vieillissement chez le Rotifère Brachionus plicatilis: Nage, Cils et Battements Ciliaires, Métabolisme Energétique. Thèse doctorat spécialité, 1211, Univ. Lyon I, 86 pages.
Luciani, A., Chassé, J.L. & Clément, P. (1983) Aging in Brachionus plicatilis. The evolution of swimming as a function of age at two different calcium concentrations. Hydrobiol. (In press)
Menzel, R. & Roth, F. (1972) Spektrale Phototaxis von Planktonrotatorien. Experientia 28: 356-357.
Nørrevang, A. (1974) Photoreceptors of the phaosome (hirudinean) type in a pogonophore. Zool. Anz. 193: 297-304.
Osafune, T. & Schiff, J.A. (1980) Stigma and flagellar swelling in relation to light and carotenoids in Euglena gracilis var. bacillaris. J. Ultrastruct. Res. 73: 336-349.

Pourriot, R. (1963) Influence du rythme nycthéméral sur le cycle sexuel de quelques Rotifères. C.R. Acad. Sci. Fr. 256: 5216-5219.

Pourriot, R. (1965 a) Recherches sur l'Ecologie des Rotifères. Vie et Milieu, Banyuls s/Mer, Fr., suppl. 21, 224 p.

Pourriot, R. (1965 b) Sur le déterminisme du mode de reproduction chez les Rotifères. Schweiz. Z. Hydrobiol. 27: 76-87.

Pourriot, R. (1977) Food and feeding habits of Rotifera. Arch. Hydrobiol. Beih. 8: 243-260.

Pourriot, R. (1980) Workshop on culture techniques of Rotifers. Hydrobiol. 73: 33-35.

Pourriot, R. & Clément, P. (1972) Influence du photopériodisme sur l'apparition de phases de reproduction sexuée chez Notommata copeus (Rotifère): étude du spectre d'action de la lumière visible. C.R. Acad. Sci. Fr. 274: 106-109.

Pourriot, R. & Clément, P. (1973) Photopériodisme et cycle hétérogonique chez Notommata copeus (Rotifère Monogononte). II - Influence de la qualité de la lumière. Spectres d'action. Arch. Zool. Exp. Gén. 114: 277-300.

Pourriot, R. & Clément, P. (1975) Influence de la durée de l'éclairement quotidien sur le taux de femelles mictiques chez Notommata copeus Ehr. (Rotifère). Oecol. (Berlin) 22: 67-77.

Pourriot, R. & Clément, P. (1976) Comparison of the control of mixis in three clones of Notommata copeus. Arch. Hydrobiol. Beih. 8: 174-177.

Pourriot, R. & Clément, P. (1981) Action de facteurs externes sur la reproduction et le cycle reproducteur des rotifères. Acta Oecol. Oecol. Gen. 2: 135-151.

Pourriot, R., Clément, P. & Luciani, A. (1981) Perception de la photopériode par un Rotifère; hypothèses sur les mécanismes. Arch. Zool. Exp. Gén. 122: 317-327.

Pourriot, R. & Snell, T.W. (1983) The resting eggs in rotifers. Hydrobiol. (In press)

Preissler, K. (1977) Horizontal distribution and "avoidance of shore" by rotifers. Arch. Hydrobiol. Beih. 8: 43-47.

Preissler, K. (1980) Field experiments on the optical orientation of pelagic Rotifers. Hydrobiol. 73: 199-203.

Remane, A. (1929-1933) Rotatoria. In: Dr. H.G. Bronn's Klassen und Ordnungen des Tierreichs, IV (Vermes), 2 (Aschelminthes), 1 (Rotatorien, Gastrotrichen un Kinorhynchen), 3, 4 vol.: p. 1-448.

Salvini-Plawen, L.V. (1982) On the polyphyletic origin of photoreceptors in visual cells in evolution: In: Visual Cells in Evolution. Ed. J.A. Westfall, New York, Raven Press, p. 137-154.

Salvini-Plawen, L.V. & Mayr, E. (1977) On the evolution of photoreceptors and eyes. In: Evolutionary Biology. Ed. M.K. Hecht, W.S. Sterre & B. Wallace. Vol. 10, p. 207-263.

Saunders, D.S. (1976) Insect Clocks. Oxford, New York, Toronto, Sydney, Paris, Frankfurt, Pergamon Press, 280 p.

Seldon, H.L. (1972) Observations on Symmetry and on Development in the Brain of the Rotifer Asplanchna brightwelli. PhD. Thesis, M.I.T., U.S.A., 104 p.

Stasko, A.B. & Sullivan, C.M. (1971) Responses of Planarians to light: an examination of Klino-Kinesis. Anim. Behav. Monographs, 4, 2, 124 p.

Stossberg, K. (1932) Zur Morphologie der Rädertier-gattungen Euchlanis, Brachionus und Rhinoglena. Zeitsch. Wiss. Zool. 142: 313-424.

Ullyott, P. (1936) The behaviour of Dendrocoelum lacteum. I - Responses at light- and dark boundaries. II - Responses in non directional gradients. J. Exp. Biol. 13: 253-278.

Vanfleteren, J.R. (1982) A monophyletic line of evolution? Ciliary induced photoreceptor membranes. In: Visual Cells in Evolution. Ed. J.A. Westfall, New York, Raven Press, p. 107-136.

Vanfleteren, J.R. & Coomans, A. (1975) Photoreceptor evolution and phylogeny. Z. Zool. Syst. Evolut. Forsch. 14: 157-169.

Verger-Bocquet, M. (1983) Photoréception et vision chez les Annélides. (This volume)

Viaud, G. (1938) Recherches Expérimentales sur le Phototropisme des Daphnies. Etude de Psychologie Animale. Thèse doctorat, Strasbourg.

Viaud, G. (1940) Recherches expérimentales sur le phototropisme des Rotifères. I - Bull. Biol. Fr. Belg. 74: 249-308.

Viaud, G. (1943 a) Recherches expérimentales sur le phototropisme des Rotifères II. Bull. Biol. Fr. Belg. 77: 68-93.

Viaud, G. (1943 b) Recherches expérimentales sur le phototropisme des Rotifères III. Stroboscopie des mouvements ciliaires. Mouvement ciliaires et phototropisme. Bull. Biol. Fr. Belg. 77: 224-242.

Wallace, R.L. (1980) Ecology of sessile rotifers. Hydrobiol. 73: 181-194.

Ware, R. (1971) Computer Aided Nerve Tracing in the Brain of the Rotifer Asplanchna brightwelli. PhD. Thesis, M.I.T., U.S.A. 213 p.

PHOTORECEPTION ET VISION CHEZ LES ANNELIDES

MARTINE VERGER-BOCQUET
Laboratoire de Biologie Animale
Laboratoire Associé au C.N.R.S. n° 148
Université des Sciences et Techniques de Lille
59655 Villeneuve d'Ascq Cédex, France

ABSTRACT

In Annelids, the photoreceptors are located not only on the head but are also present in other parts of the body. They are uni- or pluricellular organs. The sensory cells, in the majority of cases, possess numerous microvilli which are arranged either on the apical process (all prostomial Polychaete eyes) or either around an internal vacuole called phaosome or "Binnenkörper" (Oligochaete and Leech photoreceptors, some Polychaete photoreceptive structures). However a narrow connection with the extracellular space has been demonstrated for Hirudo by lanthanum impregnation. In addition to microvilli, a varying number of cilia (9 + 0) is present in Oligochaetes. Sometimes, a rudimentary cilium is also found at or close to the tip of apical processes of the prostomial Polychaete eyes. Ciliary photoreceptors have been described in the brain of Nereidae and Phyllodocidae and in the eyes of branchial filaments of Sabellidae. In the latter case, the photosensory structures are stacked lamellar "sacs" which project into an extracellular space. Every "sac" is attached to a kinetosome (9 + 0).

Absent in Oligochaete photoreceptors, the pigmented cells in the leeches form a pigment cup around the photoreceptoral cells whereas in the cerebral Polychaete eyes which have many cells, they are placed side by side of sensory cells and often called supportive cells. When the cerebral eye has only one pigmented cell, it forms a cup around the light-sensitive organelles of the receptoral cell. The shape of the supportive cells and particularly that of their apical processes changes with the family and sometimes within the same family with the species. In some cases, the apical processes constitute the lens.

Present in Polychaete cerebral eyes, the lens can be very diverse; it can be secreted by special cells (secretory cell of Vanadis, lenticular cell of Eulalia) or derived from pigmented cells which either secrete a hyaline material (Autolytus) or develop slender interdigitated apical processes (Nereis, Syllis). The branchial eyes of Sabellidae also possess a lens. In Dasychone and Potamilla the lens are secreted whereas in Branchiomma it is constituted of a single cell which is sitting at the top of the receptoral cell. It contains basally a dish-shaped nucleus and more centrally an ovoid mass of vesicles.

The electroretinograms have been recorded in Nereidae and in Alciopidae. They are cornea negative, but in T. candida (Alciopidae which has two retinas) a reversal of polarity of ERG appears with wavelength and at 560 nm with position of the active electrode on the cornea. The cornea negative response is from the main retina while the cornea positive response is from an accessory retina.

In Torrea, the primary retina peaks in sensitivity at 400 nm, the secondary retina at 560 nm. The Vanadis eye (another Alciopidae) peaks in unit at 460 to 480 nm. Nereis mediator is also maximally sensitive at 480 nm.

In the medicinal leech (Hirudo medicinalis) the absorption of light increases the conductance of the microvillar membrane. Therefore a flow of current appears; it flows inward through the microvillar membrane and outward through the external membrane.

In general, Annelids are photonegative in their behaviour but they become photopositive under some conditions. It is the case for the blood-sucking leeches, several Polychaete families which are pelagic during the breeding season and L. terrestris which crawls toward dim lights and away from strong ones.

The Annelids also react to changes in illumination (moving shadows, sudden decrease or sudden increase in light intensity) and habituate more or less rapidly to these stimulations.

Les organes photorécepteurs sont présents chez la majorité des Annélides; ils sont soit unicellulaires, soit pluricellulaires. Ils se situent généralement au niveau du prostomium; toutefois chez ces Invertébrés certaines structures photoréceptrices se rencontrent aussi dans d'autres régions du corps.

1. MORPHOLOGIE

Polychètes

Depuis plus de cent ans, l'organe photorécepteur des Polychètes a retenu l'attention des zoologistes. De nombreux travaux ont été réalisés en microscopie photonique; nous ne citerons pour mémoire que ceux de De Quatrefages (1850), de Malaquin (1893), de Hesse (1899) et de Tampi (1949).

Avec l'apparition de la microscopie électronique, les ocelles cérébraux d'un nombre relativement important de familles ont été décrits au niveau infrastructural. De ces travaux il ressort que, chez cette classe d'Annélides, les ocelles prostomiaux présentent une grande diversité de structure (vision directe ou indirecte, organe bicellulaire ou pluricellulaire, lentille de type "cellulaire" ou sécrétée).

Les structures photoréceptrices présentes dans d'autres régions du corps chez certaines familles ont également retenu l'attention de plusieurs auteurs (Hermans, 1969 b; Ermak & Eakin, 1976; Dragesco-Kernéis, 1980 a; Verger-Bocquet, 1981 b) (Fig. 1). Ces structures visuelles qui se forment assez souvent après la métamorphose de la larve présentent généralement un modèle morphologique différent de celui des ocelles cérébraux.

1. Les organes photorécepteurs localisés dans le prostomium.

a. Les yeux.

Chez les Annélides Polychètes, au niveau du prostomium, se rencontrent différents types d'yeux. Généralement peu développés et à vision indirecte chez les espèces sédentaires, ils sont à vision directe et constitués par la juxtaposition de nombreuses cellules sensorielles et de soutien chez les espèces errantes sans toutefois être composés de plusieurs unités visuelles comme les yeux des filaments branchiaux des Sabellidae (Kernéis, 1975).

L'oeil à vision indirecte des Polychètes sédentaires.

Ce type d'oeil décrit par Hermans et Cloney (1966) et par Ermak et Eakin (1976) respectivement chez Armandia brevis (Opheliidae) et chez Chone ecaudata (Sabellidae) n'est constitué que de deux cellules: une cellule de soutien et une cellule sensorielle (Fig. 2 a); la première en forme de cupule entoure le prolongement apical de la seconde. La cellule de soutien contient de nombreux granules de pigment. Selon la répartition de ces derniers, deux régions peuvent être distinguées: la cupule où se regroupent les granules pigmentaires et le diaphragme qui en est dépourvu. Chez les Sabellidae le diaphragme est mince (0,3 µm) tandis que chez les Opheliidae il est plus épais (0,3 à 0,6 µm) et contient de nombreux granules renfermant des particules denses aux électrons. La fonction de ce diaphragme n'est pas définie; toutefois Hermans et Cloney (1966) suggèrent qu'il serait l'homologue de la lentille des Nereidae. Dépourvue de granules pigmentaires, la cellule sensorielle se compose de deux parties: le corps cellulaire et le prolongement apical qui est inclus en totalité dans la cavité formée par la cellule de soutien. L'extrémité du prolongement avec ses nombreuses microvillosités rhabdomériques se dirige vers la concavité de la cupule pigmentaire. Nous sommes donc en présence d'un oeil à vision indirecte par opposition à l'oeil à vision directe chez lequel les extrémités photosensibles s'étendent vers la lumière incidente.

L'oeil des Flabelligeridae décrit par Spies (1975) est par contre constitué de plusieurs cellules sensorielles et de plusieurs cellules de soutien (Fig. 2b). Dans ce cas, la cupule pigmentaire est formée par les parties distales des cellules de soutien qui sont étroitement juxtaposées et dans lesquelles sont regroupés de nombreux granules de pigment. Comme les yeux précédents, celui des Flabelligeridae est un oeil à vision indirecte. Après avoir longé la couche des cellules de soutien, les cellules

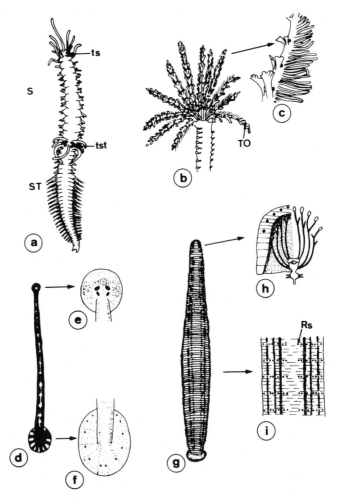

Fig. 1.: Localisation des structures photoréceptrices chez diverses Annélides.

Fig. 1 a: <u>Autolytus</u> sp. (Polychète) - Lors de la reproduction des yeux de grande taille se différencient dans le segment céphalique du stolon (d'après Dales, 1963).

Figs. 1 b et 1 c: <u>Dasychone bombyx</u> (Polychète) - Vue antérieure de l'animal (Fig. 1 b) et vue de détail d'un des filaments du panache branchial (Fig. 1 c) (d'après Carricaburu & Kernéis, 1975).

Figs. 1 d à 1 f: <u>Piscicola geometra</u> (Hirudinée) - Vue d'ensemble de l'animal (Fig. 1 d) et vues de détail des ventouses antérieure (Fig. 1 e) et postérieure (Fig. 1 f) (d'après Jung, 1963).

sensorielles subissent une torsion et se dirigent par l'ouverture de la cupule pigmentaire dans la cavité oculaire. Avant de pénétrer dans cette dernière, la région médiane des cellules photosensibles donne naissance à une zone qualifiée par Spies de "cornéenne de type lenticulaire".

L'oeil des Polychètes errantes.

Entre l'oeil de Nephtys qui n'est constitué que de quelques cellules et l'énorme oeil camérulaire des Alciopidae dont la complexité structurale rivalise avec celle des yeux des Céphalopodes et des Vertébrés, se rencontrent chez les autres espèces des yeux assez bien différenciés.

L'oeil pigmenté de Nephtys situé dans la région postérieure du cerveau n'est constitué que de quatre cellules (deux cellules sensorielles et deux cellules de soutien); par sa morphologie il s'apparente plus à l'oeil des Polychètes sédentaires qu'à celui des Polychètes errantes. En forme de cupule, les deux cellules de soutien dans lesquelles deux régions peuvent être différenciées, se font face (Fig. 2 c). Leurs régions postérieures où se regroupent les granules de pigment, donnent naissance à la cupule pigmentaire dont l'ouverture est orientée latéralement ou dorso-latéralement vers l'épiderme. Leurs régions antérieures forment par contre une structure biconvexe qui est considérée comme étant une lentille; elles sont réunies latéralement aux régions postérieures. Chaque cellule sensorielle est incluse dans la concavité d'une cellule de soutien. Chez Nephtys, les yeux antérieurs ou accessoires situés de chaque côté du prostomium, immédiatement au-dessus du cerveau, ne sont par contre constitués que d'une ou deux cellules photoréceptrices identiques à celles des yeux pigmentés.

Les yeux des autres Polychètes errantes sont constitués par la juxtaposition de nombreuses cellules de soutien pigmentaires et de nombreuses cellules sensorielles photoréceptrices qui renferment parfois des granules pigmentaires (Figs. 2 d et 2 f). A ces deux types cellulaires s'ajoutent chez les Phyllodocidae une cellule lenticulaire (Whittle & Golding, 1974) et chez les Alciopidae une cellule sécrétrice (Hermans & Eakin, 1974).

Les granules pigmentaires essentiellement localisés à l'apex du corps cellulaire des cellules rétiniennes forment la couche pigmentaire qui, à l'encontre de ce que l'on observe chez les Vertébrés ne délimite pas la totalité de l'oeil mais seulement la cavité oculaire (Fig. 2 d). Les parties nuclées des corps cellulaires situées à l'extérieur de la cupule pigmentaire constituent la couche rétinienne qui s'étend par l'intermédiaire des

---

Figs. 1 g à 1 i: Hirudo medicinalis (Hirudinée) - Vue d'ensemble de l'animal (Fig. 1 g) et vues de détail de la région céphalique (Fig. 1 h) et de la région moyenne du corps (Fig. 1 i) (d'après Kretz et al., 1976).

Rs: récepteur segmentaire; S: souche; ST: stolon; TO: tache oculaire; ts: tête souche; tst: tête stolon.

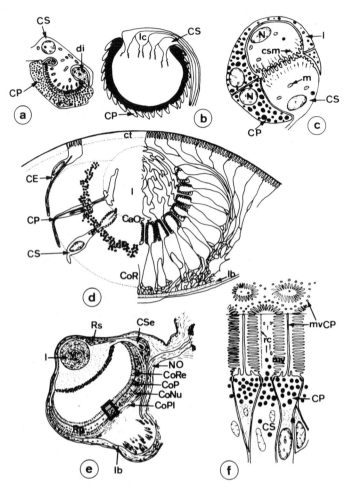

Fig. 2 : Différents types d'yeux céphaliques observés chez les Annélides Polychètes.

Fig. 2 a : Armandia brevis (Opheliidae) - L'oeil à vision indirecte n'est constitué que de deux cellules. Le diaphragme (di) prolongement apical de la cellule de soutien (CP) serait l'homologue de la lentille (d'après Hermans & Cloney, 1966).

Fig. 2 b : Flabelligeridae - La zone médiane des cellules sensorielles (CS) donne naissance à une zone "cornéenne de type lenticulaire" (lc) (d'après Spies, 1975).

Fig. 2 c : Nephtys (Nephthydidae) (d'après Zahid & Golding, 1974).

Fig. 2 d : Platynereis dumerilii (Nereidae) - L'oeil est formé par la juxtaposition de nombreuses cellules prolongements basaux jusqu'à la région nerveuse sous-jacente. La cavité oculaire située entre la couche pigmentaire et la cuticule renferme les prolongements apicaux des cellules rétiniennes. Ceux des cellules sensorielles avec leurs nombreuses microvillosités se rencontrent toujours à la périphérie, tandis que la région centrale contient une lentille de structure très variable.

Chez les Alciopidae, l'oeil est plus complexe (Fig. 2 e). Les cellules sensorielles et de soutien constituent la rétine primaire qui délimite la cavité optique. Cette rétine se divise en quatre couches: les couches plexiforme, nuclée, pigmentée et réceptrice. L'oeil de Vanadis tagensis (Hermans & Eakin, 1974) possède en outre un iris pigmenté, une rétine accessoire (= rétine secondaire) et une cellule sécrétrice.

Comme nous venons de le voir, l'oeil des Polychètes errantes est essentiellement constitué par la juxtaposition de deux types cellulaires distincts: les cellules sensorielles et les cellules de soutien. La morphologie des cellules de soutien et en particulier celle de leurs prolongements apicaux diffère selon la nature de la lentille; par contre l'aspect des cellules sensorielles ne varie guère (Figs. 2 d et 2 f).

La cellule sensorielle présente trois régions: le corps cellulaire, le prolongement apical situé dans la cavité oculaire et la fibre basale, prolongement qui se dirige vers les couches nerveuses sous-jacentes. Le corps cellulaire renferme le noyau et les organites cytoplasmiques classiques. En outre il contient des corps multivésiculaires et chez certaines espèces des granules pigmentaires. Ceux-ci sont absents chez Nephtys (Zahid & Golding, 1974) et chez les Polychètes sédentaires (Hermans & Cloney, 1966; Ermak & Eakin, 1976). Ils se rencontrent plutôt chez les espèces dont l'oeil est plus complexe. Il existe cependant des exceptions: ainsi chez les Nereidae, la cellule photoréceptrice de N. virens (Dorsett & Hyde, 1968) en est dépourvue tandis que celle de

---

sensorielles (CS) et de nombreuses cellules de soutien (CP). Les prolongements apicaux de ces dernières forment la lentille (l) qui est de "type cellulaire" (d'après Fischer & Brökelmann, 1965).

Figs. 2 e et 2 f: Vanadis (Alciopidae) - Vue générale de l'oeil de V. formosa (d'après Hesse, 1899) et vue de détail de la rétine primaire (Rp) de V. tagensis (d'après Hermans & Eakin, 1974).

CaO: cavité oculaire; CE: cellule épidermique; CoNu: couche nucléaire; CoP: couche pigmentaire; CoPl: couche plexiforme; CoR: couche rétinienne; CoRe: couche réceptrice; CP: cellule de soutien; CS: cellule sensorielle; CSe: cellule sécrétrice; csm: citerne sous-microvillaire; ct: cuticule; l: lentille; lb: lame basale; m: mitochondrie; mv: microvillosités; mvCP: microvillosités de la cellule de soutien; N: noyau; NO: nerf optique; rc: racine ciliaire; Rs: rétine secondaire.

Platynereis dumerilii (Fischer & Brökelmann, 1966) et de N. vexillosa (c f. Eakin, 1972) en contiennent. La même variabilité dans la présence de ces granules pigmentaires se retrouve chez les Syllidae avec toutefois une certaine constance à l'intérieur d'une même sous-famille (Verger-Bocquet, 1981 a); absents chez les Syllinae, ils sont présents chez les Autolytinae et les Eusyllinae.

La partie distale du corps cellulaire, fortement rétrécie forme le col de la cellule, lieu de passage à travers la couche pigmentaire. Elle se poursuit dans la cavité oculaire par le prolongement apical muni de nombreuses microvillosités disposées radialement. Celles-ci sont plutôt droites chez les Alciopidae (Hermans & Eakin, 1974) alors que chez les Nereidae (Eakin, 1972) et chez les Syllidae (Bocquet & Dhainaut-Courtois, 1973) elles sont moins régulières et leurs parcours est sinueux; elles s'enchevêtrent assez souvent avec celles des cellules sensorielles adjacentes. Ces microvillosités sauf chez Nephtys se dirigent vers la lumière incidente: les yeux des Polychètes errantes sont donc à vision directe. Une formation ciliaire plus ou moins développée se rencontre fréquemment dans le prolongement photosensible. Elle est représentée soit par un cil rudimentaire, un centriole et une racine ciliaire (Fischer & Brökelmann, 1965, 1966; Eakin, 1968; Bocquet, 1976, 1977), soit par un centriole et une racine ciliaire (Whittle & Golding, 1974; Hermans & Eakin, 1974) ou soit uniquement par une racine ciliaire (Dorsett & Hyde, 1968; Singla, 1975). La racine ciliaire peut même être absente (Zahid & Golding, 1974).

Une forme particulière de réticulum qualifiée de "citernes sous-microvillaires" est généralement présente dans le prolongement photosensible (c f. discussion dans Whittle & Golding, 1974). De telles citernes se rencontrent également chez les Syllidae mais leur développement et leur disposition varient d'une espèce à l'autre (Verger-Bocquet, 1981 a).

La morphologie de la cellule de soutien et en particulier celle de son prolongement apical est étroitement liée à la nature de la lentille. Celle-ci qui a parfois été qualifiée de "Füllmasse" (Fischer & Brökelmann, 1966) ou de corps vitré (Hermans & Cloney, 1966) a une structure très variable.

Chez Eulalia et Vanadis, elle est formée par des cellules spéciales (cellule lenticulaire, cellule sécrétrice). Elle est composée de vésicules remplies d'un matériel finement granulaire de densité variable. Ces vésicules sont plus étroitement regroupées dans la région centrale de la lentille où elles présentent des signes de fusion. Dans ce cas, les prolongements apicaux des cellules de soutien sont peu développés. Chez les Alciopidae, il s'agit de fines microvillosités qui s'élèvent du corps cellulaire et constituent une charpente autour de chaque extrémité photoréceptrice (Fig. 2 f) (Hermans & Eakin, 1974). Chez les Phyllodocidae, les prolongements provenant des cellules de soutien forment par contre une étroite rangée qui s'intercale entre la lentille et la région apicale des prolongements photosensibles (Whittle & Golding, 1974).

Chez les autres espèces, les cellules de soutien contribuent à la formation de la lentille, celle-ci étant due, soit à un important développement des prolongements apicaux, soit à la sécrétion d'un matériel dense aux électrons. Les Autolytinae possèdent en effet, une volumineuse lentille sécrétée, constituée de deux parties séparées par une constriction. Les prolongements apicaux des cellules de soutien sont comme dans les deux cas précédents peu développés. Après avoir cheminé entre les microvillosités des cellules sensorielles adjacentes, leur région distale se divise et forme une sorte de manchon qui entoure la lentille (Fig. 3 c) (Bocquet, 1976). Ils ne sont donc guère différents de ceux observés chez les Phyllodocidae. Toutefois, la lentille des Autolytinae est sécrétée par les cellules de soutien. Lors de la différenciation oculaire, ces cellules présentent en effet une intense activité: le réseau ergastoplasmique très abondant renferme des granules intraciternaux et l'appareil de Golgi émet de nombreuses vésicules qui migrent vers le prolongement apical et le traversent avant de se condenser avec le matériel lenticulaire déjà formé (Verger-Bocquet, 1977).

Chez les Nereidae, la lentille est par contre formée par les prolongements apicaux des cellules de soutien (Fig. 2 d) d'où le développement accru de ces derniers par rapport aux cas précédents. Après avoir cheminé entre les extrémités photosensibles des cellules sensorielles adjacentes, le prolongement apical s'enchevêtre avec celui des cellules de soutien voisines. Un tel oeil dépourvu de lentille sécrétée se rencontre aussi chez la majorité des Syllinae (Bocquet & Dhainaut-Courtois, 1972, 1973) (Fig. 3 a). Cependant chez ces derniers les prolongements apicaux ne renferment pas de vésicules au contenu finement granulaire comme cela a été décrit chez N. vexillosa (Eakin & Westfall, 1964) et chez Platynereis dumerilii (Fischer & Brökelmann, 1966), ni de petits granules ou petits filaments comme chez N. virens (Dorsett & Hyde, 1968). Chez les Syllinae seuls quelques microtubules et quelques rares vésicules pénètrent parfois dans les différentes digitations du prolongement.

Chez les Aphroditidae, la lentille est constituée de plusieurs éléments qui renferment des tubules de réticulum endoplasmique disposés de manière à former un ensemble paracristallin. Pour Singla (1975) et pour Bassot et Nicolas (1978) chaque élément représente le prolongement apical d'une cellule de soutien tandis que pour Whittle (1976), ces éléments sont les prolongements d'un petit nombre de cellules lenticulaires (5 à 10) qui ont leur péricaryon situé sous celui des cellules rétiniennes; dans ce cas, les prolongements apicaux des cellules de soutien réduits, séparent incomplètement la périphérie de la lentille des extrémités photosensibles.

Chez les Eusyllinae et plus particulièrement chez Odontosyllis ctenostoma il est difficile de savoir quelle structure doit être qualifiée de lentille. En effet, chez les Syllinae les prolongements apicaux des cellules de soutien, bien développés, ont reçu cette dénomination tandis que chez les Autolytinae, la lentille est représentée par un volumineux corps sécrété. Or chez O. ctenostoma, ces deux éléments sont présents. Ainsi pour donner la même dénomination à toute structure identique dans l'oeil des Syllidae nous avons convenu d'appeler lentille toute formation d'origine sécrétrice, située dans la cavité oculaire. Ainsi chez cette famille d'Annélides Polychètes, nous avons pu mettre en évidence selon le degré de développement de la lentille sécrétée trois types d'organes photorécepteurs (Fig. 3) (Verger-Bocquet, 1981 a). Le premier, dépourvu de cette formation, se rencontre chez la majorité des Syllinae; seuls les prolongements apicaux des cellules de soutien contribuent à la formation

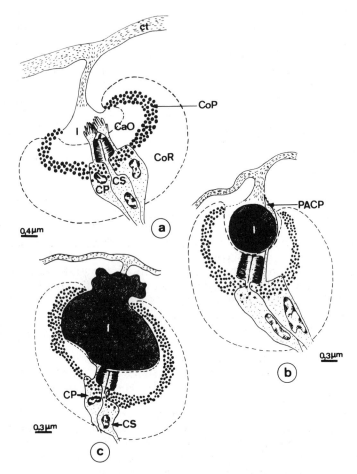

Fig. 3: Différents types d'yeux observés chez les Syllidae (Annélides Polychètes).

Fig. 3 a: Syllis amica (S/F: Syllinae) - Cet oeil est dépourvu de lentille sécrétée. Seuls les prolongements apicaux des cellules de soutien constituent une lentille (l) qualifiée de "type cellulaire".

Fig. 3 b: Odontosyllis ctenostoma (S/F: Eusyllinae) - Cet oeil possède une lentille (l) sécrétée entièrement incluse dans la cavité oculaire (CaO). Cette lentille coexiste avec les prolongements apicaux des cellules de soutien (PACP) bien développés.

Fig. 3 c: Autolytus pictus (S/F: Autolytinae) - Cet oeil possède une volumineuse lentille (l) sécrétée formée de deux parties séparées par une constriction.

CoP: couche pigmentaire; CoR: couche rétinienne; CP: cellule de soutien; CS: cellule sensorielle; ct: cuticule.

d'une lentille que nous avons qualifiée de "type cellulaire" par opposition à la précédente. Le second, s'observe chez les Autolytinae et S. krohnii; il présente une volumineuse lentille sécrétée formée de deux parties: l'une proximale, logée dans la cavité oculaire et l'autre distale, qui s'étend vers la cuticule. Le troisième, décrit chez les Eusyllinae, possède également une lentille sécrétée mais elle est entièrement incluse dans la cavité oculaire où elle coexiste avec les prolongements apicaux des cellules de soutien bien développés.

b. Les autres organes photorécepteurs céphaliques.

A côté des yeux que nous venons de décrire, il existe dans le prostomium de certaines Annélides Polychètes des structures de type photorécepteur dont la fonction n'a pas encore été démontrée. Généralement situées dans le cerveau, ces formations ont été qualifiées d'extra-oculaires par Corbière-Tichané (1977).

Chez les Nereidae, ces photorécepteurs ont particulièrement bien été étudié chez l' Heteronereis mâle. La cellule photoréceptrice se caractérise par la présence de nombreux organites ciliaires qui prennent naissance à partir des corps basaux présents à l'apex de la fibre sensorielle. La cellule de soutien très allongée, coiffe cette région et tous les organites ciliaires (Dhainaut-Courtois, 1965, 1968, 1970). Une telle structure photoréceptrice se rencontre aussi chez les Phyllodocidae (Whittle & Golding, 1974).

Chez Nephtys il s'agit par contre d'une paire de "sacs sensoriels". Ces structures sont présentes dans le ganglion cérébral de N. caecoïdes, de N. hombergi et de N. caeca. Chaque sac est formé d'une seule cellule qui semble posséder une vaste vacuole intracellulaire (phaosome) dans laquelle se projettent de nombreuses microvillosités irrégulières et très rarement un cil. En fait, la membrane limitante est en continuité avec la membrane cellulaire (Zahid & Golding, 1974).

Dans le cerveau d' Armandia brevis (Hermans & Cloney, 1966; Hermans, 1969 b) les cellules photoréceptrices sont entourées par des cellules gliales. Comme chez Nephtys, ces cellules possèdent de nombreuses microvillosités et sont dépourvues de formation ciliaire.

Les cellules présumées photoréceptrices décrites par Manaranche (1968, 1971) dans la partie dorsale du ganglion cérébral de Glycera convoluta diffèrent de celles décrites jusqu'à présent. Chez cette espèce le cytoplasme est envahi par des empilements de sacs très aplatis disposés plus ou moins parallèlement. Ces sacs résultent de l'invagination de la membrane ergastoplasmique qui est repliée plusieurs fois sur elle-même.

2. Les organes photorécepteurs localisés dans d'autres régions du corps.

Certaines espèces d'Annélides Polychètes ont la particularité de posséder des taches pigmentées supposées photoréceptrices à des endroits très divers du corps (Figs. 1 a à 1 c). Ces structures visuelles ne sont donc pas en relation avec le cerveau. Toutefois lors de la reproduction de certains Syllidae, on assiste à la différenciation d'un segment céphalique pourvu d'yeux de grande taille (Fig. 1 a) dans la région antérieure du stolon (segments postérieurs de l'individu contenant les produits génitaux qui à maturité sexuelle se détachent de la région antérieure asexuée appelée souche).

a. Les ocelles associés à un cerveau.

Ces structures oculaires différenciées secondairement soit sur un métamère précis en l'occurence le 14è pour Autolytus pictus soit sur un segment sétigère non déterminé de la région médio-postérieure chez Syllis amica, sont hypertrophiées par rapport aux yeux formés lors de l'ontogenèse (Fig. 1 a). Mise à part cette différence de taille, il existe généralement une grande analogie dans la structure fondamentale des yeux de souche et de stolon. L'hypertrophie oculaire est due à une augmentation du nombre des cellules rétiniennes et à un accroissement de leur volume en particulier de celui de leurs prolongements apicaux. De plus, chez le stolon, ces prolongements renferment de très nombreux canalicules (Bocquet & Dhainaut-Courtois, 1973). Les prolongements apicaux des cellules photoréceptrices dont le nombre de microvillosités latérales s'est accru, possède toujours une formation ciliaire; le cil rudimentaire se situe à l'apex du segment photosensible tandis que la racine ciliaire traverse ce dernier. Dès les premiers stades de l'organogenèse oculaire des formations ciliaires ont été observées dans les cellules sensorielles (Bocquet & Dhainaut-Courtois, 1973; Verger-Bocquet, 1977) mais de telles formations s'observent aussi dans les cellules non encore différenciées et dans les cellules de soutien en cours de différenciation.

Chez S. amica, des cas de polycéphalie se rencontrent assez fréquemment lors des stolonisations obtenues expérimentalement par ablation du proventricule (Durchon, 1952, 1959). Les têtes avec leur deux paires d'yeux volumineux situés de part et d'autre d'un cerveau (Bocquet, 1971) apparaissent indifféremment sur les métamères de la souche ou sur ceux du stolon. Parfois, il ne se forme qu'une demi-tête surnuméraire. Dans ce cas, aux deux yeux différenciés ne correspond qu'un demi-cerveau localisé dans la moitié correspondante du métamère (Verger-Bocquet, 1981 a).

b. Les ocelles extracérébraux.

Les taches oculaires des filaments branchiaux des Sabellidae

De nombreuses espèces de la famille des Sabellidae possèdent des taches pigmentaires oculaires à l'extrémité ou le long de plusieurs filaments du panache branchial (Figs. 1 b, 1 c).

Ces taches oculaires sont formées de nombreuses unités visuelles élémentaires. Chaque unité visuelle composée d'une cellule sensorielle toujours dépourvue de granules pigmentaires est surmontée d'une formation réfringente: la lentille; elle est séparée de la suivante par des cellules de soutien pigmentaires, une seule cellule de soutien pouvant former chez Dasychone bombyx un manchon autour de la cellule photoréceptrice (Kernéis, 1968 b).

La morphologie de la cellule sensorielle diffère quelque peu selon les espèces (Figs. 4 a à 4 c) mais dans tous les cas les structures photosensibles sont représentées par des sacs lamellaires d'origine ciliaire, superposés plus ou moins régulièrement (400 chez Branchiomma vesiculosum (Lawrence & Krasne, 1965); 70 - 100 chez Dasychone (Kernéis, 1966, 1968 b). La cellule photoréceptrice est ainsi creusée d'une cavité dans laquelle se projettent les cils (structure = 9 + 0) modifiés en expansions lamellaires. Celles-ci se situent généralement au-dessous de la lentille (Kernéis, 1968 b, 1971, 1975); mais chez Branchiomma vesiculosum une remarquable disposition de longues mitochondries s'intercale entre ces structures et la cellule lenticulaire (Fig. 4 a) (Krasne & Lawrence, 1966). Chez les autres espèces, les mitochondries, presque toujours aussi abondantes sont disposées différemment; soit autour de l'étranglement de la lentille, soit sous les sacs lamellaires ou autour d'eux; dans ce dernier cas leur nombre est plus réduit.

Si les structures purement photosensibles des taches oculaires des filaments branchiaux des Sabellidae caractérisent cette famille, les lentilles qui les surmontent diffèrent d'une espèce à l'autre par leur localisation, leur forme et leur structure (Figs. 4 a à 4 c). Il s'agit généralement de lentilles secrétées alors que chez Branchiomma vesiculosum la lentille est composée d'une seule cellule contenant une multitude de vésicules plus ou moins denses aux électrons d'une taille de 200 à 300 nm regroupées au-dessus d'un noyau aplati (Krasne & Lawrence, 1966). Le matériel lenticulaire, s'élabore au niveau du réticulum endoplasmique granulaire, transite dans les dictyosomes puis est libéré sous forme de vésicules golgiennes; il migre alors vers la partie apicale de la cellule (Dragesco-Kernéis, 1979). Chez les autres espèces la lentille n'est qu'une masse extracellulaire de texture variable en continuité avec la cuticule. Cette lentille est soit formée de granules denses aux électrons de 20 à 30 nm régulièrement répartis, soit constituée de particules tellement fines qu'il est impossible d'en mesurer la taille (Kernéis, 1975).

Les taches oculaires segmentaires et pygidiales.

Les structures oculaires situées au voisinage des parapodes apparaissent selon les espèces à des moments plus ou moins précoces du développement. Chez Dasychone lucullana les premières taches oculaires sont visibles sur l'avant dernier métamère dès le stade 5 segments sétigères (Dragesco-Kernéis, 1980 b) alors qu'il faut attendre le stade 23-24 métamères pour Armandia cirrosa (Guérin, 1972) et 26 métamères pour Armandia brevis (Hermans, 1969 b). Chez certaines espèces les taches oculaires n'apparaissent que lorsque l'individu entre en reproduction. C'est le cas de Syllis spongicola (Syllinae) qui différencie sur chaque métamère du stolon une tache pigmentaire brune à la base du cirre dorsal du parapode. C'est aussi le cas du palolo (Eunice viridis) mais chez cette espèce la tache oculaire impaire se forme sur la face ventrale de la région sexuée, sous la chaîne nerveuse (Fauvel, 1959).

. L'ocelle segmentaire d' Armandia brevis (Opheliidae)

Armandia brevis possède 11 paires d'ocelles segmentaires situées du 7è au 17è métamère. Chaque ocelle ne possède qu'une cellule sensorielle qui émet approximativement 15 extrémités photosensibles. Ces prolongements, composés d'un noyau central de neurofibrilles, d'une rangée de

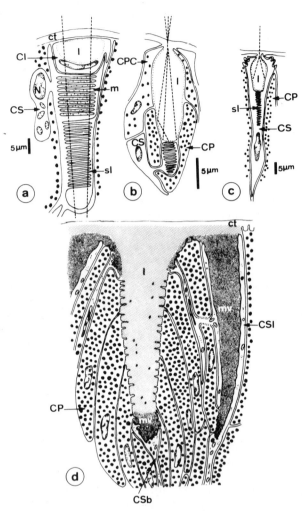

Fig. 4 : Yeux des Sabellidae (Annélides Polychètes).

Figs. 4 a à 4 c : "Unités visuelles" constituant les taches oculaires du panache branchial chez *Branchiomma vesiculosum* (Fig. 4 a) (d'après Krasne & Lawrence, 1966), chez *Dasychone bombyx* (Fig. 4 b) (d'après Kernéis, 1968), et chez *Potamilla reniformis* (Fig. 4 c) (d'après Kernéis, 1971). Les lignes pointillées indiquent le trajet probable des rayons lumineux (d'après Carricaburu & Kernéis, 1975).

Fig. 4 d : *Dasychone* - Coupe longitudinale d'une tache oculaire segmentaire (d'après Dragesco-Kernéis, 1980).

mitochondries et d'un ensemble de microvillosités, se projettent vers l'intérieur de la cavité oculaire qui est doublée par des cellules gliales: l'ocelle est à vision indirecte. La cupule pigmentaire, couche unicellulaire d'une trentaine de cellules pigmentaires d'origine mésodermique entoure la cellule sensorielle. Cet ocelle est donc d'origine ecto-mésodermique (Hermans, 1969 b).

. L'ocelle pygidial de Chone ecaudata (Sabellidae)

Les deux ocelles de Chone situés de chaque côté du pygidium, sont formés par une couche sous-cuticulaire de cellules sensorielles, pigmentaires et sécrétrices (Ermak & Eakin, 1976). La région apicale des cellules photoréceptrices émet à partir de ses bords latéraux de courtes et irrégulières microvillosités et à partir de son centre deux cils; les microvillosités et les cils s'étendent dans une petite dépression souscuticulaire de 3 μm de profondeur. Les cellules pigmentaires adjacentes, caractérisées par la présence de nombreux granules de pigment de 0,3 à 0,4 μm de diamètre, sont dépourvues de cils. Elles renferment par contre de nombreux granules muqueux identiques à ceux des cellules sécrétrices de la cuticule. Ainsi, la cellule pigmentaire continue apparemment à sécréter la cuticule tout en servant de support et d'écran aux cellules sensorielles. Les cellules sécrétrices se rencontrent plutôt à la périphérie de l'ocelle. Leur région supranucléaire contient de nombreux granules de 0,1 μm de diamètre tandis que de l'extrémité apicale s'élèvent un cil et des microvillosités plus courtes et plus droites que celles de la cellule sensorielle. Comme pour cette dernière, elles s'étendent dans une dépression peu profonde.

. La tache oculaire segmentaire de Syllis spongicola (Syllinae)

Chaque tache oculaire est composée d' "unité visuelle", une "unité" étant constituée d'une cellule sensorielle entourée de ses cellules de soutien. Deux autres types cellulaires sont également présents: il s'agit de cellules muqueuses situées à la périphérie ou entre les "unités visuelles" et de cellules chargées en grains de sécrétions localisées à la base de la tache oculaire, au voisinage de la jonction neuro-musculaire (Verger-Bocquet, 1981 b). L'extrémité distale invaginée des cellules photoréceptrices forme une cavité (phaosome) dans laquelle s'étendent de nombreuses microvillosités et un cil rudimentaire. A la périphérie de cette cavité s'observent de très nombreuses mitochondries. Les cellules de soutien se caractérisent par la présence de nombreux granules pigmentaires plutôt localisés dans la région nucléaire. Quatre sortes de granules ont pu être distingués d'après leur taille et leur texture.

CL: cellule lenticulaire; CP: cellule de soutien; CPC: cellule péricristallinienne; CS: cellule sensorielle; CSb: cellule sensorielle basale; . CSl: cellule sensorielle latérale; ct: cuticule; l: lentille; m: mitochondrie; mv: microvillosités; N: noyau; sl: sac lamellaire.

. La tache oculaire segmentaire de Dasychone (Sabellidae)

Situées entre les rames dorsales et ventrales des parapodes thoraciques et abdominaux (Kernéis, 1968 a), ces taches oculaires sont constituées de plusieurs cellules pigmentaires et de quelques cellules sensorielles regroupées autour d'une lentille granuleuse extracellulaire qui est en continuité avec la cuticule (Fig. 4 d). La morphologie des cellules photoréceptrices diffère selon leur emplacement par rapport à la lentille. Les cellules latérales digitées ont une forme complexe; par contre la cellule basale est plus simple. Une cavité extracellulaire ménagée dans ces cellules contient les nombreuses microvillosités disposées irrégulièrement, support présumé du pigment photosensible. Ces amas de microvillosités qui entrent en contact avec la lentille à différents niveaux, naissent à partir de la membrane plasmique à proximité d'un ou deux centrioles (Dragesco-Kernéis, 1980 a).

3. Les organes photorécepteurs des larves.

Les quelques études réalisées au niveau infrastructurale sur les organes photorécepteurs des larves montrent que ces structures ne sont guère différentes de celles de l'adulte. A côté des yeux pigmentés se rencontrent aussi d'autres photorécepteurs (Holborow & Laverack, 1972; Niilonen, 1980).

a. Les yeux

Les yeux pigmentés de la larve trochophore des Nereidae qui disparaissent lorsque les yeux adultes se forment (Eakin & Westfall, 1964) ressemblent à ceux de la larve trochophore des Aphroditidae (Holborow & Laverack, 1972). Il s'agit d'un organe constitué de deux cellules et dépourvu de lentille. L'extrémité distale de la cellule sensorielle munie de microvillosités se prolonge dans la concavité de la cellule de soutien pigmentée. Cette extrémité renferme également une formation ciliaire représentée chez Neanthes succinea par un centriole et une racine ciliaire (Eakin & Westfall, 1964).

Chez les Nereidae, l'oeil de la larve de trois segments pourvu d'une lentille est encore constitué d'un nombre restreint de cellules. Fischer et Brökelmann (1966) signalent la présence de deux cellules sensorielles et de deux cellules de soutien, Eakin et Westfall (1964) celle d'une seule cellule photoréceptrice dont le prolongement s'étend dans la cupule pigmentaire formée par deux cellules de soutien.

Chez S. amica (Syllinae) la présence dans l'oeil d'une larve non segmentée ou dans celui d'une larve de six métamères, d'un corps de forme arrondie, moyennement dense aux électrons, situé dans la cavité oculaire au-dessus des microvillosités rhabdomériques (Verger-Bocquet, observation non publiée), soulève le problème du devenir de cette structure. Ce corps dense qui vraisemblablement correspond à une lentille sécrétée n'a en effet jamais été observé dans l'oeil adulte.

b. Les autres photorécepteurs

Des photorécepteurs de type phaosome ont été trouvés dans la larve de Polydora ligni (Spionidae) (Niilonen, 1980). La cellule sensorielle située entre l'oeil médian et l'oeil latéral se caractérise par la présence d'une vaste vacuole qui est entièrement remplie par les microvillosités. La partie distale de cette cellule qui renferme le noyau est en contact direct avec la surface épidermique tandis que la partie proximale repose sur une membrane basale.

Chez Harmothoë, parmi les corps des cellules nerveuses sous l'apex de la trochophore âgée (14 jours) se rencontre un organe composé de cils et de microvillosités (Holborow & Laverack, 1972). Sa structure se rapproche de celle des photorécepteurs décrits dans le cerveau de Nereis pelagica (Dhainaut-Courtois, 1965).

Archiannélides

Ces Annélides qui forment un groupe hétérogène ont tantôt été considérées comme des formes voisines des Polychètes, tantôt comme des formes primitives. Selon Hermans (1969 a) elles devraient être considérées comme un ordre de la classe des Polychètes et non comme une classe séparée; néanmoins nous traiterons séparément ce groupe de Vers afin de pouvoir établir plus aisément une comparaison entre la structure des organes photorécepteurs de ces Vers et celle des yeux des Polychètes.

Merker et Vaupel Von Harnack (1967) décrivirent l'infrastructure de l'oeil de Protodrilus. Ultérieurement Eakin et al. (1977) élargirent cette étude à trois autres espèces d'Archiannélides (Saccocirrus, Dinophilus, et Nerilla). De plus un membre des Polygordiidae vient dernièrement d'être examiné (Brandenburger & Eakin, 1981; Eakin, 1982).

Les Archiannélides ne possèdent qu'une paire d'yeux, à l'exception du genre Nerilla qui en a deux paires et des individus adultes de Polygordius cf. appendiculatus qui en sont dépourvus. Les yeux sont constitués d'un nombre relativement réduit de cellules. Les plus simples sont ceux de Protodrilus et ceux de Saccocirrus. Chaque oeil est formé par la juxtaposition de deux cellules: une cellule de soutien pigmentée en forme de cupule et une cellule sensorielle dont le prolongement apical est inclus dans la cavité formée par la première (Fig. 5 c). Les yeux de Dinophilus et de Nerilla sont un peu plus complexes. Celui de Dinophilus est constitué de trois cellules (2 sensorielles et 1 de soutien) (Fig. 5 d) tandis que celui de Nerilla en possède six (2 sensorielles, 2 de soutien et 2 cornéennes) (Fig. 5 e).

La cellule sensorielle des cinq espèces possède un prolongement photosensible muni de nombreuses microvillosités. Celles-ci chez Protodrilus, Saccocirrus et Polygordius (Figs. 5 b et c) se dirigent vers la concavité de la cellule de soutien tandis que chez Dinophilus elles s'orientent vers la lumière incidente (Fig. 5 d). Celles des deux cellules sensorielles de Nerilla se font par contre face et s'interdigitent (Fig. 5 e); un court cil sans rapport avec les microvillosités a de plus parfois été observé dans l'une de ces deux cellules. Le prolongement apical renferme aussi selon les espèces un nombre plus ou moins important de mitochondries et de citernes de réticulum sous-microvillaires.

Les cellules de soutien qui entourent plus ou moins bien les extrémités photosensibles des cellules photoréceptrices possèdent de nombreux granules pigmentaires. Généralement sphériques et très denses aux électrons, ces granules sont remplacés chez Nerilla par plusieurs

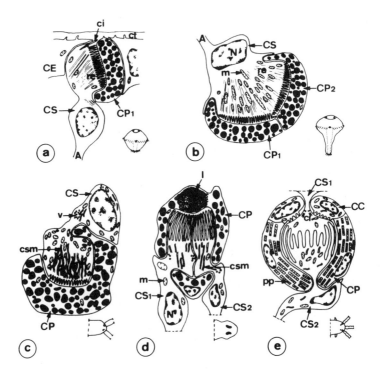

Fig. 5: Ocelles des Archiannélides

Figs. 5a et 5 b: <u>Polygordius</u> cf <u>appendiculatus</u> - Ocelle de la larve trochophore (Fig. 5 a) et ocelle de la larve segmentée (Fig. 5 b) (d'après Brandenburger & Eakin, 1981).

Figs. 5 c à 5 e: Ocelle de <u>Protodrilus</u> (Fig. 5 c), de <u>Dinophilus</u> (Fig. 5 d) et de <u>Nerilla</u> (Fig. 5 e) (d'après Eakin <u>et al</u>., 1977).

A: axone; CC: cellule cornéenne; CE: cellule épidermique; ci: cil; CP, CP1 et CP 2: cellules de soutien; CS, CS 1 et CS 2: cellules sensorielles; csm: citerne sous-microvillaire; ct: cuticule; l: lentille; m: mitochondrie; N: noyau; pp: plaquette pigmentaire; re: réticulum endoplasmique lisse (tubules); v: vésicule.

rangées de plaquettes pigmentaires vidées de leur contenu (Fig. 5 e). La morphologie de la cellule de soutien est variable d'une espèce à l'autre. Chez <u>Protodrilus</u> et <u>Saccocirrus</u> elle a la forme d'une cupule (Fig. 5 c) mais chez cette dernière espèce les extrémités distales de la cupule différencient de longues microvillosités. Chez <u>Dinophilus</u> (Fig. 5 d), la région médiane de la partie distale de cette cellule est modifiée en lentille; il s'agit d'un corps biconvexe regroupant des globules de matériel finement granulaire entourés d'une membrane.

L'oeil de la larve trochophore de Polygordius est composé de deux cellules de soutien pigmentaires formant une cupule et d'une cellule sensorielle munie d'une rangée de microvillosités (Fig. 5 a). Cet oeil n'est guère différent de celui des larves trochophores des autres Polychètes qui ne possèdent qu'une cellule de soutien (Eakin & Westfall, 1964; Holborow & Laverack, 1972).

Oligochètes

A l'exception de certains Naididae, les Oligochètes sont dépourvus d'yeux véritables mais non de cellules photoréceptrices. Ces cellules ont surtout été étudiée chez les Lumbricidae que ce soit en microscopie photonique (Hesse, 1896; Hess, 1925) ou en microscopie électronique (Hirata et al., 1969; Röhlich et al., 1970; Myhrberg, 1979). Les autres familles n'ont pas retenu l'attention des auteurs si ce n'est celle des Enchytraeidae chez laquelle Bradke (1962) a donné une brève description de cellules sensorielles épidermiques présumées photoréceptrices. Les organes photorécepteurs des Oligochètes ont récemment fait l'objet d'une intéressante synthèse (Jamieson, 1981). Ainsi nous décrirons l'aspect général des cellules sensorielles pour faciliter la comparaison entre les structures visuelles chez les diverses classes d'Annélides; pour plus de détail nous renvoyons le lecteur à cette étude.

1. Les cellules photoréceptrices des Lumbricidae

Chaque récepteur se compose d'une seule cellule. Limitée à l'épiderme chez Eisenia (Hirata et al., 1969), ces cellules sont chez Lumbricus terrestris particulièrement abondantes dans le prostomium où elles se rencontrent dans l'épiderme mais aussi dans les renflements ganglionnaires des branches des nerfs prostomiaux et dans la partie dorsolatérale du ganglion cérébral entre les cellules neurosécrétices (Röhlich et al., 1970). Ces cellules de forme très irrégulière sont soit incluses dans le tissus nerveux et entourées surtout au niveau du ganglion cérébral par des cellules gliales, soit logées profondément dans l'épiderme (Fig. 6 b); elles ne semblent révéler aucune communication avec la surface. Chez Eisenia, les cellules photoréceptrices sont beaucoup plus allongées et présentent une large ouverture sous-cuticulaire (Fig. 6 a).

Les cellules photoréceptrices se caractérisent chez Lumbricus comme chez Eisenia par la présence d'une vaste vacuole interne encore appelée phaosome ou "Binnenkörper" dans laquelle se discernent de nombreuses microvillosités parfois ramifiées et plusieurs cils sensoriels (Figs. 6 a et 6 c). Chez Lumbricus les cils de type (9 + 0) dont les tubules sont souvent désorganisés dans leur région distale, possèdent un corps basal mais pas de racine ciliaire. Ils sont indépendants des microvillosités qui chez cette espèce sont uniquement internes et remplissent la totalité du phaosome ainsi que les petites cavités adjacentes. Chez Eisenia la région centrale de la cavité est dépourvue de structure cellulaire (Fig. 6 a). D'autre part, toujours chez cette espèce, les microvillosités s'élèvent non seulement des parois du phaosome mais aussi des parois du col qui mènent au pore extérieur. Des microvillosités s'élèvent également de la surface extérieure de la cellule et traversent la cuticule. Toujours dépourvues de granules pigmentaires, les cellules photoréceptrices renfer-

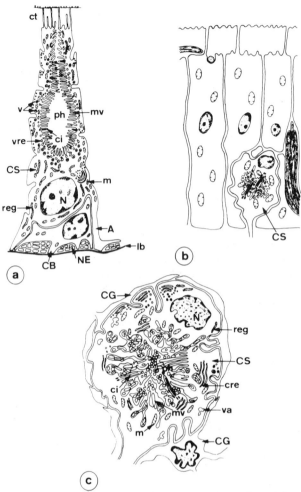

Fig. 6 : Les organes photorécepteurs des Oligochètes.

Fig. 6 a : Cellule photoréceptrice d' Eisenia foetida (d'après Hirata et al., 1969).

Figs. 6 b et 6 c : Lumbricus terrestris - Localisation de la cellule photoréceptrice (Fig. 6 b) (d'après Myhrberg, 1979) et représentation schématique d'une cellule sensorielle (Fig. 6 c) d'après Röhlich et al., 1970).

A: axone; CB: cellule épidermique basale; cre: citerne de réticulum endoplasmique; CG: cellule gliale; ci: cil; CS: cellule sensorielle photoréceptrice; ct: cuticule; lb: lame basale; m: mitochondrie; mv: microvillosité; N: noyau; NE: nerf épidermique; ph: phaosome; reg: réticulum endoplasmique granulaire; v: vésicule; va: vacuole; vre: vésicule de réticulum endoplasmique.

ment de nombreuses vésicules et vacuoles. Chez Lumbricus, autour du phaosome, s'étendent des empilements de deux à cinq citernes de réticulum qui pourraient correspondre aux citernes sous-microvillaires de Whittle (1976); chez Eisenia, seules des vacuoles et des vésicules (qui peuvent former un réticulum) occupent cette position. Des enroulements de réticulum, en continuité avec le système de vacuoles et de vésicules sous-microvillaires peuvent être observés dans le cytoplasme périphérique.

2. Les cellules photoréceptrices des Enchytraeidae.

Il s'agit d'un groupe de cellules ciliées s'étendant de la membrane basale de l'épiderme à la cuticule (Bradke, 1962). Les cils émanent de l'extrémité apicale de la cellule et pénétrent dans la cuticule qu'ils traversent. Ces cellules se caractérisent par la présence d'un vaste corps homogène composé de microtubules. Ce corps dont la longueur peut varier de 5 à 11 μm s'étend de la région moyenne à la base de la cellule ou se replie autour du noyau basal. Il serait l'équivalent du phaosome, les microvillosités étant remplacées par des microtubules. Autour de ce corps microtubulaire sont disposés concentriquement plusieurs complexes golgiens sécréteurs qui sont à leur tour entourés par du réticulum endoplasmique granulaire.

3. Les cellules sensorielles ciliées qui pourraient être photoréceptrices.

De chaque côté du cerveau de quatre oligochètes limicoles (Limnodrilus, Stylodrilus, Stylaria et Enchytraeus) il existe un groupe de cellules sensorielles ciliées; chaque groupe étant composé de 2, 4 ou 12 cellules (Golding & Whittle, 1975): ces cellules ne sont pas associées à des cellules de soutien ou des cellules pigmentaires. Chaque cellule possède de 30 à 40 cils sauf chez Stylodrilus où elles n'en ont que cinq. Dans tous les cas les cils sont du type (9 + 2). Actuellement il est impossible d'affirmer que ces cellules soient photoréceptrices mais nous ne pouvons pas pour autant écarter la possibilité d'une telle fonction.

Hirudinées

Les organes photorécepteurs des Hirudinées ont soulevé l'intérêt de nombreux cytologistes. Nous ne citerons que les travaux de Whitman (1886), de Hesse (1897) et de Hachlov (1910). Pour de plus amples références nous renvoyons le lecteur aux études de Hansen (1962) et de Röhlich & Török (1964).

Les cellules photoréceptrices se rencontrent en petit nombre dans les bourgeons sensoriels et parfois dans l'épiderme. Mais, à l'encontre des Oligochètes, les Hirudinées possèdent aussi des yeux.

1. Les yeux

a. Localisation

Les Hirudinées sont généralement pourvues d'yeux dont le nombre et la disposition varient selon la famille et le genre (Harant & Grassé, 1959). Hirudo (Hirudidae) possède cinq paires d'yeux tandis que Helobdella (Glossiphoniidae) en a une paire et Theromyzon (Glossiphoniidae) en a quatre paires. Ces yeux sont généralement localisés dans la région céphalique (Figs. 1 d, e, g, h). Cependant chez les Piscicolidae, de petits ocelles pigmentés s'observent souvent sur la ventouse postérieure (Fig. 1 f). Chez P. geometra il a été possible d'y dénombrer 16 petits ocelles (Jung, 1963).

b. Description générale:

Les yeux des Hirudinées sont constitués d'une cupule pigmentaire dans laquelle est inclus un nombre plus ou moins important de cellules sensorielles qui chez Hirudo sont séparées les unes des autres par un espace intercellulaire (Walz, 1979). Au contraire chez Helobdella les cellules photoréceptrices sont serrées les unes contre les autres; leur membrane repliée forme de fins plis de cytoplasme qui les séparent (Clark, 1967) (Fig. 7 b). Les yeux de la ventouse antérieure de P. geometra renferme de 20 à 25 cellules photoréceptrices (Jung, 1963) tandis que chez Hirudo medicinalis chaque cupule pigmentaire contient de 30 à 50 cellules sensorielles empilées verticalement pour former 6 à 10 couches de cellules réceptrices (Kretz et al., 1976). Les yeux de la ventouse postérieure des Piscicolidae ne sont guère différents. La cupule pigmentaire peu développée ne contient qu'un nombre très réduit de cellules photoréceptrices (1 à 5). Ces yeux sont cependant moins bien organisés que ceux de la ventouse antérieure; les cellules pigmentaires entourent les cellules sensorielles mais aussi d'autres groupes de cellules et des cellules isolées (Jung, 1963).

La position des fibres nerveuses est très variable (Fig. 7 a). Nous passons donc progressivement d'un oeil à vision indirecte à un oeil à vision directe.

c. Infrastructure de la cellule photoréceptrice:

Depuis les travaux de Hansen (1962) qui décrivaient la structure des cellules visuelles de sept espèces d'Hirudinées, d'autres études ont été réalisées au niveau infrastructural essentiellement chez Hirudo medicinalis (Röhlich & Török, 1964; Yanase et al., 1964; White &

Fig. 7: Les organes photorécepteurs des Hirudinées.

Fig. 7 a: Plusieurs types d'ocelles sont présents chez les Hirudinées. A gauche, type inversé, nerf optique partant à l'opposé de la cupule pigmentaire (cas de Glossiphonia: en haut et de Erpobdella: en bas); au centre, type direct à nerf optique latéral (4° et 5° paires d'ocelles d' Hirudo medicinalis); à droite type direct à nerf optique terminal (1er et 3° paires d'ocelles d' H. medicinalis) (d'après Hesse dans Harant & Grassé, 1959).

Fig. 7 b: Cellule photoréceptrice de l'oeil d' Helobdella stagnalis (d'après Clark, 1967).

Fig. 7 c: Cellules sensorielles de l'oeil de Piscicola

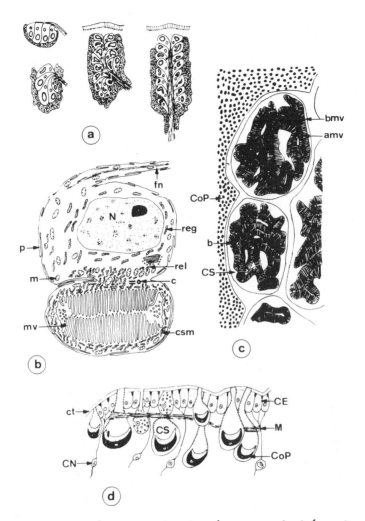

geometra montrant la complexité des "Binnenkörper" (b) (d'après Jung, 1963).

Fig. 7 d: Section à travers les parois du corps d' Haemadipsa zeylanica mettant en évidence des cellules photoréceptrices (d'après Bhatia, 1975).

amv: apex des microvillosités; bmv: base des microvillosités; c: centriole; CE: cellule épidermique; CN: cellule nerveuse; CoP: couche pigmentaire; CS: cellule sensorielle photoréceptrice; csm: citerne sous-microvillaire; ct: cuticule; fn: fibre nerveuse; M: muscle; m: mitochondrie; mv: microvillosité; N: noyau; p: pli; reg: réticulum endoplasmique granulaire; rel: réticulum endoplasmique lisse.

Walther, 1969; Lasansky & Fuortes, 1969; Walz, 1979).

Chez toutes les espèces étudiées, la cellule photoréceptrice se caractérise par la présence d'une vaste vacuole (phaosome ou "Binnenkörper"), dans laquelle se projettent de très nombreuses microvillosités. Il a été montré chez Hirudo que cette vacuole située à l'intérieur de la cellule, communique avec l'espace intercellulaire par plusieurs canaux d'un diamètre de 16-17 nm. En section transversale, ces canaux ont l'aspect de structures annulaires (Lasansky & Fuortes, 1969; White & Walther, 1969).

La cellule photoréceptrice des Hirudinées est généralement dépourvue de formation ciliaire. Seul Clark (1967) a pu observer chez Helobdella stagnalis et Placobdella rugosa une paire de centrioles au voisinage des microvillosités.

Selon les familles, la morphologie du phaosome varie. De forme assez simple chez Helobdella (Fig. 7 b) (Clark, 1967) et Hirudo (Röhlich & Török, 1964; Walz, 1979) il présente une structure complexe avec de très nombreuses circonvolutions chez Piscicola geometra (Fig. 7 c) (Jung, 1963). De plus chez cette dernière espèce aucun espace libre n'est ménagé entre les rangées de microvillosités contrairement à ce qu'a décrit Hansen chez Hirudo medicinalis, Haemopis sanguisuga et Erpobdella octoculata. Chez certaines espèces, un matériel homogène, plutôt finement granulaire a été observé dans l'espace vacuolaire (Röhlich & Török, 1964; Clark, 1967; Lasansky & Fuortes, 1969). Les microvillosités situées à la périphérie du phaosome ne sont pas toujours réparties uniformément. Chez Hirudo la densité des microvillosités est moindre autour des canaux qui entrent dans la vacuole (Walz, 1979). De forme cylindrique les microvillosités chez cette espèce sont un peu moins larges (0,08 µm contre 0,1 µm) et un peu plus longues (1 à 2 µm contre 0,4 à 1,4 um) que chez Helobdella (Clark, 1967). Chez Hirudo la membrane réceptrice (microvillaire) subit très probablement un intense mouvement membranaire (Walz, 1979).

L'étude en cryofracture des membranes réceptrices réalisées par Walz (1979) a montré que les faces-P sont riches en particules et que les faces-E sont presques lisses. Sur les faces-P les particules semblent distribuées au hasard excepté pour deux sites: les dilatations des extrémités des microvillosités où les particules sont groupées irrégulièrement et les membranes des faces-P entourant l'entrée des canaux radiaires où plusieurs rangées de particules sont disposées concentriquement. Sur la face-P des membranes réceptrices la densité des particules est de 2700 particules/μm^2. Cette densité est plus faible que celle observée chez les autres Invertébrés mais après une adaptation à l'obscurité de 28 jours la densité des particules chez Hirudo atteint 5000 particules/μm^2. Le diamètre moyen de ces particules est de 7 nm; il est compris dans les limites observées chez les autres Invertébrés. Ainsi cet auteur suggère que les particules de la face-P des membranes microvillaires représentent les sites où le photopigment est incorporé à la membrane.

Une comparaison entre les membranes réceptrices (microvillaires) et non réceptrices (externes) de la cellule sensorielle révèle que cette dernière possède environ 33% de particules intramembranaire/μm^2 de plus que la première et, que les particules adhérant sur la face-P de la membrane non réceptrice sont plus grandes (diamètre moyen 9 nm) que celles qui adhèrent sur la face-P de la membrane réceptrice.

2. Les autres organes photosensibles.

a. Les récepteurs segmentaires:

Depuis les études morphologiques de Whitman (1886), de Hesse (1897) et de Livanow (1903, 1904) on admettait généralement que Hirudo medicinalis et les membres apparentés à la famille des Hirudidae possédaient deux types d'organes photosensibles: les yeux et les récepteurs segmentaires. En 1969 Laverack a montré que des photorécepteurs existaient dans les segments du corps et en 1976 Kretz et al. ont apporté la preuve physiologique d'une telle existence.

Chez Hirudo, les récepteurs segmentaires au nombre de sept paires réparties sur la circonférence de l'anneau médian de chacun des 21 segments métamériques du corps (Fig. 1 i), sont constitués de cellules sensorielles et de cellules sphériques, la morphologie de ces dernières étant la même que celle des cellules photoréceptrices trouvées dans l'oeil.

Chez Haemadipsa zeylanica (Haemadipsidae) Bhatia (1975) signale aussi la présence dans les récepteurs dorsaux et ventraux disposés métamériquement sur le premier anneau de chaque segment du corps, de cellules avec une vaste vacuole hyaline.

b. Les cellules présentes dans les parois du corps.

Chez Haemadipsa zeylanica (Bhatia, 1956, 1975) il existe dans les parois du corps un grand nombre de cellules en forme de bouteille, à la base desquelles se rencontre une cupule pigmentaire (Fig. 7 d). Ces cellules qui ressemblent à celles présentes dans l'oeil, restent en relation avec l'extérieur.

2. PHYSIOLOGIE

Si la morphologie des organes photorécepteurs des Annélides a été largement étudiée, leur physiologie semble par contre avoir peu retenu l'attention des auteurs.

Les Polychètes.

1. Enregistrement d'électrorétinogrammes.

Les électrorétinogrammes ont été enregistrés chez trois espèces de Nereidae (N. mediator, N. diversicolor et Pl. dumerilii) et chez deux espèces d'Alciopidae (Torrea et Vanadis).

Chez Pl. dumerilii et N. diversicolor, la réponse à un flash de lumière consiste en un potentiel négatif relativement rapide, une brève déflection positive et un lent retour au niveau de base (Figs. 8 a et 8 b) (Gwilliam, 1969).

La forme générale de l'ERG de N. mediator est similaire (Yingst et al., 1972). Cependant ces auteurs signalent que les différences dans la latence des réponses et l'absence apparente de phase dynamique et

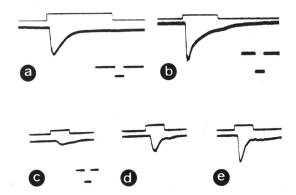

Fig. 8:   Electrorétinogrammes de Nereidae

Figs. 8 a et 8 b: Yeux antérieurs de N. diversicolor (1 mV, 20 ms)

Figs. 8 c à 8 e: Yeux postérieurs de Pl. dumerilii montrant une réponse graduée à l'augmentation de l'intensité lumineuse (2 mV, 200 ms) (d'après Gwilliam, 1969).

d'oscillations chez Pl. dumerilii et N. diversicolor pourraient être attribuées à la faible amplitude des réponses montrées par Gwilliam.

La forme de l'ERG est fonction à la fois de l'intensité (Figs. 8 c à 8 e) et de la durée du stimulus. Yingst et al. (1972) ont montré qu'elle varie aussi selon l'adaptation à l'obscurité ou à la lumière verte.

Chez Torrea candida, la polarité de l'ERG peut être inversée selon les longueurs d'onde (Fig. 9 a) (Wald & Rayport, 1977) et aux longueurs d'onde élevées (560 nm) selon la position de l'électrode active sur la cornée (Fig. 9 b). La déflection cornée-négative est obtenue pour la majorité des positions de l'électrode. Elle provient de la rétine principale; par contre celle cornée-positive n'apparaît que pour certaines positions peu faciles à trouver. Elle est due à la rétine accessoire. Dans les deux cas, la réponse du récepteur est identique; le segment externe devient négatif par rapport au reste de la cellule. L'inversion de la polarité de l'ERG est due à la situation des récepteurs par rapport à l'électrode. Chez Vanadis, espèce qui ne possède qu'une rétine accessoire, la réponse cornée-positive n'a pas pu être observée.

Wald et Rayport (1977) ont également montré que si chez ces deux espèces (Vanadis et Torrea) on stimule la rétine principale répétitivement avec des flashes, l'amplitude des ERG s'élève. Il s'agit du phénomène de "facilitation" qui a déjà été signalé dans d'autres phylums.

2. Les courbes de sensibilité spectrale.

Les yeux antérieurs et postérieurs de N. mediator présentent un maximum de sensibilité pour les longueurs d'onde de 480 nm (Fig. 10) mais sont aussi très sensibles à la lumière entre 400 et 540 nm (Yingst et al.,

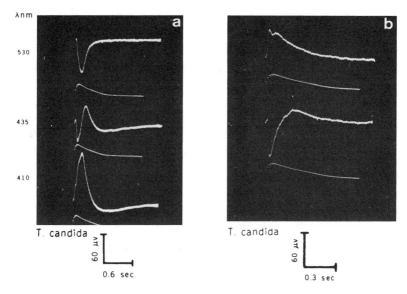

Fig. 9:   ERG de Torrea candida (Alciopidae).

Fig. 9 a: Renversement de la polarité de l'ERG avec les longueurs d'onde.

Fig. 9 b: Renversement de la polarité de l'ERG à 560 nm d'après la position de l'électrode active sur la cornée (d'après Wald & Rayport, 1977).

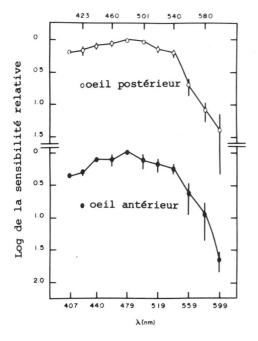

Fig. 10: Nereis mediator au repos sexuel - Courbes de la sensibilité spectrale de l'oeil antérieur et de l'oeil postérieur adaptés à l'obscurité (d'après Yingst et al., 1972).

1972). La sensibilité des récepteurs ne change pas lors de la reproduction lorsque les yeux s'hypertrophient. Les expériences d'adaptation sélective, à l'obscurité ou en lumière verte, montrent la présence d'un système multirécepteur. Un tel système doterait N. mediator de la possibilité de distinguer les couleurs et de détecter la lumière sur une large étendue spectrale (Yingst et al., 1972).

Chez Torrea, la rétine principale a un maximum de sensibilité vers 440 nm et la rétine accessoire vers 560 nm (Fig. 11 a). Les deux à la fois pourraient servir de jauge de profondeur puisque la longueur d'onde de 560 nm s'atténue plus vite dans l'eau de mer que celle de 400 nm (Wald & Rayport, 1977). Les sensibilités spectrales de la rétine primaire et secondaire, tracées linéairement ont été comparées au nomogramme de Dartnall. Ce nomogramme est basé sur l'observation que, la plupart des pigments visuels connus présentent des courbes de spectre d'absorption de forme similaire quand on les trace point par point en fonction de la fréquence. La courbe de la rétine principale s'adapte bien au nomogramme de Dartnall, celle de la rétine secondaire beaucoup moins (Fig. 11 b), surtout parce que la lumière frappe cette rétine indirectement, après transmission à travers les pigments oculaires ou réflexion sur eux.

3. Dioptrique des yeux de quelques Polychètes.

L'étude morphologique des organes photorécepteurs (Section I) a

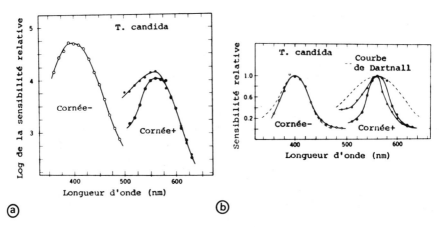

Fig. 11 : Courbes de la sensibilité spectrale de T. candida.

Fig. 11 a : Courbes de la sensibilité spectrale de la rétine primaire (cornée négative) et de la rétine accessoire (cornée positive).

Fig. 11 b : Courbes de la sensibilité spectrale de la rétine primaire et de la rétine accessoire tracées linéairement point par point et comparées à la fonction de Dartnall (d'après Wald & Rayport, 1977).

fait ressortir la grande diversité des structures qui d'après leur position par rapport aux formations photoréceptrices ont été qualifiées de lentille; mais leur rôle exact n'est pas connu. Seul Carricaburu et Kernéis (1975) ont mesuré l'indice de réfraction et essayé de déterminer le trajet des rayons lumineux (Figs. 4 a à 4 c) dans les yeux élémentaires des taches oculaires des filaments branchiaux des Sabellidae. Chez les trois espèces étudiées (Branchiomma vesiculosum, Dasychone bombyx et Potamilla reniformis) ces auteurs font remarquer que ni les cuticules, ni les lentilles ne présentent de biréfringence et trouvent que les lentilles sont constituées d'une substance dont l'indice de réfraction constant est supérieur à celui des cuticules qui sont optiquement homogènes. L'oeil élémentaire de B. vesiculosum est un système convergent fortement hypermétrope et comme les structures photosensibles interceptent les rayons contenus dans un cône de demi-angle au sommet de l'ordre de 30°, le système réfringent ne semble jouer qu'un rôle secondaire, sinon nul. Les yeux élémentaires de D. bombyx et de P. reniformis ne présentent aucune convergence et la diffraction empêche toute concentration de lumière sur l'élément photosensible (Carricaburu & Kernéis, 1975).

Ainsi comme le pensent ces auteurs, les yeux élémentaires de ces Annélides ne sont pas optiquement semblables aux ommatidies des Arthropodes. Les cuticules cornées des yeux composés d'Insectes et de Crustacés sont en effet fortement biréfringentes et sont formées de deux couches avec un indice moyen de réfraction supérieur à celui du cône cristallin (Carricaburu, 1967).

Les Hirudinées

Les cellules visuelles des sangsues sont spécialement appropriées pour analyser les changements électriques associés aux réponses à la lumière. Nous savons en effet que chez ces Annélides, la membrane microvillaire entoure une vacuole qui est reliée à l'extérieur par d'étroits canaux (Section 1 et Fig. 12). Il est donc possible d'enregistrer non seulement les potentiels entre l'intérieur et l'extérieur de la cellule mais aussi les courants à travers la membrane microvillaire.

Dès 1963, Walther a fait des enregistrements intracellulaires des cellules photoréceptrices et a trouvé que leur membrane est dépolarisée en réponse à l'éclairement. Ultérieurement cette étude a été confirmée (Walther, 1966, 1970; Lasansky & Fuortes, 1969; Fioravanti & Fuortes, 1972). Kretz et al. (1976) résument ainsi ces travaux: l'absorption de la lumière accroît la conductance de la membrane vacuolaire et de ce fait induit, une série de courants dans la cellule réceptrice. Les courants passent de l'espace extracellulaire vers la vacuole, créant à la fois un flux vers l'intérieur à travers la membrane microvillaire et un flux vers l'extérieur à travers la membrane externe (Fig. 12). Le passage de ce courant provoque une dépolarisation de la membrane de la cellule réceptrice et une polarisation négative de la vacuole par rapport au fluide extracellulaire. La dépolarisation induite par la lumière au niveau de la membrane de la cellule réceptrice est à l'origine d'impulsions qui se dirigent vers un site d'initiation axonal. Ces impulsions sont enregistrées comme des "spikes" positifs par une microélectrode introduite soit dans le cytoplasme, soit dans la vacuole de la cellule réceptrice.

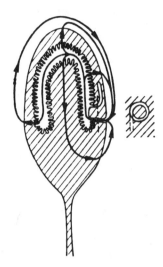

Fig. 12. Schéma d'une cellule photoréceptrice de sangsue mettant en évidence les courants induits lors de l'éclairement (flèches) (d'après Fioravanti & Fuortes, 1972).

Kretz et al. (1976) ont tenté de déterminer la manière dont l'information sensorielle primaire est acheminée des yeux et des récepteurs segmentaires au système nerveux central. Ils ont montré que les caractéristiques de la réponse de la cellule sensorielle ressemblent beaucoup à celles des courants de réponse enregistrés simultanément sur le nerf optique. En accord avec ce que laissaient présumer les données anatomiques, le courant de réponse correspond à la somme des activités élémentaires du photorécepteur correspondant. En réponse à un éclairement, les photorécepteurs produisent un train d'impulsions dont la fréquence s'élève d'abord fortement puis décroît et reste stable jusqu'à la fin de l'éclairement; l'amplitude du pic et le niveau du plateau augmentent linéairement avec le log de l'intensité de l'éclairement. Les photorécepteurs des yeux et des métamères s'adaptent à l'intensité du fond lumineux entre les périodes d'éclairement. L'adaptation des photorécepteurs oculaires est tellement importante que leur niveau d'activité de base est approximativement indépendant de l'intensité du fond lumineux. L'adaptation des récepteurs segmentaires est par contre incomplète ainsi leur activité de base s'élève avec l'intensité du fond lumineux. Les récepteurs segmentaires S_1 et S_2 pourraient, en plus de la détection des fluctuations rapides de l'intensité lumineuse, servir comme indicateur du niveau de l'éclairement ambiant. Les récepteurs segmentaires permettraient aux sangsues de détecter les variations lentes de l'éclairement: en particulier la photopériode et ses variations saisonnières.

Chez Hirudo, la réponse photosensible d'un récepteur segmentaire présente beaucoup moins de sensibilité sélective que celle d'un oeil simple. Il semblerait que la principale information que le système nerveux de la sangsue retire des informations fournies par les photorécepteurs segmentaires est de savoir qu'elles sont les parties du corps qui sont à la lumière et celles qui sont à l'obscurité. Par contre, la plus grande sélectivité directionnelle des yeux fait qu'il est possible qu'ils fournissent une vision sommaire des formes.

Chez Hirudo, la courbe de la réponse spectrale présente un maximum vers 540 nm. Les photorécepteurs oculaires et segmentaires sont presque insensibles aux lumières violette et rouge.

Des neurones sensoriels d'ordre supérieur qui reçoivent et traitent les informations photosensibles fournies par les récepteurs segmentaires ont pu être identifiés (Kretz et al., 1976).

3. PHOTOCOMPORTEMENT

L'orientation vis-à-vis de la lumière (phototactisme)

En général les Annélides recherchent les zones les moins éclairées mais dans certaines conditions elles se dirigent néanmoins vers la lumière.

Chez Lumbricus terrestris, Hess (1924) a montré que si les vers sont adaptés préalablement à l'obscurité, l'éclairement latéral du corps provoque une réaction de fuite pour des intensités supérieures à 0,001 lux tandis que pour des éclairements d'intensité inférieure les vers présentent un phototactisme positif. Toutefois si le ver est exposé au préalable à une lumière forte, le phototactisme n'est négatif que pour des intensités lumineuses supérieures à 0,3 lux. Chez cette espèce le phototactisme est donc négatif pour des éclairements forts mais devient positif pour des éclairements faibles; l'intensité critique correspondant au changement du sens de la réaction, dépendant de l'adaptation préalable de l'animal à la lumière ou à l'obscurité.

D'autres espèces d'Oligochètes telle Pheretima agrestis sont par contre complètement photonégatives et répondent en proportion de l'intensité lumineuse (Howell, 1939).

Un renversement partiel de la réponse normale à la lumière a été signalé (Hess, 1924; Howell, 1939) après ablation du ganglion cérébral ou section des commissures périoesophagiennes. L. terrestris devient photopositif en lumière modérée et P. agrestis en lumière faible. D'autre part, chez les individus (L. terrestris) dont la chaîne nerveuse ventrale a été sectionnée, la partie antérieure qui est sous le contrôle du ganglion cérébral, est photonégative tandis que la partie postérieure est photopositive (Hess, 1924). Le ganglion cérébral semblerait être le centre ultime de contrôle et de coordination (Hess, 1924; Nomura, 1926; Prosser, 1934; Howell, 1939). Cependant plus récemment Doolittle (1972) trouve que les vers privés de leurs ganglions cérébroïdes fuient la lumière comme les témoins. Blue (1976) confirme ce résultat chez L. terrestris et signale en outre que les individus opérés manifestent une plus grande capacité à fuir la lumière (100 W). Cet auteur pense que dès l'ablation du prostomium, les régions moyennes et postérieures deviennent plus sensibles à d'intenses sources de lumière; il y aurait activation de cellules sensorielles de type lenticulaire.

Les Hirudinées sont aussi fortement photonégatives. La Table I montre le résultat d'une série d'observations qui consiste à laisser à diverses sangsues le choix entre la zone sombre et la zone éclairée d'un aquarium. La position des individus est notée durant 20 jours et chaque

TABLE I: Répartition des sangsues entre les zones éclairées et les zones sombres d'un aquarium (d'après Mann, 1962).

Espèces	Une sangsue		Plusieurs sangsues		
	% à la lumière	% à l'obscurité	Nb de sangsues	% à la lumière	% à l'obscurité
Piscicola geometra	0	100	6	2.9	97.1
Hemiclepsis marginata	35	65	3	25.9	74.1
Theromyzon tessulatum (repue)	11.8	88.2	3 repues	33.3	66.7
Theromyzon tessulatum (affamée)	80.0	20			
Glossiphonia complanata	0	100	5	21.2	78.8
Glossiphonia heteroclita	0	100	8	27.2	72.8
Helobdella stagnalis	0	100	6	4.5	95.5
Hirudo medicinalis	35	65	—	—	—
Erpobdella sp.	0	100	3	15.4	84.6

jour la zone sombre est changée de côté. Il ressort que la majorité des sangsues choisissent la zone sombre. Cependant les Theromyzon affamées se rencontrent beaucoup plus souvent dans la zone éclairée que les individus repus. Ceci laisserait penser que la nécessité d'obtenir un repas chez les espèces suceuses de sang, modifie leur réactions normales qui consistent à éviter la lumière (Mann, 1962; Bhatia, 1975).

Herter puis Denzer-Melbrandt (1935) qui donnèrent le choix de quatre intensités lumineuses (éclairée, légèrement sombre, fortement sombre et obscure) à diverses sangsues montrèrent également que les espèces parasites s'établissent plus souvent dans la zone éclairée que dans l'une ou l'autre des zones partiellement sombres. Les espèces parasites peuvent être photopositives (Mann, 1962).

Les Polychètes présentent aussi un comportement photonégatif. Ceci a été établi chez les Nereidae (Bohn, 1902; Herter, 1926; Ameln, 1930) et chez Nephtys (Clark, 1956). De même, les Polychètes sédentaires retirées de leur tube se dirigent vers les zones d'éclairement minimal (Nicol, 1950; Dragesco-Kernéis, 1980 b). Mais Nicol (1950) a également montré que Branchiomma vesiculosum oriente la partie distale de son tube et en conséquence sa couronne branchiale, vers la lumière. Un tel comportement avait déjà été observé chez Sabella spallanzanii et Hydroides uncinata (Loeb, 1906, 1918).

Chez certaines espèces de Polychètes, un renversement de la réponse photique habituelle a été signalé à plusieurs reprises dans la littérature. Durant la saison de reproduction, certaines espèces deviennent pélagiques et photopositives. Clark et Hess (1942) ont montré chez L'Eunicien Leodice fucata que les parties épitoques détachées de la souche asexuée réagissent photopositivement à toutes les intensités lumineuses au-dessus d'un seuil de 0,05 à 1 lux alors que les vers immatures n'ont une réponse positive que pour de faibles éclairements (0,01 à 0,1 lux), le phototactisme étant négatif pour des éclairements supérieurs.

Chez Dasychone lucullana (Sabellidae), Dragesco-Kernéis (1980 b), a montré que les taches oculaires segmentaires et peut être aussi les ocelles prostomiaux sont responsables des réactions phototactiques négatives de cette espèce. En outre cet auteur pense que des enregistrements d'électrorétinogrammes donneraient des réponses "on" pour les structures oculaires qui appartiennent au type rhabdomérique alors que les taches oculaires du panache branchial de type ciliaire donneraient des réponses "off".

Réactions aux changements d'intensité lumineuse

Chez les Oligochètes, Unteutsch (1937) lors de l'analyse de la réponse motrice aux ombres portées a été amené à concevoir l'existence de deux catégories de photorécepteurs chez Eisenia foetida et Lumbricus rubellus. L'une dont le maximum de sensibilité est dans le bleu serait activée par l'augmentation de l'intensité lumineuse tandis que l'autre excitable par une diminution de cette intensité permettrait la perception des ombres portées, leur maximum de sensibilité étant dans le jaune.

Les Hirudinées réagissent aussi aux ombres portées mais les espèces suceuses de sang et celles qui ne le sont pas se comportent différemment. Les espèces parasites présentent souvent des mouvements d'extension et d'exploration (Stammers, 1950; Mann, 1962; Bhatia, 1975). Les espèces non parasites s'aplatissent par contre contre le substrat ou cessent brusquement les mouvements de ventilation (Mann, 1962). Dans les circonstances favorables, la totalité de l'animal ne doit pas nécessairement être assombri; Kaiser (1954) a montré que les sangsues peuvent répondre à une diminution de l'intensité lumineuse qui a lieu sur une partie de leur ventouse postérieure. D'autre part, les sangsues décapitées répondent aussi bien que les animaux entiers. La réponse aux changements d'intensité lumineuse serait donc due aux récepteurs épidermiques sensibles à la lumière, répartis sur le corps.

Les sangsues s'adaptent assez rapidement aux diminutions de l'éclairement lorsque ce stimulus est répété plusieurs fois (Stammers, 1950; Kaiser, 1954). De plus, Kaiser (1954) a montré que chez Haemopis avec une réduction de 100% de l'intensité lumineuse, la première période de latence est de 100 secondes; celle-ci tombe rapidement aux environ de 10s où elle reste durant une vingtaine d'essais puis elle disparaît rapidement. Par contre avec une chute de 30% de l'intensité lumineuse, la réponse disparaît après seulement trois essais (Fig. 13).

Chez les Polychètes, de brusques changements de l'intensité lumineuse entraîne une réaction de retrait. Celle-ci a très souvent été observée chez les Polychètes sédentaires (cf. introduction dans Nicol, 1950) mais elle existe aussi chez les Polychètes errantes (Clark, 1960 a et b; Evans, 1969 a et b; Gwilliam, 1969).

Les Polychètes sédentaires réagissent aux brusques diminutions de l'intensité lumineuse mais non aux brusques augmentations de celle-ci (Hargitt, 1906; Hess, 1914; Nicol, 1950). Lorsque la stimulation lumineuse est répétée plusieurs fois, les animaux répondent à la première

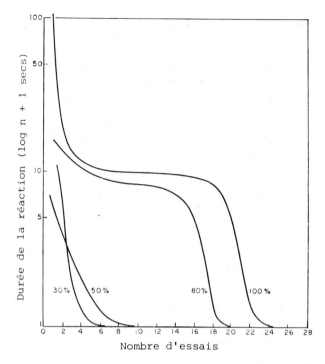

Fig. 13. Courbes d'adaptation d' Haemopis (Hirudinée) aux diminutions de l'intensité lumineuse. Le pourcentage de réduction de l'intensité lumineuse est indiqué à côté des courbes (d'après Kaiser, 1954).

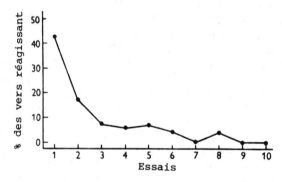

Fig. 14. Courbe d'adaptation de Branchiomma vesiculosum aux brusques diminutions de l'intensité lumineuse (d'après Nicol, 1950).

ou aux quelques premières par un retrait et ensuite ne réagissent plus (Fig. 14); ils s'adaptent rapidement aux brusques diminutions de l'intensité

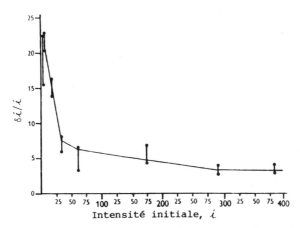

Fig. 15. Changement d'intensité minimal nécessaire pour provoquer une réponse aux intensités (i). Le point supérieur représente la valeur minimale pour laquelle une réponse se produit, le point inférieur la valeur maximale pour laquelle aucune réponse n'apparaît.

(En ordonnée: $\delta i/i$ x 100; en abscisse: intensité i en lux).
(d'après Nicol, 1950).

lumineuse (Bohn, 1902; Hargitt, 1906; Hess, 1914; Nicol, 1950). Chez Branchiomma vesiculosum un changement d'intensité relativement plus grand est nécessaire pour avoir une réponse aux faibles intensités (Nicol, 1950) (Fig. 15). Nicol (1950) a d'autre part montré que des animaux maintenus un certain temps soit à la lumière, soit à l'obscurité sont sensibles à des intensités de lumière au moins aussi faibles que 0,36 lux. Les animaux non importunés, montrent une sensibilité accrue et de bonnes réactions; par contre des stimulations fréquentes augmentent le seuil de stimulation et établissent un état d'adaptation durable. Des animaux adaptés à une brusque diminution de l'intensité lumineuse, sont encore sensibles aux ombres portées (Fig. 16).

La plupart des espèces de Nereis vivent d'une façon plus ou moins permanente dans des tubes et à beaucoup d'égard leur comportement est adapté à une vie tubicole (Clark, 1959); en particulier, elles manifestent un réflexe de retrait lors de brusques stimulations (Clark, 1960 a et b). N. pelagica réagit à une brusque diminution de l'éclairement et aux ombres portées par une contraction rapide. Toutefois le ver s'adapte rapidement à ce second stimulus et de ce fait la réponse rapide est remplacée par une contraction plus lente. N. pelagica réagit aussi à une brusque augmentation de l'éclairement mais dans ce cas les contractions sont souvent lentes et incomplètes (Clark, 1960 a). Evans (1969 a) a montré de plus chez N. diversicolor que les réponses aux brusques diminutions de l'éclairement sont d'ordinaire des retraits des extrémités antérieures tandis que de brusques augmentations de l'intensité lumineuse provoque des retraits des

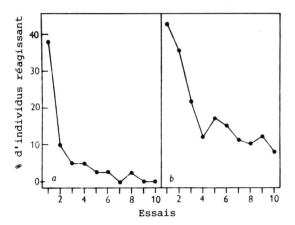

Fig. 16. Courbes d'adaptation de Branchiomma vesiculosum aux stimulations lumineuses: (a) brusques diminutions de l'intensité lumineuse; (b) ombres portées. L'essai b 1 suit immédiatement l'essai a 10 (d'après Nicol, 1950).

extrémités postérieures; les retraits dus aux premiers stimuli sont d'autre part plus rapides et ont une latence plus courte que ceux dus aux seconds stimuli.

N. pelagica s'adapte rapidement aux ombres portées, moins vite aux brusques diminutions de l'intensité lumineuse et lentement aux brusques augmentations de l'éclairement (Figs. 17 et 18) (Clark, 1960 a). Toutes les Polychètes errantes ne réagissent cependant pas de la meme façon (Evans, 1969 b). Il n'existe pas de différences nettes dans les fréquences des réponses ou les vitesses d'adaptation à de soudaines diminutions de l'éclairement et aux ombres chez N. diversicolor, N. pelagica et Pl. dumerilii. Le seul stimulus lumineux dont les vitesses d'adaptation et les fréquences de réponses différent, est une brusque augmentation de l'éclairement; Pl. dumerilii répond à peine à ce stimulus. Cette espèce est la seule des trois à être infra-littorale. Son comportement envers les brusques augmentations de lumière paraît ressembler à celui des autres polychètes infra-littorales telles que Branchiomma vesiculosum (Nicol, 1950), Hydroides dianthus (Yerkes, 1906) et Serpula vermicularis (Hess 1914) dont les réflexes aux ombres portées sont bien développés mais qui ne répondent pas aux augmentations brusques de l'intensité lumineuse. Chez N. diversicolor et N. pelagica, Clark (1960 b) a également pu montrer en adaptant des vers aux brusques diminutions de l'éclairement puis aux ombres portées et vice versa que N. diversicolor ne distingue pas entre une diminution de l'intensité lumineuse produite par des ombres et celle produite soudainement (Fig. 19). N. pelagica par contre distingue entre des ombres portées et une brusque diminution de l'éclairement. Chez cette espèce, l'adaptation au second stimulus est plus longue que l'adaptation au premier (Fig. 20) en outre il existe une interaction complexe entre les deux processus d'adaptation (Clark, 1960 a).

PHOTORECEPTION IN ANNELIDS 325

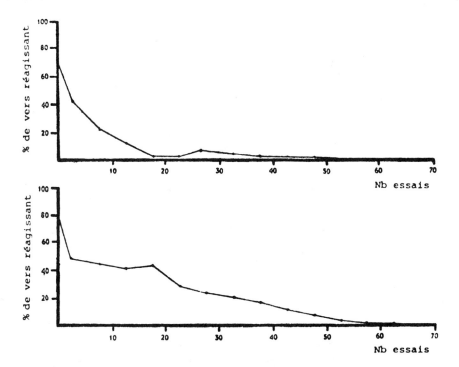

Fig. 17. Diagramme supérieur. Adaptation de Nereis pelagica aux brusques diminutions de l'intensité lumineuse. Les vers sont préalablement adaptés à la lumière. Diagramme inférieur - Adaptation de N. pelagica aux brusques augmentations de l'intensité lumineuse. Les vers sont préalablement adaptés à l'obscurité. (d'après Clark, 1960 a).

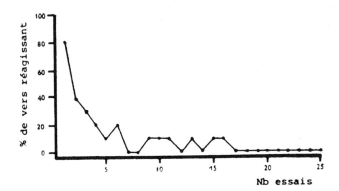

Fig. 18. Courbe d'adaptation de N. pelagica aux ombres portées. Les vers sont préalablement adaptés à la lumière. (d'après Clark, 1960 a).

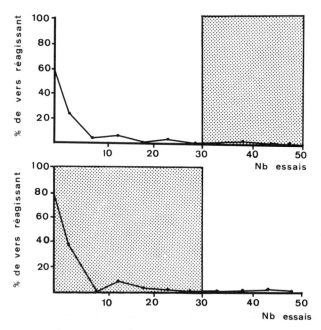

Fig. 19. Diagramme supérieur - Adaptation de N. diversicolor aux brusques diminutions de l'intensité lumineuse durant 30 essais puis aux ombres portées durant 20 essais. Diagramme inférieur - Les stimuli sont inversés. (d'après Clark, 1960 b).

CONCLUSIONS

L'étude morphologique des organes photorécepteurs des Annélides révèle l'existence de différentes structures. Les organites sensibles à la lumière sont représentés soit par des cils, soit par des microvillosités qui sont disposées autour d'un prolongement apical photosensible ou qui délimitent une vacuole interne encore appelée phaosome ou "Binnenkörper". Toutefois si le phaosome des Hirudinées est dépourvu de cils (seule une paire de centrioles a pu être observée au voisinage des microvillosités chez Helobdella), celui des Oligochètes en possède plusieurs de structure $9 + 0$. Le prolongement apical photorécepteur renferme aussi assez souvent une formation ciliaire; le cil rudimentaire lorsqu'il est présent se situant à l'apex de l'extrémité photosensible.

Une gradation dans le développement du cil s'observe donc chez les Annélides; les photorécepteurs passent insensiblement du type rhabdomérique au type ciliaire et constituent alors des exceptions à la théorie diphylétique d'Eakin (1968, 1972), théorie selon laquelle les organes photorécepteurs des Annélides appartiennent au type rhabdomérique. Devant le nombre croissant des exceptions Vanfleteren et Coomans (1976) ont avancé l'hypothèse que l'organe photorécepteur est toujours initié par une formation ciliaire; après la phase d'induction, cette formation se

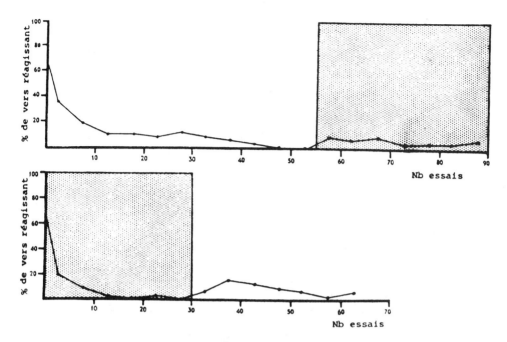

Fig. 20. Diagramme supérieur - Adaptations de N. pelagica aux brusques diminutions de l'intensité lumineuse durant 55 essais puis aux ombres portées. Diagramme inférieur - Les stimuli sont inversés (d'après Clark, 1960 b).

réduirait (type rhabdomérique) ou se développerait (type ciliaire). Par contre Salvini-Plawen et Mayr (1977) prétendent que les photorécepteurs sont issus indépendamment d'au moins 40 si ce n'est 65 ou plus lignées phylogénétiques différentes. Nous ne discuterons pas ici de la valeur de ces diverses théories car il nous paraît primordial de connaître l'ontogenèse des photorécepteurs, ce qui jusqu'à présent n'a guère été réalisé.

L'étude des organes photorécepteurs des Annélides fait ressortir plusieurs points essentiels. Avant de nous attarder sur ces derniers nous rappellerons les grandes lignes de l'évolution de ce phylum. Les Polychètes sont considérés comme le groupe le plus ancien. Ils ont donné naissance aux Oligochètes desquels ont dérivé ultérieurement les Hirudinées.

1) Si l'on excepte les cellules ciliées observées de chaque côté du cerveau de quelques Oligochètes limicoles, les Clitellates (Oligochètes + Hirudinées) ne possèdent que des cellules sensorielles photoréceptrices de type phaosome.

2) Les organes photorécepteurs cérébraux des Polychètes présentent une grande diversité de structure (organe bi- ou pluricellulaire, vision indirecte ou directe, lentille de "type cellulaire" ou sécrétée). Mais dans tous les cas, ils sont formés par la juxtaposition de deux types cellulaires (cellule sensorielle photoréceptrice et cellule pigmentaire de

soutien) auxquelles s'ajoutent parfois une cellule lenticulaire ou une cellule sécrétrice. La cellule sensorielle est de morphologie constante. Toujours pourvue de nombreuses microvillosités rhabdomériques réparties autour du prolongement apical, cette cellule renferme aussi assez souvent une formation ciliaire plus ou moins développé.

3) Chez les Polychètes, les structures visuelles localisées dans les autres régions du corps différent des organes photorécepteurs cérébraux.

4) La structure de l'ocelle de la trochophore des Polychètes est très voisine de celle de l'ocelle de la trochophore des Archiannélides.

REFERENCES

Ameln, P. (1930) Der Lichtsinn von Nereis diversicolor O.F. Müller. Zool. Jb., Abt. Allg. Zool. 47: 685-722.

Bassot, J.-M. & Nicolas, M.-T. (1978) Similar paracrystals of endoplasmic reticulum in the photoemitters and the photoreceptors of scale-worms. Experientia 34: 726-728.

Bhatia, M.L. (1956) Extraocular photoreceptors in the land leech Haemadipsa zeylanica agilis (Moore), from Nainital, Almora (India). Nature (Lond.) 176: 420-421.

Bhatia, M.L. (1975) Land leeches, their adaptation, and response to external stimuli. Zool. Poloniae 25: 31-53.

Blue, J. (1976) Effect of anterior ganglia removal on phototaxis in the earthworm (Lumbriscus terrestris). Bull. Psychonomic Soc. 7: 257-259.

Bocquet, M. (1971) Contribution à l'étude infrastructurale de l'organe photorécepteur des Syllidae (Annélides Polychètes). D.E.A. Lille, p. 1-21.

Bocquet, M. (1976) Ultrastructure de l'organe photorécepteur d'Autolytus pictus (Annélide Polychète). Etude chez la souche, le stolon parvenu à maturité sexuelle et la tête régénérée. J. Microsc. Biol. Cell. 25: 61-66.

Bocquet, M. (1977) Etude ultrastructurale de l'organe photorécepteur d'Odontosyllis ctenostoma S/F Eusyllinae (Annélide Polychète). J. Ultrastruct. Res. 58: 210-217.

Bocquet, M & Dhainaut-Courtois, N. (1972) L'infrastructure de l'organe photorécepteur des Syllidae (Annélides Polychètes). C.R. Acad. Sci. 274 D: 1689-1692.

Bocquet, M. & Dhainaut-Courtois, N. (1973) Structure fine de l'organe photorécepteur des Syllidae (Annélides Polychètes). I. Etude chez la souche et le stolon parvenu à maturité sexuelle. J. Microsc. 18: 207-230.

Bohn, G. (1902) Contribution à la psychologie des Annélides. Bull. Inst. Gén. Psychol. 2: 317-325.

Bradke, D.L. (1962) A unique invertebrate photoreceptor. In: Fifth Intern. Congr. Electron Microscopy, Vol. 2, Ed. S.S. Breese (Jr.). New York, London, Academic Press, p. R5.

Brandenburger, J.L. & Eakin, R.M. (1981) Fine structure of ocelli in larvae of an Archiannelid, Polygordius cf. appendiculatus. Zoomorphol. 99: 23-36.

Carricaburu, P. (1967) Contribution à la dioptrique oculaire des Arthropodes: détermination des indices des milieux transparents de l'ommatidie. Thèse de Doctorat ès Sciences. In: Mém. Soc. Hist. nat. Afr. N. 1968, 9: 1-146.

Carricaburu, P. & Kernéis, A. (1975) Dioptrique des yeux de quelques Annélides Polychètes. Vision Res. 15: 123-127.

Clark, A.W. (1967) The fine structure of the eye of the leech, Helobdella stagnalis J. Cell. Sci. 2: 341-348.

Clark, R.B. (1956) The eyes and the photonegative behaviour of Nephtys (Annelida, Polychaeta). J. Exp. Biol. 461-477.

Clark, R.B. (1959) The tubicolous habit and the fighting reactions of the polychaete Nereis pelagica. Anim. Behav. 7: 85-90.

Clark, R.B. (1960 a) Habituation of the polychaete Nereis to sudden stimuli. 1. General properties of the habituation process. Anim. Behav. 8: 82-91.

Clark, R.B. (1960 b) Habituation of the polychaete Nereis to sudden stimuli. 2. Biological significance of habituation. Anim. Behav. 8: 92-103.

Clark, R.B. & Hess, W.N. (1942) The reactions of the Atlantic Palolo, Leodice fucata, to light. Pap. Tortugas Lab. Carnegie Inst. Washington 33: 71-81.

Corbière-Tichané, G. (1977) Photorécepteurs extraoculaires chez les Invertébrés. Vision Res. 17: 459-462.

Dales, R.P. (1963) Annelids. Ed. H. Munro Fox. London, Hutchinson & Co. Ltd. p. 1-200.

Denzer-Melbrandt, U. (1935) Helligkeits - und Farbensinn bei deutschen Süsswasseregeln. Zool. Jhb. Abt. Physiol. 55: 525-562.

De Quatrefages, A. (1850) Etudes sur les types inférieurs de l'embranchement des Annelés. Mémoire sur la famille des Polyophthalmiens (Polyophthalmea nob.) Ann. Sci. Nat. Série III 13: 1-24.

Dhainaut-Courtois, N. (1965) Sur la présence d'un organe photorécepteur dans le cerveau de Nereis pelagica L. (Annélide Polychète). C.R. Acad. Sci. 261: 1085-1088.

Dhainaut-Courtois, N. (1968) Etude histologique et ultrastructurale des cellules nerveuses du ganglion cérébral de Nereis pelagica L. (Annélide Polychète). Comparaison entre les types cellulaires I-VI et ceux décrits antérieurement chez les Nereidae. Gen. Comp. Endocr. 11: 414-443.

Dhainaut-Courtois, N. (1970) Contribution à l'étude morphologique des processus sécrétoires dans le système nerveux central et au niveau de la glande infracérébrale des Nereidae (Annélides Polychètes). Thèse de Doctorat Sci. Nat. Lille, p. 1-191.

Doolittle, J.H. (1972) The role of anterior ganglia in phototaxis and thigmotaxis in the earthworm. Psychonomic Sci. 27: 151-152.

Dorsett, D.A. & Hyde, R. (1968) The fine structure of the lens and photoreceptors of Nereis virens. Z. Zellforsch. 85: 243-255.

Dragesco-Kernéis, A. (1979) Sur la régénération de la tache oculaire du panache, de Branchiomma vesiculosum (Montagu), Annélide Polychète. C.R. Acad. Sci. Paris 288 D: 1179-1182.

Dragesco-Kernéis, A. (1980 a) Taches oculaires segmentaires chez Dasychone (Annélides Polychètes). Etude ultrastructurale. Cah. Biol Mar. 21: 287-302.

Dragesco-Kernéis, A. (1980 b) Phototaxie chez Dasychone lucullana (Delle Chiaje). Cah. Biol. Mar. 21: 467-478.

Durchon, M. (1952) Recherches expérimentales sur deux aspects de la reproduction chez les Annélides Polychètes: l'épitoquie et la stolonisation. Ann. Sci. Nat. Zool. Biol. Animale 14: 119-206.

Durchon, M. (1959) Contribution à l'étude de la stolonisation chez les Syllidiens (Annélides Polychètes). I. Syllinae. Bull. Biol. France et Belgique 93: 155-219.

Eakin, R.M. (1968) Evolution of photoreceptors. In: Evolutionary Biology, Vol. 2. Ed. T. Dobzhansky, M.K. Hecht & W.C. Steere. New York; Appleton-Century-Crofts, p. 194-242.

Eakin, R.M. (1972) Structure of invertebrate photoreceptors. In: Handbook of Sensory Physiology, VII/I. Ed. H.J.A. Dartnall. Berlin, Heidelberg, New York, Springer-Verlag, p. 625-684.

Eakin, R.M. (1982) Continuity and diversity in photoreceptors. In: Visual Cells in Evolution. Ed. J.A. Westfall. New York, Raven Press, p. 91-105.

Eakin, R.M., Martin, G.G. & Reed, C.T. (1977) Evolutionary significance of fine structure of Archiannelid eyes. Zoomorphol. 88: 1-18.

Eakin, R.M. & Westfall, J.A. (1964) Further observations on the fine structure of some invertebrate eyes. Z. Zellforsch. 62: 310-332.

Ermak, T.H. & Eakin, R.M. (1976) Fine structure of the cerebral and pygidial ocelli in Chone ecaudata (Polychaeta: Sabellidae). J. Ultrastruct. Res. 54: 243-260.

Evans, S.M. (1969 a) Habituation of the withdrawal response in nereid polychaetes. 1. The habituation process in Nereis diversicolor. Biol. Bull. 137: 95-104.

Evans, S.M. (1969 b) Habituation of the withdrawal response in nereid polychaetes. 2. Rates of habituation in intact and decerebrate worms. Biol. Bull. 137: 105-117.

Fauvel, P. (1959) Organes des sens. In: Traité de Zoologie. Anatomie, Systématique, Biologie, tome V. Ed. P.-P. Grassé. Paris, Masson et Cie, p. 95-107.

Fioravanti, R. & Fuortes, M.G.F. (1972) Analysis of responses in visual cells of the leech. J. Physiol. 227: 173-194.

Fischer, A. & Brökelmann, J. (1965) Morphology and structural changes of the eye of Platynereis dumerilii (Polychaeta). In: The Structure of the Eye. II. Symposium. Ed. J.W. Rohen, Stuttgart, F.K. Schattauer Verlag, p. 171-174.

Fischer, A. & Brökelmann, J. (1966) Das Auge von Platynereis dumerilii (Polychaeta) sein Feinbau im ontogenetischen und adaptiven Wandel. Z. Zellforsch. 71: 217-244.

Golding, D.W. & Whittle, A.C. (1975) Secretory end-feet, extracerebral cells, and cerebral sense organs in certain limicole oligochaete annelids. Tissue Cell 7: 469-484.

Guérin, J.P. (1972) Le développement larvaire d' Armandia cirrosa Filippi (Annélide Polychète). Tethys 4: 963-974.

Gwilliam, G.F. (1969) Electrical responses to photic stimulation in the eyes and nervous system of nereid polychaetes. Biol. Bull. 136: 385-397.

Hachlov, J. (1910) Die Densillen und die Entstehung der Augen bei Hirudo medicinalis. Zool. Jb. Anat. Ontog. Tiere 30: 261-300.

Hansen, K. (1962) Elektronenmikroskopische Untersuchung der Hirudineen - Augen. Zool. Beitr. N.F. 7: 83-128.

Harant, H. & Grassé, P.-P. (1959) Classe des Annélides Achètes ou Hirudinées ou Sangsues. In: Traité de Zoologie. Anatomie, Systématique, Biologie, Tome V. Ed. P.-P. Grassé. Paris, Masson et Cie, p. 471-593.

Hargitt, C.W. (1906) Experiments on the behaviour of tubicolous annelids. J. Exp. Zool. 3: 295-320.

Hermans, C.O. (1969 a) The systematic position of the archiannelida. Syst. Zool. 18: 85-102.

Hermans, C.O. (1969 b) Fine structure of the segmental ocelli of Armandia brevis (Polychaeta: Opheliidae). Z. Zellforsch. 96: 361-371.

Hermans, C.O. & Cloney, R.A. (1966) Fine structure of the prostomial eyes of Armandia brevis (Polychaeta: Opheliidae). Z. Zellforsch. 72: 583-596.

Hermans, C.O. & Eakin, R.M. (1974) Fine structure of the eyes of an Alciopid polychaete, Vanadis tagensis (Annelida). Z. Morph. Tiere 79: 245-267.

Herter, K. (1926) Versuche über die Phototaxis von Nereis diversicolor O.F. Müller. Z. vergl. Physiol. 4: 103-141.

Hess, C. (1914) Untersuchungen über den Lichtsinn mariner Würmer und Krebse. Pflüger's Arch. ges. Physiol. 155: 421-435.

Hess, W.N. (1924) Reactions to light in the earthworm, Lumbricus terrestris. J. Morph. 39: 515-542.

Hess, W.M. (1925) Photoreceptors of Lumbricus terrestris, with special reference to their distribution, structure and function. J. Morph. Physiol. 41: 63-93.

Hesse, R. (1896) Untersuchungen über die Organe der Lichtempfindung bei niederen Tieren. I. Die Organe der Lichtempfindung bei den lumbriciden. Z. Wiss. Zool. 61: 393-419.

Hesse, R. (1897) Untersuchungen über die Organe der Lichtempfindung bei niederen Tieren. III. Die Sehorgane der Hirudineen. Z. Wiss. Zool. 62: 247-283.

Hesse, R. (1899) Untersuchungen über die Organe der Lichtempfindung bei niederen Tieren. V. Die Augen polychäter Anneliden. Z. Wiss. Zool. 65: 446-516.

Hirata, K., Ohsako, N. & Mabuchi, K. (1969) Fine structure of the photoreceptor cell of the earthworm Eisenia foetida. Rep. Fac. Sci. Kagoshima Univ. 2: 127-142.

Holborow, P.L. & Laverack, M.S. (1972) Presumptive photoreceptor structures of the trochophore of Harmothöe imbricata (Polychaeta). Mar. Behav. Physiol. 1: 139-156.

Howell, C.D. (1939) The response to light in the earthworm, Pheretima agrestis Goto and Hatai, with special reference to the function of

the nervous system. J. Exp. Zool. 81: 231-259.
Jamieson, B.G.M. (1981) The Ultrastructure of the Oligochaeta. London, Academic Press.
Jung, D. (1963) Bau und Feinstruktur der Augen auf dem vorderen und hinteren Saugnapf des Fischegels Piscicola geometra L. Zool. Beitr. N.F. 9: 121-172.
Kaiser, F. (1954) Beiträge zur Bewegungsphysiologie der Hirudineen. Zool. Jb. (Allg. Zool.) 65: 59-90.
Kernéis, A. (1966) Photorécepteurs du panache de Dasychone bombyx (Dalyell) Annélides Polychètes. Morphologie et ultrastructure. C.R. Acad. Sci. (Paris) 263 D: 653-656.
Kernéis, A. (1968 a) Ultrastructure de photorécepteurs de Dasychone (Annélides Polychètes Sabellidae). J. Microsc. 7: 40 a.
Kernéis, A. (1968 b) Nouvelles données histochimiques et ultrastructurales sur les photorécepteurs "branchiaux" de Dasychone bombyx Dalyell (Annélide Polychète). Z. Zellforsch. 86: 280-292.
Kernéis, A. (1971) Etudes histologique et ultrastructurale des organes photorécepteurs du panache de Potamilla reniformis (O.F. Müller), Annélide Polychète. C.R. Acad. Sci. (Paris) 273 D: 372-375.
Kernéis, A. (1975) Etude comparée d'organes photorécepteurs de Sabellidae (Annélides Polychètes). J. Ultrastruct. Res. 53: 164-179.
Krasne, F.B. & Lawrence, P.A. (1966) Structure of the photoreceptors in the compound eyespots of Branchiomma vesiculosum. J. Cell. Sci. 1: 239-248.
Kretz, J.R., Stent, G.S. & Kristan, W.B. Jr. (1976) Photosensory input pathways in the medicinal leech. J. Comp. Physiol. 106: 1-37.
Lasansky, A. & Fuortes, M.G.F. (1969) The site of origin of electrical responses in visual cells of the leech, Hirudo medicinalis. J. Cell Biol. 42: 241-252.
Laverack, M.S. (1969) Mechanoreceptors, photoreceptors and rapid conduction pathways in the leech, Hirudo medicinalis. J. Exp. Biol. 50: 129-140.
Lawrence, P.A. & Krasne, F.B. (1965) Annelid ciliary photoreceptors. Science 148: 965-966.
Livanow, N. (1903) Untersuchungen zur Morphologie der Hirudineen. I. Das Neuro- und Myosomit der Hirudineen. Zool. Jb., Abr. Anat. 19: 29-90.
Livanow, N. (1904) Untersuchungen zur Morphologie der Hirudineen. II. Das Nervensystem des vorderen Körperendes und seine Metamerie. Zool. Jb., Abt. Anat. 20: 153-226.
Loeb, J. (1906) Further observations on the heliotropism of animals and its identity with the heliotropism of plants. In: Studies in General Physiology. p. 89-106.
Loeb, J. (1918) Forced Movements, Tropisms and Animal Conduct. Philadelphia, J.B. Lippincott. Co.
Malaquin, A. (1893) Recherches sur les Syllidiens, Morphologie, anatomie, reproduction, développement. Mém. Soc. Sc. Agr. Arts, Lille, 4° série, 18: 1-477.
Manaranche, R. (1968) Sur la présence de cellules d'allure photoréceptrice dans le ganglion cérébro'ide de Glycera convoluta

(Annélide Polychète). J. Microscopie 7: 44 a.
Manaranche, R. (1971) Ultrastructure de cellules d'allure photoréceptrice dans le ganglion cérébroïde de Glycera convoluta K. (Annélide Polychète). J. Microscopie 3: 433-440
Mann, K.H. (1962) Leeches (Hirudinea). Their structure, Physiology, Ecology and Embryology. Inter. Ser. Monographs on Pure and Appl. Sci., Oxford, Pergamon Press. p. 1-201.
Merker, G. & Vaupel-Von Harnack, M. (1967) Zur Feinstruktur des "Gehirns" und der Sinnesorgane von Protodrilus rubropharyngaeus Jaegersten (Archiannelida). Mit besonderer Berücksichtigung der neurosekretorischen Zellen. Z. Zellforsch. 81: 221-239.
Myhrberg, H.E. (1979) Fine structural analysis of the basal epidermal receptor cells in the earthworm (Lumbricus terrestris). Cell Tissue Res. 203: 257-266.
Nicol, J.A.C. (1950) Responses of Branchiomma vesiculosum (Montagu) to photic stimulation. J. Mar. Biol. Ass. U.K. 29: 303-320.
Niilonen, T. (1980) Fine structure of the phaosomous photoreceptor in the larvae of Polydora ligni Webster (Polychaeta: Spionidae). Acta Zool. Stockl. 61: 183-190.
Nomura, E. (1926) Effect of light on the movements of the earthworm Allolobophora foetida (Sav.). Tôhoku Imp. Univ. Sci. Rep. 4th Ser. 1: 293-409.
Prosser, C.L. (1934) Effect of the central nervous system on responses to light in Eisenia foetida Sav. J. Comp. Neurol. 59: 61-92.
Röhlich, P. & Török, L.J. (1964) Elektronenmikroskopische Beobachtungen an den Sehzellen des Blutegels, Hirudo medicinalis L. Z. Zellforsch. 63: 618-635.
Röhlich, P., Aros, B. & Virágh, S. (1970) Fine structure of photoreceptor cells in the earthworm Lumbricus terrestris. Z. Zellforsch. 104: 345-357.
Salvini-Plawen, L.V. & Mayr, E. (1977) On the evolution of photoreceptors and eyes. In: Evolutionary Biology, Vol. 10. Ed. M.K. Hecht, W.C. Steere & W. Wallace. New York, Plenum Publ. Co., p. 207-263.
Singla, C.L. (1975) Ultrastructure of the eyes of Arctonoë vittata Grübe (Polychaeta, Polynoidae). J. Ultrastruct. Res. 52: 333-339.
Spies, R.B. (1975) Structure and function of the head in Flabelligerid Polychaetes. J. Morph. 147: 187-208.
Stammers, F.M.G. (1950) Observations on the behaviour of land leeches (genus Haemadipsa). Parasitol. 40: 237-246.
Tampi, P.R.S. (1949) On the eyes of polychaetes. Proc. Indian Acad. Sci. B. 29: 129-147.
Unteutsch, W. (1937) Über den Licht - und Schattenreflex des Regenwurms. Zool. Jahrb. Physiol. 58: 69-112.
Vanfleteren, J.R. & Coomans, A. (1976) Photoreceptor evolution and phylogeny. Z. Zool. Syst. Evolut. - forsch. 14: 157-169.
Verger-Bocquet, M. (1977) Ultrastructure de l'organe photorécepteur d' Autolytus pictus (Annélide Polychète). Formation de l'oeil du stolon. Biol. Cell. 30: 65-72.
Verger-Bocquet, M. (1981 a) Contribution à l'étude des organes

photorécepteurs des Syllidae. Structure, développement régénération, involution. Thèse de Doctorat Sci. Nat. Lille, p. 1-144.

Verger-Bocquet, M. (1981 b) Etude comparative, au niveau infrastructural, entre l'oeil de souche et les taches oculaires du stolon chez Syllis spongicola Grübe (Annélide Polychète). Arch. Zool. Exp. Gén. 122: 253-258.

Wald, G. & Rayport, S. (1977) Vision in annelid worms. Science 196: 1434-1439.

Walther, J.B. (1963) Intracellular potentials from single cells in the eye of the leech Hirudo medicinalis. Proc. XVI internat. Congr. Zool. Washington D.C.

Walther, J.B. (1966) Single cell responses from the primitive eyes of an annelid. In: The Functional Organization of the Compound Eye. Ed. C.G. Bernhard, Oxford, Pergamon Press, p. 329-366.

Walther, J.B. (1970) Widerstandsmessungen an Sehzellen des Blutegels, Hirudo medicinalis L. Verh. dtsch. Zool. ges. 64: 161-164.

Walz, B. (1979) A Comparison of receptive and non-receptive plasma membrane areas of photoreceptor cells in the leech, Hirudo medicinalis. Cell Tissue Res. 198: 335-348.

White, R.H. & Walther, J.B. (1969) The leech photoreceptor cell: ultrastructure of clefts connecting the phaosome with extracellular space demonstrated by lanthanum deposition. Z. Zellforsch. 95: 102-108.

Whitman, C.O. (1886) The leeches of Japan. Quart. J. Micr. Sci. N.S. 26: 317-416.

Whittle, A.C. (1976) Reticular specializations in photoreceptors: a review. Zool. Scripta 5: 191-206.

Whittle, A.C. & Golding, D.W. (1974) The fine structure of prostomial photoreceptors in Eulalia viridis (Polychaeta: Annelida). Cell Tissue Res. 154: 379-398.

Yanase, T., Fugimoto, K. & Nishimura, T. (1964) The fine structure of the dorsal ocellus of the leech, Hirudo medicinalis. Memoirs Osaka Gakugei Univ. B. 13: 117-119.

Yerkes, W.A. (1906) Modification in behaviour in Hydroides dianthus. J. Comp. Neurol. Psychol. 16: 441-450.

Yingst, D.R., Fernandez, H.R. & Bishop, L.G. (1972) The spectral sensitivity of a littoral Annelid: Nereis mediator. J. Comp. Physiol. 77: 225-232.

Zahid, Z.R. & Golding, D.W. (1974) Structure and ultrastructure of the central nervous system of the Polychaeta Nephtys, with special reference to photoreceptor elements. Cell Tissue Res. 149: 567-576.

PHOTORECEPTOR STRUCTURES AND VISION

IN ARACHNIDS AND MYRIAPODS

ARTURO   MUÑOZ-CUEVAS

Chargé de Recherche au C.N.R.S.
Laboratoire de Zoologie (Arthropodes) M.N.H.N.
61, rue de Buffon, 75005 Paris, France

> What an eye she has! Methinks it sounds a parley of provocation.
>
> Othello. Act II, Sc. 3.

ABSTRACT

Arachnids

Photoreceptive structures among Arachnids are very varied as a result of the heterogeneity of the whole group. This variability is expressed both by the number and position of the eyes in Arachnids. Their number may reach 4 to 10 in Scorpions; 8 in Holopeltids and Amblypygids; 2 to 4 in Acarids and Pseudoscorpions, and 2 in Opilionids. In addition they are totally lacking in Palpigradida, Schizopeltids (Uropygids), Ricinulei and in whole Acarids families (i.e. Gamasidae). Their position may be medial or lateral. The eyes are always simple (ocelli), and, according to the orientation of the retinular cells and the rhabdom they may be differentiated into direct and indirect eyes.

This study deals with the development of the median eyes, in conjunction with the cytodifferentiation of retina. The differentiation of the optic vesicle is considered as a center of cellular differentiation. Data are also given on the ciliary induction of the rhabdomere according to the pattern used in Opilionids, on the morphogenesis and spatial organization of the rhabdom, and on the differentiation of the pigmentation of retina with the cytological pattern of formation of the ommochrome granules. Concerning the dioptric system, the cellular differentiation of the vitreous cell and the lens-secretion are discussed.

The accessory eyes are studied according to the three Homann's main types, i.e. the accessory eyes of primitive type (PT), the canoe-shaped ones (KT) and those belonging to the grate-shaped type (RT). The ontogenesis of the eyes of primitive type (PT) are analyzed.

The postembryonic development of the eyes is discussed with regards to the growth of the dioptric system and of the retina as well as the phenomena of ontogenic regression of the dioptric system in the cave-dwelling Ischyropsalis. The organization of the retina in diurnal Salticid spiders is analyzed with relation to the adults. As for nocturnal spiders, the organization of the retina in Dinopid spiders is discussed.

The physiology of vision is at first studied through the physiological dioptric of scorpions and spiders. Electrophysiology is limited to the following problems: spectral sensitivity and cellular categories; nocturnal and diurnal sensitivity; critical frequency fusion; role of the eyes in the control of the circadian rhythm in scorpions; perception of the polarized light in spiders and extraocular photoreception in scorpions.

Finally, some problems related to vision and sexual behavior in spiders, together with some visual mechanisms in the predation of spiders, are also discussed. Attention is drawn to several unanswered questions on vision so as to attract young researchers.

Myriapods

In Myriapods, the eyes are generally grouped laterally. Their number varies and they are reduced or often lacking, as in Pauropods, Symphyla, Geophilidae (Chilopoda) and Diplopoda (Polydesmidae). They are simple eyes or ocelli except in the Scutigeromorpha (Scutigera) which show typical facetted eyes.

This study discusses the general organization and adaptive changes due to light and dark of the eye in Chilopods (Lithobius forficatus) The main features of the facetted eye (Scutigera) are mentioned. Some structural features of the eye of Diplopods (Glomeris) are also discussed.

This report is completed with data on the electrophysiology of the eye in L. forficatus.

INTRODUCTION

Except for well known papers on eye formation and structure published at the end of the 19th century and at the beginning of the 20th, the study of vision in this group, despite its fascinating interest, never received as much attention as in Insects or Crustaceans.

Nowadays, for many zoologists, Arachnids still remain relative unknown group. This may be due to several reasons but the psychological aspect cannot be ignored. These factors may be related to imagination or to ancestral fears, but reticence seems to have played a rather important part in delaying the development of our knowledge.

However, it is now possible, thanks to many studies, to correlate morphological, physiological and behavioral data. The present knowledge of behavior shows the importance of vision, neglected up to now.

On the other hand, the great morphological variety of Arachnids, and their adaptability to very different environments of the aquatic, desert and cavern types give to each group specific structures which constitute sources of interest for the present study.

Our purpose, in this study, is to synthesize research dealing with vision in Arachnids with relation to the present views concerning evolution in the Chelicerata.

GENERAL DEFINITION OF ARACHNIDS

The Arachnids are Chelicerata, most of them being terrestrial, carnivorous, nocturnal and cryptozoic, with a body typically consisting of two main parts prosoma (anterior) and opisthosoma (posterior). On the prosoma, there are simple eyes and in the adult, six pairs of appendages: one of them (the chelicerae) is pre-oral, ending in claws or fangs; this is followed by a pair of pedipalps, at the bases of which the mouth usually opens; and the last four pairs are walking legs. Development is generally achieved without metamorphosis; the sexes can be distinguished and courtship is common. Instinctive behavior is also highly developed and, in some groups, social organization exists (Millot, 1949; Savory, 1977). The Arachnids are an extremely ancient class, Silurian Scorpions being among the first known terrestrial animals. Uropygids, Amblypygids, Solpugids Spiders, Harvestmen and Acarids are known from the Carboniferous period onwards and their aspect is not quite different from that of the living species, showing an astonishing stability of organization in each of these groups (Petrunkvitch, 1955).

EYE NUMBER AND POSITION IN THE LIVING ARACHNID ORDERS

The data cited in this section are from Grenacher (1879), Sheuring (1914), André (1949), Grandjean (1928, 1958), Millot (1949) and Homann (1971).

The Arachnids usually possess eyes, and they always occur on the anterior part of the prosoma; their number varies considerably from order to order and even within the same order. Ordinarily, they are 4 to 10 in Scorpions, 8 in Holopeltids and Amblypygids, 6 or 8 in Spiders, 4 or 6 in Solpugids, 2 or 4 in Acarids and Pseudoscorpions, and 2 in the Harvestmen. They are lacking in Palpigrads, Schizopeltids (Uropygids), Ricinuleids, in some Acarid families (Gamasidae), as well as in several species belonging to different groups which are mainly cave-dwelling. These are always simple eyes or ocelli, very different from the compound eyes of Insects or Crustaceans; fundamentally the eye consists of a sensory cup-shaped retina and a dioptric system. Direct and indirect eyes can be distinguished from the orientation of the retinular cells or the rhabdom.

Scorpions have two types of ocelli: median, and lateral. Despite important differences, they all belong to the direct type. The median eyes, only one pair, lie on a small tubercle in the middle of the dorsal plate of the prosoma. The lateral ones are gathered in a small group on each side of the prosoma and depending on the species may contain 2 to 5 ocelli.

In Pseudoscorpions, when they occur, there are 2 or 4 eyes. They are always on the sides of the anterior part of the cephalothorax; nearly always sessile, rarely situated on small tubercles (Garypidae); and typically indirect.

The Solpugids have a well-developed pair of median eyes, near the anterior edge of the propeltidium, and one or two pairs of lateral eyes; the latter being hardly visible exteriorly. All are typically direct.

In Opilionids, the eyes (two in number) are on a median prosomatic tubercle. In some groups (Biantinae), they are widely separated and localized near the edge of the prosoma. In the Trogulidae, they are in front of the body on two expansions of the prosoma. They are always of the direct type.

Among Acarids, the Cryptostigmata and Mesostigmata are mostly eyeless. In the Protostigmata, the eyes are usually dorsal and lateral on the thorax: 2, 4 or 5 in number. Each pair of eyes is rarely very close to the median line (Eyleis). They are usually lateral in terrestrial and aquatic Acarids. Sometimes, as in Thrombidium, they are carried by a free peduncle. Grandjean (1926, 1958) has pointed out the presence of eyes in groups in Oribatids so far known as eyeless.

In the Uropygids, the eyes occur in the Holopeltids only; in the Schizopeltids, the atrophied ocelli are reduced to modified tegumentary regions. The forms with eyes, may be divided into two types: 2 median eyes (direct type) near the end of the cephalic peltidium; or 2 groups of 3 subcontiguous lateral eyes (indirect types).

The eyes of the Amblypygids belong to two different types: 2 median eyes (direct type) lying on the anterior part of the mediocephalic groove, and 2 triads of subcontiguous lateral eyes (indirect type).

The eyes of Spiders belong to two types: main and lateral eyes (Fig. 1, 2). The first which includes the anterior median eyes are of the direct type. The second type includes both the lateral and the posterior median eyes, with the retina being of the indirect type. The eyes lie on an area, the so-called ocular area, which is highly variable in size. They are typically 8 in number, which is never exceeded, but may drop to 6 in Sicariidae and Dysderidae, to 4 in Tetrablemma, and to 2 in Nops and Matta. In some cave-dwelling species (Telema) eyes are absent. Eyes are sessile and set in the tegument, or slightly raised and facing different directions. Sometimes, they are grouped on the same process (Hersilia), and more rarely carried on a true peduncle (Pholcus podophthalmus). More often, there are two subparallel transverse rows of 4 eyes, including anterior and posterior ones. But, depending on the family, important variations occur. The rows can run straight be strongly curved, or sometimes totally dislocated. In the Filistatidae, Oecobiidae, Ammoxenidae, Liphistiomorphae and many trap-door spiders, all the eyes are gathered into a more or less dense median group. In most Sicariidae three widely separated pairs occur. Their size varies not only among families, but also between the different pairs of the same animal. They may be fairly equal, strongly reduced, or considerably developed; for example, in the Salticidae, the anterior median eyes are more than ten times larger than the posterior median ones.

EMBRYONIC DEVELOPMENT OF EYES

Despite several classical papers dealing with the embryology of the eyes (Grenacher, 1879; Bertkau, 1886; Purcell, 1892, 1894; Patten, 1887;

PHOTORECEPTOR STRUCTURES IN ARACHNIDS AND MYRIAPODS   339

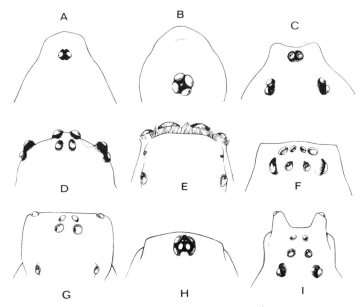

Fig. 1   Variation in the number and position of the eyes in Spiders; dorsal view
A- Nops (2)
C- Scytodes (6)
D- Aranea (8)
E- Viciria (8)
F- Heteropoda (8)
H- Liphistius (8)
I- Euphrostenops (8)
(from Millot, 1949)

Schimkewitsch, 1906) the results are generally very incomplete. It is only from the publications of Homann (1950 to 1971) that one gets the chronology and a more precise definition of eye-development. In this chapter, the general data about eye-development are based on Homann's studies on Spiders and Muñoz-Cuevas' on Harvestmen.

Median eyes

The embryological description of median eyes is mainly based on the results obtained in the Harvestmen; Pachylus quinamavidensis, Acanthopachylus aculeatus, Discocyrtus cornutus (Gonyleptidae), Ischyropsalis luteipes (Ischyropsalidae) (cf. Muñoz-Cuevas, 1981).

a)   Eye-development in P. quinamavidensis, an epigean species.

In P. quinamavidensis the embryonic development, at 20°C, lasts 37 days (Muñoz-Cuevas, 1970).
    The mean duration of each stage is as follows:
Stage I   :   Segmentation, from laying to 5 days.
Stage II :   Formation of the germ-layer, from the 5th to 10th day.
Stage III :   Metameric segmentation of prosoma, from 10th to 13th day.

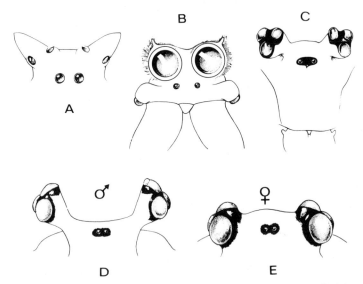

Fig. 2. Some typical dispositions of the eyes in Spiders; front view.
A- Thomisidae
B- Dinopidae
C- Pholcidae
D & E - Variation of the eyes depending on the sex, Pholcus podophthalmus
(from Millot, 1949)

Stage IV : Inversion of the embryo, from 13th to 21st day.
Stage V : Larval organogenesis, from 22nd to 29th day.
Larva : from 30th to 37th day.

During the 4th stage of embryonic development the primitive ocular rudiments are formed; they can be observed through the transparent chorion and are called ocular folds.

In P. quinamavidensis, the two ocular folds begin to differentiate on the cephalic plate, at the IV.2 stage; these crescent shaped folds are unpigmented. On histological slides, the primitive ocular rudiments occur as a deep ectodermic invagination. This invagination is at first continuous with the ectoderm; then, it divides and forms a cell cluster surrounded by the ectoderm, laterally and on its dorsal part. The ectodermic cell cluster so formed, has a ventral cavity which constitutes the primitive optic vesicle. The epidermal cells, of which it is made, discharge a material into the cavity; the shape of this material is more or less that of rolled membrans but cytomembrans are never formed. The frequent mitoses observed at the level of the layers surrounding the vesicle-cavity, are proof of the part played by the vesicle as the retinal differentiation center. The visual cells develop from the successive mitoses of the optic vesicle.

In P. quinamavidensis eye differentiation takes place during the Vth stage of the embryonic development. Retinal differentiation begins on the 22nd day. The mitotic activity of the optic vesicle results in primitive retinal cells which gather in fours to form the rhabdom. During this stage, the retina first occurs as a relatively large cupule of cells which grows in height. The growth is obvious in the apical part of the cell at the rhabdomeric level which looks like a fine net facing the epidermis. During growth and differentiation new retinal cells arise and the morphology and general disposition of this organ change. The cupule, dorsally fairly open, grows imperceptibly more and more pyriform. Changing its position in space, the base of the left retina gradually reaches the base of the right one. On frontal slides, the two retinae look like long horizontal sacs covered by the epidermis the two optic vesicles facing each other at the bottom of the retina.

During the last days of this stage, the epidermal cells which cover the retina begin to differentiate into the vitreous body. Initially it looks first like an epidermal thickening; the cells, show a basal nucleus; compared with the rhabdoms, they are vertically arranged and separated from the retina by a preretinal membrane.

The first signs of lens-differentiation can be observed at the end of the larval organogenesis period. Firstly, from the secretion of the vitreous body which includes a lamellar structure a small narrow body appears and increases rapidly in size.

In P. quinamavidensis the first signs of differentiation of the nervous fibers and the organization of these into an optic nerve can first be observed only about the 25th day. The lens quickly grows during the larval instar and at hatching, the dioptric system and the sensory part of the eye are functional.

In the larva, especially in the first nymph, the spatial arrangement of the optic vesicle undergoes changes. It is displaced towards the sagittal plane of the animal, at the bottom of the retina. This vesicular displacement is due to the retinal growth and to the orientation of the rhabdom towards the vitreous body.

b) Development of eye-pigmentation in P. quinamavidensis.

Eye pigmentation is one of the morphological features, which enables the development of the organ to be followed. In Gonyleptidae, the color of the ocular pigment does not differ from that of the pigmentary one.

The pigment, brown-colored, is first deposited at the beginning of stage V. At first, the pigment seems to be arranged irregularly in two areas on the ocular surface: a superior one, slightly pigmented, and an inferior one, more strongly colored. During the first 24 hours, only the ocular surfaces are pigmented. Twenty-four hours later, the pigmentation of the body begins and the ocular surface seems more strongly colored. Pigmentation of the superior, less pigmented ocular surface, is then reduced to form an internal clearer area.

In the prosomatic epidermis the pigment is deposited around the eyes. These are surrounded by two small pigmented areas on the anterior side of the prosoma, and two badly-delimited areas behind but on the anterior sides of the coeca. During development, the ocular pigment darkens, from light-brown at the beginning it becomes very dark. The distribution of the pigment on the ocular surface changes too and the internal half of the eye looks like a light area crossed by stripes of darker pigment.

c) Eye-development of I. luteipes, a troglophilous species.

Eye-development in this species shows the same main features as in Gonyleptidae. We shall only point out the chronology and development of pigmentation.

In I. luteipes, embryonic development, at 11,5°C, lasts 45 days; the mean duration of each stage is as follows:

Stage I : Segmentation, from laying to 6th day.
Stage II : Formation of the germ layer, from 6th to 10th day.
Stage III : Prosomatic metamerisation from 10th to 15th day.
Stage IV : Inversion of the embryo, from 15th to 25th day.
Stage V : Larval organogenesis, from 25th to 40th day.
Larva : From 41st to 45th day.

d) Development of eye-pigmentation in I. luteipes.

Eye-color of I. luteipes undergoes several changes from the first hours, when pigmentation appears, to hatching. Ocular pigmentation can be observed for the first time between the 24th and the 25th days of development (stage V 1). The ocular area is dotted on all its surface with light yellow-colored granules.

Between the 26th and 28th days (stage V 2), the ocular area divides into two irregularly pigmented regions. The anterior one, which is more densely pigmented is light reddish-brown, whereas the posterior one is light yellow, as in stage V 1.

At stage V 3 (from 29th to 31st day) the ocular surface, now more densely pigmented, shows three areas of pigmentation: the anterior and interior posterior ones are reddish-brown as in stage V 2; while the median one, lying in the middle of the ocular area along the external edge, is light brown. This area is surrounded by a narrow reddish-brown stripe.

During stage V 4 (from 32nd to 36th days) the pigmented surface is restructured and three areas occur on the eye-surface. The anterior area, reddish-brown or light brown in color, covers the anterior edge of the eye and extends beyond each side to the middle of the ocular area. The median area, quoted at stage V 3, is enlarged and is colored dark brown. The lens is differentiated towards the external edge of this area. The posterior area is triangular and shows a light reddish-brown pigmented region with a transitional color towards the median area, (reddish-brown/light brown).

At stage V 5 (37th to 40th days) the ocular surface is covered with two pigmented areas. One is dark brown and extends to the main part of the surface, it includes the lens which has moved forwards and outwards. The second area is peripheral and reddish-brown. Between these areas a small reddish-brown/light brown transitional region can be observed at the posterior end of the eye. During the larval instar (from 41st to 43rd days) pigmentation increases, and there are three areas on the ocular surface: a dark brown median area, a black area surrounding the lens, and a light reddish-brown peripheral one.

CELLULAR DIFFERENTIATION

Cellular differentiation of median eyes

Few ultrastructural studies of the retinal development have been carried on in Arthropods. The first paper was published by Waddington and Perry (1960) on Drosophila it was followed by other studies, dealing with Drosophila too, by Waddington and Perry (1963) and Perry (1968 a, b). Trujillo-Cenoz and Melamed (1973) undertook the study of the retina in two Muscid species: Phaenicia sericata and Sarcophaga bullata. More recently, Such (1975) studied the Phasme, Carausius morosus. Of note also is the study by Eakin (1966) on Peripatus (Onychophora) a group related to the Arthropods.

As for the Arachnids, the only papers published are those on Opilionids by Muñoz-Cuevas (1973 a-b, 1974, 1975 a-b, 1976 a, 1978 a, b, 1980 a-b, 1981).

a) Cellular differentiation of retina; optic vesicle.

As mentioned, the optic vesicle develops from an invagination of the cephalic ectoderm during the fourth stage of the embryonic development. The cavity of each vesicle is surrounded with one or two cell-layers. These cells belong to the epidermal type; they have relatively abundant rough endoplasmic reticulum, and a high density of free ribosomes. The plasma membrane surrounding the cavity consists of villi, comparable to the epidermal ones. The cavity is initially empty but, during development, an amorphous granular substance is secreted into it by the cells of the vesicle. Towards the 30th day of development, the main part of the cavity is filled with "membrane" structures. This course of activity of the optic vesicle is comparable to that of the neural organs during the morphogenesis of the nervous system in some Arthropods (Muñoz-Cuevas, 1973 c). The main feature of the optic vesicle is the mitotic ability of some of its cells. Cellular divisions observed in the ectodermal folds during invagination are related to the growth of the vesicular rudiment, whereas the mitoses observed in the optic vesicle after separation from ectodermis, are related to retinal differentiation.

It is relatively difficult, indeed impossible, due to the thickness of the seriated slides, to draw a curve showing the variations of mitotic frequency. An approximate evaluation of this frequency points out two periods of important mitotic activity; the first one occurs when the vesicle rudiments are formed, and the second one during the 100, or so, first hours of activity of the optic vesicle. After this second period, only a few mitoses can be observed.

Among Arthropods, the existence of a center of retinal differentiation could be observed in several Insect groups (Wolsky 1956; White 1961, 1963; Such 1975); whereas, in Arachnids, the concept of such a center is not clearly expressed.

By its ectodermal origin, its cytological structure as well as by its function as differentiation center, the optic vesicle must be considered homologous to the neural organs of the Opilionids (Muñoz-Cuevas, 1973 c).

b)   Differentiation of retinular cell and ciliary induction.

The results mentioned in this paragraph, are derived from papers dealing with the cellular differentiation of the retina in the Harvestmen, (Muñoz-Cuevas, 1973 a-b, 1975 a, b, 1981).

During the first 24 hours of stage V, the cells giving rise to the retinular cells lengthen; this lengthening is more obvious at the apical half of the cell. The size of the retinular cell approaches 13 μm; its basal nucleus is 5 to 6 μm, and its cytoplasm is highly electron-dense. This results in a sharp apical end, with two larger regions: the nucleus and the sub-apical region which will differentiate into the rhabdomeric microvilli. At this stage of the development, the presence of centrioles in the sub-apical part of the cell is easily recognizable.

In the Opilionids, the retinula consists of four cells: three peripheral and one central (Purcell 1892, 1894; Curtis 1969, 1970). The rhabdomeres of these four cells give rise to a closed rhabdom in the adult.

In the embryo of I. luteipes at stage V 1, the four cells of the retinula are arranged around a median cavity. Due to the cytoplasmic lengthening observed during the first 24 hours, these four cells adhere by means of their sub-apical halves, allowing a cavity between them, into which the rhabdomeric microvilli extend. This subapical space, made of four retinular cells, constitutes the embryonic rhabdomeric cavity.

The first indication of rhabdomeric differentiation is the occurrence of fine microvilli due to evaginations of the internal membrane of each retinular cell. The mean diameter of these microvilli is about 60 nm and their maximal length about 0,5 μm.

The microvilli which thus project into the rhabdomeric cavity during the first 24 hours (V 1) are neither similar nor regular. Some are ramified, some show slight swellings of the membrane. Inside each microvillus there is a microtubule, the size of which is 14 nm. It gives structural support during the first hours of differentiation.

Among microvilli occurring at stage V 1, there is one with a large diameter - almost 200 nm. It also projects into the rhabdomeric cavity like the thinner ones. The study of cross-sections through this large microvillus shows that it occurs in the same plane as the distal centriole. The retinular cell includes a diplosome, made of two centrioles perpendicular to each other. The distal centriole being near the internal edge of the cell is arranged along the same axis as the great microvillus.

The study of serial sections shows that microtubules from the distal centriole project into the large diameter microvillus. Long microtubules, probably the two external ones from each triplet, enter the microvillus. The mean diameter of these centriolar microtubules is 18 nm, i.e. less than that of microtubules in the cytoplasm, which is 30 nm. No ciliary root could be observed.

The centriole inside the retinular cell occurs as a small hollow cylinder, with a 0,15 to 0,20 μm diameter and 0,30 μm length. Its axoneme belongs to the type 9 + 0; the nine triplets are arranged helicoidally along the longitudinal axis. Along their whole length or so, these triplets are surrounded with an electron-dense, rather thin matrix. Nine dense, irregular masses are arranged on the periphery of the apical part of the centriole; they correspond to the satellite pericentrioles. An electron-dense, subspheric, heterogeneous mass, sub-apically arranged, can be observed.

The structural organization of the centriole as well as its relation to the enlarged microvillous lasts only 48 hours. During stage V 3 (72 hours), the diplosome position remains unchanged during the first 48 hours, the distal centriole being in contact with the enlarged interior membrane; however, due to a depolymerization of the ciliary microtubules of distal centriole C_1, the relation to the microvillus stops.

At stage V 3, size-equalization and a beginning of spatial arrangement of microvilli can be noted. The maximal length of the microvilli at stage V 3 reaches 0,9 μm with mean diameter 100 nm. The microvilli are irregular throughout their length. The structural supporting microtubules, very obvious during the first 48 hours, are rare at the 72 hour stage. Generally, these embryonic microvilli are very irregular and do not fill the rhabdomeric cavity.

c) Ciliary induction model

From the ultrastructural data observed during the first 80 hours of rhabdomeric differentiation in I. luteipes, a cytological model may be constructed (Fig. 3) (Muñoz-Cuevas, 1975 a, 1978 b, 1981).

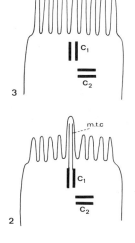

Fig. 3. Schematic representation of the ciliary pattern of development of the photoreceptor in the Opilionid Ischyropsalis luteipes. 1. 2. 3.: Cross-sections at the level of the rhabdomere of the embryonic visual cell. The three stages of the ciliary induction. C_1, distal centriole; C_2, proximal centriole; mtc, ciliary microtubules. (from Muñoz-Cuevas, 1975)

1st Period: the diplosome, perpendicularly arranged, occurs at the edge of the interior cell-membrane. The distal centriole C_1 is orientated perpendicularly to the interior plasma membrane.

2nd Period: out of the distal centriole C_1, ciliary microtubules are projected, which penetrate a microvilli, the diameter of which is greater than that of the others. This ciliary induction period occurs at a precise place of the plasma membrane and lasts 48 hours only (V 1 - V 2).

3rd Period: depolymerization of ciliary microtubules and size equalization of rhabdomeric microvilli.

d) Morphogenesis and spatial organization of the rhabdom.

Until the end of stage V, larval organogenesis stage, the rhabdomeric microvilli are not spatially arranged. From the 1st stage of larval instar L 1, the length of microvilli increases from 1 μm to 1,4 μm. This period of microvilli lengthening is the second differentiation period, the first one occurring between the first and the third days of the instar.

The second period of lenth-increase of microvilli is accompanied by shape-equalization and, especially, by spatial arrangement. During this period, the spatial disposition of microvilli and their grouping inside the rhabdomeric cavity give the peculiar honeycomb structure. In I. luteipes, geometrically perfect honeycombs can be observed from L 3 instar, and simultaneously the embryonic rhabdomeric cavity disappears.

The retinular cells, which make the retinula, show several junction points which, during rhabdomeric differentiation, enable them to keep a definite morphology and to be in accordance with morphogenetic changes. Thus, the retinular cells show two types of junctions, Macula adherens and septate junction, but during rhabdomeric differentiation and at the moment of honeycomb spatial organization of microvilli, a cellular contact of the "gap junction" type occurs between microvilli.

Macula adherens (spot desmosome). The number of retinular desmosomes is 4. They connect the four retinal cells and run along the rhabdomeric microvilli. The intermembranous space is about 20 nm; the tonofilaments are about 20 nm too. In longitudinal sections, they are 1 μm long in I. luteipes. In Acanthopachylus aculeatus, the retinular cells show desmosomes and septate junctions, with lengths varying in the same retinula from 0,3 μm to 0,6 μm. Septal junctions is about 28 nm thick. The intermembranous space is 14 nm, the septate length 12 nm and the space between septa 7 nm. These septate junctions are always placed after desmosomes.

Due to the spatial organization of microvilli within the rhabdomeric cavity, microvilli can come into contact with microvilli of the same rhabdomere or with those of other retinular cells. During the larval instar, these contacts occur giving the classic honeycomb structure. During this instar, the first contact-figures between microvilli can be observed at stage L 2. These contacts seem to belong to "gap junction" type. The whole thickness of this junction varies from 6 to 12 nm. The same contact-type can be observed between some opposite rhabdomeric microvilli.

Whereas the function of desmosomes and septate junctions in the retinula belongs to the mechanical type (Noirot-Timothée & Noirot, 1973; Eley & Shelton, 1976), the junction between microvilli belongs to the gap-type (Perrelet & Bauman, 1969). The function of gap-junctions as electrotonic synapses was proved by Satir and Gilula (1973). The importance of these junctions which participate in the ionic and metabolic coupling within the rhabdom is obvious in the retinular cells which are electrically coupled (Shaw, 1969).

e) Axon and optic nerve formation.

The cytoplasm of retinula tapers from the nuclear base, and it is also less dense than in the middle and apical part of the cell. The extensions of the four cells of each retinula come together and, with those of the neighboring cells, form bundles reaching the basal retinal membrane from which the optic nerve arises.

The first axonal extensions occur during the first 24 hours of the development. Their diameter, at the level of the nuclear base, is about 1,5 to 2 µm. Each extension includes a clear cytoplasm with numerous neurotubules (diameter 31 nm) and clear vesicles varying in size from 61 to 92 nm. These bundles may have a small number of axons arising from two or three retinulae, or they may have more numerous axons, from twenty to thirty.

During stage V, two new structures occur in the axonal extensions: dense bodies and multivesicular ones. During larval instar, granules of different size occur; they look like neurosecretory granules, they are dense and their size varies from 85 nm to 120 nm.

In the larva, the optic nerve is composed of several fiber bundles 20 to 30 µm in length and 5 µm in diameter. The diameter of the axons of which it is made varies from 0,1 to 0,3 µm. The optic nerve is surrounded by a perineural envelope, and lined by several peripheral glial cells.

f) Glial retinal cells.

Some cells resulting from mitoses of the optic vesicle differentiate into glial ones.

The cells of the glial retinal system are small and lie on the basal part of the retina. The size of their nuclei varies from 2 µm to 2,5 µm, and have very dense chromatin. The cytoplasm of glial cells shows long and narrow cytoplasmic strands, called glial processes, which reach the preretinal membrane dorsally and the retinal base ventrally; they surround the axons all along. The glial processes penetrate between the retinular and pigment cells; their lengths are about 19 µm.

The cytoplasm around the nucleus, is generally fairly dense. On the other hand, the glial processes, especially those which run towards the preretinal membrane, exhibit a clear cytoplasm with few microvacuoles, ribosomes and gliofibrils. In the apical part of the retina, these processes mix with one another and with the apical expansions of the retinular and pigment cells. The glial cells are arranged peripherically around the peripheral nerve with the glial processes surrounding it partially.

g) Pigmentary differentiation of the retina and the biochemical nature of the pigment.

In Harvestmen, as in other Arachnids, the retina is pigmented. In the Arachnids, which have been studied ultrastructurally, pigmentation seems to be due to spherical, highly-electron-dense granules. These granules are contained within the cytoplasm of the pigment-cell and the photoreceptor.

The biochemical nature of eye pigment has been widely studied in Arthropods, especially in Insects. Two groups of natural pigments are observed in the eyes of Arthropods: the pterines and the ommochromes.

Among Chelicerata, Butenandt (1959) noted the presence of ommines in the eyes and tegument of Limulus polyphemus and in the eyes of Dolomedes fimbriatus. In Harvestmen, we have recorded the presence of ommines in the eyes of I. luteipes and A. aculeatus (Muñoz-Cuevas, 1978 b, 1981).

Using the tegument of P. opilio we identified a xanthommatine with maximal absorption spectra at 235 and 440 nm in Na_2HPO_4 (0,1 M) and 243 and 475 nm in HCl (2 N) (Muñoz-Cuevas, 1978 b, 1981).

The differentiation of the pigment cell occurs very early in retinal development. Chronologically, in Harvestmen, the differentiation of the pigment cell is concommittant with that of the retinular one. The pigment cell occurs between the retinular cells; its cytoplasm is elongated and it shows the same polarity as that of the retinular cell. The morphology of the pigment granule varies widely during development, and during any one stage, several forms can be observed.

h) Cytological model for the differentiation of pigment granules in Harvestmen (Fig. 4).

1°) Formation of a more or less enlarged saccule by the endoplasmic reticulum.

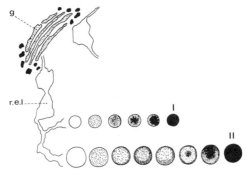

Fig. 4. Cellular differentiation of the pigment (ommochrome) of retina in I. luteipes. Formation of Type I and II granules. g, golgi; rel, smooth endoplasmic reticulum.
(from Muñoz-Cuevas, 1980)

2°) Elaboration of a matrix inside the saccule of the smooth endoplasmic reticulum.
3°) Elaboration and release of pigment by the Golgi system.
4°) Saccule filling and final granule formation.

These different phases follow one another relatively quickly and to understand the whole, it is necessary to study them during development.

Two types of granules occur during development in I. luteipes; type I occurs from the beginning of retinal differentiation; type II can be observed from the 30th day of development. In these two types of granules, the saccule filling differs.

Type I: small granules, round at final state. This type is the only one differentiating during the first 5 days; its diameter is about 0,6 µm.

Type II: its definitive mean size is 1 µm; it can be observed in retina from the 30th day of the development.

These two granule types differentiate in the pigment and retinular cells. Type I granule is the rarer one; type II is very numerous, especially in the median and apical part of the cell. In I. pyreneae and A. aculeatus, granule formation agrees with the scheme in I. luteipes. All our results dealing with differentiation of pigmentation and ommochrome function in eyes, are discussed fully in Muñoz-Cuevas (1978 b, 1981).

Cellular Differentiation of Dioptric System

The few papers dealing with cytodifferentiation of dioptric system in Arthropods are mainly with Insects. The ultrastructure of the Arachnid dioptric system, studied by several authors, has been of adults or of late stages of the postembryonic development.

The cytological process of the differentiation of the vitreous cell as well as the part it takes in lens secretion, will be described and the main cytological aspects of the structure and organization of lens and cornea will be given in accordance with the developmental chronology in Harvestmen (Muñoz-Cuevas, 1978 a, 1978 b, 1981).

a) Differentiation of the vitreous cell and lens-secretion

The differentiation of the dioptric system primarily occurs in the cells derived from ectoderm which cover the optic vesicles. In I. luteipes, between the 25th and 30th day of the development, the cells are undifferentiated. Some cells have membrane bound bodies containing a great deal of glycogen particles, α and β, occur, their size vary from 25 to 30 nm. Half the cytoplasm is filled with vacuoles. The size of the cell reaches 9 µm. The undifferentiated cells are separated from the underlying rhabdom by the preretinal membrane.

Between the 30th and the 34th days, the size of the cell increases and it becomes cubic; its height reaches 17 µm and its cytoplasm is filled with rough endoplasmic reticulum. From the 34th day on and during the whole larval organogenesis instar, this cell discharges the lens which is functional by the time of hatching (45th day). Electron-dense granules (40 nm to 60 nm in diameter) occur in the Golgi region during this period. Several microtubules, orientated along the longest cell-axis, occur with these granules; they can be observed especially within the apical third.

The apical edge of the cell is made of microvilli; their height is 110 µ m. As the secretory granules reach the apical edge of the cell, they spread along the base of the microvilli, they then cross the microvilli and become incorporated into the growing lens-mass as dense masses on the periphery of the central lens-mass. Posteriorly, the secretory product is organized in microfibrillar material and the microfibrils are arranged in parabolic curves.

The secretion of lens-material lasts during larval organogenesis and the larval instar till hatching. The lens-size of the larva is about 40 to 45 µ m in I. luteipes. Lens-secretion goes on after hatching during the whole postembryonic period until the adult lens-size be acquired.

The microfibrils which constitute its primary structure have a diameter of about 20 nm. This microfibrillar arrangement in parabolic curves gives the lens its lamellar aspect when observed under the optical microscope.

The lens is superficially covered with a cornea, which is 1,5 µm in thickness and perfectly smooth. Eakin and Branderburger (1971) note that, according to Land, the cornea of Salticidae is exuviated at each moult. The results with Harvestmen corroborate this observation. Thus the cornea is a superficial lens-specialization which is rejected at each exuviation.

During postembryonic development, some cells of the vitreous body undergo changes and show the structure of cells typical of the adult stage. Their cytoplasm is less dense, with an obvious decrease of all the organelles, especially those of the rough endoplasmic reticulum.

ACCESSORY EYES

In Spiders, the main eyes are always the antero-median ones. The accessory eyes are the lateral and postero-median, they are bilaterally symmetrical. This symmetry is the consequence of ontogenesis during which the accessory eyes differentiate and migrate in two opposite directions.

The accessory eyes show great diversity. The accessory eyes are similar in many higher taxa. Generally, they comprise of a lens, a vitreous body and a retina. Frequently, but not always, a tapetum is present. The tapetum occurs in 3 forms, enabling a description of ocular types, according to Homann's terminology (1971).

a) Accessory eyes of primitive type (P T) (Fig. 5).

The lens is usually strongly thickened with the vitreous body, a fine cellular layer, lying laterally. Due to this disposition images cannot be formed through the lens. The tapetum is transverse to the eye and occurs as a more or less thick layer. The retina belongs to the indirect type and the retinular cells possess distal nuclei; these cells extend across the tapetum through lacunae. The rhabdom differentiates laterally between the tapetum and the nuclear region. The retina also contains pigment cells. Based on their reflecting tapetum the accessory eyes of primitive type are nocturnal. This eye type occurs in Pholcidae, Urocteidae and Filistatidae Spiders (Cribellata).

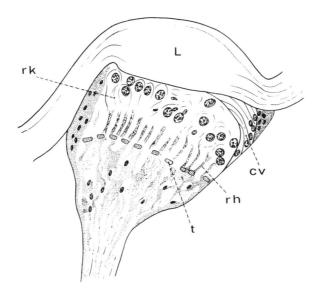

Fig. 5. Accessory eyes of primitive type (PT); sagittal section of the antero-lateral eyes in <u>Aname</u> (Spiders). L, lens; CV, vitreous cell; rh, rhabdom; rk, retinular cell. (from Homann, 1971)

b) Accessory eyes with canoe-shaped tapetum (KT) (Fig. 6).

These eyes are characterized by a two-part tapetum which face each other and form an angle opening outwards. At their base, the two parts open on a narrow slit. The rhabdom fills the space between the two parts of the tapetum; it lies transverse to retina and perpendicular to the tapetum-wings. The retinular nuclei are arranged distally to the retina. The vitreous body is made of a very fine cellular layer and the lens is more developed than in the previous type (PT). This eye-type is commonly observed in spiders, e.g. in Zodariidae, Palpimanidae, Theridiidae, Hadrotarsidae, Nesticidae, Linyphiidae, Araneidae, Agelenidae, Amaurobiidae.

c) Accessory eyes with a grate-shaped tapetum (RT) (Fig. 7).

The lens of these eyes is well-developed. The vitreous body is dorsal, ensuring that the images form on the retina. The tapetum is made of strips joined together thus giving rise to a fenestrate layer through which the retinular cells extend. The retina belongs to the indirect type, with nuclei lying distally between the tapetum and the pre-retinal membrane. The rhabdom occurs near the tapetum between this membrane and the nuclei-region. The retina is generally concave. This eye-type which provides good optic performance is found in hunting Spiders. They can be observed in Lycosidae, Oxyopidae and Senoculidae.

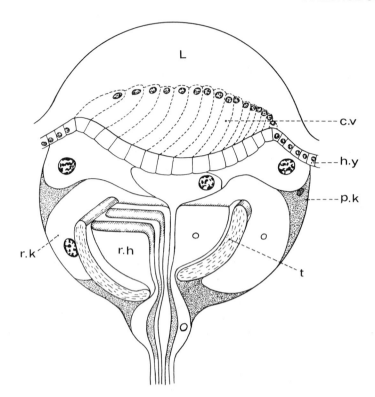

Fig. 6. Accessory eyes with canoe-shaped tapetum (KT). On the right: an horizontal section; on the left the rhabdomeric plates are schematized. The arrow indicates the saggittal plane; the circles (on the right) show part of the retinal cells. L, lens; CV, vitreous cell; hy, epidermis; rh, rhabdom; t, tapetum; pk, pigmentary cell; rk, retinular cell.
(from Homann, 1971)

d) Accessory eyes with a different structure

In addition to the three common types of accessory eyes described above there are a great many accessory eyes which do not fall under this scheme. Several eye-types can be observed in the members of the same family, like Sparassidae and Thomisidae; others occur in one genus only.

e) Lateral eyes in other Arachnid orders

Among other Arachnid orders the lateral eyes do not always belong to the direct type. The lateral eyes of Scorpions, Solpugids and Acarids belong to the direct type; whereas in Pseudoscorpions, Uropygids and Amblypygids belong to the indirect one.

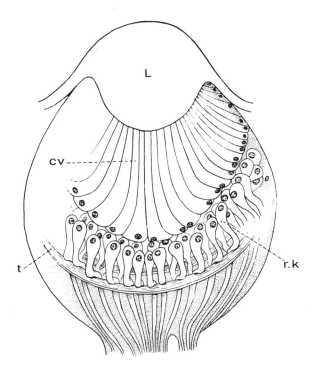

Fig. 7. Accessory eyes with a grate-shaped tapetum (RT). Horizontal section. L, lens; CV, vitreous cell; t, tapetum; rk, retinular cell.
(from Homann, 1971)

Ontogenesis of Accessory Eyes (Fig. 8).

Ontogenesis of accessory eyes have been studied by several authors; viz. Mark (1887), Hentschel (1899), Kishinouye (1891), Yoshikura (1955), Homann (1953, 1955, 1956, 1961, 1971). Despite being only light microscopic observations, Homann's studies help in the understanding of the morphogenesis of these eyes. The ontogenesis of the primitive type (PT) is described; that of KT and RT types being similar.

a) Ontogenesis of accessory eyes of primitive type (PT)

In Segestria senoculata, Homann (1971) could determine the ocular rudiment in the 5 or 6 cellular layers thick blastoderm. In the middle of this rudiment, an accumulation of relatively large nuclei with vacuolar chromatine can be observed, this constitutes the retinogen (rg). Homann noted the presence of mitoses on the superior side only; consequently, he assumed that the nuclei form here and then migrate to greater depths. In the median group, he observed two rows of nuclei; the proximal row gives rise to the pigmentogen (pg) and the distal one the tapetogen (tg). Peripherically lying nuclei give rise to the vitreous body (vitreogen, v g).

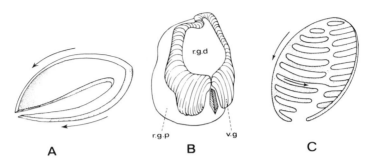

Fig. 8.
A- A canoe-shaped tapetum.
B-C. Ontogenesis of accessory eyes belonging to the grate-shaped type. Postero-median eyes of <u>Alopecosa</u>. Scheme of retinogen and vitreogen stage. The retinogen is divided by the vitreogen (v.g.) into a distal part (r.g.d.) and a proximal one (r.g.p.).
B- Frontal view
C- Formation of the tapetum. The arrow shows the orientation of the migration of the tapetogen nuclei.
(from Homann, 1971)

The tapetogen (tg) nuclei migrate around the rudiment in two rows. The cells widen in the middle of these two rows and later differentiate into tapetum crystals.

The pigmentogen (pg) - nuclei migrate round the retinogen (r g) - mass. The retinogen (r g) - nuclei maintain their position during development. The cells lengthen proximally, allowing free space between the nuclei and the tapetum. Within this space, the retinogen-cells differentiate into the rhabdomere. The cells traverse the tapetum through its narrow slits and extend into the axons which form the optic nerve.

POSTEMBRYONIC DEVELOPMENT OF EYES

In the life-cycle of Arachnids, postembryonic development is very important stage. It lasts several months during which a succession of so-called nymphal stages can be observed. During this long period, the phenomena of organic growth as well as those of ontogenic regression and of tissue-degeneration occurs in cave-dwelling animals.

a) Dioptric apparatus, lens-growth.

The dioptric apparatus differentiates during larval organogenesis and the larval stage. In Harvestmen, at hatching the first nymph shows a functional eye with a complete dioptric apparatus, although smaller in size than in the adult. The lens growth in I. luteipes (Muñoz-Cuevas, 1978 b, 1981) belongs to the continuous type. The lens-height varies from 45 µm in Ny 1 to 50 µm, 90 µm, 115 µm, 145 µm to reach 180 to 200 µm in the adult. Its height thus doubles between the first and the fourth nymphs, as well as between the fourth and the last one. The length increases faster during the second period of the postembryonic development.

b) Retinal growth in I. luteipes.

During postembryonic development the retina thickens. This is due to an increase in the height of the retinular cells and not to an increase in the number of cells. In the first nymph stage, the whole retinal height reaches 80 µm; on an average, it increases by 10 µm at each nymphal stage to reach about 150 µm in the adult. Growth of the retinular cell occurs through an increase of the rhabdomeric height rather than through the length of microvilli. At the larval stage, the rhabdomeric height is 10 µm; it increases to 17 µm in Ny 1, 20 µm in Ny 2, 22 µm in Ny 3, to reach about 30 µm in the adult. The length and diameter of microvilli do not vary during postembryonic development, i.e. the length remains 2 µm and the diameter 110 nm.

c) Ontogenic regression of the dioptric apparatus in I. strandi (Fig. 9).

Contrary to the opinion of the authors who studied the adult stage in this troglobiotic (cavernicolous) species, who assumed that I. strandi did not develop a dioptric apparatus, the study of the first nymph shows the existence of a complete eye, with a sensory and dioptric apparatus (Muñoz-Cuevas, 1976 b-c, 1978 b, 1981). As a matter of fact the vitreous cell in the first nymph reaches 18 µm in height; lens is small and biconvex, its height 34 µm and the corneal diameter is 1,8 µm; it is perfectly smooth with no peculiar cytological character.

The study of the histological reconstitution of the adult eye shows a totally regressed ocular protuberance. An histological section at the level of the apex shows the presence of an exocuticular opening outwards, this opening being more or less filled with a layer of cement. At the level of the epidermis, there is a cellular mass consisting of several nuclei of the pycnotic type; these irregular nuclei are surrounded with several lytic vacuoles of different size. In sections at the level of the epicuticle and exocuticle, a thickening of the lamellar endocuticle can be observed; it is ovoid, bulb-shaped.

The necrobiotic region in the vitreous body, the lens and the cornea, consists of degenerating organs, or tissues. In the case of I. strandi, a true ontogenic regression of the dioptric system is seen. The important degenerative modifications observed in the adult shows that this is the ultimate stage of the process of regressive evolution of an organ before its complete disappearance.

The phenomenon of ontogenic regression of the dioptric system described above can be perfectly inserted in the "complex of the phyletic rudimentation and ontogenic regression" in some cavernicolous lines.

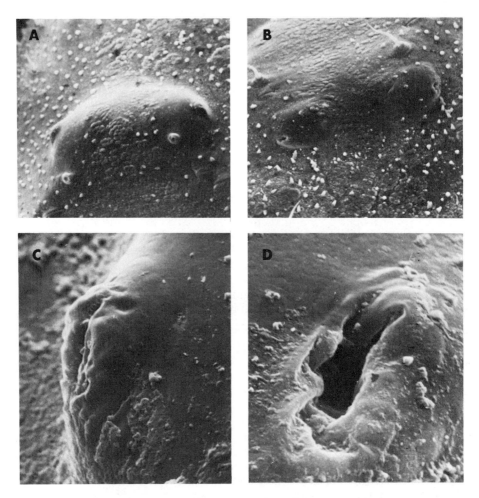

Fig. 9. Ontogenic regression
A-B Dorsal view of the eyes in two adult specimens of Ischyropsalis strandi (x 110)
C- Detail of A, cornea of the right eye, frontal view (x 1,270)
D- Detail of B lens-cavity of the left eye (x 2,170)
(from Muñoz-Cuevas, 1976)

ULTRASTRUCTURE AND ORGANIZATION OF EYES AND OPTIC CENTERS IN ARACHNIDS

Dioptric System

The ultrastructure of dioptric system of Arachnids has been observed by Baccetti and Bedini (1964), Bedini (1967), Curtis (1969-1970), Eakin and Brandenburger (1971), Gonzalez-Baschwitz (1977) and Muñoz-

Cuevas (1978 b, 1981). These observations are limited to Spiders, Scorpions and Harvestmen. However, they, in general, give an easy understanding of the dioptric system.

a) Vitreous body (Table 1)

The vitreous body in adult Arachnids varies greatly in size according to eye-type, family and ecological adaptations. Generally, the cells that form the vitreous body are cubic with a basal nucleus. The size of these cells varies from 2,5 µm in the Harvestman, Nemastoma lugubre, to 225 µm in Metaphidippus harfordi, a Salticid Spider.

The cytoplasm is characterized by its clear appearance; it is practically empty, dotted only with a granulo-filamentary substance, probably consisting of free ribosomes and glycogen particles. In the adult Opilio parietinus, the rough endoplasmic reticulum still occupies 60 to 80% of the vitreous cytoplasm (Curtis, 1970).

The large size of the vitreous body in Salticid spiders, hunting spiders, in contrast to the small size in the Harvestman, which dwells in darker layers, shows the utmost limits of morphologic adaptation of vitreous body in Arachnids.

b) Lens (Table 2)

The lens-ultrastructure of all Arachnids are similar. In adults the lens is made up of microfibrils arranged parabolically. The microfibrils are relatively homogeneous, the diameter of each being about 20 nm.

The microfibrillary arrangement in parabolic rings gives the lens the lamellar structure observed under the light microscope. As previously indicated, the lens of Arachnids is secreted by the vitreous body; as in vertebrates this secretion probably includes several proteins or "crystallines".

In Arachnids the lens is generally biconvex, nevertheless, accessory and lateral eyes of some orders show lenses which are plane or hardly curved (Pseudoscorpions). The lens size varies considerably according to eye-position, the family-considered and ecological adaptation.

c) Cornea

In Arachnids, the cornea appears as a superficial specialization of lens. Its surface is smooth; it is made up of a lamellae-piling which increases in density toward the surface. In Salticids, according to Eakin and Brandenburger (1971), the thickness of anterior-median eyes is almost 5 µm; that of posterior-lateral eyes is 3,9 µm. In I. luteipes (Opilionid) the thickness of cornea reaches 2,5 µm (Muñoz-Cuevas, 1981).

Ultrastructure and Organization of the Retina.

The understanding of the cytology and organization of the Arachnid retina has improved since 1964 thanks to several ultrastructural studies. Most of them were performed on spiders, Scorpions, Harvestmen and Acarids. The studies were conducted with diurnal or nocturnal species with different behavior; thus, they provided us with the morphological basis for the physiology and behavior of these species.

In this section, the organization of diurnal retina of Salticids will be

TABLE 1

Measurements of vitreous body in optic axis

Species	Layer/Type	Measurement	Reference
Metaphidippus harfordi (Salticidae)		225 μm, A.L.	Eakin & Brandenburger (1971)
		80 μm, P.L.	" "
Arctosa variana (Lycosidae)		70 μm, A.M.	Baccetti & Bedini (1964)
		10 μm, A.L.	" "
Lycosa bedelli (Lycosidae)		710 μm, M.	Carricaburu (1970 a)
		105 μm, P.	" "
Araneus diadematus (Argiopidae)		30 μm, P.	Gonzalez-Baschwitz (1977)
		20 μm, L.	" "
Telyphonus sepiaris (Uropygid)		20 μm, M.	Carricaburu (1970 b)
Pandinus imperator (Scorpion)		180 μm, M.	Carricaburu (1970)
Euscorpius carpathicus "		12 μm, M.	Bedini (1967)
Androctonus australis "		10 μm, M.	Carricaburu (1970)
Opilio parietinus (Opilionid)	Fieldlayer Brandes	17 μm,	Curtis (1970)
Oligolophus tridens "	Fieldlayer	6,5 μm	" "
Nemastoma lugubre "	Groundlayer	2,5 μm	" "
Ischyropsalis luteipes "	Troglophilous	17 μm	Muñoz-Cuevas (1981)
Ischyropsalis muellneri "	Troglobitic	9 μm	" "
Ceratogyrus darlingii (Therophosidae)		80 μm, M.	Carricaburu (1970 a)

TABLE 2

	Lens - diameter	
Portia fimbriata (Salticidae)	810 μm, A.	Williams & MacIntyre (1980)
Dendryphantes morsitans (Salticidae)	880 μm, A.M. 480 μm, P.L.	Carricaburu (1970 a, b) "
Ceratogyrus darlingii (Theraphosidae)	620 μm, M. 750-360 μm, A. 200 μm, P.	Carricaburu (1970 a, b) " "
Sparassus mygalinus (Sparassidae)	639 μm, A.M. 625 μm, A.L. 527 μm, P.M. 532 μm, P.L.	Carricaburu (1970 a) " " "
Lycosa bedelli (Lycosidae)	710 μm, M. 650 μm, P.M. 100 μm, A.M. 100 μm, A.L.	Carricaburu (1970 a) " " "
Arctosa variana (Lycosidae)	176-242 μm, A.M. 132-176 μm, A.L. 308-396 μm, P.M. 264-374 μm, P.L.	Baccetti & Bedini (1964) " " " " " " " " "
Dinopis subrufus (Dinopidae)	1325 μm, P.M. 300 μm, A.L. 373 μm, P.L. 232 μm, A.M.	Blest & Land (1977) " " " " " " " " "

(continued)

TABLE 2 (Continued)

	Lens - diameter	
Dolomedes aquaticus (Pisauridae)	400 μm, P.M.	Williams (1979 a, b)
Araneus diadematus (Argiopidae)	300 μm, A.M.	Gonzalez-Baschwitz (1977)
Androctonus australis (Scorpion)	500 μm, M.	Carricaburu (1968)
Telyphonus sepiaris (Urogypid)	400 μm, M. 380 μm, L.	Carricaburu (1970 b)
Telyphonus schimkewitschi (Uropygid)	445 μm, M. 400 μm, L.	Carricaburu (1970 b)
Uroproctus assamensis (Uropygid)	340 μm, M. 300 μm, L.	Carricaburu (1970 b)
Acropsopilio chilensis (Opilionid)	250 μm	Ringuelet (1959)
Tasmanopilio megalops "	280 μm	Hickman (1957)
Ischyropsalis luteipes "	150 μm	Muñoz-Cuevas (1981)
Ischyropsalis muellneri "	100 μm	"

shown as a model; it is perhaps the most evolved retina among Arachnids. On the other hand, the results obtained with Dinopidae help understand the organization in a nocturnal form having high performance. The type of eye-organization in Scorpions will also be mentioned, especially with regards to the morphological features considered as important factors of circadian adaptation.

a)   Organization of retina in diurnal Salticid Spiders.

The retina of anterior-median eyes in Salticids (Eakin & Brandenburger, 1971) consists of: the retinular and the pigmentary cells. Due to the arrangement of the retinular cells in space, the rhabdom is organized into 4 layers, the so-called 1st to 4th Land's layers (1969 a). The most basal one corresponding to the number one of Land's description. Each sensory cell may be divided into four regions: a distal region which includes the rhabdomeres, an intermediary unpigmented region crossing the layer of pigmentary cells; a basal region containing the nucleus, and an axonal region.

The rhabdomeres, Land's N°1 layer, occurs at the base of the V due to the pigmentation of retina. The rhabdomes of type 1 which in the optic axis are 2,5 μm in length and 3 μm in diameter, whereas those lying dorsally and ventrally are shorter (1,5 μm in length and 4 to 5 μm in diameter) (Fig. 10).

In the distal region, the microvilli which make up the rhabdomeres are situated at right angles to the longitudinal axis of the cell. In the type 1 cell, 1, 2, 3 or 4 rhabdomeres may occur.

The intermediary segment of the type 1 cell extends between the pigmentary cells. The basal region is outside the pigmented zone.

The rhabdomeres of the sensory layer, n°2, lie anterior to the n°1 layer; they are separated by cytoplasmic extensions of the pigmentary cells. These rhabdomeres are parallel to those of the n°1 layer and to the aperture. These rhabdoms occur in the medio-frontal plane of the eye. The rhabdoms n°2 are smaller than those of the last layer, i.e. 18 μm in the middle and 10 μm dorsally and ventrally. Due to the great number of mitochondria the cells of type 2 can be easily distinguished from those of type 1.

The rhabdomeric layer of type 3 occurs anteriorly to type 2. These rhabdoms are in the frontal-median part; they are lacking dorsally and ventrally. They are shorter and are from 10 to 14 μm. The intermediary segment is moderately dense in mitochondria.

The n°4 layer, the most anterior one, is under the vitreous body. The rhabdoms show a greater variation in size, shape and orientation of the microvilli. They lie in the central region and have the largest diameter (6 to 8 μm); their number varies from 2 to 4.

The total number of the receptors was estimated at 907 in Phidippus johnsoni (Eakin & Brandenburger, 1971). The n°1 layer contains 344 of them, arranged in 60 dorso-ventral rows. The 2nd, 3rd and 4th, respectively, have 366, 148 and 49 retinular cells. In Metaphidippus aeneolus, according to Land (1969 a), the 1st layer has 376; the 2nd layer: 302; the 3rd layer: 48 and the 4th layer: 68, i.e. total of 794 (Land's scheme n°3) retinular cells (Fig. 11).

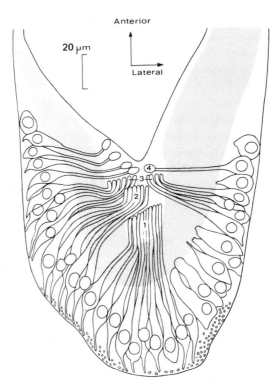

Fig. 10. Frontal (horizontal) section of the retina of the right antero-median eye of Metaphidippus aeneolus. The section is taken close to the center of the retina and shows the four layers of receptor endings (1-4). In layers 1, 2 and 3 the presumed receptive part of each receptor is the straight terminal portion, in layer 4 it is the terminal oval swelling. (from Land, 1969)

The pigmentary cells of the retina - or supporting cells according to Eakin and Brandenburger (1971) - are the 2nd cellular component of the retina of median eyes forming the black pigmented edge of this organ. The pigmentary cells contain several pigment granules with a diameter of 0,6 μm. A great number of branched cytoplasmic ramifications occur in the cell which extends between the retinular cells.

The stratified organization of the retina in M. aeneolus and P. johnsoni (Salticidae) can be easily distinguished not only from the retina of other Spiders but also from those of other Arachnids. As a matter of fact in Lycosids (Baccetti & Bedini, 1964; Melamed & Trujillo-Cenoz, 1966), Agelenids (Agelena gracilis) (Schröer, 1974 a, b) and Argiopids (Araneus diadematus) (Gonzalez-Baschwitz, 1977) the retina of main eyes is non stratified, the rhabdomeric region occurs as a relatively uniform row, under the preretinal membrane. Thus, in Salticids, the retina seems to be more complex in organization than in the other diurnal Arachnids.

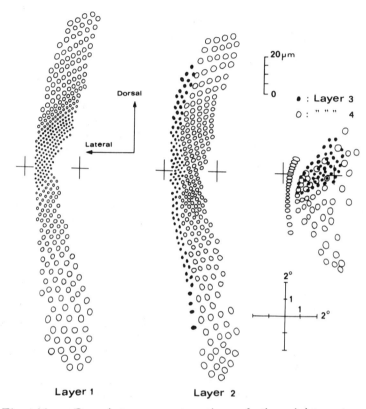

Fig. 11. Complete reconstruction of the right anteromedian retina of <u>Metaphidippus aeneolus</u> from serial transverse sections, showing the disposition of receptor endings in each of the four layers. In layer 2 the closed figures indicate those elements whose cell bodies lie on the lateral side of the retina; the open figures, those on the medial side. The small crosses are fiducial marks showing how the three drawings superimpose. The scale in degrees indicates the subtense of retinal structures at the modal point of the eyes and hence in object space; it assumes a focal length of 512 μm.
(from Land, 1969)

Based on this layered architecture Land (1969 a) put forward the hypothesis of a different spectral sensitivity at each receptor layer.

b) Organization of the retina in the nocturnal Dinopid-Spiders.

In Dinopis subrufus (Blest & Land, 1977; Blest, 1978), the organization of the retina of P.M. eyes enables us to understand the important structural changes occurring in the retina of the animals during day (diurnal retina) and night (nocturnal retina).

The topography of the diurnal retina (Blest, 1978, Fig. 1) as seen in cross-section shows hexagonal-shaped rhabdoms. Each side of the hexagon is composed of a rhabdomere which is contiguous to the adjacent receptor. At each edge, where three rhabdomeres gather, a pigmented cord can always be observed, large cytoplasmic nets of unpigmented cells may be present. A tangential scheme of the diurnal retina showing its cellular arrangement and organization is given in Fig. 12 (Blest, 1978; Fig. 2).

Distally to the cellular bodies, a region of receptor-nuclei and of nuclei of unpigmented supporting-cells can be observed; it corresponds to the so-called (SO) region, according to Blest's terminology. Blest's (o.l.m.) region occurs next, this is called external limiting membrane by Land (Blest & Land, 1977). The (rs) region or zone with rhabdomeres of 50 µm high lies between the (o.l.m.) region and the dense pigmentary layer (p.l.) which is about 12 µm in height and marks proximal limit the rhabdomeric region; this is followed by an axonal region consisting of retinular axons; the latter are surrounded by the axonal extensions of the pigmentary cells. The nuclei of the pigmentary cells are at different levels in this region. The height of this zone (a.x.) is 140 µm

The topography of the nocturnal retina (Blest, 1978) - shows considerable differences in its organization. These differences can be schematized as follows (Blest, 1978; Fig. 15).

Rhabdomeres fill less than 15% of the receptor-volume of diurnal retina fixed at noon; this configuration is maintained until night-fall; however, in the nocturnal retina, fixed one hour after sunset, the rhabdomeres fills 90% of the receptor-volume. The length of the rhabdomere varies from 50-60 µm in the diurnal state to 90-135 µm in the nocturnal state. The dense layer of pigment at the base of the rhabdomere is unaltered but a pigment-migration occurs in the proximal third of the axons.

At sunrise, the membrane of the rhabdomere is destroyed; this process is completed within two hours. The destruction of the rhabdomeric membrane occurs at the base of microvilli which are shaped like pinocytotic vesicles gathering in the neighboring cytoplasm to form the different types of multivesicular bodies (Fig. 13). In Dinopis (Blest, 1978), the cytological process corresponds to that described in Crustaceans, by Eguchi and Waterman (1976) and to the one observed in the photoreceptors of Harvestmen (Muñoz-Cuevas, 1976 a, 1978 b). Perhaps, these diurnal changes of the rhabdomeric membrane are a consequence of the photic condition, or the presence of a circadian cycle. Blest (1978) suggests that a cycle independant of the immediate changes of lighting is involved; it need not necessarily be a circadian rhythm.

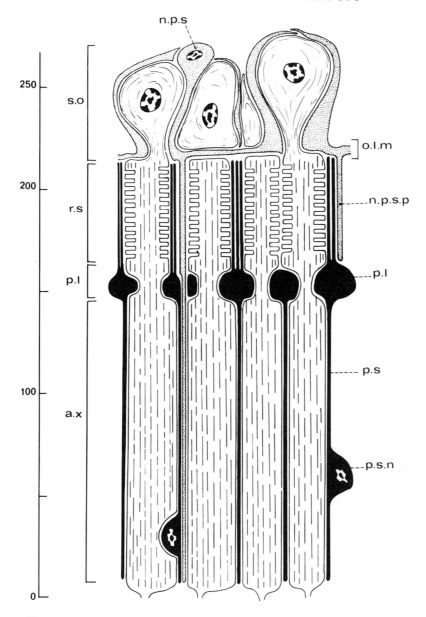

Fig. 12. Diagram of the diurnal retina of Dinopis subrufus, showing the arrangement of components as represented in a tangential section. n.p.s., non pigmented supportive cell; n.p.s.p., non-pigmented supportive cell-process; so., receptor somata; o.l.m., outer limiting membrane; r.s., receptive segments; p.l., dense pigment layer; p.s., axonal sheat formed by pigmented supportive cell-processes; p.s.n., pigmented supportive cell nucleus; ax., swollen receptor axons. (from Blest, 1978).

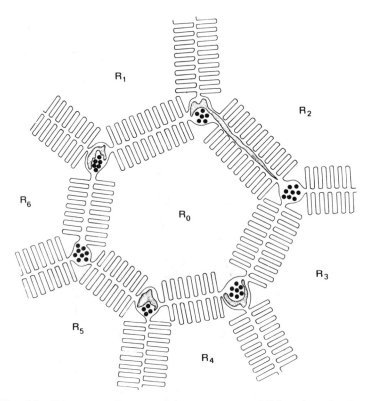

Fig. 13. Diagram of a receptive segment of Dinopis subrufus in transverse section (Ro) to show its relation to six adjacent segments (R1-6). Each face of the hexagonal profile is composed of a rhabdomere which is contiguous with one rhabdomere of an adjacent receptor. At each of the six junctions between three rhabdoms there is a strand of supportive cell processes, composed of slender extensions of pigmented supportive cells (black) which are always present, and the layer processes of non-pigmented supportive cells (stippled) which are often absent, and infrequently from sheets separating rhabdomeres which would otherwise be contiguous, as between Ro and R2. The receptors are represented in the diurnal state.
(from Blest, 1978)

The retina of posterior eyes in Spiders was described at the ultrastructural level for 3 families: Lycosidae, (Baccetti & Bedini 1964), Salticidae (Eakin & Brandenburger, 1971) and Pisauridae (Blest & Day, 1978). Lycosidae and Pisauridae have a well-structured tapetum. In Lycosidae and Salticidae the rhabdoms are wholly separated by pigmentary cells and the rhabdoms of the adjacent cell are not contiguous. In

Dolomedes (Pisauridae), the rhabdoms are linearly arranged on the tapetum and the tapetum walls are separated from the cytoplasm of the pigmentary cell. The rhabdoms are contiguous to those of the adjacent cell. In diurnal Spiders, such as Salticidae and Lycosidae, the receptors are well separated, whereas, in Dolomedes, a nocturnal genus, they are less separated and, in Dinopis, the rhabdomeres on all the sides of the photoreceptive cell is contiguous with the neighboring ones. This organization of the nocturnal retina suggests that visual acuity is sacrificed for sensitivity.

In Menneus, Dinopidae (Blest et al., 1980) the retina of posteromedian eyes shows the same organization as in Dinopis, but the receptor segments are separated from one another by glial processes, each retinular cell having 2 rhabdomeres. According to Blest et al. (1980), the specialized retina of Dinopis could have originated from that of Menneus. The lack of true tapetum in Menneus as well as in Dinopis enables these authors to consider that nocturnal behaviour is secondary in these species.

c) Ultrastructure of the tapetum.

The tapetum is made up of highly-specialized cells which are characterized by the presence of numerous crystals in their cytoplasm. In Lycosa (Baccetti & Bedini, 1964; Melamed & Trujillo-Cenoz, 1966), the cytoplasm includes a membranous system arranged parallel to the main axis of the crystals. The crystals in turn, are arranged perpendicularly to the light. Most of the crystals dissolve when the material is fixed; this leaves a space 1 to 1,5 μm in length and 0,15 μm in width. In Lycosa the nucleus of the tapetum-cell is polyhedric with light nucleoplasm and small dermatin-accumulations. In this species the tapetum occurs at the base of the rhabdoms. Between the tapetum there are spaces of 5 to 10 μm through which the retinular cells extend. In Lycosa, the tapetal strip is 5 μm thick. In this species, the retinular cells are arranged so that the basal part of the rhabdom is close to the tapetum, thus reflection of the light by the crystals induces a mirror effect on the rhabdom.

d) Organization of the retina in Scorpions.

In Scorpions, thanks to the study of Bedini (1967) on Euscorpius carpathicus and Fleissner (1972, 1974, 1977 a, b, c), and Schliwa and Fleissner (1979, 1980) on Androctonus australis, the ultrastructure, organization and physiology of these species are now better known. The retinular and pigmentary cells constitute the two cellular categories described in the median eyes. The retinula is made up of 5 retinular cells forming the rhabdom. In A. australis it may be 4 or 6. A rhabdomeric cell is added to the 5 retinular cells; therefore each retinula has 6 cells. The retinulae are partially separated by the pigmentary processes. In a cross-section of the retinula, the rhabdom occurs as a five-rayed star. The rhabdoms are not contiguous with the neighboring ones. The rhabdomeric microvilli vary from 1 to 1,5 μm in length and from 50 to 70 nm in diameter. The base of the retina is formed by the nuclear region. The rhabdomeric retinular cell lies in the basal half of the retina. Its nucleus is smaller than that of the other retinular cells. The cytoplasm of the cells belonging to these two types shows an important feature - the absence of pigment granules in the rhabdomeric cell. The apical end of this cell is made up of a long dendrite ending at the base of the rhabdom. The dendrite consists of numerous ramifications or projections extending towards the five retinular cells (Schliwa & Fleissner, 1979; Fig. 2). The size of the retinular cells varies from 150 to 180 μm in A. australis; the rhabdomeric cell is 80 μm.

During light or dark adaptation pigment migration occurs in the retinular cells of A. australis. The pigment migrates towards the proximal cellular area during the diurnal to nocturnal state; whereas, during the reverse it migrates towards the distal area facing the rhabdom. According to Fleissner, this migration is the most important mechanism expressing circadian sensitivity.

Fleissner and Schliwa (1977) described numerous neurosecretory fibers which made synaptic contacts with the retinular cells of A. australis. The number of fibers which associate with a retinular cell varies from 10 to 20. Therefore, these authors suggest that the fibers constitute the efferent way of controlling the circadian movement of pigment inside the retinular cell (Fleissner & Schliwa, 1977; Fig. 5).

The retina of the lateral eyes possesses similar categories to those described for median eyes; nevertheless, their organization differs. The rhabdomeric area appears as a net of contiguous rhabdoms. No retinulae can be easily distinguished, each lateral side of a retinular cell forms a rhabdomere. In the largest anterior eye, an arhabdomeric cell associates with two retinular cells; while in the posterior eyes it associates with one or two retinular cells (Schliwa & Fleissner; Fig. 3). There is no peculiar orientation of the rhabdomeric microvilli. As with median eyes, the retinular cell of lateral eyes shows pigment migration due to adaptation. During adaptation to light pigment granules gather in the distal region; while during adaptation to darkness, the granules migrate towards the base. In constant darkness the retinular cells show a weak circadian movement of pigment. As in median eyes, neurosecretory fibers occur in contact with the rhabdomere; there is probably one fiber per retinular cell.

e) Organization of the optic centers.

In Arachnids, the optic lobes are a part of the protocerebrum and they differ from order to order. They differ by the importance of their development as well as by the number of their optic masses. According to Hanström studies (1919, 1923, 1928, 1935), the numbers in the different orders are as follows:

Scorpions
- 3 optic masses, lateral eyes
- 2 optic masses, median eyes

Spiders:
- Lycosidae
- Thomisidae   2 optic masses, lateral eyes
- Salticidae   2 optic masses, lateral eyes
- Araneidae

Epeiridae Theridiidae	1	optic mass, lateral eyes =	<u>Uropygids</u>
Agelenidae Drossidae Dictynidae Dysderidae	2	optic masses, median eyes	<u>Solpugids</u>
Opilionids	2	optic masses	
Acarids Pseudoscorpions	1	optic mass	

In the families Epeiridae, Dysderidae and Agelenidae, the optic masses fill 3 to 5% of the whole volume of the brain, whereas in Salticidae it is 2,5%. Axons of the optic nerve synapse at the level of the first optic mass. This first synaptic junction consists of retinular fibers and the dendritic ends of the cells of the ganglionic cortex. These cells are bipolar in nature (Hanström, 1923).

The ultrastructure of the first synaptic junction was studied in <u>Lycosa</u> (Trujillo-Cenoz & Melamed, 1964; Trujillo-Cenox, 1965), in Salticids (Eakin & Brandenburger, 1971) and in Opilionids (Muñoz-Cuevas, 1978 b, 1981).

The height of the so-called synaptic area, which contains synapses between the tips of retinular and second order fibers, reaches almost 5 to 6 μm. The diameter of the second order fibers rarely reaches 0,4 μm. Around a retinular tip, several synaptic areas can be observed; the retinular tip, it may be single, bifid, or trifid. In the two last cases, it can surround the second-order fibers with its branches, and synaptic contacts may be established with them. In most cases, synaptic areas can be observed all along the retinular end, approximately 3 μm in height in <u>Ischyropsalis</u> (Muñoz-Cuevas, 1981). In the first synaptic mass, the diameter of the synaptic vesicles varies from 23 to 46 nm (Muñoz-Cuevas, 1981). The vesicles are abundant and distributed at random in the axonal tip, but at the level of the synaptic area itself, they are grouped and attached to the presynaptic membrane. The presynaptic membrane shows a highly electron dense material and is associated with the closest synaptic vesicles. The inter-synaptic space dividing the two synaptic membranes is 15,5 nm. An electron-dense material can be observed in association with the postsynaptic membrane. Generally, the synaptic area has one and sometimes two postsynaptic fibers.

Among Arachnids, in Opilionids, an optic chiasma exists, between the first and the second optic masses. This chiasma is made up of the intersection of second-order fibers arising from the neurons of the cortex (Hanström, 1923). The intersection of the superior fibers constitutes the chiasma. In <u>I. luteipes</u> (Muñoz-Cuevas, 1978 b, 1981) the second-order fibers which form the chiasma, are arranged in lamellae; their diameter varies from 0,3 to 0,7 μm. Between the fiber-bundles, light cytoplasmic strips of glial cells, rich in particles of α and β glycogen, can be observed. The area of fiber-intersection is about 6 μm high and 4 μm wide.

The present studies do not enable the projection of the axonal retinal fibers in the lobe. Thus the topography of projection of the different receptors is unknown. The only hypothesis which exists was put forward by Land (1969 a, b) for the Salticid Spider M. aeneolus.

PHYSIOLOGY OF VISION IN ARACHNIDS

Physiological Optics

With the exception of the classical works of Petrunkevitch (1907, 1911), many studies (Carricaburu (1968, 1970 a, b) on Scorpions, Spiders, Amblypygids and Uropygids; and by Blest & Land (1977), Williams (1979 a) and Williams & McIntyre (1980) on Spiders) enable us presently to better perceive the problems related to dioptrics in Arachnids.

a) Eye dioptrics in Scorpions

Carricaburu (1968) studied the median and lateral eyes of Androctonus australis.

Median eyes: The median eye is shaped like a spherical cap with a diameter varying from 0,2 to 0,3 mm. Observed through crossed Nicols it shows a black cross; indicating that the dioptric system is made up of layers gathered to form an "onion-bulb" and that it is highly bi-refringent.

According to the basic scheme of eye:

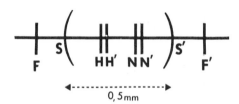

The study of the cardinal elements yield the following results: the real focus F is 90 to 17 µm from the anterior top while the virtual focus F' can be observed at a distance of 100 to 220 µm from the posterior. The virtual focal length varies from 300 to 550 µm. Carricaburu (1968), considering eye-thickness (450 to 500 µm) and the relative value of the different measurements, proposed the following means:

- Convergence   D = 3030 δ
- Real focal length   f = 333 µm
- Virtual focal length   f' = 450 µm
- Abcissa of the real focus   S F = 170 µm
- Abcissa of the virtual focus   S F' = 655 µm
- Abcissa of the real principal plane   S H = 163 µm
- Abcissa of the virtual principal plane   S H' = 205 µm
- Abcissa of the real nodal point   S N = 280 µm
- Abcissa of the virtual nodal point   S N' = 322 µm

The retina is in contact with the posterior apex S'. The eyes studied are more or less divergent from this scheme. Hence the powers observed vary from 2780 δ to 3120 δ. Sometimes, the two eyes show different powers in the same specimen. So, in an animal, in which the eyes gave clear images, Carricaburu noted 3210 and 3120δ in the right eye and 2820 and 2830δ in the left one.

Using Foucault's gratings relatively good separating power of 15' to 55' was observed. Thus aberration rather than diffraction is the limiting factor. Based on Rayleigh's formula a separating power of 500" (i.e. 8'20") can be observed for a 260 μm pupil. Astigmatism is generally absent although in A. australis a weak irregular astigmatism can be noted.

The observation of eye-sections, from the surface towards the depth, through crossed-Nicols shows, an epicuticle, a central nucleus of high refractive index, and a posterior layer with a lower refractive index. Strong birefringence and laminated structure are typical of these three layers.

The interferometry of the central nucleus shows the following values: the index in the anterior part of the central nucleus is 1,536; and it can reach 1,546 in the middle of the nucleus. As for the posterior, relatively homogenous, part of the eye, the index is 1,531. The epicuticle which is relatively heterogeneous structure has an index of 1,532 on the surface; it decreases to 1,522 in the median part and increases to 1,543 in the deep face.

Lateral eyes: A. australis has 5 lateral eyes. The three main eyes are very easily distinguishable; while the two rudimentary ones, are more difficult. The diameter of the 3 main ones varies from 0,35 to 0,38 mn. The lateral eyes give no useful image, rather they concentrate light. Thus, focuses, focal lengths or principal planes cannot be determined. Interferometry shows that there is no central nuclei with important indices. On the axis, from the surface inwards, the indices are successively 1,534, 1,520 and 1,534.

Hence, the most important point of the dioptric in A. australis is the very strong hyperopy of the median eye. Using Newton's formula, Carricaburu (1968) showed that the eye of A. australis is hyperopic by 1,200 δ. In more sensitive retinal zones, lying 20 μm behind, the hyperopy is more than 1,000 δ.

b) Eye dioptric in Spiders.

In 1970, a paper on the eye dioptric in four Spider species, (Ceratogyrus darlingii, Sparassus mygalinus, Lycosa bedelli and Dendryphantes morsitans) was published by Carricaburu (1970 a).

In C. darlingii, the optical characteristics of the median eyes are as follows :

Ceratogyrus darlingii

	D	f'	f	Δ	R
OD	2650	510	378	298	1160
	2800	485	358		1255
OG	3020	450	332	305	955
	2950	458	339		1075

Sparassus mygalinus

	D	f'	f	Δ	R
OG	3220	420	310	149	545
	2500	540	400		320
OD	3150	429	315	149	500

Anterio lateral eye in Sparassus mygalinus

D	f'	f	Δ	R
4250	320	235		200
3230	420	310	95	115

Latero posterior eye in Sparassus mygalinus

D	f'	f	Δ	R
2980	453	336	176	590
3300	410	303		775

Medio-posterior eye in Sparassus mygalinus

D	f'	f	Δ	R
2270	596	440		885
2118	623	458	312	815

Lycosa bedelli median eye

D	f'	f	Δ	R
1146	1180	875	812	100
1192	1130	840		107

Lycosa bedelli posterior eye

D	f'	f	Δ	R
1345	1005	743	675	770
1325	1020	756		740

INTERFEROMETRY INDICES

Lycosa bedelli

	Median eye	Posterior eye
Cornea	1 511	1 483
Lens	1 491	1 456
	1 439	1 446
	1 439	1 442
	1 497	1 456
	1 491	1 435

Dendryphantes morsitans

	Antero-median eye	Posterior eye
Cornea	1 371 (ext.)	1 446
	1 471 (int.)	
Lens	1 506	1 383
	1 385	1 448
	1 400	1 385
	1 385	1 403

Ceratogyrus darlingii

	Antero-median eye	Antero-lateral eye
Cornea	1 507	1 510
Lens	1 507	1 497
	1 487	1 455
	1 550	1 489
	1 536	

Sparassus mygalinus

	Antero-median eye	Antero-lateral eye	Postero-lateral eye	Median-posterior eye
Cornea	1 560	1 560	1 560	1 560
Lens	1 560	1 560	1 527	1 641
	1 546	1 560	1 560	1 614
	1 560	1 569	1 560	1 556
	1 536	1 555		

Carricaburu (1970 a) emphasizes the very strong birefringence, which varies greatly from layer to layer. These variations are also very important among individuals. Carricaburu notes that the left and right homologous eyes are generally different, the difference of power being far more important than experimental error. Finally he concludes that the eyes are always hyperopic.

In Dinopis subrufus, the physiological optics were studied by Blest and Land (1977).

Optical Dimensions of the eyes of Dinopis

eye	p.m.	a.l.	p.l.	a.m.
diameter of cornea, A	1 325	390	373	232
radius of curvature, r	660	187	185	108
focal length, f	771	277	262	164
relative aperture $(A/f)^2$	2,95	1,98	2,03	2,00
F - no	0,58	0,71	0,70	0,71
receptor separation, d	20,0	6,5	5,2	6,2
degrees (57 d/f)	1,48	1,34	1,13	2,16

A comparative table on the optical characteristics of the postero-median eyes in Dinopis and of the antero-median ones in Phidippus johnsoni was published by Blest and Land (1977).

Resolution and Light gathering Power in Dinopis and Phidippus

	D	P
Focal length, f	771	767
Receptor diameter, d	20	2,0
Receptor subtense (57 d/f)	1,48° (x 10)	0,15
Entrance pupil diameter, A	1 325	380
Relative retinal illuminance, $(A/f)^2$	2,95	0,25
Effective Receptor length, x	55	23
Relative amount of light absorbed per receptor $(A/f)^2 d^2 (1 - e - 0,01 x)$	495 (x 2360)	0,21

From this comparative study of Dinopis, a nocturnal animal, and Phidippus, a diurnal Salticid, it follows that resolution in the former is 10 times finer than in Phidippus, but light-absorption by the receptors is 2000 times less than in Dinopis.

These two examples probably constitute two extremes of eye-evolution in Spiders, the first one because of high resolution in day-light and the other one because of light-absorption in semi-darkness.

Electrophysiology

a) Spectral sensitivity in Arachnids

Spectral sensitivity in Spiders

		Maximum		
Tegenaria parietina (Agelenidae)	A.M.	460 nm 550 nm		ERG: Giulio 1962 " " "
Araneus diadematus (Argiopidae)	P.M.	350 nm 505 nm		
Lycosa baltimoriana Lycosa carolinensis Lycosa miami (Lycosidae)	A.M. P.M. P.L. A.L.	360-370 nm 510 nm 380 nm 505-510 nm		ERG; DeVoe et al. 1969 " " " " " " " " "
Lycosa baltimoriana Lycosa miami Lycosa lenta	A.M. A.L.	360-370 nm 510 nm 380 nm 510 nm		I.C.; DeVoe 1972 " " " " " "
Phidippus regius (Salticidae)	A.M.	370 nm 532 nm 370-525 nm	UV cells green cells	I.C.; DeVoe 1975 " " "
Menemerus confusus (Salticidae)	A.M. A.L., P.L.	360 nm 480-500 nm 520-540 nm 580 nm 535-540 nm	UV cells Blue cells Green cells Yellow cells	I.C. and ERG; Yamashita & Tateda 1976 " "
Argiope bruennichi	A.M.	360 nm	UV cells	ERG and I.C.; Yamashita & Tateda 1978
Argiope amoena (Argiopidae)		480-500 nm 540 nm	Blue cells Green cells	ERG and I.C.; Yamashita & Tateda 1978 " "
Plexippus validus (Salticidae)	P.L.	535 nm		I.C. and ERG; Hardie & Duelli 1978
Dinopis subrufus (Dinopidae)	P.M.	517 nm		I.C. and ERG; Laughlin et al. 1980
Plexippus validus (Salticidae)	A.M.	360 nm 520 nm	UV cells Green cells	I.C. and ERG; Blest et al. 1981

Spectral sensitivity in Scorpions

Opisthacanthus validus	M.	510 nm	ERG;	Machan 1966
Centruroides sculpturatus	L.	375 nm	"	" "
		510 nm	"	" "
Vejovis spinigerus	L.	375 nm	"	" "
		510 nm		
Buthus occitanus	M.	490 nm	ERG;	Fleissner 1968
Heterometrus fulvipes	M.	510 nm	ERG;	Geethabali & Pampapathi Rao 1973
Heterometrus gravimanus	L.	380 nm	"	"
		510 nm		

Spectral sensitivity in Acarids

Tetranychus urticae		375 nm	Exp. Phototaxy;
		530 nm	Naegele et al. 1966

Scorpions

Until now, the spectral sensitivity in Scorpions was only studied with the ERG method. From Machan's (1966, 1968) and Fleissner's (1968) studies it can be said that sensitivity in median eyes differs from that in lateral. Median eyes respond to green light only whereas lateral ones respond both to UV and green light.

Spiders

The spectral sensitivity in Spiders was better studied than in Scorpions. In spiders, members of several families with various ways of life were studied. This included not only the ERG-method, but also intracellular recording and cell-marking techniques.

Our present knowledge makes it difficult to interpret the results. Some of the results obtained from species within the same family (Salticids) contradict one other; nevertheless, a good deal of the existing data are established. Therefore, in principal eyes, sensitivity to UV and green seems to be constant. While the sensitivity to blue or yellow, as determined from intracellular recordings (Yamashita & Tateda, 1976) have presently given rise to much controversy (Blest et al., 1981). In Salticids, spectral analysis is now indefinitive and because of the morphological complexity of the retina in this family, the present data must be considered very circumspectly. At present, there is no support, for or against Land's hypothesis on spectral characteristics of the various retinal layers in Salticids.

In all the families studied (Argiopidae, Lycosidae, Salticidae and Dinopidae) the spectral sensitivity of accessory eyes shows its maximum in green, whereas UV - sensitivity can only be observed in Agelenidae and Lycosidae.

In Acarids the experiments of phototaxis on females of Tetranychus urticae showed the presence of two spikes; one at 375 nm and the other at 530 nm. These results led to the suggestion that anterior eyes have UV and green-receptors whereas UV receptors occur in the posterior ones only.

b)   Nocturnal and diurnal sensitivities.

Intracellular recordings and ERG - studies enable the comparison of nocturnal and cavernicolous Arachnids with those of diurnal animals. Laughlin et al. (1980) showed the ability of Dinopis's retina to produce bumps with very low intensity, this is probably in response to single photons. Hence, this characteristic seems to be an adaptation to nocturnal vision. A second characteristic is the extremely slow response. According to Laughlin et al. (1980), the time to reach the wave-crest in Dinopis is 150 ms, whereas it is 70 ms in Plexippus (Hardie & Duelli, 1978). Carricaburu and Muñoz-Cuevas (1978) showed that in the cavernicolous Harvestman, I. luteipes, the maximum is reached after 250 ms and that its decrease is extremely slow (about 1 second), compared to the 100 ms observed in A. aculeatus with a duration of almost 250 ms.

Based on optical physiology, Land (1969 a, b) suggested that the postero-median receptors in Dinopis receive 2000 times more photons than those of the postero-lateral eyes in Salticids. The photoreceptors of Dinopis are comparatively more sensitive, by 5 logarithmic units, than those of Plexippus (Blest & Land, 1977). In Plexippus the receptors are active all day long, whereas in Dinopis they seem to be active from nightfall to dawn (Laughlin et al., 1980).

c)   Critical fusion frequency: CFF.

Scorpions

Vejovis	M.	30	Machan 1966	
Opisthacanthus	L.		"	"
Buthus occitanus	M.	15-16	Fouchard & Carricaburu 1970	
	L.	13-14	"	"

Spiders

Tegenaria	P.	19-24	Giulio 1962	
	L.	12-15	"	"
Pardosa	L.	37	"	"
Micrommata virescens (Pisauridae)		20	"	"

Harvestmen

Pachylus quinamavidensis adult	30	Carricaburu & Muñoz-Cuevas 1981 b	
Acanthopachylus aculeatus adult	15	"	"
Acanthopachylus aculeatus N y 3	50	"	"
Ischyropsalis luteipes adult	10	"	"

The values of critical fusion frequency of the three orders studied emphasize the low frequencies obtained. These values can be considered as belonging to the so-called nocturnal animals (Figs. 14, 15, 16, 17).

Few studies have been performed on the critical fusion frequency during the postembryonic development. In the epigean Harvestman, A. aculeatus, the critical fusion frequency in Nymph 3 is three times higher than in the adult (Carricaburu & Muñoz-Cuevas, 1981 b). These observations enable a correlation to be made between eye-structure, number of retinulae, problems of ontogenic regression involved in the eye during the postembryonic development, and the way of life of the adult. Among Harvestmen, the lowest critical fusion frequencies can be observed in cavernicolous species and, of the two epigean species, the one dwelling

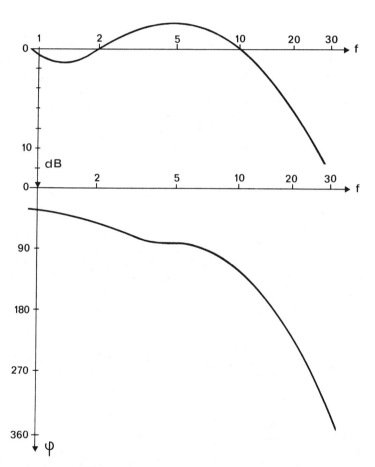

Fig. 14. Frequency transfer function (FTF) and flicker fusion frequency (FFF) in Pachylus quinamavidensis. (from Carricaburu & Muñoz-Cuevas, 1981 b)

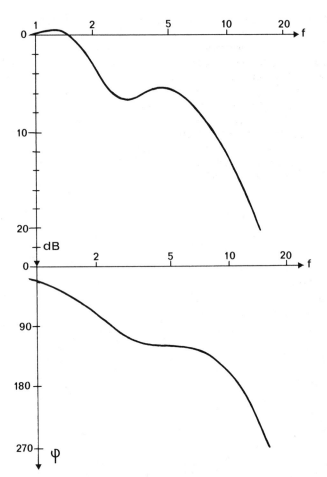

Fig. 15. Frequency transfer function (FTF) and flicker fusion frequency (FFF) in A. aculeatus, adult.
(from Carricaburu & Muñoz-Cuevas, 1981 b)

in obscure situations shows the lower frequency. The positive correlation between the number of retinulae and the critical fusion frequency corroborates the way of life (Carricaburu & Muñoz-Cuevas, 1981 a, b).

d) Role of the eyes in the control of the circadian rhythm in Scorpions

Fleissner's studies (1972, 1974, 1977 a, b, c, d) enable us to understand the contribution of eyes to the circadian rhythm in A. australis (Fig. 18). Under constant darkness the Scorpion shows a circadian change in the sensitivity of its visual system. The median eyes show periods of low sensitivity all day long; whereas periods of high sensitivity occur during the night. According to Fleissner, these periods could be called

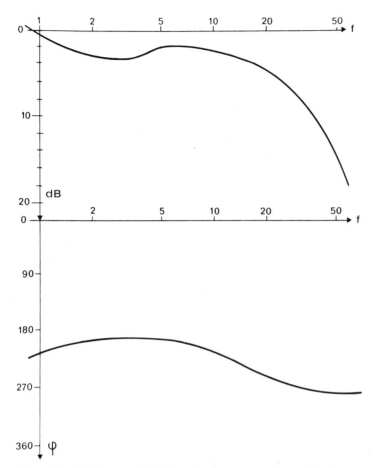

Fig. 16. (FTF) and (FFF) in A. aculeatus, juvenile Ny3. (from Carricaburu & Muñoz-Cuevas, 1981 b)

night-state or day-state. This circadian change of sensitivity was observed without exception in several tested experiments and does not change during the life-time of Scorpions. In lateral eyes, the same stimulation-test induces for greater ERG-amplitudes than in median eyes. ERG - amplitudes of the lateral eyes are constant; the oscillations are regular, of very low intensity, and synchronous with those of the median eyes. Fleissner reduced the light-intensity on the lateral eyes by two logarithmic units for this test. Lateral eyes are more sensitive than the median ones by one logarithmic unit during the night-period.

Under constant conditions, the endogenous control of lateral eye sensitivity varies vary slightly, whereas the change of sensitivity of median eyes may compensate for variations of the natural lighting during twilight and night by 3 to 4 logarithmic units. These experimental

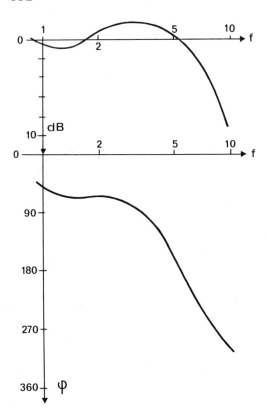

Fig. 17. Frequency transfer function (FTF) and flicker fusion frequency (FFF) in I. luteipes.
(from Carricaburu & Muñoz-Cuevas, 1981 b)

changes probably occur in nature, the Scorpion being a strictly nocturnal animal (Cloudsley-Thompson, 1956). Fleissner (1977 a, b, c, d) suggested that Scorpions have at least two receptive systems, each of them perceiving the same Zeitgeber's cycle in a different way.

e) Perception of polarized light and astronomical orientation.

The existence of a polaro-menotactic orientation in Arctosa-variana was first shown by Papi (1955 a, b). Later, similar results were obtained by Görner (1958, 1962) for Agelena labyrinthica. More recent studies on Arctosa performed by Italian researchers (Papi, 1959; Papi & Serreti, 1955; Papi et al., 1957; Tongiorgi, 1959; Papi & Tongiorgi, 1963; Magni et al., 1964, 1965), showed that the median eyes (main eyes) exhibit astronomical orientation less precise than that of all the eyes. Orientation is greatly perturbed if the secondary eyes serve as perceptors and it becomes erratic with one pair of secondary eyes. In A. variana, a short orientation with the sun can be observed. This is present even when the sum is not directly visible, or even when it is below the horizon. Bacetti and Bedini (1964) in their analysis of the retinal structure of median eyes in Arctosa, did not observe any peculiar arrangement of the rhabdom which could serve as an analyser. Arctosa shows an irregular arrangement of the receptors (pentagonal), with rhabdomeres on the 5 sides of the cell

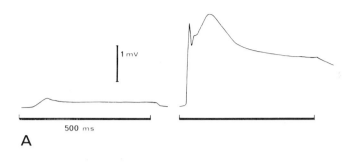

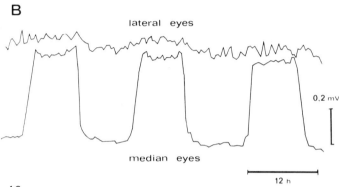

Fig. 18.
A- Electro-retinograms of a median eye in "continuous darkness" recorded during day-state of adaptation (left) and night-state (right), induced by the same stimulus. Cornea-negativity upwards. Horizontal bar: time calibration; vertical bar: amplitude calibration.
B- The north-African fat-tailed scorpion Androctonus australis L. and the circadian rhythm of its visual system. The time courses of efficiency of identical test light flashes (33 msec) simultaneously registered as ERG amplitudes of the left lateral eye and the left median eye of the same specimen are plotted for three successive circadian periods in D/D conditions. Vertical bar: calibration of ERG amplitude; horizontal bar: calibration of time scale.
(from Fleissner, 1972, 1977 a, b, c, d)

arranged randomly to the optic axis. On the contrary, A. gracilens showed retinular cells in the ventral part of the retina with rhabdomeres on the two opposite sides of the cell (Schröer, 1974 a, b). In addition, dorsal and central cells are often provided with more than 2 rhabdomeres. Schröer (1974 a, b) suggested that ventral receptors with perpendicular rhabdoms may be necessary for the analysis of polarized light. The possible rotating movement of retina would enable the ventral receptors of Agelena to analyse polarized light.

In the Scorpion, A. australis the perception of polarized light was shown by Carricaburu and Cherrak (1968). In both the median and lateral eyes, the ERG amplitude is modulated, these eyes perceiving variations of luminosity concomitant with variations in the direction of the plane of polarization.

f) Extraocular photoreception.

Zwicky (1968, 1970) showed for the first time the presence of a metasomatic sensitivity to light in the Scorpion Urodacus novaehollandiae. Quantitative variations of spectral sensitivity were obtained in the mesosomic nerve-cord between the 2nd and 3rd ganglia, a spike at 480 nm could be observed. On comparison with the results, from median eyes, obtained by Machan (1968), Zwicky noted a slight shift towards the left of the spectrum in his results. Geethabali and Pampapathi Rao (1973) and Geethabali (1974, 1976 a, b) analyzed light-perception in Heterometrus and compared the results to those obtained in Urodacus, and to the values mentioned by Machan (1968). According to these authors:

1°) An extraocular photoreception exists in the metasoma of different species of the genus Heterometrus.
2°) The telson-nerve, sensitive to light, is not the only one involved in light-perception. The response is not suppressed by cutting the nerve.
3°) Spectral sensitivity is the same in the different species of Heterometrus tested, the spike occurring at 568 nm and a smaller one at 440 nm.

Eye-Fluorescence in Spiders.

Young and Wanless (1967) studied eye-fluorescence in Spiders under visible and UV light. Adults belonging to 11 families were tested. According to these two authors, Spiders with bad sight, such as Araneus umbricatus, Ciniflo similis and Clubiona terrestris, show an important fluorescence in all their eyes. In all individuals with pearl-white eyes fluorescence is high whereas in those with dark ones (black, brown) it is low. The color of fluorescence is turquoise or blue-green under UV - light and apple-green under violet.

In Salticus scenicus postero-lateral and postero-median eyes are fluorescent whereas the antero-median and the antero-lateral ones are not. The presence or absence of tapetum is related to the intensity of fluorescence. Among all the families studied, only Thomisidae, Salticidae and Linyphiidae have non-fluorescent eyes. A fluorescent substance which may be associated with the epicuticle was observed in the cornea and lens of the animals studied.

VISION AND COMMUNICATION

Vision and Sexual Behavior.

Crane's studies (1949) on the biology of behavior in a great many of species from Rancho grande (Venezuela), and Drees' studies (1952), several solved problems related to shape-perception in Salticidae.

Thus, in Corythalia xanthopa, Crane (1949) showed that:
- "Use of the AME by the male is the single, sine qua non of display. Use of the AME only is a prime requisite in Salticid courtship".
- "Appropriate size is an important secondary sign stimulus to display. No male C. xanthopa will give any kind of a positive display reaction to a spider with a frontal area of more than five times, or less than one-third of his own. In linear measurements, positive responses may be given to stimuli measuring up to about twice the natural size and down to about one half".
- "The general shape of the spider is another secondary sign stimulus for releasing display. The primary shape requisite for releasing and directing either courtship or threat in Corythalia and Phiale is a roughly quadrilateral figure, broader than high, the lateral portions of which show some suggestions of dark and light horizontal stripes. The vertical dimension is less important than the horizontal as a factor in display".

The behavior of Salticidae in the capture of moving prey or in mating shows very intricate sequences, which can be roughly considered as an early reaction to a moving object of appropriate size and at correct distance. This may be due to lateral eyes and constitutes the "primary pattern orientation". The aim of this is to have the prosoma of the animal facing the stimulus so that its image could be projected on the retinae of the principal eyes; it is accompanied with a pursuit motion, and according to the pattern-recognition of the visual characteristics of the stimulus, the behavior may be directed towards the intricate and varied preliminaries leading up to mating, escape or prey-capture (Forster, 1977, 1982).

The retina in principal eyes of Salticidae moves. There are four types of motions (Land, 1969 a, b): spontaneous activity, saccades, tracking and scanning; all these movements are inter-connected.

"Spontaneous activity consists of a very variable, periodic side-to-side motion of the retinae. It is associated with states of high excitability, and occurs whether or not there is any structure in the field of view".

"Saccades occur when a small stimulus (e.g. a dark dot) is presented to, or moved upon, the retinae of either the principal eyes or the anterolateral eyes. In a saccade the retinae move towards the image of the target so that they come to rest with their central regions fixed on the target".

"Tracking: if the target moves, the retinae track it, maintaining central fixation".

"Scanning normally follows a saccade. It consists of an oscillatory, side-to-side movement of the retinae across the stimulus, with a period of 1-2 sec, and a simultaneous torsional movement in which the retinae partially rotate about the visual axis, through an angle of approximately 50° and with a period of 5-15 sec".

Visual Mechanisms during Prey-Capture.

Behavioral sequences of prey-capture in Salticidae are under the control of vision (Homann, 1928; Crane, 1949; Kaestner, 1950; Drees, 1952; Gardner, 1964; Forster, 1977, 1979, 1982).

Forster's (1979) most important conclusions on visual mechanisms during prey-capture in Trite planiceps are as follows:

"Jumping spiders rely on visual stimuli when catching their prey. Each of the hunting actions employed is mediated by one or more of the three pairs of eyes responsible for this behavior. No function has yet been demonstrated for the tiny fourth pair of eyes".

"After swivelling, jumping spiders engage in one of two hunting strategies:
1°) they chase and catch their prey on the run
2°) they stalk, crouch, and jump at stationary prey".

"A L eyes mediate chasing behavior, A M eyes mediate stalking behavior. Input from either two A M eyes or adjacent A L and A M eyes induce the spider to adopt a low profile as it approaches its prey, to crouch, and to jump; dual input appears to be essential. When visual input is reduced, spiders creep closer to the target before jumping. Accurate jumps of 2-3 cm towards stationary targets can be made with just the A M pair, but all four anterior eyes are used in performing accurate jumps of up to 9 cm towards moving targets".

"Depending on prey movement, spiders switch smoothly and instantly from chasing to stalking and vice versa; tracking and scanning movements of the A M eyes probably facilitate this transition".

"Short jumps are made by A M-blinded spiders towards targets whose angular velocity is greater than 15°/sec. A significant negative correlation ($P < 0,005$) is found between target velocities and jump distances, suggesting that A L eyes must process information relating to these variables".

"Blinding one A M eye or adjacent A M and A L eyes in male spiders does not affect:
    a- responses to another male,
    b- minor-image reactions,
    c- courtship behavior toward a female.
Responses occur at shorter distances than usual, i.e., not until the blinded male is about 6-8 cm from the other spider. Blinding both A M eyes prevents all displays".

Color Vision

Peckham and Peckham's (1894) studies showed variations in the color of Salticidae females lead to reduction in all the intricate activities of the male before courtship. Crane (1949) showed that the yellow color of the clypeus in C. xanthopa is essential to start mating behavior. Kästner (1949, 1950) observed in Evarcha falcota, behavior which suggests color-vision.

According to Land's studies (1969) on the stratification of retina in Salticidae, it is assumed that each retinal layer is specialized in perceiving part of the spectrum. The deepest layer may catch red-light issuing from the infinite; blue-green light should be caught by the second layer whereas violet and ultraviolet are perceived by the third one. The 4th Land's layer, the most superficial, would be a receptor enabling the analysis of the plane of polarized light. Land's hypothesis has not been supported by the studies on spectral sensitivity and more data are necessary to substantiate this hypothesis which is strictly based on morphology.

Generally data dealing with spectral sensitivity in Arachnids are positive; they indicate the presence of color vision but behavioral studies to confirm it are lacking.

CONCLUDING REMARKS AND OUTLOOK

During the past few years, several investigations have been made on circadian rhythm in Scorpions (Fleissner, 1968, 1972, 1974, 1977 a, b, c, d; Fleissner & Schliwa, 1977; Schliwa & Fleissner, 1979, 1980). Land's (1969 a, b) studies on the retina in Salticids are also commendable; as are the cytological analyses by Blest and colleagues (Blest 1978; Blest & Day, 1978; Blest & Land, 1971; Blest et al., 1980, 1981) on the retina in Dinopidae; Forster's (1977, 1979, 1982) investigations dealing with the visual mechanisms of Salticidae during prey-capture; and our own findings on cell-differentiation of the eye in Opilionids.

Several studies are presently underway, they include protein synthesis and mechanisms of membrane renewal of the rhabdom related to the way of life in Opilionids; and Ca^{++} microanalysis in the retina of Opilionids. Nevertheless, other important sectors are not receiving attention. They are: - biochemistry of visual pigments; the study of the optic centers and their retinular extensions, topography and functioning; behavioral-tests of color vision; the phyletic analysis of the origin of the inverted retina and its evolution in Chelicerata. These options remain attractive possibilities for research.

EYES OF MYRIAPODS

Myriapods, usually have two lateral groups of eyes with 40 relatively large circular lenses of different sizes. They are frequently reduced or even absent, as in Pauropoda, Symphyla, some Chilopoda (Geophilidae), and some Diplopoda (Polydesmidae). The arrangement of these lenses is never as narrow as in a facetted eye; only in the Scutigeromorpha do we find typical facetted eyes.

General Organization of Chilopod Eyes

Light microscopic investigations on the anatomy and histology of the eyes of a few species of Chilopods have been made by Grenacher (1880), Willem (1892) and Hesse (1901). According to these authors the ocellus of Chilopods does not show similar organization in all species. While it is generally agreed that there is a vitreous body in the Scolopendridae, opinions differ for Lithobius. There has also been disagreement over the presence of functionally and morphologically diverse cellular elements in the interior of the optic cup (Grenacher, 1880; Willem, 1892; Hesse, 1901).

The first ultrastructural investigation was made on Polybothrus fasciatus by Bedini (1968). She was unable to reconcile her findings on this species with existing histological descriptions of the ocelli of Lithobius forficatus. In addition electron-microscopic observations by Joly (1969) on Lithobius did not confirm the differences within the receptor layer described by Grenacher (1880), Willem (1892) and Hesse (1901).

The results of Bähr's works (1971, 1974) on L. forficatus is summarized below.

The single eye consists of a corneal lens, corneagenous cells, Hüllzellen (covering cells) and a variable number (up to 110) of visual cells. The latter, discernible by their sensory structures, contribute to either the distal or the proximal rhabdom.

Corneal lens: All lenses are biconvex (50 µm in diameter and 35 µm thick). Externally a predominantly homogeneous epicuticular layer of 2-3 µm can be distinguished from the underlying endocuticle.

The lamellated structure of the dioptric apparatus can be explained by the different alignment of microfibers running parallel in the laminae and changing their direction by approximately 180° within the interlaminar layer.

Cornegenous cells: In contrast to Bedini's (1968) observations on Polybothrus, the corneogenous layer in Lithobius are irregular in appearance. Bähr's investigation (1971, 1974) does not support Grenacher's (1880) observation of the peripherically located pigment cells. The corneagenous cells are too heterogeneous with insufficient pigment grains to serve as screening cells.

Hüllzellen (covering cells), ("cellules bordantes" - Joly & Herbaut, 1968; satellite cells - Bedini 1968); near the corneagenous cells exhibit morphological properties which are transitional between the corneagenous cells and the cells detached from the peripheral layer which cover the eye-cup. Their structural aspect and distribution suggest three possible functions to Bähr (1974):
  1- Secretion of the endocuticular part of the lenses
  2- Morphological and mechanical separation of the ocelli from one other
  3- Isolation of one or more axons.

The photoreceptor cell layer: Bähr's investigations confirm the existence of two kinds of receptor cells in L. forficatus, distinguishable by their ultrastructural properties. Bedini (1968), however, has clearly demonstrated that P. fasciatus possesses only one type of sensory cell.

The distal receptors of hair cells: The distal and greater part of the eye-cup is occupied by about 25 to 70 large sensory cells. The hair cells of the dark adapted state show a succession of sectors: the rhabdomere; the so-called Schaltzone (intercalaryzone); the main cytoplasmic part with cell organelles and the nerve fiber process. Compared with other arthropod visual cells, the size of the microvilli of Lithobius is large (1-17 µm in length and 75-250 nm in diameter). Beyond the microvilli and extending towards the periphery of the eye-cup is a clear zone, the "Schaltzone", which, depending on the state of dark adaptation, may completely surround the fused rhabdom of the distal receptors. Electron microscopic findings reveal that the Schaltzone consists of cytoplasmic bridges surrounding small cisternae of smooth endoplasmic reticulum (palisades by Horridge & Barnard, 1965).

The basal cells (retina): Longitudinal and cross sections through the proximal part of the ocellus show rhabdomeric organization quite different from that of the distal part. Whereas the microvilli of the hair cells are perpendicular to the axis of the ocellus, those of the basal cells are arranged approximately perpendicular to the long diameter of the cell body. The microvilli are inserted on all the sides of the distal and apical cell membranes, but most of them lie on opposite sides. Microvilli of neighboring basal cells are not separated by rhabdomeric radii but they interdigitate. The contents of cytoplasmic cell organelles resemble that of the distal receptors. Pigment grains are scattered throughout the cytoplasm up to the level of the desmosomes.

Light and Dark Adaptive Changes in the Visual Cells of Lithobius forficatus.

The fine structure of the visual cells in relation to their state of adaptation was investigated by Bähr (1972). In the dark-adapted state, the large distal cells show an extensive Schaltzone which takes up 35 to 42% of the cross-sectional area of the ocellus. Under conditions of illumination (250 lux) this Schaltzone is reduced to 6 to 7% within 60 minutes. An adaptive pigment migration of variable extent has been found to occur in both receptor types. Therefore it is suggested that Schaltzone and pigment migration act as a "pupil-like mechanism". Different agranular endoplasmic membrane systems which probably arise from the disintegration of multilamellated bodies are also present. These multilamellated bodies are found to increase in number during light adaptation.

Lateral Facetted Eye of Scutigera.

Scutigera is the only Myriapod representative showing a facetted eye with about 200-250 units of narrowly arranged hexagonal facets. Grenacher (1880), Adensamer (1894), Hesse (1901), Hanström (1934), Miller (1957) and Paulus (1979) have studied the eye of Scutigera. According to Paulus (1979) the cornea is biconvex with a nearly smooth surface. Just beneath the lens the very large crystalline cone looks like a vitreous body, it consists of numerous confluent parts without any visible nuclei. One segment of the vitreous body consists of a hyaline substance, with dense cytoplasmic zones, and sometimes small spheres along the cell membrane. The latter are called rudimentary nuclei. In the electron micrographs of Paulus (1979) the cone-segments seem to be extracellular secretions of the distal pigment cells surrounding the distal zone of the vitreous body. Paulus (1979) suggests that these cells secrete both the crystalline cone-segments and the cornea above them during the last ecdysis. The cone is functionally comparable with the pseudocone of some insects, but it is not homologous structurally (Paulus, 1979).

The rhabdom is two-layered. The distal retinular cells extend far around the cone laterally. In the center of the eye there are about 7-10 retinular cells, and the lateral parts (dorsally) 9-23, form a circular ring of rhabdomeres around the cone tip (Paulus, 1979). The proximal layer consists, in all cases, of four retinular cells, which form a triangular rhabdom. Between these four proximal retinular cells three or four smaller cells, which possibly function as supporting cells, one seen. In addition to the distal pigment cells, there are other distal and basal pigment cells (Paulus, 1979).

The ommatidium of Scutigera differs from that of Crustacea and Insects: there are neither typical corneagenous cells, nor a tetrapartite crystalline cone. The retinula consists of 11-16 cells, which form a closed rhabdom in two layers. Comparison of the eyes of Scutigera and Lithobius shows that the Scutigera ommatidium is apparently derived from that of the latter. Thus, the facetted eye of Scutigera is not homologous with those of Insects or Crayfishes (Paulus, 1979).

Fine Structure of the Eye in Glomeris (Diplopoda).

The ocellus of Glomeris consists of a corneal lens of epidermal, corneagenous cells and of retinal pigmented cells. The basal parts of the ocelli and the optic nerves are surrounded by a common connective sheath. Pigmented and granular cells are inserted between ocelli.

Each sensory cell consists of a distal portion, or photoreceptor process, provided with long microvilli lying parallel to the longitudinal axis of the process; a central portion or cell body; and a proximal axonal portion. The photoreceptor processes are localized in the central retinal area while the axonal cell bodies are situated peripherically. The microvilli of the photoreceptor process of adjacent cells lie parallel to the minor axis of the retina and their distal tips reach those of opposing cells. The retinal cell microvilli are about 15 µm long with a diameter of 0,1 to 0,2 µm. They are usually straight and closely packed (Bedini, 1970).

Electrophysiology of Lithobius forficatus.

The most important findings of Bähr's works (1965, 1967) are:
- ERG is negative, monophasic, with initial peak and negative plateau at long duration in a few cases. Most responses, however, consist of a negative on - and a positive off-effect.
- Response amplitudes increase with increasing stimulus intensity in a logarithmical, particularly sigmoidal relation, while latencies and negative peaks decrease.
- Wave form and amplitudes change according to the adaptative state. Dark adaptation occurs at high stimulus intensities the curve is biphasic; and seems to be complete after 20 min.
- Critical flicker fusion frequency, preceded by desynchronization-effects, reaches 10 to 36 cycles/sec.

ACKNOWLEDGMENTS

The author wishes to thank Dr. P. Carricaburu for critically reviewing the manuscript; Mrs. F. Saunier for translating it into English; and Mrs. Gh. Thibaud and Mr. M. Gaillard for technical assistance.

REFERENCES

Adensamer, T. (1894) Zur Kenntnis der Anatomie und Histologie von Scutigera coleoptrata. Verh. Zool. Bot. Ges. Wien 43: 573-578.

André, M. (1949) Ordre des Acariens. In: Traité de Zool., Vol. 6, Ed. P.P. Grassé. Paris, Masson, p. 794-892.

Baccetti, B. & Bedini, C. (1964) Research on the structure and physiology of the eyes of a Lycosid Spider. I. Microscopic and ultramicroscopic structure. Arch. Ital. Biol. 102: 97-122.

Bähr, R. (1965) Ableitung lichtinduzierter Potentiale von den Augen von Lithobius forficatus L. Naturwiss. 52: 459.

Bähr, R. (1967) Elektrophysiologische Untersuchungen an den Ocellen von Lithobius forficatus L. Z. vergl. Physiol. 55: 70-102.

Bähr, R. (1971) Die Ultrastruktur der Photorezeptoren von Lithobius forficatus L. (Chilopoda Lithobiidae). Z. Zellforsch. 116: 70-93.

Bähr, R. (1972) Licht und dunkeladaptive Aenderungen der Sehzellen von Lithobius forficatus L. (Chilopoda: Lithobiidae). Cytobiol. 6: 214-233.

Bähr, R. (1974) Contribution to the morphology of chilopod eyes. Symp. Zool. Soc. Lond. 32: 383-404.

Bedini, C. (1967) The fine structure of the eyes of Euscorpius carpathicus L. (Arachnida Scorpions). Arch. Ital. Biol. 105: 361-378.

Bedini, C. (1968) The ultrastructure of the eyes of a centipede Polybothrus fasciatus (Newport). Monitore Zool. Ital. 2: 31-47.

Bedini, C. (1970) The fine structure of the eye in Glomeris (Diplopoda). Monitore Zool. Ital. 4: 201-219.

Bertkau, Ph. (1886) Beiträge zur Kenntnis der Sinnesorgane der Spinnen. Archiv. für microskop. Anatomie 27: 289-631.

Blest, A. (1978) The rapid synthesis and destruction of photoreceptor membrane by a dinopid spider: a daily cycle. Proc. R. Soc. Lond. 200B: 463-483.

Blest, A. & Day, M. (1978) The rhabdomere organization of some nocturnal Pisaurid Spiders in light and darkness. Phil. Trans. R. Soc. Lond. 283B: 1-23.

Blest, A., Hardie, R., McIntyre, P. & Williams, D. (1981) The spectral sensitivities of identified receptors and the function of retinal tiering in the principal eyes of jumping spider. J. Comp. Physiol. 145A: 227-239.

Blest, A. & Land, M. (1977) The physiological optics of Dinopis subrufus L. Koch: a fish-lens in a spider. Proc. R. Soc. Lond. 196B: 197-222.

Blest, A., Williams, D. & Kao, L. (1980) The posterior Median eyes of the dinopid spider Menneus. Cell Tissue Res. 211: 391-403.

Butenandt, A. (1959) Wirkstoffe des Insektenreiches. Naturwiss. 46: 461-471.

Carricaburu, P. (1968) Dioptrique oculaire du scorpion Androctonus australis. Vision Res. 8: 1067-1072.

Carricaburu, P. (1970a) Dioptrique oculaire de quatre espèces d'araignées. Vision Res. 10: 943-953.

Carricaburu, P. (1970b) Les yeux des Telyphonides. Bull. Soc. Hist. Nat. Afriq. Nord. 61: 69-86.

Carricaburu, P. & Cherrak, M. (1968) Electrorétinogramme du Scorpion Androctonus australis en réponse à des éclairs périodiques. C.R. Acad. Sc. Paris, 226: 1299-1301.

Carricaburu, P. & Muñoz-Cuevas, A. (1978) L'électrorétinogramme des Opilions épigées et cavernicoles. Vision Res. 18: 1229-1231.

Carricaburu, P. & Muñoz-Cuevas, A. (1981a) Régression oculaire et électrorétinogramme chez les Opilions. C.R. Séanc. Soc. Biol. 175: 28-37.

Carricaburu, P. & Muñoz-Cuevas, A. (1981b) Fonction de transfert des fréquences et fréquence critique de fusion chez les Opilions épigées et cavernicoles. C.R. Séanc. Soc. Biol. 175: 288-294.

Cloudsley Thompson, J.L. (1956) Studies in diurnal rhythms. VI. Bioclimatic observations in Tunisia and their significance in relation to the physiology of the fauna, especially woodlice, centipedes, scorpions and beetles. Ann. Mag. Nat. Hist. 12: 305-329.

Crane, J. (1949) Comparative biology of Salticid spiders at Rancho Grande, Venezuela. Part IX. An analysis of Display. Zoologica 34: 159-214.

Curtis, D. (1969) The fine structure of photoreceptors in Mitopus morio (Phalangida). J. Cell Sci. 4: 327-351.

Curtis, D. (1970) Comparative aspects of the fine structure of the eyes of Phalangida (Arachnida) and certain correlations with habit. J. Zool. (Lond.) 160: 231-265.

DeVoe, R. (1972) Dual sensitivities of cells in wolf spider eyes at ultraviolet and visible wave lengths of light. J. Gen. Physiol. 59: 247-269.

DeVoe, R. (1975) Ultraviolet and green receptors in principal eyes of jumping spiders. J. Gen. Physiol. 66: 193-207.

DeVoe, R., Small R. & Zvargulis, J. (1969) Spectral sensitivities of wolf spider eyes. J. Gen. Physiol. 54: 1-32.

Drees, O. (1952) Untersuchungen über die angeboren Verhaltensweissen bei Springspinnen (Salticidae). Zeit. für Tierpsych. 9: 169-209.

Eakin, R. (1966) Differentiation in the embryonic eye of Peripatus (Onychophora). In: Sixth Int. Congr. Elect. Microsc. Kyoto, Japan, Ed. Maruzen 2: 507-508.

Eakin, R. & Brandenburger, J. (1971) Fine structure of the eyes of jumping spiders. J. Ultrastruct. Res. 37: 618-663.

Eguchi, E. & Waterman, T. (1976) Freeze-etch and histochemical evidence for cycling in crayfish photoreceptor membranes. Cell Tissue Res. 169: 419-434.

Eley, S. & Shelton, P. (1976) Cell junctions in the developing compound eye of the desert locust Schistocerca gregaria. J. Embryol. Exp. Morph. 36: 409-423.

Fleissner, G. (1968) Untersuchungen zur Sehphysiologie der Skorpione. Verh. Dtsch. Zool. Ges. 36: 375-380.

Fleissner, G. (1972) Circadian sensitivity changes in the median eyes of the North African scorpion, Androctonus australis. In: Information Processing in the Visual Systems of Arthropods. Ed. R. Wehner. Berlin, Springer Verlag, p. 163-202.

Fleissner, G. (1974) Circadiane Adaptation und Schirmpigment -

verlagerung in den Sehzellen der Medianaugen von Androctonus australis L. J. Comp. Physiol. 91: 399-416.

Fleissner, G. (1977a) Differences in the physiological properties of the median and the lateral eyes and their possible meaning for the entrainment of the scorpion's circadian rhythm. J. Interdiscipl. Cycle Res. 8: 15-26.

Fleissner, G. (1977b) Entrainment of the scorpion's circadian rhythm via the median eyes. J. Comp. Physiol. 118: 93-99.

Fleissner, G. (1977c) Scorpion's lateral eyes: extremely sensitive receptors of Zeitgeber stimuli. J. Comp. Physiol. 118: 101-108.

Fleissner, G. (1977d) The absolute sensitivity of the lateral eyes of the scorpion, Androctonus australis L. (Buthidae, Scorpiones). J. Comp. Physiol. 118: 109-210.

Fleissner, G. & Schliwa, M. (1977) Neurosecretory fibres in the median eyes of the scorpion Androctonus australis L. Cell Tissue Res. 178: 189-198.

Forster, L. (1977) A qualitative analysis of hunting behaviour in jumping spiders. (Aranae: Salticidae). N. Z. J. Zool. 4: 51-62.

Forster, L. (1979) Visual mechanisms of hunting behaviour in Trite planiceps, a jumping spider (Araneae: Salticidae). N. Z. J. Zool. 6: 79-93.

Forster, L. (1982) Visual communication in jumping spiders (Salticidae). In: Spider Communication. Ed. P.N. Witt & J.S. Rovner. New Jersey, Princeton University Press, p. 161-212.

Fouchard, R. & Carricaburu, P. (1970) Quelques aspects de la physiologie visuelle chez le scorpion Buthus occitanus; étude électrorétinographique. Bull. Soc. Hist. Nat. Afr. Nord. 61: 57-68.

Gardner, B.T. (1964) Hunger und sequential responses in the hunting behavior of salticid spiders. J. Comp. Physiol. Psych. 58: 167-173.

Geethabali, M. (1974) Physiology of the metasomatic photoreceptor neurons in scorpions. Proc. 6th Int. Arachn. Congr. p. 197.

Geethabali, M. (1976a) Motor excitation with reference to neural photoreception in scorpion. Life Sci. 18: 1009-1012.

Geethabali, M. (1976b) Central course of photic input in the ventral nerve of scorpion (Heterometrus fulvipes). Experientia 32: 345-347.

Geethabali, M. & Pampapathi Rao, K. (1973) A metasomatic neural photoreceptor in the scorpion. J. Exp. Biol. 58: 189-196.

Giulio, L. (1962) Sensibilita spettrale degli ocelli di Araneus diadematus. Boll. Soc. Ital. Biol. Sper. 38: 1598-1599.

Gonzalez-Baschwitz, G. (1977) Ultrastructura de los ojos de Araneus diadematus Clerk (Araneae, Arachnida). Bol. R. Soc. Espanola Hist. Nat. (Biol.) 75: 129-147.

Görner, P. (1958) Die optische une kinästhetische Orientierung der Trichterspinne Angelena labyrinthica. Z. vergl. Physiol. 41: 111-153.

Görner, P. (1962) Die Orientirung der Trichterspinne nach polarisiertem Licht. Z. vergl. Physiol. 45: 307-314.

Grandjean, F. (1928) Sur un Oribatidé pouvu d'yeux. Bull. Soc. Zool. France 53: 235-242.

Grandjean, F. (1958) Au sujet du Naso et de son oeil infère chez les

Oribates et les Endeostigmata (Acariens). Bull. Mus. Hist. Nat. Paris 30: 427-435.

Grenacher, H. (1879) Untersuchungen über das Sehorgan der Arthropoden, insbesondere der Spinnen, Insekten und Crustaceen. Ed. Vandenhoek und Ruprecht. Göttingen, Verlag von Vandenhoeck und Ruprecht, 188 p.

Grenacher, H. (1880) Ueber die Augen einiger Myriapoden. Zugleich eine Entgegnung an V. Graber. Arch. mikrosk. Anat. EntwMech. 18: 415-473.

Hanström, B. (1919) Zur Kenntnis der centralen Nervensystems der Arachnoiden und Pantopoden. Thesis, Faculty of Sciences, Stockolm.

Hanström, B. (1923) Further notes on the central nervous system of Arachnids: Scorpions, Phalangids and Trapdoor spiders. J. Comp. Neur. 35: 249-274.

Hanström, B. (1928) Vergleichende Anatomie des Nervensystems der wirbellosen Tiere. Berlin, Springer, 628 p.

Hanström, B. (1934) Bemerkungen über das Komplexaugen der Scutigeriden. Acta Univ. Lund (Adv. 2, n.f.) 30(6): 1-14.

Hanström, B. (1935) Fortgesetzte Untersuchungen über das Araneengehirn. Zool. Jahrb. 59: 455-478.

Hardie, R.C. & Duelli, P. (1978) Properties of single cells in posterior lateral eyes of jumping spiders. Z. Naturforsch. C 33: 156-158.

Hentschel, E. (1899) Beiträge zur Kenntnis der Spinnenaugen. Zool. Jb. Abt. Anat. u. Ontog. 12: 509-534.

Hesse, R. (1901) Untersuchungen über die Organe der Lichtempfindung bei niederen Tieren. VII. Von dem Arthropoden Augen. Z. Wiss. Zool. 70: 347-473.

Hickman, V. (1957) Some Tasmanian Harvestmen of the sub-order Palpatores. Pap. Proceed. R. Soc. Tasmania 91: 65-79.

Homann, H. (1928) Beitrag zur Physiologie der Spinnenaugen. I und II. Z. vergl. Physiol. 7: 201-268.

Homann, H. (1950) Die Nebenaugen der Araneen. Zool. Jb. Abt. Anat. u. Ontog. 71: 1-144.

Homann, H. (1953) Die Entwicklung der Nebenaugen bei den Araneen. I. Biol. Zbl. 72: 373-385.

Homann, H. (1955) Die Entwicklung der Nebenaugen bei den Araneen. II. Biol. Zbl. 74: 427-432

Homann, H. (1956) Die Entwicklung der Nebenaugen bei den Araneen. III. Biol. Zbl. 75: 416-421.

Homann, H. (1961) Die Stellung der Ctenidae, Textricinae und Rhoicinae im System der Araneae. Senck. biol. 42: 397-408.

Homann, H. (1971) Die Augen der Araneae. Z. Morph. Tiere 69: 201-272.

Horridge, G. & Barnard, P. (1965) Movement of palisada in locust retinula cells when illuminated. Quart. J. Microsc. Sci. 106: 131-135.

Joly, R. (1969) Sur l'ultrastructure de l'oeil de Lithobius forficatus L. (Myriapode Chilopode). C.R. Hebd. Séanc. Acad. Sc. Paris 268: 3180-3182.

Joly, R. & Herbaut, C. (1968) Sur la régénération oculaire chez <u>Lithobius forficatus</u> L. (Myriapode Chilopode). Arch. Zool. Exp. Gén. 109: 591-612

Kästner, A. (1949) Uber der Farbsinn der Spinne. Naturwissenchaften 36: 58-59.

Kästner, A. (1950) Reaktion der Hüpfspinnen (Salticidae) auf unbewegte farblose une farbige Gesichtsreize. Zool. Beitr. 1: 13-50.

Kishinouye, K. (1891) On the development of Araneina. J. Cell Sc. Tokyo 4: 55-88.

Land, M. (1969a) Structure of the principal eyes of jumping spiders in relation to visual optics. J. Exp. Biol. 51: 443-470.

Land, M. (1969b) Movements of the retinae of jumping spiders (Salticidae Dendryphantinae) in response to visual stimuli. J. Exp. Biol. 51: 471-493.

Laughlin, S., Blest, A. & Stowe, S. (1980) The sensitivity of receptors in the posterior median eye of the nocturnal spider, <u>Dinopis</u>. J. Comp. Physiol. 141: 53-65.

Machan, L. (1966) Studies on Structure Electroretinogram and Spectral Sensitivity of the Lateral and Median Eyes of the Scorpion. Ph.D. Thesis, Univ. Wisconsin.

Machan, L. (1968) Spectral sensitivities of scorpion eyes and the possible role of shielding pigment effect. J. Exp. Biol. 49: 95-105.

Magni, F., Papi, F., Savely, H. & Tongiorgi, P. (1964) Research on the structure and physiology of the eyes of a Lycosid spider. II. The role of different pairs of eyes in astronomical orientation. Arch. Ital. Biol. 102: 123-136.

Magni, F., Papi, F., Savely, H. & Tongiorgi, P. (1965) Research on the structure and physiology of the eyes of a Lycosid spider. III. Electroretinographic responses to polarized light. Arch. Ital. Biol. 103: 146-158.

Mark, E. (1887) Simple eyes in Arthropods. Bull. Mus. Comp. Zool. 13: 49-105 (Arachn. 72-99).

Melamed, J. & Trujillo-Cenoz, O. (1966) The fine structure of the visual system of <u>Lycosa</u> (Araneae). Z. Zellforsch. 74: 12-31.

Miller, W.H. (1957) Morphology of the ommatidia of the compound eye of Limulus. J. Biophys. Biochem. Cytol. 3: 421-428.

Millot, J. (1949) Classe des Arachnides (Arachnida). Morphologie générale et anatomie interne. In: Traité de Zool. Vol. 6. Ed. P.P. Grassé. Paris, Masson, p. 263-319.

Muñoz-Cuevas, A. (1970) Etude du développement embryonnaire de <u>Pachylus quinamavidensis</u> (Arachnides, Opilions). Bull. Mus. Nat. Hist. Nat. Paris 2è sér. 42: 1238-1250.

Muñoz-Cuevas, A. (1971) Contribution à l'étude du développement postembryonnaire de <u>Pachylus quinamavidensis</u> Muñoz-Cuevas (Arachnides, Opilions, Laniatores). Bull. Mus. Natn. Hist. Nat. 3è sér. Zool. 12: 629-641.

Muñoz-Cuevas, A. (1973a) Premiers stades de la différentiation du rhabdome chez l'embryon d' <u>Ischyropsalis luteipes</u> et remarques sur la théorie de Eakin. C.R. 2è Coll. Arach. Langue Fr. Montpellier p. 79.

Muñoz-Cuevas, A. (1973b) Sur la présence d'un centriole aux premiers stades de la différenciation du rhabdome chez l'embryon d' Ischyropsalis luteipes (Opilion, Arachnida). J. Microscopie 17: 81.

Muñoz-Cuevas, A. (1973c) Embryogénèse, organogénèse et rôle des organes ventraux et neuraux de Pachylus quinamavidensis (Arachnides, Opilions). Comparaison avec les Annélides et d'autres Arthropodes. Bull. Mus. Natn. Hist. Nat. Paris 3è sér. n°196 Zool. 128: 1517-1538.

Muñoz-Cuevas, A. (1974) Ultrastructure de la vésicule optique embryonnaire chez les Opilions (Arachnida). J. Microsc. Biol. Cell. 20: 72.

Muñoz-Cuevas, A. (1975a) Modèle ciliaire de développement du photorécepteur chez l'Opilion Ischyropsalis luteipes (Arachnida). C.R. Acad. Sc. Paris 280: 725-727.

Muñoz-Cuevas, A. (1975b) Aspects ultrastructuraux de la différenciation et de l'organisation de la rétine chez les Opilions (Arachnida). Proc. 6th Int. Arachn. Congr. Amsterdam, p. 129-132.

Muñoz-Cuevas, A. (1976a) Formation des corps multivésiculaires au cours du développement du photorécepteur chez l'Opilion Ischyropsalis luteipes (Arachnida). J. Microsc. Biol. Cell. 26: 20A.

Muñoz-Cuevas, A. (1976b) Structure oculaire de l'Opilion cavernicole Ischyropsalis strandi K. (Arachnida). Ann. Spéléol. 31: 203-211.

Muñoz-Cuevas, A. (1976c) Structure dégérative de l'oeil chez l'Opilion cavernicole Ischyropsalis strandi. C.R. 3è coll. Arachn. Langue Fr. Les Eyzies, p. 106-108.

Muñoz-Cuevas, A. (1978a) Différenciation cellulaire du système dioptrique chez les Opilions (Arachnida). Symp. Zool. Soc. Lond. 42: 399-405.

Muñoz-Cuevas, A. (1978b) Développement, Rudimentation et Régression de l'Oeil chez les Opilions (Arachnida). Recherches Morphologiques, Physiologiques et Expérimentales. Thèse, Faculté des Sciences, Université Pierre et Marie Curie, Paris VI, 144 p. (vol. I) et planches 117 p. (vol. II).

Muñoz-Cuevas, A. (1980a) Evolution régressive du genre Ischyropsalis et modifications ultrastructurales de la pigmentation de l'oeil. (Opilions, Arachnida). 8. Internat. Arachn. Kongress Wien, p. 319-324.

Muñoz-Cuevas, A. (1980b) Différenciation cellulaire des grains de pigment (ommochrommes) de la rétine dans la série régressive formée par quatre espèces d'Opilions trogrophiles et troglobies appartenant au genre Ischyropsalis (Arachnida). C.R. Acad. Sc. Paris 290: 57-60.

Muñoz-Cuevas, A. (1981) Développement, rudimentation et régression de l'oeil chez les Opilions (Arachnida). Recherches morphologiques, physiologiques et expérimentales. Mém. Mus. Natn. Hist. Nat. Paris Sér. A, Zool. 120: 1-117.

Naegele, J., McEnroe, W. & Soans, A. (1966) Spectral sensitivity and orientation response of the two-spotted spider mite, Tetranychus urticae Koch, from 350 mµ to 700 mµ. J. Insect Physiol. 12: 1187-1195.

Noirot-Timothée, C. & Noirot, C. (1973) Jonctions et contacts intercellulaires chez les insectes. I. Les jonctions septées. J. Microsc. 17: 169-184.

Papi, F. (1955a) Astronomische Orientierung bei des Wolfsspinne Arctosa perita (Latr.). Z. vergl. Physiol. 37: 230-233.

Papi, F. (1955b) Ricerche sull'orientamiento di Arctosa perita (Latr.) (Araneae-Lycosidae). Pubbl. Staz. Zool. Napoli 27: 76-103.

Papi, F. (1959) Sull' orientamento astronomico in specie del gen. Arctosa (Araneae Lycosidae). Z. vergl. Physiol. 41: 481-489.

Papi, F. & Serreti, L. (1955) Sull' esistenza di un senso del tempo in Arctosa perita (Latr.) (Araneae-Lycosidae). Atti Soc. tosc. Sci. Nat. Mem. 62(B): 98-104.

Papi, F., Serreti, L. & Parrini, S. (1957) Nuove ricerche sull'orientamento e il senso del tempo di Arctosa perita (Latr.) (Araneae Lycosidae). Z. vergl. Physiol. 39: 531-561.

Papi, F. & Tongiorgi, P. (1963) Innate and learned components in the astronomical orientation of wolf spiders. Ergebn. Biol. 26: 259-280.

Patten, W. (1887) Studies on the eyes of Arthropodes. J. Morph. 1: 193-226.

Paulus, H.F. (1979) Eye structure and the monophyly of the Arthropoda. In: Arthropod Phylogeny. Ed. A.P. Gupta. New York, Van Nostrand Reinhold Co., p. 299-383.

Peckham, G. & Peckham, E. (1894) The sense of sight in spider with some observations of the color sense. Trans. Wisconsin Acad. Sci. Arts Lett. 10: 231-261.

Perrelet, A. & Bauman, F. (1969) Evidence for extracellular space in the rhabdom of the honey bee drone eye. J. Cell Biol. 40: 825-830.

Perry, M. (1968a) Further studies on the development of the eye of Drosophila melanogaster. I. The Ommatidia. J. Morphol. 124: 227-248.

Perry, M. (1968b) Further studies on the development of the eye of Drosophila melanogaster. II. The interommatidial bristles. J. Morphol. 124: 249-261.

Petrunkevitch, A. (1907) Studies in Adaptation. I. The sense of sight in spiders. J. Exp. Zool. 5: 275-309.

Petrunkevitch, A. (1911) Sense of sight, courtship and mating in Dugesielle hentzi (Girard), a Theraphosid spider from Texas. Zool. fahrb. 31: 355-376.

Petrunkevitch, A. (1955) Arachnida. In: Treatise on Invertebrate Palaeontology (Arthropod 2). Ed. R.C. Moore. Lawrence, Kansas, The Univ. Press of Kansas, p. 42-162.

Purcell, F. (1892) Ueber den Bau und die Entwichlung der Phalangidenaugen. Zool. Anz. 15: 461-465.

Purcell, F. (1894) Ueber den Bau der Phalangidenaugen. Z. wiss. Zool. 58: 1-53.

Ringuelet, R. (1959) Los Aracnidos argentinos del order Opiliones. Rev. Mus. arg. Cienc. nat. "Bernardino Rivadavia". Ciencias Zool. 5(2): 127-439.

Satir, P. & Gilula, N. (1973) The fine structure of membranes and intercellular communication in Insects. Ann. Rev. Entomol. 18:

143-166.
Savory, Th. (Ed.) (1977) Arachnida. London Academic Press, 340 p.
Schimkewitsch, W. (1906) Ueber die Entwicklung von Telyphonus caudatus, L. Z. wiss. Zool. 81: 1-95.
Schliwa, M. & Fleissner, G. (1979) Arhabdomeric cells of the median eye retina of scorpions. I. Fine structural analysis. J. Comp. Physiol. 130: 265-270.
Schliwa, M. & Fleissner, G. (1980) The lateral eyes of the scorpion, Androctonus australis. Cell Tissue Res. 206: 95-114.
Schrber, W.-D. (1974a) Zum Mecanismus der Analyse polarisierten Lichtes bei Agelena gracilens C.L. Koch (Araneae, Agdenidae). I. Die Morphologie der vorderen Mittelaugen (Hauptaugen). Z. Morph. Tiere 79: 215-231.
Schrber, W.-D. (1974b) Polarised light detection in an agelenid spider, Agelena gracilens (Araneae, Agelenidae). Proc. 6th Int. Arach. Congr., p. 191-192.
Shaw, S. (1969) Interreceptor coupling in ommatidia of drone honey bee and locust compound eye. Vision Res. 9: 999-1029.
Sheuring, L. (1914) Die Augen der Arachnoideer. II. Zool. Jahrb. 37: 369-464.
Such, J. (1975) Recherches Descriptives et Expérimentales sur la Morphogénèse Embryonnaire de l'Oeil Composé du Phasme Carausius morosus Br. Mise en Place et Différenciation Ultrastructurale des Eléments Cellulaires Constituant l'Ommatidie. Thèse, Faculté des Sciences, Univ. de Bordeaux I. Vol. I, 127 p. Vol. II, planches 44 p..
Tongiorgi, P. (1959) Effects on the reversal of the rhythm of nyctemeral illumination on astronomical orientation and diurnal activity in Arctosa variana Koch (Araneae Lycosidae). Arch. Ital. Biol. 97: 251-265.
Trujillo-Cenoz, O. (1965) Some aspects of the structural organization of the Arthropod eye. Cold Spring Harb. Symp. Quant. Biol. 30: 371-382.
Trujillo-Cenoz, O. & Melamed, J. (1964) Synapses in the visual system of Lycosa. Z. Naturwiss. 19: 470-471.
Trujillo-Cenoz, O. & Melamed, J. (1967) The fine structure of the visual system of Lycosa (Araneae). Part II. Primary visual centers. Z. Zellforsch. 76: 377-388.
Trujillo-Cenoz, O. & Melamed, J. (1973) The development of the retina-lamina complex in Muscoid flies. J. Ultrastruct. Res. 42: 544-581.
Waddington, C. & Perry, M. (1960) The ultrastructure of the developing eye of Drosophila. Proc. R. Soc. 153B: 155-178.
Waddington, C. & Perry, M. (1963) Inter-retinular fibres in the eyes of Drosophila. J. Insect. Physiol. 9: 475-478.
White, R. (1961) Analysis of the development of the compound eye in the mosquito Aedes aegypti. J. Exp. Zool. 148: 223-239.
White, R. (1963) Evidence for the existence of a differentiation center in the developing eye of the mosquito. J. Exp. Zool. 152: 139-147.
Willem, V. (1892) Les ocelles de Lithobius et de Polyxenus (Myriapoda). Bull. Seances Soc. R. Malacolog. Belg. 27: 1-12.
Williams, D. (1979a) The physiological optics of a nocturnal semi-aquatic

spider, Dolomedes aquaticus (Pisauridae). Z. Naturforsch. 34c: 463-469.
Williams, D. (1979b) The feeding behaviour of New-Zealand Dolomedes species (Araneae: Pisauride). N. Z. J. Zool. 6: 95-105.
Williams, D. & McIntyre, P. (1980) The principal eyes of a jumping spider have a telephoto component. Nature (Lond.) 288: 578-580.
Wolsky, A. (1956) The analysis of eye development in Insects. Trans. N.Y. Acad. Sci. 18: 592-596.
Yamashita, S. & Tateda, H. (1976) Spectral sensitivities of jumping spider eyes. J. Comp. Physiol. 105: 29-41.
Yamashita, S. & Tateda, H. (1978) Spectral sensitivities of the anterior median eyes of the orb web spiders, Argiope bruennichii and A. amoena. J. Exp. Biol. 74: 47-57.
Yoshikura, M. (1955) Embryological studies on the Liphistiid spider: Heptathela kimurai. Part II. Kuamoto J. Sc. Ser. B 2(1): 1-86.
Young, M. & Wanless, F. (1967) Observations on the fluorescence and function of spider's eyes. J. Zool. (Lond.) 151: 1-16.
Zwicky, K. (1968) A light response in the tail of Urodacus, a scorpion. Life Sci. 7: 257-262.
Zwicky, K. (1970) The spectral sensitivity of the tail of Urodacus, a scorpion. Experientia 26: 317.

CRUSTACEA

M.F. LAND

School of Biological Sciences
University of Sussex, Falmer
Brighton BN1 9QG England

ABSTRACT

Although one usually thinks of the Crustacea as linked to the insects by the presence of an exoskeleton and compound eyes, there is actually a much greater diversity of eye types in the Crustacea than in the insects or any other invertebrate group. The copepods, for example, have no compound eyes, but a single tripartite nauplius eye. In most this is a simple structure with little in the way of optics, but in one or two groups it has developed into separate eyes equipped with complex optical systems. In the pontellids the eyes are sexually dimorphic, the males having the more complex optics, and even though the eyes contain only 6 receptors it is clear that they are used to identify the females. Nauplius eyes are present in other groups too, but the only other occasion that they attain any size is in the deep-sea ostracod, Gigantocypris, where the eye's optical system is based on two huge parabolic mirrors. The advantage here seems to be that the light gathering power is very great, in an almost lightless environment.

Apart from the copepods most groups have compound eyes of one sort or another. In the malacostraca: the isopods, amphipods, stomatopods and brachyuran crabs all have apposition eyes that are basically similar in design to those of diurnal insects, in which each lens forms a small inverted image immediately behind it. Since these are mainly aquatic animals, the corneal surface contributes little or nothing to the refracting power, and the lenses tend to be variants of the Exner "lens cylinder" in which the rays are focussed by inhomogeneities within the lens itself. Of the apposition eyes the most bizarre are found in deep-sea amphipods, many of which have double eyes and some, like Phronima have wide (10 µm) light guides up to 5 mm long joining the lens to the rhabdom.

Superposition eyes, in which the optical array forms a single erect image as in firefly and moth eyes, are found in 3 malacostracan groups:

the mysids, euphausiids (krill) and the macrurous decapods (shrimps, crayfish and lobsters). The former two groups, however, use refracting optics to produce the image while the decapods use an array of radially arranged mirrors instead. It is characteristic of the latter type of eye that the corneal facets are square, because the optical mechanism requires the mirrors to act in pairs at 90° (corner reflectors) and this makes it easy to tell the two types of superposition eye apart. It looks as though the fundamental difference between these two types of eye may shed some light on the origins of the different malacostracan groups. (My view is that it was right before 1880 but has got worse since).

Physiologically and morphologically crustacean compound eyes are not very dissimilar from those of insects, and this holds for the visual nervous system too, which has the same layout. Does this mean that there is only one sensible way of designing compound eyes and their neural backup, or should we be looking for common ancestors?

INTRODUCTION

In the class Crustacea there exist more different kinds of eye, distinct in the type of optical mechanism they use, than in any other phylum or subphylum of the animal kingdom. All the types of "simple" or single chambered eye are present, as are all the types of compound eye. There is to my knowledge only one exception, and that is the kind of eye that terrestrial vertebrates share with spiders, in which the main optical element is a curved cornea separating air from the higher refractive index substance of the eye itself. In the predominantly aquatic crustaceans this, it seems, was never a possible option. In this chapter I will concentrate more on optics than on cell morphology or retinal physiology. This is partly because there is simply a lot to be said, and partly because at the receptor level crustacean eyes are not radically different from those of insects, which are well discussed elsewhere in this book. The spectacular attribute of the crustaceans seems to be their optical inventiveness.

Crustacean eyes are of two embryologically and anatomically distinct types, and both may exist in the same animal. There are on the one hand the nauplius eyes - so called because they are best seen in a particular larval stage, the "nauplius", common to many crustaceans. These usually consist of a single centrally placed cluster of 3 small eye cups each containing a handful of receptors. At its most basic a nauplius eye is equipped with no optical arrangements other than shielding pigment, and its behavioural role is not likely to be more complicated than the mediating of simple taxes towards or away from light. In some groups, however, and especially in the copepods, these eyes have become equipped with optical mechanisms of various kinds - lenses, mirrors, or combinations of the two - and have taken on behavioural tasks, such as mate finding, of a quite demanding kind. On the other hand there are the paired lateral compound eyes, and these are found in all the major classes except the Copepoda. They may be stalked and movable, as in the mysids, euphausiids and decapods, or sessile as in the isopods and amphipods. The commonest, simplest and probably the ancestral type of compound eye is the apposition type, in which each lens throws an image onto the tip of a single rhabdom, which is usually a rodlike structure consisting of the fused contributions (rhabdomeres; microvillous structures containing photopigment) of 5 to 8 receptors. As in the eyes of most diurnal insects these eyes have an "ommatidial" structure, with each ommatidium comprising the receptor cell cluster with its own private lens. In the higher malacostraca, specifically the mysids, euphausiids and long-bodied decapods, there are two other types of compound eye which despite their apparent similarity to the apposition type are really quite different in structure and optical function. These are superposition eyes in which many lenses (or in the decapods, mirrors) contribute to the formation of a single erect image in the eye, on a layer of contiguous deep-lying receptors. In a sense, then, these eyes have "public" optics, more analogous to the single lens of a vertebrate eye than the individual lens system of apposition eyes, although there the resemblance ends. It is interesting that the mirror version of the superposition eye, which was only discovered recently (Vogt, 1975) appears to be an invention unique to the Crustacea.

These, then, are the main themes: the adaptive radiation of simple nauplius and lateral compound eyes. I shall spend most of this chapter elaborating the variations. For those unfamiliar with the Crustacea and with the types of eye they possess Table 1 and Fig. 1 will, I hope, be helpful. Partial reviews of crustacean optics can also be found in Waterman (1961, 1981) and Land (1981a, b). Rather than go through the Crustacea group by group, it seems more profitable to discuss the simple eyes (A, C, D, G in Fig. 1) and the compound eyes (E, F, H) in separate sections, and rely on Table 1 to provide the taxonomic continuity.

SIMPLE EYES

The Basic Nauplius Eye

In the copepod Calanella mediterranea the nauplius eye is straightforward and unspecialised. As Grenacher's beautiful old illustrations show (Fig. 2) the eye has three cups, two dorsal each with 8 receptors and one ventral with 10. Each cup has a backing of pigment. There are no lenses. The positions of the rhabdomeres are not clear in Grenacher's figure, but from later studies, especially the E.M. investigation of Fahrenbach (1964) on Macrocyclops, it is evident that each receptor has a rhabdomere consisting of oriented microvilli on one of its faces. The only indication of a possible image-forming structure in these eyes also comes from Fahrenbach's study, where he found that the back of the eye-cup was lined with a tapetum consisting of a multilayer of crystals with spaces between them, each layer having an optical thickness of about $\frac{1}{4}$ of the wavelength of light. Such a structure must behave as a mirror (see Land, 1972a, 1981a) and indeed if one looks carefully into nauplius eyes of unfixed copepods it is often possible to see the metallic appearance of the back of the eye. Similar multilayer structures were also found by Vaissière (1961) in his extensive review of copepod eye structure.

One cannot immediately jump to the conclusion that nauplius eyes of this kind have concave mirror optics, as in the eyes of scallops described in another chapter. For mirror imaging to be effective the rhabdomeres

TABLE 1

Crustacea: Taxonomy of Eye Types

S. Cl. Branchiopoda (800)	N present. E in adults (st, Artemia; se, Triops; fu, Daphnia).
S. Cl. Ostracoda (2000)	N present. E in Macrocypridina; G in Gigantocypris and Notodromas.
S. Cl. Copepoda (7500)	N only. A or G in most; variations of C in Copilia and pontellids.
S. Cl. Branchiura (75)	N present. E in larvae and adults (se).
S. Cl. Cirripedia (900)	N in larvae and adults. E only in cypris and larval stage.
S. Cl. Malacostraca Section. Phyllocarida (25)	N not known. E (st, Nebalia)
Section. Eumalacostraca Superord. Syncarida (35)	N in some. Probably E (st, Anaspides). Bathynellacea are blind.
Superord. Hoplocarida (300)	N present. E (large st in adult stomatopods).
Superord. Eucarida Ord. Euphausiacea (90)	N present. E or F in furcilia larva, F in adults (st). Some double.
Ord. Decapoda (8500)	N present. E in all larvae. H in all adults except paguran and brachyuran crabs which retain E (st).
Superord. Peracarida Ord. Mysidacea (450)	N absent. F in adults (st). Some double.
Ord. Cumacea (770)	N absent. E (fu) in some.
Ord. Tanaidacea (350)	N absent. E (se) in some.
Ord. Isopoda (4000)	N absent. E (se).
Ord. Amphipoda (5500)	N absent. E (se), often double in pelagic hyperiids.

Compiled from Waterman (1961), Barnes (1980) and other sources.

Classes Cephalocarida (7), Mystacocarida (3) are omitted. Numbers in brackets are number of species (from Barnes, 1980).

Abbreviations: N: nauplius eye usually a triple cup of type A, G or a combination; st: stalked; se: sessile; fu: fused. Other letters A-H see Fig. 1.

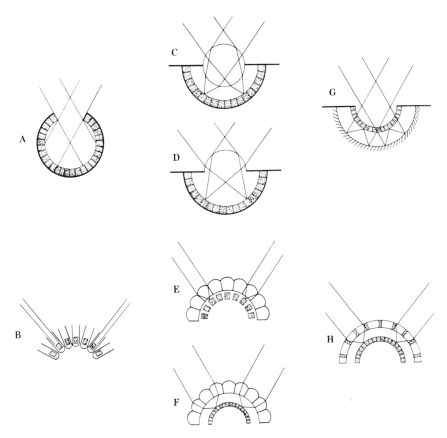

Fig. 1: Major types of optical system in animals. A, simple eye cup; B, multiple pigmented tubes; C, lens eye; D, corneal eye; E, apposition compound eye; F, refracting superposition compound eye; G, spherical mirror eye; H, reflecting superposition compound eye. A and B rely only on shadowing, C - F on refraction and G and H on reflection. All except D and perhaps B are found in the Crustacea. Based on Land (1981a).

should be arranged in a hemisphere with a radius half that of the mirror and concentric with it. If the rhabdomeres are in contact with the mirror, for example, no image will be formed on them although they will receive twice as much light as they would without the mirror. The rather ill-defined positioning of the rhabdomeres in the receptors of typical nauplius eyes does not at present let us say how much angular resolution the receptors derive from the presence of the mirror. I will return to the role of mirrors in copepod eyes when discussing lens-mirror combinations, but in the meanwhile there is one magnificent example of a nauplius eye with mirror optics, though this time in the Ostracoda.

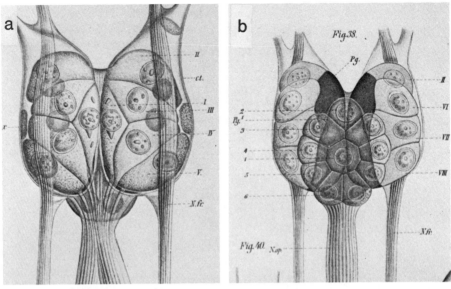
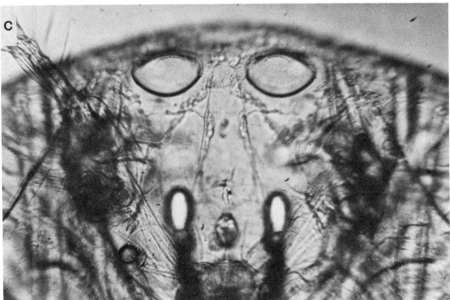

Fig. 2: Unspecialised and highly specialised nauplius eyes in copepods. **a** and **b**: 2 views of the nauplius eye of Calanella from above (**a**) and below (**b**). The whole eye is about 60 μm long (from Grenacher, 1879). **c**: photograph of the eyes of Sapphirina showing the double lens system of each lateral eye-cup. The medial eye is the object between the two proximal lenses. The separation of the distal lens centres is about 230 μm (after Land, 1981a).

Mirror Optics in Gigantocypris

The ostracods are a venerable group with fossil species going back to the Cambrian. Most living species are minute, a mm or so in body length, and whereas a few like the oceanic "chocolate drop", Macrocypridina castanea, have compound eyes, the majority have small, undistinguished nauplius eyes. The exception is Gigantocypris, which is huge (1 cm diameter) and has a pair of eyes, derived from the nauplius eye, each about 3 mm across. They are well described by Lüders (1909). Each eye consists of a large mirror, which in life reflects blue-green light, at the centre of which lies the retina consisting not of a few, but of nearly 1 000 receptors. These are very large cells, about 25 µm in diameter and up to 700 µm long, with their long axes arranged antero-posteriorly, at right angles to the mirror surface. The retina is not coextensive with the mirror, but is a vertical sausage-shaped structure lying dorso-ventrally near the centre of each mirror. Anything less like a conventional eye is hard to imagine.

Each mirror has an odd profile (Fig. 3). Seen from above they are clearly parabolic in shape, with the axes of the parabolas making an angle with each other of about 32°. However, from the side the profile is spherical, or nearly so. Given the dimensions of the mirror (measured in life; they tend to collapse when fixed) this means that the focal lengths for a horizontally oriented slit of light encountering the parabolic profile,

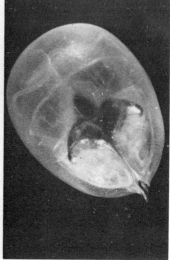

Fig. 3: Photographs of Gigantocypris mülleri by Dr. M.R. Longbottom (Institute of Oceanographic Sciences, U.K.). Note the parabolic profile of the eyes seen from above (right). The animal is about 1 cm in diameter.

and a vertically oriented one encountering the spherical profile are very different. The former focus will be about 370 µm in front of the reflector and the latter about 670 µm in front. If one imagines a slit of light rotating around the eye's axis, these observations mean that it will produce a focus that is not a point, but a line, with the focus moving in and out between these two extreme positions. What this built-in astigmatism seems to do is to produce a line focus, for a whole beam of parallel light, which lies along the length of the massive retinal receptors (Fig. 4).

This arrangement seems to ensure that all axial light passes through a single receptor, somewhere along its length. It will, however, have passed through many other receptors first so there is no chance that these eyes will have good resolution. What they do gain, however, is a massive light-gathering power, with an F-number of about 0,25! Since these animals live at depths of around 1 000 m in the ocean where almost no daylight penetrates, a trade-off in which sensitivity is gained by the sacrifice of resolution seems entirely sensible. No information is available about the uses of these eyes in the animal's life.

Lens Optics in Copepods

In a number of copepod genera the 3 cups of the nauplius eye have separated from each other to form distinct eyes, and frequently these eyes are provided with a lens apparatus. The most famous instance of this, which attracted the attention of Grenacher (1879), Exner (1891) and more

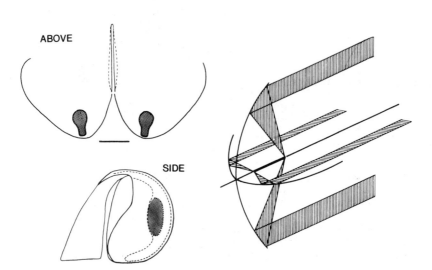

Fig. 4: Profiles of the Gigantocypris reflector and location of the retina (left). Scale: 1 mm. Diagram on the right shows how a mirror with a part spherical and part parabolic profile produces a line image.

recently Gregory et al. (1964) and Wolken and Florida (1969), are the eyes of Copilia. It is not difficult to see why. In Copilia and related genera (Sapphirina, Corycaeus) the two dorso-lateral cups are displaced laterally from the fairly normal medial cup, and are hugely elongated into what appear to be a pair of binoculars (Fig. 2c). Each lateral eye has a biconvex lens in the cuticle of the front of the animal, and then, at a considerable distance behind it a second, smaller pear-shaped lens which directs light into the elongated receptors which lie immediately behind it. There are very few receptors: 5 according to Wolken and Florida (1969), 7 according to Vaissière (1961) of which 4 are very short, and only the remaining 3 make up the long tubular structure that leads out from the rear of the eye. There is some debate as to whether the focus of the first lens lies on the second lens (Downing, 1972) or behind it (Wolken & Florida, 1969), but in either case the field of view of the receptor ensemble will be small - a few degrees at the most. Thus the animal has two extreme telephoto eyes, looking at perhaps 3° of space each. The problem, clearly, is how the animal manages to find anything worth looking at, and part of the answer may lie in the observation that the eyes make scanning movements. Gregory et al. (1964) found that the rear lenses and retinae move laterally, rapidly towards each other, and then more slowly apart at a rate varying between 0,5 and 10 Hz. Scanning can be elicited in active tethered animals with visual stimuli (Downing, 1972) so it presumably is concerned with vision and not indigestion! Nevertheless, the scanning movements are 1-dimensional and do not greatly extend the field of view; during a scan the eyes sweep across no more than about 1/1000 th of the total field around the animal. In Copilia the specialised eyes are found only in the small female and not in the larger male, which leads to the speculation that they may be used by the female for finding males. There is however, no such sex difference in Sapphirina or Corycaeus, according to Vaissière (1961), and Gophen and Harris (1981) point out that Corycaeus feeds much faster in light than dark, so the eyes may be involved in prey-capture. The observations so far made on these extraordinary animals simply do not add up, and they certainly invite further study.

Another copepod group, the Pontellidae, have eyes just as remarkable as those of Copilia, but because they mostly live in surface waters well offshore they have attracted less attention. The observations reported here are based on material collected during recent cruises of the R.R.S. Discovery, and supplement the anatomical accounts of some of the genera given by Vaissière (1961). I will concentrate on 2 genera, Labidocera and Pontella, because they seem to exemplify the way that evolution can go about the same task in quite different ways. In both genera there is a separation of the nauplius eye into three separate eyes, and in both genera there is clear sexual dimorphism (Fig. 5) with the male having the larger eyes. However, whereas in Labidocera it is the dorsal pair of eyes that is specialised, in Pontella it is the unpaired ventral eye. The pontellids are large and colourful copepods, often adorned with spots and blotches which contrast with the typically blue background colour. It thus seems worth entertaining the idea that the eyes I shall describe are concerned with detecting and recognising these patterns.

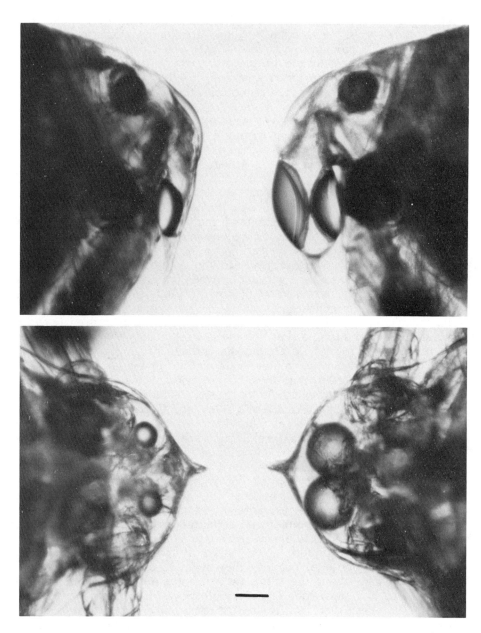

Fig. 5: Visual sexual dimorphism in two pontellid species; Pontella spinipes (top) and Labidocera acutifrons (below). The males are on the right. Scale bar: 100 μm.

In _Labidocera_ _acutifrons_ the dorsal eyes have spherical lenses which are more than twice as large in the male as in the female. The ventral eye is similar in both sexes, with no lens and 6 receptors arranged in front of a reflector. It is thus the dorsal eyes of the male that are more likely to be of special interest. They have a pair of deep (200 μm) cups beneath them, containing the receptor cells. The cups of the two eyes are joined together (Fig. 6) and, judging from the positions of the eye-cups in preserved specimens, they are capable of moving, together, through about 40° in the antero-posterior plane. This speculation is confirmed by the presence of a pair of muscles which join the tops of the eye-cups of the dorsal cuticle. It seems, therefore, that the eye-cups "scan" the image produced by the lenses, in a manner reminiscent of the scanning eye of the mollusc _Oxygyrus_ (see Land, 1983). Even more surprising was the finding that the rhabdoms in the depth of the eye-cup are arranged in a very distinct linear fashion (Fig. 7) with two long, then one short, then two long rhabdoms in the focal plane of each lens, and arranged at right angles both to the body axis, and to the presumed direction of scanning (Fig. 8). It seems then that this eye uses a nearly one-dimensional line of receptors to survey the water above the animal, by moving this line at right angles to itself. The angular subtense of the retinal line and the scan path are about the same in angular extent, so that the field of view of the eyes, as extended by the scanning movements, is about 40° square. One may speculate furiously as to what such a system might be capable of extracting from the

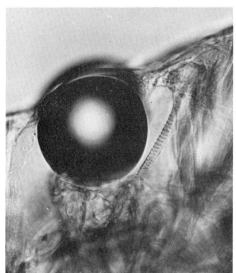 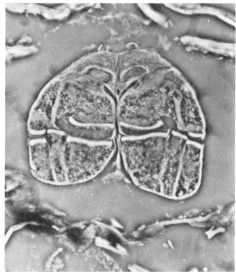

Fig. 6: _Labidocera_. **Left:** lens and top of eye-cup showing the muscle fibre that moves the eye-cup. (Lens diameter 150 μm). **Right:** section through base of eye-cup showing the 5 straight rhabdoms in each half of the cup. Anterior is upwards, and the width of the whole cup is 170 μm.

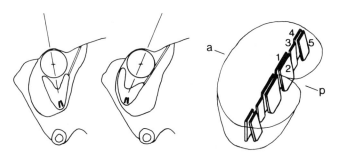

Fig. 7: **Left:** extreme positions of the eye-cup of male Labidocera, seen from the side, showing the probable extent of the scanning movements. **Right:** reconstruction of the 10 rhabdoms in the base of the eye-cup.
a, p: anterior and posterior.

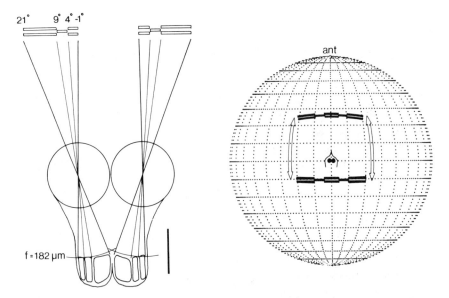

Fig. 8: **Left:** projection of the angular subtense of the rhabdoms in the eye-cup of male Labidocera onto the surroundings. Lens focal lengths were measured from image magnifications. Scale bar: 100 μm. **Right:** projection of the visual fields of the eye-cup onto a sphere, seen from above. The arrows indicate the extent of scanning movements in the antero-posterior direction. Dotted lines at 10° intervals.

surroundings, given that it only has 10 receptors to make its analysis, but an answer must surely be provided when we know more about the animal's behaviour, and in particular what kind of pattern the females present to the males during the movements that precede mating.

An interesting feature of the Labidocera lenses is that the ratio of their measured focal length to their radius is 2,5. This ratio is exactly the same as in fish lenses (Matthiessen's ratio; see Pumphery, 1961), and it has important implications for the way the lens itself is designed. The short focal length either means that the lens has an implausibly high refractive index (1,66) or that it has an inhomogeneous construction, with a high refractive index of about 1,52 falling to near that of sea-water at the periphery (Fletcher et al., 1954; see Land, 1981a). Such lenses have short focal lengths because light is bent continuously within them, and they are found in both fish and cephalopod molluscs. It seems certain that Labidocera must have this design too, and so far as I know it is the only example of such a lens in the Crustacea, although there are parallels in certain spiders (Blest & Land, 1977).

A glance at Fig. 5 shows that Pontella spinipes has adopted a quite different approach to lens construction. The lens in the ventral eye of the male is in fact a triplet (Fig. 9) and looks for all the world like a design for a high power microscope objective. Here, it seems, the animal has chosen a design based not on inhomogeneous optics, but on multiple surfaces, as in conventional lens construction in human technology. If we assume that the lenses have a refractive index of 1,53, which is typical for crustacean exoskeletal material, then the lens system brings parallel light to a focus about 30 μm behind the final surface, and the triplet has an overall focal length (posterior nodal distance) of 240 μm. It is interesting that the first surface of the lens assembly is not spherical, but clearly parabolic, at least in the profile it presents from the ventral side (in the lateral view the curvatures are rather different (Fig. 5) although the total power of the combination is the same). What the parabolic profile appears to do is to eliminate the spherical aberration of the lens system as a whole (Fig. 9b) so that the on-axis image is near perfect; without this refinement the system would produce a blur-circle at least as large as the rhabdoms of the eye-cup.

The retina in the ventral eye of Pontella contains just 6 cells, as in other pontellids. One pair lies medially, and the other two pairs are mirror images of each other, and lie latero-ventrally. The axons leave the eye-cup ventrally. What is special about Pontella is the arrangement of the two dorso-medial cells, and in particular their rhabdoms. The rhabdom of one is ball-shaped, about 25 μm across, and it is enclosed by a doughnut-shaped rhabdom in the other cell (Fig. 9c). Both ball and doughnut lie in the focal plane of the lens system. It is almost impossible to avoid the supposition that these two cells constitute a centre-and-surround spot detector. They seem to be analogous to the centre-surround ganglion cells found in vertebrate retinae, except that the latter are 3rd order neurons, but here we are dealing with 1st order receptors. It is easy to imagine ways in which the two cells might be "wired up" to produce a device for detecting a light spot on a dark background (or vice versa) or, with different photopigments in each cell, a spot of one colour against a

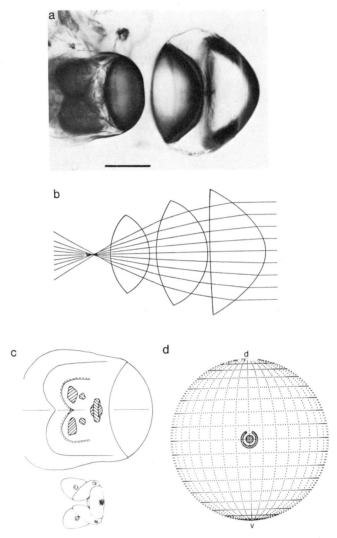

Fig. 9: <u>Pontella</u> male eye. a) view of eye from below, showing triplet lens, eye-cup with "ball" rhabdom immediately behind the 3rd lens, and the two deep eye-lobes. Scale bar 100 μm. b) position and quality of the axial image, assuming a uniform refractive index of 1,53 for the lenses. c) reconstruction of the rhabdoms (above) and cells (below) of the eye-cup. The focussed image is situated on the "ball and doughnut" rhabdom behind the lens. d) projection of the ball and doughnut rhabdom onto a sphere around the animal, seen from in front. Dotted lines at 10° intervals.

background of another. Although P. spinipes is not well described, P. securifer females have 3 lemon yellow spots on their abdomens, against a blue background (H. Roe, pers. com.), which makes sense. Behavioural studies would certainly be rewarding. Pontella has much the same problem as Copilia, in that the detector structure only has a small field of view, about 12° across in object space (Fig. 9d). However, there is a strong possibility that the dorsal eyes (Fig. 5) might act as "finders" for the ventral eye, perhaps in a similar way to the manner in which the lateral eyes of jumping spiders detect objects that are subsequently scrutinised by the narrow-field principal eyes (Land, 1972b; Forster, 1982).

The rhabdoms of the two pairs of lateral receptors in the Pontella ventral eye are very definitely not in the focal plane of the lens system (Fig. 9c), and no reasonable assumption about the refractive index of the lenses can bring them into focus. However, each cell pair is enclosed in a nearly hemispherical cup of reflecting pigment, and it is at least possible that they receive a refocussed image, approximately in the way indicated in Fig. 10:6. If this is true, then each pair of cells will have a roughly cross-shaped field of view, with the anterior cell contributing the vertical element, and the posterior one the horizontal. Their exact fields of view remain to be determined.

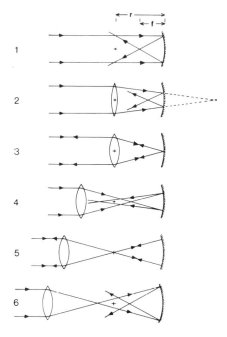

Fig. 10: Some possible combinations of lens and mirror optics in animal eyes. Explanation in the text.

Lens-Mirror Combinations

It is clear from the studies of Vaissière (1961) and Fahrenbach (1964) that many, perhaps most, copepod nauplius eyes are equipped with reflecting tapeta behind the receptors, and it is of at least passing interest to see how different combinations of lenses and mirrors might be used in the formation of images. It was suggested in the last paragraph that it might be possible for light imaged once by a lens to be re-imaged by a mirror, and since the copepods - perhaps uniquely - have exploited combined lens and mirror optics, this seems a suitable place to explore some of the possibilities (Fig. 10). 1 shows a simple spherical mirror, with a focus at a distance half the radius of curvature in front of the reflecting surface. This may turn out to be the typical optical system of lensless copepod eyes, and the example of the ostracod Gigantocypris, albeit somewhat complicated, indicates that concave mirrors are indeed used in nature. 2 is a combination of a spherical mirror and a weak lens. This is probably the kind of optical system present in the dorsal eyes of female Labidocera and the dorsal eyes of both sexes of Pontella. An excellent example of such an eye has recently been demonstrated in an ostracod Notodromas monarchus, again in the nauplius eye, by Andersson and Nilsson (1981), and in other phyla the eye of the scallop Pecten is perhaps the archetype (Land, 1983). The effect of the lens in such a system is to draw the focus of the eye closer to its rear surface, and practically this means that the rhabdoms can be set quite close to the mirror and still receive a tolerably well resolved image. In 3 the focal length of the lens is equal to the radius of curvature of the mirror, and this special case is very familiar in the tapeta of many vertebrate eyes (Walls, 1942) and also some spider eyes. The image formed by the lens lies on or near the mirror, whose function is not to refocus the light but to redirect it back through the receptors lining the mirror, thereby improving their potential photon catch. If the lens is moved back from the centre of curvature of the mirror (4), double images occur, but I know of no case of this design in nature. Its disadvantage is that the light focussed by the lens and by the mirror interfere too much, with an out-of-focus image in one place overlying an in-focus image in the other. 5 is a fine design for a projector lamp housing, with the lamp placed at the joint focus of lens and mirror, but again I know of no natural examples. 6 is perhaps the most interesting. Moving the focus of the lens yet further from the mirror gives an eye with two quite distinct foci, and unlike 4 rays passing through the first (lens) focus do not, in general, return through it after refocussing by the mirror. Two distinct images are formed at different levels, and this, I believe is the explanation for the consistent double layering of receptors in the ventral eyes of pontellids, stressed by Vaissière (1961). The anterior cells receive the direct image from the lens, and the posterior cells the image from the reflector. At least in the case of Pontella, and its relative Anomalocera the dimensions of the lens and mirror, and the positions of the various receptors, make this speculation almost a certainty. The other interesting feature of an eye of this kind in that the removal of the front lens, or its replacement with a weaker version (c f. Fig. 5), doesn't destroy the ability of the eye to focus, but converts it back into a type 2 eye. The fields of view of the receptors will be larger, but resolution will not be altogether lost. I would like to think that this is the optical basis of the difference between male and female eyes in Pontella and its relatives.

Simple Eyes: A Resume

In this brief account I have deliberately concentrated on some of the more exotic examples of nauplius eyes, to show the range of adaptation that is possible from a basically simple ground plan (Fig. 2). Whilst generally retaining an extraordinary parsimony in terms of the numbers of receptors they contain, some nauplius eyes have become specialised as target detectors by adding a range of optical devices to their eye-cups, including mirrors, lenses of two kinds (inhomogeneous fish-type and multi-surface) and combinations of lenses and mirrors. Nowhere else in the animal kingdom does there appear such an extraordinary range of optical invention.

Nauplius eyes are found in almost all crustacean groups, although it is not certain that they are strictly homologous with each other (Elofsson, 1965, 1966). Compound eyes, however, are also present in the majority of crustaceans - the great exception being the copepods - and they are usually much more important organs, giving a reasonably fine-grain image together with a wide field of view. I will now turn to their equally remarkable diversity.

COMPOUND EYES

Apposition Eyes

This kind of eye, in which each rhabdom sees a small solid angle of space, imaged by its own private lens system (Fig. 1E), is by far the most common type of compound eye. It evolved early, certainly in the trilobites in the Cambrian 500 - 600 million years ago, giving it a pedigree probably longer than any of the single chambered camera eyes, including our own. Although crustaceans and insects are the groups that everyone knows have compound eyes, they are not the only ones. The xiphosuran, Limulus, probably related to the early arachnids, has famous compound eyes, and they also crop up in a few filter-feeding annelids (Branchiomma) (see Verger, 1983), a few genera of bivalve molluscs (e.g. Arca), some specialised centipedes (Scutigera), and as compound eyespots at the tips of the tentacles of some starfish (Asterias) (see Yoshida et al., 1983). Including the insects and the Crustacea, compound eyes, always initially of the apposition type must have evolved independently at least 7 times. Even with the Crustacea it is not possible to be certain that the compound eye are all homologous; their common ancestors are buried in the radiations of the Cambrian period, and we are never likely to be certain whether the eye of (say) the water-flea, Daphnia, and a crab like Carcinus are traceable to a common origin.

In this brief essay all I will try to do is to indicate firstly the way apposition eyes are constructed optically, and then explore some of the variations of the basic design that give some eyes the sensitivity for them to be of use in the ocean depths, and others the acuity to recognise fine detail where light conditions are adequate. To explore, in other words, the limits to which the basic apposition eye design can be pushed. I will begin by discussing the eyes of Limulus, not a crustacean, but a distant relative whose eyes are similar to, and perhaps better studied than, those of any one crustacean. We can take it as a prototype, just as Exner (1891) did when he first described its mode of operation.

Limulus Eyes and Lens-Cylinders

The lateral eyes of Limulus each contain about 850 ommatidia, each with a wide corneal lens (200-300 μm across compared with 20-50 μm for brachyuran crabs). Behind the lens are 8-12 photoreceptors clustered around an eccentric cell, the whole cluster contributing to a complex rhabdom arranged like the spokes of a wheel. Exner's main concern was the lens because, in common with most crustacean eyes, it has no useful refracting outer surface in water, and must therefore have an unconventional mechanism for bending light if it is to form an image near its rear surface - as indeed it does (Fig. 11c). Exner's proposal was that it acted as a lens-cylinder, a structure which has a high refractive index along its axis, falling approximately parabolically to a lower value at the cylinder's lateral surface. If one imagines a ray of light striking the distal surface of such a cylinder, parallel to the axis but some distance from it, that ray will encounter a higher refractive index towards the axis, and a lower index away from it, and will be drawn towards the axis. This bending will continue until the ray reaches the axis, and there, provided the gradient in the structure is the correct one, it will meet all initially paraxial rays, and form a focus (Fig. 12). Although there have been recent suggestions that reflexions from the walls of the cylinders may play a part in focussing in Limulus (Levi-Setti et al., 1975), there is now no doubt that the principal focussing mechanism is as Exner described. An inverted image is formed at the proximal tip of each corneal "crystalline cone" and the refractive index gradient across the cone, measured by interference microscopy (Fig. 11b) is almost exactly what one would predict from the structure's focal length (Land, 1979). It seems certain that lens-cylinder optics will turn out to be the most common means of producing images in crustacean apposition eyes, and with the important modification that the cylinders are twice as long, this system occurs in the superposition eyes of mysids and euphausiids as well (Fig. 16).

Sensitivity and Resolution

Before going on to discuss some of the variations of the apposition design, it is useful to have some yardsticks against which to judge their performance. At low light levels the ability of an eye to see anything at all is dependent on the capacity of its receptors to collect enough photons from the surroundings. In dim light the visual world is extremely "noisy" for the simple reason that small photon numbers show large random fluctuations. Thus if a cell receives an average of 10 photons per second, the standard deviation is $\sqrt{10}$ (i.e. 32%); if on the other hand it receives 1 000 the standard deviation is $\sqrt{1\,000}$ (or only 3,2% of the mean). Clearly whatever decision an eye has to make about the presence or absence of objects, their relative brightness, size, shape or speed, these judgements

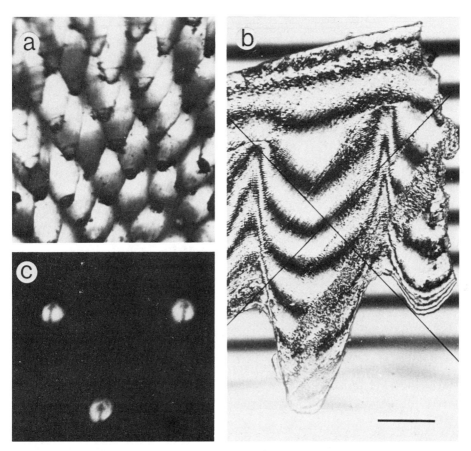

Fig. 11: Compound eye of Limulus. a) View of the crystalline cones from the rear. The spacing is about 250 µm. b) Fringe-field interference micrograph of a 130 µm section of a single Limulus crystalline cone. The distortion of the interference fringes parallels the parabolic refractive index gradient in the proximal part of the cone. c) Inverted images of an arrow (upside down in object space) photographed at the tips of three crystalline cones. The scale on (b) is 100 µm. From Land (1979).

are all going to be improved by the receptors having a high photon capture rate. This in turn depends on a number of structural features of the eye, all of which can be determined by anatomical and optical studies. Thus the capture rate is increased if the ratio of the lens diameter (**A**) to focal length (**f**) is high (i.e. if the F-number (**f/A**) is low, as in photography), if the receptors have a large diameter (**d**) and if they have sufficient length (**x**) to absorb a high proportion of photons entering them. These various considerations are summarised in a single equation which yields the sensitivity (**S**). This is the number of photons absorbed by a receptor when

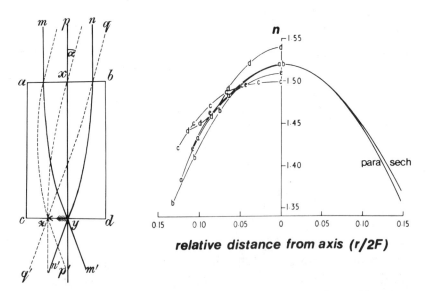

Fig. 12: **Left:** Exner's (1891) diagram of image formation in a lens-cylinder. **Right:** Measured refractive index gradients in 5 lens-cylinder eyes (a, euphausiid; b, firefly; c, moth; d, skipper butterfly; e, Limulus) compared with the theoretical gradients proposed by Exner (parabolic) and the hyperbolic secant function of Fletcher et al. (1954). Based on Land (1980).

the eye is viewing an extended source emitting 1 photon per μm^2 per steradian, or 10^{12} per m^2 per sr, assuming the dimensions on **A, f, d** and **x** are in µm. A derivation is given in Land (1981a).

$$S = (\frac{\pi}{2})^2 \cdot (\frac{A}{f})^2 \cdot d^2 \cdot (1 - e^{-kx})$$

In this equation **k** is the absorption coefficient of photopigment in the rhabdoms, and I have taken the figure derived from microspectrophotometry of lobster rhabdoms, 0,69% per µm (Bruno et al., 1977) as applying to all Crustacea. There is a useful alternative form of the equation in which (**d/f**) is replaced by $\Delta\rho$, the acceptance angle of a photoreceptor, which can sometimes be determined electrophysiologically or by ray tracing when a straightforward measure of focal length (**f**) is not available. This is legitimate because the angle that a receptor (or in general a rhabdom) tip subtends at the nodal point of the optical system is equal to $\Delta\rho$, where this is measured in radians. In this form **S** is given by:

$$S = (\frac{\pi}{2})^2 \cdot A^2 \cdot (\Delta\rho)^2 \cdot (1 - e^{-kx})$$

Applying these formulae to published data one obtains values for **S** ranging from 0,69 for the crab Leptograpsus (data from Stowe, 1980) to

6 400 for the deep-sea isopod Cirolana (data from Nilsson & Nilsson, 1981) (Fig. 14). This represents a range of photon catching power of nearly 10^4, and it means that receptors in the eyes of these two species would receive roughly equal numbers of photons when the former was viewing a scene in overcast sunlight, and the latter by streetlight at night. Alternatively, a factor of 10^4 represents a descent from the surface to a depth of about 280 m in clear ocean water. Limulus itself has a variable iris which can change the effective diameter of the rhabdom from 17 to 60 μm (Barlow et al., 1980) giving values of S between about 130 and 1 200, figures which definitely indicate a nocturnal lifestyle when compared with, say, Leptograpsus (see Table 2).

The ability of the eye to resolve detail depends on the angular separation of its receptors, and in a spherically symmetrical apposition eye this separation is the inter-ommatidal angle $\Delta\phi$. There are many ways of obtaining a measure for $\Delta\phi$, but perhaps the most accurate is to use the phenomenon of the pseudopupil, when this is visible. The pseudopupil is the small black spot, easily seen in eyes without heavy pigmentation (Fig. 13), which appears to move around the eye with the observer. It

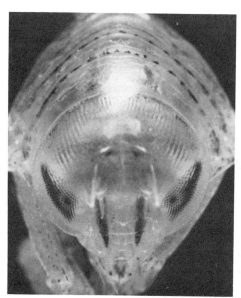 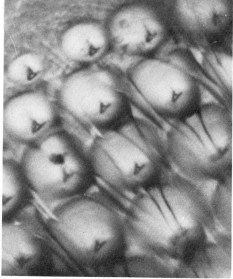

Fig. 13: Amphipod eyes: Left: Phrosina semilunata showing the 2 pseudopupils in the lower part of the eye As in Phronima the upper part has larger facets and is directed towards the sea surface. The animal is 7 mm wide. **Right**: inverted images behind the crystalline cones of Phronima, showing the apposition nature of the eye. Light-guides, originating from the cones and visible in the photograph convey the central 3,5° of each image to the retina, situated 5 mm below. Each facet is about 130 μm across.

TABLE 2

Crustacea: Resolution and Sensitivity of Compound Eyes

Species	Type of Eye	$\Delta\phi$	S
Daphnia (Branchiopoda) (Shallow-living)	Apposition (E)	38°	13,7
Artemia (Branchiopoda) (Shallow-living)	Apposition (E)	9°	0,5 (L) 2,3 (D)
Limulus (Xiphosura) (Active at night)	Apposition (E)	8°	127 (L) 1190 (D)
Cirolana (Isopoda) (Deep-sea)	Apposition (E)	15°	6410
Phronima (Amphipoda) (Mid-water)	Apposition (E)	0,4°(*)	70
Leptograpsus (Decapoda) (Shore crab)	Apposition (E)	1,5°(min)	0,7 (L) 3,7 (D)
Palaemonetes (Decapoda) (Shallow-water shrimp)	Reflecting superposition (H)	2,5°	0,2 (L) 40,2 (D)
Oplophorus (Decapoda) (Deep-sea shrimp)	Reflecting superposition (H)	8,1°	3330
Meganyctiphanes (Euphausiacea) (Mid-water krill)	Refracting superposition (F)	2,9°	266
Thysanopoda furcilia (Euphausiacea) (Surface-living larva)	Refracting superposition (F)	4°	11,5

Notes

$\Delta\phi$ and S are defined in the text.

L and D are light- and dark-adapted.

The low value of $\Delta\phi$ in Phronima dorsal eye (*) only applies to the resolution for single objects, not stripe patterns.

represents the location of those ommatidia which are looking at (and hence absorbing light from) the observer (see Stavenga, 1979; Horridge, 1978). Thus if the animal is rotated through 15°, and the pseudopupil moves across 5 facets, then the inter-ommatidial angle is 3°. Clearly, the more facets in the eye, and so, usually, the larger the eye is, the smaller $\Delta\phi$ can be. In Daphnia, with only 22 ommatidia $\Delta\phi$ is 38° (Young & Downing, 1976) whereas in the crab Leptograpsus with several thousand ommatidia the value of $\Delta\phi$ in the centre of the eye is about 1,5° (Sandeman, 1978).

The problem with using $\Delta\phi$ as a measure of resolution is that $\Delta\phi$ gets smaller as the quality of the image improves. Acuity ($1/\Delta\phi$) is better, or perhaps better still a measure of the ability of the eye to resolve the lines in a grating. Since it requires 2 receptors (one for the light and one for the dark stripe) the spatial period of a grating that an eye can just resolve will be $1/(2\Delta\phi)$. In Table 2, for simplicity, I have left the inter-ommatidial angle in its raw form ($\Delta\phi$).

In general, the rhabdom acceptance angle ($\Delta\rho$) and inter-ommatidial angle ($\Delta\phi$) are similar to each other in apposition eyes in both crustaceans and insects. If $\Delta\rho$ were substantially greater than $\Delta\phi$ there would be overlap between receptive fields, and if the converse were the case the field of view of the eye would be undersampled and have "holes" in it. Given, then, that $\Delta\phi$ and $\Delta\rho$ are similar, we find that sensitivity, which is proportional to ($\Delta\rho$)^2, and resolution, which is inversely proportional to $\Delta\phi$, are at odds. If resolution is good, sensitivity must be low and vice versa. The only cure, if both resolution and sensitivity are required is to keep $\Delta\rho$ small in the sensitivity equation but to increase the aperture (A) of each facet, and that in turn means making the eye as a whole much bigger. One gets what one pays for: high resolution means a large number of facets, and high sensitivity means that each must be large. Either way a large eye is required.

Variations on the Apposition Theme

Daphnia has a single compound eye formed by the fusion of the two lateral eyes. Although the number of ommatidia is very small (22) there appear to be no gaps in the visual field (Young & Downing, 1976). The wide acceptance angles of the rectangular rhabdoms are achieved by having under-focussed optics, that is, the focal point of the lenses is some distance behind the rhabdom, and the acceptance function is not the usual gaussian shape, but rather broad and flat-topped, with a width at half maximum of about 27°, compared with an inter-ommatidial angle of about 38°. Despite their small size, low resolution and sensitivity, these eyes are certainly important in the animal's behaviour. Von Frisch and Kupelweiser as long ago as 1913 found evidence of colour vision, in that the animals are attracted by long and repelled by short wavelengths, implying at least two visual pigments either in different parts of the eye, or in different receptors in each ommatidium. There is evidence too of polarisation sensitivity, as one might expect for a rhabdom structure with the microvilli from different receptors oriented at right angles (Waterman, 1981). Most surprisingly, perhaps, the eyes are highly mobile, Frost (1975) found four classes of eye movement: a high speed tremor (3-4° at 16 Hz), rhythmic scanning through 5-6° at 4 Hz, large fast "saccadic" eye movements, and optokinetic nystagmus in response to moving stripe patterns. The large movements can be as great as 150° in the saggital plane, and 60° in the horizontal. It is impressive to find so large a repertoire of eye movements in an animal with such an apparently "low" taxonomic position in the Crustacea. Some of the eye movements are concerned in keeping the eye in a constant orientation as the animal moves and thus providing a frame of reference for orientation of the body itself. A recurring suggestion has been that the eyes detect and "track" the outline of Snell's window, the 97° angle at the eye within which vision into the air is possible (Young, 1981).

Artemia, the brine shrimp, has separate eyes with about 300 ommatidia each, and a typical apposition structure. Resolution is better than in Daphnia ($\Delta\phi$ about 9°, although it varies somewhat across the eye). The most unusual feature of Artemia eyes concerns the changes that accompany light- and dark-adaptation. In the dark the rhabdoms enlarge, from a diameter of 5,2 to 7,4 µm, but such changes are quite a common strategy for increasing sensitivity in arthropods (e.g. Nässel & Waterman, 1979). The unique feature in Artemia is that the lens of each ommatidium, which is made of glycogen, actually changes its shape and hence focal length between dark and light. In the dark the lenses elongate, making the proximal and distal ends more curved, and so decreasing the focal length. The crystalline cone, which spaces the lens from the rhabdom, shortens proportionately. For a typical lens the light-dark focal length change is from 72 µm down to 50 µm, and since the sensitivity equation has a $(1/f)^2$ term in it, this alone will produce a sensitivity increase of 2,08 times. Together with the change in rhabdom diameter, the total increase is by a factor of 4,2. The acceptance angle necessarily increases, from 4,8 to 9,9, inevitably decreasing resolution, but this sort of trade-off is not disadvantageous (Nilsson & Odselius, 1981)

Moving to the Malacostraca, one finds apposition eyes in 4 important groups: the isopods, amphipods, brachyuran decapods (true crabs) and the stomatopods. The isopods have modest apposition eyes usually containing tens to hundreds of ommatidia. The pill-bug (or woodlouse) Armadillidium, for example, has 20 to 25 per eye, giving an inter-ommatidial angle of around 30°. One of the most interesting isopods is a deep-sea species, Cirolana borealis, which is very obviously adapted to extremely dim light, and has been studied recently by Nilsson and Nilsson (1981). The adaptations that it shows are firstly very large facets (150 µm) and very large rhabdoms consisting of 7 rhabdomeres fused distally to a single rhabdom, but somewhat unfused proximally (rather as in flies) so that there may be some degree of resolution within the rhabdom. The overall rhabdom dimensions are 80-100 µm wide by 100 µm long (Fig. 14). As in Daphnia the ommatidia are under-focussed, broadening the acceptance angle to about 45°. The partially focussed light, having passed through the rhabdoms is reflected back by a diffusing tapetum, which effectively doubles the rhabdom length for photon capture. From this data one arrives at a sensitivity figure of 6 400, which puts Cirolana top of the list, not just for apposition eyes which are not usually very sensitive, but higher even than even than any of the superposition eyes whose "raison

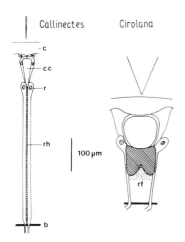

Fig. 14: Ommatida of the brachyuran crab Callinectes and a deep-sea isopod Cirolana to the same scale. Callinectes has a light-guide rhabdom, a small acceptance angle, and low sensitivity. Cirolana has a wide acceptance angle, extremely high sensitivity, and its rhabdom cannot function as a light-guide. Based on Waterman (1981) and Nilsson and Nilsson (1981).

d'être" is sensitivity (Table 2). This eye is certainly exceptional, but it does illustrate the extent to which the basic apposition plan can be bent, when a high photon catch is required.

Of all the apposition eyes in crustaceans, those of the deep-sea hyperiid amphipods are certainly the most peculiar (Fig. 13). They are large, often double with a distinct dorsal region of large facets and a ventral part with smaller ones. They have unconventional optics, usually with elongated crystalline cones containing two separated lens-cylinders and in one case (Phronima) an enormously elongated light guide (5 mm) joining the lens part of the crystalline cone to the retina (Ball, 1977; Land, 1981c). The eyes may be bizarre shapes too. Streetsia, for example, has a barrel-shaped eye occupying half the head. In Phronima, perhaps the best studied member of the group, the two upward-pointing medial eyes each contain about 400 ommatidia, but have a field of view, as measured from the pseudopupil, of only about 10°, with almost total binocular overlap between the eyes. In contrast, each ommatidium in the lateral eyes has an acceptance angle of about 10°, the same as a whole medial eye. Phronima seems to have equipped itself with a "telephoto" compound eye for looking upwards, into the reasonably bright down-welling light where it is probably looking for potential prey. Another oddity of the medial eyes of Phronima is the extent of overlap between the acceptance angles of adjacent ommatidia. $\Delta\rho$ is about 4°, but $\Delta\phi$ is about 0,5°. In most apposition eyes the $\Delta\rho : \Delta\phi$ ratio is close to one, and rarely exceeds 2, but here it is 8. I have argued elsewhere (Land, 1981c) that provided the task of the eye is to detect single small prey objects, and provided there is a neural arrangement for pooling the signals from all rhabdoms whose fields overlap - for which there is some anatomical evidence - then resolution need not be lost as a result of the high $\Delta\rho$ value. What is lost in the blurring effect of the wide acceptance angle is recovered in the improved signal-to-noise ratio that results from pooling. This conclusion should apply to other hyperiids; in Streetsia for example $\Delta\phi$ in the rostro-caudal direction is only about 0,2° (Meyer-Rochow, 1978) and from my own pseudo-pupil measurements $\Delta\rho$ is at least 4° and probably higher. It should be pointed out that a pooling strategy to recover lost resolution will only work for small objects on an uncluttered background. It would be of no value for a terresterial animal that needs to resolve background texture for optomotor orientation, a brachyuran crab for example. There is no substitute, under those circumstances, for high resolution (low $\Delta\rho$) ommatidial optics.

The true crabs have large stalked eyes with a relatively straight-forward apposition design. Each of the several thousand ommatidia consists of a cornea, which is usually flat, a distinctly conical crystalline cone which must function as a lens-cylinder since it is the only possible image-forming structure, and immediately behind this a narrow but very long rhabdom (5 by 400 µm in Callinectes; Eguchi & Waterman, 1966) (Fig. 14). The rhabdom must behave as a light guide, accepting a portion of the crystalline cone's image at its distal tip and transmitting this down its length. If this were not the case the acceptance angles would be very much wider than they actually are (e.g. $\Delta\rho$ has a minimum value of 1,6° in Leptograpsus, from intracellular recordings by Stowe (1980) which closely matches the minimum value of $\Delta\phi$ of 1,5° found by Sandeman (1978). The rhabdoms are "banded", as in most other Malacostraca, and are made up of alternating layers of orthogonally arranged microvilli, contributed by 7 retinula cells. In Callinectes there are as many as 450 1um thick bands (Eguchi & Waterman, 1966). The extent to which the orthogonal microvillous structure of the rhabdom may give rise to polarisation sensitivity is discussed by Waterman (1981). In addition to the 7 receptors of the main rhabdom, many, perhaps all, decapods have an 8th cell with a short rhabdom lying distal to, and therefore optically in front of, the other 7. In the crayfish Procambarus (a decapod, but with a superposition eye) it is now proven that the 8th cell is a violet receptor (λ_{max} 440 nm) whereas the other 7 are green receptors (530 nm) (Cummins & Goldsmith, 1981). There is strong evidence (Martin & Mote, 1982) that a similar violet/green dichromatic arrangement exists in the crabs Callinectes and Carcinus. There thus seems to be a photopigment basis for colour vision in decapods, as Wald (1968) had earlier inferred. There may, however, be other mechanisms involved. As Stowe (1980) points out there are as many as 5 types of movable screening pigment in the crab Leptograpsus, and these can in various ways passively modify the spectral sensitivities of the rhabdoms they ensheath. Such passive pigments are known elsewhere, as for example in the oil droplets of bird eyes. Colour vision in crabs is certain to exist, but may be complicated matter.

It is now clear that in addition to movements of shielding pigments, which can alter the amount of light entering and retained by the rhabdoms and so alter their absolute sensitivity, a major strategy of light/dark adaptation is the synthesis and breakdown of the structure of the rhabdom itself. This phenomenon had already been encountered in Limulus and certain spiders (Behrens & Krebs, 1976; Blest, 1978), and in 1979 Nässel and Waterman reported a massive turnover of photoreceptor membrane in the crab Grapsus. The microvilli increased in length by 154% and in number by 117% in the dark. The length increase obviously makes the rhabdom wider by a factor of about 2,5 and presumably the increased number changes the effective absorption of the rhabdom. These alterations to **d** and **x** in the sensitivity equation probably produce a total sensitivity change of about X10, although when coupled with screening pigment changes the overall figure will be larger. It cannot, however, come anywhere near the figure of 10^6 needed to make up the difference between sunlight and moonlight.

Not all crab eyes are spherically symmetrical, nor do they have equal resolution in all parts of the field of view. The ghost crab Ocypode has a nearly cylindrical eye, oriented vertically, and it is clear from the vertical elongation of the pseudopupil that the resolution is greater in the vertical than the horizontal direction (Horridge, 1978). From Horridge's eye maps one can see that there is an equatorial acute zone around the eye, about 20° high, where $\Delta\phi$ is about 0,5° along a vertical row of facets and 1° along a horizontal row. Away from this acute zone $\Delta\phi$ rapidly falls to 2,5° in both directions. Sandeman (1978) found less dramatic but basically similar variations in resolution in the eye of Leptograpsus. One can probably interpret the Ocypode result as an enhancement of vertical resolution around the horizon, which is where an active beach-living animal would have most to look at. Similarly oriented acute zones exist in many insects.

Finally, the most specialised and intriguing apposition eyes in the Crustacea occur in the stomatopods (mantis shrimps). These are very visual animals, armed with both clubs and daggers for catching prey, as well as a repertoire of social signals for dealing with conspecifics (Caldwell & Dingle, 1976). Although the stomatopods probably split off early in the radiation of the Malacostraca, the eyes are convergently similar to those of the crabs (Waterman, 1981; Schönenberger, 1977). The striking difference is the relative sizes of the eyes: a 150 mm Gonodactylus has 6 mm diameter eyes (Horridge, 1978) which are huge by any crustacean standard. Functionally, the strangest feature is the presence of 2 or 3 regions of the same eye that all look in the same direction. In Odontodactylus each eye is divided into two by a band of large facets, 6 across, which runs obliquely upwards and outwards (Fig. 15). When the eye faces the observer it presents 3 separate pseudopupils,

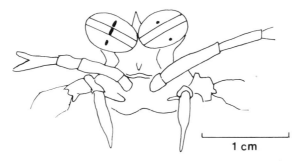

Fig. 15: The stomatopod Gonodactylus chiragra showing 3 pseudopupils in one eye and two in the other. From a photograph by Horridge (1978).

one in the central band and one on either side; and if both eyes face the observer than 6 pseudopupils are visible, which in turn means that 6 separate regions of the eye are viewing the same point in space. This sextuple vision is without parallel in the animal kingdom: binocular vision is adequate for most of us! In the central region of each eye the interommatidial angle is about 0,5°, as in Ocypode (Horridge, 1978). The eyes are very mobile and inquisitive in their activity; the nearest thing to a crustacean primate.

Refracting Superposition Eyes

The existence of compound eyes in which many facets contribute ray bundles to the formation of a single erect image was established by Exner (1891). Exner worked mainly with eyes of fireflies, but his conclusions certainly apply to the eyes of moths (Kunze, 1979) and some other beetles (Caveney & McIntyre, 1981) amongst the insects, and to the euphausiids (krill) and the mysids amongst the crustaceans (Land et al., 1979). Exner believed that the eyes of the macruran decapods (shrimps, crayfish, lobsters) also have eyes of this kind, but as we shall see he was partly mistaken in this.

Refracting superposition eyes have a morphology which is characteristically different from apposition eyes. The crystalline cones tend to be long and bullet-shaped (see Fig. 17 and Fig. 18) and they do not abut directly onto the rhabdoms. Instead, there is a wide "clear zone" separating cones and rhabdoms, and it is across this zone that rays entering many facets are brought to a focus. The rhabdom layer typically lies about halfway between the centre of the eye and the cones, and it is here that the erect image lies. There has, over the past 2 decades, been considerable controversy over the homogeneity of the clear zone, and in particular whether or not it is crossed by light-guides joining the crystalline cones to the rhabdoms (effectively converting these eyes back to apposition eyes), and it may be that in the light adapted state where in some insect superposition eyes the clear zone fills with screening pigment that cuts off oblique rays, these structures do exist and channel some of the light (see Horridge, 1975; Kunze, 1979). However, in dark adapted eyes there is now no longer any doubt that Exner's original scheme for the way the image is formed is the correct one (Land, 1980).

As can be seen from Fig. 16 the function of the crystalline cones in a superposition eye is to redirect rays across their own axes, in such a way that they emerge from the proximal end of the cone making (in general) the same angle with the axis as the incident ray. That is, if the initial angle of the ray with the cone axis was α, it is rotated through 2α. A simple lens will not do this, because it contains a nodal point through which rays pass undeviated, and forms an image on the opposite side of the axis to the object; central rays are not rotated. Exner realised that the cones must behave like inverters: inverting telescopes with unity magnification. In conventional optics such devices require two lenses of equal power, with an internal focus halfway between them. Realising that the surface of the crystalline cones did not have enough power to constitute such a telescope, Exner proposed instead that the cones acted as lens cylinders, not unlike those of Limulus (Fig. 12) but twice as long

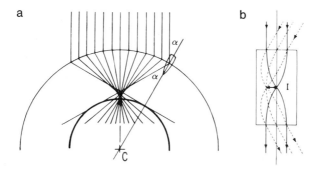

Fig. 16: Refracting superposition optics. a) Paths of rays contributing to the image on the receptor layer. Note that each crystalline cone bends light across its axis. C, centre of curvature of the eye. b) Ray paths within a double-length lens-cylinder. Both figures modified after Exner (1891).

(Fig. 16). Recalling that Limulus cones produce an inverted image at their proximal tip, two such cones placed end to end will first focus and then unfocus a light beam, and as in the inverting telescope redirect it across the axis as well. For this mechanism to work, the crystalline cone must have an appropriate refractive index gradient. This has been measured in euphausiid eyes by interference microscopy (Land & Burton, 1979), and it is indeed present.

The eye of a euphausiid (Meganyctiphanes) is shown in Fig. 17, together with the ray path entering and leaving a single cone, and an interference micrograph of a section through a cone which shows the contour lines of equal path difference that represent the refractive index gradient. The figure illustrates all three features of classical superposition eyes: the presence of a wide clear zone, the bending of rays across the cone axes, and the refractive index gradient in the lenses. That they operate as superposition eyes is no longer in dispute. The convergence between this design, and that of a moth eye is quite extraordinary, even down to the detailed shape of the crystalline cones (Land et al., 1979). Mysid eyes have the same layout as those of euphausiids, and although the refractive index gradient in the cones has yet to be measured it is hard to believe that they operate differently.

The reason that some animals employ a superposition mechanism is undoubtedly that it offers a much brighter image than that given by apposition optics. If each crystalline cone can accept light up to an angle of 30° from its axis, then the total diameter of the image-forming ray bundle that contributes to the image at any point will be equal to the radius of the eye, and as the focal length of a superposition eye is approximately half the radius, the F-number (f/A) of the eye will be 0,5. This, as photographers will recognise, is astounding. Typical F-numbers for the lenses of apposition eyes are around 2, indicating a factor of 16 in

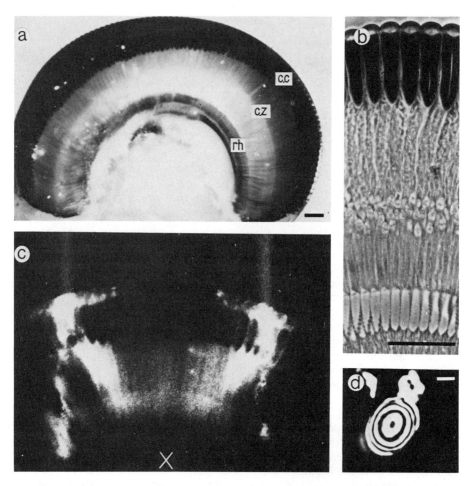

Fig. 17: Eye of the euphausiid <u>Meganyctiphanes norvegica</u>. a) Hemisection of an eye, showing crystalline cones (**c.c**), clear zone (**c.z**) and rhabdoms (**rh**). Scale 100 μm. b) Detail from a semi-thin section, showing receptor nuclei in the clear zone, and the banded rhabdoms. Scale 100 μm. c) Two beams of light directed at the surface of a hemisected eye are bent into the fixed clear-zone as indicated in Fig. 16a. They would meet at the **X** in the rhabdom layer. d) Interference micrograph of a section of a cone. The contour lines indicate contours of refractive index (see Fig. 12). Scale 100 μm. After Land (1981a).

in the light capturing power of the different types of eye. Coupled with wide receptors (17 μm in <u>Meganyctiphanes</u>) this produces a very high sensitivity figure (see Table 2). This fits in with the deep-sea life of the krill, many of which live at depths where the light levels are very much lower than at the surface (even the clearest oceanic water attenuates light by a factor of 10 for every 70 m).

Not all euphausiid eyes are spherically symmetrical (Fig. 18). As in the hyperiid amphipods with apposition eyes many species of both euphausiids and mysids have double eyes with an upward-pointing region of higher resolution. In Stylocheiron maximum the receptors are spaced 1,2° apart in the upper eye, but 2,6 apart in the lower. Presumably the upper eyes are exploiting the greater intensity of the down-welling light in order to boost their resolution. The shapes of double eyes are interesting, and the upper eyes in particular tend to depart substantially from concentric spherical arrangement of Meganyctiphanes. I have suggested elsewhere that these differences probably improve resolution by correcting the spherical aberration inherent in the symmetrical kind of superposition eye (Land et al., 1979). Not all euphausiid eyes exploit the full light-gathering potential of the superposition design, and eyes with reduced numbers of crystalline cones (and hence reduced effective pupil apertures) are quite common. Stylocheiron suhmii, for example, has only 7 cones in its upper eye, and in S. microphthalma there are only 4. Similarly, the furcilia larva of Thysanopoda tricuspidata has only 7 cones although about 90 rhabdoms

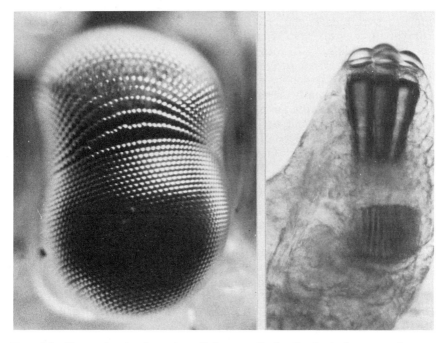

Fig. 18: Non-spherical euphausiid eyes. **Left:** Stylocheiron maximum. **Right:** Furcilia larva of Thysanopoda tricuspidata, showing cornea, 7 cones, clear zone and retina. Stylocheiron eye is 2,3 mm top to bottom; furcilia cones are 110 µm long.

(Fig. 18). Its adult, however, has a conventional spherical eye (Land, 1981a). Notice that the mismatch between crystalline cone and rhabdom numbers means that it is not really possible to speak of an "ommatidium" when discussing superposition eyes.

Reflecting Superposition Eyes

The long-bodied decapods have eyes which are similar in many ways to those of euphausiids. There is a superficial layer of optical elements, a wide clear-zone which is empty of pigment in the dark, and a deep-lying layer of rhabdoms composed, as in crabs, of contributions from 7 receptors plus a small distal 8th cell (see Waterman, 1981). The fact that they were importantly different arose out of the observation by Kuiper (1962) who found that the crystalline cones (or better, pyramids since they have a square cross-section) had a low and, worse, homogeneous refractive index. They thus could not possibly act as Exner-type lens cylinders, nor, it seemed, was superposition image formation possible. Kuiper proposed instead that these were really apposition eyes, with the "tails" of the crystalline pyramids acting as light guides. So disastrous did these observations seem for Exner's original ideas that by the end of the 1960's there was serious doubt as to whether he had been correct at all about both lens-cylinders and superposition optics (see Horridge, 1975 for a discussion).

The situation was rescued by Vogt in 1975 who observed that the crystalline pyramids in the crayfish Astacus were silvered: coated with a multilayer of crystals that Vogt thought to be guanine, but are more likely to be a pteridine, isoxanthopterin (Zyznar & Nicol, 1971). Vogt's observation, which was independently confirmed by myself (Land, 1976) on a deep-sea shrimp Oplophorus, changed the situation completely (Fig. 19a). A series of radially arranged plane mirrors can perform an almost identical function to the array of inverting lens-cylinder telescopes in refracting superposition eyes, for the simple reason that a mirror has precisely the same property of rotating a ray of light making an angle α with the mirror through an angle of 2α. These eyes are thus still superposition eyes, as their overall anatomy indicates, but the optical elements are mirrors, not lenses (Land, 1978). A full account of the optics of the crayfish eye is given by Vogt (1980).

The other interesting feature of all the macruran eyes (carid and pennaeid shrimps, crayfish, lobsters and the anomuran galatheids) is that the facets are square, rather than hexagonal as in almost all other compound eyes. It is not immediately obvious that the mirrors in these eyes must be arranged at right angles to each other, but there turns out to be an intriguing reason why this must be the case (Vogt, 1977), and Fig. 19b is an attempt to explain it. Although, in Fig. 19a, which represents an idealised section through an eye along a row of facets, light falling on each facet is brought to a common focus, it is not clear what would happen to light falling on facets in front of or behind the plane of the page. Such rays would, in general, be reflected from 2 faces of each crystalline pyramid rather than one, provided the mirrors are of the right length, and this means that what they encounter is a "corner-reflector". It is easy to demonstrate that a pair of mirrors at right angles has the

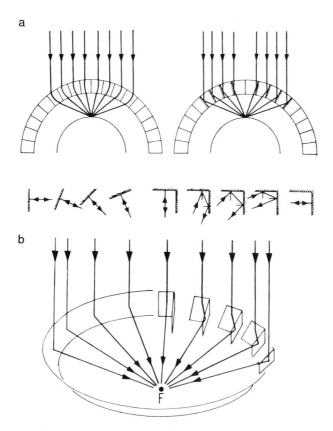

Fig. 19: a) Diagram showing the identity of ray paths in a refracting (left) and reflecting (right) superposition eye. b) Image formation in a reflecting superposition eye. The "corner-reflector" construction (right) gives similar ray paths to a single circular reflecting strip (left), but has the advantage that a system of corner-reflectors has no single axis (see text). The inserts above show ray paths as seen from normals to the eye surface for the two cases. Based on Land (1981a).

property of reflecting all rays back in a plane parallel to their incident path (one often encounters corner mirrors in clothing shops, and they have the alarming property that one cannot escape from one's image by walking around them). The direction from which the rays are incident does not matter: the corner behaves as though it were a single mirror oriented normally to the incident ray. This simplifies things considerably, because it means that the "single mirror" diagram in Fig. 19a is also appliciable to rays not in the plane of the page, because 90° corners behave like single mirrors! Since the 120° corners that a hexagonal lattice would produce do not have this useful property it follows that reflecting superposition eyes must have a square facet lattice. It is probably usually also true that eyes with a wholly square facet lattice are of the reflecting superposition type.

A further interesting finding is that the larvae of the macrurans do not have square facets, but hexagonal ones (Land, 1981a, b). They have a small distinct pseudopupil, and are clearly of the apposition type (Fig. 20). Only rather late in the development, around stage 15 in Palaemonetes, do the crystalline cones elongate and square off and the eye attains the superposition optics of the adult. Even the adult eye retains the capacity for reverting to the apposition type during light adaptation. Bryceson (1981) found in crayfish that in addition to the mirror boxes each facet has a long focus lens in the cornea which forms an image at the level of the retina. In the light screening pigment migrates into the clear zone and isolates the individual lens-rhabdom combinations from each other, and renders rays reflected from the mirror boxes ineffective. In this condition the eyes are for all practical purposes apposition eyes (like those of crabs) with the facet lenses forming small inverted images on the rhabdoms. In the dark, with the mirrors operating, the lenses would seem to have the beneficial effect of pre-focussing the beam that is reflected so that it reaches the retina as a narrow pencil, rather than a beam as wide as the facet it entered (typically about 50 µm). Nevertheless, resolution in the dark-adapted (superposition) condition seems to be poor. In the crayfish

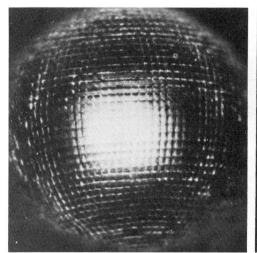
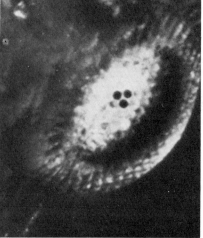

Fig. 20: **Left:** Eye of Palaemonetes varians, dark adapted. Note the square facets, and the patch of eye-shine reflected from the tapetum behind the receptors. It is similar in size to the superposition "pupil" of the eye itself. Eye diameter 1,3 mm, facets 30 µm. **Right:** Larval eye of Palaemon serratus showing hexagonal facets, and a 3 facet pseudopupil (see Fig. 13). The eye is of the apposition type. The facets are about 22 µm across (Land, 1981a).

Cherax Walcott (1974) found acceptance angles ($\Delta\rho$) as great as 24° from single cell recordings in dark-adapted animals, and these reduced to 4° in light adaptation. However, what is lost in resolution should be more than made up in sensitivity. If, in a 1 mm diameter eye with 50 µm facets (e.g. Palaemonetes), the light-adapted pupil is confined to one facet, the F-number will be about 5. If this decreases to 0,5 in the dark the retinal brightness ratio between dark and light will be $(5/0,5)^2$, or 100. Compared to the brightness ratio achieved by the human pupil (about 1:16) this is impressive.

CONCLUSIONS

The only safe generalisation that one can make about crustacean eyes is that they are usually of two kinds. One cannot be sure, for example, that the compound eye of Daphnia is related to the eyes of Palaemonetes by common lineage. Nor can all the nauplius eyes be clearly related to each other (Elofsson, 1965). The diversity is such that it may be that each type of eye evolved a number of times, and certainly the compound eyes of insects and crustacea are most unlikely to have a common ancestor. This makes the extent of convergence between the eyes of krill and those of moths all the more remarkable, and one can only wonder at the fact that there seem to be only a few ways of building optical structures properly. That particular example is at least as impressive to me as the convergence between lens eyes in fish and cephalopods.

REFERENCES

Andersson, A & Nilsson, D.-E. (1981) Fine structure and optical properties of an ostracode (Crustacea) nauplius eye. Protoplasma 107: 361-374.

Ball, E.E. (1977) Fine structure of the compound eyes of the midwater amphipod Phronima in relation to behaviour and habitat. Tissue Cell 9: 521-536.

Barlow, R.B., Chamberlain, S.C. & Levinson, J.Z. (1980) Limulus brain modulates the structure and function of the lateral eyes. Science 210: 1037-1039.

Barnes, R.D. (1980) Invertebrate Zoology. Fourth Edition. Philadelphia, Saunders College, 1089 pages.

Behrens, M. & Krebs, W. (1976) The effect of light-dark adaptation on the ultrastructure of Limulus lateral eye retinular cells. J. Comp. Physiol. 107: 77-96.

Blest, A.D. (1978) The rapid synthesis and destruction of photoreceptor membrane by a dinopid spider. I. The daily cycle. Proc. R. Soc. Lond. 200B: 463-483.

Blest, A.D. & Land, M.F. (1977) The physiological optics of Dinopis subrufus L. Koch: a fish-lens in a spider. Proc. R. Soc. Lond. 196B: 197-222.

Bruno, M.S., Barnes, S.N. & Goldsmith, T.H. (1977) The visual pigment and visual cycle of the lobster Homarus. J. Comp. Physiol. 120: 123-142.

Bryceson, K.P. (1981) Focusing of light by corneal lenses in a reflecting superposition eye. J. Exp. Biol. 90: 347-350.

Caldwell, R.L. & Dingle, H. (1976) Stomatopods. Sci. Am. 234 (1): 80-89.

Caveney, S. & McIntyre, P. (1981) Design of graded-index lenses in the superposition eyes of scarab beetles. Phil. Trans. R. Soc. Lond. 294B: 589-635.

Cummins, D. & Goldsmith, T.H. (1981) Cellular identification of the violet receptor in the crayfish eye. J. Comp. Physiol. 142: 199-202.

Downing, A.C. (1972) Optical scanning in the lateral eyes of the copepod Copilia. Perception 1: 247-261.

Eguchi, E. & Waterman, T.H. (1966) Fine structure patterns in crustacean rhabdoms. In: The Functional Organization of the Compound Eye. Ed. C.G. Bernhard. Oxford, Pergamon Press, p. 105-124.

Elofsson, R. (1965) The nauplius eye and frontal organs in Malacostracea (Crustacea). Sarsia 19: 1-54.

Elofsson, R. (1966) The nauplius eye and frontal organs of the non-Malacostraca (Crustacea). Sarsia 25: 1-128.

Exner, S. (1891) Die Physiologie der facettirten Augen von Krebsen und Insecten. Leipzig und Wien: Deuticke.

Fahrenbach, W.H. (1964) The fine structure of a nauplius eye. Z. Zellforsch. 62: 182-197.

Fletcher, A., Murphy, T. & Young, A. (1954) Solutions of two optical problems. Proc. R. Soc. Lond. 223A: 216-225.

Forster, L. (1982) Visual communication in jumping spiders. In: Spider Communication. Ed. P.N. Witt & J.S. Rovner. Princeton, University Press, p. 161-212.

Frisch, K.v. & Kupelwieser, H. (1913) Über den Einfluss der Lichtfarbe auf die phototaktischen Reaktionen niedere Krebse. Biol. Zentr. 33: 517-552.

Frost, B.J. (1975) Eye movements in Daphnia pulex (De Geer). J. Exp. Biol. 62: 347-350.

Gophen, M. & Harris, R.P. (1981) Visual predation by a marine cyclopoid copepod, Corycaeus anglicus. J. Mar. Biol. Ass. U.K. 61: 391-399.

Gregory, R.L., Ross, H.E. & Moray, N. (1964) The curious eye of Copilia. Nature (Lond.) 201: 1166.

Grenacher, H. (1879) Untersuchungen über das Sehorgan der Arthropoden, insbesondere der Spinnen, Insecten und Crustaceen. Göttingen; Vandenhoeck und Ruprecht.

Horridge, G.A. (1975) Optical mechanisms of clear zone eyes. In: The Compound Eye and Vision of Insects. Ed. G.A. Horridge. Oxford, Clarendon Press, p. 255-298.

Horridge, G.A. (1978) The separation of visual axes in apposition compound eyes. Phil. Trans. R. Soc. 285B: 1-59.

Kuiper, J.W. (1962) The optics of the compound eye. Symp. Soc. Exp. Biol. 16: 58-71.

Kunze, P. (1979) Apposition and superposition eyes. In: Handbook of Sensory Physiology. Vol. VII/6A. Ed. H. Autrum. Berlin, Springer, p. 441-502.

Land, M.F. (1965) Image formation by a concave reflector in the eye of the scallop Pecten maximus. J. Physiol. (Lond.) 179: 138-153.

Land, M.F. (1972a) The physics and biology of animal reflectors. Prog.

Biophys. Mol. Biol. 24: 75-106.
Land, M.F. (1972b) Mechanisms of orientation and pattern recognition by jumping spiders (Salticidae). In: Information Processing in the Visual Systems of Arthropods. Ed. R. Wehner. Berlin, Springer, p. 231-247.
Land, M.F. (1976) Superposition images are formed by reflection in the eyes of some oceanic decapod crustacea. Nature (Lond.) 263: 764-765.
Land, M.F. (1978) Animal eyes with mirror optics. Sci. Am. 239 (6): 126-134.
Land, M.F. (1979) The optical mechanism of the eye of Limulus. Nature (Lond.) 280: 396-397.
Land, M.F. (1980) Compound eyes: old and new optical mechanisms. Nature (Lond.) 287: 681-686.
Land, M.F. (1981a) Optics and vision in invertebrates. In: Handbook of Sensory Physiology. Vol. VII/6B. Ed. H. Autrum. Berlin, Springer, p. 471-592.
Land, M.F. (1981b) Optical mechanisms in the higher Cructacea with a comment on their evolutionary origins. In: Sense Organs. Ed. M.S. Laverack & D.J. Cosens. Glasgow, Blackie, p. 31-48.
Land, M.F. (1981c) Optics of the eyes of Phronima and other deep-sea amphipods. J. Comp. Physiol. 145: 209-226.
Land, M.F. (1983) Mollusca (This volume).
Land, M.F. & Burton, F.A. (1979) The refractive index gradient in the crystalline cones of the eyes of an euphausiid crustacean. J. Exp. Biol. 82: 395-398.
Land, M.F., Burton, F.A. & Meyer-Rochow, V.B. (1979) The optical geometry of euphausiid eyes. J. Comp. Physiol. 130: 49-62.
Levi-Setti, R., Park, D.A. & Winston, R. (1975) The corneal cones of Limulus as optimised light concentrators. Nature (Lond.) 253: 115-116.
Lüders, L. (1909) Gigantocypris agassizii. Z. wiss. Zool. 92: 103-158.
Martin, F.G. & Mote, M.I. (1982) Color receptors in marine crustaceans: A second spectral class of retinular cell in the compound eyes of Callinectes and Carcinus. J. Comp. Physiol. 145: 549-554.
Meyer-Rochow, V.B. (1978) The eyes of mesopelagic crustaceans. II. Streetsia challengeri (Amphipoda). Cell Tissue Res. 186: 337-349.
Nässel, D.R. & Waterman, T.H. (1979) Massive diurnally modulated photoreceptor membrane turnover in crab light and dark adaptation. J. Comp. Physiol. 131: 205-216.
Nilsson, D.-E. & Nilsson, H.L. (1981) A crustacean compound eye adapted to low light intensities (Isopoda). J. Comp. Physiol. 143: 503-510.
Nilsson, D.-E. & Odselius, R. (1981) A new mechanism for light-dark adaptation in the Artemia compound eye (Anostraca, Crustacea). J. Comp. Physiol. 143: 389-399.
Pumphrey, R.J. (1961) Concerning vision. In: The Cell and the Organism. Ed. J.A. Ramsay & V.B. Wigglesworth. Cambridge University Press, p. 193-208.
Sandeman, D.C. (1978) Regionalization in the eye of the crab Leptograpsus variegatus: eye movements evoked by a target moving in different parts of the visual field. J. Comp. Physiol. 123: 299-306.

Schönenberger, N. (1977) The fine structure of the compound eye of Squilla mantis (Crustacea, Stomatopoda). Cell Tissue Res. 176: 205-233.
Stavenga, D.G. (1979) Pseudopupils of compound eyes. In: Handbook of Sensory Physiology. Vol. VII/6A. Ed. H. Autrum. Berlin, Springer, p. 357-439.
Stowe, S. (1980) Spectral sensitivity and retinal pigment movement in the crab Leptograpsus variegatus (Fabricius). J. Exp. Biol. 87: 73-99.
Vaissière, R. (1961) Morphologie et histologie comparées des yeux des crustacés copépodes. Arch. Zool. Exp. Gén. 100: 1-125.
Verger-Bocquet, M. (1983) Photoreception et vision chez les annelides (This volume).
Vogt, K. (1975) Zur Optik des Flusskrebsauges. Z. Naturforsch. 30c: 691.
Vogt, K. (1977) Ray path and reflection mechanism in crayfish eyes. Z. Naturforsch. 32c: 466-468.
Vogt, K. (1980) Die Spiegeloptik des Flusskrebsauges. The optical system of the crayfish eye. J. Comp. Physiol. 135: 1-19.
Walcott, B. (1974) Unit studies on light adaptation in the retina of the crayfish, Cherax destructor. J. Comp. Physiol. 94: 207-218.
Wald, G. (1968) Single and multiple visual systems in arthropods. J. Gen. Physiol. 51: 125-156.
Walls, G.L. (1942) The Vertebrate Eye and Its Adaptive Radiation. New York, Hafner, 785 pages.
Waterman, T.H. (1961) Light sensitivity and vision. In: The Physiology of Crustacea. Vol. II. Ed. T.H. Waterman. New York, Academic Press, p. 1-64.
Waterman, T.H. (1981) Polarization sensitivity. In: Handbook of Sensory Physiology. Vol. VII/6B. Ed. H. Autrum. Berlin, Springer, p. 281-496.
Wolken, J.J. & Florida, R.G. (1969) The eye structure and optical system of the crustacean copepod, Copilia. J. Cell Biol. 40: 279-285.
Yoshida, M., Takasu, N. & Tamotsu, S. (1983) Photoreception in echinoderms (This volume).
Young, S. (1981) Behavioural correlates of photoreception in Daphnia. In: Sense Organs. Ed. M.S. Laverack & D.J. Cosens. Glasgow, Blackie, p. 49-63.
Young, S. & Downing, A.C. (1976) The receptive fields of Daphnia ommatidia. J. Exp. Biol. 64: 185-202.
Zyznar, E.J. & Nicol, J.A.C. (1971) Ocular reflecting pigments of some malacostraca. J. Exp. Mar. Biol. Ecol. 6: 235-248.

NOTE

Shortly after this manuscript was written, another important and authoritative review of photoreception in the Crustacea was published (Shaw, S.R. & Stowe, S. (1982) Photoreception. In: The Biology of the Crustacea, Vol. III. Series Ed. D. Bliss. Eds. H.L. Atwood & D.C. Sandeman. New York, Academic Press, p. 292-367). This review concentrates relatively more on electrophysiological aspects of vision than on optics, and is thus complementary in emphasis to the present paper.

RETINAL MOSAIC OF THE FLY COMPOUND EYE

NICOLAS FRANCESCHINI

Institut de Neurophysiologie et Psychophysiologie, C.N.R.S.
31 Chemin Joseph-Aiguier
13277 Marseille Cedex 9
France

> Nurse
> Hie you, make haste, for it grows very late.
>
> Romeo and Juliet, Act III, Sc. 3.

INTRODUCTION

In contrast to humans who, like squids, are equipped with camera eyes, three quarter of the animal species on this earth are endowed with compound eyes the beauty of which is shared by the arthropod world. In recent years the compound eye of insects in particular has become a valuable tool for studying basic principles in vision and their neuronal implementation in the living organism. Several comprehensive reviews of this subject have been published recently, in particular in the three volumes of the Handbook of Sensory Physiology (Autrum, 1979, 1981 a, b).

The visual system of higher dipteran insects has occupied a major place among the analyses conducted over the last 25 years and it certainly has become one of the best known insect visual systems. Several advantages make this system particularly amenable to experimental analysis:

1) Flies display a rich repertoire of visually guided behaviors. In observing a fly's aerobatic chase for instance, Rochon-Duvigneaud's comment about birds comes to mind: "a wing guided by an eye, something which requires the precision and speed of retinal functions". As is the case for many insects (rev. Wehner, 1981), many visually-guided behaviors in flies are quite reproducible and quantifiable. Since they constitute the final performance of the nervous system, they offer an invaluable means of analysis,

especially when elicited by precise optical stimuli (rev. Buchner, 1983).

2) However complex and challenging they may appear, the visually-guided behaviors encountered in flies are accomplished by a limited population of a few hundred thousand neurons whose most salient aspect is their modular organization (rev. Strausfeld, 1976, 1983). The very repetitivity of the compound eye persists throughout the three optic ganglia, a situation which is of great help for the fine analysis of the neuronal circuits at the single cell level, both with neuroanatomical (rev. Meinertzhagen, 1983) and electrophysiological techniques (rev. Laughlin, 1983). In the vertebrate visual system single neurons cannot be recognized from individual to individual, whereas here single neurons can be identified and found again and again in several individuals. This provides the experimenter with the invaluable opportunity of accumulating data about one single visual neuron (rev. Hausen, 1983).

3) A worthy representative of higher dipterans is the little fly Drosophila melanogaster whose many hundreds visual mutants offer an interesting tool for exploring visual functions in the framework of what is now called neurogenetics (rev. Heisenberg, 1979; Pak, 1979; Hall, 1982; Fischbach, 1983).

4) In each ommatidium of the fly compound eye, the light-gathering organelles, the rhabdomeres, are physically separated from one another (Fig. 1 a) instead of being intermixed into a single, central rhabdom as is the case in most insect and crustacean compound eyes. This allows single receptor cells to be analysed individually without too much interference from their direct neighbors. Moreover, the receptor cells in these eyes are separated from the outside world by very thin transparent components (corneal facet lens, crystalline cone) whose total thickness is less than 0,1 mm. In recent years we have taken advantage of this situation and devised several microscopic techniques which permit single cell analysis and single cell stimulation to be performed in the living, intact fly (Rev. Franceschini, 1975).

Today fly photoreceptor cells distinguish themselves by the fact that they can be studied individually under physiological conditions with no less than three different methods: intracellular recordings of the light induced receptor potential, transmission microspectrophotometry and microspectrofluorimetry. There is no other animal so far, among invertebrates and vertebrates, in which all three methods can be applied in vivo. It is thanks to this trio of non-invasive methods that we have been able to gain detailed knowledge about the spectral properties of the highly organized fly retinal mosaic whose 50 000 receptor cells distributed between both eyes constitute as many inputs to the nervous system. In the following account I have attempted to highlight some of the new aspects which could be of relevance to the study of other photoreceptor systems.

SENSITIZING PIGMENT IN A PHOTORECEPTOR CELL

The first determination of the spectral sensitivity of insect

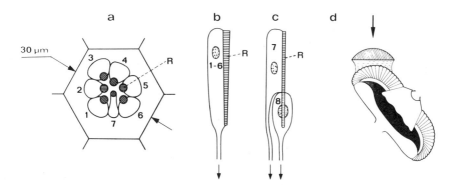

Fig. 1. Schematic, distal cross section of an ommatidium in the compound eye of the fly Musca domestica (a) (after Trujillo-Cenoz & Melamed, 1966; Boschek, 1977). Behind the hexagonal facet lenslet six receptor cells R1-R6 are found (a longitudinal section of which is schematized in b) surrounding two smaller central cells R7 and R8 which lie on top of each other (c). Hatching of the rhabdomere (R) in each cell symbolizes the direction of microvilli, the membrane of which houses the visual pigment(s). (d) = scheme of the method used to visualize the distal rhabdomere endings in vivo. A living fly is mounted at the center of a goniometric stage and its eye is observed with epi-fluorescence microscopy (arrow = blue excitation). The corneal lenslets are "optically neutralized" (Franceschini & Kirschfeld, 1971) with a drop of water and the observation is made with a water immersion objective (25 x, numerical aperture 0,65). The viewing aperture is deliberately reduced with respect to the excitation aperture, as described in Franceschini (1983 a). Focusing through the "optically neutralized" cornea (which nearly acts as a transparent parallel plate) reveals the colorful autofluorescence pattern of the retinulae within about 30 ommatidia.

photoreceptor cells to be carried out by means of intracellular electrical recordings revealed that the majority of fly receptors exhibited a dual peak spectrum (Burkhardt, 1962). Recovery of such cells after intracellular injection of a fluorescent marker has shown that this kind of spectral sensitivity was to be ascribed to the 6 peripheral receptors R1-6 (Fig 1 a) of the ommatidium (McCann & Arnett, 1972).

Recent studies concerning this odd dual-peak spectral sensitivity have led to an unexpected result. The blue-green maximum ($\lambda_{max} \simeq 490$ nm) reflects the absorption properties of the rhodopsin (Hamdorf et al., 1973; Stavenga et al., 1973; Kirschfeld & Franceschini, 1975; Stavenga & Schwemer, 1983) while the ultra-violet peak apparently stems from a photostable UV-pigment which plays the role of a sensitizing pigment (Kirschfeld et al., 1977; Stark et al., 1977) and is reminiscent of the light-harvesting pigments found in some photosynthetic membranes (Goodheer, 1969). Energy transfer from the UV donor pigment to the rhodopsin acceptor is likely to occur via "induced resonance" according to Foerster's model (1951). New supporting evidence for the UV sensitizing pigment hypothesis has been provided recently by high resolution measurements of the spectral sensitivity in the UV (Gemperlein et al., 1980). The latter is characterized by a conspicuous fine structure (one main peak at $\lambda \simeq 350$ nm) not found in any rhodopsin and reminiscent of the absorption properties of some carotenoids (Franceschini, 1983 a). This fine structure meanwhile has been confirmed both by intracellular recordings and microspectrophotometry of single receptor cells R1-6 in Musca (Kirschfeld et al., 1983).

Although many carotenoid-related substances could account for such a three-fingered spectrum (they are listed in Franceschini, 1983 a) the most plausible candidate is a retinol-protein complex (Franceschini, 1983 a; Kirschfeld et al., 1983). Retinol itself (vitamin A alcohol) displays a typically diffuse absorption spectrum peaking at 325 nm. However, upon binding to a protein the spectrum undergoes a red shift by 25 nm and becomes vibrationally resolved with 3 peaks near 330, 350 and 370 nm. This occurs when retinol binds to many model proteins (Futterman & Heller, 1972; Hemley et al., 1979; Fugate & Song, 1980) but also to bacterio-opsin (Schreckenbach et al., 1977) and to the Cellular Retinol Binding Protein (CRBP) which is a putative mediator of vitamin A action at the cellular level (Chytil & Ong, 1979). The photostable retinol chromophore is possibly bound to the rhodopsin molecule itself at a site close to the retinol chromophore (Franceschini, 1983 a). Such a proximity would explain the high efficiency of energy transfer which is reflected in the high sensitivity of the cell within the UV range of the spectrum. Energy transfer efficiency has been assessed at $\sim 0,75$, and a ratio of two molecules of sensitizing pigment per one rhodopsin has been proposed (Vogt & Kirschfeld, 1983).

It has been known for some time that the UV-peak of the spectral sensitivity was of variable height with respect to the blue-green peak (Horridge et al., 1975) and that vitamin A-deficiency in the larval diet would give rise to a weak UV-sensitivity in the adult (Goldsmith et al., 1964; Stark et al., 1977; Kirschfeld et al., 1977, 1983). In such a case however the relative sensitivity to UV versus blue-green is known to be age-dependent in the adult. In the eye of wild-type flies whose larvae have been reared on a vitamin A-deficient medium, receptor cells R1-6 not only become more and more sensitive with time (Goldsmith & Fernandez, 1968; Guo, 1980 a, b) but they also display a progressive increase in their relative sensitivity to UV which reaches the blue-green sensitivity only after 8-10 days (Guo, 1980 a, b). Thus, it would appear that the adult animal can progressively compensate for some deficiency resulting from an incomplete larval diet and restore the appropriate pigmentary outfit of the microvillar membrane over some days or weeks, including the UV-pigment or chromophore. In the context of the aforementioned hypothesis of a double chromophore bound to the opsin, this would mean that shortly after eclosion a large population of rhodopsin molecules would lack one or both chromophores.

From a functional standpoint one beneficial consequence of this UV-sensitization is the extension of the light-harvesting capability of the photoreceptor cell. In becoming "panchromatic" the cell improves its quantum catch, for the sake of a higher absolute sensitivity and lower quantum noise on the receptor potential. Although evidence for a sensitizing pigment has so far been presented exclusively for the fly, it should be kept in mind as a putative mechanism underlying the odd spectral sensitivity sometimes observed in single photoreceptors (see also De Voe et al., 1969).

SCREENING PIGMENT IN A PHOTORECEPTOR CELL

Photoreceptor cells in some vertebrates like reptiles, birds and even mammals (monotremes) sometimes harbor photostable pigments in the form of colored oil droplets (Schultze, 1866; Walls & Judd, 1933). In acting as a selective screen in the path of incoming light destined for the photosensitive pigment these colored droplets contribute towards modifying the cell's spectral sensitivity which thus does not merely reflect the absorption properties of the photosensitive pigment (Baylor & Hodgkin, 1973; Ohtsuka, 1978). A different kind of screening pigment has been found recently in some central photoreceptors R7 (see Fig. 1 a, c) of the fly retina (Kirschfeld & Franceschini, 1977; Kirschfeld et al., 1978). In contrast to the vertebrate colored oil droplets which are lumped organelles located in the inner segment of the cone, the fly screening pigment appears to be distributed throughout the rhabdomere. In Musca, 70% of the R7 receptors are equipped with this accessory pigment which strongly absorbs blue light and makes the rhabdomere appear yellow in transmitted white light (hence the name R7 yellow, R7y, given to these receptors; Kirschfeld & Franceschini, 1977). The absorption spectrum of this pigment, which has been measured under microspectrophotometry, exhibits a carotenoid "fingerprint" with a main maximum at $\lambda \simeq 460$ nm and 2 secondary peaks at ~ 430 nm and ~ 485 nm (Kirschfeld et al., 1978; McIntyre & Kirschfeld, 1981). Although screening may not be the sole "use" of this pigment in the photoreceptor (see Kirschfeld, 1979), evidence for a screening function stems from both microspectrophotometry and intracellular electrophysiological recordings. The unexpected finding that such a cell, with a violet absorbing rhodopsin (Kirschfeld, 1979) and a blue-absorbing screening pigment, seems to be mainly sensitive to ultra-violet light (Hardie et al., 1979) has led us to postulate the existence of a similar UV-sensitizing pigment to that found in R1-6 receptors (see previous section on SENSITIZING PIGMENT). Reexamination of the spectral sensitivity of such a cell with high resolution in the UV has recently confirmed this view in revealing a UV photostable pigment whose absorption spectrum displays a fine structure similar to that of R1-6 but red-shifted by ~ 5 nm (Hardie & Kirschfeld, 1983). Again it is striking that the binding of retinol to a protein leads under certain conditions (excess of retinol) to the "retro" conformation of the chromophore (for a more detailed discussion, see Franceschini, 1983 a) which is manifested by a 5 nm red-shift in its absorption spectrum with respect to that of a "normal" retinol-protein complex (Hemley et al., 1979; Fugate & Song, 1980).

The receptor R8 whose rhabdomere lies in the prolongation of R7y is called R8y for the sake of simplicity. It is maximally sensitive to green light ($\lambda_{max} \simeq 540$ nm) due to its (postulated) 520 nm rhodopsin and the screening action of the overlying R7y (Hardie et al., 1979). However its spectral sensitivity also displays a characteristic UV-fine structure whose three peaks, rather than corresponding to those of R7y coincide precisely with those of R1-6 (Hardie & Kirschfeld, 1983).

Hence some receptor cells like the R7y's of the fly retina may contain at least three pigments in their rhabdomere: a rhodopsin ($\lambda_{max} \simeq 420$ nm), a sensitizing pigment ($\lambda_{max} \simeq 355$ nm) and a screening pigment ($\lambda_{max} \simeq 460$ nm) which all contribute to shaping the spectral sensitivity of the cell so that little remains from the original rhodopsin absorption spectrum. The remarkable fact is that the rhodopsin is masked from direct photic excitation because of the presence of the blue absorbing screening pigment and it is mainly excited in an indirect way, via energy transfer from the ultraviolet sensitizing pigment. The fact that the screening pigment itself (whose molar ratio with respect to rhodopsin has been estimated at $\sim 10:1$; McIntyre & Kirschfeld, 1981) does not dramatically jeopardize the energy transfer can be easily explained by the assumption that the rhodopsin and the sensitizing pigment are built from two closely spaced chromophores within the same opsin molecule. Such a situation could indeed prevent the screening pigment from approaching the UV-chromophore closely enough to become a competitive quencher.

The "need" behind this sophisticated shaping of a UV-sensitive cell appears all the more enigmatic as UV-rhodopsins are available in insects (rev. Hamdorf, 1979) and even in flies which actually have equipped the remaining 30% of their R7's with such a "conventional" UV rhodopsin (devoid of a fine structure in its absorption spectrum, Kirschfeld, 1979; Hardie & Kirschfeld, 1983). This type of R7, called R7pale (R7p) is consequently a pure UV receptor having less than 10% sensitivity at wavelengths beyond 400 nm (Hardie et al., 1979). By contrast, R7y exhibits a long tail of sensitivity between 400 nm and 500 nm which has even been described as a genuine peak of sensitivity (Smola & Meffert, 1979). This last discrepancy is now known to be due to the rearing conditions of the animal. As in the case of R1-6 cells, and for the same reason, a carotenoid deficiency in the larval diet selectively reduces the UV-peak of R7y and artificially raises the blue tail into a blue peak (Guo, 1980 a, b).

To conclude this section we can state that the class of central cells R7y of the fly retina are certainly the most complex receptor cells ever found in an animal, as far as the intimate mechanism of their spectral sensitivity is concerned. This unexpected degree of complexity found within a single photoreceptor cell is perhaps a warning sign for the student of invertebrate photoreception. It may in particular cause some qualms in those whose endeavor is to dismantle spectral sensitivity mechanisms in cells which, to confuse the matter even more, have perversely intermixed their rhabdomeres into a "fused rhabdom", a situation so commonly encountered in the eye of invertebrates.

SPECTRAL ORGANIZATION OF THE RETINAL MOSAIC

Epi-fluorescence microscopy has recently shed new light upon the chromatic organization of the fly retina. By optically neutralizing the dioptrics of the corneal facet lenslets with a drop of water, it is possible to focus a microscope onto the rhabdomere distal endings in the living animal (Franceschini & Kirschfeld, 1971) and discover their colorful mosaic under blue excitation (Franceschini, 1977; Franceschini et al., 1981 a). Rhabdomeres R1-6 invariably fluoresce red throughout the eye whereas rhabdomeres R7's display an amazing variety since they appear either green, black or red, depending on the ommatidium. By good fortune these autofluorescence colors are tightly linked to the properties of the visual pigments (photosensitive or photostable) so that they constitute a genuine color-code for the various spectral types of receptors. Deciphering this code has involved the combination of the novel technique of "ommatidial fundus fluoroscopy" with more classical methods of single receptor analysis like transmission microspectrophotometry, intracellular recordings of the receptor potential and electron microscopy

In connection with the intracellular recordings (Hardie et al., 1979), a useful ancillary technique was developed which makes it possible to recover an intracellularly stained cell within a few minutes on the living animal, hence avoiding tedious histological processing (Franceschini & Hardie, 1980; Franceschini, 1983 b). Also it is now possible to record the spectral sensitivity of a single receptor cell within only a few seconds by using either Fourier spectroscopy (Gemperlein et al., 1980) or the less involved technique of "voltage clamp by light" (Franceschini, 1979, 1983 b).

Table 1 summarizes the spectral properties of the Musca photoreceptor cells and their appearance with epi-fluorescence and transmission microscopy. The "Spectral sensitivity" column shows that at least 6 different spectral classes of photoreceptors are encountered in the fly retina and, most strikingly, all 6 are represented in the 2 central cells R7 and R8. Depending upon the ommatidium, R7 and R8 build either a "yellow pair", a "pale pair" or a "red pair" which can be easily recognized in vivo from their characteristic fluorescence color (green, black and red, respectively). Together with its associated set of 6 receptor cells (type R1-6) which view the same point in the environment (Kirschfeld, 1967) each central pair actually builds a genuine "chromatic module" which appears to be trichromatic in two cases: "yellow" and "pale". Any such module is composed of 8 receptors (up to 10 in the equatorial region of the eye: Horridge & Meinertzhagen, 1970; Boschek, 1971) having parallel visual axes (Kirschfeld, 1967). The elegance of the situation stems from the fact that each module takes up three chromatic items of information in parallel, for the same picture element, hence without loss in spatial resolution.

Extensive mapping of the retinal mosaic using the auto-fluorescence colors of the receptor cells as a distinguishing marker have revealed that

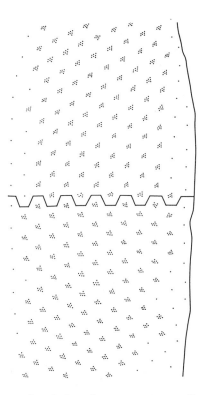

Fig. 2. Actual mosaic of the photoreceptor cells in the right eye of a housefly (white-eyed female; frontal border to the right). This map (which encompasses about 6% of the whole eye) was constructed from several, adjacent and largely overlapping color photographs obtained with the non-invasive method described in Fig. 1 d (epi-fluorescence microscopy combined with "optical neutralization of the cornea"). It reveals the exact shape of the distal rhabdomere pattern (comp. Fig. 1 a) as it appears in the focal plane of each corneal facet. Note the symmetry between dorsal and ventral rhabdomere patterns on each side of a zig-zag line which symbolizes the "equator" of the eye (Dietrich, 1909). Note also the exact orientation of the various rhabdomere patterns and the relatively small number of "flaws" in the crystal (disorganized or misaligned rhabdomere patterns). Under the blue excitation of the epifluorescence microscope, all rhabdomere R1-R6 emit red light. By contrast, the R7/8 are either non-fluorescing (R7p; 30%) or green-fluorescing (R7y; 70%). Both populations of R7/8 appear randomly distributed. A third population of (red-fluorescing) R7/8 is encountered exclusively in the frontal-dorsal part of the male retina (Franceschini et al., 1981 b).

TABLE 1. Spectral properties of the 6 peripheral cells (R1-6) and 2 central cells (R7 and R8) present in each ommatidium (see Fig. 1 a, b) of the fly compound eye (Musca domestica). R1-6 cells appear to have the same properties throughout the retinal mosaic. By contrast, the central cells R7 and R8 whose rhabdomeres lie in tandem (Fig. 1 c) show an amazing variety. They make up 3 different spectral pairs called "yellow", "pale" and "red", displaying a total of 6 different spectral sensitivities. The "red pair" is found only in the male and the R7r cell has the same dual-peak spectral sensitivity as its 6 neighbors R1-6 in the ommatidium.

	Autofluorescence colors (under blue excitation) (1,2,3)	Appearance in transmitted white light (4,5)	Spectral sensitivity λ_{max} [nm] (1,2,3,6,7,8)	Rhodopsin λ_{max} [nm] (9,10,11)	Sensitizing pigment λ_{max} [nm] (8,12,13)	Screening pigment λ_{max} [nm] (5,14)
R1-6	red	pale	350 + 490	490	350	–
yellow { 7y	green	yellow	355	430	355	456
yellow { 8y	?	?	540	520	350	?
pale { 7p	black	pale	330	330	–	–
pale { 8p	?	?	460	460	–	–
red { 7r	red	pale	350 + 490	490	350	–
red { 8r	?	?	360, 450, 520	490	?	–

References: (1) Hardie et al. (1979); (2) Franceschini et al. (1981 b); (3) Franceschini et al. (1981 a); (4) Kirschfeld & Franceschini (1977); (5) Kirschfeld (1983); (6) Hardie (1979); (7) Smola & Meffert (1979) (Calliphora); (8) Hardie & Kirschfeld (1983); (9) Kirschfeld & Franceschini (1975); (10) Kirschfeld (1979); (11) Hardie et al. (1981); (12) Kirschfeld et al. (1977); (13) Kirschfeld et al. (1983); (14) McIntyre & Kirschfeld (1981).

the proportion of R7y - R7p is usually 70% - 30% (Franceschini et al., 1981 a, b; Hardie et al., 1981). These two populations appear to be randomly distributed and do not make obvious geometric patterns. Nor is the distribution exactly the same in two different Musca flies.

The proportion given above holds for the female eye and also for the ventral part of the male eye. In the frontal-dorsal part of the male retina, surprisingly, both populations of R7p and R7y give way to the third type: the R7red (R7r) which starts investing this retinal area a few facets above the equator (Franceschini et al., 1981 a; Hardie et al., 1981). The occurrence of specific central receptors in that area once more underscores its peculiarity. It was already known that the angular resolution is finest and the binocular visual field largest in the frontal-dorsal region of the male eye (Beersma et al., 1975; Collett & Land, 1975; Franceschini, 1975) which thus resembles a kind of "fovea". Also this retinal region seems to play a strategic role in target fixation during the complex chasing behavior of the male (Wehrhahn, 1979). Finally it is known to drive male specific neurons like the MLG1 (male lobula giant neuron n°1), which is located in the third optic ganglion and is likely to be involved in the neural machinery underlying the chasing behavior (Hausen & Strausfeld, 1980).

Although the ultrastructure of R7/R8 was thought to be more or less uniform (e.g. Melamed & Trujillo-Cenoz, 1968) the striking inhomogeneity recently found in their spectral properties has led to reexamination of the matter. Two morphologically different tandem configurations have been described for the central R7's and R8's of Calliphora. One of them is composed of a relatively short R7 on top of a relatively long R8 whereas in the other one the situation is reversed (Smola & Meffert, 1979). Comparison of the frequency of occurrence of these two configurations with that of the various spectral sensitivities encountered in central cells suggests that the first configuration corresponds to the "yellow pair" and the second one to the "pale pair" (Wunderer & Smola, 1982 a).

The "red pair" too is characterized by a peculiar morphology. R7r receptors have larger and longer rhabdomeres than usual and resemble the peripheral receptors R1-6. But the most interesting feature is that their axon projects to the same neuropile (the lamina) as R1-6 receptors (Franceschini et al., 1981 b; Hardie et al., 1981), instead of projecting to the medulla as do the other R7's and R8's of the eye.

The R7r axon has a shallow terminal with a characteristic "kink" in the underlying lamina cartridge to which it conveys a seventh, spatially and spectrally matched input (Franceschini et al., 1981 b). More details about this peculiar axonal projection is given by Strausfeld (1983) and Hardie (1983). One relevant aspect of this peculiar connectivity may lie in the slight improvement of signal-to-noise ratio which arises from averaging 7 input signals instead of 6 in a lamina cartridge. As a consequence the improved smoothing of quantal noise would endow the lamina monopolar cells with an improved capacity for the detection of weakly contrasted objects like a moving female fly. However, the peculiar lamina organization in this area of the male eye may hide other biologically relevant mechanisms which could involve in particular the unusually large L_3 monopolar neuron (Braitenberg & Hauser-Holschuh,

1972), onto which the male R7r makes profuse synapses (Hardie, 1983).

It should be noted that the axon of the underlying R8r still retains its long visual fiber input to the medulla and thus does not follow the unusual connectivity of its R7 companion, even though its visual pigment is apparently the same. This very feature incidentally shows that neuronal connectivity of a photoreceptor cell cannot be predicted on the sole basis of visual pigment content and vice versa.

CONCLUSION

The spectrally homogeneous class of receptor cells R1-6, with their large and long rhabdomeres, UV-extended spectral sensitivity and neural pooling seem to be refined for high sensitivity vision (cf. Kirschfeld & Franceschini, 1968). These receptors are known to drive in particular the neural machinery responsible for movement detection (Eckert, 1971; McCann & Arnett, 1972; Heisenberg & Buchner, 1977; Kirschfeld, 1972).

In the frontal dorsal part of the male eye, participation of a seventh cell to the averaging process which apparently takes place within a cartrige (Scholes, 1969) may be considered as a specific refinement improving the ability of the male to detect the low contrast provided by a possible partner entering its frontal visual field.

Receptor cells R7 and R8, with their impressive gamut of spectral sensitivities (Table 1) seem to offer a good prerequisite for some kind of color processing. Although too little is known concerning this matter at the behavioral level, Hanstrom's conjecture (1927) that "the insect eye lags behind the vertebrate eye as far as differentiation of color shades is concerned" may have been imprudent.

Movement detection, female chasing and color processing are but three examples for which division of labor among various photoreceptor cells is apparent in flies. Another example of regionalization of the retinal mosaic has been provided by the discovery of special, marginal ommatidia (Wada, 1974) whose central receptors R7 and R8 seem to be predestinated to a function in polarized light detection (Wunderer & Smola, 1982 b). Such a feature may be widespread in the insect world and has also been observed in several hymenopterans (Schinz, 1975; Herrling, 1975; Labhardt, 1980).

The possibility we now have to assign a spectral sensitivity to any receptor cell observed under epi-fluorescence in a living animal and to stimulate that same cell selectively makes the fly a unique model for studying information processing in a highly organized visual neuropile. Several accessory techniques have emerged as by-products of the analyses conducted at the most peripheral level of the visual system. Combination of these techniques with electrophysiology (e.g. Riehle & Franceschini, 1983) may offer a promising approach to unravelling the operation of such a tiny but strong-minded piece of neural tissue.

REFERENCES

Autrum, H. (Ed.) (1979) Handbook of Sensory Physiology. Vol. VII/6A.
    Vision in Invertebrates. A: Invertebrate Photoreceptors. Berlin,

Heidelberg, New York, Springer-Verlag, 729 pages.
Autrum, H. (Ed.) (1981 a) Handbook of Sensory Physiology. Vol. VII/6B. Vision in Invertebrates. B: Invertebrate Visual Centers and Behavior I. Berlin, Heidelberg, New York, Springer-Verlag. 629 pages.
Autrum, H. (Ed.) (1981 b) Handbook of Sensory Physiology. Vol. VII/6C. Comparative Physiology and Evolution of Vision in Invertebrates. C: Invertebrate Visual Centers and Behavior II. Berlin, Heidelberg, New York, Springer-Verlag. 663 pages.
Baylor, D.A. & Hodgkin, A.L. (1973) Detection and resolution of visual stimuli by turtle photoreceptors. J. Physiol. 234: 163-198.
Beersma, D.G.M., Stavenga, D. & Kuiper, J.W. (1975) Organization of visual axes in the compound eye of the fly Musca domestica and behavioural consequences. J. Comp. Physiol. 102: 305-329.
Bernard, G.D. & Stavenga, D.G. (1979) Spectral sensitivities of retinular cells measured in intact living flies by an optical method. J. Comp. Physiol. 134: 95-107.
Boschek, B. (1971) On the fine structure of the peripheral retina and lamina ganglionaris of the fly Musca domestica. Z. Zellforsch. Abt. Histochem. 118: 369-409.
Braitenberg, V. & Hauser-Holschuh, H. (1972) Patterns of projections in the visual system of the fly. II. Quantitative aspects of second order neurons in relation to models of movement perception. Exp. Brain Res. 16: 184-209.
Buchner, E. (1983) Behavioural analysis of spatial vision in insects. (This volume).
Burkhardt, D. (1962) Spectral sensitivity and other response characteristics of single visual cells in the arthropod eye. Symp. Soc. Exp. Biol. 16: 86-109.
Burkhardt, D. & Motte, I. de la (1972) Electrophysiological studies on the eyes of Diptera, Mecoptera and Hypenoptera. In: Information Processing in the Visual Systems of Arthropods. Ed. R. Wehner. Berlin, Heidelberg, New York, Springer. p. 147-153.
Chytil, F. & Ong, D.E. (1979) Cellular retinol and retinoic acid-binding proteins in vitamin A action. Fed. Proc. 38: 2510-2514.
Collett, T.S. & Land, M.F. (1975) Visual control flight behaviour in the hover-fly Synetta pipiens L. J. Comp. Physiol. 99: 1-66.
DeVoe, R.D., Small, R.J.W. & Zvargulis, J.E. (1969) Spectral sensitivities of wolf spider eyes. J. Gen. Physiol. 54: 1-32.
Dietrich, W. (1909) Die facettenaugen der Dipteren. Z. wiss Zool. 92: 465-539.
Eckert, H. (1971) Die Spektralempfindlichkeit des Komplexauges von Musca domestica. Kybernetik 9: 145-156.
Fischbach, K.F. (1983) Neurogenetik am Beispiel des visuellen systems von Drosophila melanogaster. Habilitation work, Würzburg.
Foerster, T. (1951) Fluoreszenz organischen Verbindungen. Vandenhoeck and Ruprecht, Göttingen.
Franceschini, N. (1975) Sampling of the visual environment by the compound eye of the fly: fundamentals and applications. In: Photoreceptors Optics. Ed. A.W. Snyder & R. Menzel. Berlin,

Heidelberg, Springer. p. 98-125.
Franceschini, N. (1977) In vivo fluorescence of the rhabdomeres in an insect eye. Proc. Int. Union Physiol. Sci. XIII. XXVIIth Int. Congr., Paris, p. 237.
Franceschini, N. (1979) Voltage clamp by light. Invest. Ophtalmol., Suppl. may, p. 5.
Franceschini, N. (1983 a) In vivo microspectrofluorometry of visual pigments. In: The Biology of Photoreceptors, Symp. Soc. Exp. Biol. Ed. D.J. Cosens. Cambridge, Cambridge University Press. (In press).
Franceschini, N. (1983 b) Chromatic organization and sexual dimorphism of the fly retinal mosaic. In: Photoreceptors. Ed. A. Borsellino & L. Cervetto. New York, Plenum Press. (In press).
Franceschini, N. & Hardie, R. (1980) In vivo recovery of intracellularly stained cells. J. Physiol. (Lond.) 301: 59.
Franceschini, N., Hardie, R., Ribi, W. & Kirschfeld, K. (1981a) Sexual dimorphism in a photoreceptor. Nature (Lond.) 291: 241-244.
Franceschini, N. & Kirschfeld, K. (1971) Etude optique in vivo des éléments photorécepteurs dans l'oeil composé de Drosophila. Kybernetik 8: 1-13.
Franceschini, N., Kirschfeld, K. & Minke, B. (1981b) Fluorescence of photoreceptor cells observed in vivo. Science 213: 1264-1267.
Fugate, R.D. & Song, P.S. (1980) Spectroscopic characterization of lactoglobulin-retinol complex. Biochem. Biophys. Acta 625: 28-42.
Futterman, S. & Heller, J. (1972) The enhancement of fluorescence and the decreased susceptibility to enzymatic oxidation of retinol complexed with bovine serum albumine, lactoglubulin and the retinol binding protein of human plasma. J. Biol. Chem. 247: 5168-5172.
Gemperlein, R., Paul, R., Lindauer, E. & Steiner, A. (1980) UV-fine structure of the spectral sensitivity of flies visual cells revealed by FIS (Fourier Interferometric stimulation). Naturwiss. 67: 565-566.
Goldsmith, T.H. (1980) Hummingbirds see near ultraviolet light. Science 207: 786-788.
Goldsmith, T.H., Barker, R.J. & Cohen, C.F. (1964) Sensitivity of visual receptors of carotenoid-depleted flies: A vitamin A deficiency in an invertebrate. Science 146: 65-67.
Goldsmith, T.H. & Fernandez, H.R. (1968) The sensitivity of housefly photoreceptors in the mid-ultraviolet and the limits of the visible spectrum. J. Exp. Biol. 49: 669-677.
Goodheer, J.C. (1969) Energy transfer from carotenoids to chlorophyll in blue-green, red and green algae and greening bean leaves. Biochim. Biophys. Acta 172: 252-265.
Guo, A. (1980 a) Elektrophysiologische Untersuchungen zur Spektral- und Polarisationsempfindlichkeit der Sehzellen von Calliphora erythrocephala I. Sci. Sin XXIII: 1182-1196.
Guo, A. (1980 b) Elektrophysiologische Untersuchungen zur Spektral- und Polarisationsempfindlichkeit der Sehzellen von Calliphora erythrocephala II. Sci. Sin XXIII: 1461-1468.
Guo, A. (1981) Elektrophysiologische Untersuchungen zur Spektral- und Polarisation sempfindlichkeit der Sehzellen von Calliphora erythro-

cephala XXIV: 272-286.
Hall, J.C. (1982) Genetics of the nervous system in Drosophila. Quant. Rev. Biophys. 15: 223-479.
Hamdorf, K. (1979) The physiology of invertebrate visual pigments. In: Handbook of Sensory Physiology. Vol. VII/6A. Ed. H. Autrum. Berlin, Heidelberg, Springer. p. 145-224.
Hamdorf, K., Paulsen, R. & Schwemer, J. (1973) Photoregeneration and sensitivity control of photoreceptors of invertebrates. In: Biochemistry and Physiology of Visual Pigments. Ed. H. Langer. Berlin, Heidelberg, Springer. p. 155-166.
Hanström, B. (1927) Uber die Frage, ob Fuktionell versciedenen Zapfen- und Stäbchenartige Sehzellen im Komplexauge der Arthropoden vorkommen. Z. wiss. Biol., Abt. C: Physiol. 6: 566-597.
Hardie, R.C. (1979) Electrophysiological analysis of fly retina. I. Comparative properties of R1-6 and R7 and R8. J. Comp. Physiol. 129: 19-33.
Hardie, R.C. (1983) Morphology of sex-specific photoreceptors in the compound eye of the male housefly (Musca domestica). Cell Tissue Res. (In press).
Hardie, R.C., Franceschini, N. & MacIntyre, P. (1979) Electrophysiological analysis of fly retina. II. Spectral and polarization sensitivity in R7 and R8. J. Comp. Physiol. 133: 23-39.
Hardie, R.C., Franceschini, N., Ribi, W. & Kirschfeld, K. (1981) Distribution and properties of sex-specific photoreceptors in the fly Musca domestica. J. Comp. Physiol. 145: 139-152.
Hardie, R.C. & Kirschfeld, K. (1983) Ultraviolet sensitivity of fly photoreceptors R7 and R8: evidence for a sensitizing function. Biophys. Struct. Mech. 9: 171-180.
Hausen, K. (1983) The lobula-complex of the fly: structure, function and significance in visual behaviour. (This volume).
Hausen, K. & Strausfeld, N. (1980) Sexually dimorphic interneuron arrangements in the fly visual system. Proc. R. Soc. Lond. 208B: 57-71.
Heisenberg, M. (1979) Genetic approach to a visual system. In: Handbook of Sensory Physiology. Vol. VII/6a. Ed. H. Autrum. Berlin, Springer. p. 665-679.
Heisenberg, M. & Buchner, E. (1977) The role of retinula cell types in visual behavior of Drosophila melanogaster. J. Comp. Physiol. 117: 127-162.
Hemley, R., Kohler, B.E. & Siviski, P. (1979) Absorption spectra for the complexes formed from vitamin-A and lactoglobulin. Biophys. J. 28: 447-455.
Herrling, P.L. (1975) Topographische Untersuchung zur funktionellen Anatomie der Retina von Cataglyphis bicolor (Formicidae Hymenoptera). Dissertation. Universität Zürich.
Horridge, G.A. (1978) The separation of visual axes in apposition compound eyes. Phil. Trans. R. Soc. Lond. 285B: 1-59.
Horridge, G.A. & McLean, M. (1978) The dorsal eye of the mayfly Atalophlebia (Ephemeroptera). Proc. R. Soc. Lond. 200B: 137-150.
Horridge, G.A. & Meinertzhagen, I.A. (1970) The accuracy of the

patterns of connexions of the first- and second-order neurons of the visual system of Calliphora. Proc. R. Soc. Lond. 175B: 69-82.

Horridge, G.A., Mimura, K. & Tsukahara, Y. (1975) Fly photoreceptors II. Spectral and polarized light sensitivity in the drove fly Eristalis. Proc. R. Soc. Lond. 190B: 225-237.

Kirschfeld, K. (1967) Die Beziehung swischen dem Raster der Ommatidien und dem Raster der Rhabdomere im Komplexauge von Musca. Exp. Brain Res. 3: 248-270.

Kirschfeld, K. (1972) The visual system of Musca: studies on optics, structure and function. In: Information Processing in the Visual System of Arthropods. Ed. R. Wehner. Berlin, Springer. p. 61-74.

Kirschfeld, K. (1979) The function of photostable pigments in fly photoreceptors. Biophys. Struct. Mech. 4: 117-128.

Kirschfeld, K., Feiler, R. & Franceschini, N. (1978) A photostable pigment within the rhabdomere of fly photoreceptor n°7. J. Comp. Physiol. 125: 275-284.

Kirschfeld, K., Feiler, R., Hardie, R., Vogt, K. & Franceschini, N. (1983) The sensitizing pigment of fly photoreceptors: properties and candidates. Biophys. Struct. Mech. 10: (In press).

Kirschfeld, K. & Franceschini, N. (1968) Optische Eigenschaften der Ommatidien im Komplexauge von Musca. Kybernetik 5: 47-52.

Kirschfeld, K.& Franceschini, N. (1969) Ein Mechanismus zur steierung des Lichflusses in der Rhabdomeren des Komplexauges von Musca. Kybernetik 6: 13-22.

Kirschfeld, K. & Franceschini, N. (1975) Microspectrophotometry of fly rhabdomeres. Conf. on Visual Physiology, Günzburg (Germany).

Kirschfeld, K. & Franceschini,N. (1977) Photostable pigments within the membrane of photoreceptors and their possible role. Biophys. Struct. Mech. 3: 191-194.

Kirschfeld, K., Franceschini, N. & Minke, B. (1977) Evidence for a sensitizing pigment in fly photoreceptors. Nature (Lond.) 269: 386-390.

Kirschfeld, K. & Wenk, P. (1976) The dorsal compound eye of Simuliid flies: an eye specialized for the detection of small, rapidly moving objects. Z. Naturforsch. 31c: 764-765.

Labhart, T. (1980) Specialised photoreceptors at the dorsal rim of the honeybee's compound eye: polarisation and angular sensitivity. J. Comp. Physiol. 141: 19-30.

Land, M.F. & Collett, T.S. (1974) Chasing behaviour of house flies (Fannia canicularis): a description and analysis. J. Comp. Physiol. 89: 331-357.

Laughlin, S. (1983) The roles of parallel channels in early visual processing by the arthropod compound eye. (This volume).

Laughlin, S.B. & McGiness, S. (1978) The structures of dorsal and ventral regions of a dragonfly retina. Cell Tissue Res. 188: 427-447.

McCann, G.D. & Arnett, D.W. (1972) Spectral and polarization sensitivity of the dipteran visual system. J. Gen. Physiol. 59: 534-558.

McIntyre, P. & Kirschfeld, K. (1981) Absorption properties of a photostable pigment (P 456) in rhabdomere 7 of the fly. J. Comp. Physiol. 143: 3-15.

Meinertzhagen, I.A. (1983) The rules of synaptic assembly in the developing insect lamina. (This volume).
Melamed, J. & Trujillo-Cenoz, O. (1968) The fine structure of the central cells in the ommatidia of dipterans. J. Ultrastruct. Res. 21: 313-334.
Ohtsuka, T. (1978) Combination of oil droplets with different types of photoreceptor in a freshwater turtle, Geodemys reevesii. Sensory Proc. 2: 321-325.
Pak, W.L. (1979) Study of photoreceptor function using Drosophila mutants. In: Neurogenetics. Genetic Approaches to the Nervous System. Ed. X. Breakefield. North Holland, New York, Elsevier. p. 67-99.
Praagh, J.P. van, Ribi, W., Wehrhahn, C. & Wittmann, D. (1980) Drone bees fixate the queen with the dorsal frontal part of their compound eyes. J. Comp. Physiol. 136: 263-266.
Riehle, A. & Franceschini, N. (1983) Movement detection in flies: lateral gain-control over ON-OFF pathways. (In prep.).
Schinz, R.H. (1975) Structural specialisation in the dorsal retina of the bee Apis mellifera. Cell Tissue Res. 162: 23-34.
Scholes, J. (1969) The electrical responses of the retinal receptors and the lamina in the visual system of the fly Musca. Kybernetik 6: 149-162.
Schreckenbach, T., Walckhoff, B. & Oesterhelt, D. (1977) Studies on the retinal-protein interaction in bacterio-rhodopsin. Eur. J. Biochem. 76: 499-511.
Schultze, M. (1866) Zur Anatomie und Physiologie der Retina. Arch. Mikrosk. Anatomie 2: 175-286.
Schwind, R. (1983) Zonation of the optical environment and zonation in rhabdom structure within the eye of the backswimmer Notorecta Glauca. Cell Tissue Res. (In press).
Shaw, S.R. (1981) Anatomy and physiology of identified non-spiking cells in the photoreceptor-lamina complex of the compound eye of insects, especially Diptera. In: Neurones Without Impulses. Ed. A. Roberts & M.B. Bush. Cambridge, New York, Cambridge Univ. Press.
Smola, U. & Meffert, P. (1979) The spectral sensitivity of the visual cells R7 and R8 in the eye of the blowfly Calliphora erythrocephala. J. Comp. Physiol. 133: 41-52.
Stark, W.S., Ivanyshyn, A.M. & Greenberg, R.M. (1977) Sensitivity and photopigments of R1-6, a two-peaked photoreceptor, in Drosophila, Calliphora and Musca. J. Comp. Physiol. 121: 289-305.
Stark, W.S. & Tan, K.E. (1982) Ultraviolet light: photosensitivity and other effects on the visual system. Photochem. Photobiol. 36: 371-380.
Stavenga, D., Franceschini, N. & Kirschfeld, K. (1983) Fluorescence of visual pigments studied in the eye of intact flies. (Submitted).
Stavenga, D. & Schwemer, J. (1983) Visual pigments of invertebrates. (This volume).
Stavenga, D.G., Zantema, A. & Kuiper, J. (1973) Rhodopsin processes and the function of the pupil mechanism in flies. In: Biochemistry and

Physiology of Visual Pigments. Ed. H. Langer. Berlin, Heidelberg, Springer. p. 175-180.

Strausfeld, N. (1976) Mosaic organizations, layers, and pathways in the insect brain. In: Neural Principles in Vis on. Ed. F. Zettler & R. Weiler. Berlin, Heidelberg, New York, Springer.

Strausfeld, N. (1983) Functional neuroanatomy of the blowfly's visual system. (This volume).

Trujillo-Cenoz, O. & Melamed, J. (1966) Electron microscopial observetions on the peripheral and intermediate retina of dipterans. In: The Functional Organization of the Compound Eye. Ed. C.G. Bernhard. London, Pergamon. p. 339-361.

Vogt, K. & Kirschfeld, K. (1983) Sensitising pigment in the fly. Biophys. Struct. Mech. 9: 319-328.

Wada, S. (1974) Spezielle randzonale Ommatidien der Fliegen (Diptera Brachycera): Architektur und Verteilung in den Komplexaugen. Z. Morphol. Tiere 77: 87-125.

Walls, G.L. & Judd, H D. (1933) The intra-ocular colour-filters of vertebrates. Br. J. Ophtalmol., Nov. 1933: 641-675.

Wehner, R. (1981) Spatial vision in arthropods. In: Handbook of Sensory Physiol. Vol. VII/6C. Ed. H. Autrum. Berlin, Springer. p. 287-616.

Wehrhahn, C. (1979) Sex-specific differences in the chasing behaviour of free flying houseflies (Musca). Biol. Cybern. 32: 239-241.

Wunderer, H. & Smola, U. (1982 a) Morphological differentiation of the central visual cells R7/8 in various regions of the blowfly eye. Tissue Cell 14: 341-358.

Wunderer, H. & Smola, U. (1982 b) Fine structure of ommatidia at the dorsal eye margin of Callyphora erythrocephala Meigen: an eye region specialized for the detection of polarized light. Int. J. Insect Morphol. Embryol. 11: 25-38.

THE ROLES OF PARALLEL CHANNELS IN EARLY VISUAL PROCESSING BY THE ARTHROPOD COMPOUND EYE

SIMON LAUGHLIN

Department of Neurobiology
Australian National University
Canberra City, A.C.T. 2601
Australia

ABSTRACT

Both the lamina and the medulla are arrays of parallel neural subunits, each primarily receiving an input from a single retinal sampling element. Every subunit contains several parallel channels capable of monitoring different aspects of the incoming signal while the medulla contains abundant lateral pathways for local and global processing interactions between units. Parallel channels in lamina subunits segregate different components of the receptor input, such as wavelength or polarisation sensitivity. In addition one or more channels sums the majority of receptor inputs to provide a highly sensitive contrast coding pathway, typified by cells L1 and L2 in the fly. The sensitivity of the contrast channel is matched to the quality of incoming signals so as to improve the efficiency with which they transmit information to the medulla. Non-linear synaptic amplification ensures that all levels of graded response are fully utilised, while intensity dependant lateral and temporal inhibition reduces the range of signal amplitudes to be coded. Two further lamina channels have been recorded from, but these have not been identified or characterised to the same degree. The medulla is far more complicated than the lamina and its cells are smaller and less accessible to recording. With the exception of movement detection, these technical difficulties are compounded by a lack of hypotheses about the procedures required at this level of processing. However the medulla's ordered anatomy raises hopes that its exploration may one day bridge the gaps between anatomy and function, and lamina and lobula complex.

INTRODUCTION

Vision, involving as it does the simultaneous manipulation of intensity measurements made at a large number of points, is a parallel process, performed by arrays of receptors and interneurons. In many vertebrate visual systems it is proving difficult to define these neuronal arrays and hence to place responses within the context of parallel channels. For example, the neuronal complement and connectivity of cortical columns is poorly described while the number of types of bipolar cells in any one retina is still unresolved. Although cat retinal ganglion cells have recently been seen to form a regular array of parallel channels for sampling the retinal image (Wässle et al., 1981; Hughes, 1980), this anatomical order apparently dissolves among the wide branchings of their central terminals (Bowling & Michael, 1980).

By comparison, the visual systems of insects and crustacea are generally dominated by the clear delineation of parallel channels. The retina is a geometrical lattice of ommatidia (rev. Stavenga, 1980) and its order is imposed upon four successive layers of neurons (rev. Strausfeld & Nässle, 1981; Strausfeld, 1983) by orderly axonal projections. This retention of a parallel unit structure facilitates the anatomical and physiological characterisation of neural processing. Each neuron can be placed in its correct spatial context, so enabling one to reconstruct the neural images formed by the arrays of parallel channels at each level. Taken together with the well defined receptor optics and visual behaviour, this delineation of parallel channels is one of the reasons why compound eyes are a useful "model for the uptake, transduction and processing of optical data in the nervous system" (Reichardt, 1970).

The following brief account of neural processing by the parallel channels of the lamina and medulla concentrates primarily on the lamina, partly because it has been my field of study, and partly because it has proved much easier to investigate. The lamina provides an important lesson in early visual processing. One commonly thinks of transduction being followed by neural processes directed towards some form of feature extraction. The lamina tells us that before information can be processed for features, it must be properly coded into parallel channels to provide a robust signal that is not easily lost amid the hurly-burly of higher centres. This packaging of information (Barlow, 1961) requires some interactions between parallel channels but is fairly simple. The more complex and challenging tasks of making sense of the world probably commence in the medulla and are largely unresolved.

THE PHOTORECEPTORS' NEURAL IMAGE

The characteristics of the photoreceptors' responses to natural stimuli have a profound effect upon neural processing in the lamina. The eye's optics maps the spatial distribution of intensity onto the orderly array of receptors. Each receptor transduces intensity to amplitude of graded depolarisation (Fig. 1), with a waveform that is remarkably similar in many arthropods, and a characteristic spectral and polarisation sensitivity. A large body of evidence (rev. Laughlin, 1981a) shows that, in

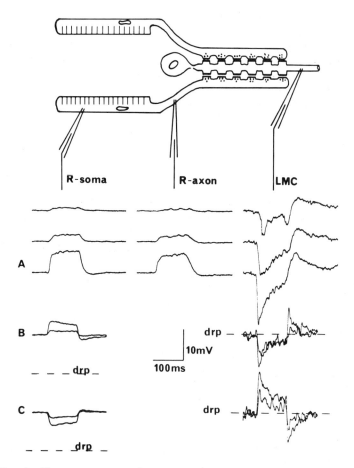

Fig. 1: The responses of a receptor soma, receptor axon, and a large monopolar cell to identical stimuli. A) Three responses each to 100 msec flashes delivered to the dark adapted retina (intensity increases from top to bottom). Note the **on** and **off** transients in the LMC, and their absence in the receptor response. B) Responses to 100 msec pulses of light superimposed on a strong steady background, producing contrast steps of 0,2 and 1,0: **drp** designates the level of the cells' resting potential to total darkness. Note the sustained depolarisation of the receptors. C) As in B but the pulses are dimming, with contrasts of -0,2 and -0,5.

addition to spectral and polarisation sensitivity, the receptors' neural image of the spatial and temporal distribution of light intensity exhibits the following important properties.

1. The neural image is blurred before being coded in the discrete receptor channels. In space this blurring results from the acceptance

of light from a finite solid angle, while in time it is caused by the finite duration of the impulse response (the response to a very brief pulse of light (rev. Laughlin, 1981a). Blurring reduces the amplitudes of responses to small or rapidly moving objects; the smaller or faster the object, the worse the attenuation.

2. The photoreceptor signal is noisy. Photon noise, resulting from the random arrival of light quanta, and transducer noise, generated during transduction, make the receptor responses unreliable. At low intensities the contributions of photon and transducer noise are equal, but at high intensities transducer noise predominates (Lillywhite & Laughlin, 1979; Howard & Snyder, 1983). As also expected of photon noise alone, the signal to noise ratio of receptor signals rises with intensity but it reaches an asymptote set by transduction (Howard & Snyder, 1983). Noise and blurring combine to give an upper limit to the information available from the receptors' neural image. Noise sets the amplitude of the smallest detectable signal and blurring determines the finest details capable of generating detectable responses.

3. The absolute amplitude of the receptor signal is low. In addition to blurring, two factors depress the amplitude of the receptor's neural image:- non-linear response summation and the restricted contrast range of natural signals. The relationship between intensity and receptor membrane depolarisation is non-linear, its slope decreasing with intensity (Fig. 2), and this non-linearity principally results from the self-shunting of parallel conductance channels (Martin, 1955). Self-shunting forms the basis of the retinal gain control (Glantz, 1972), reducing the incremental sensitivity of photoreceptors as intensity rises and, in collaboration with light adaptation mechanisms (Fig. 2), self-shunting enables the eye to operate with an appropriate sensitivity over a wide intensity range (rev. Laughlin, 1981a). By using non-linear summation for gain control the eye pays the penalty of small voltage signals. To demonstrate this point we must consider the properties of naturally generated intensity patterns.

The majority of natural objects are reflecting and absorbing, consequently the intensities of the optical signals they produce are proportional to the intensity of illumination. Although absolute intensities vary daily over many orders of magnitude, the relative intensities of one object compared to another do not: e.g. print has one-eighth the intensity of paper, no matter what the illumination. Hence contrast, a measure of relative intensity, is appropriate both for measuring intensity distributions in natural scenes, and for encoding scenes in neurons. Using a scanner with the same angular and spectral sensitivity as a fly's eye I have found that the mean contrast, averaged from several habitats devoid of human artefacts, is 0,4 and 3/4 of the points sampled have contrasts of less than 0,5 (Laughlin, 1981b).

By combining contrast measurements with measurements of the sensitivity of fly photoreceptors to modulations in contrast, one derives the mean amplitude of the signal generated in a fly's photoreceptors as

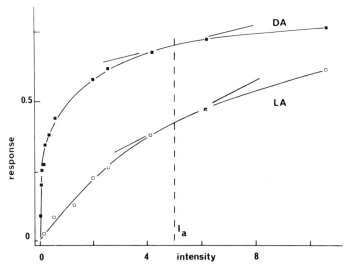

Fig. 2: Linear plots of intensity-response functions for a fly photoreceptor R1-6 under dark adapted conditions (DA), and when light adapted (LA) by a sustained stimulus of intensity (I_a). Response amplitude is normalised to a proportion of a maximum response from the dark adapted cell. Note that the slope of the LA curve about the background intensity (I_a) is greater than DA's. Plotted from the data of Matic and Laughlin (1981).

they monitor a natural scene (Fig. 3). This signal is only $\pm$ 8% of the maximum saturated response and, as the best cells generate responses of up to 70 mV, the voltage fluctuation is about 6 mV. The range of contrasts containing over three quarters of all signals lies within an envelope of 15 mV. Because these figures do not take into account temporal blurring they may slightly overestimate the values occurring in life. For receptors other than fly's R1-6, differences in spectral and angular sensitivities will further modify the range of amplitudes but the signals will still be small because the receptor's dynamic range encompasses a large span of intensities.

In summary, the visual system must retrieve information from a flickering mosaic of undulating receptor potentials. These undulations average $\pm$ 6 mV, rarely exceed 15 mV, and ride up and down upon a mean depolarisation, the DC-bias, which depends upon intensity (Fig. 3). To handle relatively small signals that are often superimposed upon a larger bias, requires a set of neuronal interactions which are, at present, only described in visual systems. These are typified by one of the sets of parallel channels running through the insect lamina, the large monopolar cells.

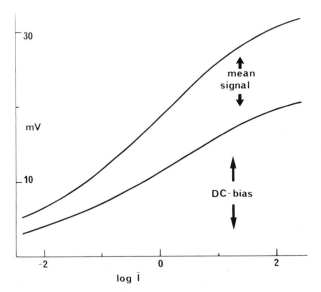

Fig. 3: The mean signal amplitude in a fly photoreceptor produced by the average environmental contrast (0,4), as a function of log background intensity, log I. For most of the operating range a relatively small signal is superimposed on a larger standing potential, the DC-bias. Log I = 0 corresponds to the intensity producing a half-maximal 35 mV response in the dark adapted cell.

THE ANATOMY OF PARALLEL PATHWAYS THROUGH THE LAMINA

The neuroanatomy of the first optic neuropile, the lamina, has been extensively studied in a variety of arthropods. In the fly, and to a lesser extent the bee, crayfish, dragonfly and Daphnia, wiring diagrams of unparallel completeness have resulted. As these studies have been extensively reviewed (Strausfeld & Nässle, 1981; Shaw, 1981; Laughlin, 1981a; Meinertzhagen & Armett-Kibel, 1982), we need only consider the basic points which establish the fundamentals of parallel processing in the lamina.

A. Processing is by Parallel Spatial Units

The lamina is a mosaic of cartridges arranged in register with the ommatidia (e.g. Strausfeld, 1983). An orderly projection of receptor axons maps the receptors' neural image onto the lamina so that each cartridge receives direct inputs from receptors sharing the same field of view. Thus each lamina unit is primarily concerned with processing the information from one spatial point and this is reflected in the narrow receptive fields of several lamina neurons (Järvilehto & Zettler, 1973).

B. Information is Directed Centrally

The majority of synapses are from receptors to the second order interneurons that connect lamina and medulla cartridges. Feedback pathways, lateral projections and intrinsic lamina circuits are, in terms of the number of synapses and neurons involved, much weaker.

C. Information is Segregated into Parallel Neurons

The eight or nine photoreceptors sharing the same field of view are potentially able to code for colour and polarisation at a single point. The projections to the lamina and medulla are able to segregate these different qualities in each cartridge although the precise pattern of segregation and the relevance of the resulting parallel channels has not been established. In general, two or three photoreceptor axons are long visual fibres, running through the lamina to the corresponding medullary cartridge and these pathways have been associated with sensitivity to polarised light and short wavelengths (Menzel & Snyder, 1974; Menzel & Blakers, 1976; Wehner, 1976; Hardie, 1979). Within each cartridge the different types of monopolar cells often receive inputs from sub-sets of photoreceptors. In the crayfish, the receptor projection to monopolars probably segregates the two receptor types with orthogonal sensitivities to polarised light. (Nässle & Waterman, 1977). In a lamina cartridge of the worker bee, the monopolar cells L1-3 receive a set of distinctive receptor inputs from a single ommatidium (Ribi, 1981). L1 sums all receptor inputs, L2 sums all receptors that terminate in the lamina, and also feeds back onto two of these receptors, while L3 just connects to one type of visual fibre (Fig. 4). The long visual fibres are probably UV receptors; consequently L1 sums blue, green and UV inputs from each ommatidium, L2 sums blue and green and may, by virtue of its feedback onto one class of axons show some opponency, while L3 receives signals from just one receptor class, probably green (Ribi, 1981). Although we still await a definitive correlation of receptor spectral sensitivity with receptor fibre type in Golgi (c.f. Menzel & Blakers, 1976) this anatomical study suggests that the bee lamina segregates parallel colour channels. Finally, in the fly Musca the pattern of segregation shows a sexual dimorphism. In the foveal lamina of males one of the long visual fibres fails to project to the medulla and terminates in the lamina. This shorter cell is modified to have the same sensitivity and probably the same correlations as the other short visual fibres, R1-6. A parallel channel has been traded for an extra input to a summation system, perhaps to enhance sensitivity while tracking females (Hardie et al., 1981; Franceschini, 1983).

D. Pathways Converge in the Lamina

In the crayfish, fly, bee and dragonfly at least one of the large monopolar cells receives inputs from all short visual fibres and sometimes from the long one too, e.g. L1 and L2 in bee, L1, L2, L3 and T1 in fly, MI and MII in dragonfly (Ribi, 1981; Strausfeld & Nässel, 1981; Meinertzhagen & Armett-Kibel, 1982). This summation enhances the signal to noise ratio in monopolars by averaging out uncorrelated noise in the individual receptors. In bee and dragonfly summation sacrifices spectral sensitivity.

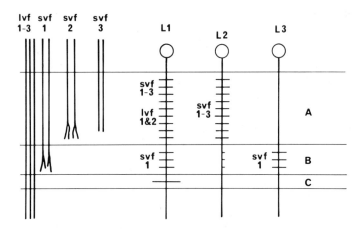

Fig. 4: The segregation of the receptor axon inputs from one ommatidium to three monopolar cells L1, L2 and L3, in the worker bee lamina (after Ribi, 1981). **lvf** 1-3 are the three long visual fibres (receptor axons) passing through the three layers of the lamina. A, B, and C, to the medulla. **svf** 1, 2 and 3 are the three pairs of short visual fibres terminating in the lamina. The presynaptic contacts from receptor axons to L1-3 are shown alongside each fibre type. Further discussion is given in the text.

Because the receptors projecting to a cartridge have the same field of view, this type of summation does not lead to a broader angular sensitivity so that the resulting channel is ideally suited for accurately registering the spatial distribution of contrast.

In muscoid flies the 8 receptors projecting to the same cartridge have the same field of view but lie in 7 different ommatidia. Summation in the lamina achieves a neural superposition of retinal images and since every cartridge receives inputs from receptors lying behind seven separate lenses, the signal is derived from about seven times more light. Thus the signal to noise ratio is inevitably better than in the equivalent apposition eye (Kirschfeld, 1973; Franceshini, 1983).

In addition to summation within one lamina cartridge, there is evidence for summation between cartridges. In bee the monopolars L2 and L4 extend dendrites to neighbouring cartridges and these may be involved with summation at low intensities and/or lateral inhibition (Ribi, 1981). In the laminas of nocturnal insects the projections of lateral dendrites into other cartridges are very prominent, leading to an apparent breakdown in cartridge structure, and again, this may be associated with the necessity to pool signals from many receptors so as to obtain a satisfactory signal:noise ratio at low intensities (rev. Laughlin, 1981a). In fly there is electrophysiological evidence that, at low intensities, many of the single photon signals seen in L1 and L2 originate from receptors outside the normal neural superposition projection, and again this wider field could be advantages for vision at low levels (Dubs et al., 1981; Dubs, 1982).

THE FUNCTION OF PARALLEL CHANNELS IN THE LAMINA

Of all the parallel channels running through the lamina, only three have been characterised physiologically. These are the large monopolar cells of fly, dragonfly, and bee, referred to as LMC's and the 'on-off' and 'sustaining' fibres of the fly lamina (Arnett, 1972). The combination of intense anatomical and physiological study has rendered the LMC's, and in particular the monopolar cells L1 and L2 of muscoid flies, one of the best understood set of second order interneurons in any sensory system. These cells provide an example of parallel spatial channels, tailored to fit the properties of the signal they carry in order to enhance neuronal information capacity, and so increase the detail they transmit centrally. This better understood pathway will be reviewed first.

A. The Function of LMC's

As suggested by their anatomy, the LMC's have a broad spectral sensitivity, resulting from the summation of receptor inputs, and are devoid of spectral opponency (Zettler & Autrum. 1975; Laughlin, 1976). Here we will describe briefly the transformation of the signal as it passes from receptor to LMC, and discuss the functional implications of this transformation in terms of contrast coding. The discussion of the cellular mechanisms responsible for these transformations is minimal, and for details of this aspect readers are referred to three recent reviews (Shaw, 1979, 1981; Laughlin, 1981a).

B. Signal Transformation

Several techniques have been applied to the description of signal transformation at the level of the synapse between photoreceptors and LMC's. These have compared receptor signals in the retina to LMC responses in the lamina or first chiasm, using a number of stimuli, including pulses of light delivered in darkness (Shaw, 1968; Zettler & Järvilehto, 1971; Järvilehto & Zettler, 1971; Laughlin, 1973), sudden dimming or brightening of sustained illumination at a number of background intensities (Laughlin & Hardie, 1978; Laughlin, 1981b), and sinusoidal (Järvilehto & Zettler, 1971) or white noise (French & Järvilehto, 1978) modulation of light. These methods have described the following set of transformations.

1. The signal is inverted. Whereas receptors respond to light with a graded depolarisation, increasing in amplitude with intensity, the corresponding LMC signal is a graded, intensity dependent hyperpolarisation and is associated with a resistance decrease, suggestive of the synaptic activation of potassium channels. (Shaw, 1968; Laughlin, 1974c).
2. The responses are transient, even under conditions where the receptor input is sustained, (Fig. 1). During sustained constant illumination the LMC membrane potential returns to the dark level within 0,5 to 1,0 sec, (Laughlin & Hardie, 1978). Clearly there is a powerful antagonism of the light induced signal, generated in the

lamina. Regardless of whether the antagonism is opposing stimulus dimming (i.e. inhibition hyperpolarises the LMC) or brightening (it depolarises the cell) we shall refer to this antagonism as inhibition because we are considering function, not mechanism. Single facet stimulation shows that much of this inhibition is generated within a single cartridge (Shaw, 1979; Laughlin, unpublished observations), but a component comes from other cartridges (Zettler & Järvilehto, 1972; Laughlin, 1974b; Mimura, 1976). This lateral inhibition is relatively weak, is not recurrent, and has a slower time constant (Shaw, 1981; Dubs, 1982). The available evidence suggests that inhibition acts pre-synaptically on the photoreceptor terminals and subtracts away the steady voltage representing the mean intensity - the DC-bias in Fig. 3. The extracellular lamina field potential is a prime candidate for performing this subtraction (Shaw, 1975, 1979, 1981; Dubs, 1982; Laughlin, 1974a, 1981a; Laughlin & Hardie, 1978) but its contribution has not been determined quantitatively. The rapid temporal transients and oscillations seen at the onset of bright stimuli suggest additional fast mechanisms, perhaps mediated by the feed-back synapses onto receptor axons by synapses onto L1 and L2, and there is evidence for a fast feed-back onto the receptors (Shaw, 1981). Finally it should be emphasised that inhibition is intensity dependent (Figs. 1 & 7). In the fly LMC's both lateral and temporal inhibition are weak when the retina is fully dark adapted and inhibition is not manifest by the responses to single photons (Dubs et al., 1981) but it increases in strength with light adaptation (Srinivasan et al., 1982; Dubs, 1982).

3. The signal is amplified. The transient LMC response is always much larger in peak amplitude than the change in receptor potential that drove it (Fig. 1). Estimates of the amplification vary from species to species and with the method and conditions of measurement. These latter complications result from the fact that the LMC behaves as a frequency dependent amplifier (Zettler & Järvilehto, 1971), principally through the action of inhibition. Thus the measured gain depends upon the temporal characteristics of the stimulus and the time of measurement. My own recent measurements in Calliphora, comparing the rates of rise of receptor and LMC responses, or the amplitudes of responses to step stimuli at the time of the peak LMC response, show that the gain is between 5,7 - 8,4. This range of values was derived over a five log unit span of background intensities, and the differences depended more upon the method of evaluation and the time course of the response, than they did upon adaptation state. These measurements confirm that the gain is largely unaltered by light adaptation - a finding that speaks strongly in favour of subtractive inhibition (Laughlin & Hardie, 1978).

C. A Simple Overview of LMC Contrast Coding

The combination of inhibition and amplification enables the LMC's to code the neural image from the receptors in a manner that preserves the fine detail in the contrast signal. We have already seen that almost all contrast information is contained within a narrow band of receptor responses that rides up and down with the intensity dependent DC-bias (Fig. 3). Subtractive inhibition removes the bias, and amplification expands the contrast signal (and the accompanying noise) to fill the dynamic range of the LMC's graded response. However much the mean intensity varies, the strength of inhibition is adjusted to keep the LMC response centred upon the band of contrast signals (Fig. 5). Thus the LMC is like a magnifying glass, expanding a small window of receptor voltage response, and kept centred on the receptor contrast signal by inhibition.

If the brain were a perfect device there would be little point in magnifying the receptor response. The tiniest receptor potential fluctua-

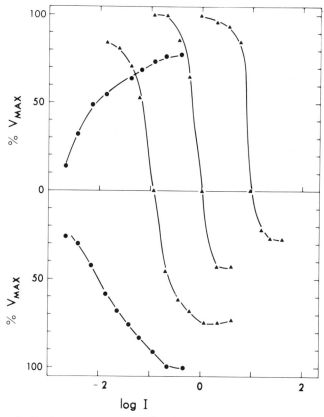

Fig. 5: Peak response amplitudes as a function of intensity for four different adaptation states of a single fly LMC. For the left hand curve the cell is dark adapted, while for the remainder it is light adapted at background intensities corresponding to the curves' intersections with the zero response level. Note that the LMC responds to a small band of intensities on either side of the background (after Laughlin & Hardie, 1978).

tions would be discriminable at a higher level and the amount of incoming information would be entirely limited by the noise in the receptor responses. But neurons are far from perfect. The random opening of conductance channels and their high resistances generate a noisy environment into which signals are launched by the essentially random processes of transmitter vesicle release and channel activation. This intrinsic noise must inevitably set a lower limit to the amplitude of the smallest signal resolved by a neural network. Amplification of the incoming signal helps minimise the limitation imposed by this later intrinsic noise and, as noise is generated at all stages of transmission, it is best to amplify as early as possible in sensory processing. Because the receptors' non-linear summation is being used as a gain control, one cannot fully amplify the signal during transduction and the visual system's first synapse is used instead. In the next section we will discuss how the processes of amplification and inhibition are carefully matched to the statistical properties of the incoming receptor signal, so as to enhance coding efficiency.

D. Coding Efficiency in a Single Channel

The efficiency with which LMC's encode contrast can be appreciated by attempting to answer two simple operational questions. Precisely what gain should the signal be amplified by, and how should one compute the DC-bias in order that it may be subtracted away?

1. Matched coding - an amplification strategy

The problem of choosing the correct gain for a neuron is similar to the problem of selecting the gain for displaying a signal on a oscilloscope. Both devices have a limited range over which to display or code the signal, both set a lower limit to the amplitude of coded or displayed signal that can be resolved, and in both cases one selects a gain which depends upon one's expectation of signal amplitude. If the gain is too low, the signal will rarely fill most of the response range and it would be advantageous to expand the signal so as to resolve it better. Conversely too high a gain leads to frequent clipping of the signal when the response saturates or disappears off the screen. The best gain is a compromise between the under-utilisation of the full response range and signal saturation. Unlike most oscilloscopes, neurons can fully utilise the entire range of response and still avoid saturation by using a gain that varies as a function of the amplitude of incoming signal, and is proportional to the expectation of encountering the signal (Laughlin & Hardie, 1978).

The argument is relatively straightforward. Consider (Fig. 6) a signal whose amplitude is normally distributed about the mean (i.e. signals set up by contrast fluctuations about the background). If the gain was uniform across the response range and was chosen to accommodate the entire signal range, the responses would cluster near the mean and the outlying response levels would be under-utilised. Using a high gain for responses close to the mean, the commoner events are better resolved and, by reducing the gain when the signal approaches the limits of its excursion the possibility of saturation is avoided but the rarer, larger signals are monitored with a lower sensitivity. The optimum trade-off between the

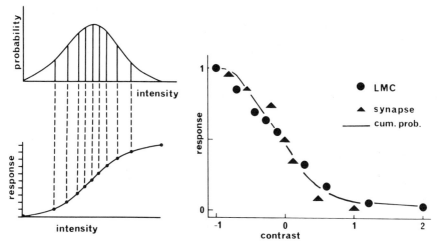

Fig. 6: The strategy for the matched amplification of a normally distributed signal (LHS) and its implementation in the fly lamina by a chemical synapse (RHS). Coding is most efficient when all neuronal response states are used equally often and to achieve this requires an intensity response curve corresponding to the cumulative probability function of the input (LHS). Comparison between the cumulative probability distribution for contrast in natural scenes and the response of fly LMC to contrast (RHS) shows that coding is matched. The sigmoidal response function is typical of a chemical synapse, as illustrated by normalising the input - output function for a chemical synapse (Blight & Llinas, 1980) to an input excursion of -1 to +1 and an output excursion of 0 to 1. Data from Laughlin (1981b) and unpublished results.

better resolution of common events and the worse resolution of the rare is attained when the gain is adjusted so that all response levels are used equally often. This condition also satisfies the dictates of information theory. Shannon demonstrated that code symbols are used most effectively when all symbols occur with equal frequency (Shannon & Weaver, 1949). As each discriminable response level in a neuron's output is equivalent to a transmitted symbol, the neuron's information capacity can be maximised when all output levels occur equally often, i.e. each output level does its fair share of the work.

A way of matching a neuron's input-output function to the statistical distribution of outputs so as to achieve an equal utilisation of all response is shown in Fig. 6. Equal voltage intervals of response encompass equal areas under the probability distribution of inputs, and the resulting coding function corresponds to the cumulative probability function for the inputs. Note that on the cumulative probability function, equal increments along the abscissa are equal probabilities.

In light adapted Calliphora LMC's the contrast-response function follows the cumulative probability distribution for contrast almost exactly (Laughlin, 1981b). This optimal <u>matched amplification</u> is achieved by the chemical synapses connecting receptors to LMC's (Fig. 6). A chemical synapse is pre-adapted to code normally distributed potentials optimally because its voltage input-output function has the necessary sigmoidal form (Fig. 6). However, for the synaptic transfer function to match the cumulative probability function both must have the same slopes in their mid-regions.

In common with other chemical synapses, the release of transmitter from receptor to LMC is apparently an exponential function of pre-synaptic receptor voltage (Shaw, 1979). In this case the slope of the synaptic transfer function in its mid-region depends on the sensitivity of the exponential transmitter release mechanism to pre-synaptic voltage. A number of other factors, including the post-synaptic cell's input resistance and the number and area of synapses, determine the absolute level of receptor voltages over which the transfer function operates (Shaw, 1981). The number and area of synapses is well regulated (Nicol & Meinertzhagen, 1982; Meinertzhagen, 1983), presumably to ensure that the LMC's operate over the appropriate receptor input level. The possibility that a part of the shift in the L-cell operating range during light adaptation is brought about by a decrease in the number of effective synapses, a depletion of the available transmitter release pool or a change in post-synaptic load resistance, has not been tested experimentally.

In summary, the contrast coding characteristics of light adapted LMC's are matched to the expected levels of signal but it is not known if this matching is hard-wired or is amenable to modification on the basis of the previous history of stimulation. Matched amplification obviously has wider implications for neuronal interactions in both sensory and motor systems because it is the informational equivalent of impedance matching. It has been independently suggested that the human psychophysical contrast transfer is matched to natural intensity signals in an identical manner, by exploiting the non-linear properties of neurons (Richards, 1982). Note that matched amplification is simply another method of adhering to a basic principle of sensory physiology, a principle which is exhibited by the wavelength maxima of fish visual pigments (Lythgoe, 1979) or the spatial distribution of retinal ganglion cells (Hughes, 1977) - one puts sensitivity where one expects the signal to be.

2. Computing the DC-bias, predictive coding

To determine the DC-bias one must average the receptor response over an area of space and/or an interval of time. Many scenes contain significant areas at different mean intensities and this produces a serious operational problem. Consider a scene divided into an area of light and an area of shade. Movement of the eye across this scene will set up corresponding temporal variations in mean intensity. If one computes the global mean over the entire scene and uses this as the estimate of DC-bias, one will subtract away a value that is too small for the light part and too large for the shade. Obviously one takes a local measure (Laughlin & Hardie, 1978), but if the area or interval is too restricted, one begins to subtract away the local variations in contrast that one wishes to encode.

We have recently shown that the fly LMC's use a simple procedure, predictive coding, to achieve the appropriate compromise (Srinivasan et al., 1982). Predictive coding exploits the fact that the receptors' neural image contains spatial and temporal correlation. Spatial correlation arises because nearby points are more likely to have the same intensity than distant ones, and temporal correlations are introduced by the finite duration of the receptor response. From these correlations one can predict the signal one expects at a given point in space or time. By subtracting away this prediction, one reduces the amplitude of the signal to be encoded. By definition these predictable components told us nothing new about the image, so that no information is lost, signal amplitude has been reduced, and the signal preserved. This predictive coding, which was first developed for the compression of video signals (Oliver, 1952), is similar to subtractive retinal inhibition. The lateral inhibitory field removes a weighted mean from surrounding areas while temporal inhibition removes a component based upon the previous history of stimulation. By exploiting the intensity dependence of the predictive coding strategy we were able to show that the lateral inhibitory fields of LMC's and the temporal inhibitory transients in their responses had the properties necessary for predictive coding (Fig. 7). At high intensities inhibition is derived from nearby points in space or confined to narrow time intervals because neighbouring values are always the most closely related. As intensity falls, photon and intrinsic noise in the receptors decorrelates the inputs and one must increase the span of inhibition to obtain a reliable signal (Fig. 7). Thus inhibition in the lamina removes predictable components that carry no information so that the unpredictable can be more fully amplified. It is these predictable components which form the most local estimate of the DC-bias that is still compatible with the retention of the contrast signal.

Our analysis of predictive coding and matched amplification shows that the interactions at the first synapse are carefully tailored to maximise the amount of information transmitted by each LMC. The coding procedures follow the basic rules of maximising signal to noise ratio, removing redundancy and exploiting the full response range. As previously suggested (Barlow, 1961) these types of coding principle may typify early processes. In support of this suggestion it is worth emphasising that the properties of vertebrate photoreceptor cells, although less well documented, are remarkably similar to the analogous insect monopolar cells (Laughlin & Hardie, 1978; Laughlin, 1981a; Srinivasan et al., 1982), suggesting that all sophisticated visual systems adopt common solutions to the operational problems associated with small receptor signals and intrinsic noise. This similarity demonstrates how the fly retina, with its well defined neuroanatomy, accessibility to recording techniques and potential to associate responses quantitative measures of behaviour, is rapidly superceding the Limulus lateral eye as an invertebrate model suitable for the discovery of general principles of neural processing in vision.

Nonetheless, some important questions about LMC's remain unanswered. We still do not know the uses to which LMC responses are put because none of their unique filter properties have been shown to

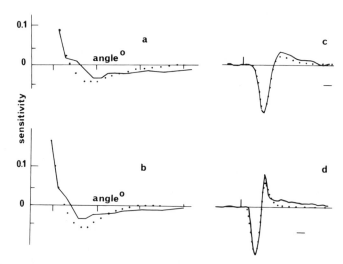

Fig. 7: The intensity dependence of inhibition in fly LMC's and the implementation of predictive coding. a & b:-portions of angular sensitivity curves at two different mean intensities. c & d:- impulse responses generated by brief intensity increments at two different mean intensities. Intensity in (b) is 0,9 log units greater than (a), and (d) is 2 log units brighter than (c). The experimental measurements (solid curves) are compared with the inhibitory interactions required for predictive coding (dotted curves). Signal to noise ratios are (a) = 0,58; (b) = 1,45; (c) = 1,1; (d) = 9,4. Data and analysis from Srinivasan et al. (1982). Timescale in (c) & (d) = 10 ms.

determine the sensitivities of higher order cells. Furthermore the two fly LMC's, L1 and L2, generate responses that are, by present critera, indistinguishable. It is perplexing that a pathway which is so obviously designed to maximise the information capacity is rendered redundant by duplication. Perhaps the fly has recuited an extra LMC to enhance the realiability of the signal in the contrast channel. My own preliminary measurements suggest that much of the noise generated in LMC's is synaptic in origin. By duplicating the number of channels the effects of such synaptic noise are reduced because the number of synapses transmitting the signal has been doubled without altering the characteristics of synaptic transmission. Because the cells L1 and L2 receive slightly different patterns of input in the lamina and terminate at different levels in the medulla, the second possibility is that L1 and L2's sensitivities are subtly tailored to their destinations. By comparison, in bee, dragonfly and crayfish, the LMC's receiving inputs from the majority of receptors are not duplicated. In these cases it has not been proved that their sentivities are matched to the expected levels of incoming signals, although, where documented, their responses and sensitivities are very similar to those of fly (Menzel, 1974; Laughlin & Hardie, 1978).

E. The function of the Remaining Parallel Channels within each Cartridge

The pathways accompanying the LMC's from lamina to medulla are, by comparison, poorly characterised, principally because the remaining cells are less accessible to recording than LMC's. Dye injections and preliminary explorations of response and receptive fields show that the remaining spiny monopolar cell of fly L3, and the basket fibre, T1, have properties very similar to L1 and L2 (Järvilehto & Zettler, 1973), and this is in agreement with the anatomical data suggesting that they have similar patterns of synaptic input (Laughlin, 1981a; Shaw, 1981). The significance of this further replication of contrast coding pathways is not known.

In fly, each lamina cartridge contains two smooth monopolar cells, L4 and L5. These smaller cells project to the medulla but are 3rd order interneurons devoid of receptor input and post-synaptic to lamina interneurons. Several lines of reasoning (Shaw, 1981; Laughlin, 1981a) suggest that L4 and L5 correspond to the sustaining and ON-OFF fibres recorded extracellularly by Arnett (1972). This supposition is supported by one dye injection of L4 (Hardie, 1978). The sustaining unit has a receptive field centre corresponding to one cartridge, in which illumination generates a sustained spike discharge. This is flanked horizontally by two inhibitory areas (Fig. 8) which antagonise the centre's response and produce responses at light off. In the dark adapted neuron the threshold for the centre response is 1,3 log units below the surround's, which suggests that inhibition is ineffective at the lowest intensities. The function of this pathway has not been established but there is evidence to

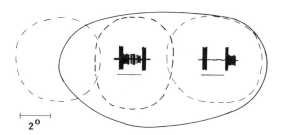

Fig. 8: The responses and receptive fields of the ON-OFF units (large spikes, solid line) and sustaining units (small spikes, dashed lines) recorded extracellularly between the lamina and medulla of the fly. ON-OFF unit's responses are generated throughout its receptive field. The sustained unit produces an ON response in the central field, and OFF responses when the stimulus is applied to the two flanking areas. Stimulus:- small spot flashed for 460 msec at position indicated by responses. Data from Arnett (1972).

support the original suggestion that it is a major input to the fly's optomotor system (McCann & Arnett, 1972). A recent analysis of the optomotor neuron H1, a large field horizontal movement detector in the lobula, suggests that the movement signal is prefiltered early in processing, before it is fed into the movement detecting circuits. Prefiltering is optimised to reject signal components that are unreliable and this involves a weak, horizontally directed lateral inhibition, similar to that exhibited by the sustaining unit (Srinivasan & Dvorak, 1980). Thus L4 may be a specialised channel carrying information that has been optimally prefiltered for elementary movement detectors in the medulla, but this attractive suggestion requires confirmation by a more searching electrophysiological analysis of L4's transfer properties, particularly in the light adapted state. Arnett's ON-OFF unit (L5?) has a receptive field with a single ON-OFF area (Fig. 8) and the cell is less sensitive than the sustaining fibre. There are no clues to the function of this channel.

NEURAL PROCESSING IN THE MEDULLA

Information has been gleaned from the medulla in three ways. The first is to study the neuroanatomy, the second is to deduce the functions of the medulla from the differences between the signal entering and leaving the medulla, and the third is to record from the neurons. The medulla is truly a frontier and only the brave have ventured into its near inpenetrable thickets. The tales brought back by these pioneers are full of promise but provide scant knowledge of commerce or currency. We will concentrate on the generalities that suggest principles, rather than reviewing a mass of detail that either generates, or obscures, hopes.

A. The Neuroanatomy of the Medulla

As in the lamina, the medulla is dominated by a columnar structure with parallel spatial channels corresponding to ommatida. Within each channel there can be up to 46 identifiable neuronal elements (Campos-Ortega & Strausfeld, 1972) so there is far more scope here for the partitioning and segregation of different stimulus qualities into separate lines. The medulla is banded by layers of laterally directed fibres and there are probably more than 70 types of these lateral neurons, each with a recognisable pattern of lateral projection (Campos-Ortega & Strausfeld, 1972). This complex pattern has the capability to draw together common features from different spatial points, correlate the responses from moving patterns or simply compare local signals to establish the signs and curvatures and directions of edges. There are easily enough degrees of freedom in the medulla to support a multitude of models for the higher level processing of information.

B. Inferences from Input and Output

These inferences are dangerous because we have a circumscribed knowledge of both the input and output channels. As seen above, several of the parallel input channels from the lamina remain undefined and our knowledge of neuronal functions central to the medulla is primarily limited to the large movement detectors and novelty detectors found in the lobula or lobula plate. The large movement detecting neurons of the fly lobula plate integrate patterns of movement over sizeable fields of the retina, to provide the optomotor system with information on retinal flow patterns required for flight stabilisation (Hausen, 1983). Despite an intensive search there is no convincing evidence for movement detection in the lamina; the optomotor system and the lobular neurons behave as if they integrate the inputs from local or elementary movement detectors, associated with pairs of facets (Kirschfeld & Lutz, 1974; Buchner, 1976, 1983). It is likely that the elementary movement detectors are associated with medullary columns. A second type of neuron integrating local information over a large area, by virtue of its extended dentritic coverage of the lobula, is the Lobula Giant Movement Detector (LGMD) of the locust. This neuron responds to a local change in intensity (a contrast modulation) but repeated responses lead to habituation, localised to the area of the retina stimulated. This site specific habituation is a direct consequence of the organisation of the medulla into a large number of parallel spatial channels because habituation probably takes place near the endings of medulla neurons in the outer lobula (O'Shea & Fraser-Rowell, 1976). In addition, the LGMD is only activated and habituated by small stimuli because stimulation of large retinal areas leads to an inhibition of the outputs of these medulla cells (Fraser-Rowell et al., 1977).

C. The Responses of Medulla Neurons

Not only is the medulla tangled, the neurons are small, so that our knowledge of responses, and in particular those of identified cells, is sparse. The available data suggests the following generalisations.

1. In the fly medulla there are neurons of the outer medulla responding to simple patterns of stimulation with graded potentials. One cell, identified as a transmedullary fibre projecting to the lobula, responds to illumination over a narrow receptive field with a depolarisation much like the photoreceptor. Other neurons hyperpolarise and the patterns of graded response can be phasic or tonic (DeVoe & Ockleford, 1976; DeVoe, 1980).
2. The majority of cells reported are spiking neurons.
3. The receptive field sizes of medulla neurons vary over a wide range. Although there are cells of the outer medulla with narrow fields, the majority of cells encompass inputs from many ommatidia. These wide fields suggest that some cells are tangential elements but the dye injection studies of Hertel (1980) indicate otherwise. Hertel proposes that arrays of neurons with wide overlapping fields could code spatial locations as combinations and permutations of responses across the array.
4. Spatial antagonism between different areas within the receptive fields of medulla neurons are by no means universal. In cricket, some neurons leaving the medulla often have concentric centre-surround RF's, reminiscent of retinal ganglion cells (Honneger, 1978, 1980), but this familiar pattern has not been encountered in fly or bee (DeVoe, 1980; Hertel, 1980; Kien & Menzel, 1977a, b).

5. Many receptive fields and sensitivity patterns are complicated. This has been illustrated by analyses of colour coding in bee medulla (Kien & Menzel, 1977a, b; Hertel, 1980). In addition to broad band neurons whose activity derives from all photoreceptor colour types, there are a wide variety of colour coding cells principally green sensitive, many of which show spectral opponency. In contrast to vertebrate retinal ganglion cells (where the 'on' centre is driven by one colour mechanism and the concentric 'off' surround by another), this opponency is not coupled to a spatial antagonism; rather the opponency is often distributed across the entire receptive field. In addition, different phases of a neuron's response have different spectral sensitivities, suggesting complex interaction between inhibitory and excitatory mechanisms with dissimilar chromatic dependencies. As well as opponent cells there are neurons with peak sensitivities falling in regions of overlap between receptor colour types (e.g. in the violet between the blue and UV receptors) and the mechanism driving this unique class is unknown. Finally space and colour can be coupled together in receptive fields to provide information of behavioural significance. Hertel (1980) describes several examples and one in particular responds to ON and OFF in the UV upper field and OFF in the green lower. The division between the fields is horizontal and equatorial which suggests that the neuron may be used as a horizon detector during flight stabilisation.
6. A small proportion of medulla neurons are multimodal, responding to stimuli such as scent or air movement, in addition to light (e.g. Hertel, 1980).
7. Assuming that all the appropriate lamina neurons are accounted for, the medulla is undoubtly the seat of motion sensitivity. The small field movement detectors have been encountered rarely and their receptive fields have not been measured with a detail sufficient to compare them with the elementary movement detectors expected from the behavioural analyses of higher order responses (DeVoe & Ockleford, 1976; Hertel, 1980). The majority of movement detecting neurons encountered so far and characterised intracellularly, are non-directional, have large fields and are extremely sensitive to rapid reversals in movement as opposed to sudden stops in movement (DeVoe, 1980; Hertel, 1980). The fine structure of their depolarising responses, and in particular the oscillations that are driven by grating movement past the eye, can provide clues as to the elementary processes underlying movement detection (DeVoe, 1980).

In summary, the medulla has the complexity expected of a sophisticated eye, capable of performing to a high level of accuracy, but there are serious methodological problems associated with analysing its function. These problems are illustrated by comparing the function of lamina and medulla in general terms. The lamina seems to be primarily concerned with packaging the receptor signals in a form that preserves most of the information contained in the necessarily small receptor signals. Packaging involves the segregation of colour or PS channels, receptor summation into a contrast channel, and the use of signal amplification and inhibition to improve coding efficiency. The constraints governing the packaging of signal and their transmission to higher levels are parameters such as input signal amplitude, channel response range and contaminating noise levels and these are familiar to users of analogue electrical equipment. Furthermore, the strict retinotopic projection of receptors to lamina interneurons preserves the retinal sampling mosaic and the receptive fields, at least of the LMC contrast channels, are narrow enough to ensure that the neural image of LMC responses looks like the object. This strict spatial correspondance is compatible with the lamina's role as an interface between receptors and interneurons and it is precisely because we know both the spatial distribution of the neural image in the lamina and the constraints upon neural coding efficiency. that we can confidently infer the lamina's function.

By comparison the forms of neural images in the medulla are unknown and there is no compelling a priori reason for neural images to look like objects. Ultimately the outputs of the visual system are motor commands and these are based upon the organisation of muscle rather than the mapping of objects in space.The relationship between 'muscle space' and visual space is still relatively unexplored but there is every reason to expect transitional forms in which information is mapped out according to an intermediate coordinate system, and this may explain why some columnar medulla neurons have large receptive fields. On the other hand, the large number of interneurons in the columns of the outer medulla suggests that a retinotopic segregation of parallel channels is taking place at each sampling point on the mosaic and this would suggest the computation of large numbers of local parameters such as direction, velocity, orientation and contrast or contrast gradient. Clearly the medulla presents three problems. We have insufficient knowledge of neuronal receptive fields to be able to reconstruct a neural image of objects and so assess the degree to which the neural image has been transformed from the object. With the exception of movement detection (Hausen, 1983; Buchner, 1983), we have little idea of how local parallel channels should analyse the incoming signals to extract the relevant information and, unlike the analogue communication systems on which we have based our assessment of the lamina, we have no satisfactory models against which to assess the roles of the neuronal responses we observe. Finally the transformations required to bridge the gap between retinal coordinate and muscular coordinate systems are obscure. Nonetheless, if we can overcome some of the technical problems associated with recording from tiny medulla neurons, this neuropile may provide significant insight into the ways in which visual systems undertake these higher order processes, just as the lamina has illustrated the principles of early signal processing, and the lobula plate the organisation of neurons closer to the output (Hausen, 1983).

REFERENCES

Arnett, D.W. (1972) Spatial and temporal integration properties of units in first optic ganglion of Dipterans. J. Neurophysiol. 35: 429-444.
Barlow, H.B. (1961) The coding of sensory messages. In: Current Problems

in Animal Behaviour. Ed. W.H. Thorpe & O.L. Zangwill. Cambridge, Cambridge University Press, p. 331-360.

Blight, A.R. & Llinas, R. (1980) The non-impulsive stretch receptor complex of the crab: a study of depolarization-release coupling at a tonic sensorimotor synapse. Phil. Trans. R. Soc. Lond. 290B: 219-276.

Bowling, D.B. & Michael, C.R. (1980) Projection patterns of single physiologically characterised optic tract fibres in the cat. Nature (Lond.) 286: 899-902.

Buchner, E. (1976) Elementary movement detectors in an insect visual system. Biol. Cybern. 24: 85-101.

Buchner, E. (1983) Behavioural analysis of spatial vision in insects (This volume).

Campos-Ortega, J.A. & Strausfeld, N.J. (1972) Columns and layers in the second synaptic region of the fly's visual system: the case for two superimposed neural architectures. In: Information Processing in the Visual Systems of Arthropods. Ed. R. Wehner. Berlin, Springer, p. 31-36.

DeVoe, R.D. (1980) Movement sensitivities of cells in the fly's medulla. J. Comp. Physiol. 138: 93-119.

DeVoe, R.D. & Ockleford, E.M. (1976) Intracellular responses from cells of the medulla of the fly, Calliphora erythrocephala. Biol. Cybern. 23: 13-24.

Dubs, A. (1982) The spatial integration of signals in the retina and lamina of the fly under different conditions of luminance. J. Comp. Physiol. 146: 321-344.

Dubs, A., Laughlin, S.B. & Srinivasan, M.V. (1981) Single photon signals in fly photoreceptors and first order interneurons at behavioural threshold. J. Physiol. 317: 317-334.

Franceschini, N. (1983) The retinal mosaic of the fly compound eye. (This volume).

Fraser-Rowell, C.H., O'Shea, M. & Williams, J.L.D. (1977) The neuronal basis of a sensory analyser, the acridid movement detector system. IV. The preference for small field stimuli. J. Exp. Biol. 68: 157-188.

French, A.S. & Järvilehto, M. (1978) The transmission of information by first and second order neurons of the fly visual system. J. Comp. Physiol. 126: 87-96.

Glantz, R.M. (1972) Visual adaptation: a case of non-linear summation. Vision Res. 12: 103-109.

Hardie, R.C. (1978) Peripheral Visual Function in the Fly. Ph. D. Thesis. Canberra, Australian National University

Hardie, R.C. (1979) Electrophysiological analysis of fly retina. I. Comparative properties of R1-6 and R7 and 8. J. Comp. Physiol. 129: 19-33.

Hardie, R.C., Franceschini, N., Ribi, W. & Kirschfeld, K. (1981) Distribution and properties of sex specific photoreceptors in the fly Musca domestica. J. Comp. Physiol. 145: 139-153.

Hausen, K. (1983) The lobula-complex of the fly: Structure, function and significance in visual behaviour (This volume).

Hertel, H. (1980) Chromatic properties of identified neurons in the optic lobe of the bee. J. Comp. Physiol. 137: 215-231.

Honneger, H.-W. (1978) Sustained and transient units in the medulla of the cricket, Gryllus campestris. J. Comp. Physiol. 125: 259-266.

Honneger, H.-W. (1980) Receptive fields of sustained medulla neurons in crickets. J. Comp. Physiol. 136: 191-210.

Howard, J. & Snyder, A.W. (1983) Transduction as a limitation on compound eye function and design. Proc. R. Soc. Lond. B. (In press).

Hughes, A. (1977) The topography of vision in mammals with contrasting life styles - comparative optics and retinal organization. In: Handbook of Sensory Physiology, Vol. VII/5. Ed. F. Crescitelli. Berlin, Springer, p. 614-756.

Hughes, A. (1980) Cat retina and sampling theorem: the relation of transient and sustained brisk-unit cut-off frequency to   and   cell density. Exp. Brain Res. 40: 250-257.

Järvilehto, M. & Zettler, F. (1971) Localised intracellular potentials from pre- and postsynaptic components in the external plexiform layer of an insect retina. Z. vergl. Physiol. 75: 422-440.

Järvilehto, M. & Zettler, F. (1973) Electrophysiological-histological studies on some functional properties of visual cells and second order neurons of an insect retina. Z. Zellforch. 136: 291-306.

Kien, J. & Menzel, R. (1977a) Chromatic properties of interneurons in the optic lobes of the bee. I. Broad band neurons. J. Comp. Physiol. 113: 17-34.

Kien, J. & Menzel, R. (1977b) Chromatic properties of interneurons in the optic lobes of the bee. II. Narrow band and colour opponent neurons. J. Comp. Physiol. 113: 35-53.

Kirschfeld, K. (1973) Das neurale Superpositionsauge. Fortschr. Zool. 21: 229-257.

Kirschfeld, K. & Lutz, B. (1974) Lateral inhibition in the compound eye of the fly, Musca. Z. Naturforsch. 29C: 95-97.

Laughlin, S.B. (1973) Neural integration in the first optic neuropile of dragonflies. I. Signal amplification in dark-adapted second order neurons. J. Comp. Physiol. 84: 335-355.

Laughlin, S.B. (1974a) Neural integration in the first optic neuropile of dragonflies. II. Receptor signal interactions in the lamina. J. Comp. Physiol. 92: 357-375.

Laughlin, S.B. (1974b) Neural integration in the first optic neuropile of dragonflies. III. The transfer of angular information. J. Comp. Physiol. 92: 377-396.

Laughlin, S.B. (1974c) Resistance changes associated with the response of insect monopolar neurons. Z. Naturforsch. 29C: 449-450.

Laughlin, S.B. (1976) Neural integration in the first optic neuropile of dragonflies. IV. Interneuron spectral sensitivity and contrast coding. J. Comp. Physiol. 112: 199-211.

Laughlin, S.B. (1981a) Neural principles in the peripheral visual systems of invertebrates. In: Handbook of Sensory Physiology, Vol. VII/6B. Ed. H. Autrum. Berlin, Springer, p. 133-280.

Laughlin, S.B. (1981b) A simple coding procedure enhances a neuron's information capacity. Z. Naturforsch. 36C: 910-912.

Laughlin, S.B. & Hardie, R.C. (1978) Common strategies for light adaptation in the peripheral visual systems of fly and dragonfly. J.

Comp. Physiol. 128: 319-340.
Lillywhite, P.G. & Laughlin, S.B. (1979) Transducer noise in a photoreceptor. Nature (Lond.) 277: 560-572.
Lythgoe, J.N. (1979) The Ecology of Vision. Oxford, Clarendon Press.
Martin, A.R. (1955) A further study on the statistical composition of the end-plate potential. J. Physiol. 130: 114-122.
Matic, T. & Laughlin, S.B. (1981) Changes in the intensity-response function of an insect's photoreceptor due to light adaptation. J. Comp. Physiol. 145: 169-177.
McCann, G.D. & Arnett, D.W. (1972) Spectral and polarization sensitivity of the Dipteran visual system. J. Gen Physiol. 59: 534-558.
Meinertzhagen, I.A. (1983) The rules of synaptic assembly in the developing insect lamina (This volume).
Meinertzhagen, I.A. & Armett-Kibel, C. (1982) The lamina monopolar cells in the optic lobe of the dragonfly Sympetrum. Phil. Trans. R. Soc. Lond. 297B: 27-49.
Menzel, R. (1974) Spectral sensitivity of monopolar cells in the bee lamina. J. Comp. Physiol. 93: 337-346.
Menzel, R. & Blakers, M. (1976) Colour receptors in the bee eye-morphology and spectral sensitivity. J. Comp. Physiol. 108: 11-33.
Menzel, R. & Snyder, A.W. (1974) Polarised light detection in the bee, Apis mellifera. J. Comp. Physiol. 88: 247-270.
Mimura, K. (1976) Some spatial properties in the first ganglion of the fly. J. Comp. Physiol. 105: 64-82.
Nässel, D.R. & Waterman, T.H. (1977) Golgi EM evidence for visual information channeling in the crayfish lamina ganglionaris. Brain Res. 130: 556-563.
Nicol, D. & Meinertzhagen, I.A. (1982) An analysis of the number and composition of the synaptic populations formed by photoreceptors of the fly. J. Comp. Neurol. 207: 29-44.
Oliver, B.M. (1952) Efficient coding. Bell System Tech. J. 31: 724-750.
O'Shea, M. & Fraser-Rowell, C.H. (1976) The neuronal basis of a sensory analyser; the acridid movement detector system. II. Response decrement, convergence, and the nature of the excitatory afferents to the fan-like dendrites of the LGMD. J. Exp. Biol. 65: 289-308.
Reichardt, W. (1970) The insect eye as a model for the uptake, transduction and processing of optical data in the nervous system. In: The Neurosciences: Second Study Programme. Ed. F.O. Schmitt. New York, Rockefeller University Press, p. 494-511.
Ribi, W.A. (1981) The first optic ganglion of the bee. IV. Synaptic fine structure and connectivity patterns of receptor cell axons and first order interneurons. Cell Tissue Res. 215: 443-464.
Richards, W.A. (1982) Lightness scale from image intensity distributions. Appl. Optics 21: 2569-2582.
Shannon, C.E. & Weaver, W. (1949) The Mathematical Theory of Communication. Urbana, University of Illinois Press.
Shaw, S.R. (1968) Organization of the locust retina. Symp. Zool. Soc. Lond. 23: 135-163
Shaw, S.R (1975) Retinal resistance barriers and electrical lateral inhibition. Nature (Lond.) 255: 480-483.

Shaw, S.R. (1979) Signal transmission by slow graded potentials in the arthropod peripherial visual system. In: The Neurosciences: Fourth Study Programme. Ed. F.O. Schmitt. Cambridge, MIT Press, p. 275-295.

Shaw, S.R. (1981) Anatomy and physiology of identified non-spiking cells in the photoreceptor-lamina complex of the compound eye of insects, especially Diptera. In: Neurones Without Impulses. Ed. A. Roberts & B.M.H. Bush. Cambridge, Cambridge University Press, p. 61-116.

Srinivasan, M.V. & Dvorak, D. (1980) Spatial processing of visual information in the movement-detecting pathway of the fly. J. Comp. Physiol. 140: 1-23.

Srinivasan, M.V., Laughlin, S.B. & Dubs, A. (1982) Predictive coding: a fresh view of inhibition in the retina. Proc. R. Soc. Lond. 216B: 427-459

Stavenga, D.G. (1980) Pseudopupils of compound eyes. In: Handbook of Sensory Physiology, Vol. VII/6A. Ed. H. Autrum. Berlin, Springer, p. 357-440.

Strausfeld, N.J. (1983) Functional neuroantomy of the blowfly's visual system (This volume).

Strausfeld, N.J. & Nässel, D.R. (1981) Neuroarchitecture serving compound eyes of crustacea and insects. In: Handbook of Sensory Physiology, Vol. VII/6B. Ed. H. Autrum. Berlin, Springer, p. 1-132.

Wässel, H., Peichl, L. & Boycott, B.B. (1981) Dendritic territories of cat retinal ganglion cells. Nature (Lond.) 292: 344-345.

Wehner, R. (1976) Structure and function of the peripheral visual pathways in Hymenopterans. In: Neural Principles in Vision. Ed. F. Zettler & R. Weiler. Berlin, Springer, p. 280-333.

Zettler, F. & Autrum, H. (1975) Chromatic properties of lateral inhibition in the eye of the fly. J. Comp. Physiol. 97: 181-188.

Zettler, F. & Järvilehto, M. (1971) Decrement-free conduction of graded potentials along the axon of a monopolar neuron. Z. vergl. Physiol. 75: 402-421.

Zettler, F. & Järvilehto, M. (1972) Lateral inhibition in an insect eye. Z. vergl. Physiol. 76: 233-244.

FUNCTIONAL NEUROANATOMY OF THE BLOWFLY'S VISUAL SYSTEM

N.J. STRAUSFELD

European Molecular Biology Laboratory
Postfach 10.2209
6900 Heidelberg
West Germany

ABSTRACT

Outputs from the optic lobes arise mainly in the lobula (Lo) and lobula plate (LP), two neuropils each with distinct architectures. Columnar neurons, comprising the relatively large and multilayered Lo, receive the majority of axons originating in the peripheral medulla (Me). In contrast, the LP is a thin tectum of neuropil characterized by large field tangential cells thought to be involved in visual stabilization of flight. In common with other areas, relays from the retina into the LP are retinotopically organized. Neuroanatomical studies show that pathways destined for the Lo and LP segregate peripherally at the level of synapses between receptors and interneurons. The final input to the LP is carried by four identical neurons (the T4-pair and T5-pair), each quartet representing a point in the visual field which overlaps six surrounding points. The terminals of the quartet define two functional layers in the LP neuropil (horizontality and verticality). T-cell endings are presynaptic to horizontal (HS) and vertical (VS) motion sensitive neurons. Their synaptology suggests that computation of direction (as opposed to motion) is performed in the lobula plate. HS and VS cells input to separate channels leading out of the brain (sets of descending neurons, the DNHS and DNVS). The DNHS receive additional inputs from the antennae whereas the DNVS receive additional inputs from the ocelli. It is proposed that the DNVS set carries information about pitch, yaw and roll whereas the DNHS set carries information about angular acceleration and apparent speed of the visual surround.

FOREWORD

There is no shortage of reviews on the general organization of the insect optic lobes (Strausfeld, 1976 a, b; Strausfeld & Nässel, 1981) and no need to provide another one here. I shall instead discuss the structural organization of certain columnar neurons that occur in the blowfly and other orders of Diptera (Musca, Drosophila (Fischbach, 1983), Eristalis (Strausfeld, 1970) and Tabanus (Cajal & Sanchez, 1915)). These cells comprise a simple and direct pathway from the eye into the brain. Possibly they are ubiquitous amongst flying insects. Certainly they allow a precise description of visual maps, functional layers and synaptic connections. They also provide examples of neurons that in part conform to the topology of models of motion detector pathways in the fly brain (Torre & Poggio, 1978). The cells to be described are: the "short" and "long" visual receptors, R1-6 and R7 & 8, respectively; the monopolar cells L1 & L2; the medulla neurons SUB, TM1 & TM5; and the medulla-to-lobula plate and lobula-to-lobula plate T-cells, T4 and T5. The latter terminate on, and are functionally connected to, giant horizontal (HS) and vertical (VS) cells (Pierantoni, 1976; Hausen, 1981; Hengstenberg et al., 1982).

Since I shall more or less ignore all other relays from the retina I should at least mention that these comprise 40 or so cell types. As most of these occur as multiples of the number of lenses I shall be ignoring at least one hundred thousand neurons not to speak of their filigree of synaptic connections. However, by the end of this chapter I hope that it will be obvious to the reader that the strategies for opening this Pandora's box already exist and that future research in this field is going to be rather exciting

GENERAL MORPHOLOGY

Receptor Organization

The blowfly retina consists of some 3500 - 4000 ommatidia. Each ommatidium is a functional unit consisting of an inert cornea, a pseudocone and its attendant Semper cells, eight visual (retinula) cells, and two primary and twelve secondary screening-pigment cells (Trujillo-Cenoz & Melamed, 1966; Boschek, 1971). Cornea and cone together comprise the dioptric apparatus which focuses light onto the caps of seven of the receptors (Franceschini, 1975).

In many insect orders the rhabdomeres beneath a lens are fused together and share a common optical axis (Ribi, 1975; Meinertzhagen, 1976; Strausfeld & Nässel, 1981): the axons from one ommatidium project coherently to a column of neurons (optic cartridge) in the first optic neuropil (the lamina). This is not so in the fly. Its unique property is that receptor cells and their rhabdomeres are separate. The cap of each rhabdomere inserts into a separate point in the base of the pseudocone and each cell has a unique optical axis. A point source focussed on one receptor falls outside the field of view of the other six (Franceschini, 1975). In other words, each ommatidium is wall-eyed.

The cell bodies of six of the receptors extend from the Semper cells down to the basement membrane. These are called the short receptors (R1-R6) the axons of which penetrate the basement membrane and pass through several layers of glia cells before ending in the lamina (Boschek, 1971). The seventh rhabdomere is composed of two cells whose cell bodies occupy, respectively, the outer (R7) and inner (R8) halves of the ommatidium and whose rhabdomeres are placed one over the other (Melamed & Trujillo-Cenoz, 1968; Boschek, 1971). This tandem arrangement is seen across the whole female eye. The microvilli of R7 are at right angles to R8 and in most eye regions the diameters of the R7-R8 rhabdomeres are about half the size of R1-R6 (Boschek, 1971). This is not true, however, for a small zone in the upper eye, either side of the ocelli (Wunderer & Smola, 1982 a). There, R7 and R8 are very large, whereas R1-R6 are diminutive. Unlike R7/R8 rhabdomeres elsewhere which twist down their length (McIntyre & Snyder, 1978) these dorsal receptors are virtually untwisted and are thus ideally disposed to detect polarized light (Wunderer & Smola, 1982 b). The significance of this dorsal region will be mentioned later in the context of lobula plate organization. The axons of R7 and R8 plunge directly through the basement membrane and through the underlying neuropil of the lamina, finally ending in the second visual neuropil, the medulla, via an optic chiasma (Strausfeld, 1971; Campos-Ortega & Strausfeld, 1972 a, b). In contrast, the axons of short visual cells undergo a peculiar and highly regular pattern of crossing over between the retina and lamina. Six optically divergent R1-R6 cells from an ommatidium send axons to six different optic cartridges arranged more or less as a ring beneath the ommatidium of origin. A single cartridge receives endings from the reverse pattern of six ommatidia above it (Trujillo-Cenoz, 1966).

The significance of this projection was correctly interpreted by the French anatomist Vigier, in 1907 (Vigier, 1907 a, b, c). He wrote that if the rhabdomeres beneath a lens are separate then so are their optical axis. To preserve the resolving power of the retina in the brain, receptors with different axes should go to different follower neurons (monopolar cells). He observed that axons from different ommatidia converged to a single monopolar cell where they structurally summated on its dendritic spines (Vigier, 1908) and correctly concluded that they shared the same optical axis. What happens is shown in Fig. 1. A small area of the eye (indicated in a scanning electron micrograph of, in this case, the housefly) shows diagrammatically how different receptors in different ommatidia, but sharing the same optical axis, converge to the same optic cartridge. Three receptors are shown projecting to cartridge a whose receptor outer segments view point A through three lenses. Two of these lenses also focus light from source B to receptors that project to cartridge b. Points A and B are thus represented by units a and b in the neuropil. The Lamina La is connected to the medulla (Me) by the first optic chiasma. Horizontal rows of optic cartridges are simply reversed whereas vertical rows remain the right way up (Strausfeld, 1971). However, the relationship between optic cartridges is repeated exactly between units of neuropil in the medulla (called medulla columns). This is the principle of retinotopy and I shall return to it in more detail later. The significance of

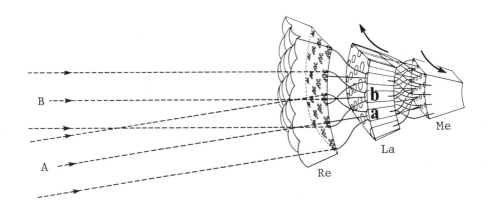

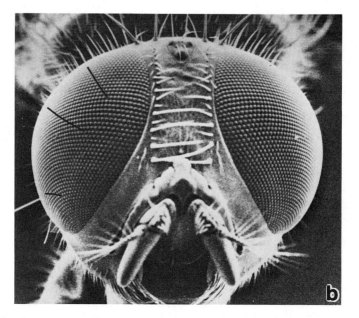

Fig. 1.

a: Projection of optical axes of receptors into space (the visual sampling units) and the convergence of their axons to optic cartridges. Further explanation in text.

b: Scanning electron micrograph of the head of Musca domestica (female). "Horizontal" axes of the retinal lattice are titled upwards dorsally and downwards ventrally. The strip of ommatidia in the top diagram is 90° from the horizontal.

the receptor optics and the projection patterns between the retina and lamina were examined with more sophistication by Kirschfeld (1967) using methods to optically locate like-oriented receptors, and by Trujillo-Cenoz (1966) and Braitenberg (1967) using electron and light microscopy to structurally locate the receptors. The projections of the R7 and R8 neurons to the medulla were described by Campos-Ortega and Strausfeld (1972 b).

The results of these studies are summarized in the inset of Fig. 2. Two patterns of projections are superimposed under the same set of dioptrics as if looking through the cornea into the eye. One pattern is carried by the R7 and R8 cells and this is identical to the pattern of projections seen in apposition eyes of bees and dragonflies, for example. The other projection pattern is carried by R1-R6. It is truly unique to many diptera and comprises the "neural superposition" eye serving to increase the light-gathering power of the retina (actually, through the follower neurons) and to improve the signal-to-noise ratio at low light intensities (Scholes, 1969).

In Fig. 2, inset Re, 20 lenses are viewed from above. Each of the small black dots, one per lens, represents the tandem R7/R8 whose central optical axis is approximately coincident with the geometrical axis of the cornea and pseudocone. Their axon cross-sections are represented by pairs of spots beside each ring of large axon cross-sections (the cartridges) in the lamina (La). The R7 and R8 axons project straight down into the lamina from their ommatidium of origin.

R1-R6 cells are depicted only in seven ommatidia in order to emphasize their peculiar distribution. Six short receptors sharing the same optical axis look out from a typical asymmetric pattern of six ommatidia centered around a seventh. Each is accompanied by five other receptors (open circles) whose optical axes diverge. The six optically parallel receptors converge to a single cartridge in the lamina which is shown as a ring of filled profiles in La. Its satellite R7/R8 axons are derived from the middle ommatidium of the set of seven. In this unit R7/R8 rhabdomeres are shown slightly apart for clarity. The six retinula cells (cross-hatched) surrounding this pair of long visual fibres look in six different directions and diverge to six different optic cartridges in the lamina (cross-hatched in La). The arrows in Re and La indicate the horizontal and two oblique axes of the retina and lamina mosaic. Other coordinates have been used referring to horizontal and vertical (Braitenberg, 1967; Stavenga, 1979). Although out of the scope of this chapter it is worth remarking that Stavenga's system ideally describes retinal tesselation and relates this to predictions about the distribution of optical axes into space and the packing of visual sampling points (Beersma et al., 1975; Stavenga, 1979).

Lamina Synaptology

Golgi impregnations of the optic lobes reveal an amazing variety of nerve cell shapes (Cajal & Sanchez, 1915; Strausfeld, 1970, 1976 b). Does each shape of cell represent a functional class of neuron? How many types of nerve cells are there? How are these connected and what is their functional significance?

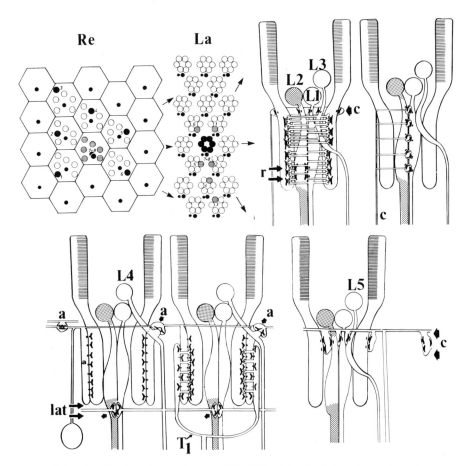

Fig. 2. Projections of long (R7/R8; black dots at center of each facet) and short visual fibres (R1-R6; larger cross-sections in facets) in Re sharing the same optical axes to the same cartridge in La. R1-R6 from a single ommatidium (cross-hatched in Re) having divergent optical axes go to six different optical cartridges in La. Explanation in text. The rest of this figure summarizes synaptic connections from receptors (R1-R6) onto lamina neurons. Parts of the L2 monopolar cells are shown stippled to differentiate them from L1 and L3. L1, L2, L3 spines are diagrammed only in the upper left cartridge, for clarity. Synapses are indicated by gull-shaped profiles. Detailed explanation in text.

The only region of neuropil yet to be examined with reference to all these questions is the lamina. Single neuron staining has revealed 12 cell types in addition to the endings of short receptor cells and R7/R8 axons (Strausfeld, 1976 a, b). These cells are: three monopolar cells (L1, L2 and

L3); a basket-shaped dendrite from the T1 cell; four species of centrifugal neurons originating in the medulla; two types of amacrine cells (one rarely stained); and two other relay cells, L4 and L5, which appear to get no input from the receptors. The synaptic connections of all these neurons were worked out by combining Golgi staining with electron microscopy (Strausfeld & Campos-Ortega, 1977) and are summarized in Fig. 2.

Optic cartridges are represented by pairs of short receptor endings from outer segments (the rhabdomeres) sharing the same optical alignment. L1 and L2 receive about the same number of synaptic inputs from receptors (in the order of 300-400 inputs; Strausfeld, 1976 b) and their paired dendrites share the same apposing presynaptic area of the receptor cell. A third monopolar cell (L3) gets about a third as many inputs from all the receptors. Two types of small field centrifugal cells provide a reafferent input back into the lamina from the medulla. One synapses onto the trunks of L1 and L2 (c, upper right) the other synapses on the necks of L1, L2, L3 and the receptor endings (c, upper left cartridge). A local pathway within the lower part of the receptor endings (r, double arrows, into the left cartridge) provides local feedback from one of the monopolars (L2) back onto receptors and onto its "twin" L1. In all the cartridges L2 is shown partly stippled, to distinguish it from the other cells. This is partly propaganda since, as will be seen later, the dual channel retina-to-lobula (via the medulla) and retina-to-lobula plate (also via the medulla) seems already to begin at this level with the divergence of the R1-R6 input onto L1 and L2. For clarity, L1-L3 are shown without their spines in the remaining cartridges. In the bottom row, the two cartridges to the left are shown with an amacrine cell (a), an L4 monopolar, and a T1 neuron. T1 is postsynaptic to receptors and provides a fourth second-order pathway from the optic cartridge, in parallel with L1, L2 and L3, to the medulla. T1 dendrites climb up the outside of receptor endings accompanied by six alpha-processes that hang down around the cartridge from amacrine cells. Amacrines are postsynaptic to receptors and pre- and postsynaptic to T1 cells. Amacrines are also pre- and postsynaptic to each other providing an extensive network of lateral connections across the top of the lamina neuropil.

The most regular pattern of connections is shown by the L4 neurons (Fig. 3). These cells receive a distal input from amacrines, and their collaterals beneath the lamina provide a regular network of reciprocal inputs to outgoing L1 and L2 monopolar cell axons (Strausfeld & Braitenberg, 1970).

What is the significance of these connections which occur more or less uniformly across the lamina? Almost all our knowledge about the physiology of lamina monopolar cells comes from studies by Laughlin (1973, 1974, 1975, 1976) (on Odonates) and Zettler and Järvilheto (Zettler & Järvilheto, 1972; Järvilheto & Zettler, 1973) (on Calliphora eyes) with some earlier contributions by Arnett (1972) and Hardie (cited by Shaw 1981). Most of these studies are reviewed by Laughlin (1981) and Shaw (1981) in the context of retinal physiology and lamina electrophysiology and structure. The barest of interpretations is repeated here with respect to the L1/L2 output.

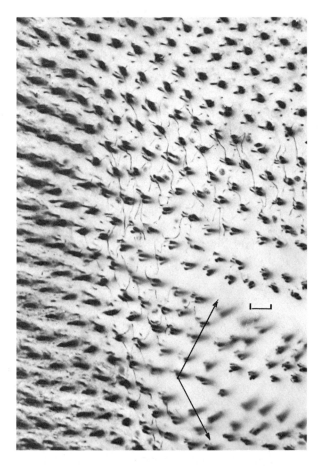

Fig. 3. A regular network of L4 collaterals connecting L1/L2 axons, organized along the two diagonal axes of the retinotopic mosaic. Scale = 10 μm.

Flies can manoeuvre as efficiently in the darkest recesses of the stable as in a summer-lit room. Their short visual cells operate over a wide range of intensity, and in dim light receptors looking at the same point in space from beneath six different lenses are ideally situated to maximize photon capture and to improve the signal-to-noise ratio (Scholes, 1969; Laughlin, 1973). This amplification is also boosted by parallel synapses between the receptor terminal and the monopolar follower neurons. Depolarization in a receptor will cause the simultaneous release of transmitter from many presynaptic sites. This ensures that synaptic noise is not confused with the multiple excitation of monopolar postsynaptic sites arising from a transient receptor signal (Laughlin, 1981). A problem for the fly, as in all high acuity eyes, is to distinguish contrast against an ever-changing background. Possibly this need is acute in a rapidly manoeuvring animal where background intensities may fluctuate rapidly. Also important for an insect that relies strongly on visual feedback to maintain stabilized flight is the ability to detect fast-moving objects. Laughlin has suggested that the smearing of voltage response to a moving point source (Srinivasan & Bernard, 1975) may be minimized by increasing the response speed of the receptor (Laughlin, 1981). However, this may still be insufficient at high velocities and possibly some local wide-field channels from, say, the R7/R8 endings in the medulla, relayed back out to the lamina, provide an adaptation pool that would compensate for movement-induced smear.

Lamina monopolars (L1, L2) have a high sensitivity and integrate contrast signals over a wide spectrum of wave lengths and intensity (Laughlin, 1975, 1976). They are also highly sensitive to intensity fluctuations relative to background illumination; in dragonflies and flies detection of change in contrast by lamina monopolars is 10 times more efficient than by the receptors (Laughlin, 1976; Laughlin & Hardie, 1978). Laughlin (1981) summarizes the properties of lamina (L1/L2-type) monopolar cells as follows: 1) they amplify the receptor signal and 2) they have a narrow dynamic range at any one mean intensity. This range is shifted to a new one by rapid light-adaptation in order to suit the new background intensity. This change is local, concerning a small area of the retina. Possibly it may be effected by neural feedback either by local adaptation pools, such as the amacrine cells, or by intracartridge adaptation by reciprocal feedback synapses. However, Shaw (1981) has pointed out that lateral inputs to L1/L2 in the lamina are highly circuitous reaching the monopolars via the sparse presynaptic connections in L4 collaterals. Shaw argues that these would be insufficient to generate conductance changes that would override the full activation of L1/L2 channels by R1-R6. He suggests that L4 collaterals are more likely to provide a network giving rise to a type of center-on flanking-off surround seen in spiking (morphologically unidentified) neurons recorded in the first optic chiasma (Arnett, 1972). Shaw suggests that the ambient background signal is subtracted from the receptor's presynaptic signal by means of extracellular field potentials which have been identified physiologically (Shaw, 1975, 1981). In any event, the L1/L2-type monopolar cell has the property of being a high-pass filter which extracts information about contrast, through lateral inhibitory and self-inhibitory networks (Laughlin, 1981). Significantly, only the L1 and L2 monopolar cells share the same neuronal synaptology with respect to amacrines and L4 cells. L3 seems to be immune from neuronal lateral-interaction networks and reciprocal feedback. Possibly it relays some qualities of the world to the medulla that are filtered out from the L1/L2 system. L4 and L5 are clearly local interneurons which, in the lamina, are designed to receive wide-field inputs via amacrines or tangentials. Shaw (1981) has suggested their possible modes of action with respect to spiking transmission to the medulla. This is in contrast to the apparent "non-spiking" transmission of L1 and L2 cells (Zettler & Järvilheto, 1971, 1972; Järvilheto & Zettler, 1973). In the medulla L5 is interposed between the L1 and L2 terminals and certain second-order interneurons. L4 endings connect adjacent medulla columns, presumably synapsing onto relay neurons there. The role of T1 is enigmatic, particularly since it appears to be pre-and postsynaptic in the medulla. We shall return later to the nature of the L2 pathway onto third-order relays.

The Retinotopic Mosaic

The basic principle of the mosaic has already been illustrated by the R1-R6 projections into the optic cartridges. The general rule is that one column of neuropil represents one group of identically aligned receptors. However, columnar cells arising in the lobula actually coarsen this mosaic, one cell representing a circular patch of nine groups of aligned receptors (Strausfeld, 1976 a; Strausfeld & Hausen, 1977; Strausfeld & Nässel, 1981).

Successive relays carry the mosaic through successive layers of synaptic connections. The principle of layering can already be seen in the lamina starting peripherally with amacrine-amacrine cell connections followed by the outer centrifugal inputs onto L1, L2 and L3 which are then succeeded by receptor-to-monopolar connections. At the base of each cartridge there are reciprocal synapses between L2 and receptors and between L2 and L1. Lastly, beneath the lamina, there is the regular network of L4 axon collaterals which are reciprocally presynaptic amongst themselves and presynaptic onto L1/L2 axon hillocks (Strausfeld & Braitenberg, 1970; Braitenberg & Debbage, 1974). This network is illustrated in Fig. 3; the plane of section cuts tangentially across the curved inner surface of the lamina to reveal bundles of monopolar cell axons connected by L4 collaterals. This picture demonstrates one of the most conservative features of the system. Namely, that interconnections between columns usually follow the diagonal axes rather than a "vertical" or "horizontal" axes in the retinotopic mosaic (an exception is the lamina amacrine; Campos-Ortega & Strausfeld, 1973).

Possibly, this rectilinear pattern reflects the development of protoclusters of ommatidia and interneurons. During eye-disc development cells divide alternately horizontally and then vertically to give alternating vertical and horizontal mitoses (Ready et al., 1976). Proliferation is from the posterior eye margin towards the anterior. The lamina and retina grow from the back to the front, whereas the medulla grows in the opposite direction, its anterior neuropil representing the posterior visual field (Meinertzhagen, 1973). Long visual fibers from the developing lamina grow out to the corresponding position of the developing medulla. Presumably, they act as guidelines for the subsequent outgrowth of monopolar axons from the optically relevant cartridge. Pulse-labelling with ^{3}H-thymidine during various stages of third larval instar development in Drosophila shows characteristic bands of differentiated neurons representing the time of isotope incorporation and, hence, the progression of a mitotic wave across the developing retina and optic lobe neuropils. The progressive proliferation of lamina cells is matched by an equivalent wavefront of proliferation of columnar neurons that are native to the lobula and lobula plate (Campos-Ortega, 1982).

Reduced-silver sections through the imago's optic lobes illustrate the progressive change of patterns of connections from one layer to the next (see Fig. 6). In contrast, the pattern of connections between columns at any one layer is highly similar across the whole eye. In a general sense, synaptology changes in depth rather than laterally across the system. There are certain exceptions to this, particularly with respect to local cell assemblies in the male and female lobula and with respect to sexual dimorphism in the upper front lamina through the medulla and into the lobula which, in part, reflect differences of retinal morphology (Strausfeld, 1979, 1980; Hausen & Strausfeld, 1980; Franceschini et al., 1981). However, these need not concern us at present and, significantly, dimorphism has not yet been substantiated in the lobula plate. Figure 4 illustrates cross-sections of the retinotopic mosaic, through the lamina (each point being represented by a cartridge and its satellite pair of long visual fibres, Fig. 4A), the outer medulla surface (4B), the outer medulla at the level of L2 endings (4C), the inner medulla at the origin of T4 cells (4D), the second optic chiasma (4F) and the outer surface of the lobula plate at T5 cell endings (4G) and deeper in the lobula (4E). Note that in Fig. 4C many line-amacrines weave vertically across many columns. However, they are always aligned along one or other of the diagonal axes of the mosaic rather than along true "vertical" axes. The inner medulla is, in comparison, a relatively "noisy" structure, where each column is divided into two units: these are axons already segregated into bundles destined for the lobula plate or for the lobula.

Beneath the medulla, horizontal rows of axon bundles converge so that sets of three horizontal rows of medulla columns contribute axons to two sheets of fibers. One sheet projects uncrossed onto the outer surface of the lobula plate, where fibers separate out and again confer an accurate representation of horizontal rows of medullary columns onto its neuropil (Fig. 4G). The other row of axon bundles twists through 180° so that posterior medulla columns map into the anterior (lateral) margin of the lobula. This crossed projection constitutes the second optic chiasma, reversing the order of fibers that was reversed once before between the lamina and medulla by the first optic chiasma.

The outer surface of the lobula plate faces the outer surface of the lobula. Inputs from the same medullary column into the lobula plate and lobula face each other. Columns in the lobula plate and lobula are also connected by axons that span the gap between them.

The deep neuropils of the lobula plate and lobula look highly disordered (Fig. 4E). This is because terminals from the medulla branch extensively amongst relay neurons whose dendrites are arranged either as wide-field tangential cells or as wide-field columnar cells spaced one to every three input bundles every third row. The original 1:1 mapping of columns to visual units is thus coarsened to a ratio of about 1:9.

<u>Retinotopic Maps and the Visual World</u>

The far-field radiation pattern of sets of identically aligned receptors describes the packing of "visual units" into space. This pattern has been directly demonstrated by Franceschini (1975) by means of a special telescope that looks at the antidromically illuminated eye and captures a large area of its visual field. Franceschini showed that the world of the fly is sampled by an orthogonal pattern of visual units, each of which is represented in the neuropil by a retinotopic column (the R1-R6 terminals). Their spacing is also a function of eye curvature. Where the

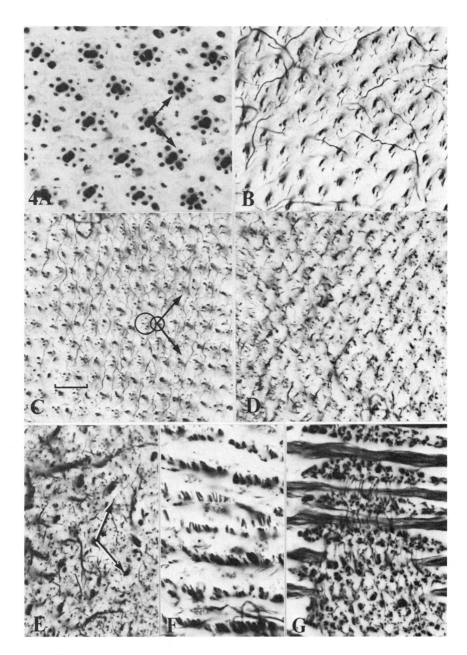

Fig. 4. Retinotopic organization of columns in the lamina (A); at the outer medulla at the level of L1, L2, L3 and R7/R8 inputs (B); at the level of line amacrines forming a plexus of interconnections immediately beneath the L2 endings (C): see also Fig. 6F; at the level of T4 cell

eye is flattest, the lenses are large and the interommatidial angle is small. The divergence angle between adjacent receptors is reduced. Local differences in eye curvature and sampling points are important when we come to consider areas of binocular overlap, sex-specific fovea, and equatorial specializations. At present, however, it is sufficient to treat the eye as a uniform structure and mention these local differences later. Figure 5 summarizes the representation of the receptor mosaic in the neuropil and out into the visual world (W). The orthogonal pattern of visual units is described along axes that are diagonal to the retina's vertical and horizontal axes (the diagonals Dd and Dv). The large spot in the visual unit array represents the six short receptors and the R7/R8 pair that have like-orientation and which are distributed in seven ommatidia (R) as in Fig. 2. This sampling point in W is recombined at a cartridge in the lamina in La (as in Fig. 1) and in a column in the medulla (Me). The

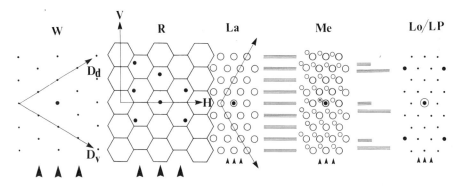

Fig. 5. The mapping of a single point in the visual field (left) through the matrices of the retina (R), lamina (La), medulla (Me) and lobula/lobula (Lo/LP). Further explanation in text. Three vertical rows are indicated at each level (arrow heads).

dendrites (D; see also the layer indicated in Fig. 6H, showing a section cut parallel to the columns). Inputs to the lobula and lobula plate appear somewhat disorganized (E) after collecting in horizontal sheets of chiasmal fibers (F). Large columnar neurons are arranged along the diagonal axes but coarsen the retinotopic mosaic. In G, small-field terminals can be seen spreading vertically either side of axon bundles. They map an accurate representation of the retina into third order neuropil. Scale = 10 µm. Double rings (C) indicate two groups of axon bundles in a column. The smaller goes to the lobula plate, the larger to the lobula. Further explanation in text.

medulla is shown with two groups of axons (represented by a small and a large circle) which separately project to the lobula and lobula plate terminating amongst columnar output cells spaced 1 to every 9 input bundles.

Structural Analysis of Functional Pathways:

1. Giant neurons of the lobula plate

One strategy for finding out what the optic lobes do is to record from as many neurons as possible and interpret their activity in the context of their synaptology. However, gone are the days of optimistic probings amongst an undetermined quantity of nerve cells. We have since learned that 1) there are rather a lot of nerve cells (a rough estimate for each lobe is in the order of 130 000 cells (Strausfeld, 1976 b), 2) the synaptology of only 12 of them (in the lamina) is complex and its analysis consumed 4 man-years (Strausfeld & Campos-Ortega, 1977), 3) most cells are so small that stable long-term recordings may be quite difficult. The obvious tactic is to probe the large neurons in the lobes or visual neurons that leave the brain. This has been done in two cases: for the locust LGMD-DCMD system (Rowell et al., 1977) and large tangential cells in the fly lobula plate.

The lobula plate tangentials are the most accessible of all neurons. They occupy characteristic layers of a thin tectum that receives a retinotopic input of columnar cells arising in the medulla and lobula. The most prominent lobula plate tangentials are the "horizontal" and "vertical" giant neurons, so called because they are, literally, giants amongst visual interneurons (some with axon diameters of 20 µm) and have dendritic trees that map into the retinotopic mosaic horizontally or vertically (Braitenberg, 1972; Pierantoni, 1976). Their visual physiology has been extensively researched (Dvorak et al., 1975; Hengstenberg, 1977, 1982; Eckert & Bishop, 1978; Eckert, 1981, 1982; Hausen, 1981, 1982 a, b; Hengstenberg et al., 1982). To summarize a long, complex, and sometimes confused story horizontal cells respond to horizontal wide-field motion over the upper, middle and lower parts of the visual field. Hausen (1981, 1982) has shown that their sensitivity is highest in the frontal part of the visual field, decreasing laterally and towards the rear. They respond ipsilaterally to progressive front-to-back motion of the visual panorama and are inhibited by the reverse movement. Their preferred directions are truely horizontal at the equator of the eye where the orientation of the horizontal axes of the ommatidial lattice is parallel to the horizon. Elsewhere this lattice skews progressively from the equator towards the poles, as can be seen in the scanning electron micrograph of Figure 1. The preferred direction is tilted off-horizontal, dorsally in the dorsal hemisphere and ventrally in the ventral hemisphere (Hausen, 1982 a, b). This suggests that if "elemental motion detectors" synapse onto horizontal cells then they are precisely oriented according to the retinal lattice and the viewing directions of visual sampling points. Sequential illumination of certain patterns of neighboring visual units of Drosophila's retina result in an appropriate "optomotor" turn. This suggests too that "elemental motion detectors" supplying pathways involved in the turning motor are precisely oriented in the lattice (Buchner, 1976). Another wide-field neuron, the HI cell, is uniquely represented in each lobula plate and provides connections between left and right lobes. Its contralateral ending covers the field of the upper (North = HSn) and middle (Equatorial = HSe) of the three horizontal cells, but omits the lower part of the represented visual field. Hausen (1982 a) found that only the upper two horizontal cells, representing the upper and middle part of the eye, are functionally influenced (depolarized) by contralateral back-to-front (regressive) movement of vertical gratings presumably relayed to them by the H1 cell. Rotational movement (regressive contralateral and progressive ipsilateral) is responded to best by the HSe neuron (Hausen, 1981 b). However, all three HS respond almost as efficiently to front-to-back streaming motion of vertical gratings presented to both eyes (Hausen, 1982 a, b).

The vertical cell system (VS) comprises an assembly of 11 neurons whose shapes and dendritic domains vary in a specific fashion across the lobula plate (Hengstenberg et al., 1982). Comparisons between individually stained VS neurons discount them as a homogenous system. Instead they comprise at least three groups of heteromorphic cells. The preferred directions of most vertical cells is ipsilateral movement of horizontal gratings downwards (Hengstenberg, 1982). However, certain vertical cells also respond to regressive or progressive motion of vertical gratings presented to the dorsal part of the field (Hengstenberg, 1977; Eckert & Bishop, 1978; Hausen, 1981; Eckert, 1982). In this they are similar to horizontal cells. Thus, as well as being heteromorphic the VS neurons are subdivisible into vertical motion sensitive neurons and "Zwitter" cells, responding to vertical and/or horizontal motions depending on which part of the eye is stimulated.

2. Layers in the lobula plate

HS and VS neurons basically occupy two layers in the lobula plate, the HS cells residing near its outer surface, the VS cells being deepest. Other neurons are known to link the left and right lobula plates providing heterolateral inputs to either the vertical cell system or the horizontal cell system (Hausen, 1981). Certain neurons (CH neurons) provide centrifugal input to the horizontal or vertical cell layer, and others provide wide-field outputs from the vertical cell layer into the dorsal contralateral neuropil of the deutocerebrum (Hausen, 1981). In all cases, any wide-field neuron that has a preferential response to horizontal motion of vertical gratings resides near the outer surface of the lobula plate whereas cells responding to vertical motion of horizontal gratings reside at/or near the inner layers. Certain neurons which prefer oblique movement lie between these two layers (Hengstenberg et al., 1983). The situation is highly reminiscent of the mammalian visual cortex where neurons having different preferred orientations are encountered at different depths in columns (Hubel & Wiesel, 1974). In the fly we have an accessible group of neurons which are a paradigm of orientation and direction selective systems.

Research on the horizontal and vertical cell systems poses the following questions:

1. What is the identity of the inputs to the VS and HS?
2. Are these motion detectors or is motion and direction (orientation) computed by synaptic interactions in the lobula plate itself?
3. If there are "elemental motion detectors" of the kind thought to exist in Drosophila (Buchner, 1976) can these be corroborated neuroanatomically?

3. Small-field relays to the lobula plate begin in the lamina.

Radioactive deoxyglucose accumulation in active cells demonstrates that movement of gratings in the visual field induces activity at specific levels in the medulla columns (Buchner et al., 1979) and in two layers of the lobula plate, each of which is divided into two narrowly separated strata (see Buchner & Buchner, 1983). None of the layers, however, actually correspond to HS or VS neurons, nor, indeed, to any other known type of wide-field cell there. What, then, do they represent?

Looking at Golgi preparations, many neurons offer themselves as possible candidates for primary inputs (motion-sensitive or otherwise) into the lobula plate. This is because they are small and, like L4, appear to connect columns in a regular fashion. Such cells include T-shaped neurons linking the medulla to the lobula plate, columnar neurons between the lobula and lobula plate and many forms of transmedullary cells whose Y-shaped axons penetrate first the lobula plate and then project across into the lobula. A useful approach, and one which narrows down the search image, is to compare neurons between species or, even better, between orders. Three insects orders have lobula plates: Coleoptera, Lepidoptera and Diptera (Strausfeld & Blest, 1970; Strausfeld, 1976 b). Certain small-field medullary cells and medulla-to-lobula plate neurons can be recognized not only amongst different Diptera species but also in butterflies, moths and beetles (Strausfeld & Blest, 1970; Strausfeld, 1976 b). Five types of neurons are common to these three insect orders. They are L1 and L2 - type monopolars, the type SUB, TM1 & TM5 medulla cells and four types of T-cells. Two of these, T2 and T3, link the medulla to the lobula. However, the bushy T-cells (T4 and T5), arising in the inner medulla and the outer stratum of the lobula, both project to the lobula plate. One significant feature is that in all species examined the SUB neurons end at the level of T4 dendrites and TM1 cells terminate in a superficial layer of the lobula in which originate dendrites of T5s. It is this layer, and the T4 dendrite layer in the medulla, which accumulate deoxyglucose after visual motion stimulation.

What features of these cells suggest that they relay information to the HS and VS neurons? And, what is the significance of their structural relationships? First, I should emphasize that their distinction from other cells in the highly complex medulla neuropil (Campos-Ortega & Strausfeld, 1972 a) is not arbitrary. In flies, only two wide-field Y-cells, for example, extending to the lobula plate and the lobula, have homologues in Lepidoptera. But they have not been found in Coleoptera. Second, SUB, TM1 & TM5 are the only medullary cells to have dendrites restricted to one group of terminals arising from a single optic cartridge. They thus preserve an "unsmudged" representation of the mosaic of visual sampling units. Third, each SUB and TM1 neuron is clearly apposed to the L2 terminal rather than the bistratified ending from the L1 monopolar cell. L2 is the simplest of the monopolar cell endings. L2 is also the only monopolar cell to provide its own reciprocal feedback onto receptors in the lamina. Fourth, T4 and T5 cells are unique in that each is duplicated in each column. One T4 terminates at the horizontal level in the lobula plate and the other terminates at the vertical level. Likewise, one T5 terminates at the horizontal level, the other at the vertical level. Assuming SUBs and TM1s output to T4s and T5s, respectively, then each visual unit will be represented in the lobula plate by duplicate inputs that segregate out to two main levels. Fig. 6 illustrates these neurons and demonstrates the precise layering of terminals, dendrites and outputs in the medulla. Fig. 6 also shows that dendritic domains of SUB and TM1 are narrow, presumably restricted to the width of a single column. In contrast, T4 dendrites spread vertically through groups of three columns. T5 cells are demonstrated as ending at two main layers in the lobula plate, one of which is subdivided into two strata (Fig. 6J).

Seen from above, as in Fig. 10A, T4 dendrites map into a vertical strip of columns and their terminals in the lobula plate spread this area out to a group of 3-4 columns. These therefore effectively represent 9-12 retinotopic units. Fig. 8 (top) illustrates the assumed receptive field of a T4 neuron extrapolated out into a matrix of visual units that stretch out to infinity. In Fig. 8, visual sampling units are depicted as a contracting pattern with increasing resolution from back to front due to the expansion of the retinal mosaic over the flat front part of the eye. The ommatidia involved in "seeing" the T4 field are shown stippled. The retinal pattern shown here incorporates the transition from the anterior A-type lattice - (using Stavenga's, (1979) annotation) -to the densely packed C-type lattice posteriorly. The retinotopic mosaic in the lobula and lobula plate is shown as an expanding pattern from back to front as columns representing the front part of the visual field become progressively more separated and contain a greater abundance of synapses. The field mapped onto the lobula plate mosaic describes the representation of the T4 receptive field amongst columns invaded by its terminal arborization. Extrapolations of wide-field cells in the lobula plate onto the retinal mosaic or out into visual space have to take this kind of smudging into consideration in order to be meaningful.

From Fig. 7 it is obvious that only the L2 monopolar cells could input to SUB & TM1. Thus, it is reasonable to say that the segregation of a direct pathway into lobula plate neuropil already begins at the divergence of receptor inputs onto the two parallel channels, L1 and L2. Fig. 8 schematizes these findings. Receptors a and a' representing R1-R6 and R7/R8 look at the point source A. Receptors b and b' look at B. The channels a, b each diverge to the two parallel pathways in the outer medulla (OMe) one of which ends on SUB & TM1 cells (represented in OMe by A and B). The other pathway from a and b ends amongst numerous neurons destined for the lobula (stippled column). The R7/R8 pathway (a', b') also diverges onto SUB & TM1 cells (A and B in the outer medulla, OMe) and onto pathways destined for the lobula. The SUB cell contacts T4 dendrites in the inner medulla (IMe) and TM1 contacts T5 cells in the outer lobula (Olo). A pair of T4 cells arises from each column (aT4, bT4)

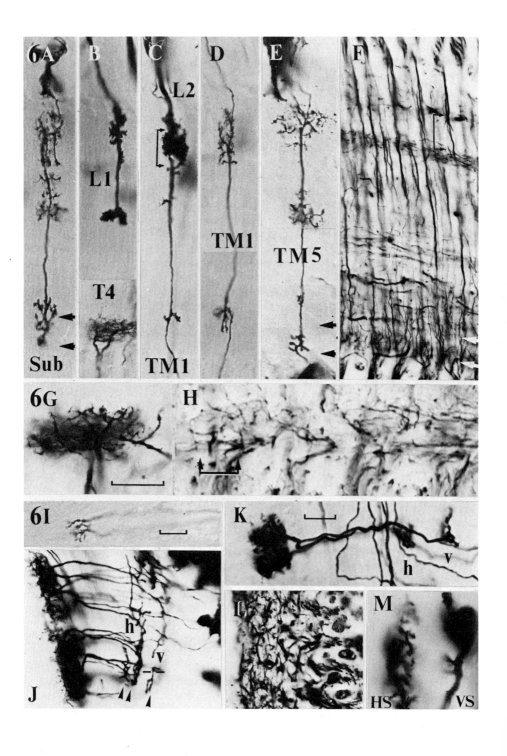

Fig. 6. Small-field neurons linking the medulla to the lobula plate (Calliphora erythrocephala).

Fig. 6A-E: Small-field relay neurons from monopolar cell endings. SUB is an intrinsic short-axoned relay neuron that does not leave the medulla. TM cells connect the medulla to the lobula. TM1 terminates in the outer lobula stratum (Fig. 6I) at the level of T5 dendrites (Fig. 6J). The double arrows in Fig. 6A-F indicate the level of T4 dendrites: SUB and TM5 are ideally disposed to interact with them.

Fig. 6F: Bodian-stained medulla showing spacing of columns. The level of L2 monopolar cell endings is bracketed at arrows and is also shown in Fig. 6C depicting L2 endings stained with TM1 of the same column.

Fig. 6G & H: High-power of a pair of T4 dendrites arising in the same medulla column (Fig. 6G) and the periodic array of T4 cells, one pair in each column (Fig. 6H: Bodian stain).

Fig. 6I & J: TM1 endings are at the same level as T5 dendrites. T5 cells project to two levels in the lobula plate (h and v). Terminals at the h layer define two strata (arrowheads).

Fig. 6K: A pair of T5 neurons which arise at the same position in the lobula and terminate at two levels in the lobula plate (h, v).

Fig. 6L: Vertically oriented connections between columns at the outer lobula surface. These are presumably T5 neurons, the dendrites of which have an orderly arrangement within the retinotopic mosaic.

Fig. 6M: The main dendrites of an HS neuron and VS neuron in the lobula plate lying at the same level as the T5 terminals in the h- and v-layer, shown above in Fig. 6M.

Scale bars. 6I (also for 6A-F, J, L) = 10 μm; 6G, H = 10 μm; 6K, M = 10 μm

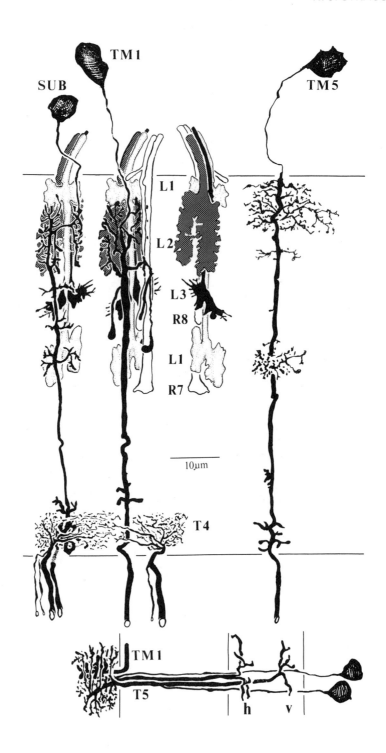

and their dendrites extend vertically to adjacent columns. Likewise, a pair of T5 cells (aT5 and bT5) arise from each TM1 terminal in the lobula. The medulla to lobula pathways do not interact with the T5 layer but make their contacts in the inner lobula (ILo). T5 pairs end at two levels in the lobula plate as do the T4 pairs.

The periodic pattern at the top is merely to remind the reader of the proposition that these pathways are the most obvious candidate for an elemental small-field motion detecting system. None of the cells have yet been recorded, something that I eagerly await. They also happen to be rather nicely corroborated by Buchner and Buchner's motion-sensitive deoxyglucose-labelled strata (Buchner & Buchner, 1983).

4. Synaptic connections of T-cells in the lobula plate

Brains containing cobalt-filled and silver-intensified HS and VS neurons can be counterstained by the Golgi method to reveal small-field inputs. Using appropriate fixation, double-marked brains have revealed T4 cells, marked with silver chromate, ending on cobalt-silver stained HS and VS neurons. Both markers are electron dense and can be readily distinguished in the electron microscope. Synapses in T4 terminals appose the HS and VS cells (Strausfeld et al., 1983). T4s are also postsynaptic to another form of small-field ending, possibly a T5 terminal at the same level. A computer reconstruction of the serial synapse between what is possibly T5 onto the identified T4 and VS7 cell is shown in the upper part of Fig. 9. In other preparations we have seen small-field endings converging onto the same postsynaptic site in the VS cell so that two inputs face a common postsynaptic membrane (Strausfeld & Bassemir, in preparation). These types of connections meet the requirements of Barlow and Levick-type small-field directionally selective, motion-sensitive units (i.e., serial inhibiting synapses and asymmetric connections (Barlow & Levick, 1965) and Torre and Poggio-type converging inputs to a common postsynaptic site (Torre & Poggio, 1978). Again, we do not know how the signals in the T4 or T5 cell interact, but it seems that they do not have simple relationships with HS and VS dendrites.

Fig. 7. Reconstruction of cell relationships amongst small field neurons linking monopolar cell inputs to the lobula plate.

Two neurons in the medulla, SUB and TM1, clasp the L2 terminal. SUB ends amongst T4 dendrites, TM1 ends amongst T5 dendrites (lower inset) which terminate at two levels in the lobula plate (lower right; h- and v-levels). TM5 probably receives its main input from the L1 monopolar cell.

T4 neurons have asymmetric dendrites which are eligned dorso-ventrally along the diagonal axes of the medulla mosaic. Other abbreviations: L3 = L3 monopolar cell ending, R7, R8 = endings of long visual fibres from the retina.

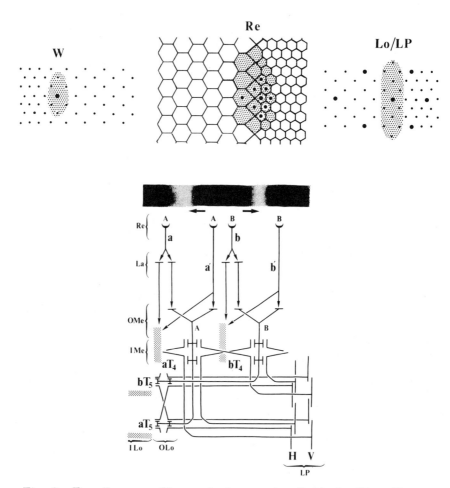

Fig. 8. Top diagram: Theoretical receptive field of a T4 or T5 cell assuming its dendrites visit a vertical surround of six retinotopic columns and receive inputs there. Possibly dendritic asymmetry (Fig. 7) reflects a synaptic asymmetry between columns. Ommatidia containing receptors inputting to a symmetric T4 receptive field are stippled in Re. Since T4/T5 endings go to several lobula plate columns their "fields-of-view" are smudged vertically onto VS-HS neurons. The lower diagram illustrates "elemental" pathways from the retina to H and V layers in the lobula plate via T4/T5 pairs. Further explanation in text.

Is the computation of directionality achieved at the level of T-cell dendrites, or amongst the T-cell endings or by the interaction of T-cells onto follower neurons? Possibly, T-cells are motion-sensitive to movement in any direction sufficient to stimulate adjacent inputs along their diagonal axes in the mosaic and are the small "elemental detectors"

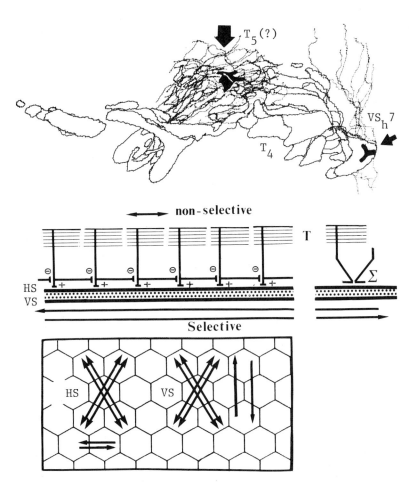

Fig. 9. Top: Computer reconstructed serial E.M. sections through a labelled T4 presynaptic onto a labelled VS7 dendrite. T4 is also postsynaptic to a tetrad of inputs, one possibly belonging to a neighboring T4 or T5 at the same layer. Middle: Suggested modes of connections of non-selective "elemental motion sensitive" neurons, amongst themselves and onto HS and VS cells, giving directionally sensitive Barlow & Levick-type or Torre & Poggio-type interactions. Bottom: Suggested "elemental movement detectors" arranged along the diagonal axes of the retinotopic mosaic compute information about preferred motion in four directions at the lobula plate; these are upwards, downwards, progressive (front-to-back) and regressive.

(Buchner, 1976). The direction and orientation selectivity could be achieved by appropriate asymmetric connections amongst their terminals as suggested in the middle part of Fig. 9. For this to occur we must propose that the weighting of synapses between T-cells at four successive levels achieves the appropriate asymmetrical connections or postsynaptic delays for computing vertically, horizontally and directionally. Observations on the patterns of HS and VS dendrites, or the pattern of T-cell endings, do not bring us any closer to an answer. What is shown is that even amongst wide-field neurons the general orientation of dendritic branches is now biased along axes vertically (Fig. 10) rather than horizontally or obliquely. Terminals of T-cells are generally oriented up-down, but when several of them are stained together hardly any pattern can be discerned. This level of resolution tells us little about how these neurons work and we have now to analyse the system using serial electron micrographs of labelled T-cells

5. Structural correlates to function of the VS and HS system.

Ten of eleven VS cells (VS 2-11) are shown in Fig. 11. The most posterior neurons (VS 7-11) are characterised by branches that occupy the horizontal layer of the lobula plate. The remaining cells, and the lower portions of VS 7-11 occupy the vertical layer of the lobula plate. The middle diagram illustrates a lobula plate surface map. Its columns expand from the posterior towards the front and represent the retina (see Fig. 8). The density of vertical cell dendrites mapped against the columnar packing is given in terms of isodensity lines (right diagram). Unity equals the area of highest density. We may expect that the heteromorphic vertical cell assembly has a graded sensitivity from the front (where it is highest) to the back of the eye (isodensity lines of the horizontal components of VS 7-11 in the top half of the lobula plate are not shown here). The mappings of vertical cell dendrites into the retinotopic mosaic is shown in Fig. 12. Here the vertical cells have been divided into different sets, or local assemblies, Hengstenberg's VS 1 neuron being a truely unique element. VS 1, and VS 7 through 11 all have dorsal dendrites that invade the horizontal layer (thus VS n_h) with lower components invading the vertical level (VS n_v). Together the VS 2-6 neurons cover the dorsal mosaic at the vertical layer where it is omitted by VS 7-11. There is however a "blind spot" where VS axons leave the lobula plate. This is depicted as zero density in Fig. 11, and a black area of the retinal map in Fig. 13. The field "NP" comprises a group of "North Pole" neurons (Fig. 14) which subtend the most dorsal part of the mosaic. This corresponds in part to a special area of the retina containing large diameter R7/R8

Fig. 10A. Top view looking on two T4 dendritic trees (columnar spacing indicating). See Fig. 8, top left.

Fig. 10B, C. On-top views of T4/T5 terminals at the horizontal (B) and vertical (C) layers in the lobula plate. Their branches spread approximately vertically and visit only one or two columns horizontally. Groups of endings comprise a "noisy" pattern of telodendria with no obvious axis (cf. Fig. 3).

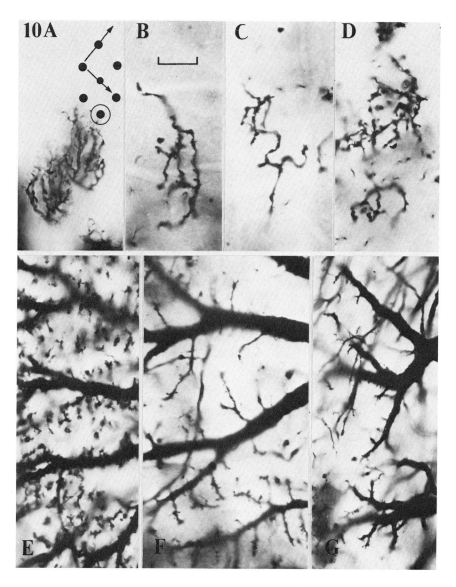

Fig. 10D, E, F, illustrate, respectively, swollen collaterals of wide-field centrifugal (CH) neurons ending in the H layer, dendrites of the equatorial HS neuron and dendrites of the VS 3 neuron. Although the main trunks of these cells are oblique or horizontal their processes are aligned vertically. Scale for all figures = 10 μm and is oriented horizontally.

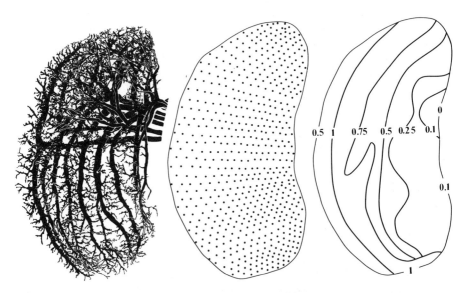

Fig. 11. Left: The heteromorphic assembly of VS 2-11. Note the horizontally oriented processes of VS 7-11 in the H-layer of the upper lobula plate (anterior towards the left). Middle: The distribution of every third retinotopic point every third row of the mosaic. Right: Isodensity lines illustrating relative packing of VS dendrites amongst V-layer retinotopic columns. Further explanation in text.

receptors specialized for the detection of polarized light (Wunderer & Smola, 1982 b). The fields extrapolated out to the planar retina map represent the visual world of all T4 and T5 cells ending on the vertical cells of the lobula plate (Fig. 13).

Part of the HS, VS and NP systems are shown in Fig. 14, omitting the VS 1 neuron and several other features that are out of the scope of this chapter (Strausfeld, Seyan & Bassemir, in preparation). The HS assembly, and VS 3-11 of the VS assembly shown here were all filled with cobalt transsynaptically (Strausfeld & Bassemir, 1983) from their respective descending neurons. Like VS 1, VS 2 is filled only indirectly and its axon is here shown cut off. In fact this cell projects with VS 1 to descending neurons of the horizontal cell system (DNHS). The cells VS 3-8 project coherently to the Y-shaped assembly of descending neurons of the vertical system (DNVS). Both the DNHS and DNVS groups consist of several uniquely identifiable descending neurons each having a common input. It will be noted that DNHS neurons have many dendrites that do not contact HS cells. In part these receive mechanoreceptor inputs from the antennae amongst which are branches of receptors arising in Johnston's organ relaying information about angular acceleration and deceleration (Gewecke, 1974). Mechanosensory axons in fact go to several descending systems but not to DNVS. Certain assemblies of small-field

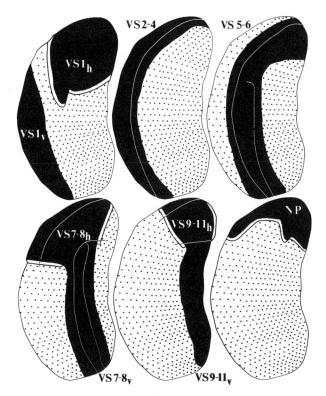

Fig. 12. Mappings of VS assemblies over lobula plate retinotopy. The suffix h denotes processes of VS 1, and VS 7 through VS 11, at the H level. White lines indicate overlap from adjacent cell sets. Maps are a smudged representation of the VS dendritic arbors appropriately connected to T4/T5 cells whose dendritic fields encompass six visual units (Fig. 8 & 10A). Anterior is towards the left. Further explanation in text.

visual neurons originating in the lobula also end on DNHS cells. Some parts of the HS axons do not contact the DNHS but impinge onto another set of DNs (not shown here) receiving inputs from VS 9 through VS 11. An important difference between VS and HS organization is the convergence of the ocellar interneurons onto certain dendrites of DNVS neurons. To add to the complexity, the NP system originating in the horizontal layer of the lobula plate, and accompanying the upper dendrites of VS 7 and VS 8, also terminates in neuropil of the DNVS output.

What do these structures mean?

First, the HS and VS system are not a set of orthogonally arranged neurons converging onto a common output, although certain of them project to the same DNs. Second, the mappings of VS dendrites into the

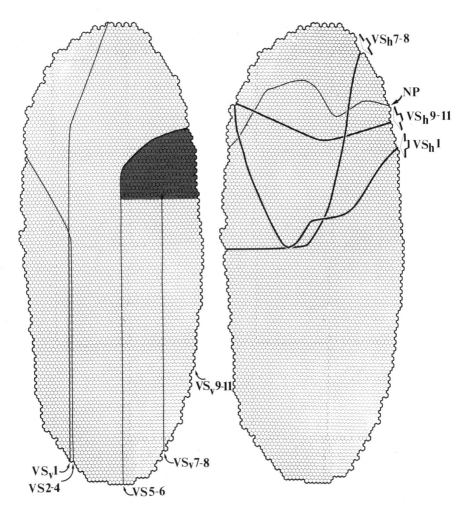

Fig. 13. Extrapolated maps of the VS T4/T5 fields at the retina incorporating the convergence of R1-R6 to single optic cartridges. VS components of the V layer are shown in the left map. VS components of the H layer are shown in the upper half of the right map. The dark area is the "blind spot". Anterior is towards the left for both maps.

retinotopic mosaic would be sufficient for the VS system alone to monitor yaw, roll and pitch, seen as displacements of the visual panorama. (This assumes that ocellar relays register displacement of the horizon as they do in locusts and flies (Wilson, 1978; Taylor, 1981). We can further speculate that the NP system is polarization-sensitive and converges to DNVS outputs. In comparison, HS cells converge onto DNs that also receive a mechanoreceptor input. Does this impart an entirely new aspect to this system? One suggestion is that although the HS responds to motion simulating yaw and pitch, the speed of the moving panorama combined with mechanosensory data about acceleration may be more relevant to the multimodal DNHS pathway. It is conceivable that the VS system (VS3-11) with the ocelli confers roll-pitch-yaw information to flight motor centers omitting any other modality, whereas the DNHS output provides data about apparent ground speed using mechanoreceptive and progressive motion inputs. Clearly these propositions, which are suggested by anatomical connections, need to be physiologically verified. A diagram summarizing this working hypothesis is shown in Fig. 15.

6. <u>Independently processed visual information convergences onto output pathways.</u>

In the previous section I described the segregation of verticality and horizontality into different layers of neuropil and showed how certain neurons from each layer converge, or diverge, to outputs from the brain. Another example of layering and regional variation in the mosaic is the sex-specific lobula neurons (Strausfeld, 1980; Hausen & Strausfeld, 1980). Three of these uniquely identifiable tangentials are shown in Fig. 16 (LMG 1-3). They subtend special regions of the visual field of the male blowfly, representing the dorso-frontal fovea and a zone of binocular overlap (Strausfeld, 1979) (Fig. 17). They are consistently filled transsynaptically after backfilling descending neurons with cobalt chloride. In females, these neurons have inconspicuous homologues (FLGS) which are difficult to spot amongst palisades of Col A cells common to both sexes (Fig. 18). Again, as in the lobula plate, different visual neurons, representing specific fields of view or specific layers in the system converge onto descending neurons. In the upper part of Fig. 18 the columnar Col A cells represent the entire visual field of one eye. They are shown inputting onto the Giant Descending Neuron (GDN) with which they share gap junctions (Strausfeld & Bassemir, 1983). The GDNs eventually lead into the jump-initiation circuit of the midleg (Strausfeld & Bacon, 1983; Strausfeld et al., 1983). Another terminal wraps around the axon of the GDN. This is the ending of the MLG 1 from the contralateral lobula. The trajectories of the contralateral MLG 1 and ipsilateral MLG 2 are shown below. Both interact with multimodal pathways from the brain.

Clearly, a high level of integration is occurring at dendritic trees of DNs. It is insufficient only to record from visual neurons in order to determine their relevance in behaviour. The anatomy of the system warns of the importance <u>not</u> to deprive the brain of other sensory inputs which may influence the activity of visual interneurons. Two strategies are suggested. One is to record from descending neurons in order to determine just how much information in, say, the HS or VS neuron is used by the follower neurons. Second, we need to know the functional significance of layers in the medulla if its filter properties are to be understood. These are presumably resident in its synaptic layers. It would be sufficient to record from small-field neurons leading to the lobula whose inputs, like the T4 and T5 cells to the lobula plate, are recognized as arising at a specific level of the medulla whose relationships with lamina outputs can be discerned. If nothing more, functionally identified neurons arising in well-documented strata will give greater impetus to those obdurated enough to unravel their synaptic connections.

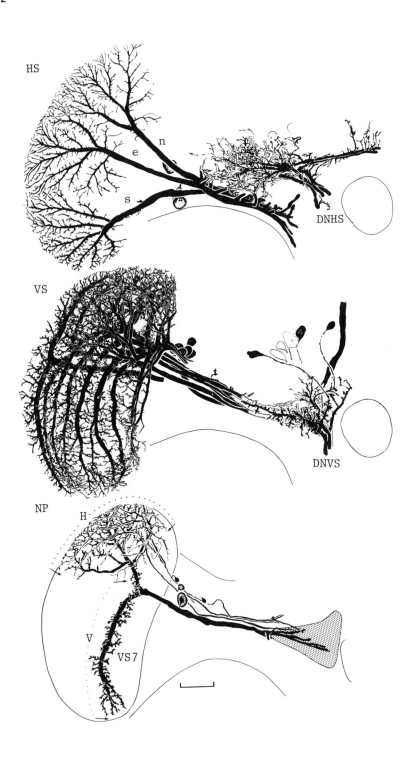

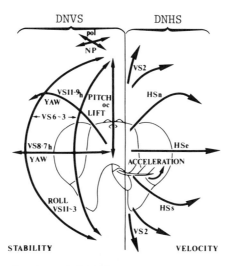

Fig. 15. Suggested roles of DNVS and DNHS neurons based on their connections with VS, HS, NP, ocellar interneurons and primary antennal mechanoreceptors. It is proposed that DNHS integrates bimodal inputs that inform the flight motor about angular acceleration (wind detection) and the front-to-back movement of the visual panorama to determine apparent ground speed (velocity). The VS neurons are ideally mapped retinotopically and into functional (H, V) layers to provide, with ocellar and some HS (e.g. HSs) inputs, the DNVS outputs data about the slip of the visual panorama incurred by movement through three degrees of freedom (pitch, roll, yaw). NP neurons may provide additional information about the angle of polarized light.

Fig. 14. Drawings of cobalt-silver stained HS and VS assemblies. HS neurons were filled transsynaptically from the five DNHS whose axons end in pro-, meso-and metathoracic neuropils (Strausfeld & Bacon, in preparation). Note that the north (n) HS cell has rather sparse branches in the dorsal part of the lobula plate and omits the most dorsal retinotopic columns. These are occupied by certain dendrites of VS 7-VS 11 and the NP system (shown with the VS 7). VS 3-11 in the middle picture were filled transsynaptically from DNVS neurons whose axons also end in the pro-, meso- and metathoracic ganglia (Strausfeld et al., 1983). Certain ocellar interneurons also terminate on DNVSs. VS 2 is shown with its axon truncated. The VS 2 axon normally projects with that of the "Zwitter cell" VS 1 (Eckert, 1962; Hengstenberg et al., 1982) (which is not shown) onto DNHS outputs. The axons of VS 9-11 are also cut. These normally converge with branches of certain HS neurons (HSe, HSn) to a third group of DNs (not shown (Strausfeld & Seyan, in preparation)). Further explanation in text. Scale = 50 µm

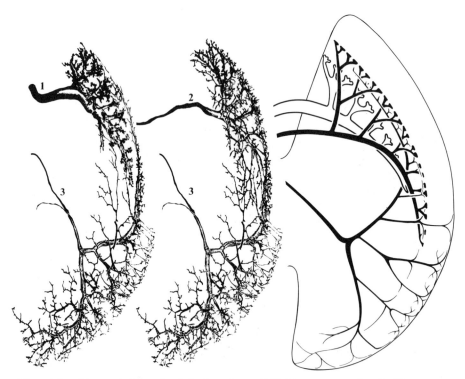

Fig. 16. Type 1-3 male lobula giant (MLG) neurons in Calliphora. Like HS and VS neurons, their dendrites in the upper and front of the mosaic occupy discrete layers where they locally interact with segregated inputs of lobula channels from medulla columns

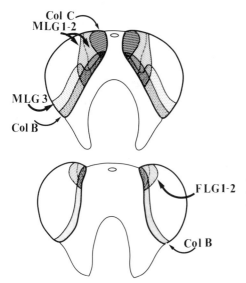

Fig. 17. MLG 1, 2, 3 and local assemblies of columnar neurons (Col B and Col C) are visited by relays representing the front and upper retina. Female homologues (FLG 1, 2 and Col B cells) subtend smaller areas of the retinal projection. Col C cells are absent.

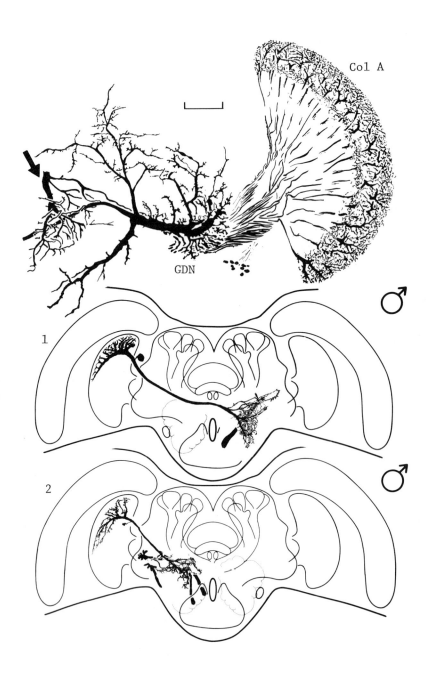

Fig. 18. Top: Camera lucida of the sexually unspecific Col A assembly (one Col A to every nine columns of medulla inputs) whose axons map onto certain dendrites of the giant descending neuron (GDN). This cell also receives a number of other types of visual, mechanosensory and olfactory inputs (Strausfeld & Bacon, 1983). Its axon receives branches of the contralateral MLG 1 neuron (arrow). Lower figures: MLG 1 through to MLG 6 terminate on DNs, usually at their axons (Strausfeld, in preparation). MLGs confer sex-specific outputs onto DNs whose dendritic inputs are identical in males and females. Scale = 50 µm. Further explanation in text

ACKNOWLEDGEMENTS

I am grateful to Renate Weisskirchen for printing the photographs and to Marianne Remy for typing the manuscript. The final version of this chapter benefited from discussions with Dr. Karl Fischbach.

REFERENCES

Arnett, D.W. (1972) Spatial and temporal integration properties of units in the first optic ganglion of dipterans. J. Neurophysiol. 35: 429-444.

Barlow, H.B. & Levick, W.R. (1965) The mechanism of directionally sensitive units in rabbit's retina. J. Physiol. 178: 477-504.

Beersma, D.G.M., Stavenga, D.G. & Kuiper, J.W. (1975) Organization of visual axes in the compound eye of the fly Musca domestica L. and behavioural consequences. J. Comp. Physiol. 102: 305-320.

Boschek, C.B. (1971) On the fine structure of the peripheral retina and lamina ganglionaris of the fly Musca domestica. Z. Zellforsch. 118: 369-409.

Braitenberg, V. (1967) Patterns of projections in the visual system of the fly. I. Retina-lamina projections. Exp. Brain Res. 3: 271-298.

Braitenberg, V. (1972) Periodic structures and structural gradients in the visual ganglia of the fly. In: Information Processing in the Visual System of Arthropods. Ed. R. Wehner. Berlin, Heidelberg, New York, Springer, p. 1-15.

Braitenberg, V. & Debbage, P. (1974) A regular net of reciprocal synapses in the visual system of the fly Musca domestica. J. Comp. Physiol. 90: 25-31.

Buchner, E. (1976) Elementary movement detectors in an insect visual system. Biol. Cybern. 24: 85-101.

Buchner, E. & Buchner, S. (1983) Neuroanatomical mapping of visually induced nervous activity in insects by ^{3}H-deoxyglucose. (This volume)

Buchner, E., Buchner, S. & Hengstenberg, R. (1979) 2-deoxy-D-glucose maps movement-specific nervous activity in the second visual ganglion of Drosophila. Science 205: 687-688.

Cajal, S.R. & Sanchez, D. (1975) Contribution al conocimiento de los céntros nerviosos de los insectos. Parte I, retina y céntros opticos. Trab. Lab. Invest. Biol. Univ. Madrid 13: 1-168.

Campos-Ortega, J.A. (1982) Development of the nervous system. In: Handbook of Drosophila Development. Ed. R. Ransom. Amsterdan, New York, Oxford, Elsevier Biomedical.

Campos-Ortega, J.A. & Strausfeld, N.J. (1972 a) Columns and layers in the second synaptic region of the fly's visual system: the case for two superimposed neuronal architectures. In: Information Processing in the Visual System of Arthropods. Ed. R. Wehner. Berlin, Heidelberg, New York, Springer.

Campos-Ortega J.A. & Strausfeld, N.J. (1972 b) The columnar

organization of the second synaptic region of the visual system of Musca domestica L. I. Receptor terminals in the medulla. Z. Zellforsch. 124: 561-585.

Campos-Ortega, J.A. & Strausfeld, N.J. (1973) Synaptic connections of intrinsic cells and basket arborizations in the external plexiform layer of the fly's eye. Brain Res. 59: 119-136.

Dvorak, D.R., Bishop, L.G. & Eckert, H.E. (1975) On the identification of movement detectors in the fly optic lobe. J. Comp. Physiol. 100: 5-23.

Eckert, H. (1981) The horizontal cells in the lobula plate of the blowfly, Phaenicia sericata. J. Comp. Physiol. 143: 511-526.

Eckert, H. (1982) The vertical-horizontal neurone (VH) in the lobula plate of the blowfly, Phaenicia. J. Comp. Physiol. 149: 195-205.

Eckert, H. & Bishop, L.G. (1978) Anatomical and physiological properties of the vertical cells in the third optic ganglion of Phaenicia sericata (Diptera, Calliphoridae). J. Comp. Physiol. 126: 57-86.

Fischbach, K.F. (1983) Neural cell types surviving congenital sensory deprivation in the optic lobes of Drosophila melanogaster. Dev. Biol. 95: 1-18.

Franceschini, N. (1975) Sampling of the visual environment by the compound eye of the fly: fundamentals and applications. In: Photoreceptor Optics. Ed. A.W. Snyder & R. Menzel. Berlin, Heidelberg, New York, Springer, p. 97-125.

Franceschini, N., Hardie, R., Ribi, R. & Kirschfeld, K. (1981) Sexual dimorphism in a photoreceptor. Nature (Lond.) 291: 241-244.

Gewecke, M. (1974) The antennae of insects as air current sense organs and their relationships to the control of flight. In: Experimental Analysis of Insect Behaviour. Ed. L. Barton-Browne. Berlin, Heidelberg, New York, Springer, p. 100-113.

Hausen, K. (1981) Monocular and binocular computation of motion in the lobula plate of the fly. Verh. Dtsch. Zool. Ges. 1981: 47-70.

Hausen, K. (1982 a) Motion sensitive interneurons in the optomotor system of the fly. I. The horizontal cells: structure and signals. Biol. Cybern. 45: 143-156.

Hausen, K. (1982 b) Motion sensitive interneurons in the optomotor system of the fly. II. The horizontal cells: receptive field organization and response characteristics. Biol. Cybern. 46: 67-79.

Hausen, K. & Strausfeld, N.J. (1980) Sexually dimorphic interneuron arrangements in the fly visual system. Proc. R. Soc. Lond. 208B: 57-71.

Hengstenberg, R. (1977) Spike responses of "non-spiking" visual interneurones. Nature (Lond.) 270: 338-340.

Hengstenberg, R. (1982) Common visual response properties of giant vertical cells in the lobula plate of the blowfly Calliphora. J. Comp. Physiol. 149: 179-193.

Hengstenberg, R., Bülthoff, H. & Hengstenberg, B. (1983) Three-dimensional reconstruction and stereoscopic display of neurons in the fly visual system. In: Functional Neuroanatomy. Ed. N.J. Strausfeld. Springers series in Experimental Entomology. Heidelberg, Berlin, New York, Springer. (In press)

Hengstenberg, R., Hausen, K. & Hengstenberg, B. (1982) The number and structure of giant vertical cells (VS) in the lobula plate of the blowfly Calliphora erythrocephala. J. Comp. Physiol. 149: 163-178.

Hubel, D.H. & Wiesel, T.N. (1974) Sequence, regularity and geometry of orientation columns in the monkey striate cortex. J. Comp. Neurol. 158: 267-294.

Järvilehto, M & Zettler, F. (1973) Electrophysiological-histological studies on some functional properties of visual cells and second order neurons of an insect retina Z. Zellforsch. 136: 291-306.

Kirschfeld, K. (1967) Die Projektion der optischen Umwelt auf das Raster der Rhabdomere im Komplexauge von Musca. Exp. Brain Res. 2: 248-270.

Laughlin, S.B. (1973) Neural integration in the first optic neuropil of dragonflies. I. Signal amplification in dark-adapted second-order neurons. J. Comp. Physiol. 84: 335-355.

Laughlin, S.B. (1974) Neural integration in the first optic neuropile of dragonflies. III. The transfer of angular information. J. Comp. Physiol. 99: 377-396.

Laughlin, S.B. (1975) Receptor and interneurone light adaptation in the dragonfly visual system. Z. Naturf. 30c: 306-308.

Laughlin, S.B. (1976) Neural integration in the first optic neuropile of dragonflies. IV. Interneurone spectral sensitivity and contrast coding. J. Comp. Physiol. 122: 199-211.

Laughlin, S.B. (1981) Neural principles in the visual system. In: Handbook of Sensory Physiology. Vol. VII/6B. Ed. H. Autrum. Berlin, Heidelberg, New York, Springer.

Laughlin, S.B. & Hardie, R.C. (1978) Common strategies for light adaptation in the peripheral visual system of fly and dragonfly. J. Comp. Physiol. 128: 319-340.

McIntyre, P. & Snyder, A.W. (1978) Light propagation in twisted anisotropic media: application to photoreceptors. J. Opt. Soc. Am. 68: 149-157.

Meinertzhagen, I.A. (1973) Development of the compound eye and optic lobes of insects. In: Developmental Neurobiology of Arthropods. Ed. D. Young. Cambridge, London, New York, Cambridge University Press, p. 51-104.

Meinertzhagen, I.A. (1976) The organization of perpendicular fibre pathways in the insect optic lobe. Phil. Trans. R. Soc. Lond. 274B: 555-596.

Melamed, J. & Trujillo-Cenoz, O. (1968) The fine structure of the central cells in the ommatidia of dipterans. J. Ultrastruct. Res. 21: 313-334.

Pierantoni, R. (1976) A look into the cock-pit of the fly: the architecture of the lobular plate. Cell Tissue Res. 171: 101-122.

Ready, D.F., Hanson, T.E. & Benzer, S. (1976) Development of the Drosophila retina. A neurocrystalline lattice. Dev. Biol. 53: 217-240.

Ribi, W.A. (1975) The neurons of the first optic ganglion of the bee Apis mellifera. Adv. Anat. Embryol. Cell Biol. 50: 1-43.

Rowell, C.H.F., O'Shea, M. & Williams, J.L.D. (1977) The neuronal basis of a sensory analyser, the acridid moment detector system. IV. The preference for small field stimuli. J. Exp. Biol. 68: 157-185.

Scholes, J. (1969) The electrical responses of the retina receptors and the lamina in the visual system of the fly Musca. Kybernetik 6: 149-162.

Shaw, S.R. (1975) Retinal resistance barriers and electrical lateral inhibition. Nature (Lond.) 255: 480-482.

Shaw, S.R. (1981) Anatomy and physiology of identified non-spiking cells in the photoreceptor-lamina complex of the compound eye of insects, especially diptera. In: Neurons without Impulses. Ed. A. Roberts & B.M.H. Bush. Cambridge, London, New York, Cambridge University Press.

Srinivasan, M.V. & Bernard, G.D. (1975) The effect of motion of visual acuity of the compound eye: a theoretical analysis. Vision Res. 15: 515-525.

Stavenga, D.G. (1979) Pseudopupils of compound eyes. In: Handbook of Sensory Physiology. Vol. VII/6A. Ed. H. Autrum. Heidelberg, Berlin, New York, Springer, p. 357-439.

Strausfeld, N.J. (1970) Golgi studies on insects. Part II. The optic lobes of diptera. Phil. Trans. R. Soc. Lond. 258B: 175-223.

Strausfeld, N.J. (1971) The organization of the insect visual system (light microscopy). II. The projection of fibres across the first optic chiasma. Z. Zellforsch. 121: 442-454.

Strausfeld, N.J. (1976 a) Mosaic organizations, layers, and visual pathways in the insect brain. In: Neural Principles in Vision. Ed. F. Zettler & R. Weiler. Berlin, Heidelberg, New York, Springer.

Strausfeld, N.J. (1976 b) Atlas of an Insect Brain. Berlin, Heidelberg, New York, Springer.

Strausfeld, N.J. (1979) The representation of a receptor map within retinotopic neuropil of the fly. Verh. Dtsch. Zool. Ges. 1979: 167-177.

Strausfeld, N.J. (1980) Male and female neurons in dipterous insects. Nature (Lond.) 233: 381-383.

Strausfeld, N.J. & Bacon, J.P. (1983) Multimodal convergence in the central nervous system of insects. In: Multimodal Convergence in Sensory Systems. Fortschr. Zool. 28. Ed. E. Horn. Stuttgart, New York, Gustav Fischer.

Strausfeld, N.J. & Bassemir, U.K. (1983) Cobalt-coupled neurons of a giant fiber system in Diptera. J. Neurocytol. (In press)

Strausfeld, N.J., Bassemir, U.K., Singh, R.N. & Bacon, J.P. (1983) Organizational principles of outputs from dipteran brains. J. Insect Physiol. (In press)

Strausfeld, N.J. & Blest, A.D. (1970) Golgi studies on insects. Part I. The optic lobes of Lepidoptera. Phil. Trans. R. Soc. Lond. 258B: 81-134.

Strausfeld, N.J. & Braitenberg, V. (1970) The compound eye of the fly (Musca domestica): connections between the cartridges of the lamina ganglionaris. Z. vergl. Physiol. 70: 95-104.

Strausfeld, N.J. & Campos-Ortega, J.A. (1977) Vision in insects: pathways possibly underlying neural adaptation and lateral inhibi-

tion. Science 195: 894-897.
Strausfeld, N.J. & Hausen, K. (1977) The resolution of neuronal assemblies after cobalt injection into neuropil. Proc. R. Soc. Lond. 199B: 563-476.
Strausfeld, N.J. & Nässel, D.R. (1981) Neuroarchitecture of brain regions that subserve the compound eyes of crustacea and insects. In: Handbook of Sensory Physiology. Vol. VII/6B. Ed. H. Autrum. Berlin, Heidelberg, New York, Springer.
Taylor, C.P. (1981) Contribution of compound eyes and ocelli to steering of locusts in flight. 1. Behavioural analysis. J. Exp. Biol. 93: 1-18.
Torre, V. & Poggio, T. (1978) A synaptic mechanism possibly underlying directional selectivity to motion. Proc. R. Soc. Lond. 202B: 409-416.
Trujillo-Cenoz, O. (1966) Some aspects of the structural organization of the intermediate retina of dipterans. J. Ultrastruct. Res. 13: 1-33.
Trujillo-Cenoz, O. & Melamed, J. (1966) Electron microscopical observations on the peripheral and intermediate retina of dipterans. In: The Functional Organization of the Compound Eye. Ed. C.G. Bernhard. London, Pergamon.
Vigier, P. (1907 a) Méchanisme de la synthèse des impressions lumineuses recueillies par les yeux composés des Diptères. C.R. Acad. Sci. (Paris) 63: 122-124.
Vigier, P. (1907 b) Sur les terminations photoréceptrices dans les yeux composés des Muscides. C.R. Acad. Sci. (Paris) 63: 532-536.
Vigier, P. (1907 c) Sur la réception de l'excitant lumineux dans les yeux composés des insectes, en particulier chez les Muscides. C.R. Acad. Sci. (Paris) 63: 633-636.
Vigier, P. (1908) Sur l'existence réelle et le rôle des appendices piriform des neurones. La neurone périoptique des Diptères. C.R. Soc. Biol. (Paris) 64: 959-961.
Wilson, M. (1978) Functional organization of locust ocelli. J. Comp. Physiol. 124: 297-316.
Wunderer, H. & Smola, U. (1982 a) Morphological differentiation of the central visual R7/R8 in various regions of the blowfly eye. Tissue Cell 14: 341-358.
Wunderer, H. & Smola, U. (1982 b) Fine structure of ommatidia at the dorsal eye margin of Calliphora erythrocephala Meigen (Diptera: Calliphoridae): an eye region specialized for the detection of polarized light. Int. J. Insect Morphol. Embryol. 11: 25-38.
Zettler, F. & Järvilheto, M. (1971) Decrement-free conduction of graded potentials along the axon of a monopolar neuron. Z. vergl. Physiol. 75: 402-421.
Zettler, F. & Järvilheto, M. (1972) Lateral inhibition in an insect eye. Z. vergl. Physiol. 76: 233-244.

THE LOBULA-COMPLEX OF THE FLY:
STRUCTURE, FUNCTION AND SIGNIFICANCE IN VISUAL BEHAVIOUR

KLAUS HAUSEN

Max-Planck-Institut für biologische Kybernetik
Spemannstrasse 38, D 7400 Tübingen
Federal Republic of Germany

ABSTRACT

The lobula-complex of flies consists of the highest order visual neuropils of the optic lobe, the lobula plate and the lobula. Anatomical and electrophysiological investigations of the lobula plate have revealed that it contains a system of large directionally selective motion sensitive interneurons. The structure, response characteristics and synaptic interactions of these interneurons are described. There is strong evidence that the lobula plate is the main motion computation centre of the optic lobe controlling the optomotor responses of the fly. Additional functions in the visual fixation and tracking behaviour and in the figure-ground discrimination seem likely. The lobula has so far been studied only anatomically but not physiologically. Rather indirect evidence suggests that it computes visual signals initiating escape behaviour. The existence of male specific interneurons in this neuropil indicates that it is additionally involved in the control of chasing behaviour typical of males.

INTRODUCTION

Processing of information in the visual system of the fly is currently investigated at two different levels. The analysis of visually induced behaviour reveals general operating rules of the system and provides the theoretical framework for an understanding of the underlying neural mechanisms. Electrophysiological and anatomical investigations elucidate the actual wiring of the visual neuropils and the functional properties of their single interneurons. Both approaches are complementary and influence each other strongly: only behavioural observations enable us to interpret the responses of interneurons in terms of their functional significance and only electrophysiological studies can prove neuronal models derived from behavioural investigations as right of wrong.

The uptake and peripheral processing of visual signals in the compound eye and lamina of the fly have been treated in previous chapters of this volume (Franceschini, 1983; Laughlin, 1983). The present account deals with the structural and functional properties of the highest order visual neuropils in the optic lobe and their significance in visual behaviour.

GENERAL ARCHITECTURE OF THE VISUAL SYSTEM OF THE FLY

The neuroanatomy of the fly has been thoroughly studied in recent years and its compound eyes, optic lobes and optic centres in the brain (Fig. 1) may presently be regarded as the best known insect visual system (Strausfeld, 1983). This section summarises some optical and structural

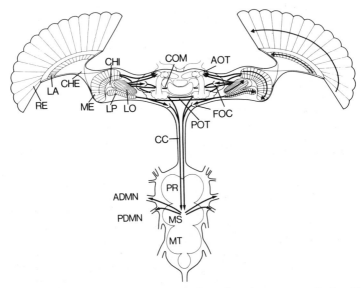

Fig. 1. Basic organisation of the visual system of the fly. The two compound eyes (RE) are connected to the brain by optic lobes, each containing successive visual neuropils (LA: lamina, ME: medulla, LP: lobula plate, LO: lobula) and two chiasms (CHE: external chiasma, CHI: internal chiasma). The visual neuropils consist of columns (indicated by thin lines) which are retinotopically arranged (indicated by arrows in the neuropils of the right optic lobe). Main optic tracts in the brain and to the thoracic ganglion are marked by arrows. Further abbreviations: ADMN, PDMN: anterior and posterior dorsal metathoracic nerve; CC: cervical connective; COM: central commissure; FOC: anterior and posterior optic foci; POT: posterior optic tract; PR, MS, MT: pro-, meso-, metathoracic neuromeres of the thoracic ganglion.

features of this system which are particularly relevant with respect to the higher order visual interneurons discussed below.

The compound eye

The visual field of the compound eye comprises about one hemisphere of the surrounding space and overlaps frontally and caudally with the field of the contralateral eye (Beersma et al., 1977; Franceschini, 1975; Franceschini et al., 1979). Positions within the fields are described in terms of the coordinates ψ and θ (Fig. 2 a), denoting horizontal angular distance with respect to the vertical symmetry plane of the head and vertical angular distance with respect to the horizontal plane through the frontal crossection of both eye equators, respectively.

Each ommatidium in the compound eye contains 2 central receptors R7/8 and 6 peripheral receptors R1-6. Central receptors are characterised by narrow angular sensitivities, low absolute sensitivities and spectral sensitivities which peak in the UV (R7) and blue-green range (R8) of the spectrum; peripheral receptors have larger angular and higher absolute sensitivities and are UV-green receptors. All receptors show polarisation sensitivity (Hardie, 1979; Hardie et al., 1979). Central receptors share a common optical axis which represents the optical axis of the ommatidium.

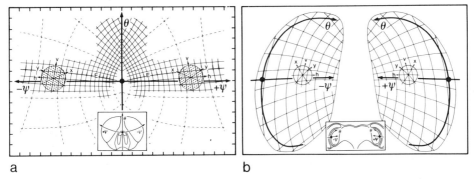

Fig. 2. Lattice orientations in the frontal parts of the two compound eyes and orientations of the columnar lattices in the lobula plates.

a) Telescopic inspection of the eyes reveals hexagonal lattices of ommatidial axes or sampling stations as demonstrated in the encircled areas as examples; the lattice axes are termed **h, v, x** and **y**. The orientations of lattice axes **h** and **v** in the frontal fields of the eyes are plotted with respect to the angular coordinates ψ and θ (see inset). Solid lines were drawn after optical measurements, broken lines are extrapolations. Scale unit on frame of figure: 10°. The plot demonstrates the gradual changes of lattice orientations in the dorso- and ventrofrontal eye regions as compared to the regions of the eye equators (E). Drawn after measurements of Franceschini, 1975, and Franceschini, unpubl. results. Calliphora ♂

b) Schematic representation of the orientations of the columnar lattice axes **h** and **v** and the coordinates ψ and θ in the left and right lobula plate (frontal view; see inset).

The axes of the peripheral receptors in each ommatidium diverge, the divergence angles being identical to those between neighbouring ommatidia. As a consequence of this arrangement the optical axes of six different peripheral receptors lying in a characteristic group of six ommatidia surrounding a central one are parallel. The axons of these receptors converge and terminate in one column of the lamina, where their signals are pooled and which can thus be understood as a neuroommatidium (neural superposition: Kirschfeld, 1967; Braitenberg, 1967). Since the polarisation sensitivities of the peripheral receptors differ with respect to the preferred orientation of the E-vector of the incident light, the first order interneurons postsynaptic to all six receptors are polarisation insensitive. The axons of the central receptors bypass the lamina columns and project into the second visual neuropil, the medulla.

The optical axes of the ommatidia are arranged in a hexagonal lattice which represents the sampling stations of the eye in visual space. Spatial relationships between lattice points are described using the horizontal, vertical and oblique lattice axes **h, v, x** and **y** as reference (Braitenberg, 1970; Stavenga, 1975; Beersma et al., 1975, 1977). Optical measurements have revealed that the orientation of the lattice is not constant with respect to the coordinates ψ and θ but changes gradually over the eye (Fig. 2a; Franceschini, 1975, 1983). This is particularly evident in the frontal (and very likely also in the caudal) eye region: whereas the horizontal lattice axes **h** are approximately horizontally aligned near the eye-equator, they become vertically tilted towards the dorsal and ventral margins of the eye. The orientations of the lattice axes **v, x** and **y** change respectively. A second characteristic of the eye also becomes apparent from Fig. 2a. The angular distances between adjacent lattice points, i.e. the divergence angles between the single ommatidia are smallest in the fronto-equatorial part of the eye and increase gradually in dorsoventral and in lateral direction (Beersma et al., 1975). This indicates an inhomogeneous angular resolution of the eye, being optimal frontally.

Recent studies have revealed a remarkable sexual dimorphism in the compound eyes of flies which is mainly expressed by the existence of a specific type of central receptors in the dorsofrontal eye region of males. These receptors have the same spectral sensitivity as the peripheral receptors and terminate in the lamina, presumably contributing to the neural superposition of signals from the receptors R1-6 (Franceschini et al., 1981; Franceschini, 1983; Hardie et al., 1981).

The optic lobe

The optic lobe subserving the compound eye consists of discrete subsequent visual neuropils (Braitenberg, 1967, 1970, 1972; Strausfeld, 1976a, b, 1983; Strausfeld & Nässel, 1981): the above mentioned lamina, the medulla and the lobula-complex, which is subdivided into the anterior lobula and the posterior lobula plate (Fig. 1). All visual neuropils are composed of columns of basically identical sets of small field neurons (columnar cells) and larger tangential and amacrine cells which invest plexiform layers oriented orthogonally to the columns. The number of columns in each neuropil is identical to the number of ommatidia in the compound eye. The visual neuropils show a strict retinotopic order: each lamina column can be correlated to a corresponding ommatidium and is connected via the external chiasma to a corresponding medulla column. The complicated internal chiasma connects each medulla column to corresponding columns of both, the lobula and lobula plate. Hence, each visual neuropil can be interpreted as an anatomical map of the mosaic of sampling stations in the ipsilateral visual field. Due to the fibre crossings in the chiasms, this map is horizontally inverted between lamina and medulla and rotated again between medulla and lobula-complex. The retinotopic order, however, remains unchanged. In the lobula-complex, columns at the distal and proximal margins of both neuropils represent sampling points in the frontal and caudal regions of the ipsilateral visual field; dorsal and ventral columns represent sampling points in the dorsal and ventral visual field, respectively (Fig. 2b).

The visual neuropils show characteristic anatomical gradients which indicate the predominant importance of the fronto-equatorial eye region (Braitenberg & Hauser-Holschuh, 1972; Strausfeld & Nässel, 1981; Strausfeld, 1983). In the lobula complex of females, the columns subserving this part of the eye are enlarged (Fig. 2b). In males the situation is slightly different, since larger columns are also present in the dorsolateral part of the neuropil which corresponds to the dorsofrontal eye-region.

Visual tracts

The sequence of visual neuropils connected by the chiasms represents the main visual pathway from compound eye to brain. It terminates in the anterior and posterior optic foci, which are the projection centres of interneurons originating from the lobula and the lobula plate, respectively. Additional visual tracts connect both medullae (posterior optic tract), both lobulae (giant commissure) and both lobula plates, and mediate direct binocular interactions between the optic lobes. Visual side tracts are the anterior optic tracts emerging from each medulla and lobula and terminating in the anterior optic tubercles near the dorsofrontal surface of the brain. The ventrolateral protocerebrum and, in particular, the optic foci are densely invested by dendrites of descending neurons which project through the cervical connective into the motor control centres of the fly in the thoracic ganglion.

The thoracic ganglion

This part of the nervous system is a dense neuropil fused from three pairs of neuromeres (pro-, meso- and metathoracic ganglia) each of which gives rise to one pair of ventral nerves, two pairs of dorsal nerves (anterior and posterior dorsal nerves) and one pair of accessory nerves (Power, 1948). Most important in this context are the anterior dorsal mesothoracic nerves which contain the main bulk of fibres innervating the direct flight muscles, and the posterior mesothoracic nerves which lead to the tergo-trochanter muscles. The direct flight muscles are the steering muscles of the wings which are activated during visually induced flight manoeuvres (Heide, 1982). The tergo-trochanter muscles mediate the jump-movement of the mesothoracic leg during flight initiation and escape (Levine & Tracey, 1973; Tanouye & Wymann, 1980).

THE LOBULA COMPLEX: INPUT ORGANISATION

As pointed out above, each column in the lobula and each column in the lobula plate receive their inputs from one corresponding medulla column. In addition, corresponding columns of both neuropils of the lobula complex are directly coupled. The cellular architecture of the internal chiasma mediating these connections is not yet fully analysed. Golgi studies (Strausfeld, 1976a, b, 1983) have demonstrated, however, that each medulla column contains at least one centripetal Y-cell which bifurcates in the internal chiasma, and terminates in both the lobula and the lobula plate. Direct connexions between medulla and lobula are established by columnar TM-cells. In addition, each medulla column contains two T4 cells, one of which terminates in anterior layers of the lobula plate while the other one projects into posterior layers. Similarly, each lobula column contains a pair of T5 cells, which project into the lobula plate and terminate at different levels near the anterior and posterior surface of this neuropil.

The functional properties of the columnar input elements to the lobula complex are unknown and only few data are available about signal processing in the medulla. Anatomically the medulla is the most complex structured region of the optic lobe (see Strausfeld, 1983) and the variety of functionally different units which have been recorded in this neuropil seems to mirror its anatomical complexity (Bishop et al., 1968; McCann & Dill, 1969; Mimura, 1971, 1972; DeVoe & Ockleford, 1976, DeVoe, 1980).

Medulla units with small receptive fields (diameters 10° - 40°) which might represent columnar cells can be roughly classified into (a) "simple" units (phasic or tonic excitatory or inhibitory responses to luminance changes: on-, off-, on/off-, on-sustained units), (b) more "complex" units (on-sustained units with preference for horizontal or vertical orientation of contrast borders within their fields), (c) non-directional movement sensitive units (which are probably not motion- but flicker-sensitive) and (d) directionally selective movement sensitive units. Large field units (receptive fields: ipsilateral, contralateral or binocular; field sizes: up to one hemisphere or total visual field of the animal) which presumably represent tangential or amacrine cells were in most cases found to be sensitive to motion. In general, the described units were not identified histologically and their functional characteristics remain to be investigated in detail. The fact, that motion detection takes place already at the level of the medulla is, however, firmly established electrophysiologically. This is of considerable importance with respect to the function of the lobula complex.

THE LOBULA PLATE: ANATOMY OF THE TANGENTIAL CELLS

In contrast to all other neuropils of the optic lobe, the anatomical appearance of the lobula plate is dominated by an array of large tangential neurons. These cells arborise within distinct, thin vertical layers of the neuropil and have characteristic shapes which allow their unambiguous identification in histological preparations of different animals. At present 21 anatomical classes of tangential cells are known, more than 10 of which have been studied physiologically. These studies have revealed that all tangential cells of the lobula plate are large-field, motion sensitive elements responding selectively to horizontal or vertical motions in the visual field of the fly (Hausen, 1981).

Structures and preferred directions

Examples of physiologically studied tangential cells from the lobula plate of the female blowfly (Calliphora erythrocephala) are compiled in Fig. 3. The cells were reconstructed from frontal sections of dye-injected or cobalt-impregnated material and are arranged according to the sequence of their arborisation layers within the neuropil from anterior to posterior. This sequence may be subject to minor changes in the future since the evaluation of the precise relative positions of the different cells is technically complicated and not satisfactory in all cases. In addition, some cells (marked with an asterisk) have bistratified arborisations which renders the specification of their position difficult. The main preferred and null directions of the cells are indicated on the right side of the figure by filled and open arrowheads, respectively. A number of tangential cells show additional responses to motion in the contralateral visual-field, which results from interactions between both lobula plates. These interactions and the resulting binocular sensitivities of tangential cells will be discussed later.

Anatomically, the tangential cells of Fig. 3 can be classified into centripetal elements, centrifugal elements and heterolateral connexion elements projecting to the contralateral lobula plate. Major centripetal (output) cells are represented by two classes of giant neurons termed the horizontal and vertical systems. The horizontal system (HS) consists of 3 cells (north, equatorial and south horizontal cell: HSN, HSE, HSS) residing near the anterior surface of the lobula plate and projecting into the posterior, ventrolateral protocerebrum (Fig. 4). The cells respond preferentially to progressive (front to back) motion in the ipsilateral visual field (Braitenberg, 1972; Pierantoni, 1973, 1976; Dvorak et al., 1975; Hausen, 1976a, b, 1981, 1982a, b; Hengstenberg, 1977; Eckert, 1978, 1981). The vertical system (VS) consists of 11 cells (VS1 - VS11) arranged serially at the posterior surface of the lobula plate and projecting into the posterior perioesophageal region of the protocerebrum. Vertical cells respond in general to vertical motion but are a functionally inhomogenous group: whereas the cells VS2-6 respond exclusively to ipsilateral downward motion, the bistratified cells VS1 and VS7-9 show this preferred direction only in the ventral parts of their receptive fields. Dorsally, the VS1 responds to regressive (back to front) motion whereas the VS7-9 respond to progressive motion. The directional selectivities of the VS10-11 are supposed to resemble those of the VS7-9. (Braitenberg, 1972; Pierantoni, 1973, 1976; Hausen, 1976a, b, 1981; Hengstenberg, 1977, 1981, 1982a, b; Hengstenberg et al., 1982; Eckert &Bishop, 1978; Soohoo & Bishop, 1980). Heterolateral output elements of the lobula plate are the H2, H3 and H4 cells which show dendritic arborisations in deeper layers of the neuropil and which terminate in the contralateral protocerebrum or exhibit terminal arborisations in both sides of the protocerebrum. The H2 and the H3 respond selectively to ipsilateral regressive

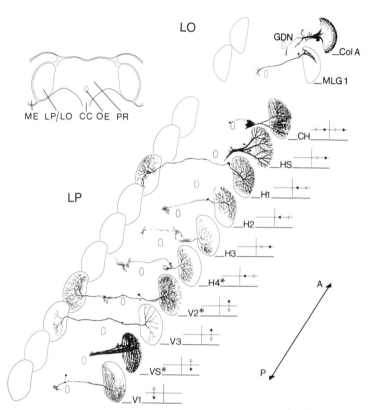

Fig. 3. Neuronal arrangements in the lobula (LO) and lobula plate (LP). The margins of both neuropils in the left and right optic lobe are shown in frontal projection (see frontal representation of protocerebrum and proximal optic lobes in inset; ME: medulla, LP, LO: lobula plate and lobula, CC: cervical connective, OE: oesophagus, PR: protocerebrum). Cells of the right lobula-complex are arranged according to the sequence of their arborisation layers from anterior A to posterior P. The demonstrated cells of the lobula plate are motion sensitive. Arrows on the right side indicate schematically their main preferred and null directions (filled and open arrowheads, respectively) in the right or left visual field (thin crossed lines represent ψ/θ coordinates). Asterisks mark cells having bistratified dendritic fields and complex preferred directions (see text). The cells were reconstructed after intracellular dye-injections or cobalt impregnations from frontal serial sections. Reconstructions of H2, H4, V3: Courtesey of R. Hengstenberg. <u>Calliphora</u> ♀

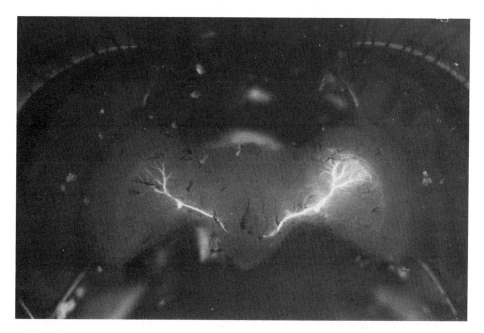

Fig. 4. Lucifer-injected north horizontal cells (HSN) of the right and left lobula plate in a wholemount preparation of the head. The rearside of the head capsula was opened and the photograph was taken from behind. Calliphora ♀

and progressive motion, respectively (Hausen, 1976 a, b; 1981, 1983b; Eckert, 1981; Hengstenberg & Hengstenberg, 1980). The bistratified H4 is activated by regressive motion but responds also to upward motion in the ipsilateral visual field (Hengstenberg, unpubl. results).

Centrifugal elements include the centrifugal horizontal cells (CH), a class consisting of two cells (dorsal and ventral centrifugal horizontal cell: DCH, VCH) having small dendritic arborisations which surround the terminals of the horizontal cells. Their extended terminal arborisations cover the entire frontal surface of the lobula plate. CH-cells respond binocularly to contralateral regressive and ipsilateral progressive motion. (Hausen, 1976a, b, c; 1981, 1983a; Eckert, 1979). A second centrifugal element is the V1 cell, which projects from the terminal region of the vertical system into the contralateral lobula plate where it terminates on the posterior surface. The V1 responds to downward motion (Hausen, 1976a, b, 1981; Eckert & Bishop, 1978).

The H1, V2 and V3 cells are heterolateral cells connecting both lobula plates directly. The H1 is sensitive to ipsilateral regressive motion (Hausen, 1976a, b; 1981, 1983b; Eckert, 1980) and the V3 responds to upward motion (Hengstenberg, pers. comm.). Like the H4, the bistratified V2 shows a complex response behaviour: the main preferred direction is ipsilateral upward, additional responses can be elicited also by regressive motion (Hausen, 1976a, b). It should be noticed that in contrast to the horizontal, centrifugal horizontal and vertical cells, the H1-H4 cells and the V1-V3 cells are in fact unique cells in the sense that only one of them exists per lobula plate.

In previous studies employing extracellular recording techniques (Bishop & Keehn, 1966; Bishop et al., 1968; McCann & Dill, 1969), units of the lobula plate were characterised physiologically but could not be identified anatomically. From comparisons of response characteristics it is evident that the former unit II-1 is identical to the H1; V1 and V2 probably represent the units II-2 and II-4, respectively. The unit II-3 is still unidentified and is termed below as U.

The above given list of tangential cells is based on studies in the blowfly Calliphora erythrocephala. It is assumed, however, that the basic architecture of the lobula plate in different species and even in different families of flies is essentially the same. This is indicated by the fact, that horizontal and vertical cells have been observed also in other species (Drosophila: Heisenberg et al., 1978; Musca: Pierantoni, 1976; Fannia: Hausen, unpubl. observations; Phaenicia: Dvorak et al., 1975; Eckert & Bishop, 1978) and that, in addition, very similar physiological results were obtained from recordings in Calliphora (reviewed by Hausen, 1981); Phaenicia (Dvorak et al., 1975; Eckert, 1980, 1981; Eckert & Bishop, 1978; Soohoo & Bishop, 1980) and Musca (Bishop et al., 1968). Since significant species dependent differences with respect to the tangential cells of the lobula plate have not been reported so far, it seems justified to neglect this problem in the following sections.

Synaptic connexions

The functional discrimination between dendritic and telodendritic arborisations of the tangential cells described in the previous section results mainly from light microscopical and electrophysiological evidence. Ultrastructural studies in order to determine the location and polarity of synaptic structures have been performed so far only on the giant horizontal and vertical cells and the centrifugal horizontal cells. These investigations have revealed that both classes of giant cells are purely postsynaptic in their dendritic regions and that synaptic sites are located on the dendritic main branches and, in highest density, on the fine higher order profiles (Pierantoni, 1976; Hausen et al., 1980; Bishop & Bishop, 1981; Eckert & Meller, 1981); synaptic contacts are made onto the above mentioned columnar T4 cells of the medulla and T5 cells of the lobula (Strausfeld & Bassemir, in prep.). The axons of the giant cells are surrounded by multilamellar glia sheaths, reminiscent of the myelin sheaths of vertebrate neurons, which may play a role in signal transmission of these neurons (Hausen et al., 1980; Hausen, 1982a). The terminal arborisations of both classes of giant cells show pre- and postsynaptic specialisations, indicating that they are not simply output regions but also receive inputs from other cells (Pierantoni, 1976; Hausen et al., 1980; Eckert & Meller, 1981). In the case of the vertical system synaptic output connexions to a large descending neuron have been demonstrated which consist of parallel chemical and gap-junction-like structures (Strausfeld & Bassemir, in prep.). More indirect evidence suggests that the terminals of the horizontal system are also synaptically coupled to descending neurons (Eckert, 1981). Finally, it was found that the dendrites of the centrifugal horizontal cells are postsynaptic to the horizontal cell terminals (Eckert & Meller, 1981) which is in accordance with previous electrophysiological evidence (Hausen, 1976a, 1983a).

Dendritic constancy and variability

The tangential cells of the lobula plate are particularly well suited for comparative studies of their dendritic geometries, since they reside in thin, nearly planar vertical layers of the neuropil. This allows almost distortion-free reconstructions of the cells from well oriented frontal serial sections of the brain. In addition, at least the horizontal and vertical systems of both optic lobes can be routinely and reliably revealed in silver intensified cobalt impregnations by employing simple backfilling techniques (Strausfeld & Obermayer, 1976; Hausen et al., 1980; Hausen & Wolburg-Buchholz, 1980; Wässle & Hausen, 1981).

Detailed studies of the dendritic arborisations of horizontal and vertical cells (Hausen, 1982 a; Hengstenberg et al., 1982) and further tangential cells (Hausen, 1981, 1983b) have demonstrated that in general the branching pattern of a certain cell in the left and right lobula plate of the same animal and in different animals is highly variable, although its characteristic gestalt can always be recognised. Vertical cells are exceptional in this respect since branching variability is mainly expressed in their second and higher order dendrites while the first order dendritic main trunks are always very similar in shape. Interestingly, the same tangential cells from the two optic lobes of one brain do not show a stronger similarity of branching patterns than the respective cells from different animals.

The dendritic arborisations are highly constant, on the other hand, with respect to (a) the neuropil layer invested by the tips of the fine branches. (b) the position and extent of the invested field within this layer, and (c) the branching density distribution within the field (Fig. 5). In addition, the tangential cells are highly constant with respect to the number of individual elements constituting one anatomical class (e.g. inspection of horizontal systems from more than 100 animals revealed only one system consisting of only two instead of three cells; Hausen, unpubl. observation).

These findings are interesting for two different reasons. First, the constancy of certain dendritic features suggests that they are of particular functional relevance to the cells. That this is indeed the case will be shown below. Second, the analysis of the branching geometries may give some insight into the mechanisms governing the outgrowth of the dendritic trees during development. It seems very likely that dendritic growth is rigidly controlled with respect to the neuropil area that a given cell has to invest with a certain branching density in order to establish the proper distribution of synaptic connexions to its input cells. The actual route of the single dendritic profiles within this area, however, seems to be less important and less accurately determined. This becomes particularly evident from the fact that the dendritic patterns of the same cells in the two lobula plates of one animal differ significantly al-

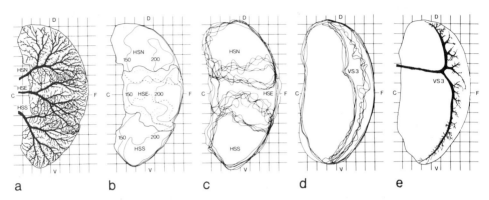

Fig. 5. Anatomical characteristics of the dendritic fields of the three horizontal cells HSN, HSE, HSS, and the vertical cell VS3. a) Graphical reconstruction of a cobalt-stained horizontal system of the right lobula plate obtained from frontal serial sections. b) Dendritic density of the horizontal system shown in a). Dendritic density was measured as dendritic length per square unit and is plotted in arbitrary units. Dendritic densities are highest in the lateral lobula plate and decline in proximal direction. c) Constancy of dendritic fields. The dendritic field margins of horizontal cells from 6 lobula plates were obtained from reconstructions as shown in a) and were superimposed after normalization of the lobula plates with respect to their dorso-ventral extent. The plot demonstrates the extraordinary spatial constancy of the dendritic fields of the horizontal cells. d, e) Reconstruction of a cobalt-filled vertical cell VS3 and dendritic field plots of 4 VS3-cells. Field constancy in the VS cells is similar to the HS cells and represents a general feature of tangential cells in the lobula plate. Calliphora ♀

though they are genetically identical.

Studies of branching geometries in sensory neurons of other insects (Goodman, 1974; O'Shea et al., 1974; Altman & Tyrer, 1977) demonstrate that in general the overall shape of the cells and the areas invested by dendritic or terminal arborisations are rather constant whereas the patterns of fine branches are more or less variable. The degree of variability in sensory neurons however is consistently found to be much smaller than in motoneurons (Burrows, 1973). This suggests again that constant features in the branching geometries play a significant functional role in sensory neurons.

Dendritic stability under visual deprivation

In a recent deprivation study (Hausen, in prep.) adult flies were kept after emergence from pupae for 7-10 days in (a) complete darkness, and (b) under illumination with short stroboscope flashes which provided a practically motion-free visual input to the animals (flash duration: 3 msec; flash frequency 2 Hz). The horizontal and vertical cells of the deprived animals were cobalt impregnated and their structure compared with the cells of control animals from the same stock, kept for the same period of time under normal rearing conditions (12 h daylight and 12 h darkness per day), and with cells of animals which were impregnated directly after emergence. Light microscopical analysis of the cells from the four groups revealed no obvious differences with respect to their dendritic patterns. Hence the dendritic morphologies of the giant cells, and presumably those of the other tangential cells in the lobula plate also, are fully developed at emergence and seem to remain constant irrespective of visual experience or deprivation of the animal. This is in contrast to the marked deprivation effects found in the vertebrate visual system (Wiesel & Hubel, 1963) and it remains to be investigated whether the synaptic connexions and the physiological properties of tangential cells from deprived animals are also unaffected. Such functional alterations have been observed in visual interneurons of locusts (Bloom & Atwood, 1980) and in the mechanosensory system of crickets (Murphey et al., 1977).

THE LOBULA PLATE: PHYSIOLOGY OF THE TANGENTIAL CELLS

The tangential cells of the lobula plate described in the last section represent a functionally homogeneous set of motion sensitive interneurons differing from each other mainly with respect to their receptive fields and preferred directions. In general, all tangential cells investigated are non-habituating elements responding tonically to motion even under sustained stimulation over long periods. The responsiveness of the cells seems to be rather independent of the overall state of responsiveness of the fly and there are no indications that they are affected by the restraints resulting from the recording situation. Hence the lobula plate can be regarded as a neural centre computing motion in the visual surrounding of the animal in a highly stereotype fashion, largely unaffected by the ongoing activity in the brain and probably also unaffected by the behavioural activity of the animal.

Although the tangential cells are often called motion detectors it is presently unknown whether they actually represent parts of motion detection circuits or whether they simply integrate the signals of motion sensitive input elements. Motion detection relies on correlation of signals from at least two spatially separated sampling stations in the eye, and a simple model of an elementary motion detector (EMD) consists of two input channels, one having high pass filter characteristics, the other one having low pass filter characteristics, the signals of which are multiplied. A detailed discussion of the different models of motion detectors described in the literature is given in another chapter of this volume (Buchner, 1983). Since the actual site of motion detection with regard to the tangential cells is not known, it is assumed for simplicity in the following sections that motion detection takes place between columnar cells in the medulla and that at least four classes of motion sensitive columnar cells (preferred directions: upward, downward, progressive, regressive) project retinotopically into the lobula plate where they contact synaptically the various tangential cells. The fact that motion sensitive small field neurons have been demonstrated in the medulla favours this view. It must be emphasised, however, that there is no evidence at present that excludes the possibility of a separate motion detection stage in the lobula plate.

Signal transmission

The dendritic trees of tangential cells are depolarised under stimulation with motion in their preferred directions and hyperpolarised by motion in their null directions. The heterolateral tangential cells H1-H4 and V1-V3 which are characterised anatomically by long and rather thin axons (length up to 1200 μm; diameter of about 5 μm) transmit information about their dendritic potentials to their terminals in the usual way by modulating the frequency of action potentials travelling down their axons (Dvorak et al., 1975; Hausen, 1976a, b, 1981; Eckert, 1980).

Homolateral cells having relatively short and thick axons (lengths up to 500 μm; diameters up to 25 μm) behave differently. The centrifugal horizontal cells are non-spiking elements operating exclusively with graded potentials (Hausen, 1976a, b, 1981, 1982 c). The giant horizontal and vertical cells were originally considered to respond to ipsilateral stimulation also with graded potentials (Hausen, 1976 a, b). Later, it was demonstrated that they have active membranes (Hengstenberg, 1977; see also Eckert & Bishop, 1978; Eckert & Hamdorf, 1981) and recent recordings demonstrate that they generate complex signals under ipsilateral stimulation consisting of both graded potentials and superimposed action potentials (Hausen, 1981, 1982a). The axonal length constants λ of the giant cells have so far not been determined, but estimations in the horizontal cells (λ = 700 μm - 1500 μm; Eckert, 1981; Hausen, 1982a) show that they exceed by far the actual axon lengths and hence that graded dendritic potentials can be transmitted into the terminals without significant decrement. This is in accordance with electrophysiological observations (Hausen, 1982a). The dense glial sheaths enclosing the axons of both the vertical and the horizontal cells may play a supporting role in graded signal transmission by reducing current loss over the axonal membranes (Hausen et al., 1980). Examples of all three types of signals are given in Fig. 6 showing responses of the HSE, DCH, H1 and V1 to movement in the respective null and preferred directions.

Directional selectivities

The directional selectivity of the tangential cells is demonstrated in Fig. 7a using the south horizontal cell HSS, the V1 and the H1 as examples. The responses of the cells are maximal for motion in their particular preferred directions and decline when the direction of motion is gradually changed. The shape of the obtained response curves can be approximated by cosine-functions. Stimulation with motion in null direction is inhibitory to the tangential cells, the inhibition depending again in a cosine-function on the direction of motion. Preferred and null direction are precisely antiparallel for each cell. Examination of the directional sensitivities of various tangential cells have demonstrated that only four preferred directions are obviously represented in the lobula plate, namely progressive, regressive, upward and downward (some

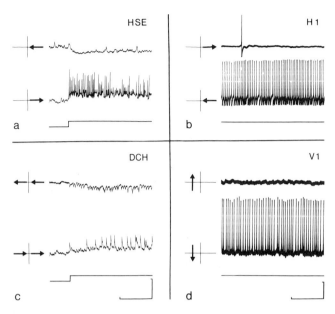

Fig. 6. Axonal signals of the tangential cells of the right lobula plate under stimulation with motion in their respective preferred and null directions. The H1 and V1 respond to motion in preferred direction with regular trains of action potentials. The responses of the HSE consist of graded potentials and superimposed irregular action potentials. The DCH is non-spiking and generates exclusively graded responses resulting from summation of clearly visible epsps and ipsps. Stimulus parameters in a, c: circular periodic pattern of dark and bright stripes; diameter: 39°, spatial wavelength: 13°, mean luminance: 70 cd m^{-2}, contrast: 80%, contrast frequency of motion: 1,5 Hz; position of pattern-centre: $\psi = +40°$ / $\theta = 0°$ (and $\psi = -40°$ / $\theta = 0°$ in c). Stimulus parameters in b, d: rectangular periodic pattern (54° x 72°) of dark and bright stripes; wavelength: 18°, luminance: 70 cd m^{-2}, contrast: 78%, contrast frequency of motion: 1,6 Hz, positions of pattern-centres: $\psi = +50°$ / $\theta = 0°$ in b, and $\psi = -50°$ / $\theta = 0°$ in d. Resting potentials: -60 mV (a), -51 mV (b), -55 mV (c), -53 mV (d). Calibration: 20 mV/200 msec (a, c), 25 mV/100 msec (b), 16 mV/100 msec (d). Calliphora ♀

exceptions are discussed below). Interestingly, these preferred directions are not constant but change considerably in the frontal part of the visual field: horizontal (vertical) preferred directions are tilted vertically (horizontally) in the dorsal and ventral parts of the visual field as

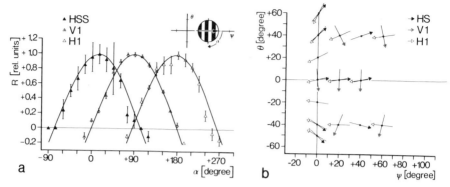

Fig. 7. Directional selectivity of the tangential cells in the lobula plate. a) Motion induced responses R of a HSS, V1 and H1 cell as function of direction α of motion (definition: see inset). The response curves are approximated by cosine-functions; peak values indicate the preferred directions. The HSS, V1, H1 respond maximal to progressive, downward and regressive motion respectively. Stimulus parameters as specified in Fig. 6a. Positions of pattern centres: $\psi = +40°/\theta = -40°$ (HSS), $\psi = +40°/\theta = 0°$ (V1, H1). Stimulus induced responses R (V1, H1: spikes/sec, HSS: mV) are normalised with respect to the peak values and represent means and standard deviations of 3 measurements. b) Preferred and null directions of the three cells of the horizontal system HS, the H1 and the V1, evaluated in different test-positions of the frontal visual field and plotted in the ψ/θ - coordinate system. The resulting pattern of preferred directions resembles the pattern of the **h** and **v** lattice axes shown in Fig. 2a. <u>Calliphora</u> ♀

demonstrated in Fig. 7b. The gradual changes of the preferred directions are parallel to the orientation changes of the horizontal and vertical lattice axes **h** and **v** in the regions of the compound eye scanning these parts of the visual field (compare Figs. 2a and 7b). This indicates that the effective sampling stations of the single EMDs are arranged according to **h** and **v** axes of the eye lattice (Hausen, 1981, 1982b, 1983b). Hence, cells with "horizontal" preferred directions show in fact a characteristic distribution of local preferred directions which is to some extent similar to the flow field seen by an animal during straight flight in a stationary visual surrounding, whereas cells with "vertical" preferred directions are also sensitive to horizontal motion components in the dorso- and ventrofrontal part of the eye which are generated on the retina during roll movements of the animal around its longitudinal body axis.

Examination of the directional sensitivities and the spatial arrangement of the tangential cells in the lobula plate (Fig. 3) reveals a striking correlation between their preferred directions and their dendritic layers: all cells responding to horizontal motion reside in the frontal part of the neuropil whereas those having vertical preferred directions are located posteriorly (Hausen, 1976a, b). Closer inspection suggests that a further subdivision into four directionally layers is possible. The most frontal layer contains elements selectively responding to progressive motion, followed in posterior direction by three layers with cells having regressive, upward and downward preferred directions, respectively. As pointed out previously, the precise determination of the relative positions of the cells to each other is very complicated. The H1 for example, which was previously thought to reside in front of the CH-cells (Hausen, 1981) was only very recently shown to lie posterior to the HS (Hausen, 1983b). This fits the above given concept. In contrast, the position of the H3 is inconsistent with the concept; it seems possible however, that further analysis may reveal that the relative positions of the H2 and H3 have to be changed. The hypothesis of four directionality layers is supported by the results of recent deoxyglucose-studies which have demonstrated directionally selective motion induced activity in four subsequent layers of the lobula plate according to the above given scheme (Buchner, 1983).

It is tempting to assume that the proposed four layers represent discrete synaptic regions established by the termination levels of different classes of columnar input elements which the dendrites of the various tangential cells contact in order to gain their specific directional selectivity. The bistratified dendritic field organisations of the vertical cells V1 and VS 7-9 are particularly suggestive in this respect: these cells show dorsal dendritic fields in frontal layers of the lobula plate and ventral fields in the posterior layer which is also invested by the other vertical cells VS 2-6; correspondingly, VS1 and VS 7-9 respond to horizontal motion in the dorsal visual field and - like the other vertical cells - to vertical motion in the ventral field (Hengstenberg, 1981, 1982b). Similar preferred directions are expected in the cells VS 10 and VS 11, the dendrites which are also bistratified. In addition, the bistratified cells H4 and V2 which give rise to more anterior and more posterior dendritic profiles, respond to both horizontal and vertical motions and have, hence, resulting oblique preferred directions.

Although the effective preferred directions of the tangential cells are in general parallel to the **h** and **v** axes of the eye lattice, this does not exclude the possibility, that the spatial arrangement of their inputs is in fact more complicated: e.g. sensitivity for horizontal motion could result from output summation of arrays of EMDs the sampling stations of which are obliquely arranged according to the **x** and **y** directions in the eye lattice. The problem of the alignment and spacing of the sampling stations has been studied in a number of investigations on the H1 cell using different experimental approaches and leading to contradictory results. Whereas previous studies led to the conclusion that the detection of horizontal motion is in fact mediated by pairs of EMD's having sampling stations in adjacent lattice points in **x** and **y** direction (Zaagman et al., 1977; Masteborek et al., 1980; Srinivasan & Dvorak, 1980; see also McCann & Arnett, 1972) a very recent study, based on direct optical stimulation of single receptors in the eye, demonstrates clearly that motion detection also takes place between adjacent lattice points in the h-direction (Riehle & Franceschini, 1982). This discrepancy remains to be clarified by further investigations. Concerning EMD's detecting vertical motion, there is electrophysiological evidence that the sampling stations are located in adjacent lattice points in the in **v**-direction (McCann, 1973). Behavioural studies addressed to the problem of the spatial arrangement of input channels to EMDs will be discussed below.

Spectral and polarisation sensitivities

Measurements of the response dependency of the tangential cells on stimulus parameters like mean luminance, contrast, spatial and spectral composition, and the orientation of the E-vector of polarised light reveal further information about the cellular architecture and the functional characteristics of the input channels to the lobula plate. Studies of this kind have been mainly performed on the H1 cell, but the obtained results may be valid also for the other tangential cells.

The spectral sensitivity of the H1 is basically identical to that of the UV-green receptors R1-6 in the retina indicating that these retinula cells constitute the main input elements to the EMDs subserving the tangential cells (Bishop, 1969; McCann & Arnett, 1972; Lillywhite & Dvorak, 1981). Only under very specific conditions, designed to stimulate mainly the central receptors R 7/8 (moving periodic patterns of very small spatial wavelength), a different spectral sensitivity was found which peaks in the blue range of the spectrum. In addition, only under this condition the H1 shows polarisation sensitivity. Since R7 and R8 were originally thought to be blue receptors, and since their polarisation sensitivities are not abolished by signal pooling in the lamina as is the case in R1-6, these findings were taken as evidence that the central receptors are additional inputs to the EMDs (McCann & Arnett, 1972). This inference is in contrast to the other studies cited above and disagrees partially with the recently measured spectral sensitivities of R 7/8 (see above). Hence, participation of the central receptors in the movement detection circuitry seems questionable but cannot be excluded with certainty (this problem is discussed at length with respect to behavioural experiments by Heisenberg & Buchner, 1977; see also Buchner, 1983).

Response dependencies on mean luminance

The absolute threshold sensitivity of the H1 is extraordinarily high. Single photons, which are known to elicit quantum bumps in the receptor cells (Hardie, 1979) are also sufficient to induce detectable responses in the H1 (Lillywhite & Dvorak, 1981). Under the usual experimental conditions, i.e. stimulation with moving periodic gratings of mean spatial frequency (0,05 - 0,1 cycles/degree) and high contrast, the response range of the H1 covers about 2 log units of mean luminance from threshold to saturation as shown in Fig. 8 (Bishop et al., 1968; McCann & Foster, 1971; McCann, 1973; Hausen, 1981). This is different from the situation in receptor cells (response range: 4 log units of mean luminance) but agrees well with the characteristics of the first order visual interneurons (response range: 2 log units of mean luminance) and reflects the peripheral processing of visual signals in the lamina (Laughlin, 1981, 1983).

Response dependencies on spatial frequency

Responses of the H1 depend strongly on the contrast and on the

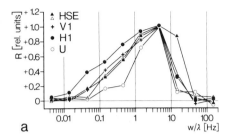 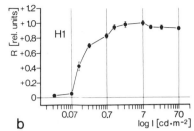

Fig. 8. Dependency of motion induced responses R of tangential cells (HSE, V1, H1 and U) on the contrast frequency w/λ (a) and the mean luminance I (b) of a periodic pattern used for stimulation. The cells respond maximally to a contrast frequency of 4,5 Hz; the response range covers 2 log units of mean luminance. Pattern parameters as specified in legend of Fig. 6a. Values plotted are normalised with respect to maximal responses obtained and are means of 3 measurements. Calliphora ♀

spatial frequency content of moving stimuli. At high mean luminances the sensitivity of the cell is highest for spatial frequencies of about 0,05 cycles/degree and declines steeply above this value towards an upper cutoff frequency of about 0,5 cycles/degree. Qualitatively, this decline is to be expected since the optics in front of each input channel limit the transfer of high spatial frequencies. Interestingly, the sensitivity of the H1 is also reduced in the low spatial frequency range. This can be explained by lateral inhibition in the input circuitry of the EMDs, and it has been speculated that the on-sustained units of the external chiasma (Arnett, 1972), the receptive fields of which consist of a central on-field and two lateral inhibitory off-fields, are neuronal candidates mediating the connection between receptors and EMDs. Low spatial frequency suppression is interpreted as an optimization for the extraction of the most reliable motion cues from the incoming visual signal (Srinivasan & Dvorak, 1980; Dvorak et al., 1980).

Velocity characteristics

Elementary movement detectors as sketched above are in fact not velocity but contrast frequency detectors (contrast frequency = angular velocity/spatial wavelength). The responses of all tangential cells tested so far depend in a characteristic fashion on the contrast frequency of periodic patterns used as stimuli: Lower and upper response thresholds lie at 0,01 - 0,05 Hz and 20 - 50 Hz respectively; the response-peaks are consistently found at 1 - 5 Hz (McCann & Foster, 1971; Hengstenberg, 1977, 1982a; Zaagman et al., 1978; Eckert, 1980; Mastebroek et al., 1980; Soohoo & Bishop, 1980; Hausen, 1981, 1982b, 1983b). In logarithmic plots the response amplitudes increase nearly linearly from lower threshold to peak and fall off sharply above this value. This response behaviour is shown in Fig. 8 for the cells HSE, H1, V1 and U. The similarity of the plotted curves again emphasises the functional homogeneity of the input network of all tangential cells.

Response characteristics as described above are in full agreement with the correlation model of motion detection (Reichardt, 1957; Poggio & Reichardt, 1973) and it has been shown that this model predicts the responses of the H1 even quantitatively if an additional memory-term is implemented (Mastebroek et al., 1982; see also Zaagman et al., 1978).

Receptive fields and spatial sensitivity gradients

With regard to the size of their dendritic fields one can discriminate between tangential cells extending over the entire lobula plate (H1, V2) and cells covering only particular regions of the neuropil (HSN, HSE, HSS: dorsal, equatorial and ventral fields respectively; V2-V11: narrow vertical fields arranged in series from distal to proximal; H2: lateromedial field, V3: medial field; H3, H4: proximal fields). Comparisons between the receptive fields of tangential cells (Bishop et al., 1968; Eckert & Bishop, 1978; Hausen, 1981, 1982b, 1983b; Lillywhite & Dvorak, 1981) and their structures reveal a general rule: the position and extent of the receptive field of a cell in the ipsilateral visual field is determined by the position and size of its dendritic field within the lobula plate. The receptive field of the HSE lies, for example, equatorially, whereas the H1 scans the whole ipsilateral hemisphere (Fig. 9). The close correlation between dendritic and receptive fields demonstrates that information about local motion components is conveyed onto the lobula plate without major deterioration and agrees well with the notion that the dendrites of the tangential cells integrate the signals of columnar input cells to the lobula plate.

A striking functional property of the tangential cells is a gradual change in their spatial sensitivity, which is maximal in the frontal eye region and declines drosoventrally and laterally (Fig. 9). As mentioned above, optical and anatomical gradients in the visual system show the same slope indicating that the frontal eye region is of special significance to the animal. Analysis of the dendritic branching densities of the H1 cell and the three horizontal cells (Fig. 5; Hausen, 1981, 1982a, 1983b) has revealed that they are also highest in the mediolateral part of the lobula plate subserving the fronto-equatorial region of the eye. It can therefore be assumed that the gradients in the spatial sensitivity distributions of the tangential cells are caused anatomically by gradual changes in their dendritic density and, hence, in the density of their input synapses.

Spatial integration properties

It has been discussed above that stimulation of a single EMD is sufficient to elicit a response in a tangential cell. Since large numbers of EMDs feed into these neurons (e.g. at least 5000 in the H1 of Calliphora), one should expect that they are immediately driven into saturation under large-field stimulation. Studies of the spatial integration properties of the HSE (Hausen, 1982b) and H1 (Egelhaaf, unpubl. results) show in fact that their responses increase in highly nonlinear fashion with increasing stimulus size (Figs. 10a, b, curves 1): after a steep, initial increase, the response amplitudes become nearly constant and hence almost independent of the size of stimulus. Further measurements with reduced stimulus

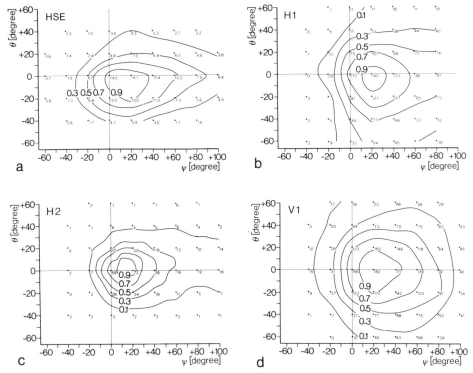

Fig. 9. Receptive fields and spatial sensitivity distributions of 4 tangential cells of the right lobula plate (HSE, H2, H1, V1). Responses to motion of a periodic pattern in the respective preferred directions were measured in all field positions marked by a dot. Small numbers below dots indicate local motion responses (HSE: mV; H1, H2, V1: spikes/sec), lines delineate spatial sensitivity profiles of the cells, large numbers indicate fractions of maximum responses. Stimulus parameters are identical to those specified in Fig. 6a. The receptive fields of the HSE and H2 are restricted to the equatorial region of the ipsilateral visual field. The H1 and V1 are whole field cells. All cells show maximal sensitivity to motion in the fronto-equatorial visual field. Calliphora ♀

strength (e.g. reduced mean luminance: curves 2 and 4 in Fig. 10a, reduced contrast frequency of the stimulus: curve 3 in Fig. 10a, reduced velocity: curves 2 - 4 in Fig. 10b) however demonstrate clearly that this response behaviour cannot be attributed simply to saturation since, again, nearly constant but significantly lower response levels are reached under large field stimulation. Spatial integration properties of this kind can be explained by a recently proposed gain control mechanism which relies basically on large-field inhibitory interactions within the input circuitry of the tangential cells (Poggio et al., 1981; Reichardt et al., 1983). It is

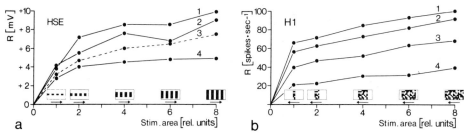

Fig. 10. Spatial integration properties of the tangential cells of the lobula plate. Stimulus induced responses R of a HSE and a H1 to pattern motion in preferred direction are plotted as function of stimulus size. a) The HSE was stimulated by progressive motion of a periodic pattern, the vertical extent of which was increased in 4 steps (full extent of pattern: 32° x 50°; spatial wavelength: 12,5°, contrast: 80%). Curves 1 - 4 were obtained for different mean luminances of the pattern and contrast frequencies of motion (strongest stimulus (1): 70 cd m^{-2}; 1,5 Hz; stimulation with reduced mean luminance (2): 20 cd m^{-2}, (4): 7 cd m^{-2}; stimulation with reduced contrast frequency (3): 0,45 Hz). Values plotted are means of 3 measurements. b) The H1 was stimulated with a random dot pattern, the horizontal extent of which was increased in 4 steps and which was moved sinusoidally in horizontal direction with a constant frequency of 2,5 Hz (full extent of pattern: 48° x 84°). Pattern velocities were altered by changing the oscillation amplitudes (1: $\pm 10°$, 2: $\pm 5°$, 3: $\pm 3°$, 4: $\pm 1°$). Responses were measured during phases of regressive motion (preferred direction of H1); values plotted are means of 95 measurements. In both cells the response amplitudes increase sharply initially but become nearly independent of stimulus size for extended patterns. See text for further discussion. Calliphora ♀

assumed in this model that the signals of all EMDs are integrated by large pool neurons which, in turn, inhibit the output channels of the EMDs before the latter contact the tangential cells. It has not been possible yet to identify neuronal correlates of the proposed inhibition network. One of its major advantages is, however, reflected in the measurements plotted in Fig. 10: the tangential cells are highly sensitive to small stimuli and yet still capable of coding the velocity (or contrast frequency, luminance etc.) of large stimuli since they are protected from saturation.

Synaptic interactions

The receptive fields of the tangential cells in the ipsilateral visual field are determined by their dendritic connexions to columnar input elements. A number of tangential cells exhibit, however, a pronounced binocular sensitivity to motion which results from synaptic contacts to elements of the contralateral lobula plate. Some of these connexions have

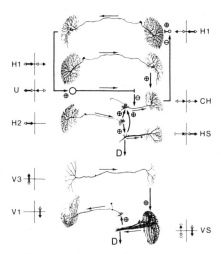

Fig. 11. Synaptic interactions between tangential cells of both lobula plates and resulting binocoular sensitivities to motion. The single connexions are described in the text. Calliphora ♀

been established physiologically and anatomically by means of double recordings and subsequent dye-injections and are demonstrated in Fig. 11.

Examples of binocular elements responding to horizontal motion are the HSN, HSE, H1 and the unidentified unit U. The HSN and HSE which respond to progressive motion in their ipsilateral receptive fields and receive excitatory input from the regressive sensitive H2 of the contralateral lobula plate via synapses at their terminal regions (Hausen, 1976a, b, c, 1981, 1983c; Eckert, 1981). Hence, rotatory motions around the dorsoventral axis of the animal in a clockwise direction are optimal stimuli to the HSN and HSE of the right lobula plate; the respective cells of the left lobula plate are preferentially activated by counterclockwise motions.

The regressive sensitive heterolateral H1 is excitatory to the progressive sensitive unit U in the contralateral lobula plate. This connexion leads to an enhanced sensitivity of U again for rotatory stimuli. Unit U, in turn, is inhibitory to the contralateral CH-cells which also receive excitatory input from the terminals of the H1, the H2 and the horizontal cells. There is evidence that the CH-cells inhibit the input elements of the ipsilateral H1. In short, these interactions lead to a response depression in both H1 cells during binocular regressive motion and a release from inhibition during horizontal rotatory motions. In addition, progressive motion has an inhibitory effect on the H1 via activation of the H and CH cells (Hausen, 1981, 1983c).

Concerning neurons sensitive to vertical motion, it is likely that the V3 cell which responds selectively to upward motion in the medial part of the visual field is excitatory to the medial vertical cells (probably VS5 and VS6) of the contralateral lobula plate. This connexion has not been demonstrated physiologically; recordings have established, however, the resulting binocular sensitivity of medial vertical cells to rotatory motion around the longitudinal body axis of the fly (Hengstenberg, 1981, 1982b). Finally, the anatomy and response characteristics of the V1 cell strongly suggests a synaptic connexion to the vertical system (Hausen, 1976 a, b; Eckert & Bishop, 1978); the output connexions of the V1 in the lobula plate are still unclear.

These interactions can be summarised by stating that in general, binocular interactions between tangential cells serve to enhance their sensitivity to rotatory motion (or to suppress their responses during translatory motion) whereas monocular interactions lead to inhibition between elements having opposite preferred directions. The latter may be regarded as a mechanism to enhance directional "contrast" in the lobula plate. It should be mentioned in this context, that the complex field organisation in the VS1 and the VS7-9, described previously, also represents a specialisation to achieve particular sensitivity to rotation: these cells respond selectively to rotations of the visual world around the transverse headaxis of the fly.

Synopsis

The physiological findings discussed in the last sections can be summarised as follows: (1) Each neuroommatidium represents an input channel to EMDs with different preferred directions. Detection of vertical motion presumably takes place between sampling stations aligned along the v-axes of the eye lattice; detection of horizontal motion is achieved by EMDs having adjacent sampling stations on the h axes or results from summation of two EMDs with adjacent inputs on the x and y axes. (2) EMDs receive predominantly input from receptors R1-6. (3) There is some evidence that the input channels of EMDs underly lateral inhibition and that they are constituted, at least in part, by the on-sustained units of the external chiasma. (4) The multiplication stage of the EMDs is located either in the medulla or the lobula plate; the resulting motion information is represented retinotopically in the lobula plate, separated into four directionality layers for progressive, regressive, upward and downward motion (sequence from anterior to posterior). (5) The tangential cells of the lobula plate integrate the outputs of the EMDs. The directional selectivities of the tangential cells depend on the directionality layer(s) invested by their dendrites; size and position of their dendritic fields determine the spatial arrangements of their receptive fields; the spatial sensitivities of the tangential cells depend on their dendritic density distributions. (6) The spatial integration properties of the tangential cells indicate an inhibitory gain control mechanism within their input circuitry. (7) Many, if not most, tangential cells are particularly sensitive to rotatory motions. Binocular sensitivities to rotation result from direct synaptic interactions between cells of both lobula plates. (8) At least the vertical and horizontal cells, probably also the H3, H4 and V2 cells are coupled to descending neurons projecting into the thoracic ganglion.

THE LOBULA PLATE: FUNCTIONAL SIGNIFICANCE

The lobula plate is considered to play a role in three different types of visually induced behavioural responses, namely the optomotor responses, the fixation and tracking of stationary and moving targets and the discrimination of textured figures from a textured background (for reviews see Hausen, 1977, 1981).

Optomotor responses

The optomotor responses of flies have been thoroughly studied and will be treated in detail in another chapter of this volume (Buchner, 1983; for review see also Poggio & Reichardt, 1976; Reichardt & Poggio, 1976). In short, flying and walking flies transform the visually perceived apparent rotatory motion of their stationary surrounding into torque around their dorsoventral, longitudinal and transverse body axes (yaw, roll and pitch responses). These torque responses are appropriate in sign and amplitude to minimise rotations of the fly relative to its surrounding and thus serve to stabilise its course. Vertical translatory motions induce lift and thrust responses (Götz, 1968; Wehrhahn, 1978; Wehrhahn & Reichardt, 1975). The motion patterns seen by the two compound eyes during locomotion of the animal consist of local motion components differing in direction and velocity. The tangential cells of the lobula plate will be stimulated by such patterns according to their preferred directions and the particular motion components present within their receptive fields. A simple scheme depicting the tangential cells of both lobula plates which are activated by the apparent motions resulting from pure translatory movements (lift, thrust, side slip) and rotatory movements (yaw, roll, pitch) of the head of the fly with respect to a stable surrounding is shown in Fig. 12. It demonstrates that in all cases specific and different groups of tangential cells are activated. Respective groups of other cells, which are not listed in Fig. 12, will be inhibited or remain unstimulated in the different situations. Hence, it seems conceivable that even under complex stimulus conditions, as encountered by the animal e.g. during flight manoeuvres, the motion-information necessary to generate optomotor responses is represented unambiguously in the lobula plate in terms of the relative excitation levels of all tangential cells. A number of investigations have yielded accumulating evidence that the lobula plate indeed controls motion induced torque responses: (a) Responses of tangential cells (H1, HSE), visually induced activity in direct flight muscles (steering muscles), and yaw torque responses studied under stimulation with moving periodic gratings depend similarly on various stimulus parameters like contrast frequency, mean luminance, orientation, size and position (McCann & Foster, 1971; Zaagman et al., 1977, 1978; Mastebroek et al., 1980; Eckert, 1980; Hausen, 1981, 1982b, 1983b; Hausen & Wehrhahn, 1983; Heide, 1975, 1982; Spüler, 1980; Spüler & Heide, 1978). (b) The inputs to both the tangential cells and the optomotor system are mediated predominantly by receptors R1-6. Behavioural experiments concerning the spacing and orientation of the sampling stations of EMDs underlying optomotor responses led to similar - and in the same sense contradictory - results as the above discussed

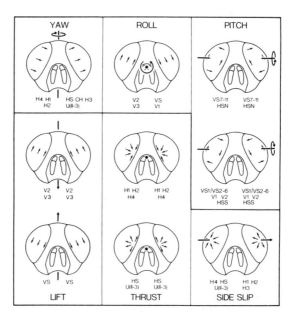

Fig. 12. Diagram of retinal motions induced by translations (lift, thrust, side slip) and rotations (yaw, roll, pitch) of the head of a fly in a stable visual surrounding. For each situation, the tangential cells of the lobula plate excited selectively by the sketched retinal motion pattern are listed. The diagram demonstrates that in all cases different groups of cells are activated and that the relative excitation levels of all tangential cells may code adequately even complex motion patterns seen by the animal during flight.

electrophysiological studies on the EMDs of the lobula plate (Kirschfeld, 1972; Kirschfeld & Lutz, 1974; Buchner, 1976; Buchner et al., 1978; Heisenberg & Buchner, 1977; Götz, 1982; Götz & Buchner, 1978). (c) Electrical stimulation of the lobula plate leads to motor responses (Blondeau 1977; 1981). (d) Optomotor yaw torque responses are reduced in flies, the giant cells of which are reduced or absent due to mutations or due to laser beam ablation during larval development (Heisenberg et al., 1978; Geiger & Nässel, 1981, 1982). (e) Behavioural tests with flies, the horizontal cells of which have been unilaterally lesioned by microsurgery show that the yaw torque responses to binocular stimulation with horizontal pattern motion are seriously changed and seem to consist only of the response component controlled by the horizontal system left intact. This is shown in Fig. 13 (Hausen & Wehrhahn, 1983).

Hence, the evidence presently available indicates clearly that the generation of optomotor torque responses is directly governed by the lobula plate and it is likely that any higher motion centres in the brain (if existing at all) are not essentially involved. It must be stressed however,

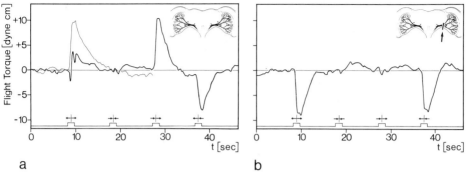

Fig. 13. Change of motion induced optomotor yaw torque by unilateral lesion of the horizontal system. a) Control animal with intact horizontal systems. Binocular stimulation with two horizontally moving periodic patterns placed symmetrically in the left and right equatorial visual field of the fly leads to yaw torque responses of the tethered flying fly which were measured by a torque compensator and which are drawn as thick curve. The sequence and timing of motion stimuli is indicated at the bottom of the figure; between the successive motion stimuli the patterns remain stationary but illuminated. Monocular stimulation (thin curve) of the right eye with the right motion component indicated at bottom leads to a strong positive torque response (intended turning of animal to the right side) for progressive motion; the response to regressive motion is insignificant. Torque responses to binocular motion result from (nonlinear) summation of the monocularly induced torque components: translatory motion stimuli (binocular progressive and regressive motion) are ineffective; rotatory stimuli (clockwise and counterclockwise motion) induce strong responses. b) Animal with lesioned right horizontal system. The torque responses consist only of the response components controlled by the left (intact) side; the response components induced by stimulation of the right eye are abolished. The experiments demonstrate the significance of the horizontal system in yaw torque generation. Response curves are averages obtained from 530 stimulation sequences in 33 flies (a), and from 410 stimulation sequences in 13 flies (b). <u>Calliphora</u> ♀

that optomotor torque generation is not determined exclusively and in a stereotype fashion by the lobula plate since visual cues other than motion and further sensory inputs (from antennae, ocelli, halteres, hair plates on the head) to the motor system play an additional role. In addition, modulatory influences of the protocerebrum on the torque generating system are likely to be present.

Information about the actual wiring between motion detectors and the torque generating motor system has been gained from behavioural studies which established the necessary functional channels between particular EMDs and particular components of the motor output (Götz,

1968, 1982; Götz et al., 1979). The neural correlates of these functional channels (i.e. the connexions between tangential cells of the lobula plate, descending neurons, interneurons of the thoracic ganglion, motoneurons and direct flight muscles) are presently being investigated.

Fixation and tracking

The second type of behavioural response of the fly to be considered in this context is concerned with the fixation and tracking of stationary and moving targets (Reichardt, 1973; see Reichardt & Poggio, 1976; Poggio & Reichardt, 1976; and Buchner, 1983, for review). In principle, these responses rely on the evaluation of the position of an object relative to the fly and a transformation of this positional information into torque responses turning the animal towards the target. Two alternative neural mechanisms underlying these responses are discussed, and both may be used by the visual system: fixation (and tracking) may be controlled by position detectors (sensitive to temporal contrast modulations of spatially stationary input signals; Pick, 1976) or may result from different efficiencies of the motion detectors for progressive and regressive motion with respect to the strength of the elicited yaw torque (Reichardt, 1973; Wehrhahn & Hausen, 1980; see also Geiger, 1981). Concerning the latter mechanism, it is evident that, again, the tangential cells of the lobula plate could be essential in mediating the correct motor output.

Figure-ground discrimination

Flies are able to detect, and to orient towards, textured figures moving in front of a textured background. This response is based on the evaluation of relative motion, and a simple neuronal model for this - rather complicated - computation has been proposed recently (Poggio et al., 1981; Reichardt et al., 1983): The output of the two arrays of EMDs for progressive and regressive motion scanning the entire visual field of the compound eye are summed by a pool cell which inhibits the output channels of the single EMDs via forward or recurrent shunt inhibition. The shunted output channels of the EMDs are synaptically coupled to a large output cell, the synapses of the EMDs sensitive to progressive (regressive) motion being excitatory (inhibitory). The output cell is, thus, a directionally selective motion sensitive large field element sensitive to progressive motion and is assumed to control yaw torque: excitation of the output cell subserving the right eye turns the fly to the right side. It is essential for this model, that the pool cell shows saturation under large field stimulation, that the inhibition is of the shunting type and that the synapses between EMDs and output cells have nonlinear input-output characteristics. Naturally, the same networks are thought to exist behind both eyes and it is further assumed that both pool cells are directly coupled with each other. The reader will have noticed that a basic outline of this model has already been described in the above section on the spatial integration properties of the tangential cells. In fact, spatial integration in e.g. the horizontal cells is identical to that in the model output cells: under large field stimulation the response amplitudes of both are nearly or completely independent of the number of EMDs activated and hence, gain controlled. This and further pieces of evidence indicate that the horizontal cells are possible neuronal correlates of the output cell of the model. It should be stressed, however, that this notion is merely a starting point for further investigations, since other properties of the horizontal cells are in conflict with the model (see discussion in Reichardt et al., 1983). Computer simulations of the model demonstrate that the output cells are highly sensitive to small figures moving in the same direction but with different velocity as compared to the surrounding field (a common situation encountered by the fly during flight when passing nearby objects in front of a more distant background) and that the circuitry can thus be used for figure-ground discrimination. The behavioural output of this model is in good agreement with the yaw torque responses of flies elicited experimentally by stimulation with relative motion of small random dot figures in front of an extended random dot panorama. In summary, there is some evidence that the lobula plate and, in particular, the horizontal system is also involved in the figure-ground discrimination behaviour of the fly.

THE LOBULA: AN UNSOLVED PROBLEM

In contrast to the lobula plate, nothing is known about the functional properties of the lobula and the discussion in this section will be limited to some speculations arising from structural observations. The neuropil of the lobula consists mainly of columnar cells (Strausfeld, 1976a; Strausfeld & Hausen, 1977; Strausfeld & Nässel, 1981), the most dominant of which have been studied anatomically and were termed Col A cells (Fig. 3; Hausen & Strausfeld, 1980). These cells invest each third column in h-direction and each third column in v-direction of the columnar lattice of the lobula, constituting a highly regular network. The Col A cells converge onto the ipsilateral giant descending neuron of the fly (GDN, see Fig. 3), which has been recently studied anatomically and physiologically (Levine & Tracey, 1973; Koto et al., 1981; Strausfeld, 1983; Strausfeld & Nässel, 1981; Strausfeld & Bacon, pers. comm.). The GDN projects through the cervical connective, and its terminal in the thoracic ganglion is synaptically coupled to the motoneuron of the tergo-trochanter muscle and, via the peripheral synapsing interneuron PSI, to the motoneurons of the dorsal longitudinal flight muscles which are the power muscles of the wings (Tanouye & Wyman, 1980). This pathway from Col A neurons to motoneurons represents the functional channel between optic lobe and motor system for the command signals mediating the jump of meta-thoracic legs and initiating flight in the visually induced escape response. The physiological properties of the Col A neurons are unknown but may be inferred from the visual responses of the GDN. This neuron is insensitive to large field motion and responds to changes in ambient luminance (Strausfeld & Bacon, pers. comm). Hence, it may be expected that the Col A cells are an array of small field on-off elements. On the basis of the anatomical and physiological evidence available it is very tempting to speculate that the Col A-GDN-system of flies resembles the LGMD-DCMD-system of Orthoptera (review: O'Shea & Rowell, 1977).

A second structural feature of the lobula leading immediately to functional speculations is the existence of sexually dimorphic interneurons (see Fig. 3) which are large in males and absent or reduced in size in females (Male Lobula Giant (MLG) cells; Hausen, 1979; Hausen & Strausfeld, 1980; Strausfeld, 1980). These neurons lie in the dorsal and lateral marginal regions of the lobula. The invested dorsal neuropil region correlates well with the dorsal eye region containing sex-specific receptors R7 in males (see above). It has been shown recently (Wehrhahn, 1979) that male-specific visual behaviour, namely the chasing of females, employs particularly the dorsal eye region, and it is indeed very suggestive, that the male receptors and the male giant neurons in the lobula play a role in this behaviour. Future studies will demonstrate whether this inference is right or wrong.

CONCLUSION

A main question in the analysis of neural circuitry underlying visual behaviour concerns the essential cues, extracted by the visual system from the incoming signals, and used by the animal for orientation. The present paper demonstrates that for a fast moving animal like the fly information about motion and relative motion is certainly essential during navigation. The importance of another feature, namely the position of stimuli relative to the animal becomes evident from the fact that the visual world is retinotopically represented throughout the successive neuropils by means of small field columnar neurons and that, hence, positional information is available in the optic lobe at all levels of computation. It is still an open question whether, in addition, the spatial orientation of visual stimuli is also an important feature for the fly as is certainly the case in vertebrates (Hubel & Wiesel, 1962). Spatially inhomogeneous centre-surround organisations as present in the on-sustained cells of the external chiasma (see above) may be interpreted as peripheral stages of orientation computation.

Although there is poor or no evidence from the medulla and the lobula, the highly stereotype response patterns of the lobula plate-cells indicate that computation of visual information in the optic lobe is performed rather independently from influences of other sense organs and the central brain. In contrast, visual behaviour is certainly influenced by other sensory modalities and its variability indicates also additional, more complex central inputs to the motor system. Thus, convergence of very different kinds of inputs must take place in the channel between optic lobe and motorneurons, probably at the level of descending neurons which show a remarkable plasticity of their response behaviour. It is challenging to analyse this convergence in detail in order to understand a bit more of the neural machinery generating complex visual behaviour in insects.

ACKNOWLEDGMENTS

I am indebted to E. Buchner, M. Egelhaaf, R. Hardie, R. Hengstenberg, W. Reichardt and C. Wehrhahn for numerous helpful discussions and for critically reading the paper. I thank T. Wiegand for drawing the figures, I. Geiss for typing the manuscript, and especially N. Franceschini and M. Egelhaaf for the permission to use their data for preparation of Figures 2a and 10b.

REFERENCES

Altman, J.S. & Tyrer, N.M. (1977) The locust wing hinge stretch receptors. II. Variation, alternative pathways and "mistakes" in the central arborizations. J. Comp. Neurol. 172: 431-439.

Arnett, D.W. (1972) Spatial and temporal integration properties of units in the first optic ganglion of dipterans. J. Neurophysiol. 35: 429-444.

Beersma, D.G.M., Stavenga, D.G. & Kuiper, J.W. (1975) Organization of visual axes in the compound eye of the fly Musca domestica L. and behavioural consequences. J. Comp. Physiol. 102: 305-320.

Beersma, D.G.M., Stavenga, D.G. & Kuiper, J.W. (1977) Retinal lattice, visual field and binocularities in flies. J. Comp. Physiol. 119: 207-220.

Bishop, C.A. & Bishop, L.G. (1981) Vertical motion detectors and their synaptic relations in the third optic lobe of the fly. J. Neurobiol. 12: 281-296.

Bishop, L.G. (1969) A search for color encoding in the responses of a class of fly interneurons. Z. vergl. Physiol. 64: 355-371.

Bishop, L.G. & Keehn, D.G. (1966) Two types of neurones sensitive to motion in the optic lobe of the fly. Nature (Lond.) 212: 1374-1376.

Bishop, L.G., Keehn, D.G. & McCann, G.D. (1968) Motion detection by interneurons of optic lobes and brain of the flies, Calliphora phaenicia and Musca domestica. J. Neurophysiol. 31: 509-525.

Blondeau, J. (1977) Electrically evoked motor activity in the fly (Calliphora erythrocephala). Dissertation, Eberhard-Karls-Universität Tübingen.

Blondeau, J. (1981) Electrically evoked course control in the fly Calliphora erythrocephala. J. Exp. Biol. 92: 143-153.

Blondeau, J. & Heisenberg, M. (1982) The three-dimensional optomotor torque system of Drosophila melanogaster. Studies on wild-type and the mutant optomotor blind H31. J. Comp. Physiol. 145: 321-329.

Bloom, J.W. & Atwood, H.L. (1980) Effects of altered sensory experience on the responsiveness of the locust descending contralateral movement detector neuron. J. Comp. Physiol. 135: 191-199.

Braitenberg, V. (1967) Patterns of projection in the visual system of the fly. I. Retina-lamina projections. Exp. Brain Res. 3: 271-298.

Braitenberg, V. (1970) Ordnung und Orientierung der Elemente im Sehsystem der Fliege. Kybernetik 7: 235-242.

Braitenberg, V. (1972) Periodic structures and structural gradients in the visual ganglia of the fly. In: Information Processing in the Visual Systems of Arthropods. Ed. R Wehner. Berlin, Heidelberg, New York, Springer-Verlag, p. 3-15.

Braitenberg, V. & Hauser-Holshuch, H. (1972) Patterns of projection in the visual system of the fly. II. Quantitative aspects of second order neurones in relation to models of movement perception. Exp. Brain Res. 16: 184-209.

Buchner, E. (1976) Elementary movement detectors in an insect visual system. Biol. Cybern. 24: 85-101.

Buchner, E. (1983) Behavioural analysis of spatial vision in insects (This volume).
Buchner, E., Götz, K.G. & Straub, C. (1978) Elementary detectors for vertical movement in the visual system of Drosophila. Biol. Cybern. 31: 235-242.
Burrows, M. (1973) The morphology of an levator and a depresser motoneuron of the hindwing of a locust. J. Comp. Physiol. 83: 165-178.
DeVoe, R.D. (1980) Movement sensitivities of cells in the fly's medulla. J. Comp. Physiol. 138: 93-119.
DeVoe, R.D. & Ockleford, E.M. (1976) Intracellular responses from cells of the medulla of the fly, Calliphora erythrocephala. Biol. Cybern. 23: 13-24.
Dvorak, D.R., Bishop, L.G. & Eckert, H.E. (1975) On the identification of movement detectors in the fly optic lobe. J. Comp. Physiol. 100: 5-23.
Dvorak, D., Srinivasan, M.V. & French, A.S. (1980) The contrast sensitivity of fly movement-detecting neurons. Vision Res. 20: 397-407.
Eckert, H. (1978) Response properties of dipteran giant visual interneurones involved in control of optomotor behaviour. Nature (Lond.) 271: 358-360.
Eckert, H. (1979) Anatomie, Elektrophysiologie und funktionelle Bedeutung bewegungssensitiver Neurone in der Sehbahn von Insekten (Phaenicia). Habilitationsschrift, Universität Bochum.
Eckert, H. (1980) Functional properties of the H1-neurone in the third optic ganglion of the blowfly, Phaenicia. J. Comp. Physiol. 135: 29-39.
Eckert, H. (1981) The horizontal cells in the lobula plate of the blowfly, Phaenicia sericata. J. Comp. Physiol. 143: 511-526.
Eckert, H. & Bishop, L.G. (1978) Anatomical and physiological properties of the vertical cells in the third optic ganglion of Phaenicia sericata (Diptera, Calliphoridae). J. Comp. Physiol. 126: 57-86.
Eckert, H.E.A. & Hamdorf, K. (1981) Action potentials in "non-spiking" visual interneurones. Z. Naturforsch. 36c: 470-474.
Eckert, H. & Meller, K. (1981) Synaptic structures of identified, motion-sensitive interneurones in the brain of the fly, Phaenicia. Verh. Dtsch. Zool. Ges. 1981: 179.
Franceschini, N. (1975) Sampling of the visual environment by the compound eye of the fly: Fundamentals and applications. In: Photoreceptor Optics. Ed. A.W. Snyder & R. Menzel. Berlin, Heidelberg, New York, Springer Verlag, p. 98-125.
Franceschini, N. (1983) The retinal mosaic of the fly compound eye (This volume).
Franceschini, N., Hardie, R , Ribi, W. & Kirschfeld, K. (1981) Sexual dimorphism in a photoreceptor. Nature (Lond.) 291: 241-244.
Franceschini, N., Münster, A. & Heurkens, G. (1979) Aquatoriales und binokulares Sehen bei der Fliege Calliphora erythrocephala. Verh. Dtsch. Zool. Ges. 1979: 209.

Geiger, G. (1981) Is there a motion independent position computation of an object in the visual system of the housefly? Biol. Cybern. 40: 71-75.

Geiger, G & Nässel, D.R. (1981) Visual orientation behaviour of flies after selective laser beam ablation of interneurones. Nature (Lond.) 293: 398-399.

Geiger, G. & Nässel, D. (1982) Visual processing of moving single objects and wide-field patterns in flies: Behavioural analysis after laser-surgical removal of interneurons. Biol. Cybern. 44: 141-149.

Goodman, C. (1974) Anatomy of locust ocellar interneurons: Constancy and variability. J. Comp. Physiol. 95: 185-201.

Götz, K.G. (1968) Flight control in Drosophila by visual perception of motion. Kybernetik 4: 199-208.

Götz, K.G. (1982) Bewegungssehen und Flugsteuerung bei der Fliege Drosophila. BIONA-report 2. Ed. W. Nachtigall. Stuttgart, New York, Gustav Fischer, p. 21-33.

Götz, K.G. & Buchner, E. (1978) Evidence for one-way movement detection in the visual system of Drosophila. Biol. Cybern. 31: 243-248.

Götz, K.G., Hengstenberg, B. & Biesinger, R. (1979) Optomotor control of wing beat and body posture in Drosophila. Biol. Cybern. 35: 101-112.

Hardie, R.C. (1979) Electrophysiological analysis of the fly retina. I. Comparative properties of R1-6 and R7 and 8. J. Comp. Physiol. 129: 19-33.

Hardie, R.C., Franceschini, N. & McIntyre, P.D. (1979) Electrophysiological analysis of the fly retina. II. Spectral and polarization sensitivity in R7 and R8. J. Comp. Physiol. 133: 23-29.

Hardie, R.C., Franceschini, N., Ribi, W. & Kirschfeld, K. (1981) Distribution and properties of sex-specific photoreceptors in the fly Musca domestica. J. Comp. Physiol. 145: 139-152.

Hausen, K. (1976 a) Struktur, Funktion und Konektivität bewegungsempfindlicher Interneuroney im dritten optischen Neuropil der Schmeissfliege Calliphora erythrocephala. Dissertation, Eberhard-Karls-Universität Tübingen.

Hausen, K. (1976 b) Functional characterization and anatomical identification of motion sensitive neurons in the lobula plate of the blowfly Calliphora erythrocephala. Z. Naturforsch. 31 c: 629-633.

Hausen, K. (1976 c) Funktion, Struktur und Konnektivität bewegungsempfindlicher Interneurone in der Lobula plate von Dipteren. Verh. Dtsch. Zool. Ges. 1976: 254.

Hausen, K. (1977) Signal processing in the insect eye. In: Function and Formation of Neural Systems. Ed. G.S. Stent. Berlin, Dahlem Konferenzen, p. 81-110.

Hausen, K. (1979) Neural circuitry of visual orientation behavior in flies: structure and function of the lobula complex. Invest. Ophthalmol. Visual Sci. (Suppl.) 18: 109.

Hausen, K. (1981) Monocular and binocular computation of motion in the lobula plate of the fly. Verh. Dtsch. Zool. Ges. 1981: 49-70.

Hausen, K. (1982 a) Motion sensitive interneurons in the optomotor system of the fly. I. The horizontal cells: structure and signals. Biol. Cybern. 45: 143-156.

Hausen, K. (1982 b) Motion sensitive interneurons in the optomotor system of the fly. II. The horizontal cells: Receptive field organization and response characteristics. Biol. Cybern. 46: 67-79.

Hausen, K. (1983 a) Motion sensitive interneurons in the optomotor system of the fly. III. The centrifugal horizontal cells. (In prep.)

Hausen, K. (1983 b) Motion sensitive interneurons in the optomotor system of the fly. IV. The H1, H2 and H3 cells. (In prep.)

Hausen, K. (1983 c) Motion sensitive interneurons in the optomotor system of the fly. V. Monocular and binocular interactions. (In prep.)

Hausen, K. & Strausfeld, N.J. (1980) Sexually dimorphic interneuron arrangements in the fly visual system. Proc. R. Soc. Lond. 208 B: 57-71.

Hausen, K. & Wehrhahn, C. (1983) Microsurgical lesion of horizontal cells changes optomotor yaw responses in the blowfly Calliphora erythrocephala. Proc. R. Soc. Lond. (In press)

Hausen, K. & Wolburg-Buchholz, K. (1980) An improved cobalt-sulfide silver-intensification method for electron microscopy. Brain Res. 187: 462-466.

Hausen, K., Wolburg-Buchholz, K. & Ribi, W.A. (1980) The synaptic organization of visual interneurons in the lobula complex of flies. Cell Tissue Res. 208: 371-387.

Heide, G. (1975) Properties of a motor output system involved in the optomotor responses in flies. Biol. Cybern. 20: 99-112.

Heide, G. (1982) Neural mechanism of flight control in diptera. BIONA-report. (In press)

Heisenberg, M. & Buchner, E. (1977) The role of retinula cell types in visual behaviour of Drosophila melanogaster. J. Comp. Physiol. 117: 127-162.

Heisenberg, M., Wonneberger, R. & Wolf, R. (1978) Optomotor-blind^{H31} - a Drosophila mutant of the lobula plate giant neurons. J. Comp. Physiol. 124: 287-296.

Hengstenberg, R. (1977) Spike responses of 'non-spiking' visual interneurone. Nature (Lond.) 270: 338-340.

Hengstenberg, R. (1981) Rotatory visual responses of vertical cells in the lobula plate of Calliphora. Verh. Dtsch. Zool. Ges. 1981: 180.

Hengstenberg, R. (1982 a) Common visual response properties of giant vertical cells in the lobula plate of the blowfly Calliphora. J. Comp. Physiol. 149: 179-193.

Hengstenberg, R. (1982 b) Characteristic visual response properties of particular giant vertical cells in the lobula plate of Calliphora. (In prep.)

Hengstenberg, R. & Hengstenberg, B. (1980) Intracellular staining of insect neurons with Procion Yellow. In: Neuroanatomical Techniques. Insect Nervous System. Ed. N.J. Strausfeld & T.A. Miller. New York, Heidelberg, Berlin, Springer-Verlag, p. 307-324.

Hengstenberg, R., Hausen, K. & Hengstenberg, B. (1982) The number and structure of giant vertical cells (vs) in the lobula plate of the

blowfly Calliphora erythrocephala. J. Comp. Physiol. 149: 163-177.
Hubel, D.H. & Wiesel, T.N. (1962) Receptive fields, binocular interaction and functional architecture in the cat's visual cortex. J. Physiol. 160: 106-154.
Kirschfeld, K. (1967) Die Projektion der optischen Umwelt auf das Raster der Rhabdomere im Komplexauge von Musca. Exp. Brain Res. 3: 248-270.
Kirschfeld, K. (1972) The visual system of Musca: Studies on optics, structure and function. In: Information Processing in the Visual Systems of Arthropods. Ed. R. Wehner. Berlin, Heidelberg, New York, Springer Verlag, p. 61-74.
Kirschfeld, K. & Lutz, B. (1974) Lateral inhibition in the compound eye of the fly, Musca. Z. Naturforsch 29 c: 95-97.
Koto, M., Tanouye, M.A., Ferrus, A., Thomas, J.B. & Wyman, R.J. (1981) The morphology of the cervical giant fiber neuron of Drosophila. Brain Res. 221: 213-217.
Laughlin, S. (1981) Neural principles in the peripheral visual system of invertebrates. In: Handbook of Sensory Physiology. Vol. VII/6B. Ed. H. Autrum. Heidelberg, Berlin, New York, Springer Verlag, p. 135-280.
Laughlin, S. (1983) The roles of parallel channels in early visual processing by the arthropod compound eye. (This volume)
Levine, J. & Tracey, D. (1973) Structure and function of the giant motoneuron of Drosophila melanogaster. J. Comp. Physiol. 87: 213-235.
Lillywhite, P.G. & Dvorak, D.R. (1981) Responses to single photons in a fly optomotor neurone. Vision Res. 21: 279-290.
Mastebroek, H.A.K., Zaagman, W.H. & Lenting, B.P.M. (1980) Movement detection: performance of a wide-field element in the visual system of the blowfly. Vision Res. 20: 467-474.
Mastebroek, H.A.K., Zaagman, W.H. & Lenting, B.P.M. (1982) Memory-like effects in fly vision: Spatio-temporal interaction in a wide-field neuron. Biol. Cybern. 43: 147-155.
McCann, G.D. (1973) The fundamental mechanism of motion detection in the insect visual system. Kybernetik 12: 64-73.
McCann, G.D. & Arnett, D.W. (1972) Spectral and polarization sensitivity of the dipteran visual system. J. Gen. Physiol. 59: 534-558.
McCann, G.D. & Dill, J.C. (1969) Fundamental properties of intensity, form, and motion perception in the visual nervous system of Calliphora phaenicia and Musca domestica. J. Gen. Physiol. 53: 385-413.
McCann, G.D. & Foster, S.F. (1971) Binocular interactions of motion detection fibers in the optic lobes of flies. Kybernetik 8: 193-203.
Mimura, K. (1971) Movement discrimination by the visual system of flies. Z. vergl. Physiol. 73: 105-138.
Mimura, K. (1972) Neural mechanisms, subserving directional selectivity of movement in the optic lobe of the fly. J. Comp. Physiol. 80: 409-437.
Murphey, R.K., Matsumoto, S.G. & Levine, R.D. (1977) Does experience play a role in the development of insect neuronal circuitry? In:

Identified Neurons and Behavior of Arthropods. Ed. G. Hoyle. New York, London, Plenum Press, p. 495-506.

O'Shea, M. & Rowell, C.H.F. (1977) Complex neural integration and identified interneurons in the locust brain. In: Identified Neurons and Behavior. Ed. G. Hoyle. New York, London, Plenum Press, p. 307-328.

O'Shea, M., Rowell, C.H.F. & Williams, J.L.D. (1974) The anatomy of a locust visual interneurone: the descending contralateral movement detector. J. Exp. Biol. 60: 1-12.

Pick, B. (1976) Visual pattern discrimination as an element of the fly's orientation behaviour. Biol. Cybern. 23: 171-180.

Pierantoni, R. (1973) Su di un tratto nervoso nel cervello della mosca. In: Atti della prima riuniore scientifica plenaria. Soc. Ital. Biofis. Pura e Applicata, p. 231-249.

Pierantoni, R. (1976) A look into the cock-pit of the fly. The architecture of the lobular plate. Cell Tissue Res. 171: 101-122.

Poggio, T. & Reichardt, W. (1973) Considerations on models of movement detection. Kybernetik 13: 223-227.

Poggio, T. & Reichardt, W. (1976) Visual control of orientation behaviour in the fly. Part II: Toward the underlying neural interactions. Quart. Rev. Biophys. 9: 377-438.

Poggio, T., Reichardt, W. & Hausen, K. (1981) A neuronal circuitry for relative movement discrimination by the visual system of the fly. Naturwiss. 68: 443-446.

Power, M. (1948) The thoracico-abdominal nervous system of an adult insect, Drosophila melanogaster. J. Comp. Neurol. 88: 347-409.

Reichardt, W. (1957) Autokorrelationsauswertung als Funktionsprinzip des Zentralnervensystems. Z. Naturforsch. 12 b: 418-457.

Reichardt, W. (1973) Musterinduzierte Flugorientierung. Verhaltens-Versuche an der Fliege Musca domestica. Naturwiss. 60: 122-138.

Reichardt, W. & Poggio, T. (1976) Visual control of orientation behaviour in the fly. Part I. A quantitative analysis. Quart. Rev. Biophys. 9: 311-375.

Reichardt, W. & Poggio, T. (1979) Figure-ground discrimination by relative movement in the visual system of the fly. Part I: Experimental results. Biol. Cybern. 35: 81-100.

Reichardt, W., Poggio, T. & Hausen, K. (1983) Figure-ground discrimination by relative movement in the visual system of the fly. Part II: Towards the neural circuitry. Biol. Cybern. 46 (suppl.): 1-30.

Riehle, A. & Franceschini, N. (1982) Response of a movement-sensitive neuron to microstimulation of two photoreceptor cells. (In prep.)

Soohoo, S.L. & Bishop, L.G. (1980) Intensity and motion responses of giant vertical neurons of the fly eye. J. Neurobiol. 11: 159-177.

Spüler, M. (1980) Erregende und hemmende Wirkungen visueller Bewegungsreize auf das Flugsteuersystem von Fliegen-Elektrophysiologische und verhaltensphysiologische Untersuchungen an Musca und Calliphora. Dissertation, Universität Düsseldorf.

Spüler, M. & Heide, G. (1978) Simultaneous recordings of torque, thrust and muscle spikes from the fly Musca domestica during optomotor responses. Z. Naturforsch. 33 c: 455-457.

Srinivasan, M.V. & Dvorak, D.R. (1980) Spatial processing of visual information in the movement-detecting pathway of the fly. J. Comp. Physiol. 140: 1-23.

Stavenga, D.G. (1975) The neural superposition eye and its optical demands. J. Comp. Physiol. 102: 297-304.

Strausfeld, N.J. (1976 a) Atlas of an Insect Brain. Berlin, Heidelberg, New York, Springer Verlag.

Strausfeld, N.J. (1976 b) Mosaic organizations, layers, and visual pathways in the insect brain. In: Neural Principles in Vision. Ed. F. Zettler & R. Weiler. Berlin, Heidelberg, New York, Springer Verlag, p. 245-279.

Strausfeld, N.J. (1980) Male and female visual neurones in dipteran insects. Nature (Lond.) 283: 381-383.

Strausfeld, N.J. (1983) Functional neuroanatomy of the blowfly's visual system. (This volume)

Strausfeld, N.J. & Hausen, K. (1977) The resolution of neuronal assemblies after cobalt injection into neuropil. Proc. R. Soc. Lond. 199 B: 463-476.

Strausfeld, N.J. & Nässel, D. (1981) Neuroarchitectures serving compound eyes of crustacea and insects. In: Handbook of Sensory Physiology. Ed. H. Autrum. Vol. VII/6B. Berlin, Heidelberg, New York, Springer Verlag, p. 1-138.

Strausfeld, N.J. & Obermayer, M.L. (1976) Resolution of intraneuronal and transsynaptic migration of cobalt in the insect visual and nervous system. J. Comp. Physiol. 110: 1-12.

Tanouye, M. & Wyman, R.J. (1980) Motor outputs of giant nerve fibre in Drosophila. J. Neurophysiol. 44: 405-421.

Wässle, H. & Hausen, K. (1981) Extracellular marking and retrograde labelling of neurons. In: Techniques in Neuroanatomical Research. Ed. Ch. Heym & W.G. Forssmann. Berlin, Heidelberg, New York Springer Verlag, p. 317-338.

Wehrhahn, C. (1978) Flight torque and lift responses of the housefly (Musca domestica) to a single stripe moving in different parts of the visual field. Biol. Cybern. 29: 237-247.

Wehrhahn, C. (1979) Sex-specific differences in the chasing behaviour of houseflies (Musca). Biol. Cybern. 32: 239-241.

Wehrhahn, C. & Hausen, K. (1980) How is tracking and fixation accomplished in the nervous system of the fly? Biol. Cybern. 38: 179-186.

Wehrhahn, C. & Reichardt, W. (1975) Visually induced height orientation of the fly Musca domestica. Biol. Cybern. 20: 37-50.

Wiesel, T.N. & Hubel, D.H. (1963) Effects of visual deprivation of morphology and physiology of cells in the cat's lateral geniculate body. J. Neurophysiol. 26: 978-993.

Zaagman, W.H., Mastebroek, H.A.K., Buyse, T. & Kuiper, J.W. (1977) Receptive field characteristics of a directionally selective movement detector in the visual system of the blowfly. J. Comp. Physiol. 116: 39-50.

Zaagman, W.H., Mastebroek, H.A.K. & Kuiper, J.W. (1978) On the correlation model: Performance of a movement detecting neural element in the fly visual system. Biol. Cybern. 31: 163-168.

BEHAVIOURAL ANALYSIS OF SPATIAL VISION IN INSECTS

ERICH BUCHNER

Max-Planck-Institut
für biologische Kybernetik
Spemannstrasse 38
7400 Tübingen, West Germany

Oh, for a life of sensations rather than of thoughts!
John Keats to Benjamin Bailey, 22 November 1817.

ABSTRACT

Two aspects of spatial visual orientation in insects constitute the central theme of this chapter: The detection of movement and the evaluation of the position of contrast elements in the visual world. In the first section the visual stimulus situation of an insect moving freely in its natural surround is described. The received "flow field" can be decomposed into three components which result from rotatory and translatory self-movement of the animal and from moving objects (e.g. birds). The next section on basic behavioural phenomena outlines the techniques of open- and closed-loop experiments and describes a few simple experiments on visual movement and position detection in flies. On the basis of these experiments an expression is derived which describes the rotatory component of a fly's flight path through space. The equation is equivalent to the phenomenological equation of Reichardt and Poggio (1976). The third section investigates the basic principles underlying movement and position detection. Comparison of two schemes for movement detection, the gradient scheme and the correlation scheme, with measured behavioural responses demonstrates that the visual system of flies utilises a correlation-based mechanism for the detection of large-field movement. For position detection again two schemes are discussed which are based on flicker detection and movement detection. A final decision on which of the two schemes might be more relevant for fixation and tracking of moving objects by flies seems not yet possible. In the fourth section interactions between elementary movement and position detectors are deduced from behavioural experiments. In the discussion various more sophisticated aspects of visual spatial orientation behaviour of insects are reviewed with emphasis on recent literature (1979 & later).

INTRODUCTION

Observing a bee approaching a flower, a housefly escaping a flyswatter or a fruitfly circling a beer bottle focuses our curiosity on the two central questions of insect visual physiology: What does an insect see when it walks or flies around? And how does its visual system abstract the relevant information from the light signals in the photoreceptors? Both these problems, the what and the how, can be investigated by behavioural analysis although rather different approaches are involved. Observation of free animals in their natural habitat usually is the method of choice when the first problem is attacked, i.e. when visual capabilities of insects are to be characterised. Clearly, a systematic survey on visual functional capacities throughout the class of insect species would be well beyond the scope of a chapter in this book. Such a survey has however recently been compiled in three comprehensive reviews (Menzel, 1979; Waterman, 1981; Wehner, 1981; see also Waterman, 1983). The behavioural approach to the second question defines the field of behavioural physiology. It views the animal essentially as a "black box" and tries to determine input-output characteristics by quantitatively measuring well-defined responses to well-defined stimuli under unnaturally restricted conditions. If these measurements are compared with predictions from theoretical systems analysis, they allow conclusions on basic principles of information processing, on logical interactions between components of the system and on properties of exposed components like the input elements (receptors) or the output structures (motor system). It is this aspect of behavioural analysis that I shall concentrate on in the present chapter. Visual orientation of flies will be chosen as a model to illustrate the power of the approach. Consequently, the seminal work of Reichardt and Poggio will play a central role. In order to avoid redundancy with available reviews on that work (Reichardt & Poggio, 1976, 1981; Poggio & Reichardt, 1976, 1981 a; Wehrhahn, 1980) emphasis will be on understandability without background knowledge rather than on mathematical rigour or completeness. The limitations of this approach for the interpretation of free-flight behaviour under natural conditions will become apparent in the discussion. The literature survey is in part to be considered an update to the extensive list of references compiled by Wehner (1981) and therefore mainly includes accessible publications on behavioural aspects of spatial vision in insects from 1979 to September 1982.

COMPONENTS OF NATURAL VISUAL STIMULATION

During free flight an insect simultaneously undergoes rotatory and translatory motion which lead to a highly complex visual input to its compound eyes: Each contrast element of the visual surround will generate a motion vector of a particular magnitude and direction on a certain region of the retina. These motion vectors may originate either from stationary or from moving objects and constitute a so-called "flow field" (Fig. 1 a). Such a complex flow field can be decomposed into three components which result from the insect's rotatory and translatory movements in space and from the movements of non-stationary objects, e.g. birds. It has been pointed out in a previous chapter (Hausen, 1983) that this conceptual separation appears to have its structural and functional counterpart in the neuronal wiring of insect visual systems.

Rotatory movement

Pure rotation of a stationary visual world will be experienced by an insect which hovers and, at the same time, turns either due to an air turbulence or to a turning force (torque) effected by the wings. In such a situation the flow field will be very simple. If for example the insect rotates about its vertical body axis all contrast elements move horizontally along minor circles, and their angular velocities on the retina depend on the elevation angle θ according to a simple $\cos \theta$ law (Fig. 1 b). Since all contours on a given minor circle (same θ) will move at the same speed regardless of the object's distance from the insect we may call such a field a homogeneous rotational field. It has two "poles" of no motion along the axis of rotation

For a freely flying insect it must be of utmost importance to keep close control of its rotations about all three body axes if it wants to stay on a purposeful course. Small angular deviations e.g. induced by turbulences may lead to large errors in final position as the angular error is multiplied by the distance flown after the deviation occurred. For control of rotatory movements during flight, vision is but one possible mechanism for course stabilisation, mechanical systems or orientation to the earth's magnetic field may also contribute. In dipteran flies the importance of the mechanical flight control system is emphasised by the fact that stable flight is not possible when the halteres are incapacitated (Derham, 1713). Basic flight on the other hand can be maintained in complete darkness. Yet vision certainly plays a major role in rotatory flight control of diurnal insects as will become apparent below.

Translatory movement

If an insect moves on a straight course through a stationary world without turning its body or head, the eyes will receive an inhomogeneous translational field. Such a field again will have two poles which in this case are lying along the direction of flight. All contrast elements move along great circles through the poles and, most importantly, their retinal velocities depend on their distance from the insect. Very distant objects like the sun or stationary clouds do not contribute to the translational field at all, while close objects move over the retina at high speed (Fig. 1 c).

Control of translatory movement during flight may not be quite as critical as rotations because errors are not multiplied by the distance flown. On the other hand, in the presence of unknown wind velocities vision must be of great importance because a mechanical system can detect only acceleration and deceleration but not constant translatory

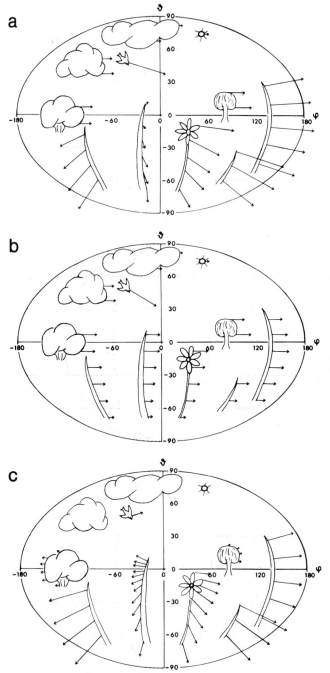

Fig. 1. Schematic illustration of the "flow-field" seen by an insect as it flies through visual space which contains stationary objects close to the animal (blades of grass,

movement (note that constant rotatory movement can be detected mechanically either due to its intrinsic radial acceleration, the cause for centrifugal forces, or by evaluating Coriolis forces (Fraenkel & Pringle, 1938)). As was just pointed out, the translational flow field is "inhomogeneous" with respect to the individual speed components since they depend on the distance of the object to the insect. Conversely, if the animal knew its own velocity, it could in principle reconstruct from the distribution of the velocity vectors on the retina a faithful representation of the 3-dimensional arrangement of the visual objects in space (Longuet-Higgins & Prazdny, 1980). This information about depth in space can be extracted from the motion parallax (Helmholtz, 1867). A foreground object could be identified due to its movement relative to the background. Humans certainly make use of the information contained in flow patterns both for shape and depth evaluation. This was demonstrated using random dot displays of 3-dimensional surfaces under condition of optical flow (Rogers & Graham, 1979). Detection of shapes by relative movement may be illustrated when the dotted outline of Fig. 2a is transfered to a transparent foil and superimposed on Fig. 2b. The pretty woman (Sanyo, 1982) disappears but can be recovered when the foil is moved relative to the background visual noise pattern. In flies it has been demonstrated that contours in fact can be detected when they move relative to a background pattern (Virsik & Reichardt, 1974; Poggio & Reichardt, 1976; Reichardt & Poggio, 1979). The proposed underlying mechanisms will be discussed below. Whether in addition these mechanisms are also used for depth perception during normal flight has not yet been clarified. Distance estimation by means of motion parallax has been demonstrated for locust before a jump (Collett, 1978; Eriksson, 1980; Goulet et al., 1981) and mantids before a prey catching strike (Mittelstaedt, 1957; Maldonado & Rodrigues, 1972; rev. Collett & Harkness, 1982). The importance of such

---

flower), at some intermediate distance (bush, tree) and very far remote (clouds, sun) as well as a moving object (bird). The diagrams should be considered as retinal snapshots of animals looking down onto the paper with the ϕ-axis aligned to the horizontal plane and the θ-axis aligned to the sagittal plane of the head. The arrows illustrate direction and speed (arrow length) of the movement of the contours over the retina during a pure counterclockwise rotation of the animal about its vertical body axis (b), during a pure translatory forward movement in the direction $\phi = 0$, $\theta = 0$ (c) and during the simultaneous occurrence of the rotation in b) and the translation in c) (a). Note that in b) all stationary contours of the same elevation move in the same direction at the same speed regardless of their distance from the animal while in c) close objects move very fast, intermediately remote objects move slowly, whereas very far stationary objects cause no retinal movement during translatory motion.

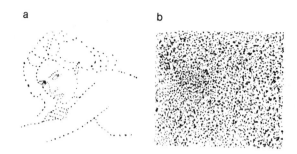

Fig. 2. Demonstration of the use of relative movement information for pattern recognition in humans. When the dotted outline in a) is copied onto a transparent foil and placed over the random dot background pattern (b) the outlines of the young female disappear. As soon as the foil is moved with respect to the background pattern the picture of the woman vividly reappears.

mechanisms for insects is emphasised by the fact that binocular depth perception through disparity analysis could only have a very limited spatial range due to the small separation of the two compound eyes and their limited angular resolution. During landing which represents an extreme flight situation characterised by most rapidly approaching objects, it appears that in fact certain flow field parameters rather than disparity are responsible for triggering deceleration in houseflies (Wagner, 1982). On the other hand, this does not prove that the fly has a percept of spatial depth since landing responses are easily elicited by purely 2-dimensional stimuli like light dimming or planar stripe movement (Taddei-Ferretti, 1973; Taddei-Ferretti & Perez de Talens, 1973; Eckert, 1980, 1982; Eckert & Hamdorf, 1980; Wehrhahn et al., 1981; Chillemi & Taddei-Ferretti, 1981; Fischbach, 1981). Depth perception through binocular disparity analysis has been suggested for several insect species including carabid beetles (Bauer, 1981, 1982, pers. comm.), and stalked-eye flies like Cyrtodiopsids in which the interocular distance is considerably increased due to the long eye stalks (Fig. 3) (Burkhardt & de la Motte, 1983). However, disparity analysis has been demonstrated conclusively only for mantids by the use of prisms in front of the eyes (Rossel, 1980, 1983).

Moving objects

The distinctive feature of objects moving with respect to the stationary surround (e.g. the bird in Fig. 1) is the fact that their retinal images neither move coherently within the rotational field (Fig. 1 b) nor does the direction of their motion (in general) fit into the translational field (Fig. 1 c). Detection of moving objects in insects appears to be achieved by neuronal subsystems which are quite distinct from the large-field movement detection network responsible for visual course control during self-movement of the animal (Palka, 1969; Geiger & Poggio, 1975;

Fig. 3. Drawing of a female stalked-eye fly, Cyrtodiopsis whitei (Diopsidae, Diptera). Recent behavioural experiments suggest distance perception within a range from a few millimetres up to one metre. The conspicuously increased interocular distance is likely to be of fundamental importance for this capacity (Burkhardt & de la Motte, 1983). (Redrawn after Fig. 1 of that work).

Rowell et al., 1977; Srinivasan & Bernard, 1977; Pinter, 1979; Olberg, 1981; Bülthoff, 1982 b; Geiger & Nässel, 1982; Heisenberg & Wolf, 1984).

BASIC BEHAVIOURAL PHENOMENA

In this section we shall demonstrate by a few simple thought experiments that visual orientation and tracking in flies are based on two fundamental mechanisms which detect the movement and the position of a contrast element on the retina. The corresponding real experiments essentially were done by Virsik and Reichardt (1976) and the conclusions were incorporated into a phenomenological theory by Reichardt and Poggio (1976). It should be pointed out that the simple presentation given here entirely builds on the thorough work of Reichardt and his co-workers (reviews: Reichardt & Poggio, 1976, 1981; Poggio & Reichardt, 1976, 1981 a; Wehrhahn, 1980). A fly's translatory and rotatory movement in space may be described by the instantaneous vectors of linear and angular velocity, respectively. The angular velocity vector points along the axis of rotation, its length signifies the angular speed. Both vectors can be decomposed into three orthogonal components along the main body axes of the fly (vertical, horizontal-longitudinal and horizontal-transverse axis) (Fig. 4). The translatory components denote velocities of upward, forward and sideward movements (Fig. 4 a), the rotatory components denote angular velocities of yaw, roll and pitch turns (Fig. 4 b), respectively. For the demonstration of the basic phenomena of visual orientation we shall consider only yaw turns about the vertical body axis as this component has been investigated most thoroughly.

At this point we have to introduce some very basic concepts of flight dynamics. Changes in velocity can only be effected by forces, as is expressed by the well-known Newton's law: force = mass x acceleration. The only other physical law we shall need describes the every-day experience that the frictional force (drag) on a body in still air increases as its velocity increases. The simple proportionality relation between frictional force and velocity used in the following is valid as long as laminar flow prevails (Stokes' law). Turbulences and thus a quadratic relationship must be expected for large insects at high cruising speeds but

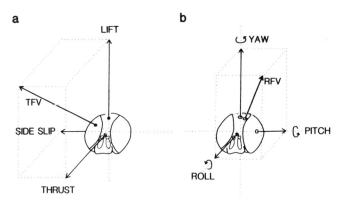

Fig. 4. Illustration of the vectorial decomposition of the translatory flight velocity (TFV) (linear velocity) into three orthogonal components of thrust, lift and side slip velocity (a) and of the rotatory flight velocity (RFV) (angular velocity) into roll, yaw and pitch turning. (Angular velocity is represented by a vector pointing along the axis of rotation in the direction in which a right-handed screw would advance). In the same fashion the translatory and rotatory force-of-flight vectors can be decomposed into thrust, lift and side-slip forces and roll, yaw and pitch torques, respectively. Note that with the usual convention clockwise rotation of the animal (as seen from above) would carry a positive sign. A yaw vector pointing upward as in the example in b) thus corresponds to a negative torque.

probably not for rotatory movements in the physiological range. Also we have to recollect that velocity means change of position with time or first time derivative of position, acceleration means change of velocity with time or first time derivative of velocity, or second time derivative of position. All this holds both for translatory and for rotatory movement. The corresponding relations have been summarised in Table 1.

<u>Closed-loop and open-loop experiments</u>

If in a free-flight situation a visual stimulus, e.g. a stationary contour seen in passing, elicits a motor response, e.g. a yaw turn, this turning will instantaneously modify the visual stimulus, i.e. the movement of the contour over the retina will be changed. Since the "feedback control loop": visual stimulus → motor response → modified visual stimulus → modified motor response etc., is closed, such conditions are called "<u>closed-loop</u>". If on the other hand the insect is attached to a physical instrument such that a certain component of the instantaneous motor output, e.g. yaw torque, can be measured <u>without</u> allowing the animal to turn, the visual stimulus will not be <u>modified</u> by the elicited motor response. Thus the above feedback control loop is opened, and the experimental conditions are termed "<u>open-loop</u>". It is only in this case

Table 1: Fundamentals of Flight Dynamics

	TRANSLATION [LINEAR COORDINATES]	ROTATION [ANGULAR COORDINATES]
POSITION	x [m]	α [°]
VELOCITY	$\dot{x} = \frac{dx}{dt}$ [m/s]	$\dot{\alpha} = \frac{d\alpha}{dt}$ [°/s]
ACCELERATION	$\ddot{x} = \frac{d\dot{x}}{dt} = \frac{d^2x}{dt^2}$ [m/s²]	$\ddot{\alpha} = \frac{d\dot{\alpha}}{dt} = \frac{d^2\alpha}{dt^2}$ [°/s²]
INERTIA (NEWTON'S LAW)	$F = m \cdot \ddot{x}$ [N=kg·m/s²] FORCE = MASS × ACCELERATION	$R = \Theta \cdot \ddot{\alpha}$ [N·m] TORQUE = MOMENT OF INERTIA × ANG. ACCELERATION
FRICTION IN AIR (STOKES' LAW)	$F = c \cdot \dot{x}$ [N] FORCE ∝ VELOCITY	$R = k \cdot \dot{\alpha}$ [N·m] TORQUE ∝ ANG. VELOCITY
"EQUATION OF MOTION"	$m \cdot \ddot{x} + c \cdot \dot{x} = F$	$\Theta \cdot \ddot{\alpha} + k \cdot \dot{\alpha} = R$
SOLUTION ($F = F_0$; $R = R_0$; $\dot{x}(0) = 0$; $\dot{\alpha}(0) = 0$)	$\dot{x} = \frac{F_0}{c} \cdot (1 - e^{-t/\tau})$	$\dot{\alpha} = \frac{R_0}{k} (1 - e^{-t/\tau})$
TIME CONSTANT	$\tau = m/c$	$\tau = \Theta/k$

that the experimenter has full control over the actual visual stimulus. He may choose it to be simple enough that theoretical predictions on the system can be compared to measured responses. We shall have ample occasion to realise the importance of this feature of open-loop experiments later in the review.

However, such severe restraints on the animal as attaching it to a torque meter raise a question which is of fundamental importance for the entire approach of behavioural physiology: Are the responses to a given retinal stimulus the same under closed- and open-loop conditions? The interpretation of natural free-flight behaviour from open-loop measurements makes sense only when this question has found a positive answer. The question may also be put in a somewhat modified form: Does a particular visual response depend on the "internal state" of the animal, and is this internal state influenced by the highly unnatural condition of open-loop measurements? We shall on several occasions come back to this question later in the chapter. Closed-loop conditions may be realised by three different experimental arrangements. Obviously any free-flight experiment is closed-loop, as just described. In many situations it is advantageous to let the insect fly on the spot allowing it only to turn about a single predefined axis. Fig. 5 illustrates the simple device which we shall use for the thought experiments below but which actually has been employed in student laboratory courses and some qualitative optomotor experiments (Heisenberg & Wolf, 1979). A fly is attached to a small piece of mu-metal wire which is suspended on a very thin thread. The pointed magnet below the fly stretches the thread and keeps the

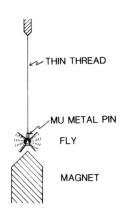

Fig. 5. Simple experimental arrangement for the isolated observation of rotatory movements of a fly about its vertical body axis. The small mumetal pin to which the fly is attached by its thorax is strongly attracted by the pointed magnet. In this way the fly is kept in place and the thread is stretched. Rotations about the vertical axis are almost unimpeded due to the fact that both the tortional modulus of the thread and the moment of inertia of the pin are negligibly small. The experimental situation of the fly may be characterised as "one-parameter closed-loop".

animal in place without impeding its rotations about the vertical axis. Thus yaw turns can be carried out essentially as in free flight, all other movements are however blocked. We may call such a condition "one-parameter closed-loop". It has the advantage that simple stimuli can be presented and the qualitative behaviour of the fly can be easily observed. The third set-up which was developed by Reichardt and Wenking (1969) is more sophisticated and less imaginable for the inexperienced, it has, however, the great advantage of allowing quantitative closed-loop experiments under controlled visual conditions: As in the open-loop situation described above, the fly is attached to a torque meter which measures the instantaneous yaw torque without permitting the animal to turn. The torque signal in this case is used to control the speed of a cylindrical panorama rotating about the fly's vertical axis in such a way as to simulate the visual situation of the "one-parameter closed-loop": When the fly intends to make a right turn it generates a corresponding yaw torque. This signal is measured by the torque meter and it is electronically amplified to drive the cylindrical panorama to the left at about the angular speed by which the fly's torque would have turned the fly to the right if it were free to turn. Note that under these "artificial closed-loop" conditions only the visual but not the mechanosensory feedback control loop is closed whereas in the situation of Fig. 5 both loops are closed. Again one may inquire about the "internal state" of the fly under these two unnatural closed loop conditions.

Movement detection as a component of visual orientation

Consider a fly attached to the set-up of Fig. 5. Let it be surrounded by a cylindrical panorama which can be rotated about the vertical body axis of the fly. Fig. 6 shows the situation as seen from above: The angles α_f and α_p describe the angular position of the fly's long axis and a fixed point ("O") on the panorama with respect to space "O", respectively. To simplify the situation further we rigidly attach the fly's head to the thorax to prevent any head movements. In this case the position of the panorama

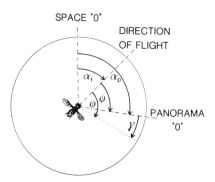

Fig. 6. Definition of the angles between an arbitrarily chosen "zero" direction in space and the fly's long body axis (α_f) as well as a "zero" position on the panorama cylinder (α_p). The fly can freely rotate as in free flight (or the set-up of Fig. 5), and the panorama cylinder can be rotated by the experimenter. If a visual object (e.g. a black stripe) is mounted on the cylinder at position panorama "0", its horizontal position on the fly's retina will be given by $\psi = \alpha_p - \alpha_f$, whereas an ommatidium at horizontal position ϕ will observe the environment at the panorama-fixed coordinate value γ.

with respect to the fly's retina is given by

$$\psi = \alpha_p - \alpha_f \qquad (1)$$

Let us start with no visual stimulus, e.g. a homogeneous white panorama, and interpret ("solve") the fly's "equation of motion" (cf. Table 1)

$$\Theta \cdot \ddot{\alpha}_f + k \cdot \dot{\alpha}_f = R \qquad (2)$$

Θ is the fly's moment of inertia about its vertical body axis, k is its aerodynamic friction constant. We assume that a constant torque ("rotatory force") R_o is externally applied to the fly at time t = 0 (this might be achieved by twisting the thread several times and releasing the fly at time 0):

$$R = \begin{cases} 0 & \text{for } t \leq 0 \\ R_o & \text{for } t > 0 \end{cases}$$

If we could neglect friction (k = 0) the angular acceleration would be constant ($\ddot{\alpha}_f = R_o/\Theta$), and the angular velocity would increase linearly ($\dot{\alpha}_f = (R_o/\Theta) \cdot t$) as shown by the straight line in Fig. 7. As $k \neq 0$ due to

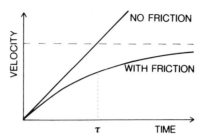

Fig. 7. Illustration of the increase of the fly's translatory (rotatory) velocity with time under the action of a constant force (torque) in the absence and presence of the frictional influence of still air. The time after which the fly has reached 63% of its final speed when friction is taken into account is denoted by τ. Note that in the situation illustrated in Fig. 5 the value of τ for the housefly <u>Musca</u> is only about 0,008 s (Reichardt & Poggio, 1976).

the frictional influence of the air, angular acceleration in fact will decrease as angular velocity increases. As a consequence, the angular velocity of the fly $\dot{\alpha}_f$ will level off and approach a constant value according to the equation $\dot{\alpha}_f = (R_o/k) \cdot (1-e^{-t/\tau})$ as illustrated in Fig. 7. The time constant $\tau = \Theta/k$ denotes the time it takes for the angular velocity $\dot{\alpha}_f$ to reach ~ 63% of its final value. For the housefly <u>Musca domestica</u> τ has been determined to be about 0,008 s (Reichardt & Poggio, 1976). This means that Θ is so small compared to k that after some 20 milliseconds $\dot{\alpha}_f$ essentially has reached its final value. If we are not interested in very rapid changes of $\dot{\alpha}_f$ we may therefore neglect Θ and approximate (2) by

$$k \cdot \dot{\alpha}_f = R \qquad (3)$$

Thus the instanteneous angular velocity $\dot{\alpha}_f$ is proportional to the instanteneous torque R generated by the fly.

In our first visual experiment we mount a low-contrast visual noise pattern on the panorama cylinder. The visual panorama thus provides no prominent contours in which the fly might be particularly interested. Yet when we rotate the cylinder at a constant speed $\dot{\alpha}_p$ the fly will try to follow the movement, it will rotate in the same direction at a speed

$$\dot{\alpha}_f = G_p \cdot \dot{\alpha}_p \qquad (4)$$

G_p relates input ($\dot{\alpha}_p$) and output ($\dot{\alpha}_f$) of the fly's rotatory movement control system under closed-loop conditions. Thus G_p represents the closed-loop gain of the system. The more efficient the control system operates the closer will G_p approach 1. The subscript p is used to indicate that the value of G_p depends on the particular visual pattern on the panorama.

The actual visual stimulus in this experiment of course is the speed at which the pattern moves across the fly's retina:

$$\dot{\psi} = \dot{\alpha}_p - \dot{\alpha}_f \tag{5}$$

$\dot{\psi}$ is called "retinal slip speed". It constitutes the "error signal" of the movement control system and is in fact the parameter that in the present experiment induced the torque which made the fly rotate at speed $\dot{\alpha}_f$. $\dot{\psi}$ may be small but never can be zero for a movement control system, otherwise no torque would be induced. (Non-zero error signal is a general characteristic of any (non-integrating, cf. below) feedback control system). According to (3) the induced torque must be

$$R_{ds} = k \cdot \dot{\alpha}_f \tag{6}$$

The index ds stands for <u>d</u>irection <u>s</u>ensitive as the sign of R_{ds} and therefore $\dot{\alpha}_f$ is determined by the direction of pattern rotation. The relation between the torque R_{ds} and the slip speed $\dot{\psi}$ can be investigated directly by a simple open-loop experiment: We stimulate the fly by the same panorama as before but now attach the animal to a torque meter instead of the thread such that it cannot turn any more. Various kinds of torque meters have been described in the literature (Kunze, 1961; Götz, 1964; Srinivasan & Bernard, 1977; Blondeau, 1981 a; Geiger et al., 1981). Since $\dot{\alpha}_f = 0$ in this case, the pattern speed $\dot{\alpha}_p$ is the same as the slip speed $\dot{\psi}$. If we use low values of $\dot{\psi}$ the measured torque signals will be proportional to $\dot{\psi}$:

$$R_{ds} = r_p \cdot \dot{\psi} \tag{7}$$

Like G_p the constant r_p carries the subscript p to indicate that its value depends on the particular pattern on the panorama. Using eq. (6) we see that under open-loop conditions the output of the fly's movement control system $\dot{\alpha}_f$ and its input, the pattern speed $\dot{\alpha}_p = \dot{\psi}$, are related by the equation $\dot{\alpha}_f = (r_p/k) \cdot \dot{\psi}$. Thus r_p/k represents the open-loop gain of the system.

The closed- and open-loop gain factors G_p and r_p/k are not independent. Their interrelationship brings up the question on the "internal state" of the fly during the "one-parameter closed-loop" condition of Fig. 5 and the open-loop experiment just described. Under the assumption that the different conditions in the two experiments have no influence on the response of the fly to the retinal movement, one can easily derive a simple relation between G_p and r_p/k by using eqs. (6), (7), (5) and (4) (in this order):

$$k \cdot \dot{\alpha}_f = r_p \cdot \dot{\psi} = r_p \cdot (\dot{\alpha}_p - \dot{\alpha}_f) = r_p \cdot \dot{\alpha}_f \cdot \left(\frac{1}{G_p} - 1 \right)$$

$$G_p = \frac{1}{1 + k/r_p} \qquad (8)$$

Interestingly eq. (8) has never been tested experimentally in insects. When the value of G_p is derived from measurements of $\dot{\alpha}_f$ and $\dot{\alpha}_p$ in eq. (4) one has to be aware of two problems. First, in any real closed-loop experiment non-visual cues (mechanical, magnetic) may influence the control system. Values on G_p found in the literature (Collett & Land, 1975 a; Collett, 1980 a) were obtained under conditions of free locomotion such that the visually induced turning may well be counteracted by mechanical systems whose function it is to keep the animal on a straight course. These G_p-values can therefore not be used in the present discussion. This problem can be eliminated by employing the "artificial closed-loop" conditions described above. The second problem is more difficult to handle: There is no movement information without position information, such that one has to rule out that a position control mechanism contributes to the measured responses. This is especially difficult as a movement control loop employing an integration stage may become almost indistinguishable from a true position control system. Only when this problem has been settled can a discrepancy between the theoretical relation (8) and measured values of G_p and r_p be taken as an indication of different "internal states" of the animal under closed- and open-loop conditions.

Position detection as an element of visual orientation

Returning to our experimental set-up of Fig. 5 we now replace the visual noise panorama by a homogeneous white cylinder carrying a single prominent object, e.g. a vertical black stripe which we may assume to be positioned at panorama "O" of Fig. 6. As above we first perform a closed-loop experiment by turning the panorama with the stripe at constant speed $\dot{\alpha}_p$ and observe the response of the fly. We notice a qualitatively different behaviour: the fly now follows at the angular speed of the pattern with zero net slip

$$\dot{\alpha}_f = \dot{\alpha}_p \qquad (9)$$

such that the mean value of $\dot{\psi}$ is zero. From what has been said about movement control systems it is unlikely that such a system can achieve this effect. Zero slip speed would require infinite open-loop gain in eq. (8) which may not be realistic. Thus we conclude that in this situation the fly most likely employs a position control mechanism rather than a movement control mechanism. Accordingly, in order to follow the stripe, the fly must generate a torque which depends on the position of the visual stimulus, so we may call it position sensitive torque R_{ps}. Using again eq. (3) we have (with eq. (9)):

$$k \cdot \dot{\alpha}_f = R_{ps} = k \cdot \dot{\alpha}_p \tag{10}$$

Just as there had to be a non-zero slip speed $\dot{\psi}$ for the (non-integrating) movement control system to function, there now must be a non-zero position error signal between the direction of flight and the rotating stripe. We can measure this error angle $\psi = \alpha_p - \alpha_f$ as a function of the panorama speed $\dot{\alpha}_p$. In fact these measurements have been carried out by Virsik and Reichardt (1976) in the "artificial closed-loop" arrangement described above. The results in Fig. 8 demonstrate a linear relation between ψ and $\dot{\alpha}_p$ over the range measured according to the equation

$$\psi = \frac{k}{\beta_p} \cdot \dot{\alpha}_p \tag{11}$$

The new constant β_p which again depends on the particular pattern used can be determined from the slope of the line in Fig. 8. Using eqs. (10) and (11) we realise that the position-induced torque R_{ps} must be proportional to the position error angle:

$$R_{ps} = \beta_p \cdot \psi \tag{12}$$

The corresponding open-loop experiment which in fact verifies this relation will be described later.

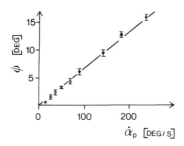

Fig. 8. Dependence of the angle between the fly's long body axis and a black stripe moving at constant speed $\dot{\alpha}_p$ around the fly in an experimental situation equivalent to that illustrated in Figs. 5 and 6. The fly follows the stripe at the same speed ($\dot{\alpha}_f = \dot{\alpha}_p$) but lags behing by a certain "tracking angle" ψ. In the velocity range up to $\pm$ 250°/s there is a linear relationship between ψ and $\dot{\alpha}_p$. (The actual data were obtained in the "artificial closed-loop" situation). (after Virsik & Reichardt, 1976).

We have now derived the basic elements of visual course control, viz. a movement-induced direction-sensitive torque (eq. (7)) and a position-sensitive torque (eq. (12)). We can transfer this knowledge to a natural free-flight situation when we restrict ourselves to considering rotations about the vertical body axis. In a natural free-flight situation the fly would not be interested in a black stripe moving at constant angular speed but rather it might chase for example another insect. The angular speed $\dot{\alpha}_p$ of the leading insect will of course not be constant such that the instantaneous slip speed $\dot{\psi}$ will not be zero. Accordingly, there will be both a movement stimulus and a position stimulus at the same time, and each component will induce a torque response, R_{ds} and R_{ps}, respectively. It was shown by Virsik and Reichardt (1976) that the two components of the torque signal add linearly:

$$R = R_{ds} + R_{ps}$$

It should be pointed out at this point that the conceptual decomposition of the torque response into a direction-sensitive and a position-sensitive component is always possible (Geiger & Poggio, 1975) and must not be taken as evidence for the existence of two separate neuronal subsystems for movement and position detection. Possible neuronal mechanisms underlying these responses will be discussed in detail below.

A final torque component has yet to be considered: In the absence of any visual stimulus the fly will not fly on a perfectly straight line forever. Rather there is some internal torque generated by the fly causing course deviations that are not related to the visual stimulus. Such an internal torque component R_{int} includes whatever one likes of the following phenomena (or others as well): random fluctuations, e.g. due to noisy signals in the visual, nervous and motor system, search strategies, "initiation" (Heisenberg, 1983), free-will decisions, etc. In summary, the total torque (about the vertical body axis) will be

$$R = R_{ds} + R_{ps} + R_{int} \qquad (13)$$

With eqs. (3), (7) and (12) we obtain an equation that predicts the time course of the angular velocity of the fly $\dot{\alpha}_f(t)$ from the time course of position and movement of visual contours on the fly's retina:

$$k \cdot \dot{\alpha}_f = r_p \cdot \psi + \beta_p \cdot \dot{\psi} + R_{int} \qquad (14a)$$

When we do not neglect the fly's moment of inertia Θ in eq. (2) this becomes

$$\Theta \cdot \ddot{\alpha}_f + k \dot{\alpha}_f = r_p \cdot \psi + \beta_p \cdot \dot{\psi} + R_{int} \qquad (14b)$$

Eqs. (14 a, b) were essentially derived from experiments in which the flies were severely restrained by attaching them to a torque meter and forcing them to fly on the spot in still air. This raises the question whether these equations are of any value for the analysis of free flight in a visually structured surround and for such complex behavioural episodes like pursuits of other insects. It was therefore very satisfying to realise that an expression essentially identical to eq. (14 a) (with $R_{int} = 0$ and a slightly delayed response) may be used to describe free-flight behaviour of one insect chasing another and even allows faithful simulation of such episodes (Land & Collett, 1974; Collett & Land, 1975 b; Collett, 1980 a; Bülthoff et al., 1980; Poggio & Reichardt, 1981 b; Reichardt & Poggio, 1981). At first sight this is rather striking as the simple theory outlined above is based on some very crude approximations. Both proportionality relations for the movement response (eq. (7)) and for the position response (eq. (12)) hold only for small slip speeds $\dot{\psi}$ and small error angles ψ, respectively. We shall come back to this point in the discussion. More realistic relations must include the experimental finding that for large values of $\dot{\psi}$ the torque R_{ds} and for large values of ψ the torque R_{ps} both return to zero (Poggio & Reichardt, 1981 b). This can be taken into account when $r_p \cdot \dot{\psi}$ and $\beta_p \cdot \psi$ are replaced by some pattern-dependent functions $r_p^*(\psi, \dot{\psi})$ and $D_p(\psi)$, respectively, which will be discussed below. Eq. (14 b) then reads:

$$\Theta \cdot \ddot{\alpha}_f + k \cdot \dot{\alpha}_f = r_p^*(\psi, \dot{\psi}) + D_p(\psi) + R_{int} \qquad (15)$$

If we try to express what we have derived in this section by a single sentence, we may state the following: For a particular visual panorama, measure how the fly responds to angular <u>movement</u> of the panorama ($r_p^*(\psi, \dot{\psi})$), measure how the fly responds to the angular <u>position</u> of the panorama ($D_p(\psi)$), and measure the properties of the <u>endogeneous</u>, pattern-independent signal (R_{int}); one can then determine the rotatory component of the fly's trajectory in space from eq. (15). Due to the indeterministic component R_{int} this prediction will not provide the precise trajectory itself but only a probability distribution that delimits a particular spatial "corridor" in which real trajectories are likely to be found. Explicit calculations in general will require numerical methods on digital computers. A description of special cases that have been solved analytically (Reichardt & Poggio, 1976) would go well beyond the scope of the present review. Rather, we shall outline in the next sections what has been learned from behavioural analysis about the neuronal networks that are involved in the two visual functions responsible for orientation behaviour, movement detection and position detection.

As a final remark on this section we may mention that in the "artificial closed-loop" set-up as it was described above, it is mathematically more convenient to express eq. (15) in the fly-centred coordinate ψ. According to eq. (1) $\dot{\alpha}_f$ can be replaced by $\dot{\alpha}_f = \dot{\alpha}_p - \dot{\psi}$. By reordering one obtains from (15) with $S(t) = \Theta \cdot \ddot{\alpha}_p + k \cdot \dot{\alpha}_p$

$$\Theta \cdot \ddot{\psi} + k \cdot \dot{\psi} = S(t) - D_p(\psi) - r_p^*(\psi, \dot{\psi}) - R_{int} \qquad (16)$$

This is the "phenomenological equation" of Poggio and Reichardt (1973 a) except for the notation $R_{int} = -N(t)$ and a slightly less restrictive approximation ($r_p^*(\psi, \dot{\psi}) \cong r(\psi) \cdot \dot{\psi}$ as long as $\dot{\psi}$ is not too large). For further applications of this theory to problems that go beyond the basic elements of orientation behaviour the reader is referred to the reviews by Reichardt and Poggio (1976; 1981), Poggio and Reichardt (1976; 1981 a) and Wehrhahn (1980).

BASIC PRINCIPLES OF MOVEMENT AND POSITION DETECTION

It was demonstrated in the previous section that spatial visual orientation of insects is mainly based on the ability to abstract from the light signals in the photoreceptors information about the <u>movement</u> and about the <u>position</u> of contrast elements in the visual world. In the present chapter we shall outline two schemes that have been proposed for movement detection, the gradient scheme and the correlation scheme, and discuss the experimental evidence on the question which of the two schemes describes the neuronal networks in insects more realistically. For the position detection system again two alternate mechanisms will be outlined, the flicker-based system and the movement-based system, and the experimental evidence will be reviewed.

Movement detection

Movement consists of direction and speed, and it should be pointed out that only mechanisms that are selective for the direction of a moving contrast element are of interest in the present context. All visual systems of higher organisms employ a large number of spatially separated input elements (receptors) and parallel processing of their signals. For the understanding of the basic principles of movement detection it seems appropriate to reduce the complexity of such a system and ask what is the minimum configuration of an assembly of receptors, interconnexions and interactions that can detect directional movement. Very simple considerations show that a movement detector must at least have two spatially separated receptors, the network connected to the receptors must be asymmetric, and the interaction of the two signals must be nonlinear. While the first property is obvious, the asymmetry becomes clear when one realises that in a symmetric 2-input system interchanging the order of the receptors has no effect. But this interchanging is entirely equivalent to inverting the direction of the movement. Therefore a symmetric network cannot detect the direction of movement. The necessity of a nonlinear interaction is less easily illustrated. A simple mathematical proof has been given (Buchner, 1976, Appendix A). It should be noted that any nonlinear operation such as a threshold, clipping, multiplication, division or rectification after the convergence of the two receptors signals may suffice to produce directional selectivity. Examples will be given below.

Consider a pattern whose intensity varies along the horizontal, pattern-fixed coordinate γ (cf. Fig. 6) according to an arbitrary (positive) function $I(\gamma)$ but is constant along the vertical coordinate. If this pattern moves horizontally across a receptor whose horizontal position is given by the eye-fixed coordinate ϕ, the light signal in the receptor will be $I(\phi,t) = I(\gamma(t))$. The function $\gamma(t)$ will depend on the pattern velocity and the starting position. For example, if the pattern starts at time $t = 0$ at position $\phi = 0$ and travels at constant velocity w, then $\gamma(t)$ will be given by

$$\gamma(t) = \phi - w \cdot t \qquad (17)$$

The sign of w determines the direction of the movement. Note that ϕ, ψ and γ are interrelated by (cf. Fig. 6)

$$\phi = \psi + \gamma$$

such that for any fixed eye position (ϕ = const.)

$$\dot{\psi} = w = -\dot{\gamma}$$

(The use of w instead of $\dot{\psi}$ in the present context has historical reasons and is retained to ease comparison with the literature). As mathematical systems analysis becomes most simple for sinusoidal input signals we choose as a stimulus pattern a sinewave grating of mean luminance $\bar{I}$, modulation amplitude ΔI, spatial frequency $1/\lambda$ (spatial wavelength λ). $I(\gamma)$ will then be (Fig. 9):

$$I(\gamma) = \bar{I} + \Delta I \cdot \sin\left[\frac{2\pi}{\lambda} \cdot \gamma\right] \qquad (18)$$

When the two receptors of a minimal movement detector configuration are separated along the horizontal coordinate by the angle $\Delta\phi^*$ and are positioned in the eye at ϕ and $\phi + \Delta\phi^*$, the light signals in the receptors will be

$$I(\phi, t) = \bar{I} + \Delta I \cdot \sin\left[\frac{2\pi}{\lambda} \cdot (\phi - w \cdot t)\right] \qquad (19)$$

$$I(\phi + \Delta\phi^*, t) = \bar{I} + \Delta I \cdot \sin\left[\frac{2\pi}{\lambda} \cdot (\phi + \Delta\phi^* - w \cdot t)\right] \qquad (20)$$

Here we have assumed that the receptors have a needle-shaped spatial sensitivity distribution. The effect of a more realistic bell-shaped sensitivity distribution will be discussed later. Note that the "effective" receptor spacing $\Delta\phi^*$ depends on the orientation of the stripes with

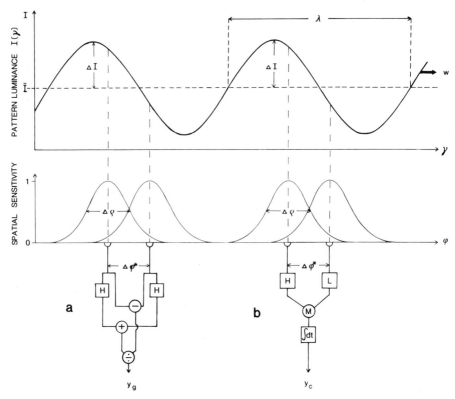

Fig. 9. Spatial luminance distribution of a sinewave grating and two models of "elementary movement detectors". The gradient model (a) involves the division ($\div$) of two gradient signals derived from the two input signals by temporal differentiation (high-pass filtering (H)) and by spatial differentiation (substraction (-)). For mathematical simplicity the mean (+) of the two temporally differentiated input signals was used in the numerator of the division operation. The correlation model (b) employs high- and low-pass filtering (H, L), a multiplication (M) and temporal averaging ($\int dt$). The stimulus situation is characterised by the mean luminance $\bar{I}$, the modulation amplitude ΔI (or contrast $\Delta I/\bar{I}$), the spatial wavelength λ (or spatial frequency $1/\lambda$) and the speed w of the pattern as well as the "effective" sampling distance $\Delta\phi^*$(cf. Fig. 10) and the half-width of the bell-shaped spatial sensitivity distribution (acceptance angle $\Delta\rho$) of the photoreceptors. γ represents the pattern-fixed horizontal angular coordinate while the ϕ-coordinate is fixed with respect to the eye of the animal.

respect to the position of the two receptors in the ommatidial array of the eye. For example, when the hexagonal array is oriented as in the horizontal eye region of flies (Franceschini, 1975), $\Delta\phi^*$ for vertical stripes

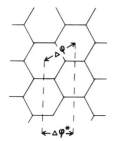

Fig. 10. Relation between the interommatidial angle $\Delta\phi$ and the "effective" sampling distance $\Delta\phi^*$ in the case of vertical stripes moving horizontally across a hexagonal array of ommatidia.

and nearest-neighbour interaction is $\sqrt{3}/2$ times the interommatidial angle $\Delta\phi$ (Fig. 10).

The gradient scheme

Gradient schemes for movement detection have been proposed for computer image processing (Limb & Murphy, 1975; Ullman, 1981). They were also incorporated as models in biological systems to explain movement responses of cortical simple cells (Marr & Ullman, 1981). Gradient models are based on the mathematical identity (chain rule of differentiation):

$$\frac{dI}{dt} = \frac{dI}{d\gamma} \cdot \frac{d\gamma}{dt} \qquad (21)$$

With eq. (17) one obtains

$$\frac{d\gamma}{dt} = -w \qquad \frac{dI}{d\gamma} = \frac{dI}{d\phi}$$

and therefore

$$w = - \frac{\frac{dI}{dt}}{\frac{dI}{d\phi}} \qquad (22)$$

The velocity w of the pattern may be computed as the quotient of the <u>temporal</u> gradient measured by a single receptor at location ϕ, and the <u>spatial</u> gradient measured by two receptors, one at ϕ, the other at $\phi + d\phi$. While temporal differentiation (high pass filtering) is a familiar process in nervous systems, the inevitable approximation of a spatial gradient computation due to the spatial separation of neighbouring receptors (sampling) has two important consequences. It renders a true velocity detector sensitive to the spatial frequency content of the pattern and it limits the spatial resolution of correct direction detection to the spatial frequency range as it is set by the sampling theorem: $0 < 1/\lambda < 1/(2\Delta\phi^*)$ (The spatial wavelength must be larger than twice the sampling interval, $\lambda > 2\Delta\phi^*$). With the input signals of eqs. (19) and (20) one can easily calculate the output y_g of the simple gradient-based movement detector shown in Fig. 9 a

$$y_g = w \cdot \frac{\frac{\pi \cdot \Delta\phi^*}{\lambda}}{\tan \frac{\pi \cdot \Delta\phi^*}{\lambda}} \qquad (23)$$

The dependence (or lack of dependence) of this output on the stimulus parameters contrast $\Delta I/\bar{I}$, speed w, spatial frequency $1/\lambda$ and temporal frequency w/λ is plotted in Fig. 11 a - d, respectively. These predicted output signals will be compared to measured responses below.

The correlation scheme

Originally this scheme for movement detection had been proposed somewhat intuitively to explain the experimental data on the visual movement responses of the beetle Chlorophanus (Hassenstein, Reichardt, Varju, 1951 - 1967). Later Poggio and Reichardt (1973 b) demonstrated that decomposition of an arbitrary multi-input system into a functional power series leads to the correlation model as the simplest general scheme with the property that its mean output depends on the direction of movement. The minimum configuration of the correlation model is shown in Fig. 9 b (Kirschfeld, 1972). It involves two different linear filters, a multiplication and temporal averaging. If we choose the filters H and L to be a high- and a low-pass filter of equal time constant τ and use eqs. (19) and (20) as input signals, the output y_c will be (Buchner, 1976)

$$y_c = \Delta I^2 \cdot \left[\frac{\pi \cdot \tau \cdot \frac{w}{\lambda}}{1 + (2\pi \cdot \tau \cdot \frac{w}{\lambda})^2} \right] \cdot \sin \frac{2\pi \cdot \Delta\phi^*}{\lambda} \qquad (24)$$

As shown in Fig. 11 e - h the predicted output signals of a correlation detector as a function of contrast, $\Delta I/\bar{I}$, speed w, spatial frequency $1/\lambda$ and contrast frequency w/λ are quite different from those of the gradient detector.

Comparison with behavioural movement responses

It is technically difficult to measure the behavioural responses of an insect as a function of pattern contrast, velocity, spatial frequency and contrast frequency when only a single pair of photoreceptors is stimulated (cf. Kirschfeld, 1972). Therefore one generally applies large-field stimulation and assumes that all stimulated movement detectors respond in a qualitatively similar fashion such that the compound response still reflects the properties of the individual detectors. Since the position-sensitive component of the response does not depend on the direction of the movement it can be eliminated by computing the differences of responses to movement in opposite directions. If in addition, the noise-like "internal" component (cf. eq. 13) is attenuated by averaging, one obtains the isolated direction-sensitive component

$$\overline{R}_{ds} = \overline{R}(w) - \overline{R}(-w)$$

which can be compared to the expressions in eqs. (23) and (24).

Fig. 11 i shows the dependence of direction-sensitive yaw torque and turning responses of flying Musca (McCann & MacGinitie, 1965) and walking Drosophila (Heisenberg & Buchner, 1977) on the pattern contrast $\Delta I/\overline{I}$. At contrast values below 3 - 4% the responses are well approximated by a parabola indicating a dependence on ΔI^2 as predicted by the correlation model (Fig. 11 e). Above this value responses increase more slowly, probably due to a gain control mechanism (cf. Discussion). The output of the gradient model does not depend on ΔI since this factor cancels at the division stage. Of course it is unrealistic to expect the model to respond with maximum output at extremely low contrast values (Fig. 11 a) since eventually photon fluctuations will mask the movement information. As signal-to-noise calculations for the gradient model are not presently available it cannot be excluded that a realistic predicted output of this model might look similar to the measured responses in Fig. 11 i. So we may conclude that the behavioural data on contrast dependence are in agreement with the correlation model but are not necessarily incompatible with the gradient scheme.

For spatial wavelengths large compared to the sampling interval $\Delta\phi$ * the gradient scheme represents a true velocity detector, its output as a function of speed approaches a line of slope 1 independent of the particular value of λ (Fig. 11 b). This is not so for the correlation scheme. Although at low velocities the denominator in the square brackets of eq. (24) may be approximated by 1 such that again we have a linear dependence of y_c on the velocity w, the slope of this function will approach zero as λ becomes large. Since appropriate behavioural data at low pattern velocities are not available, we include in Fig. 11 k extracellular recordings (Bülthoff, unpublished results) from the movement-sensitive "H1" cell of Calliphora. This cell has been described in detail in a previous chapter of this volume (Hausen, 1983). It is obvious from these recordings which were obtained under conditions of slow linear increase of the speed over the range covered by the solid lines, that the movement detection system in flies is not based on velocity detection but rather complies nicely with the theoretical expectations for the correlation model (Fig. 11 f).

A very clear distinction of the two models under consideration is also possible when the dependence of the fly's turning response on the spatial frequency of the pattern is measured. Any model whatsoever must

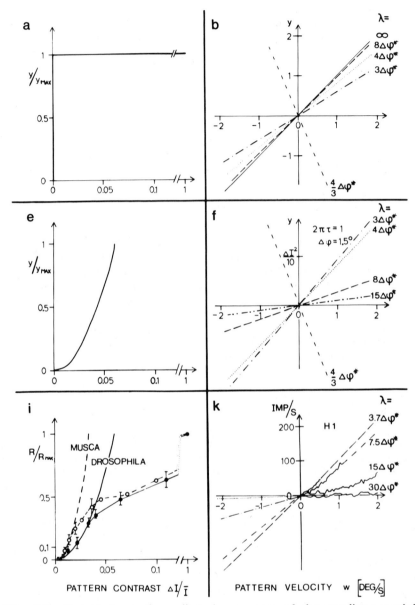

Fig. 11. Comparison of predicted responses of the gradient model (a-d) and the correlation model (e-h) with measured behavioural or electrophysiological responses of flies (i-m) as a function of the contrast (a, e, i), the velocity (b, f, k), the spatial frequency (c, g, l) and the contrast frequency (d, h, m) of a moving sinewave grating. The convincing agreement of the experimental data with the

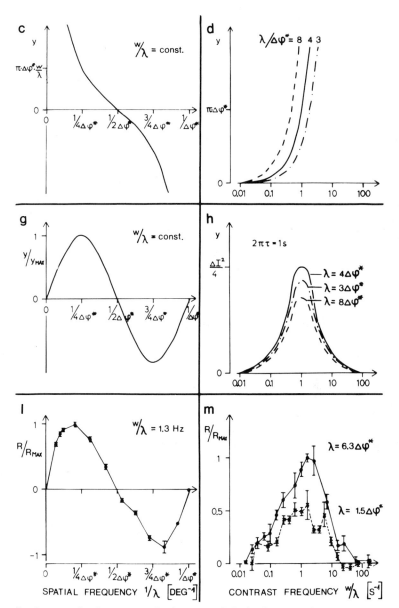

predictions of the correlation model indicate that movement detections in flies is based on correlation-like interactions rather than a gradient-computing mechanism. ((i) after McCann & MacGinitie, 1965 (Musca) and Heisenberg & Buchner, 1977 (Drosophila); (k) from H. Bülthoff, unpublished results; (l) after Buchner, 1976; (m) after Buchner, 1974).

show the sampling artefact which inverses the perceived direction of the movement in the spatial frequency range:

$$\frac{1}{2\Delta\phi^*} < \frac{1}{\lambda} < \frac{1}{\Delta\phi^*}$$

This phenomenon has long been known (v. Gavel, 1939) and has been clearly illustrated as "geometric interference" between the periodic pattern and the sampling stations (Hassenstein, 1951; Götz, 1964). The output of the correlation model becomes particularly simple when the temporal frequency of the light signals in each receptor is constant such that the filters H and L always have the same effect. This temporal frequency is called "contrast frequency" w/λ and denotes the number of luminance cycles passing each receptor per second. As is obvious from eq. (24) y_c becomes a pure sine function of spatial frequency $1/\lambda$ if w/λ is held constant (Fig. 11 g). On the other hand, since the gradient scheme approximates a true velocity detector, its output increases rapidly as λ, and therefore w, increases under these conditions (Fig. 11 c). The measured turning responses of walking Drosophila have been corrected for the effect of contrast attenuation due to the finite width of the receptor acceptance function (remember that for the derivation of eq. (24) we assumed a needle-shaped sensitivity distribution). After this correction the response curve does in fact closely resemble a sine-function of spatial frequency $1/\lambda$ (Fig. 11 ℓ). The shift of the maximum to smaller values of $1/\lambda$ will be interpreted below. Even if one takes into account that for very low spatial frequencies a signal-to-noise argument similar to the one discussed above will prevent the actual output of the gradient model from becoming infinitely large, the experimental data provide strong evidence that movement detection in flies is accomplished by a correlation-like mechanism rather than by a gradient-detecting scheme.

In the last column of Fig. 11 theoretical calculations for the two models are compared with movement-induced turning responses of walking Drosophila measured as a function of the contrast frequency w/λ (Buchner, 1974). When w/λ is held constant the shape of the curves should be the same for all values of λ if a correlation-like mechanism is in operation (Fig. 11 h). Again, even if a common high-frequency cut-off is assumed for the gradient model the experimental data favour the correlation model. The responses at $\lambda = 1,5\Delta\phi^*$ in Fig. 11 m were negative due to geometric interference as discussed above, and have been plotted with inverted sign. Contrast frequency and wavelength of such a pattern are equivalent to a pattern of $\lambda' = 3\Delta\phi^*$ according to the relation

$$\frac{2\pi \cdot \Delta\phi^*}{\lambda'} = 2\pi - \frac{2\pi \cdot \Delta\phi^*}{\lambda} \qquad (25)$$

The peak of the curve at $\lambda = 1,5\Delta\phi^*$ is certainly not shifted to the right compared to the peak of the curve at $\lambda = 6,3\Delta\phi^*$ as would be predicted for a gradient detector. These data are only suggestive, they are however supported by a wealth of experimental evidence from the literature

(Fermi & Reichardt, 1963; Götz, 1964; McCann & McGinitie, 1965; Eckert & Hamdorf, 1981). Thus from the combined evidence of Fig. 11 we may safely conclude that visual movement detection in flies is accomplished by a neuronal network which employs multiplicative two-channel interactions similar to the model of Fig. 9 b. This basic unit may therefore be called an "Elementary Movement Detector" in the visual system of flies.

Further models for movement detection

Various mechanisms apart from the gradient and the correlation scheme have been proposed as a basis for visual movement detection in animals, man and machines (van Doorn & Koenderink, 1976; rev. Ullman 1981). A scheme that has been used to model directionally selective ganglion cells in the rabbit retina (Barlow & Levick, 1965) is illustrated in Fig. 12 a. The similarity to the correlation model is apparent, a AND NOT operation is in fact a digital multiplication. In addition, a constant delay Δt_0 would work properly only over a very limited velocity range of about a factor 2 (cf. Fig. 12 c) while biological movement detectors operate over 4 $\log_{10}$ units of velocity (Fig. 11 m). Although the phenomenon of temporal interference is immediately obvious when one considers periodic patterns as a stimulus, it has not been emphasised so far. To avoid this problem the delay has to be assumed to depend on the temporal frequency of the stimulus as is the case for the delaying phase shift of a low pass filter for example. The similarity to the correlation model becomes then even more obvious. A simple physiological realisation of the AND NOT operation would consist of a linear inhibitory interaction followed by a threshold operation. Since thresholds represent the essential nonlinearity of a number of further movement detector models that have been proposed (Srinivasan & Bernard, 1976; Stegmann, 1982) it seems appropriate to outline the interrelationship of threshold and multiplication. Fig. 13 shows the transfer characteristic of a threshold operation and its approximation by a decomposition into a power series. The linear component cannot detect movement as was pointed out above. The parabolic characteristic in fact is equivalent to a squaring operation. If the input consists of the sum (or difference) of two signals $x_1 + x_2$ the output of this component will be $(x_1 + x_2)^2 = x_1^2 + 2x_1 x_2 + x_2^2$. Only one term in this expression contains information about the signals in both receptors which is a prerequisite for detecting movement. This term involves a multiplication of the two signals. Thus if a threshold operation leads to directional selectivity it will be primarily due to this term. A physiologically more plausible synaptic mechanism was proposed by Torre and Poggio (1978). It involves a nonlinear interaction in the form of a shunting inhibition. The movement-specific component of the output signal again involves a multiplication-like interaction between the two input signals as in the correlation model (Thorson, 1966 a, b). In fact, it may be true that the essential operation involved in any movement detection model proposed so far falls into either of the two schemes outlined above, the gradient scheme or the correlation scheme.

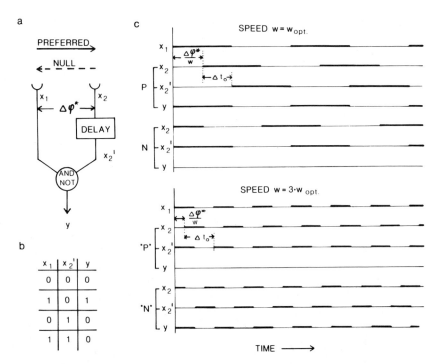

Fig. 12. a) Elementary unit of the scheme proposed by Barlow and Levick (1965) to account for directional selectivity of retinal ganglion cells in the rabbit. b) "Truth table" for the input-output relations in a). The output y is different from zero only when there is a signal in channel 1 (x_1) AND NOT in the delayed channel 2(x_2'). c) Signal flow for the scheme in a) under stimulation by a square wave grating of period λ moving at speed w either in the preferred (P) or the null (N) direction. At the optimal speed $w = w_{opt}$ the output will be maximal at preferred stimulation and zero at null stimulation. At 3-fold higher speed $w = 3\, w_{opt}$ temporal interference leads to "wrong" responses which are zero at preferred and maximal at null stimulation. This flaw in the detector can be removed if the fixed delay Δt_o is replaced by a delaying phase shift which depends on the temporal stimulus frequency w/λ. The relation to the correlation model is explained in the text.

Position detection

While, as we have seen, movement detection requires some non-trivial computation, in principle this is not true for position detection. A simple model of the fly's visuo-motor system that would be appropriate to produce a position-sensitive torque response according to eq. (12) ($R_{ps} = \beta_p \cdot \psi$) is illustrated in Fig. 14 a. The "weight" by which a stimulated receptor contributes to this response is proportional to its position ψ in

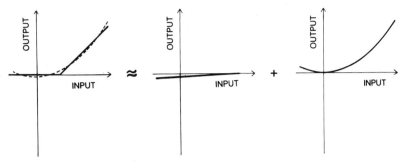

Fig. 13. Decomposition of a threshold characteristic into a power series terminated after the quadratic term. In the interval shown the approximation (dashed curve) has been optimized for least square error. The lowest order non-linearity corresponds to a squaring operation which for a sum of two inputs implies a multiplicative interaction between the two signals as in the correlation model (cf. text).

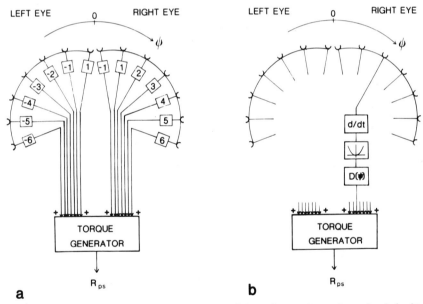

Fig. 14. Two simple schemes for position detection. In principle it is sufficient to "weight" the input signals by a factor proportional to their position ψ in the horizontal plane of the eye (a). In flies two further operations, temporal differentiation and squaring, are proposed to make the model comply with basic properties of the position-sensitive response R_{ps}. The experimentally determined weighting function $D(\psi)$ is shown in Fig. 15.

the eye. (If a single object is located at panorama position $\gamma = 0$, then $\psi \equiv \phi$ and we may continue to use ψ as is done in the original literature). However, when we carry out the open-loop experiment to verify eq. (12) we realise that the system is more intricate. A fly attached to the torque meter and stimulated by a stationary panorama carrying a single black stripe at position ψ will not generate a torque signal according to eq. (12). Rather, only the fluctuating "internal" torque signal can be measured under these conditions (Reichardt, 1973). It was soon realised that small oscillating movements of the stripe around the position ψ are capable of inducing a torque signal that is described by eq. (12) as long as the stripe remains in the very frontal eye region ($-15° < \psi < +15°$). The shape of the response over the entire range ($-180° < \psi \leq +180°$) measured under open-loop conditions with a randomly oscillating stripe is shown in Fig. 15 a (Reichardt & Poggio, 1975). The slope of the line that approximates this so-called $D(\psi)$ function in the range $-15° < \psi < +15°$ leads to the same value of β_p that had been determined from the closed-loop tracking experiment (Fig. 8) by eq. (11). This agreement can be taken as evidence that the experimental situation has no dramatic influence on the "internal state" of the animal (or the internal state does not modify the position detection system). Again it should be pointed out that free-flight tracking can be described and simulated when parameters in accordance with those obtained by the open-loop experiments are used in the theory outlined above (see also Discussion).

The finding that a stabilised retinal image does not induce a position sensitive torque while an oscillating stripe does, suggests two possible mechanisms that may underlie position detection. Oscillation consists of alternating episodes of front-to-back and back-to-front movement. It was found by open-loop experiments that front-to-back movement of a stripe e.g. on the right eye induces a torque signal to the

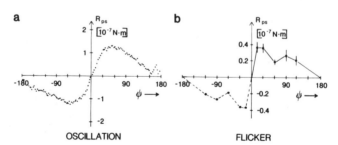

Fig. 15. Position-sensitive yaw torque responses of houseflies Musca domestica elicited by a randomly oscillating (a) or a sinusoidally flickering (b) stripe located at azimuthal position ψ (open-loop experiments). The origin $\psi = 0$ corresponds to the forward sagittal plane, positive responses at $\psi > 0$ (right hemisphere) and negative responses at $\psi < 0$ (left hemisphere) indicate a torque response towards the position of the stripe. Note the different scales in (a) and (b). For a quantitative comparison of responses to oscillation and flicker cf. Fig. 16. (after Reichardt & Poggio, 1975 (a) and Pick, 1974 (b)).

right that is larger than the torque to the left induced by the corresponding back-to-front movement. Summation of these two antagonistic torque signals over a full period of oscillation will therefore result in a net torque to the right. If this front-to-back vs. back-to-front difference is parametrised along the ψ coordinate according to the function $D(\psi)$, then this <u>movement</u>-dependent mechanism could explain the position-sensitive torque responses (Reichardt, 1973).

The alternate mechanism has been investigated by the late B. Pick whose work has profoundly influenced the discussion that follows. Any visual movement inherently contains a flicker stimulus. When the black stripe in the above open-loop experiment is replaced by a vertically oriented filament lamp of the same extensions, a position-sensitive torque response can be elicited by sinusoidally modulating the luminance of the stripe without moving it (<u>flicker</u> stimulus) (Pick, 1974; 1976). Fig. 15 b shows the $D(\psi)$ function as determined by a flickering stripe at position ψ. The similarity to the movement-induced $D(\psi)$ curve is obvious. A quantitative comparison of oscillation- and flicker-induced position-sensitive responses is possible from Fig. 16. In the same set-up and with the same flies a stripe at position $\psi = 20°$ was oscillated and flickered in alternate order. For a range of peak-to-peak oscillations and stripe widths up to at least 6° (which corresponds to 3 $\Delta\phi^*$ for <u>Musca</u>) there is no significant difference between the two curves (Pick, 1976; Reichardt, 1979). Also, in an apparent-movement experiment there was no indication of a difference between front-to-back and back-to-front movement of small amplitude (Pick, 1974). The conclusion from these experiments was that no lateral interactions between spatially separated receptors is required for detecting the position of small objects or edges and consequently, that position detection in flies under stationary conditions is accomplished by a flicker-detection mechanism rather than by an asymmetry of front-to-back vs. back-to-front movement detection. As was described above for the movement detection system one can obtain information about the basic mechanism of position detection by investigating the dependence of the position-sensitive torque response on

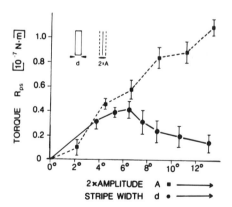

Fig. 16. Position-sensitive yaw torque responses to a 1,5° wide black bar oscillating around the position $\psi = 20°$ (right eye) with variable amplitude A (abscissa 2 x A, dashed curve), and to a stationary bar of sinusoidally flickering luminance at the same position. The width d of the bar is varied (abscissa d, solid curve). Oscillation and flicker frequency 3 Hz. (after Pick, 1976).

contrast at low values of this parameter. The corresponding experiments demonstrated a roughly quadratic contrast dependence, indicating that a nonlinear operation (squaring of input signals) is involved in position detection (Reichardt, unpublished results, cited in Poggio & Reichardt, 1976). A simple position detection scheme compatible with the results mentioned so far is shown in Fig. 14 b. It consists of a high-pass filter (to suppress responses to stabilised images), a squaring operation and a position-dependent weighting factor. Such a functional unit might be called an "Elementary Position Detector". High-pass filtering and squaring may be expected to produce neuronal activity characterised by transient spike discharges both at light-on and light-off. In fact, such responses have been found by extracellular recordings from the internal chiasm of Musca ("ON-OFF" units of Arnett, 1972). In the most simple scheme of position detection the weighted signals from these units would just be summed to produce the fly's response.

On the basis of transient-stimulation experiments the relevance of this scheme for object fixation and tracking under conditions of free flight has recently been questioned (Wehrhahn & Hausen, 1980; Wehrhahn, 1981). These authors rightly argue that mean open-loop responses averaged over 2 min of stimulation at constant eye position may not be appropriate for the interpretation of free-flight phenomena for which changes of the stimulus position within milliseconds are characteristic. They present evidence that under transient stimulation position-sensitive torque signals may not be based on flicker detection but rather on an asymmetric movement detection mechanism as described above. However, the issue is not settled since recent results from the same laboratory seem to favour a flicker detection mechanism even under transient stimulus conditions (Strebel-Brede, 1982). Unfortunately, the parameters of the visual stimuli used by the various authors were not quite comparable, such that further work is required to clarify the situation. A further problem for the unequivocal identification of the position detection mechanism is the fact that any flicker stimulus induces at its edges a specific form of apparent movement that has been recently investigated in flies and in man (Bülthoff & Götz, 1979). Since the importance of this effect in the above flicker-experiments is difficult to assess the issue must at present remain open.

FUNCTIONAL INTERACTIONS

After we have obtained a rather distinct view on the basic principles underlying visual movement and position detection we may now ask what can be learned from behavioural experiments on the way in which the corresponding basic functional subunits, the elementary movement detectors and the elementary position detectors, interact to generate the compound responses of the fly. These interactions were termed "functional" to stress the point that behavioural analysis certainly is too coarse a method to allow the identification of the actual neuronal interconnexions responsible for these interactions in a system of such complexity as the insect visual ganglia (cf. Hausen, 1983; Strausfeld, 1983).

Movement detection

If a movement detection network that is known to employ multiplicative 2-channel interactions derives its input from a 2-dimensional array of several hundred photoreceptors there is a vast number of possibilities how the pairs of interacting receptors may be distributed in the array. A priori anything between the two extremes of only nearest-neighbour interaction and interaction of every receptor with each of the rest seems possible. So one would like to find "selection rules" that determine which pairs of photoreceptors in the array interact in a movement-specific manner and which do not. The procedure on this question involves a kind of 2-dimensional Fourier analysis which is beyond the scope of this review. A detailed account has been given (Buchner, 1976). Here we can only briefly outline the principles of the analysis and state the results. Consider first a single row of equally spaced receptors. We suppose that any possible paired combination of receptors may be connected to an elementary movement detector. If we make the simplifying assumption that detectors with equal receptor separation are identical, we can classify the detectors as type 1, type 2, type 3 etc. with receptor spacings of 1 $\Delta\phi^*$, 2 $\Delta\phi^*$, 3 $\Delta\phi^*$ etc., respectively (Fig. 17). Inspection of eq. (24) shows that output signals of the various types of detectors differ significantly in their spatial frequency response as shown in Fig. 18 a. Here we have assumed that type 1 detectors contribute with an amplitude b_1, type 2 with b_2, type 3 with b_3 etc. The total output of a system with n_1 detectors of type 1, n_2 of type 2 etc. simply becomes

$$y = n_1 \cdot b_1 \cdot \sin \frac{2\pi \cdot \Delta\phi^*}{\lambda} + n_2 \cdot b_2 \cdot \sin \frac{2\pi \cdot 2\Delta\phi^*}{\lambda} + \ldots$$

Thus the output function is recognised to be a Fourier sine series. As a consequence, if one can determine the function y by behavioural experiments, it is possible to calculate the weight factors b_1, b_2, b_3 etc.

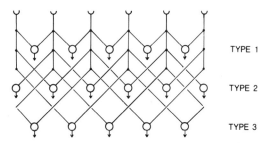

Fig. 17. Elementary movement detectors connected to a row of input elements (receptors). Qualitatively different responses are expected from detectors connected to nearest-neighbours (type 1), but-one-neighbours (type 2), but-two-neighbours (type 3) etc. (cf. Fig. 18).

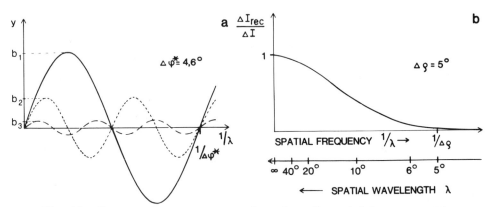

Fig. 18. Expected responses as a function of spatial frequency $1/\lambda$ of correlation-based elementary movement detectors of type 1 (a, solid curve), type 2 (dotted curve) and type 3 (dashed curve) (cf. Fig. 17) under the assumption of idealised photoreceptors with needle-shaped spatial sensitivity distributions. Arbitrary amplitudes b_1, b_2 and b_3 were chosen for the contributions of the three types of detectors. If real receptors with bell-shaped spatial sensitivity distributions are considered, the responses of (a) will be attenuated at high spatial frequencies due to the decline of the "contrast transfer characteristic" of the receptors (b). At high spatial frequencies the modulation amplitude in a receptor ΔI_{rec} will be only a fraction of that of the moving pattern due to spatial integration of luminance within the range of the spatial sensitivity distribution of the receptor. The curves have been plotted using the interommatidial angle $\Delta\phi^* = 4{,}6°$ and the acceptance angle $\Delta\rho = 5°$ as found for the compound eye of the fruitfly Drosophila.

by Fourier analysis. A significantly non-zero weight factor proves the existence of an elementary movement detector of the corresponding type. This analysis can be extended to a two-dimensional array of receptors if the output function y is determined for sine-wave gratings moving not only horizontally across the eye but also vertically and at four oblique orientations.

Before such an analysis can actually be carried out a further property of any real visual system has to be taken into account. In our theoretical considerations we so far always assumed that the photoreceptors had a needle-shaped spatial sensitivity distribution and were looking effectively at a single point of the pattern. Any real receptor has an approximately bell-shaped spatial sensitivity distribution which is shown in cross-section in Fig. 9 a, b. The width of this curve at half maximum is called the acceptance angle $\Delta\rho$. Theoretical considerations of this acceptance function must take into account diffraction phenomena at the cornea lenslet and the light guide properties of rhabdomeres. A detailed treatment has recently been published (Snyder, 1979) (For most recent experimental data cf. Dubs, 1982). In the present context it is sufficient to realise that a real receptor spatially integrates the luminance distribution located within its acceptance range and therefore represents a kind of spatial low-pass filter. As a consequence, only low spatial frequencies (gratings with broad stripes of $\lambda \gg \Delta\rho$) are "seen" by the receptors with unattenuated contrast. For sine-wave gratings the modulation amplitude in the receptors ΔI_{rec} is attenuated compared to the pattern amplitude ΔI by the factor (e.g. Götz, 1964)

$$\frac{\Delta I rec}{\Delta I} = e^{-3.56(\frac{\Delta\rho}{\lambda})^2} \qquad (26)$$

Thus for narrow gratings with $\lambda \ll \Delta\rho$ modulation in the receptors is so small that such patterns cannot be detected by the visual system. The "contrast transfer" curve of eq. (26) has been plotted in Fig. 18 b at the $1/\lambda$ scale using values of $\Delta\phi^*$ and $\Delta\rho$ as they have been determined for Drosophila (Götz, 1964; Heisenberg & Buchner, 1977). This contrast attenuation explains why contrary to the theoretical prediction (Fig. 18 a) the measured negative responses of flies (Fig. 19) in the range $2 \Delta\phi^* > \lambda > \Delta\phi^*$ are smaller than the positive responses in the range $\lambda > 2 \Delta\phi^*$. Before the Fourier analysis can be applied it is necessary to calculate the responses as they would be measured if the modulation in the receptors ΔI_{rec} could be kept at a constant value for all spatial frequencies. Such a calculation requires the knowledge of $\Delta\rho$ to determine the actual contrast in the receptors at each value of λ. Also one needs to know how the response of the fly depends on the modulation amplitude ΔI in order to determine the effect of the contrast attenuation on this response. The detailed procedure of determining $\Delta\rho$ by behavioural experiments has been described previously (Buchner, 1976; Heisenberg & Buchner, 1977). The

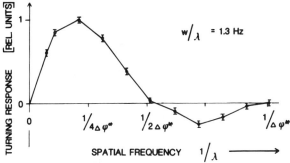

Fig. 19. Behavioural responses of walking fruitflies as a function of spatial frequency $1/\lambda$. The attenuation at high spatial frequencies is evident (cf. Fig. 18 b). Fourier analysis on these data demonstrates significant contributions of type 1 and type 2 elementary movement detectors (cf. Fig. 17). (after Buchner, 1976).

curve in Fig. 11ℓ actually had been calculated from the measured data of Fig. 19 in this manner. Within experimental error an antisymmetric curve is obtained in compliance with the theoretical prediction

The elementary movement detector analysis as outlined above was carried out for the visual system of Drosophila. Since the analysis requires precise alignment of the stripes of the sine-wave grating with the rows of optical axes of the compound eye, it was necessary to make use of the optical techniques developed by Franceschini (1975, 1983) which render the optical axes visible in a telescopic view of the eye. Also, as large amounts of behavioural data were required for this analysis, an automatic device for recording the turning tendency of the walking fly under microscopic stimulation had to be employed. It consists essentially of an air-supported styrofoam sphere whose rotations about two axes (for forward walk and rotatory turning) could be recorded by an electrooptic device (for details cf. Buchner, 1976). The responses of the flies strongly indicated that at the high pattern luminance used, wide-angle interactions do not significantly contribute. Analysis of 18 different types of elementary movement detectors including all possible interactions of receptors with neighbouring, but-one neighbouring and but-two neighbouring optical axes showed that at least 3 different types of detectors contribute significantly to the turning of the fly. They interconnect receptors whose optical axes correspond to neighbouring ommatidia along the horizontal row and two oblique rows oriented in the hexagonal array at +30 and -30° with respect to the horizontal. These 3 types of detectors are illustrated in Fig. 20 a, the thickness of the arrows represents the weight factor for each detector. The two oblique detectors which employ the smallest possible sampling separation for horizontal movement detection and therefore provide the highest spatial resolution, contribute

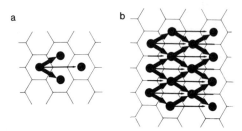

Fig. 20. Elementary movement detectors as found by a two-dimensional Fourier analysis of behavioural turning responses for the visual system of Drosophila (a). The thickness of the arrows illustrates the "weight" by which each type of detector contributes to the response (b-values in Fig. 18). If a homogeneous distribution of the detectors over the ommatidial array is assumed, a network of interactions as shown in (b) must be responsible for direction-sensitive movement detection in Drosophila. (after Buchner, 1976).

more than 70% of the total response. The horizontal detector which operates only at half the optimum spatial resolution contributes about 10%, the residual 20% being distributed among detectors whose contributions may or may not differ from zero by a statistically significant amount. (For further discussion cf. Buchner, 1976). Fig. 20 b illustrates the scheme of interactions that results for high pattern luminance under the assumption that all detectors of a given type are identical.

At low light levels the situation is quite different. By a similar analysis it was shown for Musca that interactions between receptors having one, two and three times the minimum angular separation contribute with about equal strength to movement responses elicited by very dim patterns. These differences in the interaction schemes at low and high luminance were interpreted as a manifestation of an adaptation mechanism that provides a gain in absolute sensitivity at the cost of spatial acuity (Pick & Buchner, 1979). Interestingly, a similar conclusion on optimum performance of the movement detection system at all light levels was drawn independently on the basis of contrast sensitivity measurements on direction-sensitive movement-detecting neurons in the lobula plate of the blowfly (Srinivasan & Dvorak, 1980; Dvorak et al., 1980).

When combined with appropriate optical and/or genetic techniques the power of behavioural analysis can be extended to provide information about the input elements to the visual movement detection system. The visual input of the fly is represented by two different subsystems: one is characterised by high absolute sensitivity mediated by receptors called R1-6, the other exhibits lower absolute but higher angular sensitivity mediated by receptors called R7 and 8 (Kirschfeld & Franceschini, 1968; see also Franceschini, 1983; Stavenga & Schwemer, 1983). The analogy to the photopic-scotopic dichotomy of the human visual system originally had led to the hypothesis that movement detection at low light levels was mediated by the receptor system R1-6 while at high ambient illumination system R7 and 8 would take over (Kirschfeld & Reichardt, 1970; Eckert, 1971, 1973; Heisenberg & Götz, 1975). By studies of the Drosophila mutants which lack either receptors R1-6 ("outer rhabdomers absent" ora^{JK84}, "receptor degeneration B" rdgB^{KS222}) or receptors R7 ("sevenless" sev^{LY3}) in each ommatidium as well as by bleaching experiments in which the receptors R1-6 could be reversibly blocked, it was shown that movement detection is in fact mediated predominantly if not exclusively by receptors R1-6 both at low and at high luminance (Heisenberg & Buchner, 1977). These findings have recently been supported by selective colour adaptation experiments in Musca. While for a visually induced startle-reflex the participation of two different types of photoreceptors could be demonstrated by this method, only one type of receptors apparently contributed to movement detection (Kirschfeld & Sarka, unpublished results). It must be concluded that in Drosophila the movement detectors illustrated in Fig. 20 are connected - probably via the monopolar cells of the lamina - to the receptor system R1-6 (see also Laughlin, 1983). There is ample evidence from behavioural work (Eckert, 1973; Pick & Buchner, 1979) and electrophysiological data (Mastebroek et al., 1980) that a similar scheme of elementary movement detectors should be found for the visual system of the housefly Musca and the blowfly Calliphora when periodic gratings are used for stimulation.

A more direct test for movement-specific interactions has been carried out with help of a sophisticated optical set-up by which it is possible to present a movement stimulus to only two individual receptors of the Musca eye. The response of the visual system may be examined either by measuring behavioural turning responses (Kirschfeld, 1972) or by recording from the directionally selective "H1" cell (Riehle & Franceschini, 1982) that was already mentioned above. The scheme of elementary movement detectors determined by stimulation of only two receptors is shown in Fig. 21. It significantly differs from the one discussed above. Electrophysiologically so far only the horizontal nearest-neighbour interaction has been verified which is common to both schemes (Fig. 20 a and 21) (Riehle & Franceschini, 1982). If these latter experiments should reproduce the lack of contributions from the two oblique nearest-neighbour detectors under single receptor stimulation, then this must be considered a consequence of the highly unnatural distribution of light signals on the eye, since from experiments using sine-wave gratings aligned to the receptor array under microscopic control there can be no doubt as to the existence of an oblique nearest-neighbour interaction in Musca both at high and at low light levels (Pick & Buchner, 1979). That non contiguous movement stimuli on a dark-adapted eye like those employed under single receptor stimulation may in fact lead to unexpected responses is illustrated in Fig. 22: A movement stimulus presented behind two vertical slits which are separated by a zone of 6 - 8 vertical rows of unstimulated ommatidia leads to inverted turning responses. No manifestation of inverted elementary movement detectors can be found when sine-wave gratings are used, and, in fact, these inverted responses are abolished if a constant illumination is applied to the ommatidia between the two slits (Pick & Buchner, 1979). It may be speculated that these awkward interactions are part of an inhibitory preprocessing network that has been postulated to operate distally to the movement

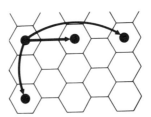

Fig. 21. Elementary movement detectors as determined by behavioural experiments on Musca under precise stimulation of individual pairs of photoreceptors (Kirschfeld, 1972). The horizontal nearest-neighbour interaction (straight arrow) has been verified by electrophysiological experiments (Riehle & Franceschini, 1982). For comparison with the results on Drosophila (Fig. 20) cf. text (after Buchner, 1976).

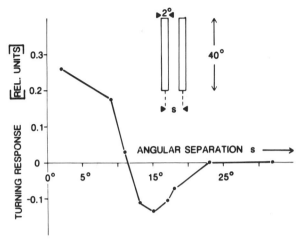

Fig. 22. Turning responses of walking Musca under stimulation by "apparent" motion (sequential flickering of two slits, cf. inset). At an angular separation of the two slits between 12° and 18° unexpected negative responses (turning against the direction of motion) are consistently obtained which are however abolished if the ommatidia between those illuminated by the two slits receive a constant background illumination. (after Pick & Buchner, 1979).

detection interaction (Srinivasan & Dvorak, 1980; Dvorak et al., 1980; see also Laughlin, 1983).

Position detection

In the simple scheme of Fig. 14 b we assumed linear spatial summation of the flicker detectors to generate the position-dependent torque response R_{ps}. The data of Fig. 16 (solid curve) agree with this assumption only over a very narrow range of lateral interactions: the response increases linearly with stripe width as long as no more than 3 rows of ommatidia are stimulated synchronously by the flickering stripe. However, beyond 6° stripe width the response not only does not increase further, it even significantly declines. On the other hand, increasing peak to peak amplitude of the oscillating stripe beyond 6° results in further approximately linear increase in the response (dashed curve in Fig. 16). Thus it was concluded that the temporal relationship of two flicker stimuli separated by more than 6° along the horizontal coordinate was of critical importance. This hypothesis was verified by experiments in which two narrow stripes could be individually flickered: Position-sensitive torque responses are inhibited if two adjacent 4,6° wide stripes are flickered in synchrony or 180° out of phase (Fig. 23). This inhibition is released if the two stripes are flickered independently, e.g. at slightly differing frequencies ("Phase drifting" in Fig. 23) (Pick, 1976). It was pointed out in

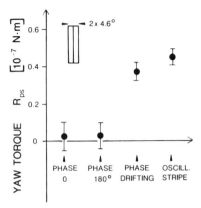

Fig. 23. Manifestations of nonlinear interactions between elementary position detectors. Position-sensitive yaw torque responses of Musca are suppressed when two 4,6° wide stripes are flickered in phase ("PHASE 0°") or in antiphase ("PHASE 180°") while strong responses are obtained when the two stripes are flickered at slightly different frequencies (3,0 Hz and 3,25 Hz) ("PHASE DRIFTING") or when a 1,5° wide black stripe is oscillated ("OSCILL. STRIPE"). Both types of stimuli were centred at ψ = +20° (right eye). (after Pick, 1976).

that work that these non-linear interactions between the (non-linear) flicker detectors are prerequisite for any form of pattern recognition and may be related to the phenomena of relative movement detection (cf. below). It also was mentioned that these interactions may well be responsible for the pattern-dependent position-sensitive responses derived from movement experiments (Geiger, 1974; Geiger & Poggio, 1975). As no manifestations of interactions between vertically separated flicker detectors can be detected (Pick, 1978) a tentative scheme of interactions between elementary position detectors as shown in Fig. 24 may be postulated. It avoids the domain of nearest-neighbour interactions which is occupied by the direction-sensitive interactions described above.

Again it was shown by genetic receptor elimination and by selective colour adaptation experiments that the position-sensitive responses probably are mediated by the receptor system R1-6 (Heisenberg & Buchner, 1977; Morton & Cosens, 1978) although the issue is still controversial (cf. Wehrhahn, 1976).

DISCUSSION

In the preceeding sections we have described movement and position detection as two fundamental components of the nervous network involved in spatial vision of insects. On this basis we can now turn our attention to

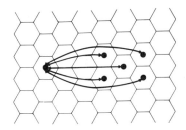

Fig. 24. Tentative scheme of nonlinear interactions between elementary position detectors. Inhibitory interactions (double-head arrows) between each detector in the hexagonal array of the compound eye with a group of five horizontally displaced detectors as shown could explain the responses of Figs. 16 and 23. Similar interactions with vertically displaced position detectors apparently do not exist (Pick, 1978).

more sophisticated aspects of visual orientation. These will provide us with a more realistic picture of the complexity of insect visual systems, and the necessary oversimplifications of the above outline on basic principles and logical interactions will be placed into the appropriate perspective. The selection of aspects to be discussed certainly reflects the author's background and interest and should be seen as a supplement and update to the extensive review by Wehner (1981).

Regional specialisation

As was described in earlier chapters of this book, the fronto-dorsal region of the compound eyes of male houseflies exhibits a conspicuous recruitment of central receptors to contribute to neural superposition at the level of the lamina (Franceschini, 1983). This region of the eye projects to a zone of the third optic ganglion (lobula) where a male-specific giant interneuron appears to collect visual information (Hausen, 1983; Strausfeld, 1983). At the behavioural level this sexual dimorphism has a likely correlate in a specific property of visually guided pursuit: During chasing episodes in free flight males try to keep the leading fly in the visual field corresponding to this fronto-dorsal eye region and may be capable of controlling the distance to their target by adjusting the forward speed. Females apparently lack this capability and try to keep their target in the lower hemisphere of the visual field (Wehrhahn, 1979; Wehrhahn et al., 1982). This latter result correlates with earlier "one-parameter closed-loop" experiments (Reichardt, 1973) and recent open-loop experiments (Wehrhahn, 1978 a; Geiger, 1981; Strebel-Brede, 1982) which show that an object in the upper hemisphere induces no appreciable position-sensitive response in female houseflies. Sexual dimorphism of visual tracking is a well-known phenomenon found also in various other insect species (rev. Wehner, 1981; Kirschfeld & Wenk, 1976; Collett & Land, 1978; Cook, 1979, 1980, 1981; Willmund & Ewing, 1982; Praagh et al., 1980; Zeil, 1981; Burkhardt & de la Motte, 1983). Regional differences are also noted for direction-sensitive movement responses and for landing reactions (Eckert, 1973, 1980; Jeanrot et al., 1981; Wehrhahn et al., 1981). These specialisations probably do not affect the "elementary movement detectors" but only their recruitment for a specific task since lift/thrust responses of Musca and Drosophila are weak or missing in the fronto-dorsal eye region where strong torque responses are recorded (Wehrhahn, 1978 a, b; Buchner et al., 1978).

Gain control

The visual system is confronted with stimulus amplitudes which vary over a range of as much as eight $\log_{10}$ units. Since the information capacity of a single neuron under reasonable conditions is limited by endogenous noise to probably less than two $\log_{10}$ units, a specific mechanism must operate to "adapt" the system to an appropriate sensitivity by adjusting the gain of the system. These mechanisms are well investigated in the periphery (cf. Laughlin, 1983) and are capable of allowing e.g. movement detection at a light flux as low as 10^{-1} photon per receptor per second without incapacitation due to saturation at 10^7 photons per receptor per second (Reichardt et al., 1968; Eckert, 1973; Lillywhite & Dvorak, 1981). A similar problem is encountered by the movement detection system in the spatial domain: How can a network which responds to a movement stimulus on a single pair of photoreceptors (cf. Fig. 21) (Kirschfeld, 1972; Riehle & Franceschini, 1982) avoid saturation when several thousands of pairs are stimulated simultaneously? The properties of this system have been incorporated into a model neuronal circuitry Poggio et al., 1981) that has been described in an earlier chapter of this book (Hausen, 1983). This model essentially reduces the gain of the individual movement detectors whenever a large number of detectors are stimulated simultaneously. A consequence of such a network will be discussed in the next paragraph. A network of somewhat related structure and function has been proposed to account for even more efficient suppression of nervous activity under large-field movement stimulation in various insect species (Rowell et al., 1977; cf. Wehner, 1981). In this case the circuitry is designed to suppress whole visual field movement as it is elicited by self-motion of the animal but to respond vigorously to small moving objects at any location in the visual field (Palka, 1969; Pinter, 1979; Olberg, 1981; Catton, 1982).

Detection of relative movement

This component of the fly's visual behaviour is presently being investigated by Reichardt and his colleagues. The basic open-loop experiment that demonstrates the essential properties of this response is depicted in Fig. 25. The fly (Musca) is suspended on a torque meter and flies inside a panorama cylinder which carries a random dot pattern (visual noise). This "background" panorama is oscillated at 2,5 Hz with ± 1° amplitude such that it moves 2° clockwise and 2° counterclockwise in alternate order 2,5 times a second. Under these conditions the fly will produce a direction-sensitive torque which also oscillates around zero.

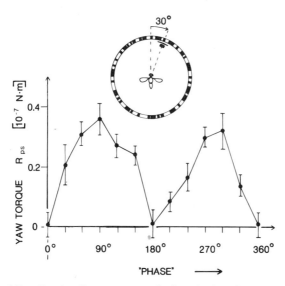

Fig. 25. Basic figure-ground discrimination experiment of Reichardt and co-workers. Houseflies (Musca domestica) flying suspended on a torque meter in still air are surrounded by a panorama carrying a random dot pattern (ground) which is oscillated horizontally at 2,5 Hz with an amplitude of ± 1° around the fly. The figure consists of a 3° wide black stripe which is oscillated by an independent drive around the position $\psi = +30°$ (right eye) at the same frequency and amplitude but at various values of the phase with respect to the ground oscillation ("PHASE"). Between the stripe and the ground a 12° wide stationary white screen has been inserted to prevent simulataneous stimulation of individual receptors by both figure and ground. Under these conditions the fly turns towards the figure (positive mean position sensitive responses) whenever the phase is different from 0° (= 360°) and 180°. When figure and ground move in phase (0°) or in anti-phase (180°) the fly does not respond to the stripe, i.e. the stripe is not "seen" by the fly. The interpretation of these results by a gain control mechanism is discussed in the text. (after Reichardt & Poggio, 1979).

Since position stimuli are equally distributed over both eyes no net position-sensitive torque is measured. If a 3° wide black stripe is mounted on a second motor and is oscillated in front of the background pattern at the same frequency and amplitude around the position $\psi = 30°$, the influence of the "phase" between the two oscillatory movements of stripe and background (x - coordinate in Fig. 25) can be measured. At 0° phase the stripe and the background move in perfect synchrony, and the fly does not respond to the position of the stripe (zero mean position-sensitive torque). This may be interpreted as an indication that the fly does not "see" the stripe. With a 90° or 270° phase difference between the oscillations of stripe and background the fly responds strongly by turning towards the stripe. At 180° phase the stripe and the background always move in opposite directions, and the fly again does not "see" the stripe. As was already mentioned in the chapter by Hausen (1983) these phenomena can be explained and simulated on a digital computer by the very same gain-control model network that was mentioned above. The relation of a mechanism which reduces the gain of movement detectors in the presence of large-field motion to the present experiment can be understood if it is realised that during inversion of the background movement there is for brief periods of time no large-field motion. Consequently, during these brief periods the gain will rise sharply and any small-field motion will cause vigorous responses. Stripe motion occurs during these periods only at phases different from 0° and 180° and is maximal at 90° and 270°. Since stripe motion alternates in direction, only the position-sensitive response sums up to give a net average torque. This is reflected by the responses of Fig. 25 (Reichardt & Poggio, 1979; Poggio et al., 1981; Reichardt et al., 1983; cf. also Bülthoff, 1981). It should be noted that the inhibitory interactions between position detectors described above (Pick, 1976) (Fig. 24) can also account for the responses of Fig. 25. It is not yet clear whether the two models are mutually exclusive or whether they represent manifestations of one and the same neuronal interaction that is not yet fully understood. It remains to be shown that these interactions are in fact of significance for the evaluation of relative movement during free locomotion. It is interesting to note in this context that motion parallax information during peering for distance estimation in locusts does not appear to be evaluated by relative movement detection since ground-movement behind the object is ignored during this behavioural episode (Collett, 1978).

Optokinetic equilibrium

The primary function of direction-sensitive responses to large-field movement ("optomotor responses") during free locomotion appears to be course stabilisation: Involuntary rotations of the animal are counteracted by the visually induced response. However, this stable equilibrium at zero rotatory movement may in fact become unstable when the animal undergoes sufficiently large spontaneous rotations during translatory locomotion (Götz, 1975, 1980). This phenomenon is a consequence of the decline of movement responses at high velocities w of the retinal image (Fig. 11 m) and can be explained by referring to Fig. 1 a. In this figure the retinal velocities caused by forward flight (Fig. 1 c) and counterclockwise rotation (Fig. 1 b) are vectorially added. It is seen that under this condition the velocity of a close object in the lateral visual field of the right eye (near $\phi = +90°$) becomes very large front-to-back while a close object on the left lateral eye (near $\phi = -90°$) moves at a smaller front-to-back velocity. If both these velocities are large enough the optomotor response to the larger velocity on the right eye will be a smaller torque to the right than the torque to the left induced by the smaller velocity on the left eye. Thus a net torque to the left results which <u>increases</u> the animals counterclockwise rotation instead of counteracting it. This positive feedback leads to instabilities of course control which in walking Drosophila can in fact be observed under suitable conditions (Götz, 1975). Note that here we have a rather informative example that simple characteristics of open-loop responses can predict complex behaviour observed under natural conditions.

Do flies measure velocity?

It was pointed out above that the gradient scheme for movement detection directly measures velocity (eq. 23). Its output signal does not depend on the contrast or the spatial frequency content of the pattern (at least as long as the spatial frequencies are low compared to the sampling frequency $1/\Delta\phi*$). However, the data of behavioural and electrophysiological open-loop experiments have made it quite clear that the visual system of flies employs a correlation-like mechanism for direction-sensitive movement detection. As is evident from eq. (24) and Fig. 11 the response of such a system strongly depends on the contrast $\Delta I/\bar{I}$ and the spatial frequency $1/\lambda$ of the pattern, and the peak response when the speed w is varied occurs for all spatial frequencies at the same value of the temporal frequency w/λ ("contrast frequency") (Fig. 11 m). How can this system be useful for visual orientation? There are several possibilities to be considered. During rotatory self-movement of the animal e.g. about the vertical body axis, the speed of any contrast element in the surround is constant at a given elevation angle (cf. Fig. 1 b) and does not depend on the spatial frequency or the distance of the contour from the animal. Evaluation of the actual speed by a correlation-based system could only be accomplished if the spatial frequency content and the contrast of the pattern would be determined independently by a different subsystem of the fly's brain. Even then the response would be unambiguous only if the contrast frequencies of all pattern components lie below the value of peak response (1Hz in Fig. 11 m). Another possibility would be that the system basically evaluates only the direction of movement and therefore effectively operates around zero rotatory speed In this case the control loop would be largely independent of the parameters of the panorama. A sensible mode of locomotion for an animal with such a system would be to advance essentially along straight line segments which are joined by saccadic turns that are not under visual movement control. Some traits of Drosophila and Syritta flight behaviour appear to resemble such a scheme, and saccadic suppression of visual input has been demonstrated in several insects (Collett & Land, 1975 b; Heisenberg & Wolf, 1979; Zaretsky & Rowell, 1979). A third possibility would be that velocity measurements are carried out by a position detection mechanism in conjunction with visual "attention" focussed on prominent contours or edges of the visual world (cf. below). Such a mechanism might be equivalent to "efferent motion perception" in humans (Dichgans et al., 1969).

During translatory self-movement of the animal the retinal velocity depends on the distance between the contour and the animal (Fig. 1 c). Since the perceived spatial frequency content of an object also depends on the distance in an inverse manner, the product, retinal velocity times spatial frequency w/λ (= contrast frequency) is invariant with distance. In this case a correlation-like mechanism may be advantageous even if the problems of contrast dependence and spatial frequency dependence remain. For a true velocity measurement the fly again would have to measure the contrast of an object and "know" its real size or measure its distance. Distance invariance in detection threshold and velocity estimation has been observed for humans in psychophysical experiments (Diener et al., 1976; Burr & Ross, 1982) which may indicate that correlation-like mechanisms are also involved in directional movement perception of man. In insects the problems of pure velocity detection have recently been investigated in free-flight experiments in a wind tunnel in combination with various stationary and moving stimuli (David, 1979 a, b, 1982 a, b; Kuenen & Baker, 1982). In some situations evidence for true velocity measurement are obtained. However, since striped patterns of high contrast were used the insect may well employ position control mechanisms rather than velocity measurements to achieve the described response. This question clearly deserves further investigation.

Context dependence and visual attention

Insect orientation behaviour as we have described it so far can be understood as a more or less complex reflex that may involve only visual and motor periphery of the nervous system and would not require interference by other sensory modalities or higher centres of the brain. The question of an influence of the "internal state" of the fly on the gain factor of movement and position responses has already been dealt with above. An interesting example of "context" dependence has recently been observed for the neuronal interactions underlying vertical movement detection in Drosophila (Fig. 26) (Götz, 1983). The recruitment of elementary movement detectors for thrust/lift responses was shown to depend on air stream input to the antennae. When the tethered animal flies in still air, two unidirectional detectors contribute most strongly which are not antiparallel but split up between two pairs of neighbouring input elements as shown in Fig. 26 a (Götz & Buchner, 1978; cf. also Wehrhahn, 1978 b). If however a stream of air is blown at the antennae, the scheme of interactions changes to the arrangement of Fig. 26 b, i.e. no evidence for contributions from non-vertical detectors is found. A plausible interpretation of these results relates the experimental situation with and without antennal air stream to forward flight and hovering, respectively. During hovering the animal's long body axis is inclined, the head is pointing obliquely upwards (David, 1978). In this posture space vertical is no longer aligned with the vertical rows of the ommatidial array but may correspond to the resultant preferred direction of the detectors of Fig. 26 a. During forward flight there may be a reasonable alignment between space vertical and the vertical ommatidial rows. The antennal input thus appears to be interpreted as a body posture signal which is used to align the preferred direction of the vertical movement detection system to space vertical. For the horizontal movement detection system no changes in detector recruitment were observed (Götz, 1983).

Apart from this capability of minor adaptive modifications the

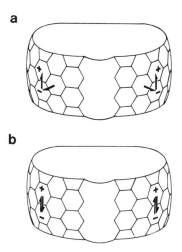

Fig. 26. Plasticity of recruitment of elementary movement detectors for thrust/lift-responses in Drosophila. When the animals fly in still air two unidirectional nearest-neighbour detectors of different orientation in the hexagonal array are found to contribute most strongly to the responses (a). In a wind tunnel at airspeeds above 0,4 m/s the main contribution appears to come from two antiparallel, vertically oriented unidirectional detectors (or from one bidirectional detector) (b). (after Götz, 1983).

movement detection system appears to be rather hard-wired. Insects cannot adjust to an artificial inversion of the direction of rotatory movement perceived during locomotion. Such an inversion can be achieved either by rotating the animal's head by 180° and fixing it to the thorax in this upside-down position (Mittelstaedt, 1949), or by electronically inverting the torque signals which drive the panorama in the "artificial closed-loop" apparatus described above (Heisenberg & Wolf, 1984.). Also, among some 60 000 mutagenised Drosophila flies tested, not a single mutant with inverted direction-sensitive movement responses was found (Heisenberg & Götz, 1975).

Position-sensitive responses are considerably less reliable. Several Drosophila mutants have been isolated which show inverted position-sensitive responses under appropriate experimental conditions (Bülthoff, 1982 a, b). Also, nervous plasticity both with respect to the gain and the sign of these responses has been observed in normal flies. In a particular experimental situation walking Drosophila flies show recurrent inversion of the sign of position-sensitive responses (Bülthoff et al., 1982). This phenomenon may well affect pattern discrimination in freely moving insects (Anderson, 1979; Mimura, 1982). An example of temporal context dependence is illustrated in Fig. 27 (Geiger, 1975). As a consequence of the position-sensitive torque response of eq. (12) a fly will be induced to

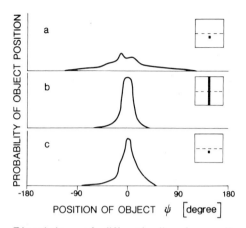

Fig. 27. Plasticity of "fixation" of small objects in "artificial closed-loop" measurements with Musca. The histograms show the proportion of the measuring time during which the visual object (cf. insets) was positioned at each value of the azimuthal angle ψ. An extended bar is most of the time "fixated", i.e. observed by the most frontal ommatidia, as indicated by the pronounced peak at $\psi = 0$ in (b). If a small stripe segment is presented to a "naive" fly which had no previous experience with the experimental situation, a rather flat histogram is obtained (a). If however the same segment is tested immediately after the bar experiment in b) the fly apparently has "learned" how to handle the situation and now is able to "fixate" even the small segment (c). (after Geiger, 1975).

"look at" a visual object, e.g. a black stripe, that is, it will align its body axis ($\psi = 0°$) with the position of the stripe ("fixation"). This behaviour can be quantitatively investigated in the "artificial closed-loop" set-up by recording a histogram of the stripe position. A pronounced peak near $\psi = 0°$ in such a histogram means that the fly looked at the object most of the time with its frontal ommatidia, i.e. the object was "fixated" by the fly (Fig. 27 b). If a small stripe segment is presented to a "naive" fly which has not been in the experimental set-up before, it will only poorly "fixate" the stripe as indicated by the broad, rather flat histogram in Fig. 27 a. If however the small stripe segment is presented to the fly immediately after the long-stripe experiment, relatively good fixation is now obtained (Fig. 27 c). Thus the long stripe apparently has aroused the fly to pay more "attention" to the position of the small segment.

Three further examples of such context dependence have been investigated by Heisenberg and his colleagues. If in an open-loop experiment fruitflies are confronted with an ambiguous position stimulus of two identical oscillating stripes, one on each eye, such that there should be no net position-sensitive torque, they seem to focus their "attention" for extended periods of time (tens of seconds) to only one stripe ignoring the other. Long-term symmetry is reestablished by occasional switching attention to the other stripe (Wolf & Heisenberg, 1980). Very recently it was demonstrated that in Drosophila the gain of the position-sensitive response does depend on the experimental condition being closed-or open-loop (Heisenberg & Wolf, 1984). Finally, a rather complex phenomenon of adaptive "learning" has been described which enables the fly to invert the sign of the position-sensitive torque if in an artificial stimulus arrangement this is prerequisite for successful visual fixation. These experiments lead to the concept of "initiation" as an important component of insect behaviour (Heisenberg, 1983). They will be described in detail in another review (Heisenberg & Wolf, 1984) and are therefore only briefly mentioned here.

Free flight phenomena

It was mentioned above that the simple equation (13) derived under rather crude approximations can be used to describe and stimulate free-flight pursuit episodes of one fly chasing another. After what has just been said about some of the more sophisticated aspects of orientation behaviour it is evident that spatial vision of insect is far more than a simple reflex involving position and movement detection. Since visually guided pursuit can even be realized by systems as primitive - compared to insect brains - as electronic devices, it may not be too surprising that selected pursuit episodes can be simulated by a variety of different mechanisms (Land & Collett, 1974; Wehrhahn & Hausen, 1980; Wehrhahn 1981; Poggio & Reichardt, 1981 b; Geiger & Poggio, 1981) all of which are essentially described by eq. (13). While the sensitive parameters of the simulation, a delay between stimulus and response and the slope β_p of the position-sensitive response (eq. 12), had to be chosen close to values obtained by open-loop measurements, it appears that neither the actual shape of the position dependence of the response nor the specific detection mechanism underlying the response are critical for a good simulation. Further aspects which illustrate that free flight has little resemblence to the simulation pertain to the various additional components of aerobatic performance. As may be seen from high-speed films of freely flying insects (Collett & Land, 1975 a, b, 1978; Collett, 1980 a, b; Zeil, 1981; Wagner, 1982) and from measurements on tethered animals (Blondeau, 1981 a, b; Blondeau & Heisenberg, 1982) forces and torques along all three axes can be generated by most insects and allow a variety of manoeuvres that are not accounted for by the simulation. Also, for houseflies it is not quite clear yet, whether only a "smooth" tracking system according to eq. (13) exists or whether, as in Syritta or Drosophila, "body saccades" occur during tracking (Collett & Land, 1975 b; Heisenberg & Wolf, 1979). Furthermore, head movements certainly play a role in visual orientation (Geiger & Poggio, 1977; Kien & Land, 1978; Wehrhahn, 1980; Varju & Bolz, 1980; Bolz & Varju, 1980; Grieger et al., 1981; Moore et al., 1981; Moore & Rankin, 1982; Hengstenberg & Sandeman, 1982). Therefore, resolving the insect's body orientation in filmed pursuit episodes certainly is not sufficient to define the position of the leading object on the retina. Contributions to visual orientation behaviour from the other visual sense organ, the ocelli, may not be negligible (Hu & Stark, 1980; Stange, 1981; Taylor, 1981) even if in most open-loop experiments described above covering the ocelli with black paint had no measurable influence on the responses. Habituation, after-effects and learning may also modify free-flight behaviour (Fischbach, 1981; Götz & Wenking, 1973; Srinivasan & Dvorak, 1979; Mastebroek et al., 1982; Folkers & Spatz, 1981; Folkers, 1982; Erber, 1982). Finally, the relation between torque generation during tethered flight in still air and turning during free flight under consideration of wing beat and body posture (Götz et al., 1979) has not yet been investigated. The concurrence of equally important mechanosensory and visual control mechanisms for spatial orientation during flight was already mentioned above (Hengstenberg & Sandeman, 1982).

As a result of this discussion it seems evident that simulations which ignore all these aspects of biological reality can only be of limited value. A good fit of such a simulation should therefore not be over-emphasised as an indication that the mechanisms and parameters of movement and position detection used in the simulation might actually reflect properties of the real system. Yet there can be no doubt that the basic characteristics of the visual component of spatial orientation and tracking behaviour of freely flying insects are in fact accounted for by movement and position detection as described by eq. (13).

<u>Concluding remarks</u>

A central theme of the present review was the illustration of value and limitations of mathematical modelling of complex function in visual behaviour of insects. Numerous models that are found in the literature make no effort to be quantitative and can at best convey a general "feel" of how a nervous system might go about to realise a particular function. The importance of the general approach of quantitative modelling of insect visual function as it was developed by Reichardt and his colleagues lies in the fact that such models not only describe the measured responses on the basis of which the models were developed but allow prediction of responses to new stimuli, usually by numerical simulation on a digital computer. These predictions can then be subjected to experimental tests which may well falsify a given model. I would like to stress the point that a model that is not specific enough to be falsified experimentally is only of very limited value. Another recent example of mathematical modelling and testing in insect visual behaviour has provided information on the basic strategy of how insects might use visual landmarks to guide their return to a food source (Cartwright & Collett, 1982).

The extensive application of this approach to the visual system of flies has provided deep insights into the basic mechanisms of movement and position detection and has yielded a wealth of information on functional interactions within this neurobiological model system. Admittedly, a problem arises when the model becomes so complex that unknown parameters can always be adjusted in such a way as to fit any experimental result. Thus behavioural analysis by itself is insufficient to unravel highly complex networks as they are found in the visual system of insects. This deficiency can only be overcome if behavioural analysis and mathematical modelling is combined with electrophysiological probing of single cells (cf. Hausen, 1983), with genetic or surgical manipulation of the system (Heisenberg, 1979; Fischbach & Heisenberg, 1981; Bülthoff, 1982 a, b; Hausen & Wehrhahn, 1983; Hausen, 1983) and with new techniques such as the deoxyglucose method for neuroanatomical functional mapping (cf. Buchner & Buchner, 1983). The technique of pharmacological manipulation of the visual system of insects has so far been almost entirely ignored as is evident from the present volume, and very little is known about the neurotransmitters involved in this system (Evered et al., 1982). Certainly with the advent of immunocytochemical techniques major progress may be expected in this field over the next few years. It seems not entirely hopeless that by incorporating these old and new techniques into the behavioural approach basic visual functions like movement and position detection in insects can eventually be understood at the level of identified neurons and their synaptic interconnexions.

ACKNOWLEDGEMENTS

I would like to thank H. Bülthoff, T. Collett, C. David, M. Egelhaaf, N. Franceschini, K. Götz, R. Hardie, K. Hausen, M. Heisenberg, R. Hengstenberg, K. Kirschfeld, I. Nicod, T. Poggio, W. Reichardt, V. Rodrigues, J. Strebel-Brede, K. Vogt, H. Wagner and C. Wehrhahn for critically reading and discussing the manuscript. I am grateful to C. Straub for preparing the figures and to C. Hanser for typing the manuscript. I am particularly indebted to H.Bülthoff, D. Burkhardt, M. Heisenberg, K. Kirschfeld and S. Rossel for the permission to use unpublished material from their work.

REFERENCES

Anderson, A.M. (1979) Visual scanning in honey bees. J. Comp. Physiol. 130: 173-182.
Arnett, D.W. (1972) Spatial and temporal integration properties of units in first optic ganglion of dipterans. J. Neurophysiol. 35: 429-444.
Barlow, H.B. & Levick, W.R. (1965) The mechanism of directionally sensitive units in rabbit's retina. J. Physiol. 178: 477-504.
Bauer, T. (1981) Prey capture and structure of the visual space of an insect that hunts by sight on the litter layer (Notiophilus biguttatus F., Carabidae, Coleoptera). Behav. Ecol. Sociobiol. 8: 91-97.
Bauer, T. (1982) Predation by a carabid beetle specialized for catching Collembola. Pedobiol. 24: 169-179.
Blondeau, J. (1981 a) Aerodynamic capabilities of flies, as revealed by a new technique. J. Exp. Biol. 92: 155-163.
Blondeau, J. (1981 b) Electrically evoked course control in the fly Calliphora erythrocephala. J. Exp. Biol. 92: 143-153.
Blondeau, J. & Heisenberg, M. (1982) The three-dimensional optomotor torque system of Drosophila melanogaster. J. Comp. Physiol. 145: 321-329.
Bolz, J. & Varju, D. (1980) Head movements of the mealworm beetle

Tenebrio molitor. II. Responses to rotating panoramas. Biol. Cybern. 36: 117-124.

Buchner, E. (1974) Bewegungsperzeption in einem System mit gerastertem Eingang. Ph.D. Thesis, University of Tübingen. 86 pages.

Buchner, E. (1976) Elementary movement detectors in an insect visual system. Biol. Cybern. 24: 85-101.

Buchner, E. & Buchner, S. (1983) Neuroanatomical mapping of visually induced nervous activity in insects by ^{3}H-deoxyglucose. (This volume)

Buchner, E., Götz, K.G. & Straub, C. (1978) Elementary detectors for vertical movement in the visual system of Drosophila. Biol. Cybern. 31: 235-242.

Bülthoff, H. (1981) Figure-ground discrimination in the visual system of Drosophila melanogaster. Biol. Cybern. 41: 139-145.

Bülthoff, H. (1982 a) Drosophila mutants disturbed in visual orientation. I. Mutants affected in early visual processing. Biol. Cybern. 45: 63-70.

Bülthoff, H. (1982 b) Drosophila mutants disturbed in visual orientation. II. Mutants affected in movement and position computation. Biol. Cybern. 41: 71-77.

Bülthoff, H. & Götz, K.G. (1979) Analogous motion illusion in man and fly. Nature (Lond.) 278: 636-638

Bülthoff, H., Götz, K.G. & Herre, M. (1982) Recurrent inversion of visual orientation in the walking fly, Drosophila melanogaster. J. Comp. Physiol. 148: 471-481.

Bülthoff, H., Poggio, T. & Wehrhahn, C. (1980) 3-D analysis of the flight trajectories of flies (Drosophila melanogaster). Z. Naturforsch. 35c: 811-815.

Burkhardt, D. & de la Motte, I. (1983) How stalk-eyed flies eye stalk-eyed flies: Observations and measurements of the eyes of Cyrtodiopsis (Diopsidae, Diptera). J. Comp. Physiol. (In press).

Burr, D.C. & Ross, J. (1982) Contrast sensitivity at high velocities. Vision Res. 22: 479-484.

Cartwright, B.A. & Collett, T.S. (1982) How honey bees use landmarks to guide their return to a food source. Nature (Lond.) 295: 560-564.

Catton, W.T. (1982) Effects of stimulus area and intensity on the on/off ratio of some locust visual interneurons. J. Insect. Physiol. 28: 285-292.

Chillemi, S. & Taddei-Ferretti, C. (1981) Landing reaction of Musca domestica. VI Neurones responding to stimuli that elicit the landing response in the fly. J. Exp. Biol. 94: 105-118

Collett, T.S. (1978) Peering - a locust behaviour pattern for obtaining motion parallax information. J. Exp. Biol. 76: 237-241.

Collett, T.S. (1980 a) Angular tracking and the optomotor response. An analysis of visual reflex interaction in a hoverfly. J. Comp. Physiol. 140: 145-158.

Collett, T.S. (1980 b) Some operating rules for the optomotor system of a hoverfly during voluntary flight. J. Comp. Physiol. 138: 271-281.

Collett, T.S. & Harkness, L.I.K. (1982) Depth vision in animals. In: Analysis of Visual Behaviour. Ed. D.J. Ingle, M.A. Goodale & R.J.W.

Mansfield. Cambridge, Mass. MIT Press, p. 111-176.
Collett, T.S. & Land, M.F. (1975 a) Visual spatial memory in a hoverfly. J. Comp. Physiol. 100: 59-84.
Collett, T.S. & Land, M.F. (1975 b) Visual control of flight behavior in the hoverfly, Syritta pipiens. J. Comp. Physiol. 99: 1-66.
Collett, T.S. & Land, M.F. (1978) How hoverflies compute interception courses. J. Comp. Physiol. 125: 191-204.
Cook, R. (1979) The courtship tracking of Drosophila melanogaster. Biol. Cybern. 34: 91-106.
Cook, R. (1980) The extend of visual control in the courtship tracking of D. melanogaster. Biol. Cybern. 37: 41-51.
Cook, R. (1981) Sex-specific tracking identified in single mosaic Drosophila. Naturwiss. 68: 267-268.
David, C.T. (1978) The relationship between body angle and flight speed in free-flying Drosophila. Physiol. Entomol. 3: 191-195.
David, C.T. (1979 a) Height control by free-flying Drosophila. Physiol. Entomol. 4: 209-216.
David, C.T. (1979 b) Optomotor control of speed and height by free-flying Drosophila. J. Exp. Biol. 82: 389-392.
David, C.T. (1982 a) Competition between fixed and moving stripes in the control of orientation by flying Drosophila. Physiol. Entomol. 7: 151-156.
David, C.T. (1982 b) Compensation for height in the control of groundspeed by Drosophila in a new, "barber's pole" wind tunnel. J. Comp. Physiol. 147: 485-493.
Derham, W. (1713) Physico-Theology. Boyle Lecture for 1711, London.
Dichgans, J., Körner, F. & Voigt, K. (1969) Vergleichende Skalierung des afferenten und efferenten Bewegungssehens beim Menschen: Lineare Funktionen mit verschiedener Anstiegssteilheit. Psychol. Forsch. 32: 277-295.
Diener, H.C., Wist, E.R., Dichgans, J. & Brandt, T. (1976) The spatial frequency effect on perceived velocity. Vision Res. 16: 169-176.
Doorn, A.J. van, & Koenderink, J.J. (1976) A directionally sensitive network. Biol. Cybern. 21: 161-170.
Dubs, A. (1982) The spatial integration of signals in the retina and lamina of the fly compound eye under different conditions of luminance. J. Comp. Physiol. 146: 321-343.
Dvorak, D., Srinivasan, M.V. & French, A.S. (1980) The contrast sensitivity of fly movement detecting neurons. Vision Res. 20: 397-407.
Eckert, H. (1971) Die spektrale Empfindlichkeit des Komplexauges von Musca (Bestimmung aus Messungen der optomotorischen Reaktionen). Kybernetik 9: 145-156.
Eckert, H. (1973) Optomotorische Untersuchungen zum visuellen System der Stubenfliege Musca domestica. Bestimmung des optischen Auflösungsvermögens, der Kontrastempfindlichkeit und der Lichtflüsse in den Rezeptoren der Komplexaugen als Funktion der mittleren Umweltleuchtdichte. Kybernetik 14: 1-23.
Eckert, H. (1980) Orientation sensitivity of the visual movement detection system activating the landing response in the blowflies,

Calliphora and Phaenicia: A behavioural investigation. Biol. Cybern. 37: 235-247.
Eckert, H. (1982) Radial pattern expansion drives the landing response of the blowfly, Calliphora. Naturwiss. 69: 348-349.
Eckert, H. & Hamdorf, K. (1980) Excitatory and inhibitory response components in the landing response of the blowfly, Calliphora erythrocephala. J. Comp. Physiol. 138: 253-264.
Eckert, H. & Hamdorf, K. (1981) The contrast frequency dependence: A criterion for judging the non-participation of neurones in the control of behavioural responses. J. Comp. Physiol. 145: 241-247.
Erber, J. (1982) Movement learning of free flying honeybees. J. Comp. Physiol. 146: 273-282.
Eriksson, E.S. (1980) Movement parallax and distance perception in the grasshopper. J. Exp. Biol. 86: 337-340.
Evered, D., O'Connor, M. & Whelan, J. (Eds.) (1982) Neuropharmacology of Insects. Ciba Foundation symposium 88. London, Pitman. 330 pages.
Fermi, G. & Reichardt, W. (1963) Optomotorische Reaktionen der Fliege Musca domestica. Abhängigkeit der Reaktion von der Wellenlänge, der Geschwindigkeit, dem Kontrast und der mittleren Leuchtdichte bewegter periodischer Muster. Kybernetik 2: 15-28.
Fischbach, K.F. (1981) Habituation and sensitization of the landing response of Drosophila melanogaster. Naturwiss. 68: 332.
Fischbach, K.F. & Heisenberg, M. (1981) Structural brain mutant of Drosophila melanogaster with reduced cell number in the medulla cortex and normal optomotor yaw response. Proc. Natl. Acad. Sci. USA 78: 1105-1109.
Folkers, E. (1982) Visual learning and memory of Drosophila melanogaster, wild type C-S and the mutants dunce, amnesiac, turnip and rutabaga. J. Insect. Physiol. 28: 535-539.
Folkers, E. & Spatz, H.C. (1981) Visual learning-behavior in Drosophila melanogaster wildtype CS. J. Insect Physiol. 27: 615-622.
Fraenkel, G. & Pringle, J.W.S. (1938) Halteres as gyroscopic organs of equilibrium. Nature (Lond.) 141: 919.
Franceschini, N. (1975) Sampling of the visual environment by the compound eye of the fly: Fundamentals and applications. In: Photoreceptor Optics. Ed. A.W. Snyder & R. Menzel. Berlin, Heidelberg, New York, Springer Verlag, p. 97-125.
Franceschini, N. (1983) The retinal mosaic of the fly compound eye. (This volume).
Gavel, L. von (1939) Die kritische Streifenbreite als Maß für die Sehschärfe bei Drosophila melanogaster. Z. vergl. Physiol. 27: 80-135.
Geiger, G. (1974) Optomotor responses of the fly Musca domestica to transient stimuli of edges and stripes. Kybernetik 16: 37-43.
Geiger, G. (1975) "Short-term learning" in flies. Naturwiss. 62: 539.
Geiger, G. (1981) Is there a motion-independent position computation of an object in the visual system of the housefly? Biol. Cybern. 40: 71-75.
Geiger, G., Boulin, C. & Bücher, R. (1981) How the two eyes add

together: Monocular properties of the visually guided orientation behaviour of flies. Biol. Cybern. 41: 71-78.

Geiger, G & Nässel, D.R (1982) Visual processing of moving single objects and wide-field patterns in flies: Behavioural analysis after laser-surgical removal of interneurons. Biol. Cybern. 44: 141-149.

Geiger, G. & Poggio, T. (1975) The orientation of flies towards visual patterns; on the search for the underlying functional interactions. Biol. Cybern. 17: 1-16.

Geiger, G. & Poggio, T. (1977) On head and body movements of flying flies. Biol. Cybern. 25: 177-180

Geiger, G. & Poggio, T. (1981) Asymptotic oscillations in the tracking behaviour of the fly Musca domestica. Biol. Cybern. 41: 197-201.

Götz, K.G. (1964) Optomotorische Untersuchung des visuellen Systems einiger Augenmutanten der Fruchtfliege Drosophila. Kybernetik 2: 77-92.

Götz, K.G. (1975) The optomotor equilibrium of the Drosophila navigation system J. Comp. Physiol. 99: 187-210.

Götz, K.G. (1980) Visual guidance in Drosophila. In: Development and Neurobiology of Drosophila. Ed. O. Siddiqi, P. Babu, L.M. Hall & J.C. Hall. New York, Plenum Press, p. 391-407.

Götz, K.G. (1983) Bewegungssehen und Flugsteuerung bei der Fliege Drosophila. In: BIONA Report 2 Ed. W. Nachtigall, G. Fischer, Stuttgart, p. 21-34.

Götz, K.G. & Buchner, E. (1978) Evidence for one-way movement detection in the visual system of Drosophila. Biol. Cybern. 31: 243-248.

Götz, K.G., Hengstenberg, B. & Biesinger, R. (1979) Optomotor control of wing beat and body posture in Drosophila. Biol. Cybern. 35: 101-112.

Götz, K.G. & Wenking, H. (1973) Visual control of locomotion in the walking fruitfly Drosophila. J. Comp. Physiol. 85: 235-266.

Goulet, M., Campan, R. & Lambin, M. (1981) The visual perception of relative distances in the wood-cricket Nemobius sylvestris. Physiol. Entomol. 6: 357 367.

Grieger, B., Bolz, J. & Varju, D. (1981) On the visually evoked head nystagmus of Tenebrio molitor and other beetles. Biol. Cybern. 41: 1-3.

Hassenstein, B. (1951) Ommatidienraster und afferente Bewegungsintegration. Z. vergl. Physiol. 33: 301-326.

Hassenstein, B. & Reichardt, W. (1956) Systemtheoretische Analyse der Zeit-, Reihenfolgen- und Vorzeichenauswertung bei der Bewegungsperzeption des Rüsselkäfers Chlorophanus. Z. Naturforsch. 11b: 513-524.

Hausen, K. (1983) The lobula-complex of the fly: Structure, function and significance in visual behaviour. (This volume)

Hausen, K. & Wehrhahn, C. (1983) The role of horizontal cells in the optomotor yaw torque response in flies. (In prep.)

Heisenberg, M. (1979) Genetic approach to a visual system. In: Handbook of Sensory Physiology. VII/6A. Ed. H. Autrum, Berlin, Springer Verlag, p. 665-679.

Heisenberg, M. (1983) Initiale Aktivität und Willkürverhalten bei Tieren. Naturwiss. 70: 70 78.

Heisenberg, M. & Buchner, E. (1977) The role of retinula cell types in visual behavior of Drosophila melanogaster. J. Comp. Physiol. 117: 127-162.

Heisenberg, M. & Götz, K.G. (1975) The use of mutations for the partial degradation of vision in Drosophila melanogaster. J. Comp. Physiol. 98: 217-241.

Heisenberg, M. & Wolf, R. (1979) On the fine structure of yaw torque in visual flight orientation of Drosophila melanogaster. J. Comp. Physiol. 130: 113-130.

Heisenberg, M. & Wolf, R. (1984) Vision in Drosophila. Berlin, Springer. (In press).

Helmholtz, H. von (1867) Handbuch der Physiologischen Optik. Leipzig: L. Voss.

Hengstenberg, R. & Sandeman, D.C. (1982) Compensatory head-roll-movements of flies. Verh. Dtsch. Zool. Ges. 1982: 313. G. Fischer, Stuttgart.

Hu, K.G. & Stark, W.S. (1980) The roles of Drosophila ocelli and compound eyes in phototaxis. J. Comp. Physiol. 135: 85-95.

Jeanrot, N., Campan, R. & Lambin, M. (1981) Functional exploration of the visual field of the wood-cricket, Nemobius sylvestris. Physiol. Entomol. 6: 27-34

Kien, J. & Land, M.F. (1978) The fast phase of optokinetik nystagmus in the locust. Physiol. Entomol. 3: 53-57.

Kirschfeld, K. (1972) The visual system of Musca: Studies on optics, structure and function. In: Information Processing in the Visual System of Arthropods. Ed. R. Wehner, Berlin, Springer, p. 61-74.

Kirschfeld, K. & Franceschini, N. (1968) Optische Eigenschaften der Ommatidien im Komplexauge von Musca. Kybernetik 5: 47-52.

Kirschfeld, K. & Reichardt, W. (1970) Optomotorische Versuche an Musca mit linear polarisiertem Licht. Z. Naturforsch. 25b: 228.

Kirschfeld, K. & Wenk, P. (1976) The dorsal compound eye of simuliid flies: An eye specialized for the detection of small, rapidly moving objects Z. Naturforsch. 31c: 764-765.

Kuenen, L.P.S. & Baker, T.C. (1982) Optomotor regulation of ground velocity in moths during flight to sex pheromone at different heights. Physiol. Entomol. 7: 193-202.

Kunze, P. (1961) Untersuchung des Bewegungssehens fixiert fliegender Bienen. Z. vergl. Physiol. 44: 656-684.

Land, M.F. & Collett, T.S. (1974) Chasing behaviour of houseflies (Fannia canicularis): A description and analysis. J. Comp. Physiol. 89: 331-357.

Laughlin. S.B. (1983) The roles of parallel channels in early visual processing by the arthropod compound eye. (This volume).

Lillywhite, P.G. & Dvorak, D.R (1981) Responses to single photons in a fly optomotor neurone. Vision Res. 21: 279-290.

Limb, J.O. & Murphy J.A. (1975) Estimating the velocity of moving objects in television signals. Computer Graphics and Image Processing 4: 311-327.

Longuet-Higgins, H.C. & Prazdny, K. (1980) The interpretation of a moving retinal image. Proc. R. Soc. Lond. 208B: 385-397.
Maldonado, H. & Rodriguez, E. (1972) Depth perception in the praying mantis. Physiol. Behav. 8: 751-759.
Marr, D. & Ullman, S. (1981) Directional selectivity and its use in early visual processing. Proc. R. Soc. Lond. 211B: 151-180.
Mastebroek, H.A.K., Zaagman, W.H. & Lenting, B.P.M. (1980) Movement detection: Performance of a wide-field element in the visual system of the blowfly. Vision Res. 20: 467-474.
Mastebroek, H.A.K., Zaagman, W.H. & Lenting, B.P.M. (1982) Memory-like effects in fly vision. Spatio-temporal interactions in a wide-field neuron. Biol. Cybern. 43: 147-155
McCann, G.D. & MacGinitie, G.F. (1965) Optomotor response studies of insect vision. Proc. R. Soc. Lond. 163B: 369-401.
Menzel, R. (1979) Spectral sensitivity and color vision in invertebrates. In: Handbook of Sensory Physiology, VII/6A. Ed. H. Autrum. Berlin, Springer Verlag, p. 503-580.
Mimura, K. (1982) Discrimination of some visual patterns in Drosophila melanogaster. J. Comp. Physiol. 146: 229-233.
Mittelstaedt, H. (1949) Telotaxis und Optomotorik von Eristalis bei Augeninversion. Naturwiss. 36: 90-91.
Mittelstaedt, H. (1957) Prey capture in mantids. In: Recent Advances in Invertebrate Physiology. Ed. B.T. Scheur. Univ. Oregon Publ. p. 51-71.
Moore, D., Penikas, J. & Rankin, M.A. (1981) Regional specialization for an optomotor response in the honeybee Apis mellifera compound eye. Physiol. Entomol. 6: 61-70.
Moore, D. & Rankin, M.A. (1982) Direction-sensitive partitioning of the honeybee optomotor system. Physiol. Entomol. 7: 25-36.
Morton, P.D. & Cosens, D. (1978) Vision in Drosophila: evidence for the involvement of retinula cells 1 - 6 in the orientation behaviour of Drosophila melanogaster. Physiol. Entomol. 3: 323-334.
Olberg, R.M. (1981) Object- and self-movement detectors in the ventral nerve cord of the dragonfly. J. Comp. Physiol. 141: 327-334.
Palka, J. (1969) Discrimination between movements of eye and object by visual interneurones of crickets. J. Exp. Biol. 50: 723-732.
Pick, B. (1974) Visual flicker induces orientation behavior in the fly Musca. Z. Naturforsch. 29c: 310-312.
Pick, B. (1976) Visual pattern discrimination as an element of the fly's orientation behaviour. Biol. Cybern. 23: 171-180.
Pick, B. (1978) Visuelles Orientierungsverhalten von Fliegen: Eine nichtlineare Systemanalyse. In: Cybernetics 1977. Ed. G. Hauske & E. Butenandt. München. R. Oldenbourg Verlag, p. 301-310.
Pick, B. & Buchner, E. (1979) Visual movement detection under light-and dark-adaptation in the fly, Musca domestica. J. Comp. Physiol. 134: 45-54.
Pinter, R.B. (1979) Inhibition and excitation in the locust DCMD receptive field. J. Exp. Biol. 55: 191-216.
Poggio, T. & Reichardt, W (1973 a) A theory of the pattern induced flight orientation of the fly, Musca domestica. Kybernetik 12: 185-

203.
Poggio, T. & Reichardt, W. (1973 b) Considerations on models of movement detection. Kybernetik 13: 223-227.
Poggio, T. & Reichardt, W. (1976) Visual control of orientation behaviour in the fly. Part II. Towards the underlying neural interactions. Quart. Rev Biophys. 9: 377-438.
Poggio, T. & Reichardt, W. (1981 a) Characterization of nonlinear interactions in the fly's visual system. In: Theoretical Approaches in Neurobiology. Ed. W. Reichardt & T. Poggio. Cambridge, Mass., The MIT Press, p. 64-84.
Poggio, T. & Reichardt, W. (1981 b) Visual fixation and tracking in flies. Mathematical properties of simple control systems. Biol. Cybern. 40: 101-112.
Poggio, T., Reichardt, W. & Hausen, K. (1981) A neuronal circuitry for relative movement discrimination by the visual system of the fly. Naturwiss. 68: 443-446.
Praagh, J.P. van, Ribi, W., Wehrhahn, C. & Wittmann, D. (1980) Drone bees fixate the queen with the dorsal frontal part of their compound eyes. J. Comp. Physiol. 136: 263-266.
Reichardt, W. (1957) Autokorrelations-Auswertung als Funktionsprinzip des Zentralnervensystems. Z. Naturforsch. 12b: 448-457.
Reichardt, W. (1973) Musterinduzierte Flugorientierung. Verhaltensversuche an der Fliege Musca domestica. Naturwiss. 60: 122-138.
Reichardt, W. (1979) Functional characterization of neural interactions through an analysis of behavior. In: Neurosciences: Fourth Study Program. Ed. F.O. Schmitt & F.G. Worden. Cambridge, Mass. MIT Press, p. 81-103.
Reichardt, W., Braitenberg, V. & Weidel, G. (1968) Auslösung von Elementarprozessen durch einzelne Lichtquanten im Fliegenauge. Kybernetik 5: 148-169.
Reichardt, W. & Poggio, T. (1975) A theory of the pattern induced flight orientation of the fly Musca domestica II. Biol. Cybern. 18: 69-80.
Reichardt, W. & Poggio, T. (1976) Visual control of orientation behaviour in the fly. Part I. A quantitative analysis. Quart. Rev. Biophys. 9: 311-375.
Reichardt, W. & Poggio, T. (1979) Figure-ground discrimination by relative movement in the visual system of the fly. Part I: Experimental results. Biol. Cybern. 35: 81-100.
Reichardt, W. & Poggio T. (1981) Visual control of flight in flies. In: Theoretical Approaches in Neurobiology. Ed. W. Reichardt & T. Poggio. Cambridge, Mass., The MIT Press, p. 135-150.
Reichardt, W., Poggio T. & Hausen, K. (1982) Figure-ground discrimination by relative movement in the visual system of the fly. Part II: Towards the neuronal circuitry. Biol. Cybern. supp. 46: 1-30.
Reichardt, W. & Varju, D. (1959) Ubertragungseigenschaften im Auswertesystem für das Bewegungssehen. Z. Naturforsch. 14b: 674-689
Reichardt, W. & Wenking, H. (1969) Optical detection and fixation of objects by fixed flying flies. Naturwiss. 56: 424-425.
Riehle, A. & Franceschini, N. (1982) Response of a direction-selective,

movement detecting neuron under precise stimulation of two identified photoreceptor cells. Neursci. Lett. Supp. 10: 411.

Rogers, B. & Graham, M. (1979) Motion parallax as an independent cue for depth perception. Perception 8: 125-134.

Rossel, S. (1980) Foveal fixation and tracking in the praying mantis. J. Comp. Physiol. 139: 307-331.

Rossel, S. (1983) Binocular steropsis in an insect. Nature (Lond.). (In Press).

Rowell, C.H.F., O'Shea, M. & Williams, J.L.D. (1977) The neuronal basis of a sensory analyser, the acridid movement detector system. J. Exp. Biol. 68: 157-185.

Sanyo(1982) Time Magazine 120 (2): 48.

Snyder, A.W. (1979) The physics of vision in compound eyes. In: Handbook of Sensory Physiology, VII/6A. Ed. Berlin, Springer, p. 225-313.

Srinivasan, M.V. & Bernard, G.D. (1976) A proposed mechanism for multiplication of neural signals. Biol. Cybern. 21: 227-236.

Srinivasan, M.V. & Bernard, G.D. (1977) The pursuit response of the housefly and its interaction with the optomotor response. J. Comp. Physiol. 115: 101-117.

Srinivasan, M.V. & Dvorak, D.R. (1979) The waterfall illusion in an insect visual system. Vision Res. 19: 1435-1437.

Srinivasan, M.V. & Dvorak, D.R. (1980) Spatial processing of visual information in the movement-detecting pathway of the fly. J. Comp. Physiol. 140: 1-23.

Stange, G. (1981) The ocellar component of flight equilibrium control in dragonflies. J. Comp. Physiol. 141: 335-347.

Stavenga, D.G. & Schwemer, J. (1983) Visual pigments of invertebrates. (This volume).

Stegmann, P. (1982) Ein Modell zur Simulation des Verhaltens von richtungsselektiven Simple-Zellen im visuellen Cortex der Katze. Diplomarbeit, Universität Stuttgart. 85 pages.

Strausfeld, N. (1983) Functional neuroanatomy of the blowfly's visual system. (This volume).

Strebel-Brede, J. (1982) Eigenschaften der visuell induzierten Drehmomenten-Reaktion von fixiert fliegenden Stubenfliegen Musca domestica L. und Fannia canicularis L. Dissertation, Universität Tübingen.

Taddei-Ferretti, C. (1973) Landing reaction of Musca domestica. IV. A. Monocular and binocular vision; B. Relationship between landing and optomotor reactions. Z. Naturforsch. 28c: 579-592.

Taddei-Ferretti, C. & Perez de Talens, A.F.P. (1973) Landing of Musca domestica. III. Dependence on the luminous characteristics of the stimulus. Z. Naturforsch. 28c: 568-578.

Taylor, C.P. (1981) Contribution of compound eyes and ocelli to steering of locust in flight: 1. Behavioral analysis. J. Exp. Biol. 93: 1-18.

Thorson, J. (1966 a) Small signal analysis of a visual reflex in the locust: I. Input parameters. Kybernetik 3: 41-53.

Thorson, J. (1966 b) Small signal analysis of a visual reflex in the locust: II. Frequency dependence. Kybernetik 3: 53-66.

Torre, V. & Poggio, T. (1978) A synaptic mechanism possibly underlying

directional selectivity to motion. Proc. R. Soc. 202B: 409-416.
Ullman, S. (1981) Analysis of visual motion by biological and computer systems. Computer 14: 57-69.
Varju, D. & Bolz, J. (1980) Head movements of the mealworm beetle Tenebrio molitor. I. Their properties in stationary environments and their role during object fixation. Biol. Cybern. 36: 109-115.
Varju, D. & Reichardt, W. (1967) Übertragungseigenschaften im Auswertesystem für das Bewegungssehen II. Z. Naturforsch. 22b: 1343-1351.
Virsik, R. & Reichardt, W. (1974) Tracking of moving objects by the fly Musca domestica. Naturwiss. 61: 132-133.
Virsik, R. & Reichardt, W. (1976) Detection and tracking of moving objects by the fly Musca domestica. Biol. Cybern. 23: 83-98.
Wagner, H. (1982) Flow-field variables trigger landing in flies. Nature (Lond.) 297: 147-148
Waterman, T.H. (1981) Polarization sensitivity. In: Handbook of Sensory Physiology, VII/6B. Ed. H. Autrum. Berlin, Springer Verlag, p. 281-469.
Waterman, T. (1983) Natural polarized light and vision. (This volume).
Wehner, R. (1981) Spatial vision in arthropods. In: Handbook of Sensory Physiology, Vol. VII/6C. Ed. H. Autrum. Berlin, Springer Verlag, p. 287-616.
Wehrhahn, C. (1976) Evidence for the role of retinal receptors R7/8 in the orientation behaviour of the fly. Biol. Cybern. 21: 213-220.
Wehrhahn, C. (1978 a) Flight torque and lift responses of the housefly (Musca domestica) to a single stripe moving in different parts of the visual field. Biol. Cybern. 29: 237-247.
Wehrhahn, C. (1978 b) The angular orientation of the movement detectors acting on the flight lift response in flies. Biol. Cybern. 31: 169-173.
Wehrhahn, C. (1979) Sex specific differences in the orientation behaviour of houseflies. Biol. Cybern. 32: 239-241.
Wehrhahn, C. (1980) Visual fixation and tracking in flies. In: Mathematical Models in Molecular and Cellular Biology. Ed. L. Segel. Cambridge, Cambridge University Press, p. 568-603.
Wehrhahn, C. (1981) Fast and slow flight torque responses in flies and their possible role in visual orientation behaviour. Biol. Cybern. 40: 213-221.
Wehrhahn, C. & Hausen, K. (1980) How is tracking and fixation accomplished in the nervous system of the fly? Biol. Cybern. 38: 179-186.
Wehrhahn, C., Hausen, K. & Zanker, J. (1981) Is the landing response of the housefly (Musca) driven by motion of a flow field? Biol. Cybern. 41: 91-99.
Wehrhahn, C., Poggio, T. & Bülthoff, H. (1982) Tracking and chasing in houseflies (Musca). Biol. Cybern. 45: 123-130.
Willmund, R. & Ewing, A. (1982) Visual signals in the courtship of Drosophila melanogaster. Anim. Behav. 30: 209-215.
Wolf, R. & Heisenberg, M. (1980) On the fine structure of yaw torque in visual flight orientation of Drosophila melanogaster. II. A temporally

and spatially variable weighting function for the visual field ("visual attention"). J. Comp. Physiol. 140: 69-80.

Zaretsky, M & Rowell, C.H.F. (1979) Saccadic suppression by corollary discharge in the locust. Nature (Lond.) 280: 583-585.

Zeil, J. (1981) Sexual dimorphism in the visual system of flies. Doctoral Thesis. University of Tübingen, Germany. 80 pages.

NEUROANATOMICAL MAPPING OF VISUALLY INDUCED NERVOUS ACTIVITY IN INSECTS BY ^{3}H-DEOXYGLUCOSE

ERICH BUCHNER and SIGRID BUCHNER
MPI für biologische Kybernetik
Spemannstrasse 38
D-7400 Tübingen
Federal Republic of Germany

ABSTRACT

Physiological nervous activity can be visualized post mortem in neuroanatomical sections by the radioactive deoxyglucose technique. The visual system of flies Drosophila melanogaster and Musca domestica has been investigated by this method. By applying visual stimuli consisting of homogeneous flicker or striped patterns moving in a particular direction across the retina, movement-specific and direction-specific nervous activity has been localized in autoradiographs of semi-thin sections of Drosophila brains. Several layers in the second neuropil (medulla) respond to visual movement by enhanced uptake of radioactivity. Although directional specificity in these layers has not yet been detected by light microscopy, the corresponding nervous activity cannot simply be of the on-off type since visual flicker is considerably less effective. In the posterior part of the third visual neuropil (lobula plate) four layers can be identified, each of which responds to one particular spatial direction of movement, front-to-back, back-to-front, upward or downward. Similar direction-specific movement-sensitive labeling is found in the visual system of Musca where in addition axons and dendritic arborizations of individual labeled cells are clearly resolved. The relation of these findings to electrophysiological data on dipteran flies is discussed.

INTRODUCTION

The visual system of the fly combines a number of features that have made it an important model system in neurobiology (c f. Buchner, 1983; Franceschini, 1983; Hausen, 1983; Laughlin, 1983; Meinertzhagen, 1983; Stavenga & Schwemer, 1983; Strausfeld, 1983; - the "fly-mafia" (ed.) -). Many "intelligent" visual functions found in higher vertebrates including man, such as movement detection, object fixation and relative movement evaluation, are also implemented by the nervous circuitry of the fly's brain (c f. Buchner, 1983). Yet the number of neurons involved in these functions is much smaller, about 10^5 to 10^6 in the fly (Strausfeld, 1976) compared to 10^{10} to 10^{12} in mammals. A further most important advantage of invertebrate nervous systems for basic research in neurobiology is the identifiability of individual neurons. Structural and functional information gathered for a particular neuron in a single individual can immediately be transferred to the corresponding neuron in another individual of the same species and often even to homologous cells in other species of the same or related families. In ganglia of repetitive structural organization such as the optic lobes of flies, periodic, sub-periodic and non-periodic neurons can be differentiated (c f. Strausfeld, 1983). This greatly reduces the number of cells that have to be characterized if the nervous circuitry underlying the above functions is to be understood at the cellular level.

The feature of identifiable neurons came fully to bear with the advent of methods that allow correlation of structure and function at the cellular level. The importance of the technique of dye injection into electrophysiologically recorded cells for anatomical characterization is evident from various chapters of this book (Franceschini, 1983; Hausen, 1983; Laughlin, 1983). However, from electron microscopy it is also clear that only a small fraction of axons and dendrites in the visual ganglia of flies are large enough to be routinely impaled by micropipettes. Thus alternatives to the electrophysiological approach to the correlation of structure and function have to be sought. A first step towards an anatomical approach to this problem has been accomplished by making the metabolic activity of neurons visible in histological sections. While several techniques for functional neuroanatomical mapping have been advocated, using Ca^{++} accumulation, enzyme activity or protein synthesis (Parducz & Joo, 1976; Mark & Sperling, 1976; Fehér & Rojck, 1976), only the radioactive deoxyglucose method (Sokoloff et al., 1977; rev. Sokoloff, 1982) has triggered an avalanche of successful work. Although spatial resolution of this method has been greatly improved over the last few years (Basinger et al., 1979; Wagner et al., 1979; Buchner & Buchner, 1980; Sans et al., 1980; Sejnowski et al., 1980; Durham et al., 1981; Kai Kai & Pentreath, 1981; Wagner et al., 1981; Buchner & Buchner, 1982; Lancet et al., 1982; Witkovsky & Yang, 1982) physiologically relevant information in insects has so far been obtained only at the level of layers and strata and, in one case, of large interneurons. Functional disentanglement of the dense feltwork of visual neuropil of flies by this method seems at present out of reach.

The deoxyglucose technique is based on the finding that physiologically active nerve cells satisfy their increased needs for metabolic energy by the uptake and breakdown of glucose (Crone, 1978). While glucose itself is ill suited for metabolic mapping - its breakdown products CO_2 and H_2O diffuse rapidly out of the cell - its analogue 2-deoxyglucose participates only in the first step of glycolysis and is intracellularly phosphorylated by the enzyme hexokinase to deoxyglucose-6-phosphate (DG-6-P). Further metabolic conversion as well as diffusion across the cell membrane proceed only at a very low rate, such that DG-6-P is essentially trapped intracellularly. Since "facilitated diffusion" of deoxyglucose across the blood-brain-barrier and the neuronal cell membrane as well as its phosphorylation proceed at a rate similar to that of glucose, DG-6-P accumulation in neurons is a measure of the rate of glucose uptake and thus of the physiological activity of these cells. If radioactively labeled deoxyglucose is introduced into the blood stream, accumulation of DG-6-P can be measured locally by autoradiography of the sectioned tissue. Extension of this technique to high resolution which requires comparatively high intracellular concentration of labeled DG-6-P may critically depend on a specific feature of brain hexokinase. Due to its low substrate specificity this enzyme accepts both glucose and deoxyglucose (and other hexoses), the regulation of its activity by end-product-inhibition is however highly specific for glucose-6-phosphate. Thus intracellular accumulation of DG-6-P does not interfere with glucose metabolism (Fromm, 1981).

These metabolic pathways for deoxyglucose have so far been investigated biochemically only in vertebrates. However, it is known that glucose constitutes the favored substrate also in the brains of lower animals (rev. Wegener, 1981), and autoradiographic studies on flies with labeled glucose and 3-0-methylglucose strongly support the notion that metabolic mapping by deoxyglucose in dipteran insects is based on the same principles as in vertebrates (Buchner & Buchner, 1983 a).

METHODS

Since the adaptation of the deoxyglucose technique to insect nervous systems has been described in detail elsewhere (Buchner et al., 1979; Buchner & Buchner, 1980; 1983 a) we shall here only briefly summarize its essential steps. High resolution autoradiography of water-soluble labeled compounds requires the animal to be given a high dose of tritiated deoxyglucose and the tissue to be processed without any contact with aqueous solutions.

In a standard experiment with the fruitfly Drosophila melanogaster we inject under cold anaesthesia 10 - 20 μ Ci of ^{3}H-2-deoxy-D-glucose or ^{3}H-2-deoxy-2-fluoro-D-glucose (Amersham, New England Nuclear) dissolved in 0,05 μl H_2O into the hemocoel by inserting a broken-off micropipette attached to a microliter syringe (Hamilton) between two tergites of the abdomen. The resulting hemolymph concentration of a few mM of deoxyglucose has no obvious toxic effects on the survival time of the fly or its behavioral response in a movement detection paradigm. Even with a 50 fold higher dose of unlabeled deoxyglucose the flies survive for several hours - though shorter than control animals - and respond to movement (Buchner, unpublished results). After recovery from anaesthesia the fly is stimulated for a period between 1 and 6 h and immediately submerged into melting nitrogen ("slush"). The frozen flies are then freeze-dried for 6 days at -70°C as described previously (Buchner & Buchner, 1980). After slow warm-up to room temperature the dry tissue is fixed for 1 h in OsO_4 vapor and embedded in EPON$^{(R)}$ (Shell). Semithin sections (2-5 μm) are cut on a dry glass knife and brought into close contact with dry autoradiographic film (AR-10, Kodak) by a special procedure (Buchner & Buchner, 1983 a). For electron microscopic investigation thin sections can be cut and floated on a water surface with the understanding that undetermined fractions of the label are displaced and lost even from embedded tissue. Yet it has been demonstrated that most of the stimulus-specific label is retained (Buchner & Buchner, 1981; 1982).

While quantitative evaluation of individual autoradiographs can be accomplished by grain counts or microphotometric measurements, it appears difficult to obtain reliable values on tissue glucose uptake from the autoradiographs of insect sections by the mathematical formalism developed for vertebrates (Sokoloff et al., 1977). Neither does it seem feasable to continuously monitor hemolymph concentrations of glucose and deoxyglucose during an experiment nor can the formalism be immediately applied to animals with open blood systems. Consequently variations in the grain density of autoradiographs from different animals do not necessarily reflect different levels of glucose uptake and nervous activity, but might be caused for instance by differences in size or nutritional state of the animals. For this reason only gross differences between different experiments can be directly interpreted. For a quantitative comparison of the effectiveness of two stimuli it seems appropriate to present both stimuli simultaneously, e.g. one on each eye. This is particularly advantageous when paired structures like the optic ganglia are under study. In addition, the retinotopic organization of these ganglia can be utilized to include further control situations in a single experiment. Such a compound stimulus configuration for the identification of movement-specific nervous activity is shown in Fig. 1. The

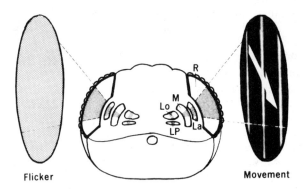

Fig. 1. Stimulus arrangement for the localization of movement-specific nervous activity in the visual system of flies. Intenseness of labeling in movement-stimulated tissue (shaded), right hemisphere) flicker-stimulated tissue (shaded, left hemisphere) and non-stimulated tissue of retina (R), lamina (La), medulla (M), lobula (Lo) and lobula plate (LP) can be compared.

movement stimulus consists of a sine-wave grating moving at constant speed behind a circular aperture that restricts the stimulated visual field to a disk of 70° diameter centered on the right eye. Since every movement stimulus will also activate on-off units that are not necessarily movement specific, a sinusoidally flickering disk is presented to the corresponding visual field of the left eye as a control stimulus. Luminance, modulation depth and frequency of the flicker stimulus are adjusted in such a way that the stimulated photoreceptors on either eye receive identical sinusoidal light signals. The movement information is exclusively coded in the phase difference between the sine-wave functions in neighboring receptors. Thus any difference in labeling between right and left hemisphere must be due to this phase difference, and therefore reflects movement-specific nervous activity. In addition, retinotopically organized stimulus-specific nervous activity can be identified in both optic lobes by comparing stimulated and unstimulated regions in each ganglion.

A second stimulus configuration appropriate for comparing the effects of the _direction_ of a moving pattern is sketched in Fig. 3 in front of the eyes. A periodic grating rotates around the fly presenting back-to-front movement to the left eye and front-to-back movement to the right eye (small arrows). Any difference in labeling between left and right hemisphere in such an experiment will reflect _direction_-sensitive nervous activity. An unstimulated control region is included by shielding a 45° wide sector in the fronto-lateral region of each eye (solid bar).

RESULTS AND DISCUSSION

The basic labeling pattern that is obtained under the stimulus conditions of Fig. 1 for a horizontal section of a _Drosophila_ head is shown in the autoradiograph of Fig. 2 a and the schematic sketch of labeled brain structures in b. Only qualitative features pertinent to the visual stimulus will be pointed out here. A detailed quantitative evaluation of these and similar experiments will be given in a forthcoming paper (Buchner & Buchner, 1983 b). By comparison of stimulated vs. non-stimulated parts of the optic ganglia it is clear that the medulla shows pronounced retinotopically organized nervous activity both on the flicker- and the movement-stimulated side. Comparing left and right hemispheres for movement-specific effects one notices significantly stronger labeling in the right medulla and a prominent band of label in the lobula plate that is not present on the flicker-stimulated side. The interpretation of the label in the medulla is still somewhat ambiguous since no direction-specific effects have been observed so far (for further details c f. Buchner & Buchner, 1983 b). However, the band in the lobula plate is movement-specific and, as is shown in Fig. 3, its fronto-caudal position depends on the direction of the movement. On the right side which had received front-to-back stimulation this band lies near the frontal surface of the lobula plate, directly adjacent to the inner chiasm, while on the left side it lies more caudally as indicated by the clear zone between the lobula and lobula-plate bands. Note that in all three visual ganglia the central zone that corresponds to the visual field shielded from the movement shows

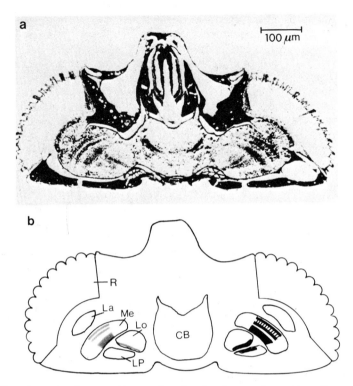

Fig. 2. a) Autoradiograph of a horizontal section through the head of a Drosophila which after injection of 20 μ Ci of ^{3}H-2-deoxy-2-fluoro-D-glucose had been stimulated as illustrated in Fig. 1 for 75 min.

b) Schematic drawing of neuropil outlines and regions of stimulus-induced accumulation of radioactive label for the autoradiograph in a) R, retina; La, lamina; Me, medulla; Lo, lobula; LP, lobula plate; CB, central brain.

reduced labeling on both sides. Experiments with vertical movement (upward and downward) have demonstrated that two further bands can be identified caudally adjacent to the two horizontal bands, an interior band responding to upward movement and a band near the caudal surface of the lobula plate responding to downward movement. While the functional segregation of frontal vs. caudal lobula plate into horizontal and vertical movement preference respectively, has long been suggested from anatomical studies (Pierantoni, 1976) and was confirmed by a wealth of electrophysiological work (c f. Hausen, 1983), the direction-specific subdivision into the four layers as shown schematically in Fig. 4 had escaped attention due to the complexity and selectivity of electrophysiological recording/staining results. However, comparison with Fig. 3 of Hausen (1983) shows that all cells except H3 and V1 do conform to the proposed scheme of directional ordering. The question whether the two

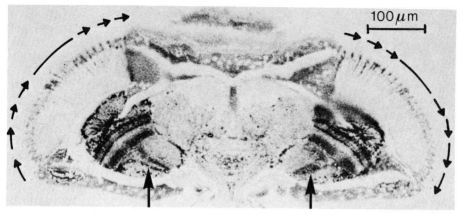

Fig. 3. Autoradiograph of a horizontal section through the head of a Drosophila which after injection of 10 μ Ci of ^{3}H-2-deoxy-D-glucose had been stimulated for 6 h in a striped drum rotating clockwise around the fly as indicated by the small arrows in front of the eyes. The solid bars on each eye indicate the retinal zones that were shielded from the movement by inserting two pieces of black cardboard into the drum. Note that the corresponding zones in the optic ganglia show reduced labeling. On the left side which was stimulated by back-to-front movement the movement-specificly labeled layer of the lobula plate lies slightly more caudally than on the right side which had received front-to-back stimulation (large arrows).

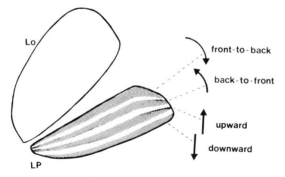

Fig. 4. Schematic diagram of a horizontal cross-section through the lobula complex with the four movement-specific directionally selective layers (shaded) of the lobula plate. For each layer the direction of visual movement that leads to enhanced labeling by deoxyglucose under ipsilateral stimulation is indicated. Lo, lobula; LP, lobula plate.

exceptions (two cells out of 23) represent a caprice of nature or whether in fact fine processes invisible in dye-injected preparations extend into the "correct" layer for synaptic interconnection must at present remain open. Recent experiments with the housefly <u>Musca domestica</u> in the rotating drum as described above have demonstrated that large heavily labeled axons and dendritic arborizations can be recognized when the present procedure is applied to larger flies. This is shown by the autoradiograph in Fig. 5 (arrows). Direction of stripe movement in the corresponding experiment was counterclockwise, and these prominent cells are labeled on the left side only. Thus they display directionally selective movement-specific labeling far more intense than the most frontal lobula plate layer described above in which they arborize. However, it is quite clear from electrophysiology (c f. Hausen, 1983) that numerous neurons in the lobula plate respond vigorously to this kind of stimulus. It will therefore be of great interest to understand why only some of these cells are heavily labeled by the deoxyglucose method while most giant neurons show weaker labeling than the surrounding neuropil irrespective of their electrical activity (Buchner & Buchner, 1980, 1983 b, in prep.).

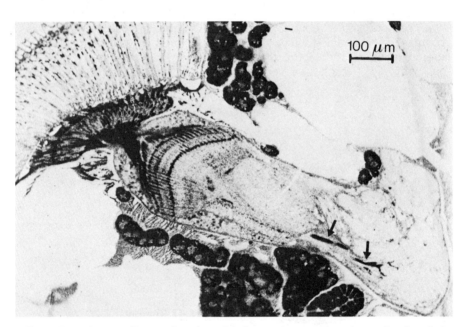

Fig. 5. Autoradiograph of a horizontal section through the left hemisphere of the head of a <u>Musca</u> fly which after injection of 150 μ Ci of ^{3}H-2-deoxy-D-glucose had been stimulated for 4 h by counterclockwise movement in the rotating striped drum. The large-diameter axons and arborizations (arrows) are heavily labeled only on this side. The reduced activity of the central medulla corresponds to the retinal region shielded from the movement as in Fig. 3.

CONCLUDING REMARKS

The comparatively brief experience with the deoxyglucose method applied to the visual system of insects has made clear that at the regional level important information can be obtained. This includes such findings as the movement-specific activity in the medulla and the direction-specific layers in the lobula plate. Equally important may however be the negative result that in the fly's brain no additional centers have been found that are obviously preoccupied with the detection of large-field movement. No other technique available today could have easily provided this information.

The main problem of identifying the labeled cells in the medulla and the lobula plate lies in the extreme complexity of these ganglia and the rather poor ultrastructural preservation of freeze-dried tissue. We do hope that comparison with silver stained or golgi impregnated preparations will provide hints on the cellular identity of the deoxyglucose layers and, in turn, on the physiological properties of structurally identified cells (Fischbach & Heisenberg, 1981; Strausfeld, 1983). However, fiber tracts appear more suitable to investigation by the deoxyglucose method at electron microscopical resolution (Buchner & Buchner, 1981, 1982). Yet it has to be pointed out that so far we have not succeeded to label movement-specific nervous activity in the cervical connective in spite of the fact that the passage of this kind of information can be verified both electrophysiologically (Hengstenberg, 1973; Strausfeld & Bacon, 1983, in prep.) and behaviorally (cf. Buchner, 1983). The reasons for this failure are at present not known, but it is tempting to speculate that it might be connected to the "internal state" of the animal or "interference by higher centers in the brain", both phenomena that are discussed at the behavioral level elsewhere in this volume (Buchner, 1983). For the investigation of these questions which may be of fundamental importance for understanding brain function and behavior of insects (Heisenberg, 1983) it will be necessary to perform "physiological" experiments on freely behaving animals. The deoxyglucose method may be uniquely suited to carry out such experiments since no physical restraints on the animal are required during the incubation period.

It finally may turn out that the combination of insects and deoxyglucose is not only fruitful for the understanding of insect nervous systems but that the work on anatomically and electrophysiologically identifiable cells as in Fig. 5 may contribute to clarify the physiological basis of the deoxyglucose method.

ACKNOWLEDGEMENTS

We would like to thank N. Franceschini, K. Götz, K. Hausen, R. Hengstenberg, I. Nicod, V. Rodrigues and N. Strausfeld for critically reading and discussing the manuscript, we are indebted to C. Straub for the preparation of the figures and to C. Hanser for the typing.

REFERENCES

Basinger, S.F , Gordon, W.C. & Lam D.M.K. (1979) Differential labeling of retinal neurones by ^{3}H-2-deoxyglucose. Nature Lond.) 280: 682-684.

Buchner, E. (1983) Behavioural analysis of spatial vision in insects (This volume).

Buchner, E. & Buchner, S. (1980) Mapping stimulus-induced nervous activity in small brains by ^{3}H-2-deoxy-D-glucose. Cell Tissue Res. 211: 51-64.

Buchner, E. & Buchner, S. (1981) The deoxyglucose method for insects: Towards electron microscopical resolution. Eur. Neurol. 20: 152-156.

Buchner, E. & Buchner, S (1983 a) Anatomical localization of functional activity in insects by ^{3}H-2-deoxy-D-glucose. In: Neuroanatomical Techniques, Vol. 2. Ed. N.J. Strausfeld. Berlin Springer Verlag (In press).

Buchner, E. & Buchner, S. (1983 b) Visual movement detection in Drosophila: A ^{3}H-deoxyglucose study. (In prep.).

Buchner, E., Buchner, S. & Hengstenberg, R. (1979) 2-deoxy-D-glucose maps movement-specific nervous activity in the second visual ganglion of Drosophila. Science 205: 687-688.

Buchner, S. & Buchner, E. (1982) Functional neuroanatomical mapping in insects by ^{3}H-2-deoxy-D-glucose at electron microscopical resolution. Neurosci. Lett. 28: 235-240.

Crone, C (1978) D-glucose - fuel for the brain. Trends Neursci. 1: 120-122.

Durham, D., Woolsey, T.A. & Kruger, L. (1981) Cellular localization of 2-^{3}H-deoxy-D-glucose from paraffin-embedded brains. J. Neurosci. 1: 519-526.

Fehér, O. & Rojik, I. (1976) Fixation of traces of excitation at the cellular level. In: Neuron Concepts Today. Ed. J. Szentagothai, J. Hamori & E.S. Vizi. Symposium, Tihany, p. 213-230.

Fischbach, K.F. & Heisenberg, M. (1981) Structural brain mutant of Drosophila melanogaster with reduced cell number in the medulla cortex and with normal optomotor yaw response. Proc. Natl. Acad. Sci. USA 78: 1105-1109.

Franceschini, N. (1983) The retinal mosaic of the fly compound eye (This volume).

Fromm, H.J. (1981) Mechanism and mode of regulation of brain hexokinase. In: The Regulation of Carbohydrate Formation and Utilization in Mammals. Ed. C.M. Veneziale. Univ. Park Press, Baltimore, p. 45-68.

Hausen, K. (1983) The lobula-complex of the fly: Structure, function and significance in visual behaviour. (This volume).

Heisenberg, M. (1983) Initiale Aktivität und Willkürverhalten bei Tieren. Naturwiss. 70: 70-78.

Hengstenberg, R. (1973) The effect of pattern movement on impulse activity of the cervical connective of Drosophila melanogaster. Z.

Naturforsch. 28c: 593-596

Kai Kai, M.A. & Pentreath, V.W. (1981) High resolution analysis of ^{3}H-2-deoxyglucose incorporation into neurons and glia cells in invertebrate ganglia: histological processing of nervous tissue for selective marking of glycogen. J. Neurocytol. 10: 693-708.

Lancet, D., Greer, C.A., Kauer, J. & Shepherd, G.M. (1982) Mapping of odor-related neuronal activity in the olfactory bulb by high-resolution 2-deoxyglucose autoradiography. Proc. Natl. Acad. Sci. USA 79: 670-674.

Laughlin, S.B. (1983) The roles of parallel channels in early visual processing by the arthropod compound eye. (This volume).

Mark, R.E. & Sperling, H.G. (1976) Color receptor identities of goldfish cones. Science 191: 487-489.

Meinertzhagen, I. (1983) The rules of synaptic assembly in the developing insect lamina. (This volume).

Parducz, A. & Joo, F. (1976) Visualization of stimulated nerve endings by preferential calcium accumulation of mitochondria. J. Cell. Biol. 69: 513-517.

Pierantoni, R. (1978) A look into the cock-pit of the fly. The architecture of the lobula plate. Cell. Tissue Res. 171: 101-122.

Sans, A., Pujol, R., Carlier, E. & Calas, A. (1980) Détection cellulaire de l'incorporation in vivo de 2-deoxyglucose tritié. Etude radioautographique dans l'oreille interne. C.R. Acad. Sc. Paris 290D: 1225-1227.

Sejnowski, T.J., Reingold, S.C., Kelley, D.B. & Gelperin, A. (1980) Localization of ^{3}H-2-deoxyglucose in single molluscan neurones. Nature 287: 449-451.

Sokoloff, L. (1982) The radioactive deoxyglucose method. Theory, procedure, and applications for the measurement of local glucose utilization in the central nervous system. Adv. Neurochem. 4: 1-82.

Sokoloff, L., Reivich, M., Kennedy, C., Des Rosiers, M.H., Patlak, C.S., Pettigrew, K.D., Sakurada, O. & Shinohara, M. (1977) The ^{14}C-deoxy-glucose method for the measurement of local cerebral glucose utilization: theory, procedure, and normal values in the conscious and anesthetized albino rat. J. Neurochem. 28: 897-916.

Stavenga, D.G & Schwemer, J. (1983) Visual pigments of invertebrates. (This volume).

Strausfeld, N J. (1976) Atlas of an Insect Brain. Springer, Berlin.

Strausfeld, N.J. (1983) Functional neuroanatomy of the blowfly's visual system. (This volume).

Strausfeld, N.J. & Bacon J.P. (1983) Multimodal convergence at the descending neurons in the central nervous system of the fly Musca domestica. In: Multimodal Convergence of Sensory Systems. Ed. E. Horn. Fortschr. Zool. 27. (In press).

Wagner, H.J., Pilgrim, C. & Zwerger, H. (1979) A system of cells in the unstimulated rat brain characterized by preferential accumulation of ^{3}H-deoxyglucose. Neurosci. Lett. 15: 181-186.

Wagner, H.J., Hoffmann, K.P. & Zwerger, H. (1981) Layer specific labelling of cat visual cortex after stimulation with visual noise: a ^{3}H-2-deoxy-D-glucose study. Brain Res. 224: 31-43.

Wegener, G. (1981) Comparative aspects of energy metabolism in non-mammalian brains under normoxic and hypoxic conditions. In: Animal Models and Hypoxia. Ed. V. Stefanovich. Pergamon Press, Oxford, p. 87-109.

Witkovsky, P. & Yang, C.Y. (1982) Uptake and localization of ^{3}H-2-deoxy D-glucose by retinal photoreceptors. J. Comp. Neurol. 204: 105-116.

THE RULES OF SYNAPTIC ASSEMBLY

IN THE DEVELOPING INSECT LAMINA

I.A. MEINERTZHAGEN

Department of Psychology, Life Sciences Centre
Dalhousie University, Halifax, Nova Scotia
Canada, B3H 4J1

INTRODUCTION

The morphologies of cells within arthropod visual systems provide very striking examples of the differentiation of neurons. Even from the earliest accounts based on impregnation by the Golgi method (Cajal & Sánchez, 1915) we have a good impression of the minute and exquisite morphological complexities of many classes of component neuron, attributes which have been eulogized generously (Cajal, 1937). Modern treatments (Strausfeld & Nässel, 1980; see also Strausfeld, 1983), moreover, considerably extend the morphological taxonomy of neurons. The major contemporary significance of these studies is the prediction that classes of neuron defined by their differences in shape also have characteristic differences in their connectivity, by which each contributes uniquely to the circuitry of the optic lobe. The analysis of this prediction is facilitated in the arthropod compound eye and optic lobe by the high degree of neuronal isomorphism that is one obvious outcome of many repeating neural pathways arranged in parallel.

Burgeoning catalogues of nerve cells and their synaptic connections create the progressively urgent need to systematize these descriptions in some way, if only for mnemonic reasons. One way is physiological, to relate each anatomical cell type and the details of its connectivity to its role in the processing of visual signals. Unfortunately electrophysiological studies lag far behind anatomical ones in the resolution of their analyses, so that in practice this most sensible flow of information is often reversed, anatomical connectivities suggesting electrophysiological interactions. A second way is developmental. An attempt can be made to define the shape and connections of neurons by the sequence and nature of the developmental decisions or rules of which they are the product. This review will attempt the second approach, drawing information from whichever arthropod has been investigated most thoroughly. Many ideas have emerged from analyses of the fly's visual system. Much of the earlier sequences of development in the following section will be glossed over, having been reviewed in a companion chapter (Mouze, 1983).

LAYING THE CELLULAR FOUNDATIONS OF THE EYE AND LAMINA

1. The Formation and Differentiation of Photoreceptors

Embryologically, the origin of arthropod photoreceptors is ectodermal, from the head epithelium. Although there are many differences between different insect groups (Meinertzhagen, 1973; Bate, 1978) all are fundamentally similar and find a counterpart in decapod Crustacea (Hafner et al., 1982). The product, ommatidial cell clusters, in all cases shows remarkable constancy of cell number despite the functional diversity to which these cells are put in eyes of different groups (e.g. Land, 1983 for Crustacea). Along with other cells of the ommatidium, the photoreceptors arise by mitoses, amongst cells of the eye margin which apparently constitute a budding zone (Bodenstein, 1953; Nowell & Shelton, 1980). In insects these mitoses commonly have two prominent waves (Egelhaaf et al., 1975; Ready et al., 1976; Mouze, 1980) passing from a postero-dorsal position across the eye field. Once born, the cells cluster, forming groups which comprise the future ommatidial complement of eight receptors. Cell identity is apparently conferred by the position of each cell within its cluster (Oncopeltus: Shelton & Lawrence, 1974; Drosophila: Lawrence & Green, 1979). In the fly, the early developmental difference in the times of origin between the groups of receptors R2-5 & 8 and R1, 6 & 7, produced by the two successive waves, stand in contrast to the receptors' later indistinguishability in the number and composition of their synapses.

2. The Formation and Differentiation of Lamina Neurons

Within the cortex of the lamina, ganglion cells are produced by proliferative mitoses of neuroblasts within the outer optic anlage (Nordlander & Edwards, 1969). Growth of the lamina cell population lags behind that of the inner neuropiles (Nordlander & Edwards, 1969; Hofbauer, 1979). The exact pattern of mitoses giving rise to the neurons of the cartridge is unknown, as are the lineage relationships, if any, between these neurons. In the adult, there is a repetitive constitution of cell bodies in the lamina cortex, identical from cartridge to cartridge (Strausfeld & Nässel, 1980). The conservative spatial relationships generally observed between insect neuroblasts and their progeny, with the latter arranged in linear columns, (Panov, 1960; Nordlander & Edwards, 1969) therefore allow the possibility of close lineage relationships between monopolar cells of a single cartridge. The existence of a double wave of mitoses in the lamina cortex of Drosophila as well as the occurrence of a later wave of cell death (Hofbauer, 1979), however, suggests that no simple, exclusive scheme of lineage relationships exists. Rather, a two-stage initial overproduction of cells is followed by loss of those that fail to incorporate into cartridge groups. In Drosophila, Hanson's observations (in Kankel et al., 1980) suggest a separate origin for the fifth of five monopolar cells L1-5 in the cartridge.

The means by which lamina ganglion cells gain their morphological and connectivity phenotypes and become incorporated into the cartridge is unknown. Some form of instructional cueing could be provided by one or several of the following: the lineage of the cells (a single ganglion mother cell could give rise to four of the monopolar cells of a cartridge); the time of cell birth (e.g. the fifth monopolar cell added late to a cartridge group in Drosophila - Hanson, in Kankel et al., 1980 - may be born at a different time to its four congeners); cell interactions between the ganglion cells dependent, for example, upon the position of each within the cartridge cluster; the sequence of innervation of cells by the photoreceptor axons of an individual ommatidium. In practice, it may be extremely difficult to disentangle which of these factors is responsible. From much that is known of the patterns of insect neuroblast mitoses in general (Panov, 1960; Goodman & Bate, 1981) cells with a fixed pattern of mitotic descent might be produced in an exact temporal sequence and hence line up in a precise spatial sequence with which they could subsequently be touched by receptor innervation; all three attributes would then co-vary, making it hard to distinguish which one assigns cell behaviour in synaptogenesis.

If lineage were conserved in evolution, then some relationship between the ancestry of a particular cell and its synaptic connectivity (and perhaps therefore morphology) should be detectable between different arthropod groups. Broad similarities between the connectivities of monopolar cells in three insects, Musca (Strausfeld & Campos-Ortega, 1977), the bee Apis (Ribi, 1981) and the dragonfly Sympetrum (Meinertzhagen & Armett-Kibel, 1982), have been discussed by the last authors but are hardly striking for some of the cells.

The sequence of innervation of monopolar cell somata by photoreceptor axons (which is most likely represented in the adult by their position in the cartridge clusters) is known in one arthropod only, the branchiopod crustacean Daphnia (Lopresti et al., 1973). One reason to believe that instructional cueing might pass from receptor axon to monopolar cell is that the stimulus eliciting monopolar neurite outgrowth comes apparently from the incoming receptor axons. Thus, the monopolar cells send out neurites in the same sequence as they themselves are touched by receptor axons (Lopresti et al., 1973); and fail to send out neurites if, following retinal lesions, receptor innervation is unforthcoming, degenerating 10-12 hrs thereafter (Macagno, 1979); or send them out late (up to 8 hrs) if receptor innervation is delayed (Macagno, 1981). Consequently, in Daphnia, neurite outgrowth from the lamina follows that from the retina in a cascade. However, this cascading sequence, although suspected, in not proven for other arthropods and is not universal for all interneurons. Some types of transmedullary cells in the medulla of Drosophila, for example, put forth axons and elaborate quite extensive dendritic trees in the eyeless mutant sine oculis, congenitally deprived of receptor innervation and an associated lamina (Fischbach, 1983). Furthermore, there is no evidence available for the stimulus for neurite outgrowth in cell types within the lamina other than monopolar cells. These include cells with a periodic distribution, one per cartridge (T1 and two types of centrifugal neuron C2 and C3) as well as cells with a supraperiodic distribution (amacrine cells and two types of tangential neuron Tan 1 and Tan 2) (Strausfeld & Campos-Ortega, 1977).

3. Innervation of the Lamina

Receptor axons gain access to their target neurons by growth in unbraided bundles, one per ommatidium. Each bundle starts to grow prior to its more anteriorly situated neighbours, so contributing to a postero-anterior sequence of lamina innervation. This rather general feature of the growth of fibre pathways between consecutive cell layers in the visual system projects upon the lamina the wave of differentiation originating in the eye. This wave is reflected in turn in the subsequent outgrowth of cartridge axon bundles between lamina and medulla, so as to give rise to the chiasma between these two neuropiles (Meinertzhagen, 1973; see Mouze, 1983). An associated feature of this sequential growth of axon bundles is that the axons are guided at all stages of their growth between neuropile layers, reaching the target neuropile as the inevitable consequence of their commencing growth. Axon outgrowth from the first photoreceptors to differentiate in many holometabolous insects is initiated around a scaffold provided by a pre-existing nerve bundle, the larval photoreceptor axon bundle of Bolwig in the case of flies (Trujillo-Cenóz & Melamed, 1973; Meinertzhagen, 1977), as if the latter possesses a pioneer function (Bentley & Keshishian, 1982).

Within the neuropile, all elements (receptor and lamina) form large expanded growth cones which make extensive contact with and explore surrounding growth cones. Differences in the timing of growth cone expansion and the presumed adhesive interactions between them are means by which the growth cones explore and subsequently connect within the neuropile differentially and so generate different morphologies, but these differences are, with but one exception, quite unknown. The exception is provided by detailed descriptive analyses of Hanson

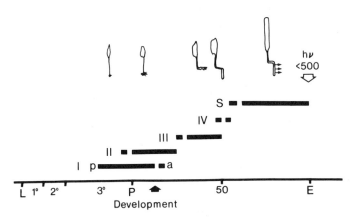

Fig. 1. Timetable for the development of the compound eye and lamina in the fly. Data are pooled from observations on Phaenicia (= Lucilia) (Trujillo-Cenóz and Melamed, 1973), Drosophila (Hanson, in Kankel et al., 1980) and Musca (Fröhlich & Meinertzhagen, 1982, 1983). Development times are expressed as proportions of pupal development, from pupation (P) to eclosion (E) (192 hr: Phaenicia; 96 hr: Drosophila; 126 hr: Musca). Four stages of axonal morphogenesis are distinguished. I. Innervation of the lamina by rapid fibre growth from the developing retina, which occurs in a sequence across the eye field from posterior (p) to anterior (a) commencing in the third instar of the larva (L). II. The formation of expanded growth cones. III. Growth cone lateral migration which occurs by slow rearrangements through predictable, shifting arrays (Hanson) to produce the adult projection pattern of receptor terminals. IV Centripetal growth of the receptor terminal after each cartridge has been correctly constituted, expanding the depth of the lamina. Although early stages of eye development reveal an anteroposterior gradient this gradually disappears, as manifested by the state of screening pigment differentiation which is synchronous across the eye field by the time synaptogenesis (S) commences. For further definition of the events occurring during synapse formation, in the last half of pupal development, see Fig. 6. The solid arrow indicates the approximate time of imaginal disc eversion; the open arrow indicates the time of first exposure, at eclosion, to light of short wavelengths (<500nm., Järvilehto & Finell, 1983).

(reported in Kankel et al., 1980) who describes the sequence of growth cone arrays by which the receptor axons R1-6 of Drosophila redistribute themselves amongst the six neighbouring bundles forming the neural superposition trapezoid (Braitenberg, 1967). This redistribution (phase III in Fig. 1) is a phase of growth, additional to that found in most insect groups, peculiar to Diptera and presumably other groups with open-rhabdomere ommatidia. It is not clear what signals the end-point, mechanistically speaking, for such growth cone manoeuvres i.e. what point has to be reached before they cease, but their result is that six receptor axons each from a separate ommatidium are finally brought into juxtaposition with target cells within a single cartridge; this occurs by about half way through the life of the pupa (Fig. 1), well before synapses are first visible at around 60% pupal development (Fröhlich & Meinertzhagen, 1982)

SYNAPTOGENESIS IN THE FLY'S LAMINA

At about 50% pupal development each receptor axon terminal grows centripetally along the length of the two axial monopolar cells L1 and L2 (growth phase IV, Fig. 1; Trujillo-Cenóz & Melamed, 1973), thus considerably increasing the area potentially available for their synaptogenesis. The membranes of cells which will enter synaptic partnerships are thus in intimate connection for 20 hr or more (in the case of Musca) before they are synaptic. If synaptogenesis is triggered by contact between intended neurons, therefore, it can only be so after a long delay which varies, moreover, in a postero-anterior direction across the eyefield so as to compensate for early differences in the sequence of innervation (Fig. 1). Starting around this time (50-60% pupal development) the receptor terminals assume a more cylindrical shape, their irregular extensions, the remnants of growth cone explorations, being gradually resorbed. Concurrently, the axons of L1 and L2 emit processes into the interstices of the surrounding receptor terminal ring. As the consequence of these explorations the processes of L1 and L2 reach the more distant sites on the receptor terminal membrane at which many of the synapses form, 200 or so of which will eventually exist in the adult (Nicol & Meinertzhagen, 1982a). This is the most numerous of the many classes of synapse formed between the elements of the cartridge (Strausfeld & Campos-Ortega, 1977).

The differentiating synaptic junctions first become distinct ultrastructurally late in development (they are only conspicuous during the last 30% of pupal development). Their increasing visibility is marked by the appearance, in section, of a small presynaptic density opposite one or usually two postsynaptic elements bearing paramembranous densities. The presynaptic density either progressively elongates or perhaps fuses with another nearby, to emerge subsequently as a bar. This bar is eventually topped off by a platform (>84% pupal development) to form the presynaptic ribbon, T-shaped in section, characteristic of Diptera (Fröhlich & Meinertzhagen, 1982). At the photoreceptor tetrad synapse, at least, the postsynaptic densities also extend and elaborate but subsequently disappear, to be replaced by flattened postsynaptic cisternae in L1 and L2.

Postsynaptic cells thus grow out to presynaptic sites during the formation of this class of synapse, a sequence which requires mental re-orientation for those who may be familiar with synaptogenesis in many other systems. The possibility of site selection for synaptogenesis (if this exists) thus devolves most apparently upon the presynaptic cell membrane, while the onus of colonizing new synaptic sites is postsynaptic, issues dealt with later on.

The criteria for site selection are quite unknown. On the one hand most of the area of the presynaptic membrane does not initiate synapse formation despite the opportunity to do so; less than 10% is occupied by synapses (Fig. 7). On the other hand, for those points at which incipient synapses do form, an eventual absolute requirement is to collect all the elements needed to complete their postsynaptic complement, which for fly photoreceptor synapses invariably comprises a tetrad (Nicol & Meinertzhagen, 1982a). At sites for which this collection is geometrically difficult, immature contacts are liable to regress. Consequently, even if selection of the sites at which synapses will form is predetermined in some way, a pattern distributed over the surface of the presynaptic terminal for instance, it takes the successful colonizing growth of postsynaptic neurites both to uncover that pattern and to rescue partially formed synapses from regression, by the stepwise completion of their tetrad.

THE COMPOSITION AND FORMATION OF INDIVIDUAL SYNAPSES

In the fly, Musca, the composition of tetrads of elements postsynaptic at photoreceptor synapses is highly predictable. Two central elements L1 and L2, with a dyadic configuration in transverse section, confer on the tetrad its elongate pattern; two polar elements are a predictable assortment of other elements, usually double α processes from an amacrine spine or, at distal synapses, double contributions from glial cells or a single contribution from L3 paired with an α or glial cell process (Burkhardt & Braitenberg, 1976; Nicol & Meinertzhagen, 1982a). Synapses with multiple-contact postsynaptic elements are a regular feature of insect visual systems (e.g. Meinertzhagen & Armett-Kibel, 1982), as well as of insect nervous systems generally (Lamparter et al., 1969; Watson & Burrows, 1982); they are perhaps both more common than has been widely acknowledged in the past and than can be appreciated from single-section electron microscopic analyses, which often miss some of the multiple contributors. Multiple-contact synapses are certainly not a freak departure from monadic normality, but a stable feature at many other classes of synapse identified in the lamina and elsewhere, for which developmental rules of assembly must exist additional to those deployed at the monads of other systems (such as the neuromuscular junction) more extensively studied as models of neural development. The rules by which the afferent synapses of photoreceptors assemble may therefore serve as a general model.

1. The Number, Symmetry and Composition of Postsynaptic Elements at Multiple-Contact Synapses

Some idea of the developmental rules controlling the assembly of specific combinations of postsynaptic element can already be gained from the chief afferent synapses described at various arthropod's photoreceptors. A broad comparative survey

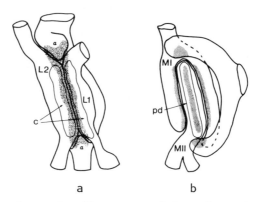

a   b

Fig. 2. Semischematic diagrams of two insect photoreceptor synapses showing the arrangement of postsynaptic elements in the plane of the presynaptic membrane; this arrangement has dyadic symmetry in the fly (a) and triadic symmetry in the dragonfly (b).
(a) In Musca, postsynaptic specializations of the processes of monopolar cells L1 and L2 are cisternae (c), situated adjacent to the textured outline of the presynaptic bar. Two polar elements shown here are double α contacts of an amacrine cell spine.
(b) In Sympetrum, postsynaptic specializations of the two monopolar cells MI and MII are shown as stippled outlines in proximity to the presumed presynaptic release site (the presynaptic density, pd, in Armett-Kibel et al., 1977). The profiles of MI are derived in this case from the double contributions of a single spine. In addition, the double polar contributions of another element with postsynaptic membrane densities is shown, although known only from preliminary analyses (Armett-Kibel & Meinertzhagen, unpublished observations).

reveals that the symmetry of these synapses can be one-, two- or three-fold (Fig. 2). Dyadic symmetry is exemplified by the fly tetrads but is seen also in the compound eyes of the cockroach (Ribi, 1977), the crustacean Artemia (Nässel et al., 1978) and also in various insect ocelli (Toh & Kuwabara, 1974, 1975; Goodman et al., 1979) or the simple eyes of spiders (Trujillo-Cenóz & Melamed, 1964; Oberdorfer, 1977) and Limulus (Whitehead & Purple, 1970). Triadic symmetry is known in the receptor synapses of the compound eyes of the bee (Ribi, 1981), crayfish (Procambarus: Hafner, 1974; Pacifastacus: Nässel & Waterman, 1977) and dragonfly (Armett-Kibel et al., 1977). Both dyads and triads have parallels in the vertebrate retina (Stell, 1972). In several of these synapses polar postsynaptic elements lying outside a transverse section plane have been described from either serial electron microscopy or skilful interpretation of different section planes (Musca: Burkhardt & Braitenberg, 1976; Nicol & Meinertzhagen, 1982a; Fig. 2; salticid spiders: Oberdorfer, 1977). They are also now known in the dragonfly Sympetrum (Armett-Kibel & Meinertzhagen, unpublished; Fig. 2) and should perhaps be suspected elsewhere. Yet other synapses are simple monads, rarely in some species e.g. Sympetrum (Meinertzhagen & Armett-Kibel, 1982) but the rule in others e.g. Daphnia (Macagno et al., 1973; Nässel et al., 1978). The somewhat close phylogenetic relationship of Daphnia (with monadic circuitry) to Artemia (with receptor dyads) suggests the relative ease with which different development rules may be selected during the evolution of the corresponding synaptic modules of equivalent neural circuits. Yet the fixity of these rules in an individual species is obvious when, for two sample populations each of about 200 synapses, all were tetrads (Nicol & Meinertzhagen, 1982 a, b).

2. The Combinations of Postsynaptic Elements

The combinations of postsynaptic element of these synapses show great selectivity. Often there are substitutions between elements so as to allow a greater number to participate than there are spaces available at one site. In the crayfish triad, the median postsynaptic element is invariably from one interneuron, M2. One lateral element comes from either of two other interneurons M3 or M4, which have processes segregated respectively in the distal and proximal lamina (Nässel & Waterman, 1977). The remaining lateral element is contributed by one of the variants of another monopolar cell M1 (Strausfeld & Nässel, 1980). In the fly tetrad, L1 and L2 are the invariant central pair, but the polar elements in the distal lamina are drawn from α (amacrine), glial or monopolar (L3) cells. Alpha and glial processes rarely occur together at one synapse but usually as both processes; L3 always contributes singly, in partnership with one of them. Often there is redundancy in the

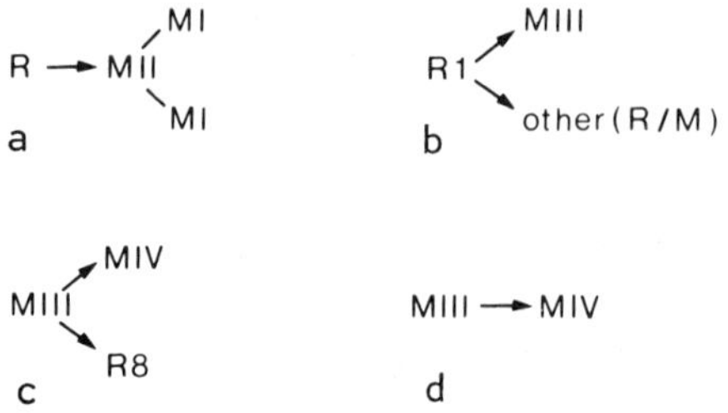

Fig. 3. Synaptic configurations in the dragonfly Sympetrum. (a) Triad A. (b) Dyad D1. (c) Dyad E3. (d) Monad F1. (All from Meinertzhagen & Armett-Kibel, 1982).

deployment, at the same synapse, of two postsynaptic elements from a single parent cell, as in these double contributions at fly tetrads or the monopolar (MI) contributions at the lateral positions of dragonfly triads. When this occurs some form of exclusion principle seems to keep these double contributions apart and thus maintain the neighbourliness of different elements. These are examples of what must constitute the rules of assembly for synapses, for which cell recognition during development is ultimately responsible. Different synapse populations at which the same photoreceptor terminal is presynaptic may have different symmetry characteristics depending on the number and identity of the postsynaptic elements. In Sympetrum, for example, certain receptor terminals are presynaptic both at triads, the chief afferent synapse upon monopolar cells MI and MII (Figs. 2, 3a), as well as at dyads, upon other elements: (Fig. 3b; Meinertzhagen & Armett-Kibel, 1982). There are even examples at which the same two elements are synaptic under different symmetry conditions, e.g. dyad E3 and monad F1 (Fig. 3c, d). These last two examples imply that it must be the identity of R8 or its interaction with MIV that confers the symmetry condition upon the synapse.

3. The Developmental Assembly Sequence of Fly Tetrad Synapses

The assembly of the tetrad of postsynaptic elements at the receptor presynaptic sites in the fly, and perhaps at all other multiple-contact synapses occurs in a sequence (Fröhlich & Meinertzhagen, 1982). The evidence for this comes from our analysis of synapse populations in animals of different age; young animals possess only dyads and triads, the number of elements postsynaptic at a synapse gradually increasing with age to the maximum of four in the adult. There is also no counter-evidence either of pre-fabrication of tetrads at sites before final synaptogenesis occurs, or of groups of growing postsynaptic neurites fasiculating and hunting in packs for a presynaptic site. Sequential assembly, then, seems to be the rule. This form of on-site assembly has two important consequences. First, the sequence in which assembly progresses may be determinate i.e. the

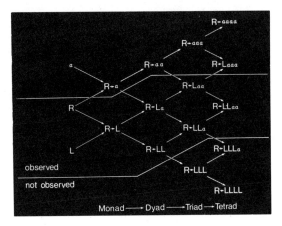

Fig. 4. An inferred sequence of assembly of the fly photoreceptor tetrad synapse by selective sequential additions of postsynaptic elements involving, in turn, monadic, dyadic and triadic contacts antecedent to a tetrad. Of the diverging possibilites of combination which exist with each postsynaptic addition, only those are actually observed which could lead from a R→L monad (involving either L1 or L2) to the single class of tetrad incorporating L1, L2, α, α. (Modified from Fröhlich & Meinertzhagen, 1983.)

synapse may not accept all of the eligible postsynaptic contenders for incorporation at a particular time, the choice being restricted by those already incorporated. Second, the synapse must await the arrival of each element, during which time it may be susceptible to some form of de-stabilization through being incomplete.

Evidence for the sequence of incorporation of postsynaptic elements comes from the analysis of combinations of postsynaptic elements in all synapses of the population (monads through tetrads) at each of a number of immature stages during the last 25% or so of pupal development (Fröhlich & Meinertzhagen, 1983). Each population is assumed to transform into the one at the next older stage and dyads gradually transform into tetrads according to the scheme in Fig. 4. Thus, starting with the basic cellular participants (receptors, L1, L2 and amacrine, α processes) the following generalities ("rules") emerge. Synapses are apparently only initiated at contacts between receptors and monopolar cell processes. Either L1 or L2 can initiate such an early contact and apparently with about equal frequency, thus demonstrating an initial equivalence which heralds their future dyadic symmetry. The remaining elements then add in any of the possible sequences which will result in the collection of just the right number and combination of elements (e.g. in the most proximal lamina to form L1, L2, α, α). Only very infrequently are combinations seen which incorporate a wrong or supernumerary element; there are in general no "mistakes". Postsynaptic elements must, therefore, carefully assay each other, recognizing not only the presynaptic site with which they will bond, but also the existing occupants. Elongation of the tetrad and the resultant development of a linear presynaptic ribbon seems to occur only when the tetrad is complete, and is accompanied by an increase in the area of apposition of the L1 and L2 processes. Substitutions for either or both the normal complement of two α processes occur in the distal zone of the lamina as previously noted, and involve L3 and glial cell processes, for which additional exclusion rules must operate. The selectivity of these events defines the performance of cell recognition, in which:

1. Recognition must initially be possible between receptor and monopolar (L1 or L2) cells. There is so far no evidence, however, for site selection at which this recognition will occur (whether the site exists before the postsynaptic element chances upon it or whether the postsynaptic element induces the formation of the site).
2. Mutual recognition must subsequently occur between L1 and L2, to exclude the incorporation of double contributions (L1, L1 or L2, L2) which have never been found.
3. Recognition must be possible between an α process and either a receptor itself or the R-L complex.
4. Recognition between L1 and L2, or between two α processes, must exclude the acceptance of a third element in any of these categories, and so preempt pentad combinations.
5. Additional recognition rules in the distal lamina normally prevent α and glial cell elements from cooperating at a single tetrad, or L3 from contributing twice (Nicol & Meinertzhagen, 1982a).

With the possible exception of rule 5, none of these features arises from differential availability of elements; all participants are present and competent to engage in synaptogenesis, which they are undertaking concurrently in different combinations at different synapses.

Incomplete tetrads are unstable in the long term, given the complete absence of dyads and triads in the adult. The time course of their destabilization must be sufficiently long to allow for an eligible postsynaptic element to stumble upon the synapse during its random growth. Otherwise, synapses would disassemble faster than they could assemble. Analysis of the dynamics of assembly of individual synapses is almost impossible from the only analyses of their populations so far available, which are carefully-timed but infrequent snap-shots from different animals of increasing age. Though it requires a certain amount of faith to accept that these static analyses truly capture the patterns of developmental transformation it seems less likely that the development of each fly follows an idiosyncratic time course, given what we know of the constancy of flies and their development rates. Nor does it seem likely that whole synaptic populations are built up and then torn down wholly within the intervals between sampling. The analysis of the numbers of dyad, triad and tetrad synapses at different ages indicates that some triads can form wholly within 14 hrs at 27°C, thus adding their elements at least as fast, on average, as one per 7 hrs or so (Fröhlich & Meinertzhagen, 1983). In other words, an incomplete synapse could be stable for a period of up to 7 hrs and perhaps longer, before needing to secure the next element(s) of its tetrad. If it fails to do so the synapse disappears quickly. Disassembly could also conceivably be sequential, however, and with the loss of each postsynaptic element the synapse may gain a moment of reprieve in which to add a fresh one, so that there could be an equivalent but more rapidly occurring scheme of synapse disassembly to that for assembly. There is no structural evidence for such different phases of synapse formation or loss, however, unlike those reported by Wernig et al. (1980), interpreted as the disestablishment of synaptic contacts in mature vertebrate muscle. A single published example (Boschek, 1971) of a receptor terminal monadically presynaptic upon a glial cell in adult Musca is conceivably interpretable as a tetrad undergoing disestablishment and having lost its postsynaptic tetrad. The strength of this interpretation is however clearly eroded by the similar report of a single synapse formed upon a glial cell from one member of the exclusive receptor axon population of an ectopic eye transplant (Trujillo-Cenóz & Melamed, 1975). Perhaps both examples represent an occasional lapse in selectivity in synapse formation.

These are some of the additional rules during synaptogenesis imposed by the formation of multiple-contact synapses, over and above those involved at their monadic counterparts in other nervous systems. Additional complexities arise when we consider the several classes of synapse formed concurrently by the same neuron.

4. The Segregation of Synaptic Populations

Where synapses of two qualitative classes are formed at which the membrane of the same neuron is both pre- and postsynaptic, the possibility exists of some form of membrane mosaicism. An example is provided by the class of monopolar feedback connection formed, reciprocally, back upon the monopolar's receptor input, as described in both the fly (Fig. 5a) and dragonfly (Fig. 5b).

The suggestion of membrane mosaicism arises naturally from the observation that a single patch of membrane is either pre- or postsynaptic but is never given over to both functions. As a mechanistic interpretation, membrane mosaicism may therefore be entirely notional, given that it is always possible to draw a

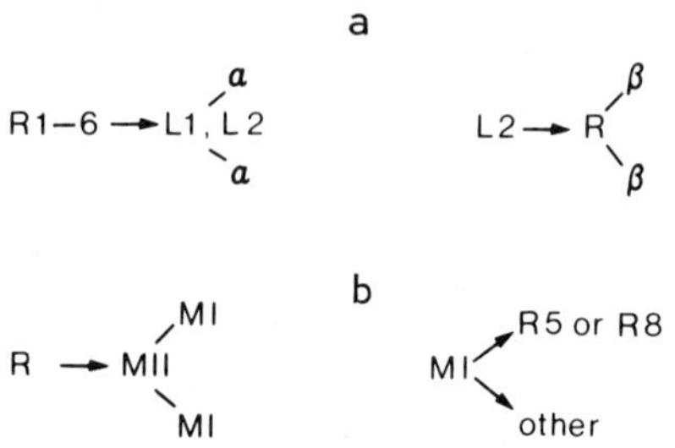

```
                    a
                   /
      R1-6 ──► L1, L2        L2 ──► R
                   \                 \
                    a                 β
                                     /
                                    β
                          b
           ,MI                    ,R5 or R8
   R ──► MII            MI
           \                     \
            MI                    other
```

Fig. 5. Reciprocal receptor-monopolar interconnections in the fly (a) and dragonfly (b).
(a) The afferent tetrad synapse at which L2 is one of the elements postsynaptic to R1-6, is described in Musca (Burkhardt & Braitenberg, 1976) and Lucilia (Trujillo-Cenóz, 1965; S.R. Shaw, personal communication). The efferent synapse at which L2 is presynaptic at synapses in the proximal lamina, first described in Musca (Strausfeld & Campos-Ortega, 1977) is now known to be triadic in symmetry with lateral β (T1 cell) elements (Shaw, unpublished observations on Lucilia, personal communication).
(b) The afferent triad synapse at which MI is one of the elements postsynaptic to receptors, most frequently R5 and R8, is reciprocated by the efferent synapse at which MI is presynaptic, at a dyad containing R5 or R8 and another element. (From Armett-Kibel et al., 1977).

line between two synapses at which one element functions both pre- and postsynaptically and that the limit is therefore set by the closeness of spacing of individual synapses. The mechanism of mosaicism may be entirely local, if participation at a synapse as the presynaptic partner, for instance, denied postsynaptic participation for some minimum area surrounding the first synapse. On the other hand, a great deal of informal observational evidence supports the impression that certain synaptic populations are carefully segregated from others on the same cell, and that this segregation often corresponds to the partition of membrane between an axon and its spines. In Sympetrum, for instance, the presynaptic site of a receptor terminal is invariably approached by the process of a postsynaptic element (Armett-Kibel & Meinertzhagen, unpublished). In the monopolar cells of Sympetrum, the distribution of mosaicism may obey different rules, with pre- and postsynaptic sites being segregated between different dendrites, for instance (Armett-Kibel & Meinertzhagen, unpublished). Similar examples may be culled from published studies on Musca but have never been systematically studied; they require extensive sampling to substantiate and are vitiated by a single contradiction. It may also be that the partition of synaptic populations between different structural components of the neuron is a coincidence,

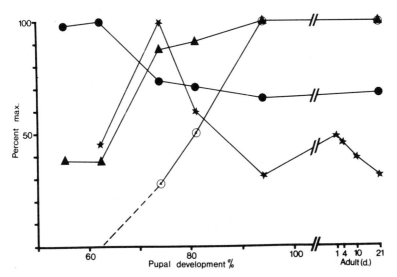

Fig. 6. The time course of four functions of synaptogenesis during the last half of pupal development and at various intervals (in days) following eclosion (E). Each function peaks at a different time; all values are expressed as a proportion of these peaks.
●——● The sum of all L1 and L2 postsynaptic neurites, per cartridge segment, peaks early in development.
▲——▲ The proportion of these neurites which extend into and terminate in the outer perimeter of the cartridge cross section (see Fig. 8) gradually increases during development as the neurites grow in length.
★——★ The number of synaptic contacts per receptor terminal rises to an early maximum; post-eclosion values disclose an unconfirmed secondary peak.
⊖——⊖ The proportion of all synapses at each stage which have complete tetrads reaches a maximum last of all during development, as synaptic contacts initiated earlier in development gradually acquire their complete complement of postsynaptic elements.

the fortuitous outcome of the regular arrangement of pre- and postsynaptic cells within the cartridge, which ensures that cells come into contact only at certain points over their surface. In the spatially less well organized neuropiles of the ventral nerve cord the case for mosaicism would seem poor. For example, the dendrites of several motor neurons in the locust are locally both pre- and postsynaptic (Watson & Burrows, 1982) with no obvious segregation of synaptic classes.

QUANTITATIVE ASPECTS OF SYNAPSE FORMATION

Rules for the qualitative composition of synapses tell us nothing about the overall selection of the size and number of synapses. These quantitative aspects of synapse formation have been investigated in a number of recent studies (Nicol & Meinertzhagen 1982, a, b: Fröhlich & Meinertzhagen, 1983) and recently reviewed (Meinertzhagen & Fröhlich, 1983).

As individual synapses collect their postsynaptic elements, they grow in size until they attain a maximum length of about 0.5μm, for the presynaptic ribbon of fully-formed tetrads. The small variation in size within the population of tetrads suggests the regulation of synapse size. The increase in size reflects primarily an increase in postsynaptic composition (Fig. 6) but also an increase simply in age, older tetrads being larger than younger ones. As synapses age and grow in size, their number also changes, increasing until about 74% pupal development and decreasing thereafter to a value about half this (Fig. 6). The total synaptic input at L1 or L2 also shows an early maximum and later decline.

The precisely regulated synapse populations of adult photoreceptors, (about 200 apiece: Nicol & Meinertzhagen, 1982a) are therefore produced by selective loss from a larger immature population of incipient synapses. To some extent the loss of synaptic sites is ineffectual, being offset by the increase in size of surviving sites, but even so the size of the summed areas of presynaptic ribbons (the collective 'corporate' synapse: Meinertzhagen & Fröhlich, 1983) declines from a maximum at 74% pupal development. This method of producing a synaptic population by selective attrition from a larger, immature population has not previously been emphasized as a specific characteristic of other developing synaptic populations. It is therefore interesting to speculate upon its significance. One obvious feature is that because the synapses have multiple postsynaptic elements there is provision for loss of those synapses which, for reason of their having a geometically less favourable location than others, fail to complete their tetrad of elements within the appropriate deadline. It may also be that the extent of synapse loss is activity dependent. In <u>Calliphora</u>, receptors can generate photoelectric responses and their synapses are functional at the developmental stages equivalent to those of anatomical synaptogenesis analysed in <u>Musca</u> (Järvilehto & Finell, 1983). In the bee, the short visual fibre (type 1: Ribi, 1979) terminals of green-sensitive receptors in UV-reared bees have fewer synapses compared with control bees (Hertel, 1983). The decrease is not attended by a decrease either in synapse size or the profiles of terminal cross sections. It therefore seems that the number of synapses surviving into adulthood depends upon the light stimulation to which the receptor is exposed, either during development or after eclosion.

In the adult, examination of the frequency of synapses in the hyperinnervated cartridges of the equatorial region of the fly's lamina (Nicol & Meinertzhagen, 1982b) demonstrates the close correlation of synapse number (N) and membrane area (A) for both the presynaptic receptor cell terminals and the spines of their postsynaptic partners, L1 and L2. This finding provides a convenient measure of synapse frequency (A/N, the membrane area allocated on average to each synapse, a reciprocal measure of synapse spacing) as well as a plausible area-dependent mechanism for distributing synapses.

During development the value of A/N is not held constant, but varies for the presynaptic receptor terminals in proportion to the size of the synapse (measured as the mean area of the presynaptic ribbon), increasing gradually through development. So, as development proceeds, a large population of small synapses transforms into a smaller population of larger synapses (Fig. 7). At each stage the average area of membrane "allotted" to each synapse increases, as if the relationship existing between the two were a regulated one. This regulation might involve, for instance, the redistribution amongst the surviving synapses of each population of presynaptic components derived from the presynaptic membrane.

As the neurons increase in age, then, the size and the spacing of their synapses gradually increases. These findings

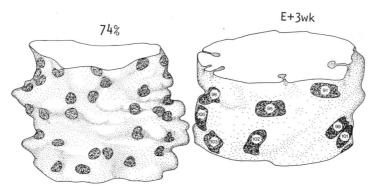

Fig. 7. Computer-aided displays of reconstructed segments of receptor terminals, showing the approximate size, number and distribution of presynaptic sites over the surface of 74% pupal development and 3-wk adult stages. Development proceeds so as to preserve a constant ratio (about 1:10) between the sums of the areas of all presynaptic ribbons (the total synaptic area) and the total membrane surface area of the terminal.

carry with them all the slender weight of correlation. They do not necessarily imply causation and their significance would be enhanced only if other parameters of the cell did not also correlate with synapse number and/or size. The volume of the neuron, for example, does not increase in proportion to increases in the number of synapses in the case of the slender, high surface area/volume ratio spines of L1 and L2 (Nicol & Meinertzhagen, 1982b). It correlates reasonably well for the much lower surface area/volume ratio of the stout receptor cell terminals, both in the adult (calculated from the data of Nicol & Meinertzhagen, 1982a) and for developmental stages (Fröhlich & Meinertzhagen, 1983). In other words, for the presynaptic receptor terminals of adults the quotient volume/N is an equally powerful predictor of synapse number as the ratio A/N. Furthermore, the relationship between synapse size and the ratio A/N is an average one, established between the receptor terminal and its synapsse population as a whole, and does not necessarily apply for the size and membrane territory of individual synapses. It might be that in the areal relationship between synapse and membrane territory, larger synapses have larger territories. However, the spacing between presynaptic ribbons observed in single sections often seems to be too close for strict observance of a particular A/N ratio within individual membrane territories, so apparently excluding this possibility.

Changes during development in the spacing of synaptic sites over the presynaptic membrane have obvious implications for the spacing and distribution of postsynaptic neurites which service these sites. In other words, as postsynaptic neurites withdraw from presynaptic sites with the loss of synapses, the morphologies of the postsynaptic neurons change.

MORPHOGENESIS OF THE POSTSYNAPTIC NEURONS

The developmental transformation of the neurons postsynaptic at receptor tetrads is best illustrated by L1 and L2, because of their invariable presence as a pair at all synapses (Nicol & Meinertzhagen, 1982a). L2 is also involved at synapses of other classes (e.g. feedback synapses upon receptor terminals, Strausfeld & Campos-Ortega, 1977) and some portion of its surface is therefore not directly involved at receptor tetrad synapses. The relative frequency of this involvement is, however, low (Hauser-Holschuch, 1975) and not in the mid-region of the lamina's depth most frequently sampled in the studies discussed here.

The loss of synaptic sites from receptor terminals during development therefore results in a reduction of the potential adult synaptic input upon L1 and L2. This loss is paralleled by a selective loss of postsynaptic spines from L1 and L2, which develop their characteristic bottle-brush morphology following exuberant neurite growth earlier in development (Figs. 6, 8).

This loss is to be compared with the end-point of their development in the adult, where variations in the size of synaptic populations between normal cartridges and the hyperinnervated equatorial cartridges (Nicol & Meinertzhagen, 1982b) are associated with hypertrophy of individual spines without an increase in their total number. Apparently, the adult morphologies of L1 and L2 are established by the selective attrition of neurites, to a fixed final number. This could perhaps occur by the adoption of a fixed spacing between consecutive dendrites of the same comb (the 'pitch' of the spiral pattern of dendrite bases, Nicol & Meinertzhagen, 1982b), still allowing some regulative capacity through variation in the size of individual dendrites.

The overall loss of neurites during development involves their selective loss in two categories identified so far (Fröhlich and Meinertzhagen, 1980), amongst late neurites and those imperfectly interleaved with the neurites of the partner L1/L2 cell. Neurites emitted after 81% pupal development never extend far, never colonize the synaptic grounds of the cartridge periphery, and subsequently regress and are resorbed. Consequently, the number of processes extending into and terminating in the outer perimeter of the cartridge gradually increases during development as a proportion of the total (Fig. 6), both by elongation of the shorter processes emitted before 81% pupal development and by loss of all of those emitted after this stage. Analysis of the neurites extending between any receptor terminal pair, down the length of the cartridge, reveals the second category of their selective loss. The strict interdigitating sequence L1, L2, L1, L2 (i.e. the absence of consecutive like-pairs L1, L1 or L2, L2) characteristic of the adult, develops only slowly from early in development when the sequence approaches randomness. With later development, the pairing ratio (the sum of all L1 and L2 pairs, as a proportion of all pairs, including L1, L1 and L2, L2) gradually increases from around 0.5 to 1.0. One consequence of neurite resorption is therefore to ensure the elimination of one or other member of each like pair, leaving only unlike pairs, L1, L2, both of which then contribute to the same synaptic sites at adjacent receptor terminals' membrane (Fröhlich & Meinertzhagen, 1980).

Initial dendrite outgrowth itself seems to be random, both in the probability of its occurrence and its direction. At around 55% pupal development, fine processes are sent from variable locations around the axon's circumference, extending not only between receptor terminals but also invaginating for short distances into the terminals themselves. The low proportion of dissimilar pairs referred to above reflects the initial randomness with which the L1 and L2 dendrites are emitted down the length of their parent axons. The morphology of a segment of L1 or L2 is at this stage highly disordered (Fig. 8a). Finally, all dendrites are directed between adjacent receptor terminals

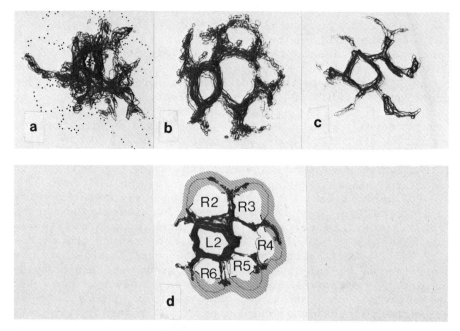

Fig. 8. Segments of L2 at different development stages, displayed by computer as superimposed profiles traced from 50-100 micrographs of transverse sections and arranged with corresponding orientations.
(a) At the 55% pupal development stage the outlines are highly disordered, with predominantly short neurites extending from the axon (asterisk) in frequently random directions. The disorder arises partially from the irregularity of the morphology of the receptor terminals (eight of which are displayed from one section only of this equatorial cartridge) around which many of the neurites grow.
(b) At 74% pupal development, neurites have grown exuberantly, most extending in length fully around each receptor terminal, so outlining the cross section of the cartridge.
(c) At 94% pupal development, many neurites have been resorbed. Those remaining have shortened, although still embracing the receptor terminals (located in the identified spaces) and still terminating in the outer perimeter of the cartridge cross section (see Fig. 8c).
(d) At the 3-wk adult stage, the branching pattern of dendrites resembles qualitatively the appearance at 94% pupal development (e). Individual spines are, however, more slender. The outlines of five of the six receptor terminals are traced from the first section of the series and the outer perimeter of the cartridge cross section, into which all dendrites extend (Fig. 6), is shown by the hatched area.

(Fig. 8b, c), those passing between the same pair interleaving with the dendrites of the partner cell. The neurites seem never to braid their positions in the depth of the cartridge, a neurite emitted at a particular level maintaining its station with respect to neurites above and below.

The final morphology of the monopolar cell seems, then, to be governed entirely by the selective pattern of neurite resorption, and thus by the rules which determine the selection of one neurite over another. There is no direct evidence concerning these rules but in the following section the possibility will be explored that a major determinant of a neurite's survival is its success in synaptogenesis.

THE RELATIONSHIP BETWEEN SYNAPTOGENESIS AND NEURONAL MORPHOGENESIS

Although there is no established link between synaptogenesis in the fly's lamina and the morphogenesis of the postsynaptic neurons, the temporal correlation between the loss of synaptic sites from receptors and the loss of dendrites (Fig. 6), encourages the hypothesis that neurites are stabilized by the very synapses that they form. Neurites not so stabilized, it is postulated, regress. The temporal resolution of static serial electronmicroscopic reconstructions of synaptic populations is of course inadequate to resolve the sequence of these losses and one could just as easily argue the reverse sequence, that the loss of a postsynaptic element, brought on by a morphogenetic rearrangement of L1 or L2, de-stabilizes the synapse at which either was previously represented. Even so, the hypothesis is economical, plausible and with explicit precedent in at least one other developing neural system (Berry et al., 1980). Neglecting for the moment the detailed mechanism by which the adhesion of pre- and postsynaptic membranes might be increased, this theory predicts that the stabilization of the postsynaptic neurites of either L1 or L2 proceeds in a cooperative, stepwise fashion, dependent upon the neurite growth of the other. In this way, dendrites of L1 and L2 compete with their like partners (L1 vs L1 and L2 vs L2) to consort with their opposite partners (L1 with L2) and adhere selectively at synaptic sites with them. Within this intricate synaptic dance, the strategy optimizing a neurite's chances at synaptogenesis would therefore be one which ensured the rapid early colonization of presynaptic membrane and the consequent deprivation of other, more slowly growing dendrites of the same cell from synaptic sites within the same territory. The threshold number of synapses just sufficient to prevent the resorption of a postsynaptic neurite is not known, and perhaps varies at different developmental ages. In the adult, dendrites have been observed which persist with only two postsynaptic sites but also as many as 16 (20 dendrites of L1 and L2, data from Nicol & Meinertzhagen, 1982b). The number of postsynaptic sites established per dendrite is thus a variable outcome, commensurate with the size of the dendrite, of a competitive interaction between sibling dendrites of the same cell. In fact, a competitive interaction can only be claimed if competition for some essential ingredient (e.g. a synaptically released trophic factor or an adhesion mediating surface macromolecule) can actually be demonstrated i.e. if a prior mechanism is understood. For the moment, therefore, a competitive interaction is inferred only. This interaction is not the sole influence, however, since the final number of dendrites persisting on adult L1 or L2 neurons is apparently regulated (Nicol & Meinertzhagen, 1982b) and the morphology of these cells (Strausfeld, 1971) reveals the regularity of the spacing of their dendrites down the axis of the cartridge. Thus, competition between sibling dendrites of L1 or L2 must fall off sharply with distance, each dendrite competing effectively over a small radius only. The mean space between consecutive dendrites ($1.29 \mu m$) approximates the distance ($1.35 \mu m$) between the centres of two hexagons of area ($1.6 \mu m^2$) given by the average value of A/N (Nicol & Meinertzhagen, 1982b). In other words, dendrite frequency and spacing is adjusted in the adult to a distribution corresponding to the spacing of the synaptic sites over the presynaptic membrane and thus minimizing the need for extensive subdivision of dendrites. Upon these postsynaptic phenomena is superimposed the gradual developmental loss of presynaptic sites discussed earlier.

Acknowledgement

Original work reported in this review was supported by grants from NSERC (A-0065) and NIH (EY-03592). I wish to thank Dr. S.R. Shaw for helpful comments on the manuscript and Drs. C. Levinthal and E.R. Macagno for use of the Computer Graphics Facility, Columbia University (supported by NIH Division of Research Resources) in obtaining Figs. 7 and 8.

REFERENCES

Armett-Kibel, C., Meinertzhagen, I.A. & Dowling, J.E. (1977) Cellular and synaptic organization in the lamina of the dragon-fly Sympetrum rubicundulum. Proc. R. Soc. Lond. 196B: 385-413.

Bate, C.M. (1978) Development of sensory systems in arthropods. In: Handbook of Sensory Physiology, Vol. IX. Ed. M. Jacobson. Berlin, Heidelberg, New York, Springer-Verlag, p. 1-53.

Bentley, D. & Keshishian, H. (1982). Pathfinding by peripheral pioneer neurons in grasshoppers. Science 218: 1082-1088.

Berry, M., McConnell, P. & Sievers, J. (1980) Dendritic growth and the control of neuronal form. In: Neuronal Development,

Part I Emergence of Specificity in Neural Histogenesis. Ed. R.K. Hunt. New York, Academic Press. Curr. Top. Dev. Bio. 15:67-101.

Bodenstein, D. (1953) Postembryonic development. In Insect Physiology. Ed. K.D. Roeder. New York, Wiley, p. 822-865.

Boschek, C.B. (1971) On the fine structure of the peripheral retina and lamina ganglionaris of the fly, Musca domestica. Z. Zellforsch. mikrosk. Anat. 118:369-409.

Braitenberg, V. (1967) Patterns of projection in the visual system of the fly. I. Retina-lamina projections. Exp. Brain Res. 3:271-298.

Burkhardt, W. & Braitenberg, V. (1976) Some peculiar synaptic complexes in the first visual ganglion of the fly, Musca domestica. Cell Tiss. Res. 173:287-308.

Cajal, S.R. (1937) Recollections of My Life (transl. E.H. Craigie). Cambridge, MIT Press.

Cajal, S.R. & Sánchez, D.S. (1915) Contribucion al conocimiento de los centros nerviosos de los insectos. Trab. Lab. Invest. Biol. Univ. Madr. 13:1-164.

Egelhaaf, A., Berndt, P. & Küthe, H.-W. (1975) Mitosenverteilung und ^{3}H- Thymidin-Einbau in der proliferierenden Augenanlage von Ephestia kuehniella Zeller. Wilhelm Roux' Archiv. 178:185-202.

Fischbach, K.F. (1983) Neural cell types surviving congenital sensory deprivation in the optic lobes of Drosophila melanogaster. Dev. Biol. 95:1-18.

Fröhlich, A. & Meinertzhagen, I.A. (1980) Neurite growth and its relationship to synaptogenesis in the fly's visual system. Soc. Neurosci. Abstr. 6:373.

Fröhlich, A. & Meinertzhagen, I.A. (1982) Synaptogenesis in the first optic neuropile of the fly's visual system. J. Neurocytol. 11:159-180.

Fröhlich, A. & Meinertzhagen, I.A. (1983) Quantitative features of synapse formation in the fly's visual system. I. The presynaptic photoreceptor terminal. J. Neurosci. (Submitted.)

Goodman, C.S. & Bate, M. (1981) Neuronal development in the grasshopper. Trends Neurosci. 4:163-169.

Goodman, L.J., Mobbs, P.G. & Kirkham, J.B. (1979) The fine structure of the ocelli of Schistocerca gregaria. Cell Tissue Res. 196:487-510.

Hafner, G.S. (1974) The ultrastructure of retinula cell endings in the compound eye of the crayfish. J. Neurocytol. 3:295-311.

Hafner, G.S., Tokarski, T. & Hammond-Soltis, G. (1982) Development of the crayfish retina: a light and electron microscopic study. J. Morph. 173: 101-118.

Hauser-Holschuh, H. (1975) Vergleichend quantitative Untersuchungen an den Sehganglien der Fliegen Musca domestica und Drosophila melanogaster. Ph.D. Dissertation:

Eberhard-Karls-Universität zu Tübingen.

Hertel, H. (1983) Change of synapse freuency in certain photoreceptors of the honey bee after chromatic deprivation. J. Comp. Physiol. (in press).

Hofbauer, A. (1979) Die Entwicklung der optischen Ganglien bei Drosophila melanogaster. Ph.D. Dissertation: Albert-Ludwigs-Universität zu Freiburg.

Järvilehto, M. & Finell, N. (1983) Development of the function of visual receptor cells during the pupal life of the fly Calliphora. J. Comp. Physiol. (in press).

Kankel, D.R., Ferrús, A., Garen, S.H., Harte, P.J. & Lewis, P.E. (1980) The structure and development of the nervous system. In: The Genetics and Biology of Drosophila, Vol. 2d. Ed. M. Ashburner & T.R.F. Wright. London, Academic Press, p. 295-368.

Lamparter, H.E., Steiger, U., Sandri, C. & Akert, K. (1969) Zum Feinbau der Synapsen im Zentralnervensystem der Insekten. Z. Zellforsch. mikrosk. Anat. 99:435-442.

Land, M.F. (1983) Crustacea. (This volume.)

Lawrence, P.A. & Green, S.M. (1979) Cell lineage in the developing retina of Drosophila. Dev. Biol. 71:142-152.

Lopresti, V., Macagno, E.R. & Levinthal, C. (1973) Structure and development of neuronal connections in isogenic organisms: cellular interactions in the development of the optic lamina of Daphnia. Proc. Natl. Acad. Sci. USA 70:433-437.

Macagno, E.R. (1979) Cellular interactions and pattern formation in the development of the visual system of Daphnia magna (Crustacea, Branchiopoda). I. Trophic interactions between retinal fibers and laminar neurons. Dev. Biol. 73:206-238.

Macagno, E.R. (1981) Cellular interactions and pattern formation in the development of the visual system of Daphnia magna (Crustacea, Branchiopoda) II. Induced retardation of optic axon ingrowth results in a delay in laminar neuron differentiation. J. Neurosci. 1:945-955.

Macagno, E.R., Lopresti, V. & Levinthal, C. (1973) Structure and development of neuronal connections in isogenic organisms: variations and similarities in the optic system of Daphnia magna. Proc. Natl. Acad. Sci. USA 70:57-61.

Meinertzhagen, I.A. (1973) Development of the compound eye and optic lobe of insects. In: Developmental Neurobiology of Arthropods. Ed. D. Young, Cambridge, University Press, p. 51-104.

Meinertzhagen, I.A. (1977) Development of neuronal circuitry in the insect optic lobe. In: Approaches to the Cell Biology of Neurons. Ed. W.M. Cowan & J.A. Ferrendelli. Bethesda, MD, Society for Neuroscience, p. 92-119.

Meinertzhagen, I.A. & Armett-Kibel, C. (1982) The lamina monopolar cells in the optic lobe of the dragonfly Sympetrum. Phil. Trans. R. Soc. Lond. 297B:27-49.

Meinertzhagen, I.A. & Fröhlich, A. (1983) The regulation of

synapse formation in the fly's visual system. Trends Neurosci. 6: (in press).

Mouze, M. (1980) Étude autoradiographique de la prolifération et de la migration cellulaires au cours de la croissance larvaire de l'appareil visuel chez Aeschna cyanea Müll. (Odonata: Aeschnidea). Int. J. Insect Morphol. Embryol. 9:41-52.

Mouze, M. (1983). Morphologie et développement des yeux simples et composés des insectes. (This volume.)

Nässel, D.R., Elofsson, R. & Odselius, R. (1978) Neuronal connectivity patterns in the compound eyes of Artemia salina and Daphnia magna. Cell Tissue Res. 190:435-457.

Nässel, D.R. & Waterman, T.H. (1977) Golgi EM evidence for visual information channelling in the crayfish lamina ganglionaris. Brain Res. 130:556-563.

Nicol, D. & Meinertzhagen, I.A. (1982a) An analysis of the number and composition of the synaptic populations formed by photoreceptors of the fly. J. Comp. Neurol. 207:29-44.

Nicol, D. & Meinertzhagen, I.A. (1982b) Regulation in the number of fly photoreceptor synapses: the effects of alterations in the number of presynaptic cells. J. Comp. Neurol. 207:45-60.

Nordlander, R.H. & Edwards, J.S. (1969) Postembryonic brain development in the monarch butterfly, Danaus plexippus plexippus, L. II. The optic lobes. Wilhelm Roux' Archiv. 163:197-220.

Nowel, M.S. & Shelton, P.M.J. (1980) The eye margin and compound-eye development in the cockroach: evidence against recruitment. J. Embryol. Exp. Morph. 60:329-343.

Oberdorfer, M.D. (1977) The neural organization of the first optic ganglion of the principal eyes of jumping spiders (Salticidae). J. Comp. Neurol. 174:95-118.

Panov, A.A. (1960) The structure of the insect brain during successive stages of postembryonic development. III. Optic lobes Ent. Rev. 39:55-68 (translation).

Ready, D.F., Hanson, T.E. & Benzer, S. (1976) Development of the Drosophila retina, a neurocrystalline lattice. Dev. Biol. 53:217-240.

Ribi, W.A. (1975) The first optic ganglion of the bee. I. Correlation between visual cell types and their terminals in the lamina and medulla. Cell Tissue Res. 165:103-111.

Ribi, W.A. (1977) Fine structure of the first optic ganglion (lamina) of the cockroach, Periplaneta americana. Tissue & Cell 9:57-72.

Ribi, W.A. (1981) The first optic ganglion of the bee. IV. Synaptic fine structure and connectivity patterns of receptor cell axons and first order interneurones. Cell Tissue Res. 215:443-464.

Shelton, P.M.J. & Lawrence, P.A. (1974) Structure and development of ommatidia in Oncopeltus fasciatus. J. Embryol. Exp. Morphol. 32:337-353.

Stell, W.K. (1972) The morphological organization of the vertebrate retina. In: Handbook of Sensory Physiology, Vol. VII/2. Ed. M.G.F. Fuortes. Berlin, Heidelberg, New York, Springer-Verlag, p. 111-213.

Strausfeld, N.J. (1971) The organization of the insect visual system (light microscopy). I. Projections and arrangements of neurons in the lamina ganglionaris of Diptera. Z. Zellforsch. mikrosk. Anat. 121:377-441.

Strausfeld, N.J. (1983) Functional neuroanatomy of the blowfly's visual system. (This volume).

Strausfeld, N.J. & Campos-Ortega, J.A. (1977). Vision in insects: pathways possibly underlying neural adaptation and lateral inhibition. Science 195:894-897.

Strausfeld, N.J. & Nässel, D.R. (1980) Neuroarchitecture of brain regions that subserve the compound eyes of Crustacea and insects. In: Handbook of Sensory Physiology, Vol. VII/6B. Ed. H. Autrum. Berlin, Heidelberg, New York, Springer-Verlag, p. 1-132.

Toh, Y. & Kuwabara, M. (1974) Fine structure of the dorsal ocellus of the worker honeybee. J. Morph. 143:285-306.

Toh, Y. & Kuwabara, M. (1975) Synaptic organization of the fleshfly ocellus. J. Neurocytol. 4:271-287.

Trujillo-Cenóz, O. (1965). Some aspects of the structural organization of the intermediate retina of dipterans. J. Ultrastruct. Res. 13:1-33.

Trujillo-Cenóz, O. & Melamed, J. (1964) Synapses in the visual system of Lycosa. Naturwissen. 51:470-471.

Trujillo-Cenóz, O. & Melamed, J. (1973) The development of the retina-lamina complex in muscoid flies. J. Ultrastruct. Res. 42:554-581.

Trujillo-Cenóz, O. & Melamed, J. (1975) Transplanted eyes of dipterans. Naturwissen. 62:42.

Watson, A.H.D. & Burrows, M. (1982) The ultrastructure of identified locust motor neurones and their synaptic relationships. J. Comp. Neurol. 205:383-397.

Wernig, A., Pécot-Dechavissine, M. & Stöver, H. (1980) Sprouting and regression of the nerve at the frog neuromuscular junction in normal conditions and after prolonged paralysis with curare. J. Neurocytol. 9:277-303.

Whitehead, R. & Purple, R.L. (1970) Synaptic organization in the neuropile of the lateral eye of Limulus. Vision Res. 10:129-132.

MORPHOLOGIE ET DEVELOPPEMENT DES YEUX SIMPLES ET COMPOSES DES INSECTES

MICHEL MOUZE
Laboratoire de Biologie Animale
Laboratoire associé au C.N.R.S. n° 148
(Endocrinologie des Invertébrés)
Université des Sciences et Techniques de Lille I
59655 Villeneuve d'ASCQ, FRANCE

> The searcher's eye,
> Not seldom finds more than he wished to find.
>
> Gotthold Ephraim Lessing, Nathan der Weise, 1779.

ABSTRACT

Compound eyes macroscopically show a certain polymorphism related to the biology, the sex or the degree of evolution of the insect.

These eyes indeed correspond to a juxtaposition of a generally very high number of functional units, the ommatidia, which shows a general structure, alike in all insects, at three levels: dioptric (cornea, crystalline lens), pigmentary (primary and secondary pigmentary cells) and photosensitive (retinal cells constituting a rhabdom) levels. Modifications of structure are registered at each level, depending on the biology (diurnal or nocturnal), the adaptation of the eye to light, or the phylogenic place of the considered insect.

Compound eyes are either present in the young larva at the time of hatching (Heterometabolous) or exist, under a functional and definitive form, only in the imago (Holometabolous). While some eye growth in Heterometabolous is accomplished by an increase in the size of its elements, most growth is due to an addition of new ommatida which continually differentiate at the level of the growth zone, a permanent, well localized structure, of a similar form in the different groups. During each larval stage, new rows of ommatidia are formed at this level and intercalate, at the next moult, between the differentiated ommatidia and the growth zone. The number of ommatidia formed during larval life varies greatly according to the species. Moreover, throughout larval development, the previously formed ommatidia increase in size: cells increase in volume, facets increase in diameter, and at the level of the differentiated and functional ommatidia, some mitose will increase the number of accessory pigment cells.

In Holometabolous, larvae are either blind or have rudimentary eyes (stemmata). The differentiation of future compound eyes can begin either at the beginning of larval life or, in contrast, at the time of metamorphosis, the stemmata generally degenerating at the end of larval life. Cellular proliferation and differentiation which give birth to the cells of the future ommatidia take place in several areas spatially separated through the presumptive eye. There exists no growth zone as specialized as in Heterometabolous. The processes of growth are nevertheless comparable, the only difference being a different chronology: in Heterometabolous, the bases of multiplication and differentiation which take place all through larval life, are very close in time and space, whereas in Holometabolous these two phases are relatively separated.

The growth zone, long considered as the level where the cephalic epidermis is transformed into an eye, would rather correspond to a localized area of stem cells which persists throughout the duration of eye development. The daughter cells constantly formed at the level of the growth zone progressively cluster, each cluster leading to the formation of an ommatidia. The differentiation of each of these components results from its position in the ommatidial bundle and perhaps also from its origin from certain hypothetical stem cells. Finally, the development of compound eyes is autonomous, independent of the presence of the optic lobe.

In addition to compound eyes, "simple" eyes also exist in insects: for example the stemmata in Holometabolous larvae, where they really play a visual role; and of the ocelli found in adult insects of numerous groups and whose function is yet not well known.

INTRODUCTION

Depuis toujours les naturalistes ont été fascinés par les yeux composés des Insectes, sans doute à cause de leurs couleurs, de leur incroyable régularité géométrique et peut-être aussi à cause de leur étrange absence de regard. Aussi ces yeux ont-ils été étudiés très tôt tant au niveau de la morphologie et de la structure, qui ont bien sûr beaucoup bénéficié de la microscopie électronique, qu'au niveau du fonctionnement (électrophysiologie et optique) et de leur utilité pour l'Insecte (performances de la vision). Mais ne serait-ce que par l'incroyable régularité géométrique de l'assemblage répétitif des ommatidies, et par la complexité même de chacun de ces petits yeux élémentaires, l'oeil composé suscite de nombreuses autres questions au niveau des mécanismes mis en jeu pour la construction d'un tel ensemble. C'est pourquoi l'oeil composé des Insectes offre également un grand champ de recherches dans le domaine de la morphogénèse. Cette revue, qui n'est surtout pas une mise au point exhaustive sur la morphologie, la structure et le développement de l'oeil composé, ce qui a été bien mieux fait par d'autres (Meinertzhagen, 1973,

1975; Autrum, 1975, Shelton, 1976), voudrait seulement rassembler les différents points d'intérêt qui peuvent être étudiés par son intermédiaire.

LES YEUX COMPOSES - STRUCTURE GENERALE

A- <u>Morphologie</u>

Les yeux composés se présentent sous la forme de deux masses globuleuses généralement ovales, et occupant une surface plus ou moins importante sur la tête de l'Insecte. La surface de chaque oeil est recouverte par une cuticule transparente, faite de protubérances généralement hexagonales et alignées, les facettes. Chaque facette est en réalité la lentille cornéenne d'une unité de vision appelée ommatidie, structuralement et fonctionnellement séparée des voisines, l'ensemble de ces yeux élémentaires constituant l'oeil composé.

Ces yeux composés, qui n'existent chez les Insectes Holométaboles qu'au stade imago, sont au contraire présents dès l'éclosion de l'oeuf chez les Hétérométaboles; leur taille va augmenter progressivement au cours du développement larvaire, par addition de nouvelles ommatidies d'une part, par augmentation de la taille des ommatidies préexistantes d'autre part.

Chez la plupart des Insectes adultes, les yeux sont plutôt volumineux, comparés à la taille de la tête; le nombre de facettes qui les constituent est très variable selon les espèces, la taille du corps, mais aussi selon le mode de vie, le sexe, la caste, ou même la répartition géographique de l'espèce:

- C'est chez les plus grandes espèces que l'on va rencontrer les yeux formés du plus grand nombre d'ommatidies (Anax, Odonate: 28600/oeil - Sphinx, Lépidoptère: 27000/oeil). - Chez les Odonates Anisoptères, les yeux hémisphériques sont même jointifs sur la face dorsale de la tête.

- Les insectes très mobiles et au vol rapide ont de grands yeux composés, contrairement aux espèces sédentaires. A l'extrême, certains insectes hypogés, ou vivant à l'obscurité ou parasites possèdent des yeux très réduits ou même en sont dépourvus (ouvrière souterraine de Solenopsis à 9 ommatidies) (Kobakhidze <u>et al</u>., 1959).

- Les insectes mâles possèdent souvent des yeux plus grands que ceux des femelles, comme chez les Apoïdae ou les Simulidae. De même, chez le <u>Lampyris splendicula</u> chaque oeil est constitué de 2500 ommatidies chez le mâle, de 300 chez la femelle.

Chez certaines fourmis les yeux sont plus petits chez les ouvrières que dans les autres castes ailées. (<u>Formica pratensis</u>: mâles 1200 ommatidies, femelles 830, ouvrières 600).

- Enfin, il a été remarqué chez <u>Apis mellifera</u> (Melnichenko, 1963) que les races vivant dans la partie septentrionale de l'aire de répartition ont des yeux comportant plus de facettes que celles vivant au sud.

- Signalons aussi la curieuse position des yeux (ainsi que des ganglions optiques et des antennes) chez les Diopsides, situés sur de longs prolongements latéraux de la tête, et leur permettant probablement d'estimer avec précision les distances.

Le diamètre des facettes n'est pas identique chez tous les Insectes (varie de 15 à 40 µm), et peut même différer d'une partie à l'autre de l'oeil du même insecte. Par exemple, chez Libellula les facettes de la moitié supérieure sont 1,3 fois plus grandes que celles de la moitié inférieure. Chez Culex au contraire les grandes facettes sont ventrales. Chez certaines espèces chaque oeil présente deux régions très différentes du point de vue morphologique et physiologique, et sont appelés "yeux doubles". Ainsi, les yeux "en turban" de l'Ephéméroptère Baëtis mâle, partie dorsale de l'oeil portée par un pédoncule, séparée des yeux ventraux. De même, chez le Coléoptère aquatique Gyrinus vivant à l'interface air/eau, dont chaque oeil est divisé en deux parties, une dorsale émergée à vision aérienne, une ventrale immergée pour la vision sub-aquatique. Enfin, chez beaucoup de Diptères les régions dorsales et ventrales des yeux, identiques à l'observation macroscopique, révèlent au microscope des différences d'organisation.

Les yeux à facettes sont de couleur noire ou ont la couleur générale du corps de l'Insecte. Parfois, la "décoration" présentée par le corps sous forme de taches de couleur se poursuit au travers de l'oeil (Carausius). Chez d'autres Insectes les yeux peuvent être iridescents ou, chez certaines souches mutantes, dépourvues en partie ou totalement de pigments (Blattes, Criquets, Drosophiles). Ces colorations de l'oeil peuvent avoir pour origine des interférences lumineuses au niveau des microstructures de la cornée, ou des réflections lumineuses variant suivant la distribution des pigments en fonction de l'adaptation de l'oeil à la lumière. Chez certains insectes hétérométaboles (Aeshna, Locusta), l'oeil présente une succession de bandes claires et de bandes foncées; ces stries, larges de 3 ou 4 facettes, correspondent aux ommatidies formées au cours des stades larvaires successifs (Volkonsky, 1938; Schaller, 1960). A la surface de l'oeil de nombreux insectes il est possible d'observer un spot, ou "pseudopupille", dont la localisation change suivant la direction de l'observation. Cette pseudopupille, qui résulte de l'absorption de la lumière par le rhabdome des ommatidies dont l'axe coïncide avec l'axe de l'observateur, permet donc de déterminer l'axe optique des ommatidies chez l'Insecte vivant. (Sherk, 1977, 1978 a, b, c).

Signalons enfin que de nombreuses espèces (abeilles, papillons diurnes) portent des soies oculaires au point de contact de 3 ommatidies voisines. Ces soies seraient en fait, des récepteurs utilisés pour l'orientation de l'animal ou l'appréciation de sa vitesse en vol.

B- Structure de l'ommatidie. (Planche I)

Une ommatidie a une forme de tronc de cône fortement allongé, de quelques centaines de microns de longueur sur 20 à 50 µm de diamètre. Schématiquement, chaque ommatidie est constituée de 3 régions fonction-nelles différentes:

- Une région dioptrique, distale, dont le rôle sera de recueillir et de concentrer les rayons lumineux.

- Une région pigmentaire qui va régler l'entrée de la lumière dans l'ommatidie et empêcher les rayons lumineux de passer d'une ommatidie dans la voisine ou en contrôler la dispersion.

Une région photosensible proximale où la lumière sera analysée, les informations recueillies étant ensuite envoyées vers les ganglions optiques et le cerveau où elles seront intégrées et interprétées.

L'oeil est limité en direction proximale par une "membrane fenestrée" formant un plancher à la base de chaque ommatidie.

1) Région dioptrique:

Cette région est constituée d'une cornée et d'un cristallin.

a) La cornée.

La cornée est un épaississement transparent de la cuticule formant une lentille convergente généralement convexe vers l'extérieur, la face interne pouvant être plane, convexe ou concave. Cette cornée, dont l'épaisseur peut atteindre 50 μm est constituée d'un certain nombre de couches cuticulaires de natures différentes (jusqu'à une 20 aine de couches), présentant chacune une épaisseur et un indice de réfraction différents des autres (Seitz, 1968). Ces couches constitueraient des filtres d'interférences, ce qui expliquerait les bandes colorées visibles dans les yeux de certains diptères (Friza, 1928; Miller et al., 1968).

La surface externe de la cornéule peut être lisse, ou présenter des microsculptures plus ou moins régulières pouvant atteindre une longueur de 0,25 μm. Toute la surface de la cornée est couverte de telles papilles très serrées chez les Trichoptères et les Lépidoptères, structures qui réduiraient la perte de lumière par réflexion (Bernhard et al., 1970).

b) Le cristallin

Il s'agit d'un corps transparent, solide ou liquide, pyriforme, situé distalement dans l'axe de l'ommatidie. Le cristallin, formé par 4 cellules cristalliniennes (ou "cellules de SEMPER") peut-être une formation extra- ou intracellulaire. On distinguera donc des yeux de différents types:

- Type acône (chez les Diptères, Hétéroptères, quelques Coléoptères et Hyménoptères) où les 4 grandes cellules cristalliniennes au cytoplasme homogène constituent par elles même un cristallin sans véritable corps vitreux.

- Type eucône (Ephémères, Odonates, Phasmoptères, Mantidés, Cicadidés, Lépidoptères, beaucoup de Coléoptères et d'Hyménoptères). Chaque cellule de SEMPER sécrète intérieurement un corps réfringent, le cristallin étant donc composé de l'ensemble de 4 parties juxtaposées. Dans chacune de ces cellules de SEMPER le cytoplasme, essentiellement constitué de granules osmiophiles, de quelques mitochondries et lysosomes, est réduit à une couche très fine entourant le corps réfringent, le noyau très aplati étant plaqué sur la face distale de celui-ci. Ces parois latérales internes des cellules cristalliniennes présentent généralement de profondes interdigitations reliées entre elles par des jonctions septées, l'ensemble assurant un lien solide entre les 4 éléments du même cône. Bien que le cytoplasme des cellules cristalliniennes se trouve empli de particules réparties de façon très homogène, la densité optique de chacune de ces cellules peut parfois être fort différente des voisines même à l'intérieur d'une même ommatidie.

La nature chimique du cristallin est variable. Ainsi, chez Pieris

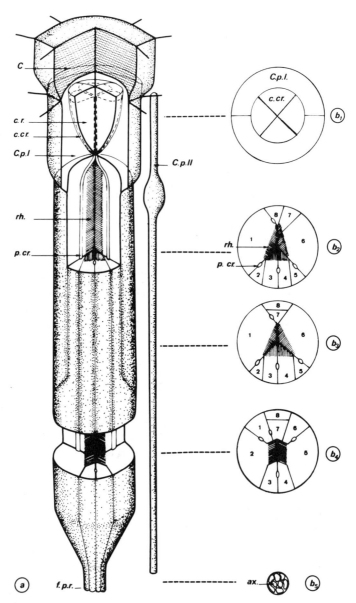

Planche 1 :

a) Représentation schématique d'une ommatidie d'Odonate Anisoptère.
b) Coupe transversale à différents niveaux de l'ommatidie : b_1 : niveau du cristallin - b_2 rétinule, niveau distal - b_3 : rétinule,

rapae, Kim (1964) a montré qu'il s'agissait de mucopolysaccharides, alors que chez Apis le cristallin serait constitué de glycogène (Fyg, 1961) ainsi que chez Carausius morosus (Such, 1969). Chez les Collemboles Entobryomorphes il s'agit de protéines (Barra, 1969).

- Type pseudocône (quelques Diptères et Coléoptères). Les cellules cristalliniennes secrètent un cristallin commun extracellulaire, plaqué à la face interne de la cornée, et jouant le rôle d'une lentille cylindrique; le cristallin forme alors une masse unique avec la cornée. Ce pseudocône est plus ou moins allongé selon les espèces, et peut également, chez certains lépidoptères, coexister avec un cône intracellulaire sous-jacent.

c) Prolongements cristalliniens: (crystalline tract-crystalline shreads).

Dans la région proximale chaque cellule cristallinienne s'effile en un long processus tubulaire intercallé entre deux cellules rétiniennes contiguës, et qui vont parcourir parallèlement au rhabdome toute la hauteur de l'ommatidie. Ces processus cristalliniens, généralement au nombre de 4, occupent toujours la même position relative entre les cellules rétiniennes, la répartition généralement asymétrique de ces prolongements par rapport aux huit cellules rétiniennes offrant la possibilité de reconnaître celles-ci individuellement. Notons que chez certaines espèces (Aedes aegypti, Brammer, 1970, Gelastocoris, Burton & Stockhammer, 1969), deux des cellules de Semper émettent deux processus cristalliniens, soit six au total pour l'ommatidie.

Ces prolongements peuvent être très longs (1000 μm chez des Lépidoptères nocturnes (Miller et al., 1968); chez certaines espèces dont les yeux s'adaptent à la lumière par déplacement proximo-distal du rhabdome, les prolongements cristalliniens peuvent également s'étirer et passer de 10 μm (oeil adapté à l'obscurité) à 120 μm (Walcott, 1969).

A l'extrémité de leur région proximale, les 4 prolongements des cellules cristalliniennes se dilatent chacun en une sorte d'ampoule qui s'accollent étroitement à la base des cellules rétiniennes. Chez Carausius (Such, 1975, 1978) ces ampoules renferment de nombreux tubules, beaucoup de glycogène, ainsi que des lysosomes, pigments et gouttelettes de sécrétion. De chaque ampoule partent plusieurs prolongements tubulaires qui plongent dans les mailles de la "membrane fenestrée" et y constituent une sorte de plexus.

Certains auteurs ont attribué à ces processus un rôle de support mécanique et de maintien des cellules rétiniennes (Horridge, 1966) ou de

---

niveau moyen - b_4: rétinule, niveau proximal - b_5: fibres post-rétiniennes.

ax.: axone - C: cornée - c. cr. cellule cristallinienne - c.r.: corps réfringent - C.p.I.: cellule pigmentaire principale - C.p.II: cellule pigmentaire accessoire - C_1 à C_8: cellules rétiniennes - f.p.r.: fibres post-rétiniennes - p. cr.: prolongement cristallinien - rh.: rhabdomère. (d'après Mouze, 1979).

guide d'ondes lumineuses (Horridge, 1968), ou encore d'intermédiaires entre cellules rétiniennes pour la transmission d'excitation grâce aux jonctions spécialisées (Burton & Stockhammer, 1969); Such (1975) se basant sur l'ultrastructure de ces processus, les considère cependant comme des neurones intervenant dans la photoréception.

2) Région pigmentaire:

Il est classique de distinguer deux types de cellules pigmentaires.

a) Les cellules pigmentaires primaires: (ou c.p. principales, ou c. cornéagènes).

Elles ont, en fait, comme fonction principale la sécrétion de la cornée. En effet, chaque facette résulte de l'activité sécrétrice de deux cellules "cornéagènes" qui, selon les espèces, seront plus ou moins chargées de pigment. Chez certains Insectes primitifs (Machilides) ces cellules cornéagènes ont uniquement cette fonction sécrétrice, et sont situées entre la lentille et l'apex du cristallin. Chez les Insectes plus évolués ces cellules acquièrent de plus une forte pigmentation. Ces cellules jouent donc en outre, le rôle d'écran optique. A leur extrémité distale, les deux C.p.I s'étalent sous la cornée, les deux minces lames cytoplasmiques, qui présentent des microvillosités sur la face en contact avec la cornée, s'unissent par l'intermédiaire d'une Zonula adherens. Le corps de la cellule pigmentaire, situé près de l'extrémité du cristallin, contient un noyau peu volumineux et de nombreux grains de pigments.

b) Les cellules pigmentaires secondaires. (C.p.II ou C.P. accessoires, ou C. interommatidiennes).

Ce sont des cellules très allongées, fortement pigmentées, s'étendant sur toute la hauteur de l'ommatidie. Leur structure est voisine de celle des C.p.I: au pôle en contact avec la cornée la membrane se plisse en microvillosités bien développées; le noyau occupe également une position distale et en direction proximale la cellule, renfermant de très nombreux microtubules et grains de pigments, s'effile en un long pédicelle très étroit (0,2 µm de diamètre chez le phasme), s'insinuant entre les cellules rétiniennes appartenant à des ommatidies voisines, comblant ainsi les interstices interommatidiens. A la base, ce pédicelle se renfle et donne naissance à de fins prolongements latéraux qui participent à l'édification du complexe structural de la membrane fenestrée.

Ces cellules constituent un manchon autour de chaque ommatidie, et sont même communes à deux ou trois ommatidies adjacentes. Leur nombre ne semble pas constant, même dans une espèce, et peut varier pour une ommatidie de 7 à 8 chez Aeshna cyanea (Mouze, 1979), 6 à 9 chez Calliphora, 8 à 12 chez Drosophila (Hofbauer & Campo-Ortega, 1976), 16 chez Locusta (Wilson et al., 1978) 25 chez Hemicordulia (Laughlin & McGinness, 1978). Ces cellules, qui peuvent être disposées en une ou plusieurs couches autour de chaque ommatidie, constituent ainsi une isolation optique des cellules photoréceptrices.

Les pigments photoprotecteurs ("screening pigment") des yeux d'Insecte appartiennent, soit au groupe des ptérines (le plus souvent de la xanthoptérine) soit aux ommochromes (xanthommatines et ommines). Ces granules pigmentaires sont trouvés indifféremment dans les deux types de cellules pigmentaires ou dans les cellules de la rétinule.

3) Région photosensible: la rétine.

Dans l'axe du cristallin, à un niveau proximal, se trouve la rétine constituée de cellules sensorielles photoréceptrices. Ces cellules rétiniennes sont très généralement au nombre de 8; on en dénombre, cependant, 9 chez les ouvrières et les mâles d' Apis (Perrelet & Baumann, 1969 a, b); 10 chez Baëtis et Ephestia, 11 chez Bombyx et 12 chez Timandra.

a) Structure de la cellule rétinienne.

Le plan structural de la cellule rétinienne paraît assez uniforme chez les Insectes (et même les Arthropodes en général).

Vers l'axe de l'ommatidie, la membrane de ces cellules rétiniennes se plisse en un très grand nombre de microvillosités parallèles les unes aux autres et empilées en un arrangement hexagonal. L'ensemble des microvillosités d'une même cellule constitue un rhabdomère, l'ensemble des rhabdomères d'une même ommatidie formant le rhabdome, qui avait autrefois été décrit en microscopie optique comme un "bâtonnet réfringent" secrété par les cellules rétiniennes. Ces microvillosités, qui ont un diamètre de 40 à 120 nm (70 nm de diamètre sur 3 µm de long chez le Phasme - Such, 1975) contiennent des pigments photosensibles appartenant aux rhodopsines, comme dans l'ensemble du règne animal; le rhabdomère est donc comparable au segment externe des bâtonnets rétiniens des Vertébrés. Les microvillosités voisines ont une base commune (par groupe de 4 microvillosités le plus souvent), et sont orientées parallèlement dans un même rhabdomère. Ces microvillosités sont fortement tassées les unes contre les autres, de telle sorte qu'en coupe parasagittale ils donnent des images en nid d'abeilles. Leur direction est généralement différente de celle des microvillosités des rhabdomères voisins; dans le cas contraire, les rhabdomères sont difficilement discernables.

Les cellules rétiniennes sont reliées les unes aux autres par des jonctions d'un type classique chez les Insectes: Les Zonula adherens, simple épaississement de deux membranes cellulaires contiguës. Elles sont situées à proximité immédiate de chaque rhabdomère, maintenant les cellules visuelles en un ensemble parfaitement cohérent autour du rhabdome central.

Le cytoplasme est très riche en ribosomes et en mitochondries, celles-ci étant surtout situées à proximité du rhabdome, perpendiculairement à celui-ci. Les dictyosomes, eux aussi très nombreux, ont une localisation proche du rhabdome. Le cytoplasme rétinien est également très riche en microtubules, orientés parallèlement à l'axe rétinien, et se prolongeant à l'intérieur de l'axone. Les cellules visuelles contiennent également en abondance des grains de pigment de taille variable groupés principalement autour du rhabdome, et en grande partie dans la région distale des cellules photoréceptrices.

Ces cellules rétiniennes contiennent aussi des "corps multivésiculaires" qui sont caractéristiques des cellules visuelles des Arthropodes. Ce sont des ensembles de petites vésicules d'un diamètre de 50 nm, étroitement tassées à l'intérieur d'une membrane en une sphère de 3 µm de diamètre.

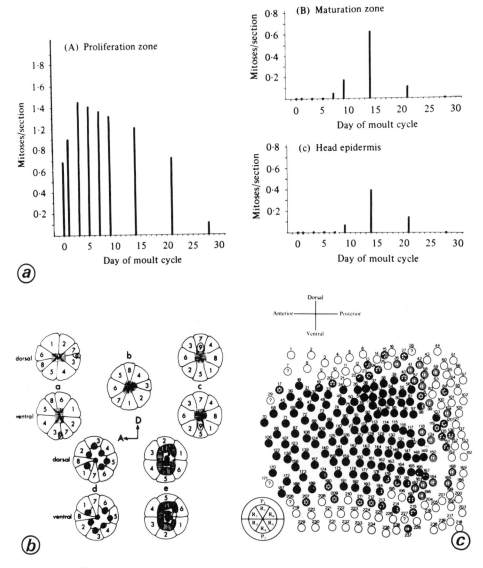

Planche II:

a) Graphiques montrant l'activité mitotique dans (A) la zone de prolifération et (B) la zone de maturation de la bordure de l'oeil, et (C) l'épiderme céphalique adjacent à la rétine de larves de Periplaneta americana. (d'après Nowel, 1981).

b) Arrangement de cellules rétiniennes dans les ommatidies de différents insectes. Toutes les cellules sont projetées sur un plan de coupe transversal unique, bien que leur représentation à différentes hauteurs puisse être variable. Les aires hachurées représentent les rhabdomères photorécepteurs. a-c sont des

Par son extrémité proximale, chaque cellule rétinienne émet un axone, les huit (ou plus) axones de chaque ommatidie se groupant en un faisceau qui, après avoir traversé la membrane fenestrée se dirige de façon centripète vers le lobe optique. Sur ces 8 axones issus d'une ommatidie, 6 courts se terminent au niveau du neuropile de la lamina (ganglion optique I) et 2 plus longs vont jusqu'au neuropile de la medulla (ganglion optique II). En plus, de l'axone chaque cellule rétinienne émet également des prolongements horizontaux qui s'insinuent à travers les mailles de la "membrane fenestrée", constituant ainsi une sorte de plexus à la base de la rétine.

b) <u>Structure de la rétinule</u>. (Planche II, b)

Morphologiquement et physiologiquement, les cellules rétiniennes sont donc des unités séparées.

La rétinule présente une certaine diversité structurale selon la disposition des cellules rétiniennes les unes par rapport aux autres.

Dans la majorité des espèces, les rhabdomères des cellules rétiniennes voisines sont contigus, l'ensemble des rhabdomères formant donc un ensemble compact dans ces ommatidies dits à "rhabdome fermé" (Grenacher, 1879). Dans un tel oeil, les rhabdomères d'une même

---

ommatidies à rhabdome ouvert, d-e à rhabdome fermé. Les axes (A, antérieur; D, dorsal) ne concernent que b, d et e-a: abeille. L'asymétrie dans l'arrangement des cellules est donnée par la position de la cellule basale 9. Près de l'appareil dioptrique, l'orientation des ommatidies diffère dans les moitiés dorsale et ventrale de l'oeil - b: le criquet <u>Schistocerca</u>. Les cellules 1-6 sont plus longues que 7 et 8; la cellule 3, distinguable par son diamètre plus étroit, apporte une asymétrie à leur arrangement - c: le papillon <u>Pieris</u>; les sections transversales sont relatives à la moitié basale de l'ommatidie, dans laquelle seules les cellules 5 à 8 contribuent aux rhabdomères, et montrent deux types avec des variations de $180°$ dans la position de la cellule basale asymétrique 9 - d: la mouche. Le complexe rhabdomérique entier des 7 rhabdomères a un pattern asymétrique avec une symétrie en miroir dans les moitiés dorsale et ventrale - e: la punaise aquatique <u>Lethocerus</u>. Les deux cellules centrales 7 et 8 ont un arrangement asymétrique. La symétrie en miroir existe entre l'arrangement habituel (en haut) et des ommatidies occasionnelles isolées (d'après Meinertzhagen, 1975).

c) Carte schématique montrant les phénotypes des cellules rétiniennes R_1 - R_6 et des cellules pigmentaires primaires P_1 et P_2 dans une rétine mosaïque obtenue expérimentalement. Les cellules de type sauvage sont noires et les cellules mutantes blanches. On remarque des ommatidies dont toutes les cellules sont du même génotype et d'autres contenant des cellules des deux génotypes. (d'après Shelton & Lawrence, 1974).

ommatidie ont un axe commun et fonctionnent optiquement comme une seule unité. Le rhabdome a un indice de réfraction plus élevé que le milieu environnant (> 1,53 chez l'abeille) si bien qu'il fonctionne comme un guide d'onde.

Au contraire, les Hémiptères aquatiques et les Diptères présentent des ommatidies à "rhabdome ouvert", les rhabdomères étant séparés les uns des autres par un espace axial extracellulaire. Chez les mouches, les axes optiques d'une même ommatidie divergent, si bien que le même champ visuel sera en commun pour 6 cellules rétiniennes périphériques appartenant à 6 ommatidies voisines. Cette différence de structure du rhabdome correspond aussi à une différence dans les projections nerveuses sur le lobe optique (Meinertzhagen, 1973, 1975).

La disposition des différentes cellules rétiniennes n'étant généralement pas symétrique autour de l'axe ommatidien, ni équivalente aux différents niveaux dans une ommatidie, la structure des yeux des Insectes présente donc une grande variabilité: chez les formes primitives, comme Lepisma, la rétinule est stratifiée avec deux couches de cellules visuelles auxquelles correspondent deux rhabdomes, l'un distal, l'autre proximal.

L'ommatidie des ouvrières d'Abeille présente le cas opposé: 8 cellules visuelles parcourent toute la longueur de la rétinule et chacune contribue à la formation du rhabdome, (plus une cellule basale).

Chez de nombreux Insectes, une ou plusieurs cellules visuelles présentent des particularités morphologiques auxquelles peuvent correspondre une spécialisation fonctionnelle. Ainsi certaines cellules rétiniennes occupent une position excentrique à la périphérie de la rétine, ou une situation très proximale proche de la membrane fenestrée, et présentent un rhabdomère réduit ou nul; par exemple, l'ommatidie de Bombyx mori se compose de 11 cellules rétiniennes, dont une excentrique qui possède un processus distal identique à celui de la limule.

Au contraire, dans les ommatidies à "rhabdome ouvert", une ou deux cellules occupent une position centrale, dans l'axe de l'ommatidie, au centre de l'espace vide extracellulaire: chez les Diptères, 6 cellules visuelles périphériques forment une colonne creuse autour de l'axe de l'ommatidie. Deux autres cellules centrales (7 et 8) sont plus courtes et n'occupent qu'une partie de la longueur de la rétinule; les 2 cellules donnent naissance à des rhabdomères bien développés et orientés à angle droit par rapport aux autres cellules réceptrices, si bien qu'on observe que 7 rhabdomères en coupe transversale.

Il est habituel de distinguer deux types morphologiques d'ommatidie liés au mode de vie. Les Insectes diurnes, actifs en grande lumière, ont des yeux à vision "à apposition" ou "photopiques", tandis que les Insectes nocturnes ou crépusculaires sont dotés d'yeux à vision "à superposition" ou "scotopiques".

Dans les yeux photopiques, l'apex du cône cristallin se termine à l'extrémité distale du rhabdome des cellules rétiniennes, mais les cellules de Semper envoient généralement un fin processus jusqu'à la membrane fenestrée. Dans les yeux scotopiques un tractus cristallin sépare le cône du rhabdome et sert de guide optique. Il peut être formé à partir du cône, ou des processus des cellules rétiniennes, ou bien être d'origine mixte.

Dans les yeux à superposition une migration importante du pigment s'effectue en réponse à l'intensité lumineuse. Dans l'oeil adapté à la lumière, les ommatidies sont isolées les unes des autres par un écran de pigment; elles répondent donc seulement à la lumière qui entre avec un angle faible par rapport à l'axe ommatidien. Mais, dans les yeux adaptés à l'obscurité, le pigment se déplace pour se concentrer distalement et aussi près de la membrane fenestrée; l'isolement optique des ommatidies est ainsi perdu.

Le rhabdome d'une ommatidie reçoit donc non seulement les rayons lumineux qui traversent le système dioptique correspondant, mais également ceux qui ont pénétré dans les facettes voisines et qui convergent vers lui.

Remarquons que chez certains groupes, comme les Ephéméroptères, les deux aires oculaires de chaque oeil diffèrent, non seulement par la taille, mais également par la structure des ommatidies: l'oeil dorsal est à superposition, le ventral à apposition.

Chez certains insectes (Drosophila) l'oeil est divisé horizontalement en deux régions. Les ommatidies de chacune d'elles présentent une structure dite "en mirroir" par rapport à la ligne équatoriale, la disposition des cellules rétiniennes étant symétrique par rapport à cette ligne de séparation.

c) Adaptation de l'oeil à la lumière. (Planche III)

Ce contrôle de la quantité de lumière qui parvient aux structures photosensibles peut-être réalisé de plusieurs manières:

Dans les yeux à superposition les granules de pigment des cellules pigmentaires secondaires présentent des migrations remarquables: lorsque le pigment est réparti tout le long de l'ommatidie et forme ainsi un écran,

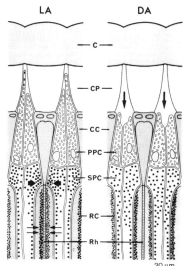

Planche III:

Représentation semi-schématique de la partie distale de la rétine de Pieris (Lepidoptère) adaptée à la lumière (LA) ou à l'obscurité (DA).

C. cornée: CC cône cristallin; CP processus cornéen; PPC cellule pigmentaire primaire; RC cellule rétinienne; Rh rhabdome; SPC cellule pigmentaire secondaire. (d'après Ribi, 1978).

le flux lumineux qui parvient au rhabdome est nettement diminué. Il peut s'y combiner des migrations de pigment dans les cellules pigmentaires primaires et dans les cellules visuelles. Chez de nombreux insectes à yeux à apposition, des déplacements de pigment ont lieu lors du passage de l'ombre au soleil. En plus de la réponse à la luminosité, de nombreux insectes montrent un rythme circadien dans les migrations pigmentaires.

Des changements morphologiques des cellules visuelles aussi bien dans les yeux à apposition qu'à superposition peuvent amener les rhabdomères à prendre une position plus distale ou plus proximale selon que la lumière est faible ou forte. Des migrations de réticulum endoplasmique, de vésicules, de mitochondries, peuvent modifier l'indice de réfraction du cytoplasme au voisinage du rhabdome et ainsi renforcer ou diminuer l'efficacité du rhabdome comme guide optique.

Des variations dans l'intensité et surtout dans la longueur d'onde de la lumière peuvent entraîner des changements beaucoup plus importants de la structure des cellules visuelles: on peut, en effet, assister à une dislocation de la structure fine du rhabdomère avec apparition corrélative d'un grand nombre de "coated vésicles" et de corps multivésiculaires lors du passage à une lumière intense (Aedes, White, 1967; Brammer & Clarin, 1976) ou vers des longueurs d'ondes courtes (Drosophila, Cosens, 1976), le rhabdome pouvant alors se raccourcir de 50 à 80 % (Gyrinus, Burghause, 1976).

d) Membrane basale et trachées.

L'oeil est limité en direction proximale par une structure très visible appelée "membrane fenestrée" ou "membrane basale" que traversent de part en part les axones des cellules rétiniennes. La résolution en microscopie optique n'étant pas suffisante pour en étudier la structure, seule une étude en microscopie électronique a permis de montrer que cette membrane n'était pas seulement une lame basale, mais qu'il s'agissait en fait d'un ensemble complexe formé de composants cellulaires et d'une partie extracellulaire. Celle-ci est effectivement une lame basale, plus ou moins développée selon les espèces, et contenant souvent des microfibrilles (cette membrane basale constitue en fait le prolongement de la lame basale de l'épiderme céphalique). Quant à la partie cellulaire, elle est formée, du côté distal de cette basale, par l'enchevêtrement de prolongements basaux de cellules cristalliniennes, de cellules pigmentaires accessoires (Odselius & Elofsson, 1981) et peut-être de cellules rétiniennes (Such, 1975); du côté proximal de la lame basale des cellules gliales constituent la face interne de cette "membrane fenestrée".

La structure plus ou moins complexe de cet ensemble reflète le degré d'évolution des Arthropodes considérés. A l'heure actuelle les rôles supposés de cette membrane basale seraient d'une part de servir de support mécanique assurant l'extension des ommatidies, d'autre part de faciliter la nutrition des cellules du cône ou l'action hormonale sur les cellules pigmentaires accessoires.

Chez la plupart des Insectes de nombreuses trachées pénètrent dans l'oeil en traversant la membrane fenestrée, puis se divisent en un très grand nombre de trachéoles qui entourent la base de chaque ommatidie. Chez les Lépidoptères nocturnes (Sphingides et Noctuides), ce réseau très développé réfléchit vers le rhabdome la lumière qui a déjà traversé l'oeil, accroissant ainsi la quantité de lumière, donc la photosensibilité de l'ommatidie.

DEVELOPPEMENT DES YEUX COMPOSÉS

A - Aspect macroscopique:

Chez les Insectes hétérométaboles, les yeux sont présents dès l'éclosion de l'oeil et comptent généralement un petit nombre d'ommatidies. Chez presque tous les Insectes étudiés, l'oeil s'accroît de l'arrière vers l'avant à partir de la région postéro-dorsale de la tête; c'est d'ailleurs tout prêt de cette région que se trouve, chez l'embryon d'Insecte, le territoire qui est à l'origine du lobe optique (Seidel, 1935), et dont la formation s'effectuera par délamination (Hétérométaboles) ou invagination (Holométaboles) de l'ectoderme (Viallanes, 1891; Schoeller, 1964; Malzacher, 1968; Such, 1975). Les yeux occupent alors une position latérale sur la tête, et vont s'accroître progressivement jusqu'au stade adulte.

1) Hétérométaboles

La croissance de l'oeil des Hétérométaboles progresse par augmentation de taille de ses éléments, mais surtout par l'addition de nouvelles ommatidies généralement le long de ses bords dorsal, antérieur et ventral.

Ces ommatidies se différencient au niveau de la "zone d'accroissement" de l'oeil (ou "growth zone"), structure permanente, bien localisée, et de structure voisine chez les différents groupes. Cette zone d'accroissement est limitée extérieurement par une frange de cuticule lisse faisant la transition entre la cuticule plissée du vertex et les rangées irrégulières de cornéules des jeunes ommatidies. Au cours de chaque stade larvaire, des nouvelles rangées d'ommatidies sont formées à ce niveau, et s'intercallent dès la mue suivante entre les ommatidies déjà différenciées et la zone d'accroissement; chez de nombreux Insectes les yeux présentent d'ailleurs des bandes d'ommatidies de teintes légèrement différentes, chacune de ces rangées correspondant aux ommatidies formées au cours d'un même stade larvaire (Volkonsky, 1938; Schaller, 1960, 1964).

Le nombre des ommatidies formées au cours de la vie larvaire varie fortement selon les espèces; ainsi la croissance oculaire, exprimée en nombre d'ommatidies, est beaucoup plus élevée chez les Odonates que dans les autres ordres: par exemple, le nombre d'ommatidies est multiplié par 4 depuis la naissance jusqu'à l'adulte dans l'oeil du Criquet et de la Notonecte (Bernard, 1937), par 12 chez Oncopeltus (Shelton & Lawrence, 1974), par 10 chez Corixa (Young, 1969), par 35 chez Periplaneta (Nowel, 1981), alors que chez Agrion virgo (Odonate Zygoptère) ce nombre est multiplié par 1200 (Richard & Gaudin, 1959) et par 1400 chez Anax (Ando, 1962). La croissance oculaire varie également selon que le stade larvaire est plus ou moins proche du stade adulte et selon la région oculaire considérée. Ainsi, par exemple, chez les Odonates Anisoptères, il a été montré expérimentalement (Lew, 1934) par l'observation de micropiqûres dont la localisation se modifie de stade en stade, que la région dorsale de l'oeil s'accroît fortement au cours des 3 derniers stades larvaires (ces ommatidies formées en dernier correspondent d'ailleurs aux grandes facettes de la partie dorsale de l'oeil de l'imago).

A chaque stade larvaire un certain nombre de nouvelles ommatidies sont donc mises en service (6 rangées d'ommatidies additionnelles par stade chez la Blatte; Nowel, 1981), intercallées entre la zone d'accroissement et les ommatidies formées au stade précédent.

Enfin, tout au long du développement larvaire, les ommatidies précédemment formées vont s'accroître en taille: les cellules augmentent de volume, les facettes de diamètre, et, au niveau des ommatidies différenciées et fonctionnelles quelques mitoses compléteront ou augmenteront le nombre des cellules pigmentaires accessoires.

2) Holométaboles

Chez les Holométaboles les larves sont aveugles, ou bien portent des yeux rudimentaires ou stemmates. La différenciation des futurs yeux composés peut commencer en début de vie larvaire (Hyménoptères et Diptères) ou au contraire au moment de la métamorphose (Lépidoptères et Coléoptères). Les stemmates larvaires dégénèrent généralement en fin de vie larvaire et ne subsistent plus que sous forme de traces pigmentées au niveau de l'épiderme ou du cerveau.

La prolifération cellulaire donnant naissance aux cellules des futures ommatidies, ainsi que la différenciation des ommatidies, ont lieu en plusieurs étapes séparées spacialement à travers l'oeil présomptif. Il n'existe pas véritablement de zone d'accroissement aussi individualisée que ce qui vient d'être décrit: l'épiderme oculaire est parcourue par une 1ère onde de mitoses qui le transforme en une placode épaisse, onde en arrière de laquelle les cellules commencent à s'organiser en groupements préfigurant les ommatidies ("preclusters"); puis vient une seconde onde de mitoses qui complètera ces préommatidies, qui pourront alors achever leur différenciation.

On voit donc que, contrairement à ce qu'il paraissait, les processus de croissance oculaire des Hétéro- et des Holométaboles sont comparables. La seule différence réside dans une chronologie différente des ondes de mitoses: chez les Hétérométaboles les deux phases de multiplication et la phase de différenciation, qui se déroulent pendant toute la vie larvaire, sont très rapprochées dans le temps et dans l'espace, alors que chez les Holométaboles ces deux phases sont relativement étalées.

B - La zone d'accroissement de l'oeil.

1) Structure

L'observation histologique montre que la zone d'accroissement de l'oeil correspond à la succession de différentes régions caractérisant chaque phase de la différenciation de l'oeil, régions qui seront retrouvées chez tous les Insectes Hétérométaboles étudiés; on pourra ainsi définir d'après l'observation en microscopie optique:

0/ - une région formée d'épiderme céphalique typique

1/ - la zone de prolifération, épaississement de l'épiderme, souvent replié en U, et où se produisent de nombreuses mitoses. Cette région présente chez presque tous les Insectes une crête interne de cuticule qui pourrait représenter chez la Blatte la limite entre l'épiderme céphalique et l'épiderme oculaire (Nowel, 1981).
2/ - Une région où les noyaux se disposent en 2 niveaux superposés, les cellules se groupant en colonnes préommatidiennes (ou "clusters").
3/ - Enfin, une région où l'on verra apparaître successivement les cristallins, d'abord fusiformes, puis s'arrondissant, les grains de pigment, les fibres post-rétiniennes etc...

Ces deux dernières régions (2 et 3) où se différencient les ommatidies sont souvent rassemblées en une "zone de maturation". A ce niveau existe encore une légère certaine activité mitotique.

2) Activité mitotique et régulation endocrinienne de la croissance de l'oeil.

La zone de prolifération oculaire est constamment le siège de multiplications cellulaires quelque soit le moment de l'intermue, contrairement à l'épiderme d'autres régions des Insectes qui subit une crise mitotique à un moment bien précis de chaque stade. La pérennité des mitoses, notée chez les Insectes tant Hétéro- qu'Holométaboles, semble être une caractéristique de cette région tégumentaire (Periplaneta, Hyde, 1972; Corixidae, Young, 1969; Sphodromantis, Yamanouti, 1933; Aedes, White, 1961, 1963; Checchi, 1969).

Cependant, le taux mitotique montre un pic de fréquence très élevé en début de chaque intermue larvaire, un niveau relativement élevé pendant une grande partie du stade, puis un arrêt total de l'activité mitotique juste avant la mue. Des courbes d'activité mitotique tout à fait comparables ont été observées chez Aeshna cyanea (Schaller, 1964), Aedes (White, 1961, 1963), Ephestia (Egelhaaf et al., 1975), Schistocerca (Anderson, 1978 a) Periplaneta (Nowel, 1981). (Planche II, a).

Par contre, les régions voisines de cette zone de prolifération, que ce soit l'épiderme céphalique adjacent ou la "zone de maturation" des ommatidies, montrent au cours de chaque intermue larvaire un pattern identique de mitoses, comparable à celui que présente l'épiderme banal de l'insecte, c'est à dire avec une brève crise mitotique au milieu du stade, les cellules ne se multipliant pas aux autres moments.

Il est généralement admis que chez les Insectes l'activité mitotique des cellules épidermiques est contrôlée par les équilibres hormonaux, en particulier par l'hormone de mue ou ecdysone. L'existence de mitoses dans la zone de prolifération tout au long du cycle de mue, suggère comme l'hypothèse a été émise (White, 1961; Schaller, 1964), que les cellules qui la constituent seraient sensibles à des quantités extrêmement faibles de cette hormone. En effet, chez des "larves permanentes" d' Aeshna cyanea, larves dont les glandes de mue contrôlant l'émission d'ecdysone ont été retirées expérimentalement, dont l'hémolymphe ne renferme plus de traces de cette hormone et qui sont donc désormais incapable de muer, les multiplications cellulaires se poursuivent néanmoins dans la zone de prolifération de l'oeil (Mouze & Schaller, 1971).

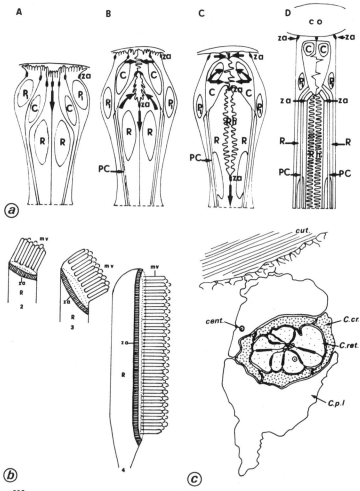

Planche IV:

a) Principales étapes de la mise en place des constituants ommatidiens. (Schémas correspondant à des sections longitudinales parallèles à l'axe de l'ommatidie).

A: Enfoncement des cellules rétiniennes (R) et début d'invagination localisée de la surface externe de la plaque optique. B: Flexion des faces apicales des cellules rétiniennes (R) et constitution de la première ébauche du rhabdome (Rh); flexion et convergence des cellules cristalliniennes (C) vers l'axe ommatidien. C: Convergence des extrémités apicales des cellules pigmentaires primaires (P_1). D: Mise en place définitive des constituants ommatidiens de la larve.
co = cornéule; PC = partie proximale de la cellule rétinienne.
(d'après Such, 1978).

Chez les Hétérométaboles, la structure typique de la zone de prolifération se maintient pendant toute la vie larvaire, puis s'estompe et enfin disparaît au cours du dernier stade larvaire. Les multiplications cellulaires deviennent moins fréquentes puis nulles, la différenciation des ommatidies s'achève dans la zone de maturation, et finalement les ommatidies différenciées de la bordure antérieure et dorsale de l'oeil sont en continuité avec les cellules de l'épiderme céphalique. Chez les Odonates, des expériences d'échanges de greffons entre larves d'âges différents, ou d'injection de mimétiques d'hormone juvénile ont montré que l'évolution de l'oeil dans le sens adulte était, comme les autres tissus des Insectes, sous le contrôle direct du taux de cette hormone juvénile (Mouze, 1971, 1978) (Anderson, 1978, b).

3) <u>Aspect dynamique de la formation de l'ommatidie.</u> (Planche IV)

Bien que l'ultrastructure des yeux composés d'Arthropodes ait fait l'objet de très nombreux travaux, il est étonnant de constater qu'à côté de l'aspect statique de ces descriptions, leur aspect dynamique, c'est à dire celui de la genèse d'une formation aussi spécialisée que l'ommatidie, n'ait pas été plus développé. Si l'on doit à Waddington et Perry (1960), Waddington (1962) les premières descriptions générales de la différenciation des ommatidies de la Drosophile, l'étude des différentes phases conduisant de la multiplication cellulaire à la construction d'un édifice aussi élaboré qu'une ommatidie, en passant par toutes les phases successives de groupement de cellules, différenciation du rhabdome etc... n'ont fait le plus souvent l'objet que de descriptions fragmentaires, les auteurs s'intéressant plutôt à l'évolution d'un organite ou d'un aspect de cette construction.

---

b) Modelage de la cellule rétinienne et origine du rhabdomere.
1: La face apicale de la future cellule rétinienne fléchit légèrement vers l'axe du groupement ommatidien.
2: Le fléchissement s'accentue; les microvilli, plus nombreux, s'orientent vers l'axe ommatidien.
3: Orientation définitive de la cellule rétinienne. Les microvilli sont désormais perpendiculaires à l'axe ommatidien. Le plan contenant la <u>zonula adhaerens</u> (très "étirée") est parallèle à cet axe.

mv: microvillosités - R: cellule rétinienne. z.a. <u>Zonula adhaerens</u> (d'après Such, 1978).

c) Coupe transversale d'un bourgeon ommatidien (coupe légèrement oblique à un niveau proche de la cuticule). On distingue les huit (futures) cellules rétiniennes (C. ret.) entourées de quatre (futures) cellules cristalliniennes (C. cr.), le tout entouré de deux (futures) cellules pigmentaires principales (C.p.I).

Cent.: centriole; cut. cuticule. (d'après Mouze, 1979).

C'est ainsi que le travail le plus complet concernant l'aspect cinétique de la formation des ommatidies est probablement l'étude du développement de l'oeil embryonnaire du phasme effectuée par Such (1975, 1978), et qui servira ici de base: une étude menée sur le développement de l'oeil de la larve d' Aeshna a confirmé et complété les interprétations de Such (Mouze, 1978, 1979), les phases successives de l'élaboration de l'ommatidie étant les mêmes. La différence essentielle concerne la chronologie des phénomènes observés: chez le Phasme Carausius les différentes phases ont été observées à un endroit donné de l'oeil sur des stades embryonnaires successifs, alors que chez Aeshna les mêmes étapes se déroulent de manière synchrone mais chez une larve à des niveaux voisins dans la zone d'accroissement.

Chez l'embryon de Carausius, l'ébauche oculaire initiale n'est représentée aux stades jeunes que par une simple plaque d'épithélium unistratifié, la "plaque optique". Les cellules qui la composent sont alors toutes identiques, leur structure étant celle de cellules ectoblastiques banales, en particulier elles participent à la genèse de la première cuticule embryonnaire par les villosités de leur surface apicale.

Le premier indice du déclenchement de la morphogénèse ommatidienne se manifeste dans la région de la placode se situant au contact du lobe optique sous-jacent qui est alors en formation: on assiste à l'apparition de groupements différenciels de cellules de la plaque optique, condensation en "bourgeons sensoriels" de 2 ou 3 cellules, qui deviendront ultérieurement des cellules rétiniennes. Ces groupes qui au moment de leur formation sont relativement éloignés les uns des autre et semblent "noyés" dans un ensemble de cellules apparemment identiques, présentent une disposition très régulière. Ils semblent en fait constituer de véritables "noyaux de condensation" auxquels s'ajouteront rapidement d'autres cellules. On aura ainsi un faisceau central de 8 futures cellules rétiniennes, chacune d'elles commençant très rapidement à bourgeonner un axone à son extrémité proximale. La face distale de ces cellules se détache alors de la cuticule, s'enfonce progressivement en direction interne et s'incline peu à peu vers l'axe ommatidien. Des microvillosités vont progressivement se développer sur les faces en vis à vis de ces cellules ommatidiennes, qui correspondent donc en réalité aux anciennes faces apicales de ces cellules. Le "caractère apical" de cette surface est d'ailleurs souligné par la Zonula adherens qui s'est déplacée au cours de l'invagination des cellules et qui forme maintenant un anneau très étiré dans un plan devenu parallèle à l'axe ommatidien. L'enfoncement des cellules ommatidiennes va entraîner, à la surface de l'épiderme oculaire, le rapprochement, puis la mise en contact de quatre autres cellules, qui seront à l'origine des cellules cristalliniennes. La poursuite de ce mécanisme de recouvrement éloignera le cristallin de la cuticule, celui-ci étant à son tour recouvert par les deux cellules cornéagènes. (Pl. IV, c). Les futures cellules pigmentaires accessoires qui ne feront jamais partie de ces bourgeons, dérivent des cellules interposées entre les différents faisceaux rétiniens. C'est à partir de ce stade, une fois la préommatidie complétée, que les différents types de cellules qui présentaient tous jusqu'alors un aspect ultrastructural assez semblable, vont progressivement se différencier et acquérir leur structure caractéristique définitive.

Signalons qu'une "ébauche ciliaire" a été observée à l'apex des cellules rétiniennes chez différents insectes (embryon de Carausius, Such, 1969 - nymphes et adultes de Coccinellidae, Carabidae, Home, 1971, 1975, 1976). Des observations comparables ont été effectuées dans un autre groupe d'Arthropodes, chez l'Opilion Ischyropsalia par Juberthie et Muñoz-Cuevas (1973), Muñoz-Cuevas (1975). Ces auteurs pensent que dans cette espèce le centriole distal joue le rôle d'organisateur morphogénétique dans l'édification du rhabdome, dont la différenciation serait de type ciliaire. Cette interprétation contredirait l'hypothèse d'Eakin (1968) selon laquelle l'évolution des photorécepteurs se serait effectuée selon deux voies bien distinctes, ciliaire et rhabdomérique. Pour l'instant, chez les Insectes, aucune observation ne permet d'affirmer que l'ébauche ciliaire joue un rôle dans le développement des microvillosités rhabdomériques.

C - Déterminisme de la croissance oculaire.

1) Relation oeil/lobe optique au cours du développement.

Les nombreux auteurs qui se sont intéressés à la morphogénèse de l'appareil oculaire des Insectes ont souvent été intrigués par les relations s'établissant au cours du développement entre l'oeil et le lobe optique et plus particulièrement par l'interdépendance qui régit la différenciation de ces deux régions. Ce problème qui est aussi un thème d'intérêt classique chez les Vertébrés, a donc suscité chez les Insectes de multiples recherches souvent effectuées par les méthodes les plus variées: si certaines conclusions ont pu être tirées de l'étude des chronologies comparées du développement de l'oeil d'une part, du lobe optique d'autre part, l'essentiel des données a été fourni par l'expérimentation. En dépit d'une approche expérimentale extrêmement variée, autant en ce qui concerne les stades de développement utilisés (embryons ou larves), les espèces étudiées (Hétéro- ou Holométaboles) ou encore les techniques employées (cautérisation, ablation chirurgicale suivie du non de transplantation) les conclusions sont longtemps restées contradictoires, comme en témoignent les revues consacrées périodiquement à ce problème (Chevais, 1937; Bodenstein, 1953; Pflugfelder, 1958; Nüesch, 1968; Edwards, 1969; Meinertzhagen, 1973, 1975).

En effet, Plagge (1936) avait vu que des disques oculaires de Ephestia, implantés avec une partie du ganglion auquel ils sont rattachés, se développent mieux que les disques détachés du système nerveux, résultats que confirment les travaux de Schrader (1938). D'autres auteurs montraient aussi que l'éloignement du système nerveux oculaire provoque des altérations dans le développement des ommatidies (Wolsky & Huxley, 1936 a, b; Wolsky, 1938; Wolsky & Wolsky, 1971 chez Bombyx; Drescher, 1960 chez Periplaneta).

Kopec (1922) avait cependant montré après extirpation du lobe optique et transplantation oculaire chez Lymantria que l'oeil présentait alors un développement normal; de la même manière chez Pentatoma (Pflugfelder, 1936, 1937), Drosophila (Chevais, 1937; Bodenstein, 1940, 1943, 1953; Ephrussi & Beadle, 1937; Steinberg, 1941; Illmensee, 1970), Calliphora (Schoeller, 1964), Periplaneta (Wolbarsht et al., 1966), Ephestia (Woolever & Pipa, 1975) chez les embryons âgés de Carausius (Pflugfelder,

1947; Such, 1975), des ébauches oculaires parfaitement normales ou des yeux privés du système nerveux optique par section ou transplantation, se développent en structures oculaires parfaitement normales. Les résultats de différents types d'opérations pratiquées chez les larves d'Odonates Anisoptères et ayant pour principe de perturber de plus en plus les relations oeil/lobe optique, aboutissaient aux mêmes conclusions, c'est à dire à une indépendance totale de la croissance de l'oeil par rapport au système nerveux sous-jacent (Mouze, 1974).

En effet, la zone d'accroissement de l'oeil, même isolée du lobe optique et greffée, par exemple, en surface d'un sternite abdominal larvaire maintient une activité normale de prolifération et de différenciation des ommatidies; (Des résultats comparables ont été retrouvés chez le criquet par Anderson, 1978 b); on remarquait cependant chez les Odonates une certaine altération des ommatidies différenciées après section des fibres post-rétiniennes.

Il semble que l'on puisse en fait accorder entre eux la majorité de ces résultats, apparemment contradictoires, en invoquant l'existence de deux phases successives au cours du développement oculaire, caractérisées chacune par des relations différentes entre l'oeil et le lobe optique:

- Tout d'abord les quelques travaux portant sur des stades embryonnaires jeunes (Seidel, 1935 sur Platycnemis; Pflugfelder, 1947 et Such, 1975 sur Carausius), ont montré que les ébauches oculaires séparées très tôt du système nerveux sous-jacent sont incapables de se différencier; de plus, lorsque ces très jeunes ébauches oculaires sont transplantées en association avec l'ébauche du lobe optique, seules les régions de la plaque optique en contact étroit avec le système nerveux vont se différencier en ommatidies. Il apparaît donc qu'une proximité immédiate, sinon un contact étroit de l'ébauche oculaire avec celle du lobe optique, est probablement nécessaire pour induire la différenciation des premières ommatidies. Notons d'ailleurs que les territoires embryonnaires très rapprochés ou confondus des ébauches de l'oeil et des deux premiers ganglions optiques, tant chez les Hétérométaboles (Viallanes, 1891; Seidel, 1935; Malzacher, 1968; Such, 1975) que chez les Holométaboles (Schoeller, 1964), seraient en mesure d'assurer une telle induction.

- Par la suite, le développement de l'oeil deviendrait progressivement autonome, une fois passée une certaine période critique dont le moment d'apparition varierait, selon les espèces (hétéro- ou holométaboles) et qui serait conditionnée par la plus ou moins grande précocité du développement oculaire. Chez Carausius, les résultats de Pflugfelder (1947) et de Such (1975) montrent que cette période critique se situe certainement au cours de la vie embryonnaire. Dans le cas des explants oculaires d'Odonates transplantés en position hétérotope, nous avons vu en effet que la zone d'accroissement oculaire persiste, et présente la même activité dans l'oeil opéré que dans l'oeil témoin. La multiplication cellulaire et la différenciation des ommatidies sont donc des propriétés intrinsèques de cette région épidermique et ne sont pas induites par quelque substance émise par le massif d'accroissement externe du lobe optique, comme cela avait précédemment été suggéré (Schaller, 1964).

- Enfin, comme nous l'avons remarqué, la section des fibres post-rétiniennes appartenant à des ommatidies différenciées, semble entraîner une dégénérescence de certains de leurs constituants, à commencer probablement par les cellules rétiniennes. Une telle altération des ommatidies explique sans doute la plupart des résultats anciens qui étaient interprétés comme une dépendance du développement de l'oeil vis à vis du lobe optique au cours de la croissance post-embryonnaire. La dégénérescence des ommatidies dont les fibres post-rétiniennes ont été lésées, phénomène retrouvé chez Calliphora (Schoeller, 1964), Drosophila (Fristrom, 1969), Locusta (Bate & Kien cités par Meinertzhagen, 1973) reste toutefois difficile à expliquer.

- Des résultats obtenus par Anderson (1978 b) sur le criquet ont cependant montré que ce serait certainement davantage la lésion de la membrane basale de l'oeil plutôt que la section des fibres post-rétiniennes qui serait à l'origine de l'altération postopératoire des ommatidies.

Enfin, Trujillo-Cenoz et Melamed (1978) pratiquant des échanges de greffons de disques oculaires entre larves de Drosophiles, ont obtenu des ommatidies anormales possédant un nombre insuffisant de cellules rétiniennes. Ces auteurs attribuent cette déficience au manque de temps qui n'a pas permis aux différentes crises mitotiques de s'effectuer dans ces disques immatures implantés dans des larves prêtes à muer en pupe.

2) Signification de la zone de prolifération.

A la suite de différents travaux relatifs aux Insectes Hétérométaboles (Odonates par Lew, 1934; Notonecte par Lüdtke, 1940; Blatte par Bodenstein, 1953), la zone de prolifération des Insectes Hétérométaboles fut longtemps considérée comme formée de cellules souches capables de proliférer pendant toute la durée du développement larvaire. Pour Bodenstein la zone de prolifération correspondait donc à un reliquat de l'ébauche oculaire embryonnaire.

Par contre, les résultats d'expériences réalisées sur des Insectes Holométaboles (Lépidoptères par Wolsky, 1938; Wolsky & Wolsky, 1971; Wachmann, 1965; Diptères Nématocères par White, 1961, 1963), expliquaient le déterminisme du développement par un mécanisme bien différent; par des expériences de cautérisations localisées, d'ablation de territoires céphaliques, ou d'échanges de greffons d'origine diverse, ces auteurs montrèrent que l'oeil des Holométaboles étudiés se formait par la transformation progressive de l'épiderme céphalique: il existerait d'abord une phase précoce de détermination conférant à l'épiderme céphalique d'un territoire donné la compétence pour être transformé en oeil; puis une "onde de multiplication", prenant son origine dans la région postérieure de la tête, traversait cet épiderme désormais compétent et le transformait en une placode épaissie renfermant de nombreuses cellules; enfin une "onde de différenciation" parcourait cette placode, onde au niveau de laquelle les cellules s'assemblaient en ommatidies. Il avait été montré que ces ondes étaient incapables de traverser un épiderme incompétent (barrage d'épiderme prothoracique greffé par exemple). Les auteurs émirent donc l'hypothèse d'un facteur diffusant à partir d'un "centre différenciateur" qui serait situé dans la tête, au niveau de la région postérieure où apparaissent les premières ommatidies, près des stemmates larvaires. La croissance de l'oeil chez les Insectes Holométaboles correspondrait donc à la progression d'une induction transformant l'épiderme céphalique en oeil.

Il paraissait donc y avoir opposition au sujet du mécanisme de croissance de l'oeil entre les Hétérométaboles (prolifération de cellules souches) et les Holométaboles (inductions successives et transformation de l'épiderme céphalique).

La nature et le fonctionnement de la zone de prolifération chez les Hétérométaboles furent profondément remises en question par les travaux de Hyde (1972) sur la Blatte. Cet auteur, à la suite d'une étude expérimentale basée sur des échanges de territoires entre animaux de type sauvage et mutants (qui différaient par la coloration des yeux) proposa en effet une nouvelle explication: la zone de croissance ne correspondrait pas à la persistance de cellules souches, reliquats embryonnaires de l'oeil larvaire, mais représenterait le niveau où l'épiderme céphalique subirait une influence issue des ommatidies différenciées, influence transformant cet épiderme "banal" en oeil par le jeux d'inductions et de divisions cellulaires successives. C'est l'hypothèse du "recrutement de l'épiderme". Comme preuve de son hypothèse, Hyde décrivait par exemple la transformation en ommatidies d'un fragment d'épiderme prothoracique greffé près de l'oeil.

Cette explication fut aussitôt testée expérimentalement par d'autres auteurs sur différents Insectes Hétérométaboles, et les résultats obtenus purent effectivement être interprétés comme un recrutement de l'épiderme (Green & Lawrence, 1975 chez Oncopeltus; Lawrence & Shelton, 1975; Mouze, 1975). Quelques corrections étaient cependant à apporter sur l'étendue des épidermes susceptibles d'être ainsi transformés en oeil, qui paraissaient beaucoup plus restreints que ce qu'avait annoncé Hyde (Mouze, 1975; Green & Lawrence, 1975). Il fut également proposé, à la suite de différents résultats expérimentaux obtenus chez Aeshna cyanea, que la zone de prolifération pourrait représenter une région de dédifférenciation épidermique causée par le contact des ommatidies et de l'épiderme, la croissance de l'oeil correspondant alors à une sorte de régénération continue de l'oeil tout au long du développement larvaire (Mouze, 1978).

Cependant, des résultats obtenus par Nowel et Shelton (1980) chez Periplaneta remettent fortement en question l'hypothèse du recrutement épidermique, pour en revenir à la notion d'une "budding zone" formée de cellules souches. Cette nouvelle interprétation repose d'une part sur le résultat d'échanges de greffons de forme choisie entre larves de génotypes différents, greffons dont l'évolution est en contradiction avec la théorie du recrutement; d'autre part, tout au long du développement larvaire, la constance du petit nombre de cellules interposées entre la zone de prolifération et les soies céphaliques en bordure de l'oeil, ne peut qu'infirmer la transformation présumée en oeil de cet épiderme. Ces auteurs pensent même que le développement de l'oeil chez les Holométaboles pourrait lui aussi être interprété par l'existence de cellules souches. En fait, les mécanismes fondamentaux du développement oculaire seraient les mêmes chez les Hétéro- et les Holométaboles, les particularités propres à chaque groupe venant de différences au niveau de la chronologie des événements ou au niveau des rapports entre l'épiderme et la cuticule céphalique (Nardi, 1977). Le front de développement des Holométaboles correspondrait en réalité à une vague stationnaire, le déplacement apparent de cette vague de prolifération étant dû à la production et à l'expansion des ommatidies d'un côté de la zone de prolifération, l'épiderme céphalique n'étant jamais transformé en oeil.

Tous les travaux expérimentaux précédents devraient donc être revus et interprétés en fonction de cette dernière explication, mais il apparaît néanmoins que certains résultats seront difficilement expliquables par cette seule hypothèse.

3) Origine des cellules de l'ommatidie.

Comme il a été vu précédemment, une ommatidie est constituée par un ensemble de plusieurs types de cellules très précisément imbriquées les unes dans les autres. En particulier, on y reconnaît un faisceau de 8 (ou plus) cellules rétiniennes, chacune de structure différente et occupant une position bien particulière; cet axe est surmonté par 4 cellules cristalliniennes et 2 cellules cornéagènes, le tout étant entouré d'un grand nombre de cellules pigmentaires accessoires. Il s'agit donc d'un ensemble complexe, orienté, et qui se répète à travers tout l'oeil en gardant la même orientation.

Comment peut s'effectuer la mise en place d'une telle structure?

On admet actuellement que le développement d'un organe peut s'expliquer par deux mécanismes différents (Garcia-Bellido, 1972):

- ou bien ce sont des séquences d'événements génétiquement programmés qui, à la suite de mitoses ségrégatives, donnent naissance à des générations de cellules dont les compétences se réduisent progressivement à l'intérieur de chacun des compartiments qui en dérivent.

- ou bien c'est par une information de position que s'explique la détermination et la différenciation de chaque cellule comme une résultante des informations reçues de son environnement immédiat.

Le mécanisme de la différenciation des ommatidies a été abordé en premier par Bernard (1937) qui décrivait chez la larve de Formicina (Hyménoptère) les différentes étapes de la construction de l'ommatidie: une cellule souche unique, par le jeu de plusieurs cycles de divisions, donnerait naissance à toutes les cellules constituant une ommatidie. Bernard décrivait une ségrégation très précoce des lignées cellulaires à l'origine d'une part des éléments dioptriques et d'autre part des cellules sensorielles. Dans la "lignée dioptrique" les divisions cellulaires produiraient successivement les différents types de cellules cristalliniennes et pigmentaires, et dans la "lignée sensorielle" les multiplications aboutiraient aux 7 (ou 8) cellules rétiniennes. L'ensemble des cellules de chaque ommatidie appartiendrait donc à une seule lignée et serait donc issu par divisions successives, d'une même cellule située au niveau de la bordure de l'oeil. Cette description fut à l'origine de la "théorie clonale" de la construction de l'ommatidie, hypothèse très séduisante qui fut longtemps admise et appliquée à l'ensemble des Insectes, d'autant que les différentes étapes décrites par Bernard étaient difficilement observables en microscopie optique. De plus, cette explication était renforcée par le fait que les différents types cellulaires qui composent de petits organes épidermiques (beaucoup plus simples que l'ommatidie cependant) viennent sans aucun doute d'une même cellule-mère (Wigglesworth, 1953). Cette explication de la genèse de l'ommatidie par la théorie clonale fut une première fois mise en doute par Yagi et Koyama (1963) qui suggérèrent que l'ommatidie de Bombyx n'était pas dérivée d'une cellule unique. Des résultats décisifs furent obtenus par la suite par des auteurs utilisant des techniques de marquage génétique, ou pratiquant des échanges de fragments d'yeux ou d'épiderme entre des souches d'Insectes aux yeux différemment colorés; (Benzer, 1973; Ready et al., 1976; Hofbauer & Campos-Ortega, 1976; Shelton et al., 1977; Anderson, 1978 b). La structure des ommatidies formées au niveau du contact entre les yeux d'origines différentes fut étudiée en microscopie électronique, et l'on s'aperçut alors qu'il pouvait quelquefois se former des "ommatidies chimères" dont toutes les cellules ne présentaient pas le même génotype; dans une même ommatidie, certaines cellules pouvant avoir le génotype du greffon, les autres celui de l'hôte (Pl. II, fig. c). Ce type d'observation infirmait donc complètement l'hypothèse de l'origine clonale des cellules de l'ommatidie puisque celles-ci pouvaient appartenir à deux lignées différentes à l'intérieur d'une même ommatidie.

En fait, il est probable que les nouveaux éléments oculaires sont constamment formés à partir d'une ligne de cellules souches localisées le long de la bordure de l'oeil, et que les ommatidies se développent à partir d'une agrégation au hasard de cellules indifférenciées plutôt qu'à partir d'un groupe de cellules clonalement dérivées; la différenciation des composants résulte de leur position propre dans le faisceau ommatidien (Stark & Mote, 1982). Des études statistiques portant sur l'analyse d'un grand nombre de ces "ommatidies chimères", ainsi que des études par autoradiographie des multiplications cellulaires dans la zone d'accroissement de l'oeil, montrèrent cependant une très forte corrélation de génotype entre certaines cellules rétiniennes. Ainsi, chez la Drosophile par exemple, la première onde mitotique en bordure de l'oeil donnait les cellules rétiniennes n° 2, 3, 4, 5, et 8, puis la seconde les cellules rétiniennes restantes (n° 1, 6, et 7) ainsi que les autres types cellulaires complétant l'ommatidie (Ready et al., 1976; Hofbauer & Campos-Ortega, 1976; Campos-Ortega & Gateff, 1976; Campos-Ortega & Hofbauer, 1977). Ces résultats tendaient ainsi à prouver qu'il existait quand même une certaine filiation reliant certaines cellules de l'ommatidie, même si toutes les cellules n'étaient pas issues d'une même cellule souche. En fait, il semble plutôt que cette fréquence élevée d'un même génotype présentée par certaines cellules voisines ne correspondrait pas à une véritable filiation, mais refléterait plutôt le synchronisme de leur mise en place au sein d'un cluster ommatidien. La position et le type d'une cellule rejoignant une ommatidie en formation seraient conditionnés par le nombre des éléments déjà groupés, les cellules utilisant l'information émise par le faisceau préommatidien préexistant pour déterminer son rôle. Ce mécanisme suppose donc l'existence d'un "noyau de condensation" autour duquel se grouperont ensuite les cellules (fonction qui pourrait, chez la Drosophile, être jouée par la future cellule rétinienne n°8).

Les premières cellules mises en place induiraient les cellules voisines à se déterminer en cellules rétiniennes, selon une séquence bien précise en accord avec leur ordre d'arrivée. Une fois le nombre de 8 cellules rétiniennes atteint, les 4 cellules suivantes recevraient l'information de leur détermination dans le sens "cristallin", puis les deux suivantes dans le sens C.p.I. Les territoires situés entre les différents bourgeons pourraient donc être considérés comme plus ou moins neutres, leurs cellules constitutives devenant les C.p.II intercalées entre les faisceaux cellulaires précédemment déterminés. Il est d'ailleurs probable que les mitoses, observées dans la zone de maturation et même dans la région dioptrique d'ommatidies différenciées et fonctionnelles, correspondent à la formation de cellules pigmentaires accessoires supplémentaires (Marullo & Mouze, 1983).

Rappelons, en outre, que les dégénérescences cellulaires au niveau de l'oeil n'ont été signalées que très rarement, et ne semblent pas faire partie des mécanismes habituels de la construction des ommatidies (Schoeller, 1964; Spreij, 1971).

Tout ceci rendrait plus compréhensible le nombre variable des C.p.II dans l'oeil d'un même insecte (8 à 12 par exemple chez la Drosophile - Hofbauer & Campos-Ortega, 1976), et fournirait un argument supplémentaire contre la théorie de l'origine clonale de l'ommatidie selon laquelle le nombre des cellules est invariablement fixé par des mitoses différentielles. Plusieurs hypothèses ont été émises pour expliquer le mode de transmission de l'information entre cellules de la future ommatidie:

- L'agrégation de cellules homologues nécessite des mécanismes de reconnaissance des cellules et ces déplacements impliquent des différences dans les propriétés d'adhésivité cellulaire. Par exemple, des cellules rétiniennes dissociées à partir de disques imaginaux de Drosophile se séparent des cellules antennaires, se réaggrègent en culture et forment de petits groupes qui pourraient préfigurer des ommatidies (Kuroda, 1970). Par ailleurs, au moment de la formation des ommatidies chez le criquet Schistocerca, Eley et Shelton (1976) ont décrit la succession de différents types de jonctions cellulaires qui apparaissent chacun à un stade particulier de la construction de l'ommatidie. Ces différents types de jonction, qui conditionnent probablement l'adhésion des surfaces et les communications intercellulaires, pourraient jouer un rôle primordial dans la formation des groupements cellulaires conduisant à l'ommatidie.

- Autre hypothèse, émise par Wilson et al., (1978): les cellules rétiniennes pourraient se déterminer en fonction de leurs positions respectives par rapport aux processus cristalliniens qui s'insinuent entre elles.

En résumé, la théorie clonale de la formation de l'ommatidie semble donc devoir être remplacée actuellement par un modèle basé sur l'information de position, où le facteur déterminant la nature des cellules serait le moment de leur incorporation dans le "cluster". On voit donc par cet exemple que même un organe (oeil composé) présentant une architecture très complexe et des structures répétitives ordonnées de façon extrêmement stricte et précise, peut être construit sans division cellulaire ségrégative. Ainsi ce mode de croissance de l'oeil d'Insecte mérite-t-il parfaitement la comparaison avec la croissance d'un cristal, telle que l'a imaginée Benzer (1973).

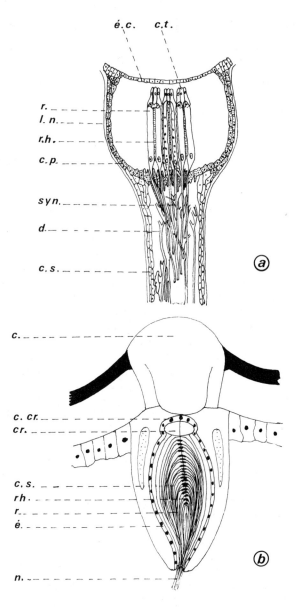

Planche V :

a) Organisation des types cellulaires dans l'ocelle latéral de Libellule (représentation schématique non conforme à l'échelle).

c.p.: cellules pigmentaires - c.s.: cellules non pigmentées de soutien - c.t.: cellules du tapetum - é.c.: épithélium cornéen - d.: dendrite de neurones synaptiques - l.n.: lamelle neurale -r.: cellule rétinienne - rh.: rhabdome - Syn.: contacts synaptiques - (d'après Ruck & Edwards, 1964).

LES YEUX SIMPLES DES INSECTES

Contrairement aux yeux à facettes, dont la structure est relativement constante, les yeux simples présentent des aspects très divers: on distingue en effet, les stemmates, qui sont des récepteurs présents uniquement chez certaines larves (essentiellement Holométaboles), et les ocelles qui existent en plus des yeux composés chez les adultes de différents groupes.

A - Les ocelles (Planche V, a)

Les ocelles sont des yeux simples, au nombre de 3 (deux ocelles latéraux et un médian) situés sur la tête, entre les yeux à facette. Il est difficile d'en donner une description générale, leur structure étant très différente d'un ordre à un autre (voir mise au point de Autrum, 1975): on peut considérer très schématiquement un ocelle sous la forme d'une coupe formée de cellules réfléchissant la lumière, coupe contenant des faisceaux parallèles de cellules photoréceptrices.

En surface, l'ocelle adulte est recouvert d'une cornée de convexité variée, épaisse et translucide. Cette cornée est formée par une couche de cellules cornéagènes plus ou moins épaisse, et d'aspect vitreux. Il n'existe jamais de cristallin dans les ocelles et les cellules pigmentaires peuvent ou non manquer. Le niveau sensoriel sous-jacent est formé par la juxtaposition de faisceau comprenant un nombre variable de cellules sensorielles (2 chez Apis, Toh & Kuwabara, 1974; 2 à 5 chez Periplaneta, Cooter, 1975), le nombre total des cellules photoréceptrices pouvant être assez élevé (800 cellules chez l'ouvrière d' Apis). L'ultrastructure des cellules sensorielles ocellaires rappelle fortement celle des cellules rétiniennes des yeux composés (en particulier on y retrouve des corps multivésiculaires). Chaque cellule forme un rhabdomère qui s'assemble à ceux des cellules voisines dans le même faisceau pour former un rhabdome. Chez Periplaneta chacun des faisceaux est formé de 5 cellules rétiniennes, mais en coupe transversale le nombre de cellules participant au rhabdome est cependant moins grand selon le niveau considéré, certaines cellules se réduisant alors à leur axone. Chez l'abeille, les cellules rétiniennes, (qui sont d'ailleurs pigmentées) étant assemblées par paires, le rhabdome présente donc une structure en deux parties. Contrairement aux cellules rétiniennes des yeux composés, chaque cellule sensorielle ocellaire n'est en contact avec les autres cellules rétiniennes que par son rhabdomère qui présente d'ailleurs en coupe transversale des desmosomes à ses extrémités.

Ces cellules photoréceptrices sont entourées par une couche de cellules renfermant des granules d'acide urique, réfléchissant la lumière:

---

b) Organisation d'un stemmate de la chenille de Gastropacha rubi.

c.: cornée - c. cr.: cellules cristalliniennes - cr.: cristallin - c.s.: cellules de soutien - é.: épithélium - n.: nerf optique - r.: cellules rétiniennes - rh.: rhabdome - (d'après Demoll, 1909).

le tapetum. Les cellules sensorielles émettent chacune un axone qui traverse ce tapetum, ces fibres formant alors un réseau synaptique très complexe avec des neurones de 2ème ordre.

Chez le criquet Schistocerca les ocelles commencent leur développement et acquièrent leur structure définitive dès la période embryonnaire, l'ocelle frontal résultant de la coalescence de deux ébauches paires (Slifer, 1960; Mobbs, 1976, 1979). Ces ocelles se remarquent parfois extérieurement dès les jeunes stades larvaires par leur coloration claire dûe à l'existence d'un tapetum encore rudimentaire, mais renfermant déjà des cristaux réfléchissant (Schistocerca, Aeshna). Ultérieurement, au cours du développement larvaire des crises de multiplication cellulaire en bordure des ocelles y ajouteront des cellules qui se disposeront aux différents niveaux: cornéagène, rétinien ou tapetal. Parallèlement, sous le tapetum, le plexus nerveux augmente progressivement en complexité, et le nombre de synapes par unité de volume s'accroît fortement. Enfin, comme pour l'oeil composé, l'accroissement volumétrique de l'ocelle est également le résultat de l'augmentation en taille de chacune des cellules (Mobbs, 1979).

B - Les stemmates (Planche V, b).

Ce sont des yeux simples, sans facette, qui n'existent que chez les larves d'Insectes Holométaboles. Ces stemmates disparaissent chez l'adulte ou ne subsistent que sous forme d'organes rudimentaires, ou peuvent même parfois persister sans modification, comme chez quelques Aphides et Nématocères.

La structure des stemmates est extrêmement variable d'un ordre à un autre, et peut aller de quelques cellules photoréceptrices situées dans des évaginations pharyngiennes (Calliphora, Schoeller, 1964), jusqu'à composer des photorécepteurs complexes constituant de véritables organes visuels (voir mise au point Goodman, 1970, 1974). Si l'on excepte les cas les plus simples, la structure générale d'un stemmate, contrairement à l'ocelle, présente de fortes analogies avec celle de l'ommatidie. En effet, on y distingue distalement un appareil dioptrique à deux niveaux: une cornée généralement bombée recouvrant un cristallin formé d'un nombre variable selon les ordres de cellules (3 à 8), cristallin lui-même entouré de cellules cornéagènes. Sous ce niveau dioptrique, le niveau photorécepteur est constitué de cellules rétinuliennes allongées formant un rhabdome stratifié, situé dans l'axe optique du stemmate, et de structure assez régulière. Chez certaines espèces comme Perga, existent plusieurs rhabdomères parallèles (Meyer-Rochow, 1974). Chez Papilio xanthus, les cellules rétiniennes, qui renferment des granules pigmentaires sont souvent disposées en plusieurs couches d'inégale importance, ce qui confère alors au rhabdome une forme différente à chacun de ses niveaux (Toh & Sagara, 1982). Ces cellules sont d'ailleurs capables d'une certains adaptation à la lumière (White & Lord, 1975). Comme dans les yeux composés, les fibres sensorielles ne subissent pas de relai au niveau des stemmates, et rejoignent directement les lobes optiques.

Du point de vue optique, contrairement aux ocelles où les images sont formées loin en dessous du niveau sensoriel, dans les stemmates l'image est focalisée sur le rhabdome. Les ressemblances existant entre ommatidies et stemmates suggèrent que ceux-ci, bien qu'étant des photorécepteurs rudimentaires et généralement groupés en petit nombre de chaque côté de la tête, peuvent être responsables de différentes fonctions visuelles comme le sont les yeux composés. En fait, les stemmates de tous les holométaboles seraient homologues et auraient évolué à partir d'un oeil composé typique présent dans les stades jeunes de quelques espèces ancestrales (Heming, 1982).

REFERENCES

Anderson, H. (1978 a) Postembryonic development of the visual system of the locust Schistocerca gregaria. I. Patterns of growth and interactions in the retina and optic lobe. J. Embryol. Exp. Morph. 45: 55-83.

Anderson, H. (1978 b) Postembryonic development of the visual system of the locust Schistocerca gregaria. II. An experimental investigation of the formation of the retina-lamina projection. J. Embryol. Exp. Morph. 46: 147-170.

Ando, H. (1962) The comparative embryology of Odonata, with special reference to a relic dragonfly Epiophlebia superstes Selys. Jap. Soc. for the Promotion of Sciences, 205-247.

Autrum, H. (1975) Les yeux et la vision des Insectes. In: Traité de Zoologie, Paris, Masson, VIII, 3: 742-853.

Barra, J.A. (1969) Les photorécepteurs des Collemboles. Nouvelles formations à structure rhabdomérique propres au genre Tomocerus (Insecte Collembole). C.R. Acad. Sci., Fr. 268: 2088-2090.

Benzer, S. (1973) Genetic dissection of behavior. Sci. Am. 229: 24-37.

Bernard, F. (1937) Recherche sur la morphogenèse des yeux composés d'Arthropodes (Développement, Croissance, Réduction). Bull. Biol. Fr. Belg. 23: 1-162.

Bernhard, C.G., Gemme, G. & Sällstrom, J. (1970) Comparative ultrastructure of corneal surface topography in insects with aspects on phylogenesis and function. Z. vergl. Physiol. 67: 1-25.

Bodenstein, D. (1940) Growth regulation of transplanted eye and leg discs in Drosophila. J. Exp. Zool. 84: 23-37.

Bodenstein, D. (1943) A study of the relationship between organ and organic environment in the postembryonic development of the yellow fever mosquito. Conn. Agr. Sta. Bull. 501: 100-114.

Bodenstein, D. (1953) Postembryonic development. In: Insect Physiology. Ed. K.D. Roeder, New York, Wiley, p. 822-865.

Brammer, J.D. (1970) The ultrastructure of the compound eye of a mosquito Aedes aegypti L. J. Exp. Zool. 175: 181-196.

Brammer, J.D. & Clarin, B. (1976) Changes in volume of the rhabdom in the compound eye of Aedes aegypti L. J. Exp. Zool. 195: 33-40.

Burghause, F. (1976) Adaptationserscheinungen in den Komplexaugen von Gyrinus natator L. (Coleoptera: Gyrinidae). Int. J. Insect Morphol. Embryol. 5: 335-348.

Burton, P.R. & Stockhammer, K.A. (1969) Electron microscopic studies of the compound eye of the Toadbug Gelastocoris oculatus. J. Morph. 127: 233-258.

Campos-Ortega, J.A. & Gateff, E.A. (1976) The development of ommatidial patterning in metamorphosed eye imaginal discs implants of Drosophila melanogaster. Wilhelm Roux Archiv. Entw. Mech. Org. 179: 373-392.

Campos-Ortega, J.A. & Hofbauer, A. (1977) Cell clones and pattern formation: on the lineage of photoreceptor cells in the compound eye of Drosophila. Wilhelm Rouw Archiv. Entw. Mech. Org. 181: 227-245.

Checchi, A.C. (1969) A Quantitative Analysis of the Compound Eye Development in the Mosquito Aedes aegypti L. Thesis, Purdue University. 85 pages.

Chevais, S. (1937) Sur la structure des yeux implantés de Drosophila melanogaster. Arch. Anat. Micros. 33: 107-112.

Cooter, R.J. (1975) Ocellus and ocellarnerves of Periplaneta americana (Orthoptera: Dictyoptera). Int. J. Insect Morphol. Embryol. 4: 273-288.

Cosens, F. (1976) The effect of short wavelength light on retinula cell structure in white-eye Drosophila. J. Insect. Physiol. 22: 497-504.

Demoll, R. (1909) Uber eine lichtzersetzliche substanz im Facettenauge sowie eine Pigmentwanderung im Appositionsauge. Arch. ges. Physiol. 129: 461-475.

Drescher, W. (1960) Regenerationsversuche am Gehirn von Periplaneta americana. Z. Morph. Oekol. Tiere 48: 576-649.

Eakin, R.M. (1968) Evolution of photoreceptors. Evol. Biol. 2: 194-242.

Edwards, J.S. (1969) Postembryonic development and regeneration of the insect nervous system. Adv. Insect Physiol. 6: 98-137.

Egelhaaf, A., Berndt, P. & Küthe, H.W. (1975) Mitosenverteilung und H^3-thymidin-Eibau in der proliferierenden Augenanlage von Ephestia kuehniella Zeller. Wilhelm Roux. Arch. Entw. Mech. Org. 178: 185-202.

Eley, S. & Shelton, P.M.J. (1976) Cell junctions in the developing compound eye of the desert locust Schistocerca gregaria J. Embryol. Exp. Morph. 36: 409-423.

Ephrussi, B. & Beadle, G.B. (1937) Revue des expériences de transplantation. Bull. Biol. Fr. Belg. 71: 54-74.

Fristrom, D. (1969) Cellular degeneration in the production of some mutant phenotypes in Drosophila melanogaster (Diptera, Drosophilidae). Mol. Gen. Genet. 103: 363-379.

Friza, F. (1928) Zur Frage der Färbung und Zeichnung des facettierten Insektenauges. Z. vergl. Physiol. 8: 289-336.

Fyg, W. (1961) Uber die Kristallkegel in den Komplexaugen der Honigbiene (Apis mellifica L.). Mitt. Schweiz. ent. Ges. 33: 185-194.

Garcia-Bellido, A. (1972) Pattern formation in imaginal disks. In: The Biology of Imaginal Disks. Ed. H. Urspung & R. Nüthinger. Berlin, Springer Verlag, p. 59-91.

Goodman, L.J. (1970) The structure and function of the insect dorsal ocellus. Adv. Insect Physiol. 7: 97-195.

Goodman, L.J. (1974) The neural organization and physiology of the insect dorsal ocellus. In: The Compound Eye and Vision of Insects.

Ed. G.A. Horridge. Oxford, Oxford University Press, p. 515-548.
Green, S.M. & Lawrence, P.A. (1975) Recruitement of epidermal cells by the developing eye of Oncopeltus (Hemiptera). Wilhelm Roux Archiv. Entw. Mech. Org. 177: 61-65.
Grenacher, H. (1879) Untersuchungen über das Sehorgan der Arthropoden. Göttingen Vanderhoeck et Ruprecht, p. 645-656.
Heming, B.S. (1982) Structure and development of the larval visual system in embryos of Lytta viridana Leconte (Coleoptera, Meloidae). J. Morphol. 172: 23-43.
Hofbauer, A., Campos-Ortega, J.A. (1976) Cell clone and pattern formation. Genetic eye mosaics in Drosophila melanogaster. Wilhelm Roux Archiv. Entw. Mech. Org. 179: 275-289.
Home, E.M. (1972) Centrioles and associated structures in the retinula cells of insect eyes. Tissue Cell 4: 227-234.
Home, E.M. (1975) Ultrastructural studies of development and light-dark adaptation of the eye of Coccinella septempunctata L., with particular reference to ciliary structures. Tissue Cell 7: 703-722.
Home, E.M. (1976) The fine structure of some carabid beetles eyes with particular reference to ciliary structures in the retinula cells. Tissue Cell 8: 311-324.
Horridge, G.A. (1966) The retina of the locust. In: Functional Organization of the Compound Eye. Ed. C.G. Bernhard. Oxford, Pergamon Press. Wenner Gren Center Internat. Symp. Series 7: 513-541.
Horridge, G.A. (1968) Pigment movement and the crystalline threads of the firefly eye. Nature (Lond.) 218: 778-779.
Hyde, C.A.T. (1972) Regeneration, post-embryonic induction and cellular interaction in the eye of Periplaneta americana. J. Embryol. Exp. Morph. 27: 367-379.
Illmensee, K. (1970) Imaginal structures after nuclear transplantation in Drosophila melanogaster. Naturwissenschaften 11: 550-551.
Juberthie, C. & Muñoz-Cuevas, A. (1973) Présence de centriole dans la cellule visuelle de l'embryon d' Ischyropsalis luteipes (Arachnides: Opilions). C.R. Acad. Sci. Paris 276: 2537-2539.
Kim, C.W. (1964) Formation and histochemical analysis of the crystalline cone of compound eye in Pieris rapae L. (Lepidoptera). Korean J. Zool. 7: 89-94.
Kobakhidze, D.N., Sicharylidze, T.A. & Svanidze, I.K. (1959) The effects of ecological conditions on the structure of the visual apparatus in orthopterans. Soobsh. AN Gruz. Z.Z.Z.E. 22: 569-579.
Kopec, S. (1922) Mutual relationship in the development of the brain and eyes of Lepidoptera. J. Exp. Zool. 36: 459-465.
Kuroda, Y. (1970) Differentiation of ommatidium-forming cells of Drosophila melanogaster in organ culture. Exp. Cell Res. 59: 429-439.
Laughlin, S. & McGinness, S. (1978) The structures of dorsal and ventral regions of a Dragonfly retina. Cell Tissue Res. 188: 427-447.
Lawrence, P.A. & Shelton, P.M.J. (1975) The determination of polarity in the developing insect retina. J. Embryol. Exp. Morph. 33: 471-486.
Lew, G.T.W. (1934) Head characters of the Odonata with special

reference to the development of the compound eye. Entomol. Americana 14: 41-73.

Lüdtke, H. (1940) Die embryonale und postembryonale Entwicklung des Agues bei Notonecta glauca (Hemiptera: Heteroptera). Zeitsch. Morph. Oekol. Tiere 37: 1-37.

Malzacher, P. (1968) Die Embryogenese des Gehirns paurometaboler Insekten. Untersuchungen an Carausius morosus und Periplaneta americana. Z. Morph. Tiere 62: 103-161.

Marullo, C. & Mouze, M. (1983) Etude du développement de l'appareil visuel chez les Odonates Zygoptères. (en préparation).

Meinertzhagen, I.A. (1973) Development of neuronal connection patterns in the visual systems of insects. In: Developmental Neurobiology of Arthropods. Ed. D. Young. London, New York, Cambridge University Press, p. 51-104.

Meinertzhagen, I.A. (1975) The development of neuronal connection patterns in the visual systems of insects. In: Cell Patterning, Ciba Foundation Symposium. Amsterdam, Oxford, New York, Elsevier, 29: 265-288.

Melnichenko, A.N. (1963) Geographic variability of the honeybee eye. Entomol. Obzr. 42: 118-126.

Meyer-Rochow, V.B. (1974) Structure and function of the larval eye of the Sawfly, Perga (Hymenoptera). J. Insect Physiol. 20: 1565-1591.

Miller, W.H., Bernard, G.D. & Allen, J.L. (1968) The optics of insect compound eyes. Science 162: 760-767.

Mobbs, P.G. (1976) Development of the locust ocellus. Nature (Lond.) 264: 269-271.

Mobbs, P.G. (1979) Development of the dorsal ocelli of the desert locust, Schistocerca gregaria Forsk (Orthoptera: Acrididae). Int. J. Insect Morphol. Embryol. 8: 237-255.

Mouze, M. (1971) Rôle de l'hormone juvénile dans la métamorphose oculaire de larves d' Aeshna cyanea Müll. (Insecte Odonate). C. R Acad. Sci. Paris 273: 2316-2319.

Mouze, M. (1974) Interactions de l'oeil et du lobe optique au cours de la croissance post-embryonnaire des Insectes Odonates. J. Embryol. Exp. Morphol. 31: 377-407.

Mouze, M. (1975) Croissance et régénération de l'oeil de la larve d' Aeshna cyanea Müll. (Odonate, Anisoptère). Wilhelm Roux Arch. Entw. Mech. Org. 176: 267-283.

Mouze, M. (1978) Contribution à l'Étude du Développement Post-Embryonnaire de l'Appareil Visuel des Odonates Anisoptères (Insectes). Thèse Doctorat Etat, Lille. 130 pages.

Mouze, M. (1979) Etude cytologique de la genèse ommatidienne chez la larve d'un Odonate Anisoptère. Rev. Can. Biol. 38: 227-248.

Mouze, M. & Schaller, F. (1971) Métamorphose oculaire de larves d' Aeshna cyanea Müll. (Insecte, Odonate) privées d'ecdysone. C.R. Acad. Sci. Paris 273: 2122-2125.

Muñoz-Cuevas, A. (1975) Aspects ultrastructuraux de la différenciation et de l'organisation de la rétine chez les Opilions (Arachnida). Proc. 6th Int. Arachn. Congr. 1974: 129-132.

Nardi, J.B. (1977) The construction of the Insect compound eye: the involvement of cell displacement and cell surface properties in the positioning of cells. Dev. Biol. 61: 287-298.

Nowel, M.S. (1981) Postembryonic growth of the compound eye of the cockroach. J. Embryol. Exp. Morph. 62: 259-275.

Nowel, M. & Shelton, P.M.J. (1980) The eye margin and compound eye development in the cockroach: evidence against recruitement. J. Embryol. Exp. Morph. 60: 329-343.

Nüesch, H. (1968) The role of the nervous system in insect morphogenesis and regeneration. Ann. Rec. Entomol. 13: 27-44.

Odselius, R. & Elofsson, R. (1981) The basement membrane of the Insect and Crustacean compound eye: Definition, fine structure, and comparative morphology. Cell Tissue Res. 216: 205-214.

Perrelet, A. & Baumann, F. (1969 a) Presence of small retinula cell in the ommatidium of the honeybee drone eye. J. Microsc. 8: 497-502.

Perrelet, A. & Baumann, F. (1969 b) Evidence for extracellular space in the rhabdome of the honeybee drone eye. J. Cell Biol. 40: 825-830.

Pflugfelder, O. (1936) Vergleichend anatomische, experimentelle und entwicklungs-Geschichtliche Untersuchungen über das Nervensystem und die Sinneorgane der Rhynchoten. Z. Wiss. Zool. 93.

Pflugfelder, O. (1937) Die Entwicklung der optischen Ganglien von Culex pipiens. Zoologischer Anzeiger 117: 31-36.

Pflugfelder, O. (1947) Die Entwicklung embryonaler Teile von Dixippus morosus in der Kopfkapsel von Larven und Imagines. Biol. Zbl. 66: 372-387.

Pflugfelder, O. (1958) Entwicklungsphysiologie der Insekten. Akad. Verlagsgesellschaft Leipzig. Geest & Portig K.G. 1-490.

Plagge, E. (1936) Transplantation von Augenimaginalscheiben Swischen der der schwarz und rotäugigen Rasse von Ephestia kühniella. Z. Biol. Zb. 56: 406-409.

Ready, D.F., Hanson, F.E. & Benzer, S. (1976) The development of the Drosophila retina, a neurocristalline lattice. Develop. Biol. 53: 217-240.

Ribi, W.A. (1978) Ultrastructure and migration of screening pigments in the retina of Pieris rapae L. (Lepidoptera Pieridae). Cell Tissue Res. 191: 57-73.

Richard, G. & Gaudin, G. (1959) La morphologie du développement du S.N. chez divers Insectes. Cas plus particulier des centres et des voies optiques. Acta Symposium de Evolutione insectorium Praha, p. 82-88.

Ruck, Ph. & Edwards, G.A. (1964) Structure of the Insect dorsal ocellus. I: General organization of the ocellus in Dragonflies. J. Morph. 115: 1-26.

Schaller, F. (1960) Etude du développement post-embryonnaire d' Aeshna cyanea. Müll. Ann. Sci. Nat. Zool. 12: 755-868

Schaller, F. (1964) Croissance oculaire au cours des développements normaux et perturbés de la larve d' Aeshna cyanea Müll. (Insecte Odonate). Ann. Endocr. Paris 25: 122-127.

Schoeller, J. (1964) Recherches descriptives et expérimentales sur la céphalogenèse de Calliphora erythrocephala au cours des développements embryonnaires et post-embryonnaires. Arch. Zool. Exp.

Gen. 103: 216.
Schrader, K. (1938) Untersuchungen über die Normalentwicklung des Gehirns und Gehirntransplantationen bei der Mehlmotte Ephestia kühniella nebst einigen Bemerkungen über das Corpus allatum. Biol. Zbl. 58: 51-90.
Seidel, F. (1935) Der Anlagenplan im Libellenei. Wilhelm Roux Arch. Entw. Mech. Org. 132: 671-751.
Seitz, G. (1968) Der Strahlengang im Appositiongauge von Calliphora erythrocephala (Meig.). Z. vergl. Physiol. 59: 205-231.
Shelton, P.M.J. (1976) The development of the insect compound eye. In: Insect Development. Ed. P.A. Lawrence. Oxford, Blackwell, p. 152-169.
Shelton, P.M.J., Anderson, H.J. & Eley, Z. (1977) Cell lineage and cell determination in the developing compound eye of the cockroach Periplaneta americana. J. Embryol. Exp. Morph. 39: 235-252.
Shelton, P.M.J. & Lawrence, P.A. (1974) Structure and development of ommatidia in Oncopeltus fasciatus J. Embryol. Exp. Morph. 32: 337-353.
Sherk, T.E. (1977) Development of the compound eyes of dragonflies (Odonata). I. Larval compound eyes. J. Exp. Zool. 201: 391-416.
Sherk, T.E. (1978 a) Development of the compound eyes of dragonflies (Odonata). II. Development of the larval compound eyes. J. Exp. Zool. 203: 47-60.
Sherk, T.E. (1978 b) Development of the compound eyes of dragonflies (Odonata). III. Adult compound eyes. J. Exp. Zool. 203: 61-80.
Sherk, T.E. (1978 c) Development of the compound eyes of dragonflies (Odonata). IV. Development of the adult compound eyes. J. Exp. Zool. 203: 183-200.
Slifer, E.H. (1960) An abnormal grasshopper with two mediam ocelli (Orthoptera: Acrididae). Ann. Entomol. Soc. Am. 53: 441-443.
Spreij, T.E. (1971) Cell death during the development of the imaginal disks of Calliphora erythrocephala. Netherlands J. Zool. 21: 221-264.
Stark, R.J. & Mote, M.I. (1982) Postembryonic development of the visual system of Periplaneta americana. I. Patterns of growth and differentiation. J. Embryol. Exp. Morph. 66: 235-255.
Steinberg, A. (1941) A reconsideration of the mode of development of the bar eye of Drosophila melanogaster. Genetics 26: 325-346.
Such, J. (1969) Contribution à l'étude histochimique et infrastructurale du "cristallin" dans l'ommatidie du phasme Carausius morosus Br. C. R. Acad. Sci., Paris 268: 948-949.
Such, J. (1975) Recherches Descriptives et Expérimentales sur la Morphogenèse Embryonnaire de l'Oeil Composé du Phasme Carausius morosus Br. Thèse de Doctorat (Etat), Université de Bordeaux, p. 1-127.
Such, J. (1978) Embryologie ultrastructurale de l'ommatidie chez le phasme Carausius morosus Br. (Phasmida: Lonchodidae): morphogenèse et cytodifférenciation. Int. J. Insect Morphol. Embryol. 7: 165-173.
Toh, Y. & Kuwabara, M. (1974) The fine structure of the dorsal ocellus of

the worker honeybee. J. Morphol. 143: 285-306.

Toh, Y. & Sagara, H. (1982) Ocellar system of the Swallowtail butterfly larva. I. Structure of the lateral ocelli. J. Ultrastruct. Res. 78: 107-119.

Trujillo-Cenoz, O. & Melamed, J. (1978) Development of photoreceptor patterns in the compound eyes of muscoid flies. J. Ultrastruct. Res. 64: 46-62.

Viallanes, H. (1891) Sur quelques points de l'histoire du développement embryonnaire de la mante religieuse. Ann. Sci. Nat., 7ème Sér. 11: 282-328.

Volkonsky, M. (1938) Sur la formation des stries oculaires chez les Acridiens. C. R Soc. Biol. 129: 154-157.

Wachmann, E. (1965) Untersuchungen zur Entwicklungsphysiologie des Komplexauges der Wachsmotte Galleria mellonella. Wilhelm Roux Arch. Entw. Mech. Org. 156: 145-183.

Waddington, C.H. (1962) Specificity of ultrastructure and its genetic control. J. Cell Comp. Physiol. 60: 93-103.

Waddington, C.H. & Perry, M.M. (1960) The ultrastructure of the developing eye of Drosophila. Proc. R. Soc. 153B: 155-178.

Walcott, B. (1969) Movement of retinular cells in insect eyes on light adaptation. Nature (Lond.) 223: 971-972.

White, R.H. (1961) Analysis of the development of the compound eye in the mosquito Aedes aegypti. J. Exp. Zool. 148: 223-240.

White, R.H. (1963) Evidence for the existence of a differentiation center in the developing eye of the mosquito. J. Exp. Zool. 152: 139-148.

White, R.H. (1967) The effect of light and light deprivation upon the ultrastructure of the larval mosquito eye. II. The rhabdom. J. Exp. Zool. 166: 405-425.

White, R.H. & Lord, E. (1975) Diminution and enlargement of the mosquito rhabdom in light and darkness. J. Gen. Physiol. 65: 583-598

Wigglesworth, V.B. (1953) The origin of sensory neurones in an insect, Rhodnius prolixus (Hemiptera). Quart. J. Microsc. Sci. 94: 93-112.

Wilson, M., Garrard, P. & Mc Guiness, S. (1978) The unit structure of the locust compound eye. Cell Tissue Res. 195: 205-226.

Wolbarsht, M.L., Wagner, H.G. & Bodenstein, D. (1966) Origin of electrical responses. In: The Functional Organization of the Compound Eye. Ed. C.G. Bernhard. Oxford, New York, Pergamon Press. Wenner Gren Symp. 7: 207-217.

Wolsky, A. (1938) Experimentelle Untersuchungen uber die Differenzierung der zusammengesetzen Augen des Seidenspinners (Bombyx mori L.). Wilhelm Roux Arch. Entw. Mech. Org. 138: 335-344.

Wolsky, A. & Huxley, J.S. (1936 a) The structure of the non-facetted region in the Bar-eye mutants of Drosophila and its bearing of the analysis of genic action upon Arthropodan eyes. Proc. Zool. Soc. Lond. 2: 485-489.

Wolsky, A. & Huxley, J.S. (1936 b) Zur Frage der Entwicklungsphysiologischen Determination des Arthropodenauges. Biol. Zentralbl. 56: 571-572.

Wolsky, A. & Wolsky, M. (1971) Phase specific and regional differences in the development of the complex eye of the mulberry silkworm (Bombyx mori L.) after unilateral removal of the optic lobe of the brain in early pupal stages. Am. Zool. 11: 679.

Woolever, P. & Pipa, R.L. (1975) Eye disk differentiation in the wax moth. Induction in vitro. J. Exp. Zool. 191: 359-382.

Yagi, N. & Koyama, N. (1963) The Compound Eye of Lepidoptera. Approach from Organic Evolution. Tokyo, Shinkys Press Ltd. 1-230.

Yamanouti, T. (1933) Waschstumsmessungen an Sphodromantis bioculata Burn. V. Bestimmung der absoluten Zenahmswerte der Facettengrosse und Facettenanzahl Anz. Akad. Wiss (Wien) 70: 7-8

Young, E.C. (1969) Eye growth in Corixidae (Hemiptera, Heteroptera) Proc. R. Ent. Soc. Lond. 44: 71-78.

MOLLUSCS

M.F. LAND

School of Biological Sciences
University of Sussex, Falmer
Brighton BN1 9QG, England

ABSTRACT

The molluscs show every type of eye design from the most simple of eye-cups in limpets to the fish-like lens eyes of squid and octopus, which are every bit as sophisticated as their vertebrate parallels. Even in the bivalves, not a group renowned for their eyesight, there have been a number of curious evolutionary experiments. Thus in scallops one finds eyes that use mirrors rather than lenses to form the image, and in Arca and its relatives the mantle bears many small compound eyes not very different from the apposition eyes of insects or sabellid tube worms. These are essentially evolutionary "one-offs" which led nowhere, but which provided their bearers with the ability to respond to moving predators before they are close enough to cast a direct shadow.

Shadow responses, which one finds in bivalve clams and some gastropods, turn out to be generated in a way that is possibly unique in the animal kingdom, in that the photoreceptors that mediate them are primary "off" receptors which hyperpolarise in the light and discharge when the light dims or is turned off. It is ironic that this was discovered in Pecten by Hartline in 1938 30 years before it was suspected that vertebrate photoreceptors might behave in a not very different way. The more usual kind of depolarising receptor is also found, usually in the true cephalic eyes of both gastropods and cephalopods. The "off" receptors tend to be highly ciliated and the "on" receptors microvillous, though what this means physiologically is still an interesting open question

In both gastropods and cephalopods the main line of evolutionary development has been in the direction of optically high quality lens eyes. Snails like Helix have only a soft, weakly refracting lens, but in the winkle Littorina this has become hard and inhomogeneous, and has a ratio of focal length to radius of about 2,5 (Matthiessen's ratio) characteristic of the aplanatic lenses of fish eyes. In the strombids this kind of eye can be quite large - a mm or more - with potentially excellent resolutions ($\frac{1}{4}°$), though no-one knows what the eyes are used for. The heteropods are even more intriguing, having fish-like eyes with long narrow retinae only a few cells wide. It now seems that they make scanning eye movements, sweeping through the visual surroundings so that the narrow retina acts as a single like on the TV scan.

Parallels between cephalopod and fish eyes are legendary. They both have Matthiessen's ratio lenses, mobile pupils and a full complement of eye-movements. The receptors, though, are quite different, and there is no suggestion of colour vision in cephalopods or any other mollusc. Nautilus is a real oddity. How did it last that long with only pinhole optics?

INTRODUCTION

The fossil record is particularly good in the molluscs, because of their hard shells, and it indicates that the 3 major classes (Bivalvia, Gastropoda and Cephalopoda) separated from each other in the Cambrian about 600 million years ago, (Yochelson, 1979). Thus, unlike the vertebrates where evolutionary changes have occurred frequently and fairly obviously over the last 400 million years, in the molluscs we are dealing with a phylum that was set in its ways much earlier, and we should not look too hopefully for signs of evolutionary progression, certainly in terms of eye design, between one class and another. I point this out because it is apparently easy to devise a conceptual series of eyes from, say, the pigmented pits of a limpet through the lens eyes of sea snails like Littorina or Strombus to the wonderfully sophisticated fish-like eyes of the cephalopods, and indeed many text-books show such diagrams. If limpets needed better eyes they have had a very long time to acquire them: we are not therefore really looking at an evolutionary progression but at solutions perfected long ago to the problems posed by the different ways of life of the animals concerned. Some eyes are undoubtedly more "advanced" than others, but existing species yield no more than the barest clues as to where or how they originated.

Fortunately there is an excellent recent review of vision in the Mollusca (Messenger, 1981) which covers all aspects of the subject thoroughly and this means that I can use this chapter to explore a small number of themes and not attempt an exhaustive review. I shall begin with a short survey of the types of eye found in molluscs, then discuss the receptor bases of vision in the phylum which are odd and possibly unique in that some molluscs possess both hyperpolarising and depolarising receptors. This leads to a section of some curious "on-off" optical experiments in the bivalves, notably the compound eyes of Arca and the mirror eyes of scallops, and then to a discussion of what probably was a major thrust of visual evolution in the molluscs - the development of high resolution lens eyes, looking first at the gastropods and then the cephalopods. When describing advanced cephalopod eyes it is impossible not to be drawn into comparisons with the eyes of their rivals in the sea the fishes, and I will conclude with a section on the possible reasons why such an extraordinary convergence should have occurred.

TYPES OF EYE

Most molluscan classes show evidence of two distinct visual systems, one based on a pair of cephalic eyes in the head, and another system distributed over the body. In bivalves there is no distinct head and in only a few genera (e.g. Mytilus) are there any signs of cephalic eyes (Rosen et al., 1978). On the other hand, the extraocular system, which in most molluscs consists of largely unidentified dermal receptors, has blossomed in a few bivalves into sets of structures that must be called eyes because they possess large numbers of receptors and optical arrangements for splitting up the light reaching them according to its direction of origin. A summary of the different types of mollusc eye is given in Table I.

Cephalic Eyes

The three least specialised existing classes of mollusc, the worm-like Aplacophora, the relict Monoplacophora (e.g. Neopilina) and the Polyplacophora (chitons) are all without cephalic eyes, although the chitons have a large number of small ocelli embedded in their shells. Most gastropods, on the other hand, have a single pair of eyes on the head, either close to the cerebral ganglion or on the end of a mobile stalk, as in the pulmonate land snails like Helix. These eyes vary enormously in their size, the extent to which their optical structure permits good image formation, and in the numbers of receptors present. The nudibranchs Hermissenda and Tritonia have only 5 receptors in each eye, and the eye of Hermissenda is only 80 µm across (Dennis, 1967). At the other extreme the eyes of the large marine prosobranch Strombus luhuanus may be 2 mm in diameter, with up to 50 000 receptors (Gillary & Gillary, 1979). The pulmonate land snail Helix is intermediate, with an eye up to 1 mm and about 4 000 receptors (for comparison, Octopus has about 20 million receptors per eye).

The optical arrangements vary markedly within the gastropods, but they all fall somewhere along a continuum from a pigmented pit without a lens of any kind (Patella, Haliotis), to forms where there is a lens, but it is a soft low refractive index structure unable on its own to form a sharp image (Helix, Fig. 1), to eyes with hard spherical lenses and undoubted image-forming capabilities. Small eyes of the last kind are found in both prosobranchs (Littorina, Newell, 1965) and pulmonates (Lymnea, Stoll, 1973). Large eyes that really look as though they would provide their owners with resolution better than 1° only occur in two gastropod groups, the strombids and the heteropods (Pterotrachea, Carinaria). The latter group have very strange eyes indeed (Fig. 2, and see Hesse, 1900), with long narrow retinae and receptors that differ from the usual rhabdomeric type in possessing numerous cilia (Dilly, 1969). The heteropods have an excuse for their large eyes; they are pelagic carnivores and presumably use them to hunt with. Strombus, however, is a browser, and so far the presence of such large high resolution eyes lacks a behavioural explanation. Visual behaviour in the gastropods in general is not remarkable: positive and negative tropotaxis are common enough, orientation in relation to the sun's position (menotaxis) is known to occur in the opisthobranchs Elydia (Fraenkel & Gunn, 1961) and Aplysia (Hamilton &

TABLE 1

Mollusca: types of eye

Class	Cephalic eyes	Other eyes
Aplacophora	None	None
Polyplacophora (chitons)	None	Up to 10^4 small eyes in shell.
Monoplacophora	None	None
Gastropoda		
Prosobranchia	Paired, pit or lens eyes. Microvillous (exc. heteropods). Large lens eyes in strombids and heteropods.	Dermal sensitivity. Behavioural "off" responses.
Opisthobranchia	Paired, small, microvillous.	Dorsal eyes (ciliary) in Onchidium.
Pulmonata	Paired pit or lens eyes.	Dermal sensitivity.
Scaphopoda	None	None known
Bivalvia	Usually absent. Microvillous in Mytilus.	"Off" responding neurones common Compound eyes in Arca and relatives. Mirror eyes in Pecten and relatives, with both ciliary and microvillous receptors.
Cephalopoda		
Nautiloidea	Large pinhole eyes, moveable, variable iris, microvillous receptors.	None known
Coleoidea	Large, highly-developed, fishlike eyes. Receptors have orthogonal microvilli.	Extra-ocular sensitivity in epistellar body.

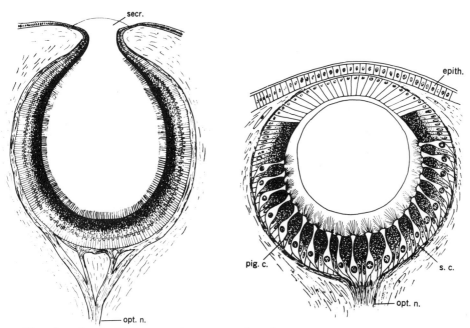

Fig. 1: Cephalic eyes of Haliotis (left) and Helix pomatia (right). After Hesse (1908). Haliotis eyes are about 1 mm long, and have no lens. Helix eye is 0,3 mm diameter, and the lens is a soft "Füllmasse" that will not form an image on its own. secr, secreted plug; opt. n, optic nerve; epith, epithelium; pig. c, pigment cell; s.c, sensory cell.

Russell, 1981), and crude form vision sometimes involving the ability to distinguish vertical structures (e.g. plant stems) from horizontal ones has been shown in Littorina (Hamilton, 1977).

The cephalic eyes of cephalopods are quite another matter. In contrast to the gastropods, vision is the most important sense in cephalopods, which are all active carnivores. I will return to the specialisations of these eyes later, but should mention here that they are of two types; the lensless "large pinhole" eyes of Nautilus which can provide only the crudest of images in spite of the presence of a fine grain retina, and the large and impressive lens eyes of the coleoids (octopus, squid and cuttlefish). The largest eyes of any animal are undoubtedly those of giant squid (Architeuthis) with diameters as great as 40 cm. Since receptor packing distances in cephalopods are typically about 4 µm, (Packard, 1972) such an eye could contain as many as 10^{10} receptors (compared with about 10^8 in the human eye). Cephalopod and gastropod eyes have two basic attributes in common: they are both single-chambered, with or without a lens (Nautilus could be compared with Haliotis in this respect, and Octopus with Strombus), and with the single exception of the heteropods the eyes of both classes have rhabdomeric receptors with densely packed microvilli. As we shall see, these

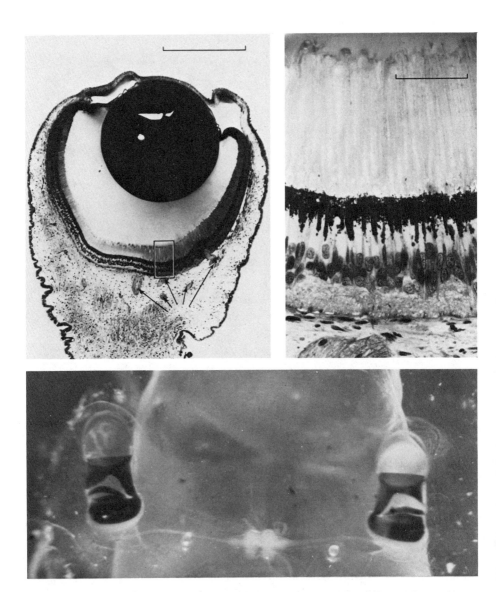

Fig. 2: Cephalic eyes with image-forming lenses. Top: <u>Strombus luhuanus</u> showing section through whole eye (left) and detail of retina (right). Scales 0,5 mm and 50 μm. Courtesy Dr. H.L. Gillary (Gillary, 1974). Arrows show branches of optic nerve. Bottom: <u>Pterotrachea coronata</u>, photograph of living eyes at base of proboscis. The optic nerves, statocysts and C.N.S. are also visible. Each eye is 1,8 mm long.

generalisations do not apply to non-cephalic eyes.

In Table 2, I have assembled estimates of the resolving power of several mollusc eyes based on anatomical measurements. Where the eye has an image-forming lens the angular receptor separation ($\Delta\phi$) is simply the receptor separation divided by the focal length, taken as the distance from the lens centre to the retinal surface. This value, in radians, gives $\Delta\phi$ in degrees when multiplied by 57,3. In lensless eyes the images of two point sources will touch on the retina if their separation is equal to angle that the pupil subtends at the retina, i.e. the pupil diameter divided by the eye's axial length. However densely the receptors are packed it is unlikely that resolution better than this angle can be achieved, so this

TABLE 2

Minimum resolvable angles ($\Delta\phi$) in some mollusc eyes

Species	Eye size (approx.)	Inter-receptor angle
Gastropoda		
Littorina littorea	f = 155 µm	4,4° (in water)
Strombus luhuanus	f = 1 mm	0,23°
Pterotrachea coronata	f = 1250 µm	0,18 x 1,14
Oxygyrus keraudreini	f = 540 µm	0,4 x 1,1°
Bivalvia		
Barbatia (compound eye)	diam: 200 µm	10°
Pecten maximus (reflecting eye)	diam: 1 mm	2°
Caphalopoda		
Nautilus (pinhole)	diam: 10 µm	2,3° (0,4 mm pupil)* 16° (2,8 mm pupil)*
Octopus	f = 10 mm	0,02° (= 1,3 min)

Eyes are lens eyes except where stated. In Nautilus (*) the value used for the minimum resolvable angle is taken to be the angular separation of non-overlapping point source images. Data from Land (1981) or references in text.

value is adopted for $\Delta\phi$ for eyes of the pinhole type. The "best" gastropod eyes (Strombus, Pterotrachea) have inter-receptor angles of 14 and 11 min which are considerably better than the minimum resolvable angle in Nautilus (2,3°) but still an order of magnitude larger than in Octopus (1,3 min). For comparison, the minimum value of $\Delta\phi$ in man is about 0,5 min, and in a bee it is 1°.

Non-Cephalic Eyes

Whilst the cephalic eyes of gastropods are principally concerned with habitat selection, and much else besides in the cephalopods, the extra-cephalic system seems to be primarily involved in defensive responses to shadow and movement (see Land, 1968). When a shadow crosses the cephalic eyes of Helix, nothing happens, but when it crosses the mantle near the base of the shell the animal withdraws (Föh, 1932). In fact, "dermal" sensitivity to dimming and sometimes to both dimming and lightening is quite common in both gastropods and bivalves; most clams, for example will close their valves when shadowed, and in the cases of Spisula and Mercenaria the activity of the neurone in the pallial nerve responsible for producing this "off" response has been studied in detail (Kennedy, 1960; Wiederhold et al., 1973). It would be of great interest to know what the detailed structure of the receptors giving rise to these neural responses is like, but as far as I know the endings have never been tracked down except at a rather gross level (Light, 1930).

In four groups the non-cephalic receptors are found in distinct eyes. These are the chitons - a class (Polyplacophora) which split off early from the three main classes - opisthobranch gastropods of the order Onchidiacea, and amongst the bivalves the scallops (Pecten and related genera) and the ark shells (Arca, Barbatia, Pectunculus) (Fig. 3). A few other

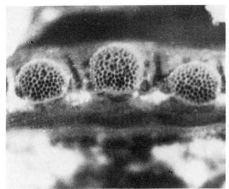 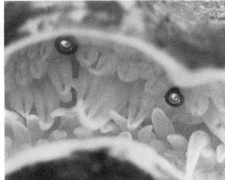

Fig. 3: Non-cephalic eyes of the arcoid Barbatia cancellaria (left) and Pecten maximus (right). Barbatia eyes are apposition compound eyes, each about 200 μm diameter. Pecten eyes are 1 mm diameter, and form an image with a concave mirror. Barbatia photograph courtesy of Thomas R. Waller, Smithsonian Institution, Washington D.C.

bivalves have eye-like structures; definite eye-spots are visible in the file shells (Lima), and cockles (Cardium) have very small eyes with reflecting cups situated around the siphons (Barber & Wright, 1969a), but unlike Pecten and Arca it is doubtful whether the latter two genera have eyes that produce usable images

The chitons have small eyes of several kinds actually embedded in the plates of their shells. These are usually divided into ocelli which have a rather flat lens 25-40 µm across, and a cup-shaped retina with up to 100 receptors, and the rather peculiarly named aesthetes which are generally smaller, have as few as one receptor in some cases and are less obviously visual in function (Boyle, 1969a, b, 1974). There may be as many as 11 500 ocelli in Acanthopleura, and Boyle (1969a) counted 1 472 on Onithochiton. Ultrastructurally the ocellar receptors certainly contain rows of oriented microvilli, forming a rhabdom, but there are whorls of lamellae derived from cilia as well, though not necessarily in cells with a receptor function. This is tantalizing because in the only really well documented case (Pecten) the rhabdomeric cells produce depolarising "on" responses and the ciliary cells hyperpolarising "off" responses. Chitons show both shadow responses and phototaxis which may be negative or positive, so it seems that they ought to possess two types of receptor, presumably in their shell eyes since they lack cephalic eyes (Bullock, 1965).

In the gastropod Onchidium the question of which eye does what is a little clearer. In addition to a pair of cephalic eyes, with microvillous receptors, Onchidium verruculatum has between 18 and 46 eye-like structures 100 to 200 µm in diameter situated in papillae in the skin of the animal's back. These eyes contain no microvilli, but instead whorls of membrane that are elaborations of numerous cilia (Yanase & Sakamoto, 1965). It has been known since a study by Arey and Crozier (1921) that the dorsal region of O. floridanum (a species without obvious dorsal eyes) was sensitive to shadow, but that the cephalic eyes were not. Fujimoto et al. (1966) found ERG's of opposite sign in the cephalic and dorsal eyes. The evidence is strong, though not yet perhaps compelling, that the ciliary receptors of the non-cephalic eyes mediate the "off" responses responsible for the behavioural shadow response. It should be pointed out that a shadow response does not logically require primary "off" receptors; in barnacles, for example, the shadow response arises from the fact that the second-order cells receive an inhibitory input from the "on" responding receptors (Gwilliam, 1963) and even in the cephalic eyes of Hermissenda, another opisthobranch gastropod, "off" responses can arise synaptically from the interactions between the 5 rhabdomeric, "on" responding receptors (Dennis, 1967, Alkon & Fuortes, 1972).

Despite these caveats, very strong support for the proposition that "on" and "off" receptors are distinct has come from studies over the years of the remarkable eyes of scallops. Their optics merit a separate section, but it will suffice here to say that they are almost the only eyes in the animal kingdom that form a good image using a concave mirror rather than a lens (Land, 1965). What is of interest here is that the retina contains two distinct retinal layers: a proximal layer (relative to the centre of the animal) containing microvillous receptors and immediately in front of it a distal layer whose receptors each contain about 100 flattened cilia and no microvilli (Miller, 1958, Barber et al., 1967). Hartline as early as 1938 had recorded from single fibres from the optic nerve of eyes of Pecten irradians, and by cutting each branch in turn managed to show that the proximal cells gave "on" responses and the distal cells "off" responses. Support for the idea that the distal cells are primary receptors, rather than second-order "off" ganglion cells comes from the fact that the image produced by the mirror lies on the distal not the proximal cells, at the level of the cilia, and that these cells do produce responses to stripe patterns that cause no overall brightness changes when they are moved, but only local changes within the image (Land, 1966a). The image is thus "seen" by the distal cells, confirming the anatomical inference that they must be photoreceptors.

Toyoda and Shapley (1967) and then McReynolds and Gorman (1970a, b) succeeded in penetrating the two cell types in the Pecten retina and found both depolarising cells, which fired during illumination, and hyperpolarising cells which fired after illumination (Fig. 4). These could only correspond to the proximal and distal cells respectively. McReynolds and Gorman (1970b) found that in both types of cell the response to illumination was an increase in conductance. In the proximal cells the

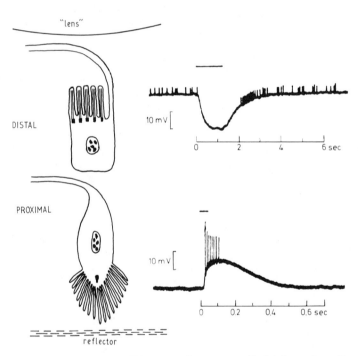

Fig. 4: Two cell types in Pecten. Structure (left) based on Barber et al. (1967) and responses (right), modified from McReynolds and Gorman (1970a). The ciliary distal cells hyperpolarise to light and give "off" responses; the microvillous proximal cells depolarise and give "on" responses.

increase results in Na^+ influx, and in the distal cells in K^+ efflux (McReynolds & Gorman, 1974). Ca^{++} is also involved in the distal cell response, but it is not entirely clear how (Gorman & McReynolds, 1978). To appreciate the importance of these findings it is useful to know what else was happening in photoreceptor research in 1970. On the one hand, Eakin had proposed that there were two lines of photoreceptor evolution (Eakin, 1972), one consisting of modified cilia and embracing the coelenterates, starfish, ascidians and vertebrates, and the other rhabdomeric line based on microvilli without cilia and including all the arthropods, some annelids and the cephalic eyes of molluscs. This classification conformed well to existing views of phylogenesis and seemed to be a promising confirmation of them. Pecten eyes, of course, straddle the division; the distal cells are very ciliated, and the proximal cells essentially rhabdomeric, although they do contain at least one ciliary base (Barber et al., 1967). On the other hand, a major bombshell had burst in vertebrate photoreceptor biophysics. Unlike Limulus, the model for 40 years, vertebrate photoreceptors did not depolarise on illumination, they hyperpolarised (Toyoda et al., 1969, Baylor & Fuortes, 1970). Could it be that Pecten proximal cells were depolarising, rhabdomeric and "arthropod", and the distal cells hyperpolarising, ciliary and "vertebrate"? The rather sad, but quite conclusive answer is no. Whereas the response of the proximal cells is like that of arthropod receptors - "typical of sense organ activity" as Hartline put it in 1938 - the basis of the hyperpolarising response is quite different in vertebrates and Pecten distal cells. In the former it is the result of the closing of Na^+ channels to light, and in the latter the opening of K^+ channels. This biophysical difference is so profound that it seems most unlikely that Pecten distal cells and vertebrate receptors have anything useful to say to each other, either about their ancestry or about their transduction mechanisms.

In spite of this heady disappointment, there remain interesting and important questions. Why do cells with large numbers of cilia respond to light by letting in K^+, not just in Pecten, but in Lima (Mpitsos, 1973) and probably also in the annelid Branchiomma, which again has ciliated receptors and a behavioural "off" response (Krasne & Lawrence, 1966)? Are the extra-ocular photoreceptors that mediate shadow responses in gastropods like Helix and bivalves like Spisula also ciliary? Do the ciliary receptors in the cephalic eyes of heteropods give "off" responses or "on" responses (Dilly, 1969)? And, perhaps most importantly, how do these primary "off" receptors really work? Authors since Hartline (1938) have all agreed that light not only inhibits, but also "charges up" the off response, so that the brighter the light and the longer it is on, the greater the "off" response at its termination. What ions are depleted? What sequestered? Does the bleaching of photopigment stimulate a pump, rather than a leak? On their own terms, the non-cephalic visual systems of molluscs present many challenges, especially to receptor physiologists.

The other bivalve eyes that would repay further study are the compound eyespots found on the mantle edge of Arca and its relatives (Fig. 3). There are no other examples of compound eyes in the molluscs, although there are a few in the annelids (sabellid tube worms like Branchiomma). Like the mirror eyes of scallops they are very much out on an evolutionary limb, with no obvious antecedents. The eyes vary in size up to a diameter of about 200 μm, and each contains between from 10 to 80 facets (Fig 3). Each "ommatidium" is a single cell with a slightly domed transparent distal region beneath which lies the nucleus and then the receptive region (Patten, 1886; Hesse, 1900). One naturally suspects that the receptors will turn out to be ciliary in nature, but Hesse's microscopy only shows a rather ambiguous looking string of blobs, and to my knowledge there is no recent EM study. The eyes are borne in large numbers: Patten counted 133 on one mantle edge, and 102 on the other, so there must be almost complete overlap between the fields of view of adjacent eyes. The function of the eyes is fairly clear; they enable the animal to respond to moving objects that do not cast a direct shadow (Braun, 1954). The advantage of having even a very simple optical arrangement for restricting the fields of view of individual receptors is that the animal can shut its valves before a potential predator is literally on top of it.

Before moving on, I should mention that photoreception in molluscs is not confined to superficial structures, whether these are eyes or free nerve endings. Several neurons in the brain of Aplysia have been identified as photoreceptive, and possibly control the diurnal rhythm of locomotion (the eye itself also has a circadian rhythm). Even in cephalopods where one would think the eyes could do all the seeing that needed to be done there are extra-ocular receptors, for example in the epistellar body of the stellate ganglion (see Messenger, 1981).

MIRROR EYES

Scallops have between 60 and 100 eyes arranged around the mantle, looking out from between the tentacles. Although their anatomical structure was known for the best part of a century (Patten, 1886, Hesse, 1900, Dakin, 1910) it had been assumed until 1965 that they were basically lens eyes, like the cephalic eyes of some gastropods. The observation that changed this view was that it is possible to see an inverted image of oneself in the eye when looking into it with a dissecting microscope (Land, 1965). If these were actually lens eyes no such image could be seen, because an object imaged onto the retina would be re-imaged at infinity by the lens, and as with the human eye one would need a telescope (or ophthalmoscope) and not a microscope to see it. The solution had to be that the image visible in the eye was not formed by the lens, but by reflection from the nearly hemispherical highly reflecting tapetum (or argentea) that lines the back of the eye (Fig. 5). There is in fact a lens, but its refractive index is so low -about 1,42 - that the image it forms would be about a mm behind the eye (Fig. 6). The reflected image lies about 140 μm in front of the reflector, which means that it lies on the region of the distal retina occupied by the ciliary receptor structures (see Fig. 4). The microvillous endings of the proximal cells are actually in contact with the reflector, and there is no image in this region. Thus visual tasks like movement detection that require a resolved image must be performed by the distal retina, and indeed the distal retina responds briskly to the movement of stripes that do not cause overall

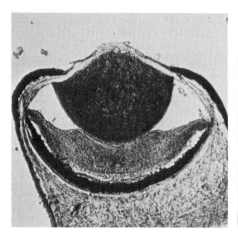 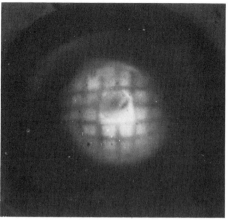

Fig. 5: Eye of <u>Pecten</u> <u>maximus</u>. Left: median frozen section of a 1 mm diameter eye, showing the large "lens", and the retina filling the whole space between the lens and the hemispherical back of the eye, which is lined inside with a reflecting tapetum. Right: reflected image of a grid photographed in the eye of <u>Pecten</u> with an ordinary microscope. Land (1965).

changes in illumination (Land, 1966a). Optic nerve responses can be obtained from <u>Pecten</u> <u>maximus</u> to 2° movements of a dark stripe. This corresponds to a movement by the image of 9 µm which is the diameter of a single distal cell. It also corresponds to the behavioural threshold of 1-2° found in an earlier study of Buddenbrock and Moller-Racke (1953). There is thus no doubt that the images produced by the concave reflector are indeed used. The role of the proximal retina, which has the same number of receptors as the distal, about 5 000, is more enigmatic. It could function as a vaguely directional intensity sensor, and there is certainly evidence that scallops will swim to particular light or dark regions of the environment (Buddenbrock and Moller-Racke, 1953, Land, 1968). The distal retina, being essentially phasic in its response, could not provide the information that would let the scallop do this. The worrying point is that the scallop has something like half a million proximal receptors altogether to perform a task that could be done by perhaps half a dozen. The distal retina raises a similar "overkill" problem, though a less severe one. Each eye has a field of view of about 90°, and since the eyes are spaced about 6° apart the images in adjacent eyes will overlap almost completely. Perhaps one has to assume that in the dim light conditions that scallops inhabit there is a need for such apparent redundancy to improve the signal to noise ratio of the detecting process, but this seems a rather clumsy explanation.

The mirror in <u>Pecten</u> is a quarter-wave multilayer reflector, made up of alternating layers of 1 µm square guanine crystals and cytoplasm (Land, 1966 b, 1972). The principle on which it works is the same as in the reflecting scales of fish and iridescent butterfly wings, and it is also the

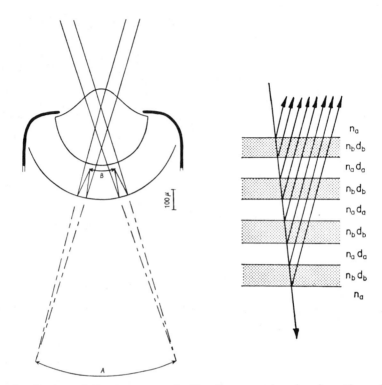

Fig. 6: Optics of Pecten eye. Left: diagram showing location of the images formed by the lens alone (A), and by the lens-mirror combination (B). The latter image lies on the ciliary endings of the distal cells (Fig. 4). Right: principle of multilayer reflection. All light emerges from the stack in phase if the optical thickness of all the layers is $\frac{1}{4}\lambda$, i.e. if $n_a d_a = n_b d_b \simeq 133$ nm. In Pecten n_a is cytoplasm (1,34) and n_b guanine crystals (1,82). Land (1966b).

way many high quality beam-splitters and colour selective mirrors are currently made. Light reaching the first water-guanine inferface is partly reflected, some of the remaining light is reflected at the next interface, and so on through the stack (Fig. 6). At each low-to-high refractive index interface the reflected light waves change phase by half a wavelength, which means that for light reflected from the next high-to-low interface to remain in phase with that from the first interface (and hence interfere constructively), that light must travel an optical path length of half a wavelength. Since it has passed through each crystal or cytoplasmic space twice, the optical thickness (thickness times refractive index) of each crystal or space must be a quarter of a wavelength. In Pecten each crystal is about 80 nm thick, and the refractive index of guanine is 1,82, so that the optical thickness is 145 nm. This should give a reflexion which has a maximum at 4 x 145, or 570 nm, which is quite close to the observed maximum in the green between 525 and 550 nm. An optical thickness of 133 nm would give a better fit. It seems that most mirrors in nature make use of the same principle; actual metallic surfaces do not exist.

Very few animals besides Pecten use mirrors for image formation Tapeta which return light back through the receptors are common in vertebrates and in spiders, but there the image is formed by a lens and the mirror is effectively in the image plane. The only other image-forming reflectors I know of are in certain crustacea - some copepods and the deep-sea ostracod Gigantocypris (see Land, 1983). The failure of the design if that is not putting it too strongly, is probably that it produces a low contrast image because light has already passed unfocussed through the receptors before returning focussed. Its merits are that the eye can be very compact, and can gather a great deal of light. The F-number of the scallop eye is about 0,6 compared with 1,25 for the eye of a fish, and the increase in image brightness is a factor of nearly 4,5.

THE DESIGN OF LENSES

In an aquatic environment the cornea is not a usable refracting surface, as it is in our eyes and those of spiders, and all the refraction in an aquatic lens eye must therefore be done by the lens itself. With few exceptions, aquatic animals with single chambered eyes have evolved spherical lenses, presumably because a sphere gives a shorter focal length than other shape, and again with few exceptions these lenses tend to have a focal length equal to about 2,5 lens radii (Matthiessen's ratio). The significance of this ratio is that all animals with lenses of this kind have hit on the same, ingenious, solution to the problem of lens design. They have all evolved lenses that are optically non-homogeneous. Most dry biological materials that lenses might be made of (protein, keratin, chitin) have refractive indices in the range 1,5 to 1,55, and if one applies the standard formulae to a homogeneous spherical lens with a refractive index of, say, 1,53 then it will have a focal length of 3,8 radii, very much longer than that of the lens of a fish, octopus or even a gastropod like Littorina. At the same time, a homogeneous lens will suffer from spherical aberration (rays off the axis are refracted too much) to the extent that it would be virtually unusable (Pumphrey, 1961). Matthiessen (1886) proposed a solution that dealt with both problems. If the lenses are inhomogeneous, with a high central refractive index falling to a value not much greater than water at the periphery, then, firstly, rays entering the lens will be bent continuously within it and not just at the surfaces, which effectively increases the power and shortens the focal length, and secondly, because the outer regions have a lower refractive index than the centre rays passing through them are bent relatively less, thereby correcting the spherical aberration (Fig. 7). To produce such a lens it is crucial that the refractive index gradient is the correct one, as there is only one right solution (Fletcher et al., 1954). This is really quite a sophisticated optical invention not yet duplicated by optical engineers, and it is therefore the more remarkable that it should have evolved so many times. (To establish that a lens is of this kind it is only necessary to

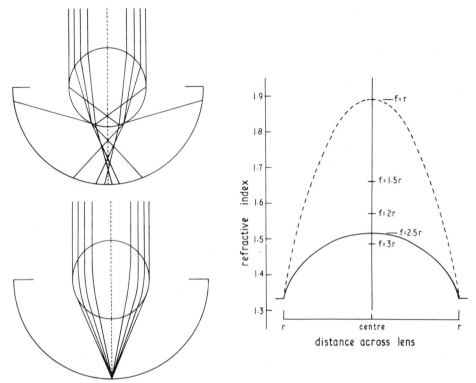

Fig. 7: Left: image produced by a homogeneous sphere of focal length equal to 2,5 radii, showing severe spherical aberration, and (below) the image formed by an inhomogeneous Matthiessen lens of the type found in fish, cephalopods and some other animals (modified from Pumphrey, 1961). Right: heavy line shows the computed refractive index gradient in a spherical aberration-free lens of focal length 2,5 radii. Dotted line shows the gradient in a lens that focuses on its rear surface. Data from Fletcher et al. (1954).

show that its focal length conforms to Matthiessen's ratio). At the last count the number of occasions that such lenses must have evolved independently is at least 6: in the fish, the coleoid cephalopods, (Sivak, 1982), in the prosobranch gastropods (Strombus, Carinaria, Fig. 2), in the pulmonate gastropods (Lymnea), in one family of polychaete worms (Alciopidae), and at least one genus of copepod crustacean (Labidocera) (see Land, 1981, 1983). There is also a similar kind of lens in certain spiders (Blest & Land, 1977). One cannot even be sure that within the prosobranch gastropods the three families known to have these lenses did not evolve them independently. It is a remarkable instance of convergent evolution, and one must conclude that there really is only one way to make a decent lens, using biological material.

Heteropod Eyes

The heteropods deserve special mention here because, although their eyes are basically of the Matthiessen type, they have a unique retinal structure whose function is just beginning to make sense. This group of prosobranchs are all active pelagic carnivores, and they all have a pair of large eyes at the base of a proboscis which houses a fearsome radula. The odd feature of the eyes is that the retina, instead of being a hemispherical cup as in, say, Strombus, is actually a long ribbon, only 3-6 cells wide, but many hundreds long. The ribbons may be linear (Pterotrachea, Oxygyrus) or curved into a horseshoe shape (Carinaria, Atlanta) (Hesse, 1900). It would seem on the face of it that such a retina could only image a very small strip of outside space, and thus not be of much value to a hunting animal.

I recently had the opportunity to study freshly caught Oxygyrus keraudreini, a small heteropod that lives close to the surface (Fig. 8). It has eyes 1,2 mm in diameter with a retina 3 receptors wide and about 410 long, covering a field of view about 3,2° by 160°. The interesting observation was that the eyes are in almost constant scanning motion.

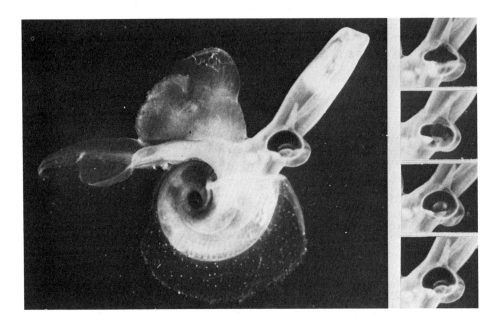

Fig. 8: Oxygyrus keraudreini, a heteropod with a scanning eye. Main photograph shows the live animal in its normal swimming position (shell down), and the inserts at the right show the eye during a downward scan, pointing horizontally (top) and downwards (bottom). The eye has a maximum diameter of 1,2 µm. Partly from Land (1982).

The whole eye swings about an axis through the lens and parallel to the retina in such a way that the retinal field of view sweeps through an arc of slightly more than 90°, covering the region below and to the side of the animal. The downward sweep is rapid (250° per s) and is followed immediately by a slower upward sweep (80° per s) to the horizontal rest position (Land, 1982). Presumably the animal is looking for food objects that glisten slightly against the dark of the water below. This "line scan" mechanism would seem to be a very compact and sensible way to operate a visual system, and I am surprised that there are few other examples. The only other similar arrangement I know of is in the pontellid copepod Labidocera where a linear array of only 10 cells scans through an arc of about 40° (see Land, 1983).

It is a pity that heteropods are oceanic and hard to work on. The retina, as Dilly has shown (1969), contains not only cilia but also plates reminiscent of vertebrate rod discs, and even microvilli as well. The affinities of these receptors to the rhabdomeric and ciliary types discussed earlier are still very much in doubt, and they are certainly unlike the receptors of the cephalic eyes of any other mollusc.

CEPHALOPOD EYES

The Enigma of the Nautilus Pinhole

Nautilus-like molluscs originated in the Cambrian with a major radiation in the Ordovician (Holland, 1979). Their modern representative, Nautilus, is not very different in structure from forms 400 million years old. The eyes of Nautilus are lensless, without a cornea, so that the interior of the eye is open to the sea (Fig. 9). This makes them similar optically to the unsophisticated eyes of a gastropod like the abalone (Haliotis) and one would be content to describe them as a primitive relic of a bygone age, if it weren't for the fact that most of their other features are far from primitive. For example, they are quite large, approaching a cm in diameter and comparable in size to an octopus eye, and larger by almost an order of magnitude than any gastropod eye. Since an eye's size is usually a good indication of its sensitivity or resolution or both (see Land, 1981), the Nautilus eye might be expected to have good capabilities. The receptor cells are narrow, 5-10 µm diameter, and long (up to 500 µm) and there may be as many as 4×10^6 in the retina (Barber & Wright, 1969b, Messenger, 1981), so that this is a large, fine-grain retina. The eye has a mobile pupil (Hurley et al., 1978) whose diameter varies from 0,4 to 2,8 mm depending on the illumination level, and again this is not a feature found in any gastropod eye. Furthermore, the eyes as a whole are mobile, and retain a vertical orientation when the animal rocks (Hartline et al., 1979), a response mediated by the statocysts and analogous to the vestibulo-ocular reflex in humans. There is also an optomotor response to wide stripes rotating around the animal (W.R.A. Muntz, personal communication). Although common in the coleoid cephalopods and in other visually advanced phyla, optomotor responses have been sought in the gastropods but not found (Dahmen, 1977).

The problem is that without a lens both the resolution and sensitivity

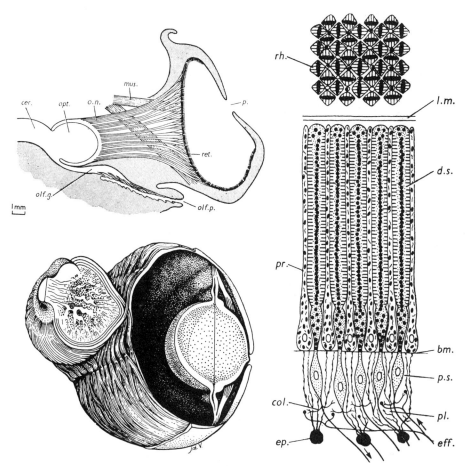

Fig. 9: Pinhole eye of Nautilus (top left), and the Matthiessen lens eye of Octopus (below). Octopus eyes vary in size, but are typically around 1 cm diameter, like the Nautilus eye. Right: structure of the Octopus retina. Note the form of the receptor lattice, in which adjacent receptors have their microvilli at right angles to each other. Abbreviations: cer, cerebral cord; mus, eye muscles; o.n., optic nerve; opt, optic ganglion; olf. p, olfactory pit; olf. g, olfactory ganglion; p, pupil; ret, retina. Col, collateral fibres; bm, basement membrane; d.s, distal segments; eff, efferent fibres; ep, epithelial (glia) cell; l. m, limiting membrane; p. s, proximal segments; pr, supporting cell processes; rh, rhabdoms. Compiled from Young (1964).

of the eye must be poor. Taking the minimum estimate of the pupil diameter (0,4 mm) the angular subtense of the blur circle on the retina, the functional value of $\Delta\phi$, will be 2,3°, and it will cover about 2 500 receptors. Octopus, with a lens, has a minimum resolvable angle equal to the receptor subtense (1,3 minutes). Nautilus, at its best, is 100 times worse. The penalty for not having a lens is even more dramatic in terms of image brightness, and hence sensitivity. A 0,4 mm pinhole in a 1 cm eye represents an F-number of 25, compared with F/1,25 for a Matthiessen's ratio squid or octopus lens. The brightness ratio is 1:400. Even with a wide open pupil (2,8 mm) the Nautilus image is 13 times dimmer than in Octopus, and at the same time $\Delta\phi$ has risen to 16°, a figure that makes anything more than simple phototaxis impossible.

The embarrassing thing about Nautilus is that almost any transparent blob put into the pinhole would improve matters. It need not initially be a perfected Matthiessen lens, a round mass of jelly of slightly higher refractive index than sea-water would help both resolution and sensitivity. When herbivorous gastropods like Strombus have evolved "proper" lenses, how is it that Nautilus, a deep-water predator, has not? How can a bad, but easily improved upon design persist for 400 million years? Is it even unthinkable, in view of the eyes' advanced features, that its ancestors possessed lenses, and that in modern Nautilus the eyes have become degenerate as olfactory senses took over their function (blind cave fish are a possible vertebrate parallel)? Either way, whether nautiloids never evolved lenses, or lost them, their eyes present a tricky evolutionary problem. They are the one eye which really should have given Darwin his famous "cold shudder".

The Scope of Convergence Between Cephalopod and Fish Eyes.

The competitors of the early cephalopods and fish would have been arthropods with compound eyes: trilobites, eurypterids, xiphosurans and crustaceans. One statement that can be made with certainty is that the single large chambered eye of the coleoid cephalopods and the vertebrates resolves better than any compound eye, and this may have been the reason for its origin. The problem with all types of compound eye is that the small size of the individual facets, rarely wider than 50 µm, makes them diffraction limited, with a minimum value for $\Delta\phi$ of about $\frac{1}{2}$°. This fundamental restriction means that they are not adaptable to a lifestyle which involves seeing small prey at a distance and lunging at it (fish) or snatching it with long tentacles (squid, cuttlefish). The single chambered eye, with a pupil several mm wide and a potential minimum resolvable angle of a minute or less imposes no such restriction, and no doubt gave the early cephalopods and fish an important competitive advantage by giving them a new niche as "long-range" carnivores. Apart from the now extinct ichthyosaurs, and the recent cetaceans, they have been the sole marine exponents of this highly successful way of obtaining food. The kind of eye they share is the right one for the job.

The similarities between the visual systems of cephalopods and fish have been reviewed extensively elsewhere (Messenger, 1981, Packard, 1972). The comparison can be made at three levels: the structure of the eyes, the ways the eyes are used in terms of the kinds of movements they make, and the uses that the nervous system makes of the visual information it obtains. I will summarise the points of convergence and difference in that order.

MOLLUSCS

Cephalopod eyes are generally similar in size to those of fish, except that the largest squid (Architeuthis) have eyes whose dimensions are outside the range of any fish (Roper & Boss, 1982). Where an animal needs both high resolution and high sensitivity it must have a large eye (see Land, 1971), and so one might expect a dim-light predator like Architeuthis to possess large eyes. The mystery, perhaps, is that they are not matched in size by those of the larger mesopelagic fish. The most striking feature the eyes have in common is the spherical lens, and since it has the same focal length to radius ratio in both groups, this ensures that the proportions of the eyes as a whole are similar. The lens, although optically similar in cephalopods and fish, is somewhat differently constructed, with distinct anterior and posterior parts in the former but not the latter group. Interestingly, both groups have developed similar specialisations for deep sea vision. Tubular eyes in particular are found in both. The tube invariably points upwards (see Locket, 1977, for fish) and it is probably best thought of not so much as a "telephoto" eye, as one in which the dark - and visually useless - part of the visual field has been excluded. In one squid genus, Histioteuthis, there are animals which have one tubular eye and one hemispherical one (Fig. 10), the latter pointing downwards (Young, 1975); I don't believe there are any parallels to this in fish. Inshore cephalopods (e.g. Octopus and Sepia) have variable pupils,

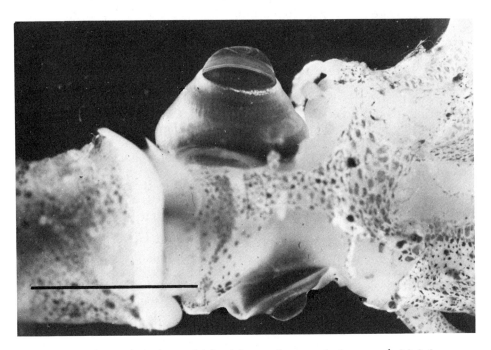

Fig. 10: Squid (Histioteuthis) with one large tubular eye (which has a yellow lens and points upwards) and one normal eye. Partially dissected; scale bar 10 mm. From Land (1981).

often W-shaped, and showing a degree of mobility not found in fish (Muntz, 1977). Pupils are circular in deeper water forms in both groups.

The most important differences in eye structure are in the organisation of the retina. In cephalopods the receptors point towards the light, whereas in vertebrates they lie behind the neural layers, apparently facing away. Since the neural layers are transparent this probably has no optical significance and reflects the peculiar way that vertebrate eyes develop as brain outgrowths. The fact that cephalopod retinae are the "right" way round means that the optic ganglia lie behind the eye, rather than partially inside it, and may also account for the absence of tapeta in cephalopods. Unlike fish eyes, there would be nowhere to put them. The receptors themselves, however, present the biggest differences. Whilst fish have the lamellar construction in their rods and cones typical of vertebrates as a whole, cephalopod receptors are exclusively microvillous. In this they are similar to the cephalic eyes of most gastropods, and indeed to the majority of invertebrates, especially the arthropods. Electrophysiologically, squid receptors depolarise on illumination, like Limulus and most other invertebrates (see Messenger, 1981) whereas fish receptors hyperpolarise. The cause is an increase in Na^+ conductance in the former, and a decrease in the latter. Perhaps the most interesting feature of cephalopod retinae is the way the microvilli are organised. The lattice of cells is a square one (Fig. 9), with the microvilli of alternating receptors at right angles to each other (Zonana, 1961). It has become well established since first put forward by Moody and Pariss (1961) that microvillous receptors are most sensitive to light polarised with its E-vector in the same plane as the long axes of the microvilli, and this in turn means that the whole retina of, say, a squid ought to behave as an "analyser" for the pattern of polarisation in the environment. The most attractive suggestion (I am not sure where it originated) is that this arrangement is to "break" the camouflage of silvery fish. Many fish, like the herring, have mirror-like sides which reflect in such a way that when seen from the side they behave as vertical plane mirrors, and thus reflect light that is similar in intensity to the light that would have passed through them (Denton, 1970). This is almost equivalent to being transparent. However, since light reflected from a non-metallic reflector is partially polarised (fish scales are constructed in the same way as the scallop tapetum), this form of camouflage will break down when viewed with a structure capable of analysing the direction of the E-vectors. Squid eat herring, and it may be that the curious arrangement of their microvilli is an adaptation to make their prey more easily visible.

Just as impressive as the structural similarities of fish and cephalopod eyes are their movements. We of course are accustomed to the ubiquitous vertebrate pattern of rapid saccades which alter the direction of gaze, alternating with periods in which the eyes are stabilised in space against movements of head or body, while they take in the scene. Whether or not we should be surprised to find that cephalopod eyes have adopted the same strategy depends on our preconceptions. Why should an animal not be able to able to follow a moving scene, keeping its eyes still relative to its body while moving around? Bees and wasps seem to do that while moving from flower to flower, or circling a jampot (unless we have missed their head movements?). Hoverflies, however, maintain rotational stability of their eyes while cruising around, just like vertebrates (Collett & Land, 1975). It would seem that both ways of seeing are possible: the moving image and the intermittently stable one. Whatever their relative merits, cephalopods have chosen the latter, as have fish. The eyes of Sepia are equipped with the analogous two methods of stabilising gaze - a statocyst mediated reflex that compensates for movements of the body, and an optokinetic reflex that prevents residual slip across the retina - as in fish and other vertebrates (Collewijn, 1970). Messenger (1981) even says that Octopus and Sepia show pursuit movements (tracking movements of a small object that are obviously in conflict with ordinary optokinetic movements which try to keep the image of the surroundings as a whole still). If this is true it is the more remarkable because such eye-movements are almost confined to primates, amongst vertebrates. This finding may need more examination, and it would be nice to know the result because a mobile predator really ought to be able to track small objects, even if the background is in relative motion. This would apply to fish as well, but their eye movements are still poorly documented. It is perhaps no surprise to find that Octopus, like fish, has six eye muscles (one antagonistic pair for each rotational degree of freedom) but it is more than a little perplexing to read (Messenger, 1981) that the cuttlefish Sepia has 13!

The upshot of this discussion is that cephalopods have not only produced a fish-like eye, but also a fish-like way of moving it in relation to the environment and to important objects in the environment. That insects, with eyes of a wholly different structure, often use the same strategy of eye movements (Land, 1975) perhaps indicates that there are greater constraints on the *process* of seeing, than there are on the physical means of acquiring an image.

For animals separated from ourselves genetically by at least 600 million years of evolution, the similarities in eye design and eye use are impressive enough for one to believe that for a large predator, at least, there is only one way of designing a system that sees well. This, though, is only the tip of the iceberg. Given a good, stable, image what should the nervous system do to turn it into an internal image, usable over time, that can distinguish between a home base and other places, between edible and inedible food, or between animals that are innocuous or predatory? Animals must learn to make these discriminations, and thanks to the efforts of J.Z. Young and his colleagues (Young, 1964, 1977, Wells, 1978) there is a massive literature on the learning capabilities of the octopus and its relatives. It must suffice here to say that these are clever, emotional animals, comparable in almost all respects with vertebrates in terms of their abilities to both learn and unlearn discrimination tasks (Sutherland & Macintosh, 1971, Packard, 1972).

As a postscript, it seems that the intellectual gulf between the cephalopods and the rest of the molluscs is enormous, although the structural similarities are undeniable. But then, if we are neotenous sea-squirts, who knows what else may have happened in the seas of the late Cambrian?

REFERENCES

Alkon, D.L. & Fuortes, M.G.F. (1972) Responses of photoreceptors in Hermissenda. J. Gen. Physiol. 60: 631-649.

Arey, L.B. & Crozier, W.J. (1921) On the natural history of Onchidium. J. Exp. Zool. 32: 443-502.

Barber, V.C. & Wright, D.E. (1969a) The fine structure of the eye and optic tentacle of the mollusc Cardium edule. J. Ultrastruct. Res. 26: 515-528.

Barber, V.C. & Wright, D.E. (1969b) The fine structure of the sense organs of the cephalopod mollusc Nautilus. Z. Zellforsch. 102: 293-312.

Barber, V.C., Evans, E.M. & Land, M.F. (1967) The fine structure of the eye of the mollusc Pecten maximus. Z. Zellforsch. 76: 295-312.

Baylor, D.A. & Fuortes, M.G.F. (1970) Electrical responses of single cones in the retina of the turtle. J. Physiol. (Lond.) 207: 77-92.

Blest, A.D. & Land, M.F. (1977) The physiological optics of Dinopis subrufus L. Koch: a fish-lens in a spider. Proc. R. Soc. Lond. 196: 197-222.

Boyle, P.R. (1969a) Rhabdomeric ocellus in a chiton. Nature (Lond.) 222: 895-896.

Boyle, P.R. (1969b) Fine structure of the eyes of Onithochiton neglectus (Mollusca: Polyplacophora). Z. Zellforsch. 102: 313-332.

Boyle, P.R. (1974) The aesthetes of chitons. II. Fine structure in Lepidochitona cinereus L. Cell. Tissue Res. 153: 383-398.

Braun, R. (1954) Zum Lichtsinn facettenaugentragender Muscheln. Zool. Jb. (Zool. u. Physiol.) 65: 91-125.

Buddenbrock, W.v. & Moller-Racke, I. (1953) Über den Lichtsinn von Pecten. Pubbl. Staz. zool. Napoli. 24: 217-245.

Bullock, T.H. (1965) The Mollusca. In: Structure and Function in the Nervous Systems of Invertebrates. Ed. T.H. Bullock & G.A. Horridge, San Francisco, W.H. Freeman, p. 1273-1515.

Collett, T.S. & Land, M.F. (1975) Visual control of flight behaviour in the hoverfly Syritta pipiens L. J. Comp. Physiol. 99: 1-66.

Collewijn, H. (1970) Oculomotor reactions in the cuttlefish, Sepia officinalis. J. Exp. Biol. 52: 369-384.

Dahmen, H.J. (1977) The menotactic orientation of the prosobranch mollusc Littorina littorea. Biol. Cybern. 26: 17-23.

Dakin, W.J. (1910) The eye of Pecten. Quart. J. Microsc. Sci. 55: 49-112.

Dennis, M.J. (1967) Electrophysiology of the visual system of a nudibranch mollusc. J. Neurophysiol. 30: 1439-1465.

Denton, E.J. (1970) On the organization of the reflecting surfaces in some marine animals. Phil. Trans. R. Soc. Lond. 258B: 285-313.

Dilly, P.N. (1969) The structure of a photoreceptor organelle in the eye of Pterotrachea mutica. Z. Zellforsch. 99: 420-429.

Eakin, R.M. (1972) Structure of invertebrate photoreceptors. In: Handbook of Sensory Physiology Vol. VII/1. Ed. H.J.A. Dartnall, Berlin, Springer, p. 625-684.

Fletcher, A., Murphy, T. & Young, A. (1954) Solutions of two optical

problems. Proc. R. Soc. Lond. 223A: 216-225.
Föh, H. (1932) Der Schattenreflex bei Helix pomatia. Zool. Jb. (Zool. u. Physiol.) 52: 1-78.
Fraenkel, G.S. & Gunn, D.L. (1961) The Orientation of Animals. New York: Dover.
Fujimoto, K., Yanase, T., Okuno, Y. & Iwata, K. (1966) Electrical responses in Onchidium eyes. Mem. Osaka Gakugei Univ. B 15: 98-108.
Gillary, H.L. (1974) Light-evoked potentials from the eye and optic nerve of Strombus: response wave form and spectral sensitivity. J. Exp. Biol. 60: 383-396.
Gillary, H.L. & Gillary, E.W. (1979) Ultrastructural features of the retina and optic nerve of Strombus luhuanus, a marine gastropod. J. Morph. 159: 89-116.
Gorman, A.L.F. & McReynolds, J.S. (1978) Ionic effects on the membrane potential of hyperpolarizing receptors in the scallop retina. J. Physiol. (Lond.) 275: 345-355.
Gwilliam, G.F. (1963) The mechanism of the shadow reflex in Cirripedia. Biol. Bull. Woods Hole 125: 470-485.
Hamilton, P.V. (1977) Daily movements and visual location of plant stems by Littorina irrorata (Mollusca: Gastropoda). Mar. Behav. Physiol. 4: 293-304.
Hamilton, P.V. & Russell, B.J. (1982) Celestial orientation by surface-swimming Aplysia brasiliana Rang (Mollusca: Gastropoda). J. Exp. Mar. Biol. Ecol. 56: 145-152.
Hartline, H.K. (1938) The discharge of impulses in the optic nerve of Pecten in response to illumination of the eye. J. Cell. Comp. Physiol. 11: 465-477.
Hartline, P.H., Hurley, A.C. & Lange, G.D. (1979) Eye stabilization by statocyst mediated oculomotor reflex in Nautilus. J. Comp. Physiol. 132: 117-126.
Hesse, R. (1900) Untersuchungen über die Organe der Lichtempfindung bei niederen Thieren. VI. Die Augen einiger Mollusken. Z. wiss. Zoll. 68: 379-477.
Hesse, R. (1908) Das Sehen der niederen Tiere. Jena: Fischer.
Holland, C.H. (1979) Early Cephalopoda. In: The Origin of the Major Invertebrate Groups. Ed. M.R. House, London, Academic Press, p. 367-379.
Hurley, A.C., Lange, G.D. & Hartline, P.H. (1978) The adjustable "pin-hole camera" eye of Nautilus. J. Exp. Zool. 205: 37-44.
Kennedy, D. (1960) Neural photoreception in a lamellibranch mollusc. J. Gen. Physiol. 44: 277-299.
Krasne, F.B. & Lawrence, P.A. (1966) Structure of the photoreceptors in the compound eyespots of Branchiomma vesiculosum. J. Cell Sci. 1: 239-248.
Land, M.F. (1965) Image formation by a concave reflector in the eye of the scallop Pecten maximus. J. Physiol. (Lond.) 179: 138-153.
Land, M.F. (1966a) Activity in the optic nerve of Pecten maximus in response to changes in light intensity, and to pattern and movement in the optical environment. J. Exp. Biol. 45: 83-99.

Land, M.F. (1966b) A multilayer interference reflector in the eye of the scallop, Pecten maximus. J. Exp. Biol. 45: 433-437.
Land, M.F. (1968) Functional aspects of the optical and retinal organization of the mollusc eye. Symp. Zool. Soc. Lond. 23: 75-96.
Land, M.F. (1972) The physics and biology of animal reflectors. Prog. Biophys. Mol. Biol. 24: 75-106.
Land, M.F. (1975) Similarities in the visual behavior of arthropods and men. In: Handbook of Psychobiology. Ed. M.S. Grazzaniga & C. Blakemore, New York, Academic Press, p. 49-72.
Land, M.F. (1981) Optics and vision in invertebrates. In: Handbook of Sensory Physiology Vol. VII/6B. Ed. H. Autrum, Berlin, Springer, p. 471-592.
Land, M.F. (1982) Scanning eye movements in a heteropod mollusc. J. Exp. Biol. 96: 427-430.
Land, M.F. (1983) Crustacea. (This volume).
Light, V.E. (1930) Photoreceptors in Mya arenaria, with special reference to their distribution, structure and function. J. Morph. 49: 1-42.
Locket, N.A. (1977) Adaptations to the deep-sea environment. In: Handbook of Sensory Physiology Vol. VII/5. Ed. F. Crescitelli, Berlin, Springer, p. 67-192.
Matthiessen, L. (1886) Ueber den physikalisch-optischen Bau des Auges der Cetaceen und der Fische. Pflügers Archiv 38: 521-528.
McReynolds, J.S. & Gorman, A.L.F. (1970a) Photoreceptor potentials of opposite polarity in the eye of the scallop, Pecten irradians. J. Gen. Physiol. 56: 376-391.
McReynolds, J.S. & Gorman, A.L.F. (1970b) Membrane conductances and spectral sensitivities of Pecten photoreceptors. J. Gen. Physiol. 56: 392-406.
McReynolds, J.A. & Gorman, A.L.F. (1974) Ionic basis of hyperpolarizing receptor potential in scallop eye: increase in permeability to potassium ions. Science 183: 658-659.
Messenger, J.B. (1981) Comparative physiology of vision in molluscs. In: Handbook of Sensory Physiology Vol. VII/6C. Ed. H. Autrum, Berlin, Springer, p. 93-200.
Miller, W.H. (1958) Derivatives of cilia in the distal sense cells in the retina of Pecten. J. Biophys. Biochem. Cytol. 4: 227-228.
Moody, M.F. & Parriss, J.R. (1961) The discrimination of polarized light by Octopus: a behavioural and morphological study. Z. vergl. Physiol. 44: 268-291.
Mpistos, G.J. (1973) Physiology of vision in the mollusk Lima scabra. J. Neurophysiol. 36: 371-383.
Muntz, W.R.A. (1977) Pupillary response of cephalopods. Symp. Zool. Soc. Lond. 38: 277-285.
Newell, G.E. (1965) The eye of Littorina littorea. Proc. Zool. Soc. Lond. 144: 75-86.
Packard, A. (1972) Cephalopods and fish: the limits of convergence. Biol. Rev. 47: 241-307.
Patten, W. (1886) Eyes of molluscs and arthropods. Mitt. Zool. Staz. Neapel. 6: 542-756.
Pumphrey, R.J. (1961) Concerning vision. In: The Cell and the

Organism. Ed. J.A. Ramsay & V.B. Wigglesworth, Cambridge, University Press, p. 193-208.

Roper, C.F.E. & Boss, K.J. (1982) The giant squid. Sci. Am. 246(4): 82-89.

Rosen, M.D., Stasek, C.R. & Hermans, C.O. (1978) The ultrastructure and evolutionary significance of the cerebral ocelli of Mytilus edulis, the bay mussel. Veliger 21: 10-18.

Sivak, J.G. (1982) Optical properties of a cephalopod eye (the shortfinned squid, Illex illecebrosus). J. Comp. Physiol. 147: 323-327.

Stoll, C.J. (1973) Observations on the ultrastructure of the eye of the basommatophoran snail Lymnea stagnalis. Proc. K. Ned. Akad. Wet. C. 76: 1-11.

Sutherland, N.S. & Macintosh, N.J. (1971) Mechanisms of Animal Discrimination Learning. New York, Academic Press.

Toyoda, J. & Shapley, R.M. (1967) The intracellularly recorded response in the scallop eye. Biol. Bull. Woods Hole 133: 490.

Toyoda, J., Nosaki, H. & Tomita, T. (1969) Light induced resistance changes in single photoreceptors of Necturus and Gecko. Vision Res. 9: 453-463.

Wells, M.J. (1978) Octopus: Physiology and Behaviour of an Advanced Invertebrate. London: Chapman and Hall.

Wiederhold, M.L., MacNichol, E.F. & Bell, A.L. (1973) Photoreceptor spike responses in the hard shell clam, Mercenaria mercenaria. J. Gen. Physiol. 61: 24-55.

Yanase, T. & Sakamoto, S. (1965) Fine structure of the visual cells of the dorsal eye in molluscan Onchidium verruculatum. Zool. Mag. (Tokyo) 74: 238-242.

Yochelson, E.I. (1979) Early radiation of Mollusca and mollusc-like groups. In: The Origin of Major Invertebrate Groups. Ed. M.R. House, London, Academic Press, p. 323-358.

Young, J.Z. (1964) A Model of the Brain. Oxford, Clarendon.

Young, J.Z. (1977) Brain, behaviour and evolution of cephalopods. Symp. Zool. Soc. Lond. 38: 377-434.

Young, R.E. (1975) Function of the dimorphic eyes of the midwater squid Histioteuthis dofleini. Pacific Sci. 29: 211-218.

Zonana, H.V. (1961) Fine structure of the squid retina. Bull. Johns Hopkins Hosp. 109: 185-205.

PHOTORECEPTION IN CHAETOGNATHA

T. GOTO
Department of Physiology
School of Medicine
Gifu University
Tsukasa-cho, Gifu 500, Japan

M. YOSHIDA
Ushimado Marine Laboratory
Kashino
Ushimado, Okayama 701-43, Japan

ABSTRACT

Chaetognaths have a pair of eyes below the epidermis on the dorsal surface of the head; an inverted type in the genus Sagitta and an everted type in the genus Eukrohnia. The inverted type is composed of capsule cells surrounding the external surface, one pigment cell at the center of the eye, photoreceptor cells extending distal processes into depressions of the pigment cell and glia cells among the photoreceptor cells. The photoreceptor cell extends a distal process via a connecting piece of ciliary nature. The distal process is composed of a conical body and a distal segment which contains a structure for photon capture. The everted type lacks the pigment cell and the photoreceptor cell is provided with a crystalline cone-like structure at the tip immediately above the photoreceptive microvilli. S. crassa show slow tactic and quick target-aiming swimming toward light. These reactions are telotactic in nature, being achieved through the structural characteristics of the lensless eye Light-dependent diurnal changes in the vertical distribution pattern in the sea are briefly discussed.

INTRODUCTION

Chaetognaths are mostly planktonic marine animals. They are small, slender and bilaterally symmetrical, possessing one or two pairs of lateral horizontal fins, and show rapid forward darts by the action of relatively well differentiated neuromuscular system.

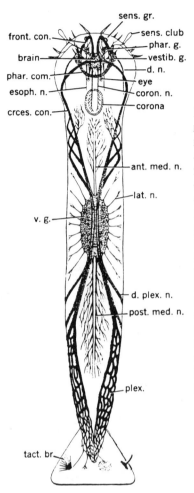

Fig. 1: General scheme of the nervous system of a chaetognatha (from Bullock, 1965).
ant. med. n., anterior medial nerve; crces. con., circumesophageal connective; coron. n., coronal nerve; d. n., dorsal nerve; d. plex. n., nerve to dorsal plexus; esoph. n., esophageal nerve; front. con., frontal connective; lat. n., lateral nerve; phar. com., pharyngeal commissure; phar. g., pharyngeal ganglion; plex., general nervous plexus of body surface; post. med. n., posterior medial nerve; sens. club, sensory club; sens. gr., sensory groove; tact. br., tactile bristles; vestib. g., vestibular ganglion; v.g., ventral ganglion.

As shown in Fig. 1, the nervous system consists of the brain with several accessory ganglia around it and a large ventral ganglion in the trunk (Hyman, 1959; Bullock, 1965). A pair of black or brown pigmented eyes (Fig. 2) are located posterior to the brain at the dorsal surface of the head and a bundle of nerves runs from the anterior end of each eye toward the brain. The ventral ganglion, on the other hand, is connected with the brain by a pair of circumesophageal connectives and gives off a number of nerves to the trunk and tail regions. Below, we will describe the structure of eyes, photic behavior and diurnal migratory patterns.

THE STRUCTURE OF EYES

Most of the anatomical studies have been carried out with the genus Sagitta (the order, Aphragmophora). The eye structure was revealed to

PHOTORECEPTION IN CHAETOGNATHA 729

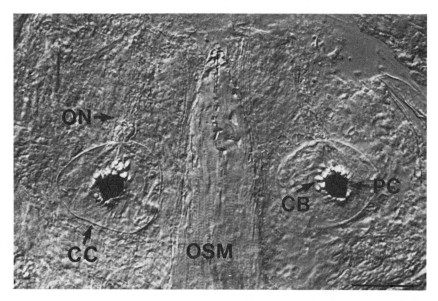

Fig. 2: Dorsal view of the eye region of <u>Sagitta crassa</u>, observed with the Nomarsky optics. CB, conical body; CC, capsule cell; ON, optic nerve; OSM, oblique superficial muscle; PC, pigment cell. Scale bar indicates 50 μm.

Fig. 3: Schematic representation of the pigment cell cut transversely (modified from Hesse, 1902). The right and left schemes are the anterior and the posterior side of the pigment cell, respectively, and the central one is the middle region of the pigment cell. Numbers 1 to 5 indicate depressed regions of the pigment cell. The depression No. 1 faces the lateral side and the others, the medial side.

the best of a light-microscope's ability by Hesse (1902) and Burfield (1927). Finding the photoreceptive endings of the visual cells of <u>Sagitta</u> within the pigment cell depressions (Fig. 3), these pioneer workers assumed that the eyes of arrow worms belong to an inverted type like those in platyhelminthes. Recently, Ducret (1975, 1978) found everted type eyes in three species of the genus <u>Eukrohnia</u> belonging to the order Phragmophora. We will describe below each type of eyes separately.

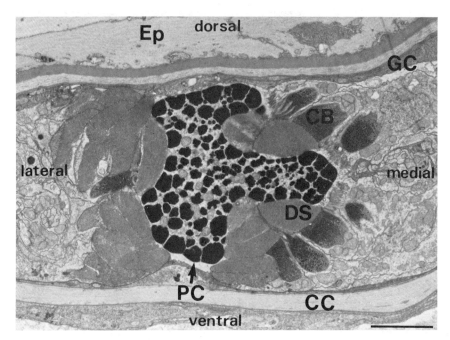

Fig. 4: An electron micrograph of transverse profile of an eye of S. crassa. CB, conical body; CC, capsule cell; DS, distal segment; Ep, epidermis; GC, glia cell; PC, pigment cell. Scale bar indicates 5 μm.

A. Inverted Eyes

Each eye is a rounded and dorso-ventrally flattened body enclosed in a capsule which is penetrated anteriorly by the optic nerve (Fig. 2). Main components of the eye are a centrally located pigment cell, numerous photoreceptor cells around it and glial cells among the photoreceptor cells (Fig. 4).

1. Pigment cell

All the species examined to date have not more than one pigment cell in each eye. As described above, the cell has several depressions (Fig. 3). It contains mitochondria, endoplasmic reticulum and cytoplasmic granules interspersed between pigment granules. The pigment granules, highly variable in shape and size (Fig. 5), appear to arise by fusion of smaller, less dense and sharply defined bodies.

It is known that the shape of the pigment cell does not change during the growth process but varies from species to species (Furnestin, 1954). Using two genera and 14 species collected from the epipelagic waters (upper 150 m) of Sagami and Suruga Bays in Japan, Nagasawa and Marumo (1976) classfied the pigment cells into three groups based on their external appearances (Fig. 6): E-shaped (upper four in the left column); star-shaped with five radiations (lower three in the left column); and T-shaped (the

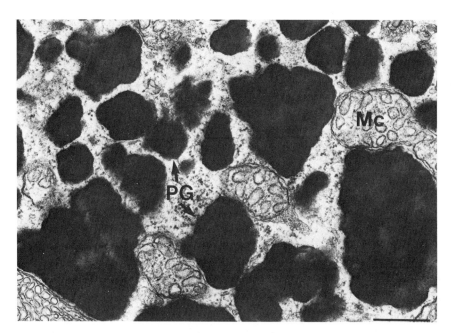

Fig. 5: A part of the pigment cell of S. crassa, showing various shape and size of pigment granules. Mc, mitochondrion; PG, pigment granule. Scale bar indicates 0,5 μm.

right column). Recently, they suggested a correlation between the shape of the pigment cell and the depth of habitat, i.e., E- and star-shaped ones occur in the species collected from the surface layer (0 - 30 m), and T-shaped ones, those from the middle (30 - 60 m) or deep (60 - 80 m) layers (Nagasawa & Marumo, 1982). By a survey of the eye size in 24 species (five genera) collected in a wider range of depths, Tokioka (1950) noticed a tendency of degeneration of the pigment cell with depth, smaller in mesoplanktonic (ca. 200 m in depth) species and none in bathyplanktonic (ca. 500 m depth) ones. On the contrary, the ratio of the eye size to the body length increases with the depth of the habitat. His interpretation is that deep-living chaetognaths develop a tendency to increase the visual function by enlarging the eye size with concomitant decrease of filtering effects due to pigment cells until they reach depths of 200 - 500 m.

Another interesting feature about the pigment cells is their ability to change in size in accordance with ambient lighting conditions. Ducret (1975) measured the size of light- and dark-adapted pigment cells in four species of Sagitta (S. enflata, S. setosa, S. minima and S. serratodentata) which were collected respectively, at noon to 1 pm and at 10 to 11 pm or maintained in the dark in a laboratory. Though no noticeable difference was found in young worms (3 - 4 mm in body length), the pigmented areas of adult worms of S. setosa (longer than 8 mm in body length) range from 0,09 - 0,15 mm^2 after light-adaptation in contrast to 0,12 - 0,20 mm^2

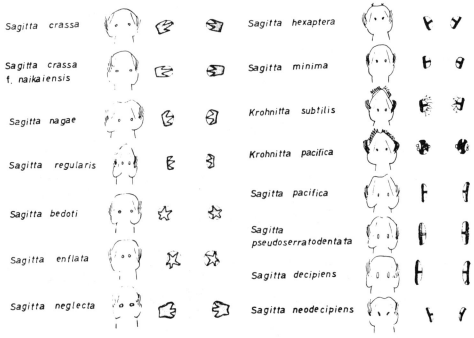

Fig. 6: Schematic representation of the head and the pigment cell in young chaetognaths (from Nagasawa & Marumo, 1976).

after dark-adaptation. It is uncertain whether this phenomenon is due to the migration of pigment granules or to morphological changes of the pigment cell itself.

2. Sensory cell

The cell body extends distally, via a connecting piece, a process which is composed of a conical body and a distal segment, and terminates proximally in an axon (Fig. 7). No synapses have been found between the eye and the brain. Assuming that the axons do not branch, the number of the sensory cells can be estimated by the number of optic nerves in transverse EM sections. The results were 500 - 600 cells in S. scrippsae (Eakin & Westfall, 1964) and 90 - 110 cells in S. crassa (Goto & Yoshida, unpublished).

a. Connecting piece

This short apparatus connects the distal process to the cell body. The cross sectional view reveals that the piece is ciliary in nature, showing nine peripheral doublets, but no central ones (Fig. 8). It can be seen that the nine ridges in the surface membrane of the connecting piece correspond in position to the nine doublets. Beneath the connecting piece lies an unpaired centriole which is different from a typical one. A striated rootlet exists normally.

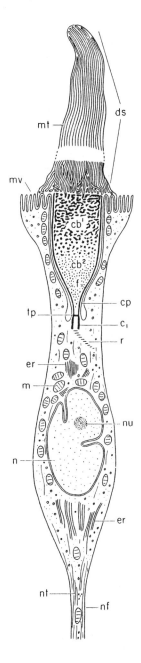

Fig. 7: Schematic representation of a sensory cell of S. scrippsae (from Eakin & Westfall, 1964). Incident light comes from below. c_1, axial centriole or kinetosome; cb^1, distal part of conical body composed of cords; cb^2, basal part of conical body composed of granules; cp, connecting piece; ds, distal segment of photoreceptor cell process; er, endoplasmic reticulum; f, two of the nine fibrils; m, mitochondrion; mt, microtubules; mv, microvilli; n, nucleus; nf, nerve fibre or axon; nt, neurotubule or neurofibril; nu, nucleolus; r, striated rootlets; tp, terminal plate.

b. Conical body

This unique structure is found only in chaetognaths. The term

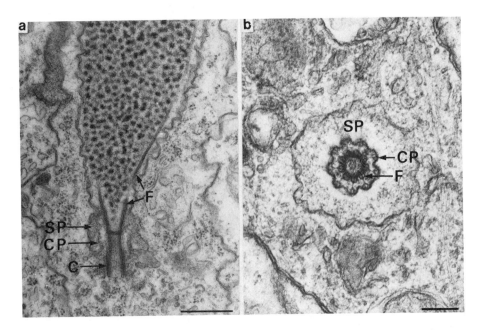

Fig. 8: Longitudinal (a) and transverse (b) section of connecting piece of S. crassa. C, centriole; CP, connecting piece; F, fibrils; SP, space between connecting piece and receptor cell proper. Scale bar indicates 0,5 μm in (a) and 0,1 μm in (b).

conical body originates from its shape, but Ducret (1978) described petaloid shape with nine "leaves" in S. tasmanica as a structure homologous to the conical body. We found in S. crassa both the round and petaloid structures co-existing in one and the same ocellus (Fig. 9). The conical body is usually divisible into two, a distal part containing irregular cords, and a basal one containing large and moderately dense granules (Fig. 7, cb^1 and cb^2, respectively). The conical body of S. tasmanica is provided distally with an additional component consisting of long filaments (Ducret, 1978).

Hesse (1902) and Burfield (1927) regarded the conical body as a dioptric apparatus from its refractive property in a fresh preparation (see Fig. 2) as well as from its location through which the incident light impinges on the receptive surface (Fig. 7). Though there is no physiological justification, recent workers such as Eakin and Westfall (1964) and Ducret (1978) seem to have no objection against such a notion. As will be described in the next section, the fact that the distal region of the conical body of a deep living worm, Eukrohnia hamata, differentiates further to a crystalline structure may be taken as an additional supporting evidence (Ducret, 1978).

Eakin and Westfall (1964) discussed other possibilities in the functional significance of the conical body. First, the conical body resembles superficially the paraboloid in vertebrate cones. However,

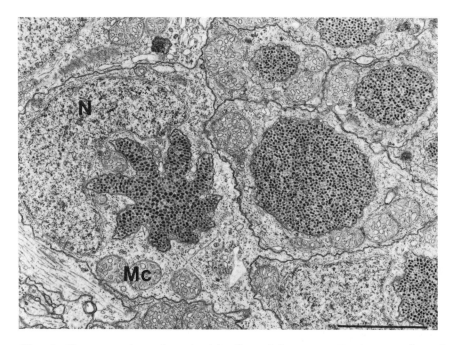

Fig. 9: Cross section of conical bodies of S. crassa. Both round (right) and petaloid (left) shapes exist in the same eye. Mc, mitochondrion; N, nucleus. Scale bar indicates 2 μm.

granules contained in the conical body are much larger than glycogen particles which are the predominant component of the cone paraboloid. This notion was supported by our unpublished histochemical studies which yielded PAS negative reactions in the region of the conical body. Secondly, based on the fact that there is an apparent connection between the base of tubule of the distal segment and a cord of the conical body (cb^1 in Fig. 7), they postulated a possibility that the cords of the conical body could be the incipience of the tubules. Unfortunately, there is no convincing evidence as regards the function of this unique structure.

c. Distal segment

This part has been considered to be the site of photon capture. As is evident from the foregoing descriptions, structures such as the conical body and the distal segment are connected to the cell body by a piece of cilium. In other words, the photoreceptor process is basically ciliary in nature. It is for this reason that the chaetognathan photoreceptors are thought to have an evolutionary relationship to chordates, rather than to platyhelminthes, as was assumed by earlier workers because of the inverted arrangement of the receptive site.

Eakin and Westfall (1964) considered that the distal segment of S. scrippsae is made up of a mass of microtubules and that these longitudinally arranged tubules do not run straight but are wavy so that

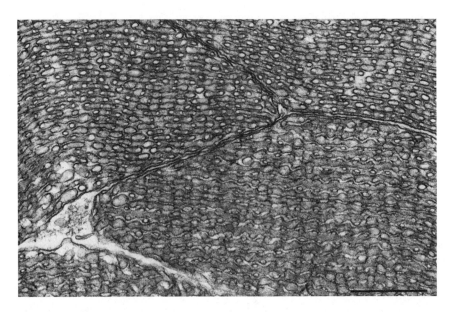

Fig. 10: Longitudinal section of the distal segment of S. crassa. Scale bar indicates 0,5 μm

they often appear as rows of short fragments when the segment is cut lengthwise. Ducret (1978), on the other hand, regarded the internal structure in the distal segment of the genus Sagitta as lamellae. As shown in Fig. 10, our observation on S. crassa revealed an even more complicated structure (Goto, unpublished). A possibility of annulate lamellae may be suggested.

As regards the orientation of the internal structure with respect to incident light, Ducret (1978) postulated an interesting hypothesis. As described above, she observed an increase in the size of the pigmented area in darkness. If such a change is due to a variation in the size of the pigment cell, the distal segment which is located in the depression of the pigment cell will be pushed out, resulting in a bend. Thus, the internal structure which oriented in parallel with the axis of the incident beam in light becomes perpendicular in dim light, a more efficient orientation for photon capture.

B. Everted Eye

Ducret (1978) reported that Eukrohnia hamata, E. fowleri and E. proboscidae have everted eyes, the distal process of each sensory cell pointing outside (Fig 11). The eyes of the latter two species have pigment cells but that of the former one lacks it. The sensory cells differ from those of the inverted type in several points. First, the connecting piece is a ciliated type but the rootlet associated with it is extremely long with no striations. Second, the conical body consists of four zones, having an

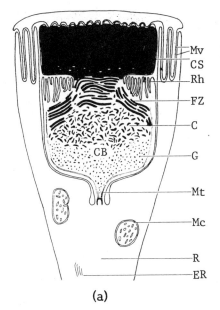
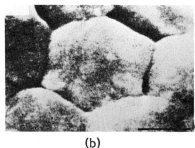

Fig. 11: Schematic representation of a sensory cell of Eukrohnia hamata (a) and the distal surface viewed by a scanning electron microscope (b) (from Ducret, 1978). C, cords; CB, conical body; CS, crystalline structure; ER, endoplasmic reticulum; FZ, fibriller zone; G, granules; Mc, mitochondrion; Mt, microtubule; Mv, microvilli; R, rootlet; Rh, rhabdome. Scale bar indicates 2 μm.

additional one to the layered type of S. tasmanica. The additional zone which is opaque is considered by Ducret (1978) to be analogous to the crystalline cone of athropods. Third, the photoreceptive site is not an internal structure but an outgrowth (microvilli) of the cilium-associated process (Fig. 11). Here again, she assumed them to be comparable to the rhabdomeres of the arthropod eyes. Her conclusion seems to be that they are analogous to the ommatida of arthropods.

The number of ommatidia of E. hamata varies with the habitat, 120 - 200 ommatidia in the Antarctic and Arctic Oceans and 80 - 120 in the Atlantic Ocean (Ducret, 1975). This tendency may depend on the depth of the habitat because the worms are bathypelagic in tropical and temperate Atlantic in contrast to epipelagic in both polar oceans.

The everted eye as in Eukrohnia is not characteristic of the order Phragmophora. Our unpublished observations on Spadella, another genus belonging to Phragmophora, revealed that their eyes rather resemble those of Sagitta, a species which belongs to the order Aphragmophora.

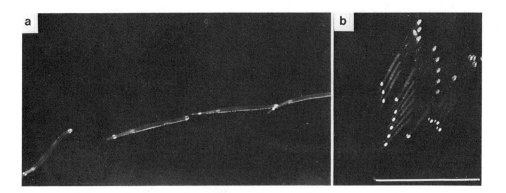

Fig. 12: Two kinds of phototactic responses of S. crassa photographed from a side with the camera shutter kept open by delivering a series of strobe flashes (five per sec) from above.
a. Quick target-aiming response induced by sudden light intensity reduction. The worm swims straight toward the light source.
b. Slow tactic response by repeatedly swimming upward and passively sinking. Scale bar indicates 1 cm.

LIGHT DEPENDENT REACTIONS

A. Phototaxis

It is reported by marine ecologists that some species of chaetognaths show positive phototaxis. But there are few laboratory experiments on photic behavior. Esterly (1919) showed that S. bipunctata is strongly positive to light ranging from that of a 15 W lamp at 50 cm to daylight which is about 3 800 times as intense. The geotropism is predominantly negative in darkness and in light that is not too bright. In a well lit room with white walls geotropism becomes strongly positive. Pearre (1973) confirmed the same tendency in S. elegans.

Goto and Yoshida (1981) analysed the behavioral sequence in positively phototactic species, S. crassa. As shown in Fig. 12, this animal exhibits two types of behavior, slow tactic and quick target-aiming. Arrow worms repeat normally upward swimming followed by passively sinking. The slow tactic movements are achieved by taking orientations of the body axis inclined more frequently toward the light source when it is optimal in intensity, rather than away from it (Fig. 13). Quick target-aiming reactions, on the other hand, are induced either upon reduction in light intensity or by water-borne vibration after a certain period of illumination. Following such a stimulation the worm steers itself in less than 0,3 sec toward a direction which is brightest around the animal and then starts swimming straight to this target at velocities greater than 14 cm/sec. The effect of light and relevant reactions induced are summarized in Fig. 14 (Goto & Yoshida, 1981). Using two-light methods of experiments and unilateral blindings, we concluded that both of the light

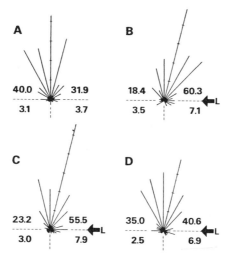

Fig. 13: Inclination pattern of the body axis of worms measured at 15-degree intervals in darkness (A: Dark control, n=160) and in a horizontal light beam of various intensities (B: 500 lux, n=141; C: 5 000 lux, n=164; D: 50 000 lux, n=160). Lengths of each radiating line show the percentages of worms inclined in that direction (calibrations on the longest line show 5 % intervals. Since upright (90°) and upside down (270°) orientations are irrelevant to the light oriented responses, the number of worms which took such orientations were summed separately. Numbers in each quadrant are total percentages in the respective direction (from Goto & Yoshida, 1981).

oriented responses are telophototaxes (Goto & Yoshida, unpublished). The morphological basis for the ability to locate a target with lenseless eyes may be sought in the sterotyped arrangement of the receptor endings assisted by shielding effects of the pigment cell.

B. Vertical Migration in the Sea

Diurnal vertical migration has been known in many zooplanktons. Some species of arrow worms show such a behavior, descending rapidly from the surface at the early post-dawn hours and rising up to the surface at dusk. Michael (1911) held that S. bipunctata seeks a depth where the light intensity is at a level of twilight intensities. Using S. setosa and S. elegans, Russel (1931) also suggested that the worms tend to gather at a depth of an optimum light intensity. Such a tendency becomes clearer with age so that larger animals leave the surface at dawn before a measurable change in light intensity occurs. Fig. 15 shows an example of diurnal changes in the vertical distribution pattern of S. crassa (Goto, unpublished).

According to Murakami (1959), the diurnal vertical migration of S. crassa depends more strongly on the light intensity than any other changes

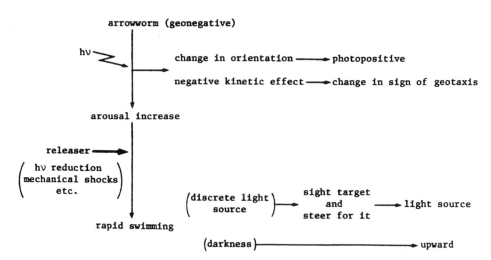

Fig. 14: Summary diagram of the results of photic reactions (from Goto & Yoshida, 1981).

in physical factors. The number of worms caught near the bottom layer at about noon is more than three times that caught from the surface or the middle layers. Here also, age differences exist, at dawn and dusk large worms are found more abundantly in the surface layer than in any other layer.

In S. crassa, Hirota (1979) also noticed an increase in the number of large worms in surface waters, at night. Also the total number of large worms caught at all depths is greater in the evening. To explain the greater number of worms caught at night, he suggested that adult animals may concentrate near the bottom during the daytime and thus be beyond the reach of the net.

In chaetognaths, the light-dependent changes in the vertical distribution pattern in the sea are evident as above, but there is no direct demonstration as yet about the controlling mechanisms of these ecologically important phenomena.

ACKNOWLEDGEMENTS

We thank Professor M.A. Ali of Université de Montréal and Dr. M. Yamamoto of Ushimado Marine Laboratory for their invaluable critism of the manuscript and Mr. W. Godo for his technical help. The work was supported in part by a grant in aid from the Ministry of Education of Japan to M.Y.

REFERENCES

Bullock, T.H. (1965) Chaetognatha. In: Structure and Function in the Nervous Systems of Invertebrates. Vol. II. Ed. T.H. Bullock & G.A.

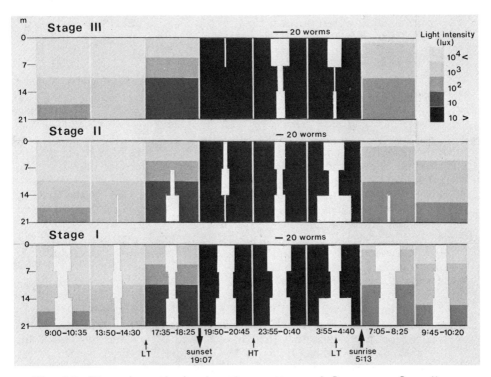

Fig. 15: Diurnal vertical migration pattern of S. crassa. Samplings were made by vertically towing a net in three separate layers (Surface - 0 to 7 m; Middle - 7 to 14 m; Deep - 14 to 21 m) on August 1 - 2, 1978 near Shibukawa coast in the Seto Inland Sea. The width of the white bar at each layer is the total number of worms collected in three trials. Stages I to III mean the degrees of maturity; the worms in stage III have long ovaries with large eggs, those in stage I are immature and those in stage II are intermediate forms between stages I and III. The background density shows the light intensity monitoring measured simultaneously at five layers with 5 m intervals in depth (calibrations shown at upper right corner). HT, high tide; LT, low tide.

Horridge. San Franciso and London, W.H. Freeman and Co., p. 1560-1564.
Burfield, S.T. (1927) Sagitta. Liverpool Mar. Biol. Comm. Mem. 28: 1-104.
Ducret, F. (1975) Structure et ultrastructure de l'oeil chez les Chaetognathés (genres Sagitta et Eukrohnia). Cah. Biol. Mar. 16: 287-300.
Ducret, F. (1978) Particularites structurales du système optique chez deux Chaetognathés (Sagitta tasmanica et Eukrohina hamata) et incidences phylogénétiques. Zoomorphol. 92: 201-215.
Eakin, R.M. & Westfall, J.A. (1964) Fine structure of the eye of a chaetognath. J. Cell Biol. 21: 115-132.
Esterly, C.O. (1919) Reactions of various plankton animals with reference

to their diurnal migrations. Univ. Calif. Publi. Zool. 19: 1-83.

Furnestin, M.L. (1954) Intérêt de certains détails anatomiques jusqu' ici peu étudiés, pour la détermination des Chaetognathes. Bull. Soc. Zool. France 79: 109-112.

Goto, T. & Yoshida, M. (1981) Oriented light reactions of the arrow worm Sagitta crassa Tokioka. Biol. Bull. 160: 419-431.

Hesse, R. (1902) Untersuchungen über die Organe der Lichtempfindung bei niederen Thieren. VIII. Weitere thatsachen. Allgemeines. Z. Wissensch. Zool. 72: 565-656.

Hirota, R. (1979) Diurnal migrations and patch formations in macroplankton in the inland sea. Mar. Sci Monthly, Tokyo 11: 627-631 (In Japanese).

Hyman, L.H. (1959) The Invertebrates. Vol. V. Smaller Coelomate Groups. The enterocoelous coelomates - phylum Chaetognatha. New York, London, Toronto, McGraw-Hill, p. 1-71.

Michael, E.L. (1911) Classification and vertical distribution of the chaetognaths of the San Diego region. Univ. Calif. Publi. Zool. 8: 21-186.

Murakami, A. (1959) Marine biological study on the planktonic chaetognaths in the Seto Inland Sea. Bull. Naikai Reg. Fish. Res. Lab. 12: 1-69 (In Japanese with English abstract).

Nagasawa, S. & Marumo, R. (1976) Identification of young chaetognaths based on the characteristics of eyes and pigmented regions. Bull. Plankton. Soc. Japan 23: 96-102 (In Japanese with English abstract).

Nagasawa, S. & Marumo, R. (1982) Vertical distribution of epipelagic chaetognaths in Suruga Bay, Japan. Bull. Plankton Soc. Japan 29: 9-23 (In Japanese with English abstract).

Pearre, S., Jr. (1973) Vertical migration and feeding in Sagitta elegans Verrill. Ecol. 54: 300-314.

Russel, F.S. (1931) The vertical distribution of marine macroplankton. X. Notes on the behavior of Sagitta in the Plymouth area. J. Mar. Biol. Assoc. U.K. 17: 391-407.

Tokioka, T. (1950) Notes on the development of the eye and the vertical distribution of chaetognatha. Nature Culture, Kyoto 1: 117-132 (In Japanese with English abstract).

PHOTORECEPTION IN ECHINODERMS

M. YOSHIDA, N. TAKASU
Ushimado Marine Laboratory
130-17 Kashino
Ushimado, Okayama 701-43, Japan

S. TAMOTSU
Physiology Department
Hamamatsu University of Medicine
Hamamatsu 431-43, Japan

ABSTRACT

The following photic reactions which are induced mostly by non-ocular receptive systems are reviewed in the first section: (1) Light-dependent aggregation of echinoplutei, (2) Amoeboid movements of directly photoresponsive chromatophores in diadematid sea urchins, (3) Reflex reactions such as spine jerkings in diadematids, contractions of pharyngeal and lantern retractor muscles in holothuroids and echinoids, respectively, and arm tip movements in asteroids, and (4) Coordinated reactions such as covering reactions in echinoids and phototaxes in echinoids and asteroids. The second part concerns the functional morphology of ocelli in holothuroids and asteroids: the ocelli in the former are flat, solitary and located at the proximal end of the tentacular nerve and those in the latter, cup-shaped, aggregated to form a compound ocellus and located at the distal end of the radial nerve. The two types of ocelli share several common features: (1) a cilium arises from the apical projection of the sensory cell, (2) the receptive sites are categorized as rhabdomeric microvilli, and (3) the sensory microvilli undergo a light-dependent turnover. In the starfish ocelli, both the density of intramembranous particles on the P-face of freeze-fractured sensory microvilli and the strength of fluorescence induced in the ocellar lumen upon reduction of lightly denatured tissue increase in darkness. The results implicate that substances incorporated in the receptive endings of the starfish ocelli resemble those in well-differentiated photoreceptors.

INTRODUCTION

A wide variety of reflex and behavioral reactions are known in adult as well as in larval forms of echinoderms. Some of those reactions are affected by the level of ambient lighting conditions and some are triggered by changes in light intensities (for reviews, see Yoshida, 1966, 1979; Binyon, 1972; Millot, 1975). Below, we shall review at first a few selected works on various photic reactions which are elicited mostly by non-ocular receptive mechanisms. In the second part, we shall describe functional morphological aspects of the ocellar photoreception, which, despite the limited occurrence in the echinoderms, may bear a broad importance for consideration of primitive ocular photoreceptive systems.

VARIOUS PHOTIC REACTIONS

A. Photoresponses in Larvae

Many embryologists are familiar with the phenomena that echinoplutei tend to gather, forming vertical streaks in the holding dish. The position of the streaks shifts with the time of day when the dish is placed by the window side, suggesting a photic effect on larval swimming activities. Neya (1965) showed, in Hemicentrotus pulcherrimus, that the speed of downward movements was always higher than that of the upward movements and that the former was affected by light to only a small extent while the latter became markedly slower under brighter conditions (Fig. 1). If a vertical, parallel light was applied from below or from above, larvae contained in a square trough aggregated near the fringe of the beam (Fig. 2c) and the postion of aggregation shifted more and more into the beam upon reduction in intensity from 138 lux down to 0,1 lux. Such a concentrated mass of larvae began to disperse again when the beam was switched off. A horizontal, parallel beam passing below the water surface was without effect in inducing the aggregation (Fig. 2a). When a beam was diverged by 5 - 20 degrees so that it transversed the water surface halfway in the trough, larvae tended to gather near the area where the water surface was illuminated by the diverged beam (Fig. 2b).

From these results in combination with some others, Neya (1965) concluded the following: In darkness, larvae swim upwards by ciliary movements. At the water surface, they move about randomly. Illumination inhibits their upward movements, and probably their ciliary activity. Since the specific gravity of larvae is higher than that of the sea water (Yasumasu, 1963), plutei will sink down when they eventually enter a brighter condition. On reaching the bottom, larvae move around and start upward movements again. Here, however, they take a path not in the beam but slightly outside of it and repetition of such movements results in the apparent aggregation near, but somewhere outside of the beam. The implication of the above interpretation is the direct effect of light on the ciliary activity but the underlying mechanism involved is yet to be discovered.

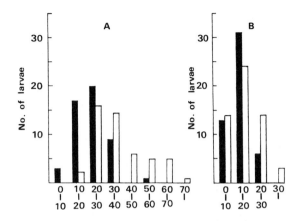

Fig. 1: Effect of light on the swimming speed of echinoplutei of Hemiceptrotus pulcherrimus. A: upward swimming. B: downward swimming. The intensities of light from above: 20 lux for black bars and 2 000 lux for white bars. The abscissae are time in sec spent by each pluteus to traverse vertically a fixed distance of 7,4 mm. Histograms are constructed by grouping the number of plutei at 10 sec intervals (from Neya, 1965).

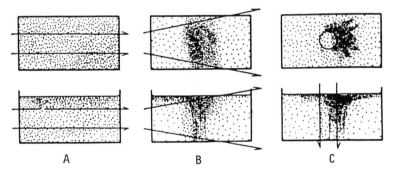

Fig. 2: The distribution pattern of echinoplutei of Hemicentrotus pulcherrimus in a square trough (5 x 10 x 5 cm), illuminated either horizontally (A, B) or vertically (C) for 30 min. The beam was made parallel, 1 - 2 cm diameter, in A and C; and diverged by 5 - 20 degrees in B. Arrows indicate light paths. The upper row shows patterns viewed from above and the lower row, from a side. The patterns are schematized from photographic recordings (from Neya 1965).

B. Color Changes

Sea urchins belonging to Diadematidae darken in light and fade in darkness, due to changes in the contour of the mass of pigment granules in chromatophores which are distributed over the epidermis (Kleinholz, 1938). The evidence for direct photoresponsiveness of the chromatophores was obtained with a Japanese species, Diadema setosum, by Yoshida (1956). If a pale region of the epidermal surface of a dark-adapted piece was illuminated by a minute spot of light, pigment granules in the stimulated branch moved to cover the illuminated spot while granules in other parts of the same chromatophore remained concentrated (Fig. 3).

The diadematid chromatophores appeared rather unusual in that those which had been heavily distorted in shape by centrifugal forces still retained the ability to respond to light (Yoshida, 1960). Weber and Dambach (1972) noted, from their electronmicroscopic studies, that in darkness, chromatophores stretch their branches actively into the intercellular space of the surrounding tissue. This property suggests an amoeboid nature. They further succeeded in isolating single cells from the epidermal tissue (Weber & Dambach, 1974). When placed in darkness for 30 min, the isolated cells took an ovoid shape, which started to transform rapidly to a stellate shape upon illumination (Fig. 4). The fact that expansion and contraction of branches were inhibited by adminstration of cytochalasin B, a substance which is known to interfere with the cytoplasmic contractile mechanisms, led Dambach and Weber (1975) to suggest involvement of microfilaments in the amoeboid movements.

The demonstration of the amoeboid nature of the diadematid chromatophores convinced Yoshida (1979) to modify his interpretation. Though the direct photoresponsiveness of the chromatophore can stand as it was, the behavior that he observed in 1956 was not due to the photosensitivity of the hyaloplasm of the cell but due to light that scattered from the illuminated spot.

C. Reflex Responses

Reflex responses that are induced by changes in photic conditions are known in asteroids, echinoids and holothuroids (for review, see Yoshida, 1966, 1979). Among these, spine jerkings of the sea urchin, Diadema antillarum, occurring in response to either an increase (Millott & Yoshida, 1959) or a decrease (Millott, 1954) in light intensities were so consistent (Fig. 5; Millott & Yoshida, 1960a) that they permitted detailed experimental analyses. Scanning over the body surface by obliterating a spot of light revealed the maximal sensitivity at ambulacral margins, next at the ambulacral center and the least at the interambulacrum. In echinoids, the superficial nervous system is formed by nerves that emerge through one of the pore pair at the ambulacral margin. Thus, the gradation in sensitivity appeared to correspond with the distribution of the superficial nervous system.

Millott (1975) has suggested that the dermal light sensitivity is a result of the superficial nerve fibres being photosensitive. Indirect evidence that appeared to support such a notion came from photic stimulation of the radial nerve cord which elicited not only the spine

PHOTORECEPTION IN ECHINODERMS

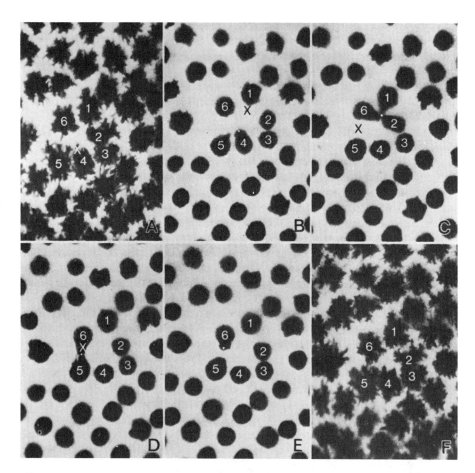

Fig. 3: Experiments in which a spot of light was projected onto various parts of dermal chromatophores of the sea urchin, Diadema setosum. A: light-adapted chromatophores (initial condition). The whole preparation was then placed in complete darkness except for the spot of light. Duration of darkness: A to B, 1,5 h; B to C; C to D; and D to E; 1 h. Crosses indicate places where the spot was projected at the beginning of dark periods. Circles indicate places where the spot had been projected until the photograph was taken. Between E and F, the spot was removed and the whole preparation was placed in darkness for 20 min and then under light for 20 min (from Yoshida, 1956).

responses at light-off in Diadema antillarum (Yoshida & Millott, 1959) but also impulse activities at light-on and -off in the radial nerve cord of D. setosum (Fig. 6; Takahashi, 1964). However, the impulses were intiated sometimes as long as several seconds after the light was turned off. It is not certain whether these photoinduced activities were in fact the result

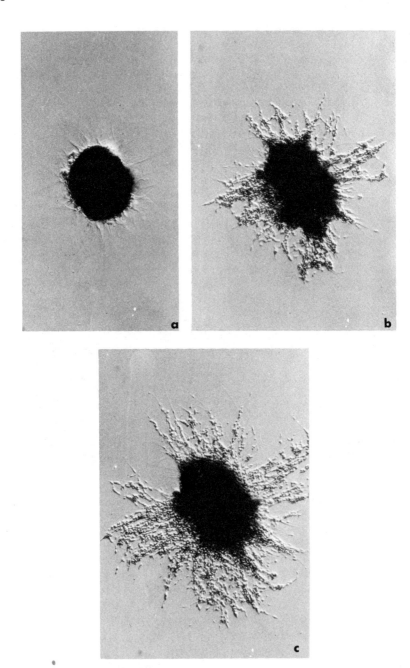

Fig. 4: Photoresponses of an isolated chromatophore of the sea urchin, Centrostephanus longispinus.

a: dark-adapted cell; b: after 11 min illumination; c: after 26 min illumination. X 400. (By courtesy of Professor W. Weber).

Fig. 5: Recordings of spine responses of the sea urchin, Diadema antillarum. The shadow of a spine cast on a slit screen was photographed by continually moving film. Traces of three separate reactions photographed at approximately hourly intervals are superimposed in the left half. The moment of light-off is shown by upward deflection in the middle trace. The bottom trace shows time signals at 1 sec intervals (from Millott & Yoshida, 1960a).

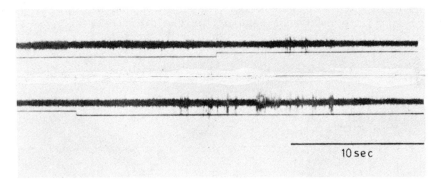

Fig. 6: Electrical responses to light-on and -off, recorded from isolated radial nerve cord of the sea urchin, Diadema setosum. Upward and downward deflection; light-on and -off respectively (from Takahashi, 1964).

of neuronal photosensitivity or whether some kinds of non-nervous structure in the radial nerve cord played a role of primary photoreceptor. The photoexcitability reported for the radial nerve cord of ophiuroids by Cobb and Stubbs (1981) is important in this respect. But, since these authors did not provide experimental details, we cannot discuss the neuronal photosensitivity in the radial nerve any further.

Whatever the primary photoreceptive site may be, the radial nerve is flat and homogenous, compared to the spiny surface of the test. Such a surface structure affords easier and more consistent manipulation of stimulating conditions. Using two spots of light (L_1 and L_2) applied in succession at spatially separated sites of the radial nerve, Millott and Yoshida (1960b) showed that the second light (L_2) applied immediately after obliteration of the preceding one (L_1) was effective in inhibiting the off-response to L_1. The position of L_2 should not necessarily coincide with that of L_1 but can be separated by some distance, 6 mm at maximum. The area within which the interaction between L_1 and L_2 occurs may be called the receptive field. These results led us to suggest that the overt response might be a result of spatial and temporal interactions between excitation and inhibition, the excitation being a rebound from inhibition due to light.

Later, Yoshida (1962, 1966) analysed characteristics of the interactions within the receptive field, by determining the extent of excitation due to the light (L_1) being off in terms of the inhibitory threshold of the second light (L_2). The summary diagram is shown in Fig. 7. The excitation at the spine base, which was termed the peripheral mechanism (surface factor) by Millott and Takahashi (1964), depends largely on spatial (d) and temporal (t) integration of excitation and inhibition occurring in the radial nerve cord. The extent of each of the centrally occurring phenomena (ambulacral factor) is determined not only by the total flux of lights (I_1,

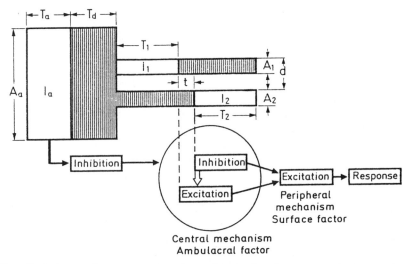

Fig. 7: Schematic representation of relation between photo-environmental factors and the mechanism involved in the shadow reaction of the sea urchin, Diadema setosum. A, size of area illuminated; I, intensity of light; T, duration of illumination or darkness. Subscript a stands for adapting; 1 for L_1 (conditioning light); 2 for L_2 (inhibitory light); and d for darkness. t, interval between extinction of L_1 and reillumination by L_2; d, spatial separation of the two lights. The terminology of ambulacral and surface factors is based on Millott and Takahashi (1963) (from Yoshida, 1966).

T_1, A for L_1; and I_2, T_2, A_2 for L_2) but also by the extent of a long lasting inhibitory effect due to pre-adaptation in a wider field (T_a, A_a, I_a), provided that T_d is sufficiently short.

A different type of interaction between excitatory and inhibitory mechanisms has been proposed to explain myoneural photosensitivities in the pharyngeal retractor muscle of a holothuroid, Cucumaria sykion (Pople & Ewer, 1954, 1958) and the lantern retractor muscle of an echinoid, Parechinus angulosus (Boltt & Ewer, 1963). These muscles respond to light with a variety of contraction patterns, which consist of two basic types, quick and delayed. The two mechanisms appeared to differ in the time course as well as in the spectral sensitivity so that a variety of contraction patterns were induced in response to repetitive, paired or prolonged stimuli by tactically manoeuvering the intensity, duration, frequency and/or spectral property of the stimulating light. For example, when two white lights were applied in succession, the muscles contracted in response to the second light as well as to the first. If, however, the second light was that of a mercury lamp, which differed spectrally from the first light of the tungsten filament lamp, the muscle relaxed rather than contracted. To explain these phenomena Boltt & Ewer (1963) postulated a group of units in the motor complex; one inhibitory citon and two excitatory citons, quick and delayed.

Another kind of reflex movements has been reported in a starfish, Asterias amurensis (Yoshida & Ohtsuki, 1966). The animal lifts up the arm tip at light-off and shifts it down at light-on. Removal of the ocellus does not abolish the reaction but raises the threshold about 10 times. On measuring the spectral sensitivity for the downward movements, we obtained two separate action spectra, peaking at 485 nm with intact ocelli and at 504 nm with eyeless preparations (Fig. 8). The shift in the peak does not necessarily imply the existence of two different photopigments, however. It is possible that a pigment which absorbs longer wavelengths in ocelli could make the blue shift in the action spectrum of the preparation with intact ocellus.

D. Coordinated Reactions

Reflex movements often proceed in a highly coordinated fashion, as revealed by jerkings in a group of spines of sea urchins, e.g., Diadema antillarum (Millott & Takahashi, 1963), at the swimming arousal in feather-like arms of a crinoid, Antedon bifida (Dimelow, 1958), etc.

The covering reaction termed by Millott (1955) is interesting in this respect. A number of regular and irregular echinoids take up opaque objects such as eel grass, shell gravel, and so forth (for review, see Yoshida, 1966). Millott (1956) observed a beautiful performance of tube feet and spines while Lytechinus variegatus was covering its body surface. These peripheral organs, characteristic to echinoids, collaborated in taking up objects from the substrata and carrying them to illuminated parts of the dermis. The covering objects were shifted over various routes, tracking a moving light source (Fig. 9). Removal of the lantern, inevitably together with the oral nerve ring, did not prevent the action of taking up gravel from the substratum. The coordination may simply be a peripheral phenomenon, being achieved through sequential reflexes which are

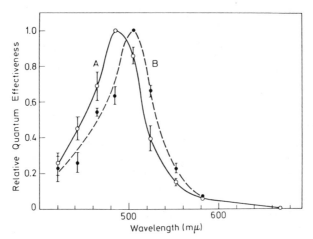

Fig. 8: Relative quantum efficiencies of photoreceptor systems of the starfish, Asterias amurensis, in inducing downward movement of the arm tip at light-on. Arms having (A) and lacking (B) intact eyespots are compared. Vertical bars; range of standard errors (from Yoshida & Ohtsuki, 1966).

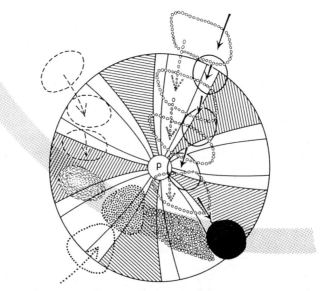

Fig 9: Analysis of the covering reaction in the sea urchin, Lytechinus variegatus. The diagram shows the placing of stones over localized, brightly illuminated areas of the body surface (stippled). Each stone, shaded by distinctive convention (lines, dots, circles or solid black) was moved approximately over the routes shown by successive outlines and arrows. The ambulacra are distinguished by cross-line shading. P, periproct (from Millott, 1956).

triggered in combination with the dermal light sense and the mechanoreceptors of the tube feet and spines, though Millott (1956) has carefully left the possibility of involvement of the nerve ring (CNS) in such highly coordinated performances shown by intact animals.

Tactic movements require coordination of locomotory organs controlled by a central commanding mechanism located in the oral nerve ring (Smith, 1950). References summarized by Yoshida (1966) as regards the polarity of taxes showed that crinoids, echinoids and holothuroids are mostly negative and asteroids, mostly positive.

Echinoids do not possess ocelli and the entire body surface has been assumed to be photosensitive (see above). Under these circumstances, an uneven illumination of the receptive surface due to the opacity of the body must be used as a cue for the animal to locate the light source. Such an asymmetrical information, after being processed in the oral nerve ring (CNS), will lead the animal to move toward or away from light by coordinated movements of tube feet.

A sea urchin, Temnopleurus toreumaticus, was found to be positively phototactic (Yoshida, 1966). Aboral halves of the urchins were cut off a little below the ambitus together with viscera, leaving one or two ambulacra intact (Fig. 10). In darkness, the majority of the operated animals moved with the intact ambulacra on their back (control, Fig. 11a). If the intact ambulacral zone was illuminated from outside, the moving direction was reversed, so that 68 % of the animals moved toward the light and none, away from it (Fig. 11b). If the inner surface was illuminated, the percentages of animals which moved away from the light source were higher than those which moved toward the light (Fig. 11c). In short, the urchins moved towards the direction where a wider area of the body surface was illuminated.

Information on pathways to the CNS was studied by leaving a sector with two radial nerve cords intact, and further by transecting either one of the of the intact nerve near the oral nerve ring. If these urchins were

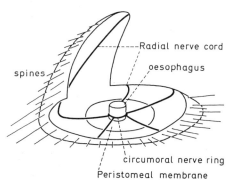

Fig. 10: The type of preparation used for analysis of tactic movements in the sea urchin, Temnopleurus toreumaticus (from Yoshida, 1979).

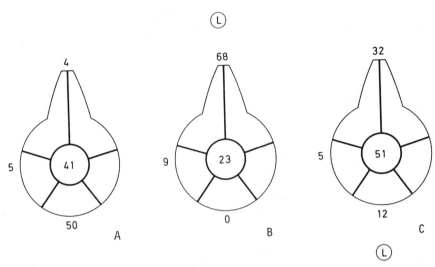

Fig. 11: Schematic representation of the direction of movements of the sea urchins, Temnopleurus toreumaticus, operated as in Fig. 10. Circumoral nerve ring and radial nerve cords are indicated by thick lines. Intact area was illuminated by a parallel beam (L) either from outside (B) or from inside (C). A, control experiments in darkness. Numbers above and below each figure show percentages of urchins which moved in the corresponding direction. Those on the left are summed percentages of urchins which moved sideways. Those at the center are percentages of urchins which did not move (from Yoshida 1966).

placed in between two light sources facing each other, they never moved toward the light source on the transected side. This experiment suggests that the information from the illuminated sector should be fed into the oral ring nerve where the motor centers are located at each junction with the radial nerve cord (Smith, 1950).

In ocellus-bearing animals, tactic movements can be demonstrated more clearly. Berrill (1966) noted in a Hawaiian synaptid holothurian, Opheodesoma spectabilis, that this animal was negatively phototactic, staying in a shaded area in the sea. The sensitivity was highest in the region of the crown of tentacles, which possessed a pair of ocelli at each proximal end. Removal of the ocelli did not completely abolish the response so that the dermal light sense seemed to co-exist.

The starfish, which also bear an ocellus at the tip of each arm, are mostly positively phototactic. Using Asterias amurensis, Yoshida and Ohtsuki (1968) determined the direction of starting movements when an arm of intact or eyeless animals was either illuminated from above or shaded by an opaque object. As shown in Fig. 12, light adaptation was essential in making animals photosensitive, so that more than 80 % of the

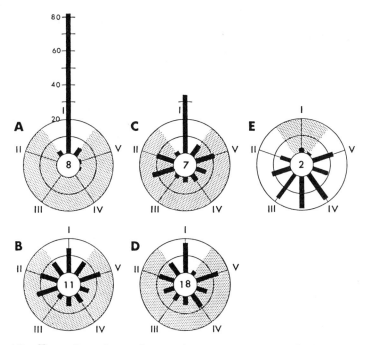

Fig 12: The direction of starting movements of the starfish, Asterias amurensis, when an arm (I) was illuminated (200 lux) from above (A - D) or shaded (E). Shaded areas are shown by stipplings. The lengths of bars and the numbers at the center represent, respectively, percentages of animals which started to move in the corresponding direction and those which did not move. Outer, middle and inner circles show, respectively, 20, 10, and 0 %. A, C and E; light-adapted animals (200 lux for 30 min). B and D; dark-adapted animals (4 h). A, B and E; intact animals. C and D; eyeless animals. Number of experiments; A, 100; B, 135; C, 90; D, 88; E, 106 (from Yoshida & Ohtsuki, 1968).

light-adapted animals started to move with the illuminated arm forward (Fig. 12a), whereas dark-adapted ones moved more or less randomly (Fig. 12b). The results obtained with light-adapted, eyeless animals (Fig. 12c) were similar to those of the intact ones (Fig. 12a), though the positivity was less evident. The extra-ocular photoreception like in the above holothurian, O. spectabilis, may play a role to some extent. If one of the arms of light-adapted specimens was shaded (Fig. 12e), a gradient occurred in the percentages of positively responding animals, increasing with distance from the shaded area. Experiments using preparations with the nerve ring transected produced such a response pattern as to suggest a blockage of sensory input and motor-commanding output at the site of transection. In any case, it is now safe to say that, both in echinoids and asteroids, the information of the environmental light conditions as perceived through their dermal and/or ocellar photosensitivity must be processed in motor centers of the nerve ring in order to achieve coordination in locomotory organs.

FUNCTIONAL MORPHOLOGY OF OCELLI

In echinoderms, organized ocelli are known only in one species of the holothurian and in most of the asteroids. The ocelli of the two groups differ as regards their location in the body, near the CNS in the former and at the farthest end of the arm in the latter, as well as in their gross appearance, flat and solitary in the former and cup-shaped and aggregated in the latter. The two kinds of ocelli share a common feature, however, so that each sensory cell is provided with a single cilium at the apex and gives rise to microvilli from the plasma membrane. Because of their having a cilium in the sensory cell as well as their sharing many embryological features with most deuterostomes, these echinoderm photoreceptors have once been placed in the ciliary line of evolution (Eakin, 1963). Based on recent findings Eakin (1982) presented a revised scheme for evolution of photoreceptors by deleting the echinoderms, though he left a possibility, from his diphyletic viewpoint, that ancestral forms could have possessed some kind of ciliary photoreceptors.

A. Gross Appearance

1. Holothuroids

Berrill (1966) described in a synaptid holothurian, Opheodesoma spectabilis, paired pigmented patches on each of the 15 tentacular nerves near the proximal end where they join the nerve ring (Fig. 13a, from Yamamoto & Yoshida, 1978), Histological preparations (Fig. 13b) reveal that the pigmented patches themselves are not the constitutive members of the ocellus but simply envelop the receptive cells. Each ocellus is composed of sensory cells, a network of supportive elements and bundles of nerve fibers. When the epineural cavity is exposed by peeling off the surrounding tentacular tissues, the ocellus is seen covered with a tuft of microvilli and a few long cilia (Fig. 13c).

The sensory cell is composed of three parts (Fig. 13e): an apical part projecting as a hillock into the epineural cavity, a slender middle part about 10 µm long and 2 µm wide, and a swollen basal part containing a nucleus. Each apical process is provided with a cilium which is associated basally with a pair of centrioles and rootlets, about 1 µm long with striation at 60 nm intervals. The microtubules in each cilium show a 9 + 2 pattern close to the centriol, but these typical axonemes usually become disordered in various ways (Fig. 13d); some cilia have only one central singlet, some lose the circular arrangement of the doublets, and some have only singlets in varying numbers. Microvilli that project into the epineural cavity (Fig. 13e) are not formed by modification of the ciliary sheath but arise as outgrowths of the plasma membrane. The nuclei (3 µm X 5 µm) are oval and less electron dense than those of the supportive cells. The proximal end of the soma gives rise to a process, about 0,5 µm in diameter, that is directed toward the underlying bundles of nerve

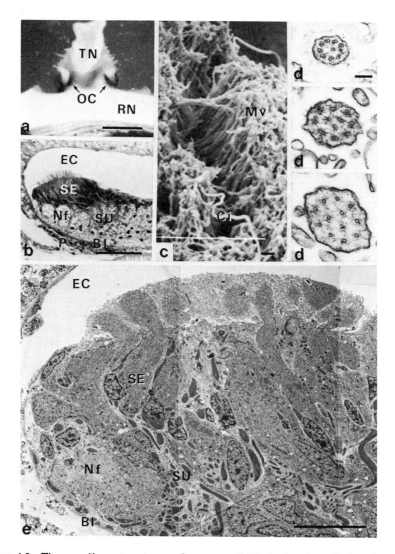

Fig: 13: The ocellar structure of a synaptid holothurian, Opheodesoma spectabilis. **a**; steroscopic appearance. The surrounding circumstomeal tissues have been peeled off. Scale, 0,5 mm. **b**; cross-section through the ocellar region of the tentacular nerve. Scale, 50 μm. **c**; scanning electron micrograph of ocellar surface in a dark-adapted specimen. Scale, 1 μm. **d**; various types of disordered arrangements of ciliary microtubules. Scale, 0,1 μm. **e**; a montage of a low power electron micrograph of the ocellar region. Scale, 10 μm. Bl, basal lamina; Ci, cilium, EC, epineural cavity; Mv, microvilli; Nf, bundle of nerve fibers; OC, ocellus; P, pigmented area; RN, ring nerve; SE, sensory cell; TN, tentacular nerve; SU, supportive cell;. (a - d, from Yamamoto & Yoshida, 1978, and **e**, by courtesy of Dr. Yamamoto).

fibers. This process contains mitochochondria and numerous microtubules and may be assumed to be an axon.

2. Asteroids

The pigmented spot situated on the oral surface of the tip end of each arm is called an optic cushion. The ultrastructure of the starfish ocelli has been studied in several species by Eakin and Westfall (1962, 1964), Vaupel von Harnack (1963), Eakin and Brandenburger (1979), Brandenburger and Eakin (1980), and Penn and Alexander (1980). Below, we shall describe briefly the gross appearance observed in a Japanese species, Asterias amurensis.

The optic cushion is composed of approximately one hundred ocelli, each being recognizable as an aggregation of pigmented cells. The term compound ocellus is sometimes used, therefore. Each ocellus (Fig. 14a) is covered with a layer of corneal cells, below which sensory and pigmented cells line to form a cup-shape (35 µm deep and 20 µm wide). Numerous microvilli arising from the apical projection of the sensory cells are found in the cup lumen (Fig. 14b). In some species such as Nepanthia belcheri (Fig. 14c), the arrangement of the microvilli is so regular as to resemble those reported in the crustacean rhabdomes (Penn & Alexander, 1980). A single cilium arises from the tip of the apical projection of each sensory cell (Fig. 14b). The sensory cells send single axons toward the nerve plexus situating below the ocellar cup. The pigmented cells also project few and short microvilli into the cup lumen. They are tapered proximally and contain tonofilaments, suggesting a supportive function

B. Microvillar Membrane

1. Light-dependent membrane turnover

Light- and dark-treatments often affect various aspects of the peripheral and/or intracellular structures of the invertebrate as well as vertebrate photosensory cells (for review, see Waterman, 1982). Such is true for the holothurian ocelli (Yamamoto & Yoshida, 1978), the most prominent changes occurring in the microvilli, which became markedly longer and more regularly arranged in dark-adapted specimens (compare Figs. 15a with 15b). Coated vesicles and invaginations of the plasma membrane became abundant with light-adaptation (Fig. 16a). The coatings over the vesicles (Fig. 16c) and those over the invaginating membranes (Fig. 16b) were much alike, possessing the characteristic basket structure. Fragments of microvilli were often present within the invaginating cavities (Fig. 16d). Such an appearance suggests an engulfing process at the cell surface. Upon being pinched off, a double-ring structure with the coat on its surface would result (Fig. 16e). The engulfed microvilli might either be digested within the vesicle leaving an empty coated vesicle (Fig. 16f and 16g) or the pinched-off structure might transform into a double-ring structure by losing the coating (Fig. 16f, double arrow). There were other variations in the shape of the coated vesicles: flat, dumbell-shaped (Fig. 16b), and horse-shoe shaped with partial coating (Fig. 16h). These observations suggest that phagocytotic activities were high in the light-adapted specimens. It is noteworthy that

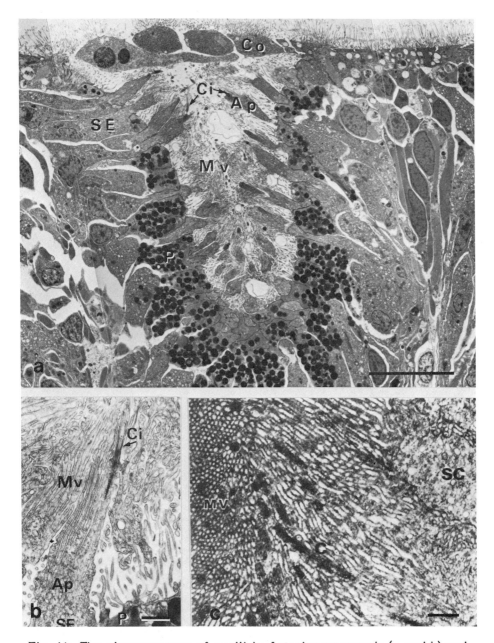

Fig. 14: The ultrastructure of ocelli in <u>Asterias amurensis</u> (a and b) and <u>Nepanthia belcheri</u> (c). **a**; a gross appearance. **b**; cilium and microvilli arise separately from the apical projection. **c**; ordered arrangement of microvilli. Scale, 10 μm in **a** and 1 μm in **b** and **c**. Ap, apical projection; C and Ci, cilium; Co, corneal cell; Mv, microvilli; P, pigmented cell; Sc and Se, sensory cell (**c** from Penn & Alexander, 1980).

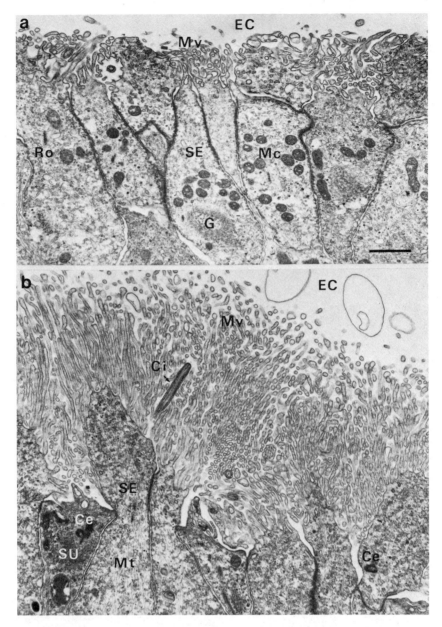

Fig. 15: The microvillar arrangements of sensory cells in holothurian ocelli (Opheodesoma spectabilis) after light- (a) and dark- (b) adaptation for 40 h. Scale, 1 μm. Ce, centriol; Ci, cilium; Ec, epineural cavity; G, Golgi complex; Mc, mitochondrion; Mt, microtubules; Mv, microvilli; SE; sensory cell; SU, supportive cell (from Yamamoto & Yoshida, 1978).

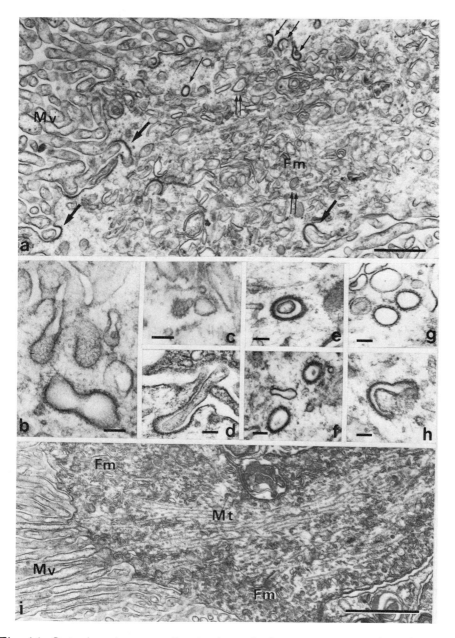

Fig. 16: Cytoplasmic organelles in the apical region of light- (a - h) and dark- (i) adapted holothurian photosensory cells (Opheodesoma spectabilis). Scale, 0,5 μm in a, 0,1 μm in b - h and 1 μm in i. Fm, filamentous fragments; Mt, microtubules; Mv, microvilli; Thin, thick and double arrows indicate, respectively, coated vesicles, invaginating plasma membrane and double ring structure with less electron dense coating (from Yamamoto & Yoshida, 1978).

we encountered no multivesicular nor lamellated bodies, which usually occur as secondary lysosomes in association with endocytotic activities.

A few more differences were noted in the distal region of the sensory cells between light- and dark-adapted specimens. Mitochondria tended to be more abundant (Fig. 15a), and membranous fragments more numerous and larger (Fig. 16a) in the light-adapted cells, whereas microtubules were more prominent in the dark-adapted cells (Fig. 16i).

Eakin and Brandenburger showed in the next year (1979) the effect of light on the sensory microvilli in three species of starfishes, Patiria miniata, Leptasterias pusilla and Henricia leviuscula. The gross appearance of similar changes observed with a Japanese species, Asterias amurensis, is shown in Fig. 17a and 17b. With more extended observations, Brandenburger and Eakin (1980) proposed a scheme for membrane recycling as shown in Fig. 18. According to them, the light-induced degradation of sensory microvilli proceeds in two separate processes; internalization as pinocytotic vesicles into sensory cells (No. 1; Resorption) and shedding as fragments (No. 2; Abscission). These sloughed fragments in the cup lumen are then removed by both corneal and pigmented cells (No. 3; Phagocytosis). The internalized materials in the three ocellar components have different fates: a either digested in (No. 5) or expelled from (No. 4; Exocytosis) the corneal cells; b transformed into phagosomes in the pigmented cells (No. 5) and then released as replenishing molecules; c transformed into smooth vesicles in the sensory cells as stored membranes; or, d digested in the sensory cells. The processes from b through d occur in darkness. Other routes for restoration of microvilli in darkness (No. 6) are also suggested: A) production of new vesicles by Golgi apparatus, and B) formation of secondary lysosomes as a result of combination of the ingested material with primary lysosomes which have been produced from Golgi-endoplasmic reticulum-lyosome complex (GERL).

The above scheme is reminiscent of that proposed in well-differentiated photoreceptive cells. In these ill-differentiated cells of starfishes, however, there is no direct proof that the microvilli arising from the so-called sensory cells are in fact the primary photoreceptive site. The following experiments were designed to tackle this problem.

2. Intramembranous particles

Recent studies on the photoreceptive membranes in vertebrates as well as in invertebrates with freeze-fracture techniques have increasingly shown that the intramembranous particles thus exposed are most probably associated with the visual protein, Rhodopsin (examples in invertebrates: Brandenburger et al., 1976; Boschek & Hamdorf. 1976; Harris et al., 1977; Schinz et al., 1982; and in vertebrates: Jan & Revel, 1974; Corless et al., 1976). The size and the density of particles on the P-face of various ocellar components in five species of starfishes were studied under light- and dark-adapted conditions (Takasu et al., 1982). As shown in Table I, the average size of particles ranged from 9 - 11 nm in all the membranous components analyzed, but the densities per unit area varied, high in the microvilli of the sensory cells. The data presented were obtained with Asterias amurensis which belongs to the order Forcipulata. Similar

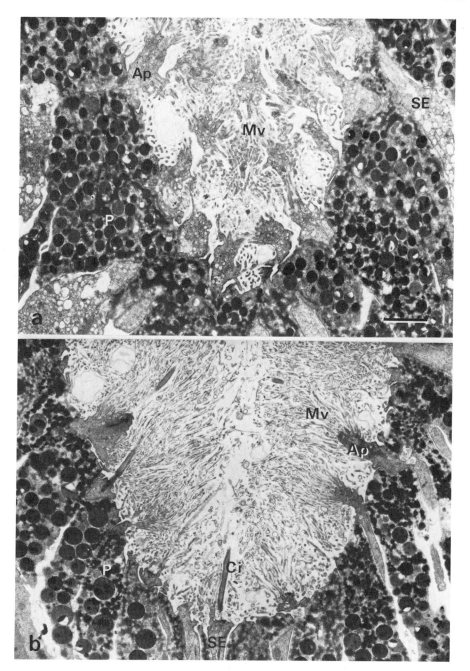

Fig. 17: Effect of light on the sensory microvilli in the ocellus of the starfish, Asterias amurensis. **a:** light-adapted for 30 h. **b:** dark-adapted for 30 h. Scale, 2 μm. Ap, apical projection; Ci, cilium; Mv, microvilli; P, pigmented cell; SE, sensory cell.

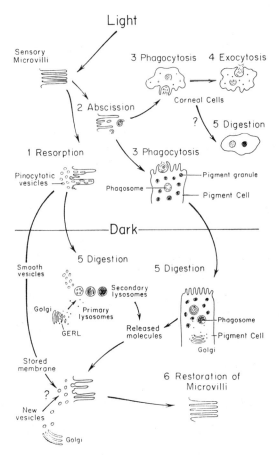

Fig. 18: Diagram of recycling of photoreceptoral membranes of the starfish, Patiria miniata. For explanation, see text (from Brandenburger & Eakin, 1980).

tendencies were obtained with other starfishes: Asteropecten scoparius, A. polyacanthus and Luidia quinaria which belong to the order Phanerozonia and Asterina pectinifera, the order Spinulosa.

The results shown in Table I reveal several important implications as regards the density of intramembranous particles (IMP's). (1) The effect of light is such as to reduce the IMP density in the sensory microvilli (Fig. 19a and 19b; 7026 $\pm$648 in DD versus 2855 $\pm$552 in LL; $P < 0,05$). This effect agrees with what has been reported for well-differentiated receptoral membranes. (2) In the case of pinocytotic vesicles, the direction of light-dependent changes in the IMP densities occurs in a reverse way, namely, higher densities in light (Fig. 19d; 2364 $\pm$795 in DD versus 3109 $\pm$1084 in LL; $P < 0,05$). These results become understandable if it is assumed that in these vesicles internalization of particle-loaded

Table I: The size and density of intramembranous particles on the P-face of freeze-fractured membrane of various ocellar components in Asterias amurensis kept under dark and light conditions for 30 h. ±, Standard deviation. *, Difference between dark and light conditions, statistically significant (P <0,05)

Membrane	Size (nm) Dark	Size (nm) Light	Density (Number/μm^2) Dark	Density (Number/μm^2) Light
SENSORY CELL				
Microvilli	9,43±1,09	9,60±1,09	7026±648*	2855±552
Plasma membrane (soma)	10,07±1,18	10,08±1,20	2351±244	2344±229
Plasma membrane (apical projection)	10,06±1,19	9,99±1,17	2219±341	1879±374
Ciliary shaft	10,40±1.37	10,03±1,06	1711±122	1568±445
Pinocytotic vesicle	9,51±1,18	9,29±1,11	2364±795*	3109±1084
SUPPORTIVE CELL				
Microvilli	10,13±1,16	9,66±1,05	1963±395	1837±274
Plasma membrane	10,76±1,23	10,61±1,23	1313±154	1303±234

membranes in light (higher IMP densities) is followed by digestion or dissolution of the ingested particles in darkness (lower IMP densities). (3) The fact that the ranges of the standard deviations in the IMP densities of pinocytotic vesicles are by far larger than those obtained from other membranous components of sensory cells may mean that these vesicles were in a dynamic phase of resorption in light and we observed, from the light- through dark-adapted states, a diverse range of particle degenerations. (4) The IMP's in the ciliary shaft, which was once considered as a candidate for photoreceptive site, are less dense (1711 ±122 in DD and 1568 ±445 in LL; difference, insignificant) than those in any other membranous structures of the sensory cells.

3. Histo-fluorescence

The pigment composition in the optic cushion of the starfish has scarcely been analyzed (Millott & Vevers, 1955). A photosensitive protein called "stellarin" was once proposed (Rockstein & Finkel, 1960; Rockstein, 1962), but no rhodopsin nor any substance which may be related to retinal has been reported in echinoderms.

During the last decade, various forms of rhodopsin have been shown to fluoresce (Ebrey, 1971; Waddel et al., 1973; Eakin & Brandenburger, 1978; Cronin & Goldsmith, 1981). Such properties may be utilized to demonstrate rhodopsin histochemically. Ozaki et al. (1982) have developed a convenient method to reveal the presence of retinal-based proteins in photoreceptive tissues. In principle, their method is to induce fluorescence after reduction of retinyl proteins which may be formed by denaturation of rhodopsin. These authors have proved the applicability of their method to retinae of four phylogenetically remote species; an octopus (Octopus

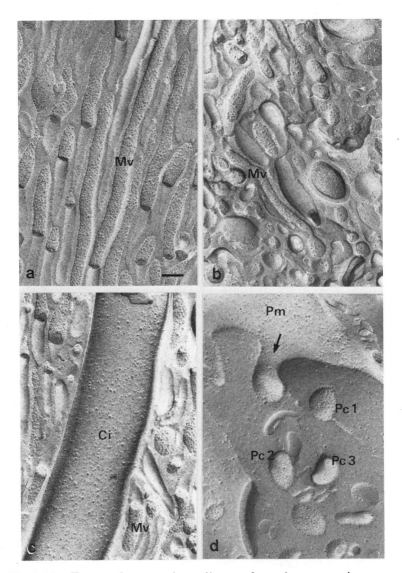

Fig. 19: Freeze-fractured replicas of various membranous components in the starfish ocelli (Asterias amurensis). Rotary shadowed with an angle of 45°. a and b, dark-and light-adapted sensory microvilli, respectively. c, dark-adapted cilium. d, light-adapted pinocytotic vesicles. Note that the density of intramembranous particles varies greatly among three pinocytotic vesicles, Pc 1 - 3. Scale, 0,1 μm. Ci, cilium; Mv, microvilli; Pc, pinocytotic vesicle; Pm, plasma membrane; Arrow, invaginating plasma membrane

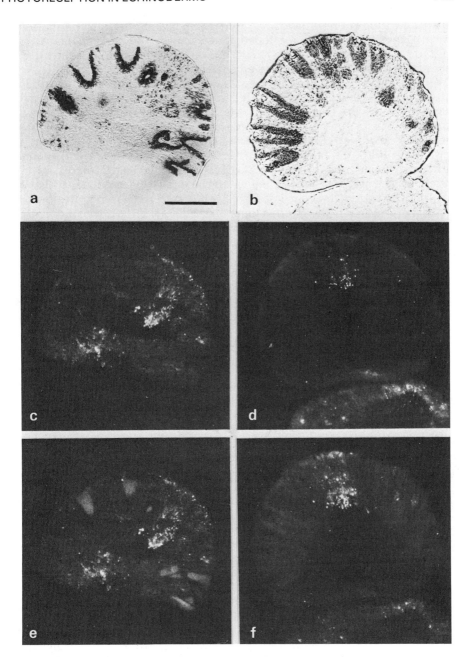

Fig. 20: Optic cushion of the starfish, <u>Asterias amurensis</u>, after dark- (a, c, e) and light- (b, d, f) adaptations. a and b; photographed with red light. c - f; photographed with epifluorescent microscope. c and d; autofluorescence (control). e and f; fluorescence induced by reduction with sodium borohydride. Scale, 100 μm.

vulgaris), a slug (Limax flavus), a crayfish (Procambarus clarkii) and a flat worm (Dugesia japonica). We found, thanks to the devoted technical help offered by Dr. Ozaki, that their method was also applicable to the starfish ocellar microvilli with small, yet essential modifications.

The results presented in Fig. 20 demonstrate that the fluorescence induced upon reduction is notable only in the ocellar lumen of dark-adapted specimens (Fig. 20e). In the starfish arm-tips, catecholamines have been shown to fluoresce in regenerating tissues of Asterina gibbosa (Huet & Franquinet, 1981). The present method definitely eliminates the possibility that the fluorescence induced is due to catecholamines Although these results should be verified by using more specific wavelength range for actinic light, the induced fluorescence, which has been demonstrated to occur in other invertebrate retinae, implies that the substance present in the ocellar lumen of the starfish is a substance which resembles that present in rhabdomes of well-differentiated cells. Conversely, the fact, that the fluorescence became very weak or was undetectable in light-adapted ocelli, may be taken as an evidence of low concentrations of the substance (most probably, visual pigment) in the cup lumen where both microvilli (Fig. 17) and intramembranous particles, which were embodied in them (Table I), decreased in quantity upon illumination.

ACKNOWLEDGEMENTS

We thank Professor M.A. Ali of Université de Montréal for his invaluable critism to the manuscript and Mr. W. Godo for his technical help. This work was supported in part by a grant-in-aid from the Ministry of Education of Japan to M.Y.

REFERENCES

Berril, M. (1966) The ethology of the synaptid holothurian, Opheodesoma spectabilis Can. J. Zool. 44: 457-482.
Binyon, J. (1972) Physiology of Echinoderms. Oxford, Pergamon Press. 264 pages.
Boltt,R E. & Ewer, D.W. (1963) Studies on the myoneural physiology of Echinodermata. IV. The lantern retractor muscle of Parechinus: Responses to stimulation by light. J. Exp. Biol. 40: 713-726.
Boschek, C.B. & Hamdorf, K. (1976) Rhodopsin particles in the photoreceptor membrane of an insect. Z. Naturforsch. 31c: 763.
Brandenburger, J.L. & Eakin, R.M. (1980) Cytochemical localization of acid phosphatase in ocelli of the seastar Patiria miniata during recycling of photoreceptoral membranes. J. Exp. Zool. 214: 127-140.
Brandenburger, J.L., Eakin, R.M. & Reed, C.T. (1976) Effects of light-and dark-adaptation on the photic microvilli and photic vesicles on the pulmonate snail Helix aspersa. Vision Res. 16: 1205-1210.
Cobb,J.L.S. & Stubbs, T.R. (1981) The giant neuron system in ophiuroids. I. The general morphology of the radial nerve cords and circumoral nerve ring. Cell Tissue Res. 219: 197-207.
Corless, J.M., Cobbs, W.H., Costello, M.J. & Robertson, J.D. (1976) On the

asymmetry of frog retinal outer segment disk membranes. Exp. Eye Res. 23: 295-324.
Cronin, T.W., & Goldsmith, T.H. (1981) Fluorescence of crayfish metarhodopsin studied in single rhabdoms. Biophys. J. 35: 653-664.
Dambach, M. & Weber, W. (1975) Inhibition of pigment movement by cytochalasin B in the chromatophores of the sea urchin Centrostephanus longispinus. Comp. Biochem. Physiol. 50C: 49-52.
Dimelow, E.J. (1958) Pigments present in arms and pinnules of the crinoid, Antedon bifida (Pennant). Nature (Lond.) 182: 812.
Eakin, R.M. (1963) Lines of evolution of photoreceptors. In: General Physiology of Cell Specialization. Ed. D. Mazia and A. Tyler. New York, McGraw-Hill, p. 393-425.
Eakin, R.M. (1982) Continuity and diversity in photoreceptors. In: Visual Cell and Evolution. Ed. J.A. Westfall. New York, Raven Press, p. 91-105.
Eakin, R.M. & Brandenburger, J.L. (1978) Autofluorescence in the retina of a snail, Helix aspersa. Vision Res. 18: 1541-1543.
Eakin, R.M. & Brandenburger, J.L. (1979) Effects of light on ocelli of seastars. Zoomorphol. 92: 191-200.
Eakin, R.M. & Westfall, J.A. (1962) Fine structure of photoreceptors in the hydromedusan, Polyorchis penicillatus. Proc. Nat. Acad. Sci. U.S.A. 48: 826-833.
Eakin, R M. & Westfall, J.A. (1964) Further observations on the fine structure of some invertebrate eyes. Z. Zellforsch. 62: 310-332.
Ebrey T.G. (1971) Energy transfer in rhodopsin, N-retinyl-opsin, and rod outer segments. Proc. Nat. Acad. Sci. U.S.A. 68: 713-716.
Harris, W.A., Ready, D.F., Lipson, E.D., Hudspeth, A.J. & Stark, W.S. (1977) Vitamin A deprivation and Drosophila photopigments. Nature (Lond.) 266: 648-650.
Huet, M. & Franquinet, R. (1981) Histofluorescence study and biochemical assay of catecholamines (dopamine and nonadrenaline) during the course of arm-tip regeneration in the starfish, Asterina gibbosa (Echinodermata, Asteroidae). Histochem. 72: 149-154.
Jan, L.Y. & Revel, J.-P. (1974) Ultrastructural localization of rhodopsin in the vertebrate retina. J. Cell Biol. 62: 257-273.
Kleinholz, L.H. (1938) Color changes in echinoderms. Pubbl. Staz. Zool. Napoli 17: 53-57.
Millott, N. (1954) Sensitivity to light and the reactions to changes in light intensity of the echinoid Diadema antillarum Philippi. Phil. Trans. 238: 187-220.
Millott, N. (1955) The covering reaction in a tropical sea urchin. Nature (Lond.) 175: 561.
Millott, N. (1956) The covering reaction of sea-urchins. I. A preliminary account of covering in the tropical echinoid Lytechinus variegatus (Lamarck), and its reaction to light. J. Exp. Biol. 33: 508-523.
Millott, N. (1975) The photosensitivity of echinoids. Adv. Mar. Biol. 13: 1-52.
Millott, N. & Takahashi, K. (1963) The shadow reaction of Diadema antillarum Philippi. IV. Spine movements and their implications. Phil. Trans. 246: 437-470.

Millott, N. & Vevers, H.G. (1955) Carotenoid pigment in the optic cushion of Marthasterias glacialis (L.) J. Mar. Biol. Assoc. U.K. 34: 279-287.
Millott, N. & Yoshida, M. (1959) The photosensitivity of the sea urchin Diadema antillarum Philippi. Responses to increase in light intensity. Proc. Zool. Soc. Lond. 133: 67-71.
Millott, N. & Yoshida, M. (1960a) The shadow reaction of Diadema antillarum Philippi. I. The spine response and its relation to the stimulus. J. Exp. Biol. 37: 363-375.
Millott, N. & Yoshida, M (1960b) The shadow reaction of Diadema antillarum Philippi. II. Inhibition by light. J. Exp. Biol. 37: 376-389.
Neya, T. (1965) Photic behavior of sea urchin larvae, Hemicentrotus pulcherrimus. Zool. Mag. (Tokyo) 74: 11-16.
Ozaki, K., Hara, R. & Hara, T. (1982) Location of retinochrome and rhodopsin in the retina as revealed by fluorescence microscopy. Jap. J. Ophthalmol. 26: 103.
Penn, P.E. & Alexander, C.G. (1980) Fine structure of the optic cushion in the asteroid Nepanthia belcheri. Mar. Biol. 58: 251-256.
Pople, W. & Ewer, D.W. (1954) Studies on the myoneural physiology of echinodermata. I. The pharyngeal retractor muscle of Cucumaria. J. Exp. Biol. 31: 114-126.
Pople, W. & Ewer, D.W. (1958) Studies on the myoneural physiology of echinodermata. III. Spontaneous actviity of the pharyngeal muscle of Cucumaria. J. Exp. Biol. 35: 712-730.
Rockstein, M. (1962) Some properties of stellarin, the photosensitive pigment of the starfish, Asterias forbesi. Biol. Bull. 123: 510.
Rockstein, M. & Finkel, A. (1960) Stellarin, a photosensitive pigment from the dorsal skin of the starfish, Asterias forbesi. Anat. Rec. 138: 379.
Schinz, R.H., Lo, M.-V.C., Larrivee, D.C. & Pak, W.L. (1982) Freeze-fracture study of the Drosophila photoreceptor membrane: Mutations affecting membrane particle density. J. Cell Biol. 93: 961-969.
Smith, J.E. (1950) Some observations on the nervous mechanisms underlying the behaviour of starfish. Symp. Soc. Exp. Biol. 4: 196-220.
Takahashi, K. (1964) Electrical responses to light stimuli in the isolated radial nerve of the sea urchin, Diadema setosum (Leske). Nature (Lond.) 201: 1343-1344.
Takasu, N., Tamotsu, S. & Yoshida, M. (1982) Morphophysiological analyses of ocellar microvilli in starfish. Proc. 4th Ann. Meet. Jap. Soc. Gen. Comp. Physiol., p. 22.
Vaupel-von Harnack, M. (1963). Über den Feinbau des Nervensystemes des Seesternes (Asterias rubens L.). III. Die Struktur der Augenpolster. Z. Zellforsch. 60: 432-451.
Waddell, W.H., Schaffer, A.M. & Becker, R.S. (1973) Visual pigments. III. Determination and interpretation of the fluorescence quantum yields of retinals, Schiff bases, and protonated Schiff bases. J. Am. Chem. Soc. 95: 8223-8227.
Waterman, T.H. (1982) Fine structure and turnover of photoreceptor membranes. In: Visual Cells in Evolution. Ed. J.A. Westfall. New York, Raven Press, p. 23-41.
Weber, W. & Dambach, M. (1972) Amöboid bewegliche Pigmentzellen im

Epithel des Seeigels Centrostephanus longispinus. Ein neuartiger Farbwechselmechanisms. Z. Zellforch. 133: 87-102.

Weber, W & Dambach, M. (1974) LIght-sensitivity of isolated pigment cells of the sea urchin Centrostephanus longispinus. Cell Tissue Res. 148: 437-440.

Yamamoto, M. & Yoshida, M. (1978) Fine structure of the ocelli of a synaptid holothurian, Opheodesoma spectabilis, and the effects of light and darkness. Zoomorphol. 90: 1-17.

Yasumasu, I. (1963) Geotaxis of swimming sea-urchin larvae. Zool. Mag. (Tokyo) 72: 160-162.

Yoshida, M. (1956) On the light response of the chromatophore of the sea urchin, Diadema setosum (Leske). J. Exp. Biol. 33: 119-123.

Yoshida, M. (1960) Further studies on the chromatophore response in Diadema setosum (Leske). Biol. J. Okayama Univ. 6: 169-173.

Yoshida, M. (1962) The effect of light on the shadow reaction of the sea urchin Diadema setosum (Leske). J. Exp. Biol. 39: 589-602.

Yoshida, M. (1966) Photosensitivity. In: Physiology of Echinodermata. Ed. R.A. Boolootian. New York, Interscience, p. 435-464.

Yoshida, M. (1979) Extraocular photoreception. In: Handbook of Sensory Physiology, Vol. VII/6A. Ed. H. Autrum. Berlin, Springer-Verlag, p. 581-640.

Yoshida, M. & Millott, N. (1959) Light sensitive nerve in an echinoid. Experientia (Basel) 15: 13-14.

Yoshida, M. & Ohtsuki, H. (1966) Compound ocellus of a starfish: Its function. Science 153: 197-198.

Yoshida, M. & Ohtsuki, H. (1968) The phototactic behavior of the starfish, Asterias amurensis Lütken. Biol. Bull. 134: 516-532.

EPILOGUE

M.A. ALI

Département de Biologie, Université de Montréal

Montréal, P.Q. H3C 3J7 Canada

> Remember our reward, when the deed is done.
> King Richard The Third, Act 1, Sc. 4.

As mentioned in the Prologue, about 97% of the animal species are invertebrate ones. If the number of individuals could even be guessed, it would probably represent over 99,5%. Such a rich source of material presents also insurmountable problems for its study. Let me first enumerate the positive aspects. The extraordinarily wide taxonomic variety offered by the invertebrates permits not only the study of evolutionary aspects but also comparative and developmental ones. The fact that they occupy every conceivable habitat makes them the material of choice for studying adaptive features. The incredible diversity that one encounters in their structures allows the choice of particular structures for investigating a particular problem in physiology or experimental biology. These same advantages also pose the "insurmountable" problems referred to above. For example, it will be impossible to study the photoreceptors of the nearly million species. Also, it will be nearly impossible to collect them from all the habitats in which they occur and keep alive in the laboratory. Thus, the solution would appear to be to select judiciously representatives of different taxa inhabiting different habitats and having different habits and structures, depending on the question that one is trying to answer or, a problem to solve. Obviously, in this context it may be superfluous to mention that the criteria for selection would and should vary according not only to the problem but also the group that one is interested in. Thus, certain investigations just simply could not be carried out using annelids, crustaceans or insects depending on the problem. Further, in the case of large groups such as the class Insecta, having not only about 3/4 of the animal species but also a remarkable similarity as far as structure and habitats are concerned it may suffice to deal with even only one representative from each order. On the other hand, in the case of Class Crustacea, a much smaller one but demonstrating considerably greater diversity of structure, function and habitats it may be appropriate to deal with representatives even from the level of families.

In the following paragraphs a gist of the discussion that took place on the last working day of the ASI (Fig. 1) is given. Six persons were asked to enumerate and discuss aspects that needed to be talked about. During and after their presentations comments or questions from the other participants were encouraged. In preparing a summary of these proceedings some liberties have been taken and, not only have I added some other points that I feel are worth including but also amplified or reduced in importance most of the comments made by others. The latter was also done in the light of reading the other chapters which were submitted after the ASI. Also, some of the points such as rhythms, sexual dimorphism etc. came up for discussion more than once during the day and in their case I have tried to combine all the aspects discussed into one section. It has also been my intention to emphasise as much as possible what appear to be promising avenues for future research.

STRUCTURE AND EVOLUTION

The premise that photoreception and vision reflect the evolutionary trend in animals can only be made once their importance, to the individual, has been established. Both photoreceptor and eye structures have been used in this classification (Fig. 2). Specific morphological charateristics are described in the chapters by Burr, Clément and Wurdak, Couillard, Goto and Yoshida, Land, Mouze, Muños-Cuevas, Verger-Bocquet, Waterman, and Yoshida et al.

Photoreceptors

Invertebrate photoreceptors are classified into two main groups based on membrane invagination - rhabdomeric and ciliary. Although this classification is generally acceptable it still remains to be established if the existence of two groups is due to the presence of two photoreceptor types or vice versa. In addition, in keeping with the laws of Mother Nature oddities exist. Ciliary and rhabdomeric photoreceptors are both found in many deuterostome and protostome phyla, even both in the same species or the same eye (Burr, 1983 a). Vertebrates are generally thought to occupy the upper levels of the evolutionary hierarchy and the presence of both rhabdomeric and ciliary type photoreceptors in invertebrates and not in vertebrates suggests the ease of switching from rhabdomeric to ciliary photoreceptors and vice versa, in one mutation. There is no definite proof for this switch but it is easy to see that such a switch in a highly evolved organ should be deleterious, thus experimentation is unlikely.

Eyes

Eye structure and optics may also be used as a guideline for evolutionary trends, however the following questions arise:

Fig. 1: A blowfly's eye view of the participants and proceedings of the ASI. Courtesy of Nick Strausfeld.

- Why do particular groups of animals select different eyes?
- **What is the basis for selection of simple eyes over compound eyes and vice versa?** Is this in any way related to the presence/absence of an exoskeleton?
- What is the practical significance of eye-size, of a scanning (tracking) eye? The latter apparently poses three primary problems (1) stabilising of the image on the retina; (2) switching, as fast as possible,

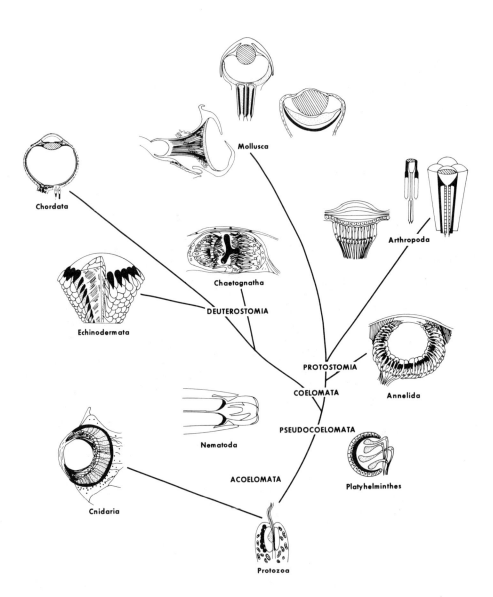

from one object to another; and (3) tracking objects they intend to track. Is this achieved by the eyes (eye stalk) or head/body movements?

This practical, humanistic approach may not provide answers. Several animals have selected apparently "unsuitable" eyes. This is examplified by the large, pin-hole camera eye of Nautilus which has all the makings of a "good, optimal" eye and yet this animal has retained an eye which according to Land is "not really good for much"! This questions the basis of selection and suggests that we should approach the subject of selective pressures from the animal's viewpoint. Does the eye perform the required functions? What can be "gained" by "improving" it and would this be in keeping with the other behavioural and physiological characteristics of the animal?

RHYTHMS

Rhythms, apparent in every aspect of biology, consist of a closed circuit of interacting systems where one event triggers another and so on. They may be observed in single cells, in tissues and organs, in the whole organism and in populations. These rhythms may originate from within the organism (endogenous rhythms e.g. transcription of molecules, synthesis of protein, turnover of enzymes) or may be governed by external periodic signals (exogenous rhythms).

Diurnal (24h) and seasonal (365 days) rhythms are of fundamental importance to most animals, while tidal rhythms are also significant to marine animals. The most common strategy in coping with the two former rhythms involves utilisation of day length (or photoperiod) as a reliable marker of phase in the environmental year. Day length (or night length) can initiate retreat strategies (e.g. hibernation, aestivation, "dormancy", diapause) in plants and animals as well as reproductive activity and growth under favourable conditions. The fact that a rhythm can be entrained by the environmental light cycle suggests the involvement of photoreceptors. Studies on photoreception have included investigations on the identification, location and properties (e.g. spectral sensitivity, intensity threshold) of the photoreceptors involved. These photoreceptors maybe "organised" (retinal), extra-retinal or multiple photoreceptors.

Fig. 2: An evolutionary tree of eyespots, light sensitive areas and eyes in the animal kingdom. The general structure only is shown. Diagrams are after the originals in:- Burr and Webster (1971), Fauvel et al. (1959), Hesse (1902), Land (1966), Pearse and Pearse (1978, Penn and Alexander (1980), Wolken (1975) and Young (1964). For a more detailed tree see Burr (1983 a). For details at the cellular level see Westfall (1982).

■ Pigment (eyespot, cup or granules).
∴∴∴ Photoreceptor (sensory cell, retinal rod or rhabdomere).

"Organised" photoreceptors are frequently bypassed in entrainment in invertebrates, however in the cockroach (Leucophaea maderae) and the cricket (Teleogryllus commodus) the compound eyes appear to be the exclusive photoreceptors for entrainment. In the scorpion (Androctonus australis) the median eyes are directly involved in entrainment, however they may also be entrained via localised, dim illumination of the lateral eyes (Fleissner, 1977 a, b). In crabs and lobsters the persistent nauplius eye is thought to function analogous to the pineal of lower vertebrates in controlling seasonal behaviour.

The primary pathway for entrainment appears to be extraretinal. This is seen in silkmoths (Hyalophora cecrophia, Antheraea pernyi), aphid (Megoura viciae), Drosophila, grasshopper (Chorthippus curtipennis) and crayfish (Procambarus clarkii). In these instances, the mid-brain and forebrain cells are directly sensitive to light and must be subjected to a particular programme before any developmental sequence will follow. In many instances of entrainment via extraretinal photoreception retinal photoreceptors can modify the behaviour "reflexively" without actually being involved (e.g. silkmoth, Truman, 1974; crayfish, Page & Larimer, 1972).

Entrainment, whether retinal or extraretinal, may be spectrum sensitive. Thus there are some insects, which are maximally sensitive to blue-green light and almost insensitive to red (e.g. Bombyx mori, Grapholitha molesta, Pieris brassicae, Antheraea pernyi, Carpocapsa pomonella, Megoura viciae). While in other there is a marked sensitivity to red light (A. rumicis, Leptinotarsa decemlineata, P. gossypiella, Nasonia vitripennis). An interesting form of entrainment is seen in Aplysia: in white light retinal photoreceptors are sufficient, however if the stimulus is red light (> 600 nm) extraretinal receptors are required.

Natural day length is a function of latitude as well as season. At equatorial latitudes there is a regular, mean photoperiod (L:D = 12:12) throughout the year. In higher latitudes (particularly polar regions) days not only increase at a greater rate after the vernal equinox but also reach a longer absolute day length at the midsummer solstice. Studies on A. rumicis have shown that for every 5° latitude the critical day length for diapause of this moth changes by 90 min. In spite of the presence of geographical clines with respect to phenological characteristics, variation may also be due to the presence of distinct races (e.g. P. brassicae, 2 races in the Soviet Union). A similar systematic study with sedentary marine invertebrates is lacking. This group of animals may provide more meaningful information on the function of day length because temperature effects will be eliminated (large water bodies have practically constant temperatures at least in the tropical and extreme northern regions).

One of the most obvious rhythms in photoreceptors is the process of membrane renewal and turnover. This process may appear wasteful, but it is nevertheless essential for the maintenance of biological functions which would otherwise be hindered by age. The extent and timing of the synthesis and breakdown of membrane seem to differ from species to species, and may be due to an endogenous rhythm, the direct effect of environmental light or a combination of the two (cf. Clément & Wurdak, 1983; Muñoz-Cuevas, 1983; Waterman, 1983; Yoshida et al., 1983). The nocturnal spider Dinopis performs a massive synthesis of rhabdomere membrane at dusk and membrane breakdown at dawn (Blest, 1978). Comparable changes are also known for some crustaceans (crabs and to a lesser extent in prawns) mosquito, dalyelliid flatworms. The interrelationship between these circadian effects and those produced by direct illumination in the control of the normal daily cycle of membrane turnover requires further analysis. It would be especially interesting in the deep sea invertebrates which exhibit diurnal vertical migration although they live in nearly constant, dim environments. Other organelles in invertebrate photoreceptors showing variation with photoperiod include vesicles, multivesicular and multilamellar bodies.

In conjunction with diurnal membrane turnover there are cellular movements associated with the adaptive state of the eye. These include migration of pigment granules and vesicules. Cytoskeletal elements (microtubules, microfilaments) present within these photoreceptive cells are thought to be implicated in these movements, however their exact function has yet to be clarified.

A further relatively unexplored area of research in this field is the method of photoperiodic induction. It is uncertain if the mechanisms of photoperiodism are set in motion at dawn or dusk, and whether day length or night length is the critical factor. Experimental results available suggest that while darkness (night length) is the governing factor day length can modify the response. In M. viciae while a critical night length of 9,75 hours is required for producing oviparae capable of producing diapausing winter eggs the period of darkness is stimulatory only after a 4-hour light period. Equally unexplored are the factors governing host-parasite relations. The dispersion of parasites, essential to the preservation of the species, is critical. The sighted, free living form of the parasite is short-lived and has to be released at an opportune time in order to find suitable hosts (Fournier, 1983). The mechanism behind the triggering-off of this response and the role of photoreceptors in the process has yet to be established.

SEXUAL DIMORPHISM

Sexual dimorphism is widespread in the animal kingdom, and in invertebrates this is extended to photoreceptive structures. Insects and copepods exhibit ocular dimorphism related to mate finding (cf. Buchner, 1983; Franceschini, 1983; Land, 1983 a; Stavenga & Schwemer, 1983; Strausfeld, 1983). Regional and sexual specialisations of insect eyes have been noted since 1909 (Dietrich) and have been described in the dragonfly, honeybee, the ant (Cataglyphis), and the male mayfly; while regional specialisation in spectral properties are known for Ascalaphus, Allograpta and Pierid butterflies. In Diptera the frontal, dorsal region of the male fly contains a unique class of central rhabdomeres (R7/R8). Although the R7 cells are similar to the peripheral R1-6 photoreceptors with respect to spectral, angular and absolute sensitivities as well as response waveform and noise characteristics, these traits are clearly distinguishable in the R8 cells. These R8 cells show a specific red autofluorescence and have been suggested to aid the chosing behaviour of the male fly while viewing dark objects (optimistically, females) against the sky. This dimorphism extends to the neuronal level where male-specific interneurons found in the lobula have been related to the tracking of females. This one neuron-one function provides a convenient model for the study of neural functions.

Sexual dimorphism of ocular structures in other invertebrates is less well known and provides avenues of study. In male nematodes, for example, the male has an ocular pigment spot while the female lacks it and contains instead a small branchial structure (Burr, 1983 b). In polychaetes, sexually mature males have apparently better differentiated eyes, which are related to the number rather than the ultrastructure of the cells. This characteristic may be related to mate finding as expressed by the swarming of Palolo (Eunice viridis) worms of Samoan coral reefs, males swimming below, head towards luminescent females swimming in a circle above. This occurs at an exact tidal and lunar period (third day after the last lunar quarter in October or November, which is the southern Spring).

BIOLUMINESCENCE AND SPECTRAL SENSITIVITY

Marine invertebrates differ from their vertebrate counterparts in that well-developed and structured photoreceptors are found in deep sea invertebrates which live in an environment of constant (practically zero) light levels throughout the year. Whether the eye in these animals is functionless or highly specialised for the perception of bioluminescent light poses an interesting question. Generally, the proportion of eyeless to eye-bearing isopods increases with depth and latitude (Menzies et al., 1973). On the other hand bioluminescent light is predominant at depths of 500-1500 m. Several marine polychaetes are luminous (e.g. members of the pelagic family Tomopteridae); while numerous pelagic crustaceans possess light organs (e.g. the deep-sea shrimp Acanthephyra purpurea possesses light organs with both reflecting layers and lenses as well as glands from which luminescent substances may be forcefully ejected). Bioluminescent light has a special significance to the behaviour of animals living below the photic zone. This light is used for camouflage, countershading, communication (identification of mate, prey and maintenance of shoals).

The relationship between the spectral sensitivity of visual pigments and the spectral emission curves of bioluminescent light of the organism is an interesting field for research. This has been shown by Lall et al. (1980 a, b) for fireflies and for pelagic invertebrates by Kampa (1955) and Kampa and Boden (1957). In the dusk-active firefly Photinus pyralis the spectral sensitivity of the compound eye is closely atuned to the species bioluminescence emission spectrum (Lall et al., 1980 a, b). In Euphausia pacifica the bioluminescence emission spectrum is located at 476 nm with the rhodopsin showing a λ_{max} of 462 nm. A survey of the spectral sensitivity of visual pigments, of both terrestrial and aquatic invertebrates, in relation to ambient bioluminescent light should prove interesting and aid in the understanding of the basic questions of communication between these animals.

VISUAL TRANSDUCTION

The process of visual transduction is initiated once the visual pigment molecule is transformed by the absorption of a photon. This aspect of photoreception and vision is basic to all organisms, and is at present enjoying a prominent place in this field as indicated by recent reviews (Miller 1981; Packer, 1982). The exact mechanism has yet to be established, indeed there is still disagreement on the intracellular messenger involved. In invertebrates measurement of receptor potentials shows that electrical events proceed within a tenth of a millisecond following the capture of photons. While the late potential, which is an order of magnitude slower (several milliseconds), takes place about the same time as the behavioural response (20-30 msec). This illustrates that the speeding mechanism is evident at the lamina and subsequent stages. Thus, once transduction is achieved, the whole neuronal machinery trips a hair trigger and a response is very rapidly elicited. Ion channel systems (at the membrane level) appear to be tied-in with the sequence of events which may be either photo (optically) regulated or rhythmically (diurnally) regulated.

INVERTEBRATES vs VERTEBRATES (Fig. 3)

Receptors

One of the most striking differences when looking at the eye of invertebrates and vertebrates is the astronomical difference in the number of receptor cells (e.g. 24×10^3 in the fly and 126×10^6 in man). The difference though significant is considerably reduced once the amount of pooling (convergence) in vertebrates is considered. Besides number, vertebrate and invertebrate photoreceptors differ in structure, nevertheless, they share a number of basic features. They both absorb photons via the membrane bound photopigments of the receptor cells; but while vertebrate photoreceptors respond by hyperpolarisation, invertebrate photoreceptors depolarise.

Vertebrates contain two types of photoreceptors (rods, cones) which provide a sensitivity range. In man the 120×10^6 rods are used for scotopic vision while the 6×10^6 cones which have blue, red and green sensitivities are used for diurnal vision. In addition, some vertebrates (e.g. birds, amphibians, reptiles, fishes) contain coloured oil droplets which fine tune sensitivity by selectively absorbing incident light. A similar wave length selectivity is provided by the R_7 and R_8 cells of the insect retina. The R_7 yellow cells contain β-carotene which acts as a screening pigment which acts as a filter for both R_7 and R_8 cells. In addition, R_7 and R_8 cells of invertebrates form a mosaic comparable to the cone mosaic of vertebrates. Furthermore regionalisation of the retina is common to both vertebrate and insect retinas with localised areas showing sensitivity to different parts of the spectrum (Bernard & Stavenga, 1979; Franceschini et al., 1981; Levine & MacNichol, 1982).

A noteworthy distinction between these two groups of animals is the

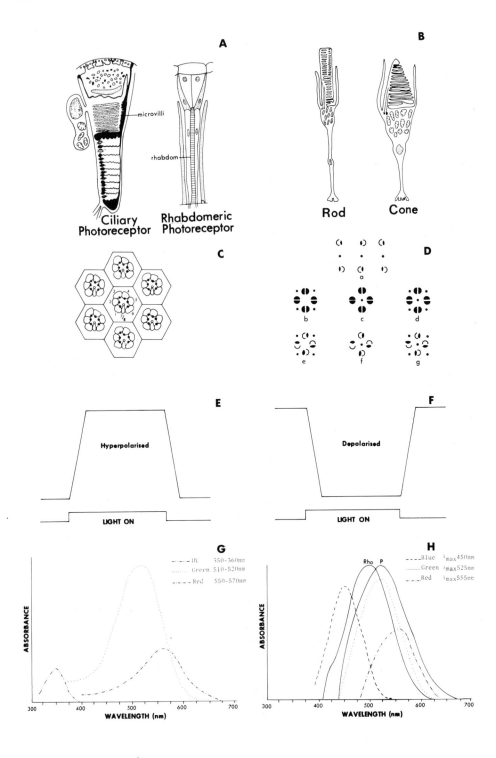

presence of ultraviolet sensitive receptors in invertebrates. Some invertebrate may also contain polarisation sensitive receptors as a result of the dichroic absorption properties of the photopigment dipoles and the geometry of the rhabdomeric microvilli (cf. Waterman, 1983). In the retina of the worker bee, the 9th retinula cell exhibits high polarisation sensitivity while the other eight, long visual cells have low polarisation sensitivity due to the twisting of their rhabdomeres. Similarly the short, distal UV-cell of the dragonfly and receptors R_7, R_8 of the fly have polarisation sensitivity. These findings suggest that the detection of polarised light, in insects, is mediated by short UV-cells which interestingly enough, in flies and bees, project directly on the second order neuropil - the medulla.

Anatomical Organisation

The compound eye of insects and the lens eye of vertebrates (and some invertebrates) have completely different optics (cf. Land, 1983 a, b) however the net result is a visual image of the external world projected onto a two-dimensional array of photoreceptors. In the insect eye the subsequent steps from photoreceptors to the brain take place in three stages: the optic lamina, the medulla and the lobula complex (see: Buchner & Buchner, 1983; Hausen, 1983; Meinertzhagen, 1983; Strausfeld, 1983). Part of the output of the lobula complex goes to the protocerebrum, this in turn projects onto the thoracic ganglion which contains motor neurons for flight muscles. The passage of information through each of these five stages may proceed by way of one or two synaptic relays. Thus in insects, optokinetic reflexes from the primary sensory input to the motor output may be mediated by pathways involving approximately ten direct synapses. Besides the direct pathway there are of course numerous lateral interactions and feedback connexions at various stages of the visual pathway. In vertebrates, the process is much more complicated. The direct pathway through the retina is of course the receptor-bipolar-ganglion cell route, however, the interactions are

Fig 3: A comparison of invertebrate (left column) and vertebrate (right column) photoreceptors.

- Structure of ciliary and rhabdomeric photoreceptors together with their lens cells (A); and vertebrate rod and cone photoreceptors (B)
- Retinal mosaics of photoreceptors (C, D) (after Franceschini et al., 1981; Wagner, 1978; respectively).
- Schematic representation of invertebrate (E) and vertebrate (F) photoreceptor response to light (Toyoda et al., 1969).
- Spectra of visual pigments present in a moth (G) (after Langer et al., 1979) and in bullfrog rods (H) (Rho - rhodopsin, P - porphyropsin; after Reuter et al., 1971) and human cones (H) (after Brown & Wald, 1964). In the moth (G) a further receptor pigment responsible for blue sensitivity (λ_{max} 460-470) is present, but it is not measurable by microspectrophotometric method.

complex. Photoreceptors are not only electrically coupled among themselves, but they also make contact at the outer plexiform layer with both horizontal and bipolar cells. The horizontal cells mediate lateral (surround) interactions antagonistic to these transmitted via the receptor-bipolar pathway by negative feedback onto the photoreceptors. Bipolar cells, on the other hand, group and transmit information from receptors to amacrine and ganglion cells at the inner plexiform layer. These interactions are even more complex. Morphological evidence suggest that reciprocal connexions exist between bipolar and amacrine cells, with ganglion cells receiving input from both these cells. In addition there is another class of cell, the interplexiform, which receives input in the inner plexiform layer and synapses onto cells within the outer plexiform layer. The ganglion cells then relay information to the dorsal lateral geniculate nucleus (dLGN), to the striate complex and beyond. Progression into visual areas away from the primary inputs leads to functional properties of cells which are increasingly complex and difficult to define. Thus the visual system of the insect will prove more amenable to thorough anatomical and physiological analyses at each stage of transmission.

Movement Perception

The mechanics of movement perception in both vertebrates and insects are relatively unexplored. Movements are detected as a result of the successive stimulation of a number of cells. In vertebrates ganglion cells are known to be movement sensitive; 50% of these cells in the frog/toad retina being movement sensitive. In the insect there are some interneurons which are sensitive to any direction across the eye. Then there are descending movement detectors (e.g. descending contralateral movement DCMD) which, although responding to movement in any direction are more sensitive to movements of small objects contrasted with the background. Besides these, other interneurons which respond to movement in preferred directions exist in the optic lobes of the insect. The unidirectional units in the medulla have a relatively narrow field, while more central units in the optic lobes have wider fields. These cells respond preferentially to either vertical or horizontal movements, thus giving information about the movement of the insect in space (cf. Buchner, 1983; Hausen, 1983; Laughlin, 1983).

Like every complex system that involves information processing the visual motor control system must be analysed and understood at different levels. Phylogeny and behaviour should be taken into account because any system under study is the outcome of an evolutionary response to some problem and it is reasonable to assume that unrelated animal species may have evolved (logically) similar solutions to common problems (convergence). Such similarities are found in the organisation of the head and body movements in flies and the eye-head coordination in monkeys (Bizzi et al., 1972; Geiger & Poggio, 1977; Land, 1983); and in saccadic eye, head, and body movements in animals with specialised central areas of vision (e.g. the fovea) and in the vestibular and optokinetic systems in widely diverse species. Thus behavioural studies (especially under natural conditions) can go a long way in providing an indication of the transmission of signals. However, due to the complexities of looking at retinal projections and the lack of coordination between neuroanatomists, neurobiologists and behavioural scientists (sometimes from the same laboratory and working on the same animal!) this is not easily achieved. Thus a gathering of individuals looking at a common problem from different angles not only provides exchange of ideas but may also provide answers to questions not readily answered should one concentrate within a specific area.

PERSPECTIVES

Lately there has been a renewed interest in invertebrates in all fields of research, including photoreception and vision. The wide diversity of environments inhabited by these animals clearly demonstrates that they are able to adapt successfully to changes. This makes them invaluable in the study of environmental pollution.

In general, very little study has been conducted on the effect of pollutants on photoreception and vision. In vertebrates heavy metals are known to elicit visual defects. Lead, mercury and cadmium have been shown to affect rod photoreceptors in frogs thus causing night-blindness; while barium caused colour-blindness by affecting cones. Recent evidence shows that methyl mercury causes an overall reduction in the spectral sensitivity of the rainbow trout (Hawryshyn et al., 1982) as well as impairing the foveal (photic) vision of monkeys (Rice & Gilberg, 1980, 1982). In addition to direct retinal involvement, pollutants may also elicit visual defects through lesions in the central nervous system. For instance, methyl mercury was found in highest concentration in areas of the brain which mediate visual sensation and motor control. Comparable studies in invertebrates along these lines are lacking. Differential responses of rod and cone photoreceptors to pollutants have been attributed to the difference in the structural organisation of their outer segment membranes. Both these photoreceptors, however, have descended from the ciliary line, thus it would be interesting to relate this to the ciliary photoreceptors of invertebrates and to determine if all ciliary photoreceptors respond similarly. It would also be interesting to verify the response of rhabdomeric photoreceptors.

The visual system of insects are less complex than that of vertebrates - approximately 10 synapses existing from the input to the output site. Thus, pollutant effects on signal transmission in insect visual systems will provide more convenient models for study. In addition, the visual pathway of insects has been more explicitly mapped out.

Invertebrates with their relatively good adaptibility and short lifecycles would prove ideal subjects for chronic pollution studies. In addition, to providing tangible evidence on the site of action, they will also show if the effects are hereditary or if these animals will be able to structurally adapt to the new environment. This versatility of the invertebrates is especially useful in the study of aquatic toxicity because water properties are affected by suspended or dissolved particles, they also form local pool of pollutants. In addition, behavioural response of invertebrates can be conveniently studied using relatively small and simple containers.

REFERENCES

Bernard, G.D. & Stavenga, D.G. (1979) Spectral sensitivities of retinular cells measured in intact living flies by an optical method. J. Comp. Physiol. 134: 95-107.

Bizzi,E , Kalil, R.E., Morasso, P. & Tabliasco, V. (1972) Central programming and peripheral feedback during eye-head co-ordination in monkeys. Bibliotheca Ophthal. 82: 220-232.

Blest,A.D. (1978) The rapid synthesis and destruction of photoreceptor membrane by a dinopid spider: A daily cycle. Proc. R. Soc. Lond. 200B: 463-483

Brown, P.K. & Wald, G (1964) Visual pigments in single rods and cones of the human retina. Science 144: 45-52.

Buchner, E. (1983) Behavioural analysis of spatial vision in insects (This volume).

Buchner, E. & Buchner, S. (1983) Neuroanatomical mapping of visually induced nervous activity in insects by ^{3}H-deoxyglucose (This volume).

Burr, A.H. (1983 a) Evolution of eyes and photoreceptor organelles in the lower phyla (This volume).

Burr, A.H. (1983 b) Photomovement behavior in simple invertebrates (This volume).

Burr, A.H. & Webster, J.M. (1971) Morphology of the eyespot and description of two pigment granules in the esophageal muscle of a marine nematode, Oncholaimus vesicarius. J. Ultrastruct. Res. 36: 621-632.

Clément, P. & Wurdak, E. (1983) Photoreceptors and photoreception in rotifers (This volume).

Couillard, P. (1983) Photoreception in Protozoa, an overview (This volume).

Dietrich, W. (1909) Die Facettenaugen der Dipteren. Z. wiss. Zool. 92: 465-539.

Fauvel, P., Avel, M., Harant, H., Grassé, P. & Dawydoff, C. (1959) Embranchement des Annélides. In: Traité de Zoologie, Vol. 5, Pt. 1. Ed. P. Grassé. Paris, Masson et Cie. p. 3-686.

Fleissner, G. (1977 a) Entrainment of the scorpion's circadian rhythm via the median eyes. J. Comp. Physiol. 118: 93-99.

Fleissner, G. (1977 b) Scorpion lateral eyes: Externally sensitive receptors of zeitgeber stimuli. J. Comp. Physiol. 118: 101-108.

Fournier, A. (1983) Photoreceptors and photosensitivity in platyhelminthes (This volume).

Franceschini, N. (1983) The retinal mosaic of the fly compound eye (This volume).

Franceschini, N., Hardie, R., Ribi, W. & Kirschfeld, K. (1981) Sexual dimorphism in a photoreceptor. Nature (Lond.) 291: 241-244.

Geiger, G. & Poggio, T. (1977) On head and body movements of flying flies. Biol. Cybern. 25: 177-180.

Goto, T. & Yoshida, M. (1983) Photoreception in Chaetognatha (This volume).

Hausen, K. (1983) The lobula-complex of the fly: Structure, function and significance in visual behaviour (This volume).

Hawryshyn, C.W., Mackay, W.C. & Nilsson, T H. (1982) Methyl mercury induced visual deficits in rainbow trout. Can. J. Zool. 60: 3127 3133

Hesse, R. (1902) Untersuchen über die Organe der Lichtempfindung bei neideren. Thieren. VIII Weitere Thatsachen. Allgemeines. Z. wiss. Zool. 72: 565-656.

Kampa, E.M. (1955) Euphausiopsin, a new photosensitive pigment from the eye of euphausiid crustaceans. Nature (Lond.) 175: 966-998.

Kampa, E.M. & Boden. B.D. (1957) Light generation in a sonic scattering layer. Deep Sea Res. 4: 73-92.

Lall, A.B., Chapman, R.M., Trouth, C.O. & Holloway, J.A. (1980 a) Spectral mechanisms of the compound eye in the firefly Photinus pyralis (Cleoptera: Lampyridae). J. Comp. Physiol. 135A: 21-27

Lall, A.B., Seliger, H.H., Biggley, W.H. & Lloyd, J.E. (1980 b) Ecology of colors of firefly bioluminescence. Science 210: 560-562.

Land, M F (1966) Image formation by a concave reflector in the eye of the scallop, Pecten maximus. J. Exp. Biol. 45: 438-447.

Land, M.F. (1983 a) Crustacea (This volume).

Land, M.F. (1983 b) Mollusca (This volume).

Langer, H. Hamann, B. & Meinecke, C.C. (1979) Tetrachromatic visual system in the moth Spodoptera exempta (Insecta: Noctuidae). J. Comp. Physiol. 129: 235-239.

Laughlin, S. (1983) The roles of parallel channels in early visual processing by the arthropod compound eye (This volume).

Levine, J.S. & MacNichol, E.F. (1982) Color vision in fishes. Sci. Am 246: 140-149

Meinertzhagen, I.A (1983) The rules of synaptic assembly in the developing insect lamina. (This volume).

Menzies R.J., George, R.Y. & Rowe, G.T. (1973) Abyssal Environment and Ecology of the World Oceans. New York, Wiley (Interscience). 488 pages

Miller, W.H. (Ed.) (1981) Molecular Mechanisms of Photoreceptor Transduction. Current Topics in Membranes and Transport, Vol. 15. New York, Academic Press. 452 pages.

Muñoz-Cuevas, A. (1983) Photoreceptor structures and vision in arachnids and myriapods (This volume).

Mouze, M. (1983) Morphologie et dévelopment des yeux simples et composes des insectes (This volume).

Page, T.L. & Larimer, J.L. (1972) Entrainment of the circadian locomotor activity in crayfish. J. Comp. Physiol. 78: 107-120.

Packer, L. (Ed.) (1982) Visual Pigments and Purple Membranes. I. Methods in Enzymology, Vol. 81 Biomembranes Part H. New York, Academic Press. 902 pages.

Pearse, J.S. & Pearse, V.B. (1978) Vision in Cubomedusan jellyfishes. Science 199: 458.

Penn P E. & Alexander, C.G. (1980) Fine structure of the optic cushion in the asteroid Nepanthia belcheri. Mar. Biol. 58: 251-256.

Reuter, T.E., White, R.H. & Wald, G. (1971) Rhodopsin and porphyropsin fields in the adult bullfrog retina. J. Comp. Physiol. 58: 351-371.

Rice, D.C. & Gilbert, S.C. (1980) Early chronic exposure to methylmercury in monkeys: Effects on spatial vision. Fed. Proc. Fed. Am. Soc. Exp. Biol. 39: 678.

Rice, D.C. & Gilbert, S.C. (1982) Early chronic low-level methylmercury poisoning in monkeys impairs spatial vision. Science 216: 759-761.

Stavenga, D.G. & Schwemer, J. (1983) Visual pigments of invertebrates (This volume).

Strausfeld, N.J. (1983) Functional neuroanatomy of the blowfly's visual system (This volume).

Toyoda, J., Nosaki, H. & Tomita T. (1969) Light induced resistence changes in single photoreceptors of Necturus and Gekko. Vision Res. 9: 435-463.

Truman, J.W. (1974) Physiology of insect rhythms. IV Role of brain in the regulation of light rhythms of the giant silkmoths. J. Comp. Physiol. 95: 281-296.

Verger-Bocquet, M. (1983) Photoréception et vision chez les annélides (This volume).

Wagner H.J. (1978) Cell types and connectivity patterns in mosaic retinas. Adv. Anat. Embroyl. Cell Biol. 55 (3): 1-81.

Waterman, T.H. (1983) Natural polarized light and vision (This volume).

Westfall, J.A. (Ed.) (1982) Visual Cells in Evolution. New York, Raven Press. 161 pages.

Wolken, J.J. (1975) Photoresponses, Photoreceptors and Evolution. New York, Academic Press. 317 pages.

Yoshida, M., Takasu, N. & Tamotsu, S. (1983) Photoreception in echinoderms (This volume).

Young, J.Z. (1964) A Model of the Brain. Oxford, Clarendon. 348 pages.

AUTHOR INDEX
(Page numbers underlined refer to the bibliography)

Able, K.P. 64, 93, 103
Abrahamsol, E.W. see Shriver, J.W. et al.
Adal, M. see Nicholls, J.G. et al.
Adensamer, T. 390, 391
Akert, K. see Lamparter, H.E.
Akiyama, J. see Eguchi, E. et al.
Alfano, R.R. see Doukas, A.G. et al.
Al-Hadthi, I. see Croll, N.A., Al-Hadthi, I.
Ali, M.A. 1-6, 28, 52, 127, 741
-- Anctil, M., Cervetto, L. 7, 8
-- Croll, R.P., Jaeger, R. 7, 8
-- Wagner, H.J. 5, 8
Alkon, D.L. 179, 211
-- Fuortes, M.G.F. 707, 730
Allen, J.L. see Miller, W.H. et al.
Allen, R.D. see Taylor, D.L. et al.
Altman, J.S., Tyrer, N.M. 534, 553
Altner, I., Burkhardt, D. 8, 82, 103
Ameln, P. 320, 328
Amsellen, J., Clément, P. 283
Amsellen, J. & Ricci, C. 254, 273
Amsellen, J. see Clément P. et al.
Anctil, M. see Ali, M.A. et al.
Anderson, A.M. 607, 611
-- Nilsson, D.E. 416, 435
Anderson, D.H. see Steinberg, R.H. et al.
Anderson, E.A. see Dusenbery, D.B. et al.
Anderson, G.L. see Dusenbery, D.B. et al.
Anderson, H. 679, 678, 682, 683, 686, 691
-- Edwards, J.S., Palka, J. 191, 211
Anderson, H.J. see Shelton, P.M.J. et al.

Ando, H. 675, 691
André, M. 337, 391
Angerer, R.C. see Davidson, E.H. et al.
Aoki, K. see Kikuchi, M., Aoki, K.
-- see Waterman, T.H., Aoki, K.
Arey, L.B. & Crozier, W.J. 707, 722
Armett-Kibel, C., Meinertzhagen, I.A., Dowling, J.E. 642, 643, 648, 656
-- see Meinertzhagen, I.A., Armett-Kibel, C.
Arnett, D.W. 465, 473, 477, 592, 612, 541, 553, 489, 491, 517
-- see McCann, G.D., Arnett, D.W.
Arnott, H.J. see Walne, P.L., Arnott, H.J.
Aros, B. see Rühlich, P. et al.
Atwood, H.L. see Bloom, J.W., Atwood, H.L.
Autrum, H. 5, 8, 35, 44, 52, 64, 103, 439, 449, 450, 663, 679, 681
-- see Zettler, F., Autrum, H.
Avel, M. see Fauvel, P. et al.
Averbach, R.A. see Song, P.S. et al

Baccetti, B., Bedini, C. 356, 358, 359, 362, 365, 366, 382, 391
Bacon, J.P. see Strausfeld, N.J. Bacon, J.P.
-- see Strausfeld, N.J. et al.
Bader, C.R., Baumann, F., Bertrand, D., Carreras, J., Fuortes, G. 41, 52
Baer, J.G., Euzet, L. 221, 234
Bühr, R. 388, 389, 390, 391

789

Baker, N.O. see Kearn, G.C., Baker, N.O.
Baker, T.C. see Kuenen, L.P.S. Baker, T.C.
Ball, E.E. 425, 435
Bannister, L.H. 643, 663
Barber, V.C., Evans, E.M., Land, M.F. 708, 709, 722
-- Wright, D.E. 708, 716, 722
Barker, R.J. see Goldsmith, T.H. et al.
Barlow, H.B. 458, 471, 477
Barlow, R.B., Chamberlain, S.C. Levinson, J.Z. 44, 51, 421, 435
Barlow, H.B., Levick, W.R. 503, 518, 597, 598, 611
Barnard, P. see Horridge, G.A., Barnard, P.
Barnes, R.D. 404, 435
Barnes, S.N. see Bruno, M.S. et al.
-- see Goldman, L.J. et al.
-- see Gorman, A.L.F. et al.
Barneveld, H.H. van see Stavenga, D.G., Barneveld, H.H. van
Barra, J.A. 667, 691
Barrell, K.F. see Pask, C., Barrell, K.F.
Basinger, S.F., Gordon, W.C., Lam, D.M.K. 624, 632
Bassemir, U.K. see Strausfeld, N.J., Bassemir, U.K.
-- see Strausfeld, N.J. et al.
Bassot, J.M., Nicolas, M.T. 298, 328
Bate, C.M. 636, 656
Bate, M. see Goodman, C.S., Bate, M.
Batschelet, E. 66, 103, 201, 211
Bauer, T. 566, 611
Baumann, F. see Bader, C.R. et al.
-- see Perrelet, A., Baumann, F.
Baylor, D.A., Fuortes, M.G.F. 709, 722
-- Hodgkin, A.L. 443, 450
Beauchamp, P. de 243, 267, 283

Becker, R.S. see Waddell, W.H. et al.
Bedini, C. 356, 358, 367, 387, 388, 390, 391
-- Ferrero, E., Lanfranchi, A. 441, 442, 448, 456, 646, 663
-- Lanfranchi, A. 224, 234
-- see Baccetti, B., Bedini, C.
Beersma, D.G.M., Stavenga, D.G., Kuiper, J.W. 448, 450, 487, 518, 525, 526, 553
Behrens, M., Krebs, W. 426, 435
Bell, A.L., see Wiederhold, M.L. et al.
Benedetti, P.A., Checcucci, A. 119, 127
-- Lenci, F. 119, 127
Bentley, D., Keshishian, H. 638, 656
Benzer, S. 686, 687, 691
-- see Ready, D.F. et al.
Berg, H.C. 179, 209, 212
-- Brown, D.A. 180, 207, 211
Berger, J. 125, 127
Bergquist, P.R. see Green, C.R., Berquist, P.R.
Bernard, F. 675, 685, 691
Bernard, G.D. 21, 32, 35, 47, 52
-- Stavenga, D.G. 35, 52, 480, 781, 786
-- Wehner, R. 25, 52, 65, 90, 103
-- see Miller, W.H., Bernard, G.D. et al.
-- see Srinivasan, M.V., Bernard G.D.
-- see Wehner, R., Bernard, G.D.
Berndt, P. see Egelhaaf, A. et al.
Bernhard, G.C., Gemme, G., Sällstrom, J. 665, 691
-- see Miller, W.H. et al.
Berrill, M. 754, 756, 768
Berry, M., McConnell, P., Sievers, J. 655, 656
Bertkau, Ph. 338, 391
Bertrand, D., Fuortes, G., Muri, R. 20, 31, 47, 52
-- see Bader, C.R. et al.

AUTHOR INDEX

Besharse, J.-C. see Papermaster, D.S. et al.
-- Pfenninger, K.H. 148, 149, 153, 171
Bavelander, G. 233, 234
Biesinger, R. see Götz, K.G. et al.
Biggley, W.H. see Lall, A.B. et al.
-- see Seliger, H.H. et al.
Birky, C.W. Jr. 243, 283
Bishop, C.A., Bishop, L.G. 532, 553
Bishop, L.G. 540, 553
-- Keehn, D.G. 532, 553
-- Keehn, D.G., McCann, G.D. 528, 532, 540, 542, 553
-- Bishop, C.A., Bishop, L.G.
-- see Dvorak, D.R. et al.
-- see Eckert, H., Bishop, L.G.
-- see Soohoo, S.L., Bishop, L.G.
-- see Yingst, D.R.
Bhatia, M.L. 310, 313, 320, 321, 328
Bizzi, E., Kalil, R.E., Morasso, P., Tabliasco, V. 784, 786
Blakers, M. see Menzel, R., Blakers, M.
Blanchot, J., Pourriot, R. 280, 283
Blest, A.D. 44, 52, 87, 103, 154, 171, 363, 364, 366, 387, 391, 425, 435, 779, 786
-- Day, W.A. 44, 52, 366, 387, 391
-- Hardie, R., McIntyre, P., Williams, D. 376, 387, 391
-- Land, M.F. 359, 363, 364, 370, 375, 378, 387, 391, 413, 445, 714, 722
-- Stowe, S., Eddey, W., Williams, D.S. 87, 103
-- Williams, D., Kao, L. 362, 377, 387, 391
-- see Couet, H.G. de, Blest, A.D.
-- see Laughlin, S. et al.
-- see Strausfeld, N.J., Blest, A.D.

-- see Carpenter, K.S. et al.
Blight, A.R., Llinas, R. 469, 478
Blondeau, J. 548, 553, 573, 609, 611
-- Heisenberg, M. 551, 609, 611
Bloom, J.W., Atwood, H.L. 535, 553
Blue, J. 319, 328
Bocquet, M. 289-334, 328, 329, 334
-- Dhainaut-Courtois, N. 296, 297, 300, 328
Boden, B.D. see Kampa, E.M., Boden, B.D.
Bodenstein, D. 636, 657, 681, 683, 691
-- see Wolbarsht, M.L. et al.
Bohn, G. 320, 323, 328
Bok, D., Hall, M.O., O'Brien, P. 153, 171
Bollerup, G., Burr, A.H. 136, 137, 171
Bolt, R.E., Ewer, D.W. 751, 768
Bolz, J., Varju, D. 609, 611
-- see Grieger, B. et al.
-- see Varju, D., Bolz, J.
Borisy, G.G. see Hyams, J.S., Borisy, G.G.
Borle, A.B. 127
Boschek, C.B. 441, 445, 450, 485, 486, 518, 647, 657
-- Hamdorf, K. 150, 171, 762, 768
Boss, K.J. see Roper, C.F.E., Boss, K.J.
Boucher, F., Leblanc, R.M. 28, 52
Boulin, C. see Geiger, G. et al.
Bound, K.E., Tollin, G. 119, 127
Bovee, E.C. see Jahn, T.L., Bovee, E.C.
Bowling, D.B., Michael, C.R. 458, 478
Bowmaker, J.K. 65, 70, 103
Bownds, D. see Hubbard, R. et al.

Boycott, B.B. see Wässle, H. et al.
Boyle, P.R. 707, 722
Brachet, P. see Darmon, M., Brachet, P.
Bradke, D.L. 307, 309, 328
Bradshaw, F. see Heelis, D.V. et al.
Braitenberg, V. 489, 496, 518, 526, 529, 553, 640, 657
-- Debbage, P. 492, 517
-- Hauser-Holshuch, H. 448, 450, 527, 553
-- see Burkhardt, W., Braitenberg, V.
-- see Reichard, W. et al.
-- see Strausfeld, N.J., Braitenberg, V.
Brammer, J.D. 667, 691
-- Clarin, B. 674, 691
-- see Stein, P.J. et al.
Brandenburger, J.L., Eakin, R.M. 154, 171, 305, 306, 328, 758, 764, 768
-- Eakin, R.M., Reed, C.T. 150, 171, 762, 768
-- Woollacott, R.M., Eakin, R.M. 152, 162, 171
-- see Eakin, R.M., Brandenburger, J.L.
Brandt, T. see Diener, H.C. et al.
Braun, R. 710, 722
Brenner, S. see Ward, S. et al.
-- see White, J.G. et al.
Briggs, M.H. 19, 52
Brehm, P. see Eckert, R., Brehm, P.
Brines, M.L. 67, 69, 97, 98, 99, 100, 101, 102, 103
-- Gould, J.L. 67, 68, 70, 98, 100, 103
Britten, R.J., Davidson, E.H. 163, 165, 167, 169, 170, 171, 172
-- see Davidson, E.H., Britten, R.J.
-- see Davidson, E.H. et al.
-- see Galau, G.A. et al.
Brükelmann, J. see Fisher, A., Brükelmann, J.

Brooker, B.E. 448, 457
Brown, D.A. see Berg, H.C., Brown, D.A.
Brown, H.M., Ito, H., Ogden, T.E. 145, 172
Brown, J.A., Nielsen, P.J. 125, 127
--see Nolte, J.E., Brown, J.A.
Brown, P.K., Brown, P.S. 16, 20, 52
--Wald, G. 783, 786
--White, R.H. 20, 52
--see Denys, C.J., Brown, P.K.
--see Kropf, A. et al.
Brown, P.S. see Brown, P.K., Brown, P.S.
Bruno, M.S., Barnes, S.N., Goldsmith, T.H. 20, 32, 53, 420, 435
--Goldsmith, T.H. 53
--see Goldsmith, T.H., Bruno, M.S.
Bryceson, K.P. 434, 435
Buchner, E. 440, 450, 475, 477, 478, 497, 498, 505, 518, 535, 539, 540, 547, 548, 550, 553, 554, 561-621, 621, 623, 631, 632, 779, 784, 786
--Buchner, S. 498, 503, 517, 611, 612, 623-634, 632, 783, 786
--Buchner, S., Hengstenberg, R. 498, 518, 625, 632
--see Götz, K.G., Buchner, E.
--see Heisenberg, M., Buchner, E.
--see Pick, B., Buchner, E.
Buchner, S. see Buchner, E., BuchnerS.
--see Buchner, E. et al.
Bücher, R. see Geiger, G. et al.
Buddenbrock, W.V., Moller-Racke, I. 711, 722
Bullock, T.H. 707, 722, 728, 740
Bülthoff, H. 567, 604, 611, 612
--Götz, K.G. 592, 612
--Götz, K.G., Herre, M. 577, 592, 607, 612
--Poggio, T., Wehrhahn, C. 577, 612

Bülthoff, H. see Hengstenberg, R. et al.
-- see Wehrhahn, C. et al.
Bundy, D.A.P. 456, 457
Burfield, S.T. 728, 734, 741
Burghause, F. 72, 104, 674, 691
Burkhardt, D. 68, 76, 103, 441, 450
-- Braitenberg, V. 641, 643, 648, 657
-- Motte, I. de la 450, 566, 567, 602, 612
-- see Altner, I., Burkhardt, D.
Burr, A.H. 131-178, 172, 179, 215, 211, 774, 777, 780, 786
-- Burr, C. 135, 136, 137, 155, 156, 172, 191, 192, 196, 211
-- Webster, J.M. 137, 172, 177, 786
-- see Bollerup, G., Burr, A.H.
Burr, C. see Burr, A.H., Burr, C.
Burr, D.C., Ross, J. 606, 612
Burrows, M. 534, 553
-- Watson, A.H.D., Burrows, M.
Bursell, E., Ewer, D.W. 182, 211
-- see Ewer, D.W., Bursell, E.
Burton, F.A. see Land, M.F., Burton, F.A.
-- Land, M.F. et al.
Burton, P.R., Stockhammer, K.A. 667, 668, 691
Butenandt, A. 348, 391
Buyse, I. see Zaagman, W.H. et al.
Bychowsky, B.E. 221, 235

Cable, R.M. see Isseroff, H. Cable, R.M.
-- see Pond, G.C., Cable, R.M.
Cajal, S.R. 635, 657
-- Sanchez, D.S. 484, 487, 517, 635, 657
Calas, A. see Sans, A. et al.
Caldwell, R.L., Dingle, H. 427, 436
Callender, R.H. see Doukas, A.G. et al.

-- Honig, B. 35, 53
Campan, R. see Goulet, M. et al.
-- see Jeanrot, N. et al.
Campbell, A. see Muller, S.W.M., Campbell, A.
Campbell, C.A. see Valentine, J.W., Campbell, C.A.
Campos-Ortega, J.A. 492, 518
-- Gateff, E.A. 686, 692
-- Hofbauer, A. 686, 692
-- Strausfeld, N.J. 474, 478, 485, 487, 492, 498, 518, 519
-- see Hofbauer, A., Campos-Ortega
-- see Strausfeld, N.J., Campos-Ortega, J.A.
Carlier, E. see Sans, A. et al.
Carlile, M.J. 116, 127
Carlson, J.D. see Stark, W.S., Carlson, J.D.
Carpenter, K.S., Morita, M., Blest, B. 154, 155, 172, 219, 226, 235
Carreras, J. see Bader, C.R. et al.
Carricaburu, P. 317, 328, 358, 359, 360, 370, 371, 375, 391
-- Cherrak, M. 384, 392
-- Kernéis, A. 292, 302, 317, 328
-- Munoz-Cuevas, A. 378, 379, 380, 381, 382, 392
-- see Fouchard, R., Carricaburu, P.
Cartwright, B.A., Collett, T.S. 97, 104, 610, 612
-- see Collett, T.S., Cartwright, B.A.
Catton, W.T. 602, 612
Caveney, S., McIntyre, P. 428, 436
Cervetto, L. see Ali, M.A. et al.
Chamberlain, S.C. see Barlow, R.B. et al.
Chamberlin, M.E. see Galau, G.A. et al.
Champ, P. 153, 157
-- Pourriot, R. 153, 157

Chapman, H.D. 231, 232, 235
Champman, R.M. see Lall, A.B. et al.
Charras, J.P. see Coulon, J. et al.
Chasse, J.L. see Coulon, J. et al.
Chassé, J.L. see Luciani, A. et al.
Checchi, A.C. 677, 692
Checcucci, A., Colombetti, G., Ferrari, R., Lenci, F. 119, 127
-- see Benedetti, P.A., Checcucci, A.
Chen, H.S., Rao, C.R.N. 68, 104
Cherrak, M. see Carricaburu, P., Cherrak, M.
Cherry, R.J. 148, 172
Chesnokova, E.G. see Gribakin, F.G., Chesnokova, E.G.
Chevais, S. 681, 692
Chillemi, S., Taddei-Ferreti, C. 566, 612
Chitwood, B.G., Murphy, D.G. 196, 211
Chrismer, K.L. see Olivo, R.F., Chrismer, K.L.
Chytil, F., Ong, D.E. 442, 450
Clarin, B. see Brammer, J.D., Clarin, B.
Clark, A.W. 310, 312, 328
-- Millecchia, R., Mauro, A. 169, 172
Clark, D. see Ware, R.W. et al.
Clark, R.B. 320, 321, 323, 324, 325, 326, 327, 329
-- Hess, W.N. 320, 329
Clément, P. 143, 156, 159, 172, 226, 228, 235, 283, 284
-- Luciani, A., Pourriot, R. 279, 284
-- Pourriot, R. 276, 277, 278, 284
-- Rougier, C., Pourriot, R. 242, 284
-- Wurdak, E. 156, 159, 172, 179, 211, 241-288, 774, 778, 786

-- Wurdak, E., Amsellen, J. 241, 242, 257, 262, 267, 284
-- see Amsellen, J., Clément, P.
-- see Cornillac A. et al.
-- see Coulon, J. et al.
-- see Luciani, A. et al.
-- see Pourriot, R., Clément, P
-- see Pourriot, R. et al.
Cloney, R.M. see Herman, C.O., Cloney, R.M.
Cloudsley-Thompson, J.L. 382, 392
Cobb, J.L.S., Stubbs, T.R. 749, 768
Cobb, N.A. 137, 172
Cobbs, W.H. see Corless, J.M. et al.
Cohen, C.F. see Goldsmith, T.H. et al.
Collett, T.S. 565, 574, 577, 604, 609, 612
-- Cartwright, B.A. 97
-- Harkness, L.I.K. 565, 612
-- Land, M.F. 93, 104, 448, 450, 577, 601, 605, 609, 613, 636, 721, 722
-- see Cartwright, B.A., Collett, T.S.
-- see Land, M.F., Collett, T.S.
Collewijn, H. 721, 722
Colombetti, G. see Checcucci, A. et al.
-- see Lenci, F., Colombetti, G.
Combes, C. 452, 457
-- Theron, A. 456, 457
-- see Fournier, A., Combes, C.
-- see MacDonald, S., Combes, C.
Condeelis, J.S. see Taylor, D.L. et al.
Cone, R.A., Pak, W.L. 38, 53
-- see Poo, M.P., Cone, R.A.
Cook, R. 601, 613
Coomans, A. 143, 172
Coomans, A., De Grisse, A. 192, 211
-- see Vanfleteren, J.R., Coomans, A.
Cooter, R.J. 689, 692

AUTHOR INDEX

Corbière-Tichané, G. 299, 329
Corless, J.M., Cobbs, W.H., Costella, M.J., Robertson, J.D. 762, 768
Corliss, J.O. see Levine, N.D. et al.
Cornillac, A. 244, 249, 254, 267, 268, 270, 274, 275, 284
-- Clément, P., Wurdak, E. 276, 268, 271, 284
-- see Coulon, J. et al.
Cornwall, M.C., Gorman, A.L.F. 20, 38, 41, 52
Cosens, D. see Morton, P.D., Cosens, D.
Cosens, F. 674, 692
Costella, M.J. see Corless, J.M. et al.
Couet, H.G. de, Blest, A.D. 87, 104
Couillard, P. 115-130, 127, 142, 172, 255, 285, 774, 786
-- see Forget, J., Couillard, P.
Coulon, J., Charras, J.P., Chasse, J.L., Clément, P., Cornillac, A., Luciani, A., Wurdak, E. 272, 285
Coulson, K.L. 68, 104
Cox, F.E.G. see Levine, W.D. et al.
Crane, J. 384, 385, 386, 392
Creutzberg, F. 179, 211
Croll, N.A. 137, 172, 191, 192, 193, 195, 196, 199, 207, 209, 211
-- Al-Hadthi, I. 192, 212
-- Evans, A.A.F., Smith, J.M. 136, 173
-- Riding, I.L., Smith, J.M. 135, 136, 173, 192, 196, 212
-- Sukhdeo, M.V.K. 135, 173
-- see Rutherfood, T.A., Croll, N.A.
Croll, R.P. see Ali, M.A. et al.
Crone, C. 624, 632
Cronin, T.W., Goldsmith, T.H. 16, 20, 23, 25, 35, 48, 49, 50, 51, 53, 86, 104, 765, 769

Crossland, K. see Ware, R.W. et al.
Crozier, W.J. see Arey, L.B., Crozier, W.J.
Cugnoli, C. see Pepe, I.M., Cugnoli, C.
Culler, G.J. see Davenport, D. et al.
Cummins, D., Goldsmith, T.H. 86, 90, 104, 426, 436
Curtis, D. 344, 356, 357, 358, 392

Dahmen, H.J. 716, 722
Dakin, W.J. 710, 722
Dales, R.P. 292, 329
Dambach, M., Weber, W. 746, 769
-- see Egelhaaf, A., Dambach, M.
-- see Weber, W., Dambach, M.
Darmon, M., Brachet, P. 179, 212
Dartnall, H.J.A. 5, 8, 27, 53, 65, 68, 69, 104
Darnell, J.E. 163, 173
Darwin, C. 131, 173
Daumer, K. 68, 104
Dauwalder, M. see Whaley, W.G., Dauwalder, M.
Davenport, D., Culler, G.J., Greaves, J.O.B., Forward, R.B. Hand, W.G. 94, 104
-- see Engel, R.E., Davenport, D.
-- see Hand, W.G., Davenport, D.
Davey, M. see Menco, B.P.M. et al.
David, C.T. 606, 613
Davidson, E.H., Britten, R.J. 165, 173
-- Galau, G.A., Angerer, R.C., Britten, R.J. 173
-- Klein, W.H., Britten, R.H. 173
-- see Britten, R.J., Davidson, E.H.
-- see Galau, G.A. et al.
Dawydoff, C. see Fauvel, P. et al.

Day, W.A. see Blest, A.D., Day, W.A.
Debbage, P. see Braitenberg, V., Debbage, P.
De Grisse, A. see Coomans, A., De Grisse, A.
Demoll, R. 689, 692
Dentler, W.L. 150, 173
Dennis, M.J. 701, 707, 722
Denton, E.J. 720, 722
Denys, C.J., Brown, P.K. 20, 53
Denzer-Melbrandt, U. 320, 329
De Quatrefages, A. 291, 329
Derham, W. 563, 613
Deroux, G. see Levine, N.D. et al.
Des Rosiers, M.H. see Sokoloff et al.
DeVoe, R.D. 376, 391, 475, 476, 478, 528, 554
-- Ockleford, E.M. 475, 476, 478, 528, 554
-- Small, R.J.W., Zvarguilis, J.E. 376, 392, 443, 450
Dhainaut-Courtois, N. 299, 305, 329
-- see Bocquet, M., Dhainaut-Courtois, N.
Dichgans, J., Körner, F., Voigt, K. 605, 613
-- see Diener, H.C. et al.
Diehn, B. 179, 181, 184, 188, 212
-- Feinleib, M., Haupt, W., Hildebrand, E., Lenci, F. 116, 127, 273, 285
-- Feinlab, M., Haupt, W., Hildebrand, E., Lenci, F., Nultsch, W. 180, 191, 192, 193, 194, 208, 209, 210, 212
-- see Doughty, M.D., Diehn, B.
Diener, H.C., Wist, E.R., Dichgans, J., Brandt, T. 606, 613
Dietrich, W. 446, 450, 779, 786
Dill, J.C. see McCann, G.D., Dill, J.C.

Dilly, P.N. 701, 709, 716, 722
Dimelow, E.J. 751, 769
Dingle, H. see Caldwell, R.L.
Dodge, F.A. see Hillman, P. et al.
Dodge, J.D. 118, 127
Donges, J. 453, 454, 455, 457
Dood, G.H. see Menco, B.P.M. et al.
Doolittle, J.H. 319, 329
Doorn, A.J. von, Koenderink, J.J. 587, 613
Dorsett, D.A., Hyde, R. 295, 296, 297, 329
Doughty, M.D., Diehn, B. 116, 119, 127
Doukas, A.G., Stefancic, V., Susuki, T., Callender, R.H., Alfano, R.R. 29, 53
Dover, G. 167, 173
Dowling, J.E. see Armett-Kibel, C. et al.
Downing, A.C. 409, 436
-- see Young, S., Downing, A.C.
Dragesco-Kernéis, A. 291, 301, 302, 304, 320, 321, 329
Draslar, K. see Schneider, L. et al.
Drees, O. 385, 386, 392
Drescher, W. 682, 692
Dubs, A. 86, 104, 465, 466, 478, 593, 613
-- Laughlin, S.B., Srinivasan, M.V. 465, 466, 478
-- see Srinivasan, M.V. et al.
Ducret, F. 729, 731, 734, 736, 737, 741
Duelli, P. see Hardie, R.C., Duelli, P.
Duniec, J. see Horridge, G.A. et al.
Durand, J.P., Gourbault, N. 219, 226, 235
Durchon, M. 300, 329, 330
Durham, A.C.H. 126, 127
Durham, D., Woolsey, T.A., Kruger, L. 624, 632

AUTHOR INDEX

Dusenbury, D.B. 179, 191, 196, 199, 200, 212
-- Anderson, G.L., Anderson, E.A. 199, 212
Dvorak, D., Srinivasan, M.V., French, A.S. 541, 554, 597, 599, 613
-- Bishop, L.G., Eckert, H.E. 496, 519, 529, 532, 536, 554
-- see Srinivasan, M.V., Dvorak, D.
-- see Lillywhite, P.G., Dvorak, D.
Dyer, F.C., Gould, J.L. 70, 97, 104

Eakin, R.M. 2, 8, 143, 156, 157, 159, 161, 169, 173, 244, 265, 285, 296, 305, 326, 330, 343, 392, 681, 692, 709, 722, 756, 769
-- Brandenburger, J.L. 152, 158, 161, 173, 189, 212, 220, 224, 235, 350, 356, 357, 358, 361, 362, 366, 369, 392, 758, 762, 765, 769
-- Martin, G.G., Reed, C.T. 305, 306, 330
-- Westfall, J.A. 138, 173, 244, 285, 297, 304, 307, 330, 732, 733, 734, 735, 742
-- see Woollacott, R.M., Eakin, R.M.
-- see Brandenburger, J.L., Eakin, R.M.
-- see Brandenburger, J.L. et al.
-- see Ermak, T.K., Eakin, R.M.
-- see Hermans, C.O., Eakin, R.M.
Ebina, Y., Nagasawa, N., Tsukara, Y. 27, 53
Ebrey, T.G. 765, 769
-- Honig, B. 14, 27, 53
Eckert, H. 75, 104, 449, 450, 496, 497, 513, 519, 529, 531, 532, 533, 536, 541, 545, 547, 554, 566, 597, 602, 613, 614

-- Bishop, L.G. 496, 497, 519, 529, 531, 532, 536, 542, 546, 554
-- Hamdorf, K. 536, 554, 566, 587, 614
-- Meller, K. 532, 533, 554
-- see Dvorak, D.R. et al.
Eckert, R., Brehm, P. 145, 150, 173
-- see Schmidt, J.A., Eckert, R.
Eddey, W. see Blest, A.D. et al.
Edgar, R.S., Wood, W.B. 191, 212
Edrich, W., Neumeyer, C., Helversen, O. von 68, 97, 104
Edwards, G.A. see Ruck, Ph., Edwards, G.A.
Edwards, J.S. 681, 692
-- see Anderson, H. et al.
-- see Nordlander, R.H., Edwards, J.S.
Egelhaaf, A., Berndt, P., Küthe, H.W. 636, 657, 677, 692
-- 72, 104
Eguchi, E. 73, 104, 157, 173
-- Goto, T., Waterman, T.H. 84, 86, 105
-- Meyer-Rochow, V.B. 68, 105
-- Waterman, T.H. 44, 53, 84, 105, 150, 173, 364, 392, 426, 436
-- Waterman, T.H., Akiyama, J. 86, 105
Ehlers, B., Ehlers, U. 224, 235
Ehlers, U. see Ehlers, B., Ehlers, U.
Eley, S., Shelton, P.M.J. 347, 392, 687, 692
Eley, Z. see Shelton, P.M.J. et al.
Ellenby, C. 136, 174
-- Smith, L. 136, 174
Elofsson, R. 417, 435, 436
-- see Odselius, R., Elofsson, R.
Engel, R.E., Davenport, D. 233, 236
Ephrussi, B., Beadle, G.B. 681, 692
Erber, J. 600, 604
Eriksson, E.S. 565, 614

Ermak, T.H., Eakin, R.M. 291, 295, 303, 330
Esterly, C.O. 738, 741
Euzet, L. see Baer, J.G., Euzet, L.
Evans, A.A.F. see Croll, N.A. et al.
Evans, E.M. see Barber, V.C. et al.
Evans, S.M. 321, 323, 324, 330
Evered, D., O'Connor, M., Whelan, J. 601, 604
Ewer, D.W., Bursell, E. 209, 212
-- see Boltt, R.E., Ewer, D.W.
-- see Bursell, E., Ewer, D.W.
-- see Pople, W., Ewer, D.W.
Ewing, A. see Willmung, R., Ewing, A.
Exner, S. 408, 418, 428, 429, 436

Fahrenbach, W.H. 403, 416, 436
Fankboner, P.V. 189, 212
Fauré-Fremiet, E. 149, 174, 255, 265, 285
-- Rouiller, C. 148, 174, 255, 265, 285
Fauvel, P. 301, 330
-- Avel, M., Harant, T., Grassé, P., Dawydoff, C.. 777, 786
Fehér, O., Rojik, I. 624, 632
Feiler, R. see Kirschfeld, K. et al.
Fein, A., Szutz, E.Z. 64, 105, 165, 174
Feinlab, M. see Diehn, B. et al.
-- see Morel-Laurens, N.M., Feinlab, M.E.
Fermi, G., Reichardt, W. 586, 613
Fernandez, H.R. 20, 53
-- Nickel, E.E. 150, 174
-- see Goldsmith, T.H., Fernandez, H.R.
-- see Yingst, D.R. et al.
-- see Waterman, T.H. et al.
Fernandez, J., Stent, G.S. 179, 212
Ferrari, R. see Checcucci, A. et al.
Ferrero, E. see Bedini, C. et al.
Ferrus, A. see Kankel, D.R. et al.
-- see Koto, M. et al.
Filippov, M.P., Ovchinnikov, A.A. 68, 105
Finell, N. see Järvilehto, M., Finell, N.
Finkel, A. see Rockstein, M., Finkel, A.
Fioravanti, R., Fuortes, M.G.F. 317, 318, 330
Fischbach, K.F. 440, 450, 484, 518, 566, 610, 613, 638, 657
-- Heisenberg, M. 611, 613, 631, 632
Fisher, A., Brükelmann, J. 295, 296, 297, 305, 330
Fisher, S.K. see Steinberg, R.H. et al.
Fitzpatrick, S.M. see Wellington, W.G., Fitzpatrick, S.M.
Fleissner, G. 367, 377, 380, 382, 383, 387, 392, 393, 778, 786
-- Schliwa, M. 387, 393
-- see Schliwa, M., Fleissner, G.
Fletcher, A., Murphy, T., Young, A. 413, 420, 436, 713, 714, 722
Florida, R.G. see Wolken, J.J.
Foerster, T. 442, 450
Füh, H. 706, 723
Folkers, E. 610, 614
-- Spatz, H.C. 610, 614
Forget, J., Couillard, P. 127
Forster, L 385, 386, 387, 393
Forward, R.B. see Davenport, D. et al.
-- see Lang W.H. et al.
Foster, K.W., Smyth, R.D. 118, 119, 121, 122, 127, 145, 174, 255, 285

AUTHOR INDEX

Foster, S.F. see McCann, G.D. Foster, S.F.
Fouchard, R., Carricaburu, P. 378, <u>393</u>
Fournier, A. 156, <u>175</u>, 217-239, <u>236</u>, 262, 281, <u>285</u>, 779, <u>786</u>
-- Combes, C. 221, <u>236</u>, 244, <u>285</u>
Fraenkel, G.S., Gunn, G.L. 5, 8, 180, 181, 182, 183, 186, 195, 200, 209, 210, 213, 219, <u>226</u>, 273, 285, 701, <u>723</u>
-- Pringle, J.W.S. 565, <u>615</u>
Franceschini, N. 31, 32, 35, 42, <u>53</u>, <u>54</u>, 439-455, <u>450</u>, <u>451</u>, 463, 464, <u>476</u>, 484, 493, <u>518</u>, 524, 525, 526, <u>554</u>, 580, 596, 597, 601, <u>614</u>, 623, 624, <u>632</u> 779, <u>786</u>
-- Hardie, R.C. 445, <u>451</u>
-- Hardie, R.C., Ribi, W., Kirschfeld, K. 73, <u>105</u>, 445, 447, 448, <u>451</u>, 526, <u>554</u>, 781, 783, <u>786</u>
-- Kirschfeld, K. 441, <u>451</u>
-- Kirschfeld, K., Minke, B. 34, <u>54</u>, 446, 447, 448, <u>451</u>
-- Münster, A., Heurkens, G. 525, <u>554</u>
-- see Hardie, R.C. et al.
-- see Kirschfeld, K., Franceschini, N.
-- see Kirschfeld, K. et al.
-- see Riehle, A., Franceschini, N.
-- see Stavenga, D.G., Franceschini, N.
-- see Stavenga, D.G. et al.
Francis, D. 122, 123, <u>127</u>
Franquinet, R. see Huet, M., Franquinet, R.
Frantsevitch, L.I.K. 101, <u>105</u>
Fraser-Rowell, C.H., O'Shea, M. Williams, J.L.F. 475, <u>478</u>
-- see O'Shea, M., Fraser-Rowell, C.H.
French, A.S., Järvilehto, M. 465, <u>478</u>
-- see Dvorak, D. et al.
Fripp, P.J. see Mason, P.R., Fripp, P.J.
Frisch, K. von 47, <u>54</u>, 64, 70, 94, 97, 100, <u>105</u>
-- Kupelweser, H. <u>436</u>
Fristrom, D. 683, <u>692</u>
Friza, F. 665, <u>692</u>
Fröhlich, A., Meinertzhagen, I.A. 639, 640, 641, 644, 645, 647, 650, 652, 653, <u>657</u>
Fromm, H.J. 625, <u>632</u>
Frost, B.J. 423, <u>436</u>
Fugate, R.D., Song, P.S. 442, 444, <u>451</u>
Fujimoto, K., Yanase, T., Okuno, Y. 708, <u>723</u>
-- see Tabasem, T. et al.

Fung, Y.M. see Kong, K.L. et al.
Fuortes, G. see Pader, C.R. et al
-- see Bertrand, D. et al.
Fuortes, M.G.F. see Aikon, D.L., Fuortes, M.G.F.
-- see Baylor, D.A., Fuortes, M.G.F.
-- see Fioravanti, R., Fuortes, M.G.F.
-- see Lasansky, A., Fuortes, M.G.F.
Furnestin, M.L. 730, 742
Furuya, M. see Watanabe, M., Furuya, M.
Futterman, S., Heller, J. 442, 451
Fyg, W. 692

Gademann, R. see Wolf, R. et al.
Gagné, H.T. see Short, R.B., Gagné, H.T.
Galav, G.A., Chamberlin, M.E., Hough, B.R., Britten, R.J., Davidson, E.H. 167, 174
-- see Davidson, E.H. et al.
Garcia-Bellido, A. 685, 692
Gardner, B.T. 386, 393
Garen, S.H. see Kankel, D.R. et al.
Garrardi, P. see Wilson, M. et al.
Gateff, E.A. see Campos-Ortega, J.A., Gateff, E.A.
Gaudin, G. see Richard, G., Gaudin, G.
Gavel, L. von, 586, 614
Gebhardt, B. see Wolf, R. et al.
Geethabali, M. 384, 393
-- Pampapathi Rao, K. 377, 384, 393
Gehrels, T. 67, 105
Geiger, G. 550, 555, 600, 601, 607, 608, 614
-- Boulin, C., Bücher, R., 573, 614
-- Nassel, D.R. 548, 555, 567, 615

-- Poggio, T. 567, 577, 600, 609, 615, 784, 786
Gelber, B. 125, 127
Gelperin, A. see Sejnowski, T.J. et al.
Gemme, G. see Bernhard, G.C. et al.
Gemperlein, R., Paul, R., Lindawer, E., Steiner, A. 42, 54, 442, 445, 451
George, R.Y. see Menzies, R.J. et al.
Gerisch, G. 123, 127
Gersen Schoorl, H.K.J. see Verheijeni, F.T.
Gewecke, M. 508, 519
Giese, A.C. 125, 127
Gifford, D. see White, R.H. et al.
Gilbert, J.J. 116, 141, 158, 159
Gilbert, S.C. see Rice, D.C., Gilbert, S.C.
Gillary, E.W. see Gillary, H.L., Gillary, E.W.
Gillary, H.L. 603, 721
-- Gillary, E.W. 701, 722
Gilula, N. see Satir, P., Gilula, N.
Giulio, L. 376, 378, 393
Glantz, R.M. 460, 478
-- see Kirk, M.D. et al.
Glas, H.W., van der 66 (W), 70 (W), 90, 99, 101, 105
Gogala, M., Handorf, K., Schwemer, J. 20, 53
-- see Hamdorf, K. et al
-- see Schneider, L. et al.
-- Schwemer, J. et al.
Golani, I. 207, 213
Golding, D.W., Whittle, A.C. 309, 330
-- see Whittle, A.C., Golding, D.W.
-- see Zahid, Z.R., Golding, D.W.
Goldman, L.J., Barnes, S.N., Goodsmith, T.H. 20, 54

AUTHOR INDEX

Goldsmith, T.H. 69, 105, 27, 42, 46, 54, 451
-- Barker, R.J., Cohen, C.F. 442, 451
-- Bruno, M.S. 23, 27, 30, 54
-- Fernandez, H.R. 442, 451
-- Wehner, R. 86, 105, 150, 174
-- see Bruno, M.S. et al.
-- see Cronin, T.W. et al.
-- see Cummins, D., Goldsmith, T.H.
-- see Hays, D., Goldsmith, T.H.
-- see Waterman, T.H. et al.
Gonzalez-Baschwitz, G. 356, 358, 360, 362, 383
Goodheer, J.C. 442, 451
Goodman, C. 534, 555
Goodman, C.S., Bate, M. 637, 657
Goodman, L.J. 690, 692
-- Mobbs, P.G., Kirkham, J.B. 643, 657
Goodsmith, T.H. see Goldman, L.J. et al.
Gophen, M., Harris, R.P. 409, 436
Gordon, W.C. see Basinger, S.F. et al.
Gorman, A.L.F. see Cornwall, M.C., Gorman, A.L.F.
-- see McReynolds, J.S., Gorman, A.L.F.
-- McReynolds, J.S. 723, 145, 174
-- McReynolds, J.S., Barnes, S.N. 145, 174
Görner, P. 382, 393
Goto, T., Yoshida, M. 727-742, 742, 774, 786-787
-- see Eguchi, E. et al.
Götz, K.G. 547, 548, 549, 550, 555, 573, 586, 595, 604, 605, 606, 607, 615
-- Buchner, E. 606, 615, 548, 555
Götz, K.G., Hengstenberg, B., Biesinger, R. 610, 615, 550, 555
-- Wenking, H. 610, 615
-- see Buchner, E. et al.
-- see Bülthoff, H., Götz, K.G.
-- see Bülthoff, H. et al.
-- see Heisenberg, M., Götz, K.G.
Gould, J.L. 64, 93, 105
-- see Brines, M.L., Gould, J.L.
-- see Dyer, F.C., Gould, J.L.
Goulet, M., Campan, R., Lambin, M. 565, 615
Gouras, P., Zrenner, E. 166, 174, 188, 213
-- see Zrenner, E., Gouras, P.
Gourbault, N. see Durand, J.P., Gourbault, N.
Goy, M.F., Springer, M.S. 179, 213
Graham, M. see Rogers, B, Graham, M.
Grandjean, F. 337, 338, 393
Grain, J. see Levine, N.D. et al.
Grassé, P. see Fauvel, P. et al.
Grassé, P.P. see Harant, H., Grassé, P.P.
Greaves, J.O.B. 94, 105, 207, 213
-- see Davenport, D. et al.
Grebecka, L. 126, 128
Grebecki, A. 126, 128
Green, C.D. 233, 236, 237, 241, 244, 251
Green, S.M. see Lawrence, P.A., Green, S.M.
Greer, C.A. see Lancet, D. et al.
Gregory, R.L., Ross, H.E., Moray, N. 409, 436
Green, C.R., Bergquist, P.R. 151, 174
Green, S.M., Lawrence, P.A. 684, 693
Greenberg, R.M. see Stark, W.S. et al.
Grenacher, H. 406, 408, 436, 337, 338, 387, 388, 389, 394, 671, 693
Grevet, C. 118, 120, 123, 128
Gribakin, F.G. 39, 54, 67, 87, 106
-- Chesnokova, E.G. 76(W), 106

-- Vishnevskaya, T.M., Polyanovskii, A.D. 87, 106
Grieger, B., Bolz, J., Varju, D. 609, 615
Griffin, D.R. 99, 106
Grundler, O.J. 80, 106
Grunn, D.L. see Fraenkel, G.S. Gunn, D.L.
Guérin, J.-P. 301, 330
Gunn, D.C. 273, 286
-- 180, 181, 195, 209, 213, 452, 458
Gunn, D.L., Kennedy, J.S., Pielou, D.P. 180, 182, 300, 213
Guo, A. 76, 106, 442, 444, 451, 452
Gwilliam, G.F. 313, 314, 321, 330, 707, 723

Haas, W. 453, 455, 458
Hachlov, J. 309, 330
Häder, D.P. see Nultsch, W., Häder, D.P.
Hafner, G.S. 643, 657
-- Tokarski, T., Hammond-Soltis, G. 636, 657
Hagins, F.M. 12, 54
Hagins, W.A., McGaughy, R.E. 38, 54
Hall, J.C. 440, 452, 74, 106
Hall, M.O. see Bok, D. et al.
Halldal, P. 115, 128
Hamacher, K.J., Kohlik, D. 20, 54
Hamann, B., Langer, H. 20, 31, 54
-- see Langer, H. et al.
Hamdorf, K. 5, 8, 16, 21, 23, 25, 36, 38, 39, 41, 54, 444, 452
-- Gogala, M., Schwemer, J. 47, 54
-- Kirschfeld, K. 36, 54
-- Langer, H. 31, 55
-- Paulsen, R., Schwemer, J. 27, 28, 29, 41, 55, 451, 452
-- Paulsen, R., Schwemer, J., Tauber, V. 20, 43, 48, 55

-- Schwemer, J. 16, 21, 41, 55
-- Schwemer, J., Tauber, V. 20, 21, 38, 55
-- see Boschek, C.B., Hamdorf, K.
-- see Eckert, H., Hamdorf, K.
-- see Gogala, M. et al.
-- see Razmjoo, Hamdorf, K.
-- see Schlecht, P. et al.
-- see Schwemer, J. et al.
Hamilton, P.V. 703, 723
-- Russel, B.J. 703, 723
Hammond-Soltis, G. see Hafner, G.S. et al.
Hand, W.G., Davenport, D. 181, 213
-- see Davenport, D. et al.
Hansen, K. 309, 310, 330
Hanson, F.E. see Ready, D.F. et al.
Hanström, B. 368, 369, 389, 394, 449, 452
Hara, R. see Hara, T., Hara, R.
-- see Osaki, K. et al.
Hara, T., Hara, R. 20, 25, 44, 45, 55
Hara, T. see Osaki, K. et al.
Harant, H., Grassé, P.P. 290, 292, 331
Harant, T. see fauvel, P. et al.
Harden Jones, F.R. 451, 458
Hardie, R.C. 42, 55, 447, 448, 449, 452, 463, 473, 478, 525, 555
-- Dvelli, P. 376, 378, 394
-- Franceschini, N., McIntyre, P.D. 20, 32, 55, 76, 77, 106 443, 444, 445, 447, 452, 525, 555
-- Franceschini, N., Ribi, W., Kirschfeld, K. 71, 76, 77, 88, 106, 447, 448, 452, 463, 478, 526, 555
-- Kirschfeld, K. 76, 77, 106, 443, 444, 447, 452
-- see Blest, A. et al.
-- see Franceschini, N. et al.
-- see Franceschini, N., Hardie, R.C.
-- see Laughlin, S.B., Hardie, R.C.

-- 540, 555
Hargitt, C.W. 321, 323, 331
Harkness, L.I.K. see Collett, T.S., Harkness, L.I.K.
Harrington, N.R., Leaning, E. 126, 128
Harris, R.P. see Gophen, M., Harris, R.-P.
Harris, W.A., Ready, D.F., Lipson, E.D., Hudpeth, A.J., Stark, W.S. 762, 769
-- Stark, W.S., Walker, J.A. 74, 106
Harte, P.J. see Kankel, D.R. et al.
Hartline, H.K. (1938) 699, 709, 723
Harline, P.H., Hurley, A.C., Lange, G.D. 716, 723
-- see Hurley, A.C. et al. 1978
Hassenstein, B. 582, 586, 615
-- Reichardt, W. 615
Haupt, W. see Diehn, B. et al.
Hausen, K. 440, 452, 475, 477, 478, 484, 496, 497, 519, 523, 560, 555, 556, 563, 583, 592, 601, 602, 604, 611, 615, 622, 623, 628, 630, 632, 783, 784, 786, 787
-- Strausfeld, N.J. 448, 452, 493, 511, 518, 552, 556
-- Wehrhahn, C. 611, 615, 547, 548, 556
-- Wolburg-Buchholz, K. 533, 556
-- Wolburg-Buchholz, K., Ribi, W.A. 532, 533, 537, 556
-- see Hengstenberg, R. et al.
-- see Paggio, T. et al.
-- see Reichardt, W. et al.
-- see Strausfeld, N.J., Hausen, K.
-- see Wässle, H., Hausen, K.
-- see Wehrahahn, C., Hausen, K.
-- see Wehrahahn, C. et al.
Hauser-Holschuch, H. 652, 657
-- see Braitenberg, V., Hauser-Holschuch, H.
Hays, D., Goldsmith, T.H. 16, 55

Hawryshyn, C.W., MacKay, W.C., Nilsson, T.H. 785, 787
Hedgecock, E.M., Russell, R.L. 197, 199, 200, 213
Heelis, D.V., Heelis, P.F., Bradshaw, F., Phillips, G.O. 119, 128
-- Heelis, P.F., Kernick, N.A., Phillips, G.O. 119, 128
Heelis, P.F. see Heelis, D.V. et al.
Heide, G. 527, 547, 556
-- see Spüler, M., Heide, G.
Heisenberg, M. 440, 452, 576, 609, 611, 615, 616, 631, 632
-- Buchner, E. 74, 75, 106, 449, 452, 540, 548, 556, 583, 585, 595, 597, 600, 603, 616
-- Götz, K.G. 597, 607, 616
-- Wolf, R. 567, 569, 605, 607, 609, 616
-- Wonneberger, R., Wolf, R. 532, 548, 556
-- see Blondeau, J., Heisenberg, M.
-- see Fischbach, K.F., Heisenberg, M.
-- see Wolf, R., Heisenberg, M.
-- see Wolf, R. et al.
Heller, J. see Futterman, S., Heller, J.
Helmholtz, H. von, 565, 616
Helversen, O. von 47, 55
-- see Edrich, W. et al.
Heming, B.S. 691, 693
Hemley, R., Kohler, B.E., Siviski, P. 442, 444, 452
Hengstenberg, B. see Götz, K.G. et al.
-- see Hengtenberg, R., Hengstenberg, B.
-- see Hengstenberg, R. et al.
Hengstenberg, R. 497, 519, 529, 536, 539, 541, 546, 556, 631, 632

-- Bülthoff, H., Hengstenberg,
   B. 497, 519,
-- Hausen, K., Hengstenberg, B.
   513, 529, 533, 556, 484, 496,
   497, 519
-- Hengstenberg, B. 531, 556
-- Sandeman, D.C. 609, 610,
   616
-- see Buchner, E. et al.
Henkart, M. Landis, D.M.D.,
   Reese, T.S. 249, 286
Henson, W.R. see Wellington, W.G.
   et al.
Hentschel, E. 353, 394
Herbault, C. see Joly, R., Herbault, C.
Hermans, C.O. 291, 299, 301, 303,
   305, 331
-- see Eakin, R.M. 160, 174, 293,
   295, 296, 331
-- see Rosen, M.D. et al.
-- Cloney, R.M. 319, 322, 323,
   324, 327, 331, 446, 458
Herre, M. see Bülthoff, H. et
   al.
Herrling, P.L. 449, 452
Hertel, H. 41, 55, 267, 268, 286,
   475, 476, 478, 650, 658
Herter, K. 320, 331
Hess, C. 307, 321, 323, 324, 331
Hess, W.N. see Clark, R.B.,
   Hess, W.N.
Hesse C. 290, 295, 307, 309, 310,
   313, 331, 387, 388, 389, 394,
   701, 710, 723, 728, 729, 734,
   742, 777, 787
Hesse, W.N. 319, 331
Heurkens, G. see Franceschini,
   M. et al.
Heuser, J.E. see Roof, D.J.,
   Heuser, J.E.
Hickman, V. 361, 395
Hildebrand, E., 145, 174, 179,
   181, 213
-- see Diehn, B. et al.
Hillman, P., Dodge, F.A., Hochstein, S., Knight, B.W., Minke, B. 20, 55

-- Hochstein, S., Minke, B. 20,
   55
-- see Hochstein, B. et al.
-- see Minke, B. et al.
Hirata, K., Ohsako, N., Mabuchi, K. 307, 308, 331
Hirota, R. 740, 742
Hitchcock, L. 126, 128
Hochstein, S., Minke, B., Hillman, P., Knight, B.W. 21,
   24, 55
Hochstein, S. see Hillman et al.
-- see Minke, B. et al.
Hodgkin, A.L. see Baylor, D.A.,
   Hodgkin, A.L.
Hodgkin, J.A. see Lewis, J.A.,
   Hodgkin, J.A.
Hofbauer, A. 636, 637, 658
-- Campos-Ortega, J.A. 342, 360,
   361, 367
-- see Campos-Ortega, J.A.,
   Hofbauer, A.
Hoffman, K.P. see Wagner, H.J.
   et al.
Holborow, P.L., Laverack, M.S.
   304, 305, 307, 331
Holland, C.H. 716, 723
Holloway, J.A. see Lall, A.B. et
   al.
Holt, E.B., Lee, F.S. 124, 128
Holtzman, E. 87, 106 106
-- Mercurio, A.M. 87, 107, 152,
   174
Homann, H. 337, 338, 351,
   353, 354, 386, 394
Home, E.M. 681, 693
Honig, B. see Callender, R.H.,
   Honig, B.
-- see Ebrey, T.G., Honig, B.
Honigberg, B.M. see Levine, N.D.
   et al.
Honneger, H.W. 475, 479
Hope, W.D. 137, 175
Horandner, H. see Kerjaschki,
   M., Horandner, H.
Horch, K.W. see Waterman, T.H.,
   Horch, K.W.

Horridge, G.A. 71, <u>107</u>, 423, 427, 428, 432, <u>436</u>, <u>452</u>, 667, 668, <u>693</u>
-- Barnard, P. 389, <u>394</u>
-- Duniec, J., Marčelja, L. 44, <u>56</u>
-- Marcelja, L, Jahnke, R., Matic, T. 69, 86, 90, <u>107</u>
-- McLean, M. 47, <u>56</u>, <u>452</u>
-- Meinertzhagen, I.A. 445, <u>452</u>, <u>453</u>
-- Mimura, K., Tsukara, Y. 442, <u>453</u>
-- see Tsukara, Y., Horridge, G.A.
-- see Tsukara, Y. et al.
Hough, B.R. see Galau, G.A. et al.
Howard, J., Snyder, A.W. 460, <u>479</u>
Howell, C.D. 319, <u>331</u>
Hu, K.G., Stark, W.S. 610, <u>616</u>
Hubbard, R., Bownds, D., Yoshizawa, T. 27, <u>56</u>
-- St-George, R.C.C. 20, 25, <u>56</u>
-- see Kropf, A. et al.
-- see Wald, G., Hubbard, R.
Hubel, D.H., Wiesel, T.N. 497, <u>520</u>, 552, <u>557</u>
-- see Wiesel, T.N., Hubel, D.H.
Hudspeth, A.J. see Harris, W.A. et al.
Huet, M., Franquinet, R. 768, <u>769</u>
Hughes, A. 458, 470, <u>479</u>
Hughes, R.L., Woollacott, R.M. 132, 133, 134, 135, 156, 166, <u>175</u>
Hurley, A.C., Lange, G.D., Hartline, P.H. 716, <u>723</u>
-- see Hartline, P.H. et al.
Huxley, J.S. see Wolsky, A., Huxley, J.S.
Hyams, J.S., Borisy, G.G. 121, <u>128</u>
Hyde, C.A.T. 677, 684, <u>693</u>
Hyde, R. see Dorsett, D.A., Hyde, R.
Hyman, L.H. <u>286</u>, 728, <u>742</u>

Illmensee, K. 681, <u>693</u>
Isseroff, H. 443, <u>458</u>
-- Cable, R.M. 221, <u>236</u>
Ito, H. see Brown, H.M. et al.
Ivanyshyn, A.M. see Stark, W.S. et al.
Iwata, K. see Fujimoto, K. et al.

Jacobs, G.H. 66, <u>107</u>
Jaeger, R. see Ali, M.A. et al.
Jahn, T.L., Bovee, E.C. 119, <u>128</u>
Jahnke, R. see Horridge, G.A. et al.
Jamieson, G.B.M. 307, <u>331</u>
Jan, L.Y., Revel, J.P. 762, <u>769</u>
Jander, R. 94, <u>107</u>, 181, <u>213</u>
-- Waterman, T.H. 64, <u>107</u>
Järvilehto, M. 36, 38, <u>56</u>
-- Finell, N. 639, 650, <u>658</u>
-- Moring, J. 75, 76, 77, <u>107</u>
-- Zettler, F. 462, 465, 473, <u>479</u>, 489, 491, <u>520</u>
-- see French, A.S., Järvilehto, M.
-- see Zettler, F., Järvilehto, M.
Jeanrot, N., Campan, R., Lambin, M. 602, <u>616</u>
Jeffreys, A.J. 167, 168, <u>175</u>
Jennings, H.S. 242, 267, <u>286</u>
Jerlov, N.G. 67, <u>107</u>
Johnson, M.A. see Stark, W.S., Johnson, M.A.
Joly, R. 388, <u>394</u>
-- Herbault, C. 388, <u>395</u>
Jones, A. see MacDonald, S., Jones, A.
Jones, F.R.H. 184, <u>213</u>
Joo, F. see Parducz, A., Joo, F.
Juberthie, C., Munoz-Cuevas, A. 681, <u>693</u>
Judd, H.D. see Walls, G.L., Judd, H.D.

Jung, D. 292, 310, 312, 331

Kai Kai, M.A., Pentreath, V.W. 624, 633
Kaiser, F. 321, 322, 332
Kalil, R.E. see Bizzi, E. et al.
Kalminjn, A.J. 99, 107
Kammon, R.L. see Kruizinga, B. et al.
Kampa, E.M. 780, 787
-- Boden, B.D. 780, 787
Kandel, E.R. 179, 213
Kaneko, H. see Naitoh, Y., Kaneko, H.
Kankel, D.R., Ferrús, A., Garen, S.H., Harte, P.J., Lewis, P.E. 637, 639, 640, 658
Kao, L. see Blest, A. et al.
Karakashian, M.W. see Karakashian, S.D. et al.
Karakashian, S.D., Karakashian, M.W., Rudzinska, M.A. 125, 128
Kästner, A. 386, 395
Katagiri, Y. see Yamaguchi, T. et al.
Kato, M., Shinzawa, K., Yoshikawa, S. 126. 128
Kauer, C.A. see Lancet, D. et al.
Kearn, G.C. 224, 226, 231, 233, 234, 236
-- Baker, N.O. 221, 224, 236
Keehn, D.G. see Bishop, L.G., Keehn, D.G.
-- see Bishop, L.G. et al.
Kelley, D.B. see Sejnowski, T.J. et al.
Kennedy, C. see Sokoloff, L. et al.
Kennedy, D. 706, 723
Kennedy, J.S. see Gunn, D.L. et al.
Kennedy, M.J. 233, 236
Kerjaschki, M., Horandner, H. 151, 175
Kernéis, A. 291, 300, 301, 302, 304, 332
-- see Carricaburu, P., Kernéis, A.
Kernick, N.A. see Heelis, D.V. et al.
Keshishian, H. see Bentley, D. Keshishian, H.
Kien, J., Land, M.F. 609, 616
-- Menzel, R. 475, 476, 478
Kikuchi, M., Aoki, K. 169, 175
Kim, C.W. 667, 693
King, C.E. 242, 286
Kirk, M.D., Waldrop, B., Glantz, R.M. 92, 107
Kirkham, J.B. see Goodman, L.J. et al.
Kirschfeld, K. 20, 41, 56, 65, 75, 76, 107, 443, 444, 445, 447, 449, 453, 464, 479, 487, 520, 526, 548, 557, 582, 598, 602, 616
-- Feiler, R., Franceschini, N. 77, 107, 443, 447, 453
-- Feiler, R., Hardie, R., Vogt, K., Franceschini, N. 442, 447, 453
-- Feiler, R., Minke, B. 32, 56
-- Franceschini, N. 42, 56, 74, 107, 441, 443, 447, 449, 453, 597, 616
-- Franceschini, N., Minke, B. 20, 31, 41, 56, 441, 447, 453
-- Lindauer, M., Martin, H. 100, 107
-- Lutz, B. 475, 479, 548, 557
-- Reichardt, W. 75, 107, 597, 616
-- Wenk, P. 453, 601, 616
-- see Franceschini, N. et al.
-- see Hamdorf, K., Kirschfeld, K.
-- see Hardie, R.C., Kirschfeld, K.
-- see Hardie, R.C. et al.
-- see McIntyre, P., Kirschfeld, K.
-- see Minke, B., Kirschfeld, K.
-- see Stavenga, D.G. et al.
-- see Vogt, K., Kirschfeld, K.
-- see Vogt, K. et al.

AUTHOR INDEX

Kishida, Y. 219, 236, 237
Kishinouye, K. 353, 395
Kito, Y., Naito, T., Nashima, K. 12, 25, 56
-- see Naito, Y. et al.
-- see Nashima, K. et al.
-- see Susuki, T. et al.
Klein, W.H. see Davidson, E.H. et al.
Kleinholz, L.H. 746, 769
Knight, B.W. see Hillman, P. et al.
-- see Hochstein, B. et al.
Kobakhidze, D.N., Sicharylidze, T.A., Svanidze, I.K. 662, 693
Kobayashi, T. see Shichida, Y. et al.
Koenderink, J.J. see Doorn, A.J. von, Koenderink, J.J.
Kohl, K.D. see Hamacher, K.J., Kohl, K.D.
Kohler, B.E. see Hemley, R. et al.
Kong, K.L., Fung, Y.M., Wasserman, G.S. 43, 56
Kopec, S. 681, 693
Körner, F. see Dichgans, J. et al.
Koshland, D.E. 93, 108
Koshland, D.E. Jr 179, 213
Koste, W. 160
Koto, M., Tanouye, M.A., Ferrus, A., Thomas, J.B., Wyman, R.J. 551, 557
Koyama, N. see Yagi, N., Koyama, N.
Krasne, F.B., Lawrence, P.A. 301, 302, 332, 709, 723
-- see Lawrence, P.A., Krasne, F.B.
Krebs, W., Lietz, R. 86, 108
-- see Behrens, M., Krebs, W.
Kretz, J.R., Stent, G.S., Kristan W.B. Jr. 293, 310, 313, 317, 318, 319, 332
Kretz, R. 78, 79, 108
Kristan, W.B. Jr. see Kretz, J.R. et al.

Kropf, A., Brown, P.K., Hubbard, R. 12, 56
Kruger, L. see Durham, D. et al.
Kruizinga, B., Kammon, R.L., Stavenga, D.G. 32, 56
-- see Stark, W.S. et al.
Kuenen, L.P.S., Baker, T.C. 606, 616
Kühn, A. 179, 213, 273, 286
Kuiper, J.W. 432, 436
-- see Beersma, D.G.M. et al.
-- see Stavenga, D.G. et al.
-- see Vries, H. de, Kuiper, J.W.
-- see Zaagman, W.H. et al.
Kung, C. see Nelson, D.L., Kung, C.
Kunze, P. 88, 108, 428, 436, 573, 616
Kuroda, Y. 687, 693
Küthe, H.W. see Egelhaaf, A.
Kuwabara, N. see Ninomiya, et al.
Kuwabara, K. see Toh, Y., Kuwabara, M.

Labhardt, T. 71, 78, 79, 108, 449, 453
-- Meyer, E.P. 81, 108
Lall, A.B., Chapman, R.M., Trouth, C.O., Holloway, J.A. 780, 787
-- Lord, E.T., Trouth, C.O. 46, 56
-- Seliger, H.H., Briggley, W.H., Lloyd, J.E. 46, 56, 780, 787
-- see Seliger, H.H. et al.
Lam, D.M.K. see Basinger, S.F. et al.
Lambin, M. see Goulet, M. et al.
-- see Jeanrot, N. et al.
Lamparter, H.E., Steiger, V., Sandri, C., Akert, K. 641, 658
Lancet, D., Greer, C.A., Kauer, J., Shepherd, G.M. 624, 633
Land, M.F. 43, 56, 71, 88, 108, 132, 140, 141, 142, 143, 175

-- 189, 190, 210, 213, 221, 238, 244, 286, 361, 362, 363, 370, 378, 385, 386, 387, 395, 401-438, 436, 437, 636, 658, 699-725, 723, 724, 774, 777, 779, 783, 785
-- Burton, F.A. 429, 437
-- Burton, F.A., Meyer-Rochow, V.B. 428, 431, 437
-- Collett, T.S. 453, 577, 609, 616
-- see Barber, V.C. et al.
-- see Blest, A.D., Land, M.F.
-- see Collett, T.S., Land, M.F.
-- see Kien, J., Land, M.F.
Landis, D.M.D. see Henkart, M. et al.
Lanfranchi, A. see Bedini, C., Lanfranchi, A.
-- see Bedini, C. et al.
Lang, W.H., Forward, R.B. Jr., Miller, D.C. 207, 214
Lange, G.D. see Hartline, P.H. et al.
-- see Hurley, A.C. et al.
Langer, H. 7, 10, 43, 56
-- Hamann, B., Meinecke, C.C. 20, 32, 56, 783, 787
-- Schlecht, P., Schwemer, J. 31, 56
-- Thorell, B. 31, 57
-- see Hamann, B., Langer, H.
-- see Hamdorf, K., Langer, H.
-- see Meinecke, C.C., Langer, H.
-- see Schlecht, P. et al.
-- see Schneider, L. et al.
-- see Schwemer, J., Langer, H.
Larrivee, D.C. see Schinz, R.H., et al.
Lasansky, A., Fuortes, M.G.F. 312, 317, 332
Laskey, R.A. 163, 175
Laughlin, S. 68, 73, 86, 88, 108, 440, 453, 457-481, 479, 489, 490, 491, 520, 523, 540, 557, 597, 599, 602, 616, 623, 624, 633, 784, 787
-- Blest, A., Stowe, S. 376, 378, 395

-- Hardie, R.C. 465, 466, 467, 468, 470, 471, 472, 479, 491, 520
-- McGinness, S. 47, 57, 73, 108, 668, 693
-- see Dubs, A. et al.
-- see Lillywhite, P.G., Laughlin, S.B.
-- see Matic, T., Laughlin, S.B.
-- see Snyder, A.W., Laughlin, S.B.
-- see Snyder, A.W. et al.
-- see Srinivasan, M.V. et al.
Laverack, M.S. 313, 332
-- see Holborow, P.L. Laverack, M.S.
Lavigne, D.M. see Wright, D.G.S. et al.
Lawrence, P.A., Green, S.M. 636, 668
-- Krasne, F.B. 334
-- Shelton, P.M.J. 684, 693
-- see Green, S.M., Lawrence, P.A.
-- see Krasne, F.B., Lawrence, P.A.
-- see Shelton, P.M.J., Lawrence, P.A.
Leaming, E. see Harrington, N.R. Leaming, E.
Leblanc, R.M. see Boucher, F., Leblanc, R.M.
Lee, F.S. see Holt, E.B., Lee, F.S.
Leedale, G.F. see Levine, N.D. et al.
Lees, A.D. 278, 286
Leggett, L.M.W. 43, 57, 87, 88, 89, 92, 108
-- Stavenga, D.G. 44, 57
Lenci, F. 8, 10
-- Colombetti, G. 115, 128
-- see Benedetti, A., Lenci, F.
-- see Checcucci, A. et al.
-- see Diehn, B. et al.
Lenting, B.P.M. see Mastebroek, H.A.K. et al.

AUTHOR INDEX

Levick, W.R. see Barlow, H.B., Levick, W.R.
Levine, J., Tracey, D. 527, 531, 557
-- MacNichol, E.F. 781, 787
Levine, N.D., Corliss, J.O., Cox, F.E.G., Deroux, G., Grain, J., Grain, J., Honigberg, B.M., Leedale, G.F., Loeblich, A.R., Lom, J., Lynn, D., Merinfeld, D., Page, E.G., Poljansky, G., Sprague, V., Vaura, J., Wallace, F.G. 115, 129
Levine, R.D. see Murphy, R.K. et al.
Levinson, J.Z. see Barlow, R.B. et al.
Lévinthal, C., Ware, R. 242, 286
-- see Lopresti, V. et al.
-- see Macagno, E.R. et al.
Levi-Setti, R., Park, D.A., Winston, R. 418, 437
Lew, G.T.W., 675, 683, 693
Lewin, B. 163, 175
Lewis, J.A., Hodgkin, J.A. 170, 175
Lewis, P.E. see Kankel, D.R. et al.
Lidington, K.J. see Pak, W.L., Lidington, K.J.
Lietz, R. see Krebs, W., Lietz, R.
Light, V.E. 706, 724
Lillywhite, P.G., Dvorak, D.R. 540, 542, 557, 602, 616
-- Laughlin, S.B. 460, 480
Limb, J.O., Murphy, J.A. 581, 616
Lindauer, E. see Gemperlein, R. et al.
Lindauer, M., Martin, H. 99, 108
-- see Kirschfeld, K. et al.
-- see Martin, H., Lindauer, M.
-- see Rossel, S. et al.
Lipson, E.D. see Harris, W.A. et al.
Lisman, J.E., Sheline, Y. 20, 57

Little, P.F.R. 167, 175
Liv, R.S.H., Matsumoto, H. 28, 57
Livanow, N. 313, 332
Llewellyn, J. 221, 236
Llinas, R. see Blight, A.R., Llinas, R.
Lloyd, J.E. see Lall, A.B. et al.
Lloyd, J.E. see Seliger, H.H. et al.
Lo, M.-V.C. see Schinz, R.H. et al.
Locket, N.A. 719, 724
Loeb, J. 273, 286, 320, 332
Loeblich, A.R. see Levine, N.D. et al.
Lom, J. see Levine, N.D. et al.
Longuet-Higgins, H.C., Prazdny, K, 565, 617
Lopresti, V., Macagno, E.R., Lévinthal,C. 637, 638, 658
-- see Macagno, E.R. et al.
Lord, E. see White, R.H., Lord, E.
Lord, E.T. see Lall, A.B. et al.
Luciani, A. 274, 275, 286
-- Chassé, J.L., Clément, P. 286
-- see Clément, P. et al.
-- see Coulon, J. et al.
-- see Pourriot, R. et al.
Luders, L. 407, 437
Lüdtke, H. 69, 108, 683, 694
Lutz, B. see Kirschfeld, K., Lutz, B.
Lynn, D. see Levine, N.D. et al.
Lyons, K.M. 224, 237
Lythgoe, J.N. 46, 57, 69, 108, 109, 470, 480

Mabuchi, K. see Hirata, K. et al.
Macagno, E.R. 638, 658
-- Lopresti, V., Levinthal, C. 179, 214, 643, 658
-- see Lopresti, V. et al.
MacDonald, S., Combes, C. 234, 237

-- Jones, A. 234, 237
Macgeshin, W.T, 233, 237
MacGinitie, G.F. see McCann, G.D., MacGinitie, G.F.
Machan, L. 377, 378, 384, 395
Macintosh, N.J. see Sutherland, N.S., Macintosh, N.J.
Mackay, W.C. see Hawryshyn, C.W. et al.
Macnab, R.M. 93, 109
MacNichol, E.F. see Levine, J.S., MacNichol, E.F.
-- see Wiederhold, M.L. et al.
MacRae, E.K. 219, 226, 237
Magni, F., Papi, F., Savely, H., Tongiorgi, P. 382, 395
Malaquin, A. 290, 332
Maldonado, H., Rodriguez, E. 565, 617
Malzacher, P. 675, 682, 694
Manaranche, R. 299, 332
Mann, K.H. 320, 321, 333
Mardia, K.V. 201, 214
Mark, E. 353, 395
Mark, R.E., Sperling, H.G. 624, 633
Marr, D., Ullman, S. 581, 617
Martin, A.R. 460, 480
Martin, F.G., Mote, M.I. 80, 86 109, 426, 437
Martin, G.G. see Eakin, R.M. et al.
Martin, H., Lindauer, M. 99, 109
-- see Kirschfeld, K. et al.
-- see Lindauer, M., Martin, H.
Marullo, C., Mouze, M. 687, 694
Marumo, R. see Nagasawa, S., Marumo, R.
Mason, P.R., Fripp, P.J. 230, 231, 237
Mast, S.O. 124, 126, 129
-- Stahler, N. 126, 129
Mastebroek, H.A.K., Zaagman, W.H., Lenting, B.P.M. 539, 541, 542, 547, 557, 597, 598, 610, 617
-- see Zaagman, W.H. et al.
Mateescu, G.d. see Shriver, J.W. et al.
Matic, T., Laughlin, S.B. 461, 480
-- see Horridge, G.A. et al.
Matsumoto, H. see Liu, R.S.H., Matsumoto, H.
Matsumoto, S.G. see Murphy, R.K. et al.
Matthiessen, L. 713, 724
Mauro, A. see Clark, A.W. et al.
Mayr, E. see Salvini-Plawen, L.V., Mayr, E.
McCann, G.D. 540, 557
McCann, G.D., Arnett, D.W. 441, 449, 453, 474, 480, 539, 540, 557
McCann, G.D., Dill, J.C. 528, 532, 557
-- Foster, S.F. 540, 541, 547, 557
-- MacGinitie, G.F. 593, 595, 596, 617
-- see Bishop, L.G. et al.
McCartney, E.J. 70, 109
McConnell, P. see Berry, M. et al.
McEnroe, W. see Naegele, J. et al.
McFarland, W.N., Munz, F.W. 69, 109
-- see Munz, F.W., McFarland, W.N.
McGaughy, R.E. see Hagins, W.A., McGaughy, R.E.
McGinness, S. see Laughlin, S.B., McGinness, S.
-- see Wilson, M. et al.
McIntyre, P., Kirschfeld, K. 20, 57, 76, 109, 443, 444, 447, 453
-- Snyder, A.W. 80, 109, 485, 520
-- see Blest, A. et al.
-- see Caveney, S., McIntyre, P.
-- see Hardie, R.C. et al.
-- see Williams, D., McIntyre, P.
McLean, M. see Horridge, G.A., McClean, M.
McReynolds, J.S., Gorman, A.L.F. 145, 175, 708, 724

-- see Gorman, A.L.F., McReynolds, J.S.
-- see Gorman, A.L.F. et al.
Meffert, P. see Smola, U., Meffert, P.
Meinecke, C.C. 72, 109
-- Langer, H. 86, 109
-- see Langer, H. et al.
Meinertzhagen, I.A. 440, 454, 470, 480, 484, 492, 520, 623, 633, 635-660, 658, 662, 671, 672, 681, 694, 783, 787
-- Armett-Kibel, C. 462, 463, 480 637, 641, 643, 644, 658
-- see Armett-Kibel, C. et al.
-- see Frühlich, A., Meinertzhagen, I.A.
-- see Horridge, G.A., Meinertzhagen, I.A.
-- see Nicol,D., Meinertzhagen, I.A.
Melamed, J., Trujillo-Cenoz, O. 362, 367, 395, 448, 454, 485, 520
-- see Trujullo-Cenoz, O., Melamed, J.
Melkonian, M., Robenek, H. 146, 147, 175
Meller, K. see Eckert, H., Meller, K.
Melnichenko, A.N. 663, 694
Menco, B.P.M., Dodd, G.H., Davey, M. 151, 175
Menzel, R. 46, 57, 66, 68, 69, 90, 109, 472, 480, 562, 617
-- Blakers, M. 82, 109, 463, 480
-- Roth, F. 267, 268, 270, 271, 286
-- Snyder, A.W. 79, 109, 463, 480
-- see Kien, J., Menzel, R.
-- see Nickel, E., Menzel, R.
-- see Snyder, A.W. et al.
Menzies, R.J., George, R.Y., Rowe, G.T. 780, 787
Mercurio, A.M. see Holtzman, E., Mercurio, A.M.
Merinfeld, D. see Levine, N.D. et al.

Merker, G., Vaupel-Von Harnack,, M. 27, 55
Messenger, J.B. 43, 46, 56, 69, 109, 700, 710, 716, 718, 720 721, 724
Meyer, E. see Wehner, R., Meyer, E.
Meyer, E.P. see Labhart, T., Meyer, E.P.
Meyer-Rochow, V.B. 84, 87, 109, 425, 437, 690, 694
-- see Eguchi, E., Meyer-Rochow, V.B.
-- see Land, M.F. et al.
Michael, E.L. 739, 742
Michaud, N.A. see White, R.H. et al.
Mikkelsen, P.M. 71, 109
Millecchia, R. see Clark, A.W. et al.
Miller, D.C. see Lang, W.H. et al.
Miller, H.M. Jr, MacCoy, O.R. 232, 237
Miller, W.H. 43, 57, 389, 395, 708, 724, 781, 787
-- Bernard, G.D. 43, 57
-- Bernard, G.D., Allen, J.L. 665, 667, 694
-- see Snyder, A.W., Miller, W.H.
Millot, J. 337, 340, 395
Millot, N. 744, 746, 751, 752, 753, 769
-- Takahashi, K. 750, 751, 769
-- Vevers, H.G. 765, 770
-- Yoshida, M. 746, 749, 770
-- see Yoshida, M., Millot, N.
Mimura, K. 466, 480, 528, 557, 577, 617
-- see Horridge, G.A. et al.
Minke, B., Hochstein, S., Hillman, P. 20, 38, 57
-- Kirschfeld, K. 20, 24, 38, 41, 57
-- see Franceschini, N. et al.
-- see Hillman, P. et al.
-- see Hochstein, S. et al.
-- see Kirschfeld, K. et al.

Mitsudo, M. see Nashima, K. et al.
Mittelstaedt, H. 565, 607, 617
Mobbs, P.G. 690, 694
-- see Goodman, L.J. et al.
Moller-Racke, I. see Buddenbrock, W.J., Moller-Racke, I.
Mollon, J.D. 66, 109
Moody, M.F., Parriss, J.R. 720, 724
Moore, D., Penikas, J., Rankin, M.A. 609, 617
-- Rankin, M.A. 101, 109, 110, 609, 617
Moore, P.L. see Taylor, D.L. et al.
Morasso, P. see Bizzi, E. et al.
Moray, N. see Gregory, R.L. et al.
Morel-Laurens, N.M., Feinlab, M.E. 121, 129
Moring, J. see Järvilehto, M. Moring, J.
Morita, M. see Carpenter, K.S. et al.
Morton, P.D., Cosens, D. 600, 617
Morton, R.A., 16, 20, 57
Mote, M.I., Wehner, R. 78, 79, 80, 110
-- see Martin, F.G., Mote, M.I.
-- see Stark, R.J., Mote, M.I.
Motte, I. de la, see Burkhardt, D., Motte, I. de la
Mouze, M. 636, 638, 659, 661-698, 694, 775, 787
-- Shaller, F 677, 694
-- see Marullo, C., Mouze, M.
Mpitsos, G.F. 709, 724
Muller, S.W.M., Campbell, A. 3, 8
Munoz-Cuevas, A. 335-399, 396, 681, 694, 778, 787
-- see Carribaru, P., Munoz-Cuevas, A.
-- see Juberthie, C., Munoz-Cuevas, A.

Münster, A. see Franceschini, N. et al.
Muntz, W.R.A. 720, 724
Munz, F.W., McFarland, W.N. 69, 110
-- see McFarland, W.N., Munz, F.W.
Murakami, A. 740, 742
Muri, R. see Bertrand, D. et al.
Muri, R.B. 16, 20, 23, 31, 58
Murphy, D.G. see Chitwood, B.G., Murphy, D.G.
Murphy, J.A. see Limb, J.O., Murphy, J.A.
Murphy, R.K., Matsumoto, S.G., Levine, R.D. 535, 557
Murphy, T. see Fletcher, A.
Myhrberg, H.E. 307, 308, 333

Naegele, J., McEnroe, W., Soans, A. 378, 397
Nagakura, S. see Shichida, Y. et al.
Nagasawa, N. see Ebina, Y. et al.
-- Marumo, R. 730, 731, 732, 742
Naito, T., Nashimo-Hayama, K., Ohtsu, K., Kito, Y. 20, 25, 28, 58
-- see Kito, Y. et al.
Naitoh, Y. 124, 129
-- Kaneko, H. 124, 129
Nardi, J.B. 685, 694
Nashima, K., Mitsudo, M., Kito, Y. 16, 58
-- see Kito, Y. et al.
Nashima-Hoyama, K. see Naito, T. et al.
Nässel, D.R. 86, 110
-- Elofsson, R., Odselius, R. 643, 659
-- Waterman, T.H. 44, 58, 87, 90, 110, 424, 426, 437, 463, 480 643, 659
-- see Strausfeld, N.J., Nässel, D.R.

Nelson, D.L., Kung, C. 179, 214
Neumeyer, C. see Edrich, W. et al.
Newell, G.E. 701, 724
Néya, T. 744, 745, 770
Nicholls, J.G., Wallace, B., Adal, M. 179, 214
Nicholson, G.L. 148, 176
Nickel, E., Menzel, R. 150, 176
-- see Fernandez, H.R., Nickel, E.E.
Nicol, D., Meinertzhagen, I.A. 470, 480, 640, 641, 643, 646, 650, 651, 652, 653, 655, 656, 659
Nicol, J.A.C. 320, 321, 322, 323, 324, 333
-- see Zyznar, E.J., Nicol, J.A.C.
Nicolas, M.-T. see Bassot, J.-M., Nicolas, M.-T.
Nielsen, P.J. see Brown, J.A., Nielsen, P.J.
Niilonen, T. 304, 333
Nilsson, D.E. Nilsson, H.L. 421, 424, 425, 437
-- Odselius, R. 424, 437
-- see Anderson, A., Nilsson, D.E.
-- see Odselius, R., Nilsson, D.E.
Nilsson, H.L. see Nilsson, D.E.
Nilsson, T.H. see Hawryshyn, C.W. et al.
Ninnemann, H. 126, 129
Ninomiya, N., Tominaga, T., Kuwabara, M. 80, 110
Nishimura, T. see Yanase, T. et al.
Niwa, H. see Tabata, M. et al.
Noirot, C. see Noirot-Timothée, C., Noirot, C.
Noirot-Timothée, C., Noirot, C., 348, 398
Nolte, J., Brown, J.E. 20, 58
Nomura, E. 319, 333
Nordlander, R.H., Edwards, J.S. 636, 637, 659

Nørrevang, A. 262, 286
Nosaki, H. see Toyoda, J. et al.
Nowell, M.S. 670, 675, 676, 677, 695
-- Shelton, P.M.J., 636, 659, 684, 695
Nowikoff, M. 58
Nüesch, H. 681, 695
Nultsch, W. 180, 181, 209, 214
-- Häder, D.P. 145, 176, 179, 214
-- see Diehn, B. et al.
Numan, J.A.J. see Stavenga, D.G. et al.

Oberdorfer, M.D. 643, 659
Obermayer, M.L. see Strausfeld, N.J., Obermayer, M.L.
O'Brien, P. see Bok, D. et al.
Ochi, K. see Yamaguchi, T. et al.
Ockleford, E.M. see DeVoe, R.D., Ockleford, E.M.
O'Connor, M. see Evered, D. et al.
Odselius, R., Elofsson, R. 674, 695
-- Nilsson, D.E. 71, 110
-- see Nässel, D.R. et al.
-- see Nilsson, D.E., Odselius, R.
Oesterhelt, D. see Schreckenbach, T. et al.
Ogden, T.E. see Brown, H.M. et al.
Ohsako, N. see Hirata, K. et al.
Ohtani, H. see Shichida, Y. et al.
Ohtsu, K. see Naito, T. et al.
Ohtsuka, T. 443, 454
Ohtsuki, H. see Yoshida, M., Ohtsuki, H.
Okuno, Y. see Fujimoto, K. et al.
Olberg, R.M. 567, 602, 617
Olive, J. 152, 176

Oliver, B.M. 471, 480
Olivo, R.F., Chrismer, K.L. 35, 58
Ong, D.E. see Chytil, F., Ong, D.E.
Oosawa, F. see Saji, M., Oosawa, F.
Osafune, T., Schiff, J.A. 255, 286
Osaki, K., Hara, R., Hara, T. 765, 770
O'Shea, M., Fraser-Rowell, C.H. 475, 480
-- Rowelle, C.H.F. 551, 558
-- Rowell, C.H.F., Williams, J.L.D. 534, 558
-- see Fraser-Rowell, C.H.F. et al.
-- see Rowell, C.H.F. et al.
Ostroy, S.E. 12, 20, 58
-- Wilson, M., Pak, W.L. 20, 58
-- see Stein, P.J. et al.
Ovchinnikou, A.A. see Filippov, M.P., Ovchinnikov, A.A.
Owens, E.D. see Prokopy, R.H., Owens, E.D.

Packard, A. 703, 716, 721, 724
Packer, L. 7, 10, 25, 58, 781, 787
Pado, R. 126, 129
Page, E.G. see Levine, N.D. et al.
Page, T.L., Larimer, J.L. 778, 787
Pak, W.L. 440, 454
-- Lidington, K.J. 38, 58
-- see Cone, R.A., Pak, W.L.
-- see Ostroy, S.E. et al.
-- see Schinz, R.H. et al.
-- see Stephenson, R.S., Pak, W.L.
Palade, G. 153, 176
Palka, J. 566, 602, 617
-- see Anderson, H. et al.

Pampapathi Rao, R. see Geethabali, M., Pampapathi Rao, K.
Panov, A.A. 637, 659
Papermaster, D.S., Schneider, B.G. 87, 110, 152, 153, 176
-- Schneider, B.G., Besharse, J.C. 153, 176
Papi, F. 382, 397
-- Serreti, L. 382, 397
-- Serreti, L., Parrini, S. 382, 397
-- Tongiorgi, P. 382, 397
-- see Magni, F. et al.
Parducz, A., Joo, F. 624, 633
Park, D.A. see Levi-Setti, R. et al.
Parker, G.H. 82, 110
Parrini, S. see Papi, F. et al.
Parriss, J.R. see Moody, M.F., Parriss, J.R.
Pask, C., Barrell, K.F. 42, 58
-- see Snyder, A.W., Pask, C.
Patlak, see Sokoloff, L. et al.
Patten, W. 338, 397, 710, 724
Paul, R. see Gemperlein, R. et al.
Paulsen, R., Schwemer, J. 12, 13, 14, 15, 58, 59
-- see Hamdorf, K. et al.
-- see Pepe, I.M. et al.
-- see Schwemer, J. et al.
Paulus, H.F. 389, 390, 397
Pearse, J.S., Pearse, V.B. 777, 787
Pearse, V.B. see Pearse J.S., Pearse, V.B.
Pearse, S. Jr. 738, 742
Peckham, E. see Peckham, G., Peckham, E.
Peckham, G., Peckham, E. 386, 397
Pécot-Dechavissine, M. see Wernig, A. et al.
Peichl, L. see Wässle, H. et al.
Penikas, J. see Moore, D. et al.
Penn, P.E., Alexander, C.G. 758, 759, 770, 777, 787, 788

AUTHOR INDEX

Pentreath, V.W. see Kaikai, M.A.,
    Pentreath, V.W.
Pepe, I.M., Cugnoli, C. 44, 59
-- Schwemer, J., Paulsen, R.
    44, 59
Perez de Talens, A.F.P. see
    Taddei-Ferreti, C., Perez de
    Talens, A.F.P.
Perrelet, A., Baumann, F. 669,
    695, 348, 398
Perry, M.M. 343, 397
-- see Waddington, C.H.
Petrunkvitch, A. 337, 370,
    397
Pettigrew, K.D. see Sokoloff, L.
    et al.
Pfenninger, K.H. see Besharse,
    J.C., Pfenninger, K.H.
Pflugfelder, O. 681, 682, 695
Phillips, G.O. see Heelis, D.V.
    et al.
Phillips. J.B., Waldvogel, J.A.
    64, 110
Pick, B. 550, 558, 590, 591, 599,
    600, 601, 604, 617
-- Buchner, E. 597, 598, 599,
    617
Piekos, W.B., Waterman, T.H.
    87, 110
-- see Waterman, T.H., Piekos,
    W.B.
Pielou, D.P. see Gunn, D.L. et
    al.
Pierantoni, R. 484, 496, 520,
    529, 532, 558, 628, 633
Pilgrim, C. see Wagner, H.J. et
    al.
Pinter, R.B. 567, 602, 617
Pipa, R.L. see Woolever, R.,
    Pipa, R.L.
Plagge, E., 681, 695
Poggio, T., Reichardt, W.
    542, 547, 550, 558, 562, 565,
    567, 578, 582, 509, 617, 618
-- Reichardt, W., Hausen, K. 543,
    550, 558, 602, 604, 618
-- see Bülthoff, H. et al.
-- see Geiger, G., Poggio, T.

-- see Reichardt, W., Poggio, T.
-- see Reichardt, W. et al.
-- see Torre, V., Poggio, T.
-- see Wehrhahn, C. et al.
Poljansky, G. see Levine, N.D.
    et al.
Polyanovskii, A.D. see Gribakin,
    F.G. et al.
Pond, G.C., Cable, R.M. 221,
    237
Poo, M.P., Cone, R.A. 148, 176
Pople, W., Ewer, D.W. 751, 770
Pourriot, R. 242, 276, 287
-- Clément, P. 242, 276, 277,
    278, 279, 280, 287
-- Clément, P., Luciani, A.,
    279, 287
-- Snell, T.W. 280, 287
-- see Blanchot, J., Pourriot, R.
-- see Clément, P. et al.
-- see Clément, P., Pourriot, R.
Power, M. 527, 558
Praagh, J.P. van, Ribi, W.,
    Wehrhahn, C., Wittmann, D.
    454, 601, 602, 618
Prazdny, K. see Longuet-Higgins,
    H.C., Prazdny, K.
Preissler, K. 242, 267, 287
Press, N. 219, 237
Pringle, J.W.S. see Fraenkel, G.,
    Pringle, J.W.S.
Prosser, C.L. 319, 333
Prokopy, R.H., Owens, E.D. 68,
    110
Pujol, R. see Sans, A. et al.
Pumphrey, R.J. 413, 437, 713,
    714, 724
Purcell, F. 337, 344, 397
Purple, R.L. see Whitehead, R.,
    Purple, R.L.

Rüber, F. 71, 82, 110
Rankin, M.A. see Moore, D., Rankin, M.A.
-- see Moore, D. et al.
Rao, C.R.N. see Chen, H.S., Rao,
    C.R.N.

Rayport, S. see Wald, G., Rayport, S.
Razmjoo, S., Hamdorf, K. 44, 59
Ready, D.F., Hanson, F.E., Benzer, S. 492, 519, 636, 659, 686, 695
Ready, D.F. see Harris, W.A. et al.
Reed, C.T. see Brandenburger, J.L. et al.
-- see Eakin, R.M. et al.
Rees, G.R. 221, 237
Reese, T.S. see Henkart, M. et al.
Reichardt, W. 458, 480, 544, 550, 556, 590, 591, 601, 618
-- Braitenberg, V., Weidel, G. 602, 618
-- Poggio, T. 93, 110, 547, 550, 558, 561, 562, 565, 567, 572, 577, 578, 590, 603, 604, 618
-- Poggio, T., Hausen, K. 543, 550, 551, 558, 602, 604, 618
-- Varju, D. 582, 618
-- Wenking, H. 570, 618
-- see Fermi, G., Reichardt, W.
-- see Hassenstein, B., Reichardt, W.
-- see Kirschfeld, K., Reichardt, W.
-- see Poggio, T., Reichardt, W.
-- see Poggio, T. et al.
-- see Varju, D., Reichardt, W.
-- see Virsik, R., Reichardt, W.
-- see Wehrhahn, C., Reichardt, W.
Reingold, S.C. see Sejnowski, T.J. et al.
Reivich, M. see Sokoloff, L. et al.
Remane, A. 243, 255, 257, 258, 287
Reuter, T.E., White, R.H., Wald, G. 783, 788
Revel, J.-P. see Jan, L.Y., Revel, J.-P.

Ribi, W.A. 80, 110, 463, 464, 480, 484, 520, 637, 643, 650, 659, 673, 695
-- see Franceschini, N. et al.
-- see Hardie, R.C. et al.
-- see Hausen, K. et al.
-- see Praagh, J.P. et al.
Ricci, C. see Amsellem, J., Ricci, C.
Rice, D.C., Gilbert, S.C. 785, 788
Richard, G., Gaudin, G. 675, 695
Richards, W.A. 470, 480
Riddle, D.L. 191, 214
Riding, I.L. see Croll, N.A. et al.
Riehle, A., Franceschini, N. 439, 454, 539, 558, 598, 602, 618, 619
Ringuelet, R. 360, 397
Robenek, H. see Melkonian, M., Robenek, H.
Robertson, J.D. see Corless, J.M. et al.
Robinson, J. see Song, P.S. et al.
Rockstein, M. 765, 770
-- Finkel, A. 765, 770
Rodriguez, E. see Maldonado, H. Rodriguez, E.
Rogers, B., Graham, M. 565 619
Rogers, J. 115, 129
Rohlf, F.J. see Sokal, R.R., Rohlf, F.J.
Rühlich, P. 149, 176, 237
-- Aros, P., Viragh, S. 307, 308, 333
-- Tar, E. 154, 176, 226, 237
-- Török, J. 154, 159, 176, 219, 226, 237, 309, 310, 312, 333
Rojik, I. see Feher, O., Rojik, I.
Ronald, K. see Wright, D.G.S. et al.
Roof, D.J., Heuser, J.E. 148, 169, 176

AUTHOR INDEX

Roper, C.F.E., Boss, K.J. 719, 725
Rosen, M.D., Stasek, S.R., Hermans, C.O. 161, 176, 701, 725
Ross, H.E. see Gregory, R.L. et al.
Ross, J. see Burr, D.C. et al.
Rossel, S. 566, 619
-- Wehner, R. 97, 98, 99, 100, 101, 111
-- Wehner, R., Lindauer, M. 97, 98, 99, 111
Roth, F. see Menzel, R., Roth, F.
Rougier, C. see Clément, P. et al.
Rouiller, C. see Fauré-Fremiet, E., Rouiller, C.
Rowe, G.T. see Menzies, R.J. et al.
Rowell, C.H.F., O'Shea, M., Williams, J.L.D. 496, 520, 567, 602, 619
-- see O'Shea, M., Rowell, C.H.F.
-- see O'Shea, M. et al.
-- see Zaretsky, M., Rowell, C.H.F.
Ruck, PH., Edwards, G.A. 689, 695
Rudzinska, M.A. see Karakashian, S.D. et al.
Ruppert, E.E. 224, 238
Russel, F.S. 739, 742
Russell, B.J. see Hamilton, P.V., Russell, B.J.
Russell, R.L. see Hedgecock, E.M., Russell, R.L.
-- see Ware, R.W. et al.
Rutherford, T.A., Croll, N.A. 197, 214

Sagara, H. see Toh, Y., Sagara, H.
Saibil, H. 144, 148, 150, 169, 176
St-George, R.C.C. see Hubbard, R., St-George, R.C.C.
Saji, M., Oosawa, F. 125, 129
Sakaguchi, H., Tawada, K. 123, 129
Sakamoto, S. see Yanase, T., Sakamoto, S.
Sakurada, O. see Sokoloff, L. et al.
Saladin, K.S. 230, 231, 232, 233, 238
Sällstrom, J. see Bernhard, G.C. et al.
Salvini-Plawen, L.V. 153, 159, 176, 265, 287
-- Mayr, E. 140, 142, 143, 156, 159, 161, 176, 255, 265, 287, 327, 333
Sanchez, D.S. see Cajal, S.R., Sanchez, D.S.
Sandeman, D.C. 423, 427, 437
-- see Hengstenberg, R., Sandeman, D.C.
Sandri, C. see Lamparter, H.E. et al.
Sans, A., Pujol, R., Carlier, E. Calas, A. 624, 633
Sanyo 565, 619
Satir, B.H. 148, 177
Satir, P., Gilula, N. 347, 397
Saunders, D.S. 278, 287
Savely, H. see Magni, F. et al.
Savory, Th. 337, 398
Schaffer, A.M. see Waddell, W.H. et al.
Schaller, F. 664, 675, 677, 682, 695
Schiff, J.A. see Osafune, T., Schiff, J.A.
Schimkewitsch, W. 339, 398
Schinz, R.H. 449, 454
-- Lo, M.-V.C., Larrivee, D.C., Pak, W.L. 177, 762, 770
Schlecht, P., Hamdorf, K., Langer, H. 32, 42, 59
-- see Langer, H. et al.
-- see Schneider, L. et al.
Schletz, K. 123, 129

Schliwa, M., Fleissner, G. 367, 368, 387, <u>398</u>
-- see Fleissner, G., Schliwa, M.
Schmidt, J.A., Eckert, R. 121, <u>129</u>
Schneider, B.G. see Papermaster, D.S., Schneider, B.G.
-- see Papermaster, D.S. et al.
Schneider, L., Gogala, M., Draslar, K., Langer, H., Schlecht, P. 71, 72, <u>111</u>
Schoeller, J. <u>675</u>, 681, 682, 683, 687, 690, <u>695</u>
Scholes, J. <u>449</u>, <u>454</u>, 487, 490, <u>520</u>
Schöne, H. 93, <u>111</u>, 189, 210, <u>214</u>
Schönenberg, N. <u>427</u>, <u>438</u>
Schrader, K. 681, <u>696</u>
Schreckenbach, T., Walckhoff, B., Oesterhelt, D. 442, <u>454</u>
Schröer, W.D. 362, 383, <u>398</u>
Schultze, M. 443, <u>454</u>
Schwemer, J. 15, 16, 20, 21, 22, 23, 24, 25, 26, 27, 44, 45, <u>59</u>
-- Gogala, M., Hamdorf, K. 20, 27, 30, <u>59</u>
-- Langer, H. 25, 26, <u>59</u>
-- Paulsen, R. 20, 30, 32, <u>59</u>
-- see Gogala, M. et al.
-- see Hamdorf, K., Schwemer, J.
-- see Hamdorf, K. et al.
-- see Langer, H. et al.
-- see Paulsen, R., Schwemer, J.
-- see Pepe, I.M. et al.
-- see Stavenga, D.G., Schwemer, J.
Schwind, R. 69, <u>111</u>
Sejnowski, T.J., Reingold, S.C., Kelley, D.B., Gelperin, A. 624, <u>633</u>
Seidel, F. 675, 682, <u>696</u>
Seitz, G. 665, <u>696</u>
Seldon, H.L. 242, 244, 247, 265, <u>287</u>
Seliger, H.H., Lall, A.B., Lloyd, J.E., Biggley, W.H. 69, <u>111</u>
-- see Lall, A.B. et al.
Serreti, L. see Papi, F., Serreti, L.
-- see Papi, F. et al.
Shaller, F. see Mouze, M., Shaller, F.
Shannon, C.E., Weaver, W. 469, <u>480</u>
Shapley, R.M. see Toyoda, J. Shapley, R.M.
Shaw, S. 87, 347, <u>398</u>
Shaw, S.R. 69, <u>111</u>, 462, 465, 466, 470, 473, <u>480</u>, 481, 489, 491, <u>521</u>
-- Stowe, S, <u>87</u>, 111
Sheline, Y. see Lisman, J.E., Sheline, Y.
Shelton, P.M.J. 663, <u>696</u>
-- Anderson, H.J., Eley, Z. 686, <u>696</u>
-- Lawrence, P.A. 636, <u>659</u>, 670, 675, <u>695</u>
-- see Eley, S., Shelton, P.M.J.
-- see Lawrence, P.A., Shelton, P.M.J.
-- see Nowel, M., Shelton, P.M.J.
Shepherd, G.M. see Lancet, D. et al.
Sherk, T.W. 664, <u>696</u>
Sheuring, L. 337, <u>398</u>
Shichida, Y., Kobayashi, T., Ohtani, H., Yoshizawa, T., Nagakura, S. 25, 27, 29, <u>59</u>
-- see Tokunaga, F. et al.
-- see Yoshizawa, T., Shichida, Y.
Shih, E. see Wolken, J.J., Shih, E.
Shinohara, M. see Sokoloff, L. et al.
Shinzawa, K. see Kato, M. et al.
Short, R.B., Gagne, H.T. 226, <u>238</u>
Shriver, J.W., Mateescu, G.D., Abrahamsol, E.W. 28, <u>59</u>

Sicharylidze, T.A. see Kobakhidze, D.N. et al.
Siddiqui, I.A., Viglierchio, D.R. 135, 136, 137, 177, 192, 214
Sievers, J. see Berry, M. et al.
Silberglied, R.E. 68, 111
Singh, R.N. see Strausfeld, N.J. et al.
Singla, C.L. 138, 139, 144, 162, 177, 296, 297, 333
-- Weber, C. 138, 144, 152, 177
Sivak, J.G. 714, 725
Siviski, P. see Hemley, R. et al.
Slifer, E.H. 690, 696
Small, R.J.W. see DeVoe, R.D. et al.
Smith, J.E. 753, 754, 770
Smith, J.M. see Croll, N.A. et al.
Smith, L. see Ellenby, C., Smith, L.
Smola, U., Meffert, P. 444, 447, 448, 454
-- Tscharntke, H. 81, 111
-- Wunderer, H. 81, 111
-- see Wunderer, H., Smola, U.
Smyth, R.D. see Foster, K.W., Smyth, R.D.
Snell, T.W. see Pourriot, R., Snell, T.W.
Snyder, A.W. 77, 80, 86, 111, 594, 619
-- Laughlin, S.B. 86, 111
-- Menzel, R., Laughlin, S.B. 41, 59
-- Miller, W.H. 42, 59
-- Pask, C. 42, 59
-- see Howard, J., Snyder, A.W.
-- see McIntyre, P., Snyder, A.W.
-- see Menzel, R., Snyder, A.W.
Soans, A. see Naegele, J. et al.
Sokal, R.R., Rohlf, F.J. 202, 214
Sokoloff, L. 624, 633
-- Reivich, M., Kennedy, C., Des Rosiers, M.H., Patlak, C.S.,

-- Pettigrew, K.D., Sakurada, O., Shinohara, M. 624, 633
Sommer, E.W. 71, 78, 81, 111, 112
Song, P.S. 124, 129
-- Walker, B., Auerbach, R.A., Robinson, J. 124, 125, 130
-- see Fugate, R.D., Song, P.S.
Soohoo, S.L., Bishop, L.G. 529, 532, 541, 558
Southgate, E. see White, J.G. et al.
Spatz, H.C. see Folkers, E., Spatz, H.C.
Sperling, H.G. see Mark, R.E., Sperling, H.G.
Spies, R.B. 291, 293, 294, 333
Sprague, V. see Levine, N.D. et al.
Spreij, T.E. 687, 695
Springer, M.S. see Goy, M.F., Springer, M.S.
Spüller, M. 547, 558
-- Heide, G. 547, 558
Srinivasan, M.V., Bernard, G.D. 491, 521, 567, 573, 587, 619
-- Dvorak, D. 474, 481, 539, 540, 558, 597, 599, 610, 619
-- Laughlin, S.B., Dubs, A. 466, 471, 472, 481
-- see Dubs, A. et al.
-- see Dvorak, D. et al.
-- see Wehner, R., Srinivasan, M.V.
Stahler, N. see Mast, S.O., Stahler, N.
Stammers, F.M.G. 321, 333
Stange, G. 610, 619
Stark, R.J., Mote, M.I. 686, 696
Stark, W.S., Carlson, S.D. 74, 112
-- Ivanyshyn, A.M., Greenberg, R.M. 32, 59, 442, 454
-- Johnson, M.A. 23, 25, 59
-- Stavenga, D.G., Kruizinga, B. 32, 33, 59
-- Tan, K.E. 454
-- see Harris, W.A. et al.

-- see Hu, K.G., Stark, W.S.
Stasek, C.R. see Rosen, M.D. et al.
Stasko, A.B., Sullivan, C.M. 180, 181, 194, 195, 208, 209, 214, 273, 288, 451, 460
Stavenga, D.G. 59, 60, 423, 438, 458, 481, 487, 499, 521, 526, 559
-- Barneveld, H.H. van 42, 60
-- Franceschini, N. 33, 34, 60
-- Franceschini, N., Kirschfeld, K. 34, 35, 60, 454
-- Numan, J.A.J., Tinbergen, J., Kuiper, J.W. 32, 60
-- Schwemer, J. 11-61, 441, 454, 597, 619, 623, 633, 779, 788
-- Zantema, A., Kuiper, J.W. 20, 31, 43, 60, 441, 454, 455
-- see Beersma, D.G.M. et al.
-- see Bernard, G.D., Stavenga, D.G.
-- see Kruizinga, B. et al.
-- see Leggett, L.M.W., Stavenga, D.G.
-- see Stark, W.S. et al.
-- See Tsukara, Y. et al.
-- see Vogt, K. et al.
Stefancic, V. see Doukas, A.G. et al.
Stegmann, P. 587, 619
Steiger, V. see Lamparter, H.E.
Stein, P.J., Brammer, J.D., Ostroy, S.E. 12, 60
Steinberg, A. 681, 696
Steinberg, R.H., Fisher, S.K., Anderson, D.H. 153, 177
Steiner, A. see Gemperlein, R. et al.
Stell, W.K. 643, 660
Stent, G.S. 170, 177
-- see Fernandez, J., Stent, G.S.
-- see Kretz, J.R. et al.
Stephenson, R.S., Pak, W.L. 38, 60
Stewart, A. 155, 177, 226, 238
Stockhammer, K.A. see Burton, P.R., Stockhammer, K.A.

Stoll, C.J. 701, 725
Stossberg, K. 257, 258, 288
Stover, H. see Wenig, A. et al.
Stowe, S. 39, 42, 44, 60, 87, 112, 420, 426, 438
-- see Blest, A.D. et al.
-- see Laughlin, S. et al.
-- see Shaw, S.R., Stowe, S.
Straud, C. see Buchner, E. et al.
Strausfeld, N.J. 440, 448, 455, 458, 481, 483-522, 521, 524, 526, 527, 528, 551, 552, 559, 623, 624, 631, 633, 635, 656, 660, 779, 783, 788
-- Bacon, J.P. 511, 516, 521, 631, 633
-- Bassemir, U.K. 508, 511, 521
-- Bassemir, U.K., Singh, R.N., Bacon, J.P. 503, 511, 513, 521
-- Blest, A.D. 498, 521
-- Braitenberg, V. 489, 492, 521
-- Campos-Ortega, J.A. 489, 496, 521, 637, 638, 640, 648, 652, 660
-- Hausen, K. 492, 522, 551, 559
-- Nässel, D.R. 88, 112, 458, 462, 463, 481, 484, 492, 522, 526, 527, 551, 559, 635, 636, 643, 660
-- Obermayer, M.L. 533, 559
-- see Campos-Ortega, J.A., Strausfeld, N.J.
-- see Hausen, K., Strausfeld, N.J.
Strebel-Brede, J. 592, 601, 619
Stubbs, T.R. see Cobb, J.L., Stubbs, T.R.
Such, J. 343, 398, 667, 668, 669, 674, 675, 678, 679, 680, 681, 682, 696
Sugahara, M. see Suzuki, T. et al.
Sukhdeo, M.V.K. see Croll, N.A. Sukhdeo, M.V.K.

AUTHOR INDEX

Sullivan, C.M. see Stasko, A.B. Sullivan, C.M.
Sullivan, C.R. see Wellington, W.G. et al.
Sundeen, C.D. see White, R.H., Sundeen, C.D.
Sutherland, N.S., MacIntosh, N.J. 721, 725
Suzuki, T., Sugahara, M., Kito, Y. 28, 60
-- Uji, K., Kito, Y. 16, 23, 27 60
-- see Doukas, A.G. et al.
Svanidze, I.K. see Kobakhidze, D.N. et al.
Szuts, E.Z. see Fein, A., Szuts, E.Z.

Tabata, M., Tamura, T., Niwa, H. 145, 177
Tabliasco, V. see Bizzi, E. et al.
Taddei-Ferretti, C. 566, 619
-- Perez de Talens, A.F.P., 566, 619
-- see Chillemi, S., Taddei-Ferretti, C.
Takahashi, K. 747, 749, 770
-- see Millott, N., Takahashi, K.
Takasu, N., Tamotsu, S., Yoshida, M. 762, 770
-- Yoshida, M. 148, 177
-- see Yoshida, M. et al.
Takeuchi, J. 20, 60
Taliaferro, W.H. 190, 191, 195, 214
Tamotsu, S. see Takasu, N. et al.
-- see Yoshida, M. et al.
Tampi, P.R.S. 290, 333
Tamura, T. see Tabata, M. et al.
Tan, K.E. see Stark, W.S., Tan, K.E.
Tanouye, M.A., Wyman, R.J. 527, 551, 559
-- see Koto, M. et al.
Tar, E. see Rühlich, P., Tar, E.
Tartar, V. 124, 130
Tateda, H. see Yamashita, S., Tateda, H.
Tauber, U. see Hamdorf, K. et al.
Tawada, K. see Sakaguchi, H., Tawada, K.
Taylor, C.P. 510, 522, 610, 619
Taylor, D.L., Condeelis, J.S., Moore, P.L., Allen, R.D. 126, 130
Theron, A. 234, 238
-- see Combes, C., Theron, A.
Thomas, J.B. see Koto, M. et al.
Thompson, J.N. see White, J.G. et al.
Thompson, N. see Ward, S. et al.
Thorell, B. see Langer, H., Thorell, B.
Thorson, J. 587, 619
Tinbergen, J. see Stavenga, D.G. et al.
Toh, Y., Kuwabara, M. 643, 660 689, 696
-- Sagara, H. 690, 697
Tokarshi, T. see Hafner, G.S. et al.
Tokioka, T. 731, 742
Tokunaga, F., Shichida, Y., Yoshizawa, T. 25, 60
Tollin, G. see Bound, K.E., Tollin, G.
Tominaga, T. see Ninomiya, N. et al.
Tomita, T. see Toyoda, J. et al.
Tongiorgi, P. 382, 398
-- see Magni, F. et al.
-- see Papi, F., Tongiorgi, P.
Török, J. see Rühlich, P, Török, J.
Torre, V., Poggio, T. 484, 503, 522, 597, 619
Toyoda, J., Nosaki, H., Tomita, T. 699, 725, 783, 788
-- Shapley, R.M. 708, 725
Tracey, D. see Levine, J.,

Tracey, D.
Trouth, C.O. see Lall, A.B. et al.
Trujillo-Cenoz, O. 369, <u>398</u>, 485, 487, <u>522</u>
-- Melamed, J. 343, 369, <u>398</u>, 441, <u>455</u>, 484, <u>522</u>, <u>638</u>, 639, 640, <u>643</u>, 647, <u>648</u>, <u>660</u>, 683, <u>697</u>
-- see Melamed, J., Trujillo-Cenoz, O.
Truman, J.W. 778, <u>788</u>
Tscharntke, H. see Smola, U., Tscharntke, H.
Tsukara, Y., Horridge, G.A., Stavenga, D.G. 41, <u>60</u>
Tsukara, Y., Horridge, G.A. 23, 39, 41, <u>60</u>
-- see Ebina, Y. et al.
-- see Horridge, G.A. et al.
Tuffrau, M. <u>130</u>
Tyrer, N.M. see Altman, J.S., Tyrer, N.M.

Uji, K. see Suzuki, T. et al.
Ullman, S. 581, 587, <u>620</u>
-- see Marr, D., Ullman, S.
Ullyot, P. 180, <u>215</u>, 229, <u>238</u>, 273, <u>288</u>
Unteutsch, W. 321, <u>333</u>
Usukura, J., Yamada, E. 151, <u>177</u>

Vaissière, R. 403, 409, 416, <u>438</u>
Valentine, J.W., Campbell, C.A. 167, <u>177</u>
Vanfleteren, J.R. 143, 144, 146, 157, 161, <u>177</u>, 265, <u>288</u>
-- Coomans, A. 143, <u>177</u>, <u>238</u>, 265, <u>288</u>, 326, <u>333</u>
Varju, D., Bolz, J. 609, <u>619</u>
-- Reichardt, W. <u>620</u>
-- see Bolz, J., Varju, D.
-- see Grieger, B. et al.
-- see Reichardt, W., Varju, D.

Vaupel-von Harnack, M. 758, <u>770</u>
-- see Merker, G., Vaupel-von, Harnack, M.
Vaura, J. see Levine, N.D. et al.
Verger-Bocquet, M. 262, 282, <u>288</u>, 417, <u>438</u>, 774, <u>786</u>
Verheijen, F.T., Gerssen Schoorl, K.H.J. 229, <u>238</u>
Vevers, H.G. see Millott, N. Vevers, H.G.
Viallanes, H. 675, 682, <u>697</u>
Viaud, G. 228, <u>238</u>, 243, <u>267</u>, 268, 270, 271, 273, <u>288</u>
Vigier, P. 485, <u>522</u>
Viglierchio, D.R. see Siddiqui, I.A., Viglierchio, D.R.
Viragh, S. see Rühlich, P. et al.
Virsik, R., Reichardt, W. 565, 567, 575, 576, <u>620</u>
Vishnevskaya, T.M. see Gribakin, F.G. et al.
Vogt, K, 43, <u>61</u>, 77, <u>112</u>, 403, 432, <u>438</u>
-- Kirschfeld, K. 76, <u>112</u>, 442, <u>455</u>
-- Kirschfeld, K., Stavenga, D.G. 37, 42, <u>61</u>
-- see Dichgans, J. et al.
-- see Kirschfeld, K. et al.
Volkonsky, M. 664, 675, <u>697</u>
Vries, H. de, Kuiper, J.W. 31, <u>61</u>

Wachmann, E. 86, <u>112</u>, 683, <u>697</u>
Wada, S. 72, 74, <u>78</u>, <u>112</u>, 449, <u>455</u>
Waddel, W.H., Schaffer, A.M., Becker, R.S. 765, <u>770</u>
Waddington, C.H. 679, <u>697</u>
-- Perry, M.M. 343, <u>398</u>, 679, <u>697</u>
Wagner, H. 566, 609, <u>620</u>
Wagner, H.G. see Wolbarsht, M.L. et al.

AUTHOR INDEX

Wagner, H.J. 783, <u>786</u>
-- Hoffman, K.P., Zwerger, H. 624, <u>634</u>
-- Pilgrim, C., Zwerger, H. 624, <u>633</u>
-- see Ali, M.A., Wagner, H.J.
Walckhoff, B. see Schreckenbach, T. et al.
Walcott, B. 667, <u>697</u>, 435, <u>438</u>
Wald, G. 426, <u>438</u>
-- Hubbard, R. 20, <u>61</u>
-- Rayport, S. 309, 310, 311, <u>334</u>
-- see Brown, P.K., Wald, G.
-- see Reuter, T.E. et al.
Waldrop, B. see Kirk, M.D. et al.
Waldvogel, J.A. see Phillips, J.B., Waldvogel, J.A.
Walker, B. see Song, P.S. et al.
Walker, J.A. see Harris, W.A. et al.
Wallace, B. see Nicholls, J.G. et al.
Wallace, F.G. see Levine, N.D. et al.
Wallace, R.L. 279, <u>288</u>
Walls, G.L. 416, <u>438</u>
-- Judd, H.D. 443, <u>455</u>
Walne, P.L., Arnott, H.J. 121, <u>130</u>
Walther, J.B. 317, <u>334</u>
-- see White, R.H., Walther, J.B.
Walz, B. 310, 312, <u>334</u>
Wanless, F. see Young, M., Wanless, F.
Ward, S. 170, <u>177</u>, 179, 191, 192, 197, 200, 206, 207, <u>214</u>
-- Thompson, N., White, J.G., Brenner, S. 191, <u>214</u>
Ware, R. 242, 244, <u>247</u>, <u>288</u>
-- see Lévinthal, C., Ware, R.
Ware, R.W., Clark, D., Crossland, K., Russell, R.L. 191, <u>215</u>

Wasserman, G.S. see Kong, K.L. et al.
Wässle, H., Hausen, K. 533, <u>559</u>
-- Peichl, L, Boycott, B.B. 458, <u>481</u>
Watanabe, M., Furuya, M. 121, <u>130</u>
Waterman, T.H. 6, <u>9</u>, 44, <u>61</u>, 63-114, <u>112</u>, <u>113</u>, 154, <u>158</u>, <u>177</u>, 188, <u>225</u>, 403, 404, 423, 425, 426, <u>427</u>, 432, <u>438</u>, 562, <u>620</u>, 758, <u>770</u>, 774, <u>778</u>, 783, <u>788</u>
-- Aoki, K. 70, <u>113</u>
-- Fernandez, H.R. 86, 87, 89, 90, <u>113</u>
-- Fernandez, H.R., Goldsmith, T.H. 70, 82, <u>113</u>
-- Horch, K.W. 89, <u>113</u>
-- Piekos, W.B. 87, <u>113</u>
-- Wiersma, C.A.G. 88, <u>113</u>
-- see Eguchi, E., Waterman, T.H.
-- see Eguchi, E. et al.
-- see Nüssel, D.R., Waterman, T.H.
-- see Piekos, W.B., Waterman, T.H.
Watson, A.H.D., Burrows, M 641, 649, <u>660</u>
Weaver, W. see Shannon, C.E., Weaver, W.
Weber, C. 138, 140, 145, <u>178</u>
-- see Singla, C.L., Weber, C.
Weber, W., Dambach, M. 746, 770, <u>771</u>
-- see Dambach, M., Weber, W.
Webster, J.M. see Burr, A.H., Webster, J.M.
Wegener, G. 625, <u>634</u>
Wehner, R. 68, 71, 72, 77, 78, 80, 81, 82, 88, 89, 93, 94, 97, 99, <u>113</u>, 439, <u>455</u>, 463, <u>481</u>, 562, 601, 602, <u>620</u>
-- Bernard, G.D. 72, 79, <u>113</u>
-- Meyer, E. 81, <u>113</u>
-- Srinivasan, M.V. 94, <u>113</u>

-- see Bernard, G.D., Wehner, R.
-- see Goldsmith, T.H., Wehner, R.
-- see Mote, M.I., Wehner, R.
-- see Rossel, S., Wehner, R.
-- see Rossel, S. et al.
Wehrhahn, C. 75, <u>113</u>, 448, <u>455</u>, 547, <u>559</u>, 562, <u>567</u>, 578, <u>592</u>, 600, <u>601</u>, 602, 609, <u>620</u>
-- Hausen, K. 550, 552, <u>559</u>, 609, <u>620</u>
-- Hausen, K., Zanker, J. 566, 602, <u>620</u>
-- Poggio, T., Bülthoff, H. 564, <u>620</u>
-- Reichardt, W. 547, <u>559</u>
-- see Bülthoff, H. et al.
-- see Hausen, K., Wehrhahn, C.
-- see Praagh, J.P. et al.
Weidel, G. see Reichardt, W. et al.
Wellington, W.G. 98, 100, <u>113</u>, <u>114</u>
-- Fitzpatrick, S.M. 64, <u>114</u>
-- Sullivan, C.R., Henson, W.R. 64, <u>114</u>
Wells, M.J. 721, <u>725</u>
Wenk, P. see Kirschfeld, K., Wenk, P.
Wenking, H. see Götz, K.G., Wenking, H.
-- see Reichardt, W., Wenking, H.
Wernig, A., Pécot-Dechavissine, M., Stover, H. 647, 660
Westfall, J.A. 6, <u>9</u>, 777, <u>788</u>
-- see Eakin, R.M., Westfall, J.A.
Whaley, W.G., Dauwalder, M. 153, <u>178</u>
Whelan, J. see Evered, D. et al.
White, J.G., Southgate, E., Thompson, J.N., Brenner, S. 191, <u>215</u>
White, J.G. see Ward, S. et al.

White, R. 343, <u>398</u>
White, R.H. 44, <u>61</u>, 674, 677, 683, <u>697</u>
-- Gifford, D., Michaud, N.A. 44, <u>61</u>
-- Lord, E. 693, <u>697</u>
-- Sundeen, C.D. <u>44</u>, <u>61</u>
-- Walther, J.B. 310, 312, <u>334</u>
-- see Brown, P.K., White, R.H.
-- see Reuter, T.E. et al.
Whitehead, R., Purple, R.L. 643, <u>660</u>
Whitman, C.O. 309, 313, <u>334</u>
Whittle, A.C. 152, <u>178</u>, 297, 309, <u>334</u>
-- Golding, D.W. 293, 296, 299, <u>334</u>
-- see Golding, D.W., Whittle, A.C.
Wiederhold, M.L., MacNichol, E.F., Bell, A.L. 706, <u>725</u>
Wiersma, C.A.G., Yamaguchi, T. 89, <u>114</u>
-- see Waterman, T.H., Wiersma, C.A.G.
Wiesel, T.N., Hubel, D.H. 535, <u>559</u>
-- see Hubel, D.H., Wiesel, T.N.
Wigglesworth, V.B. 686, <u>697</u>
Willem, V. 387, 388, <u>398</u>
Williams, D. 360, 370, <u>398</u>, <u>399</u>
-- McIntyre, P. 359, 370, <u>399</u>
-- see Blest, A. et al.
Williams, D.S. 81, 87, <u>114</u>
-- see Blest, A.D. et al.
Williams, J.L.D. see Fraser-Rowell, C.H. et al.
-- see O'Shea, M. et al.
-- see Rowell, C.H.F. et al.
Willmund, R., Ewing, A. 601, <u>620</u>
Wilson, M. 510, <u>522</u>
-- Garrard, P., McGuiness, S. 668, 687, <u>697</u>
-- see Ostroy, S.E. et al.

AUTHOR INDEX

Wilson, R.A. 226, 238
Wilson, R.S. 94, 114
Winston, R. see Levi-Setti, R. et al.
Wist, E.R. see Diener, H.C. et al.
Witkovsky, P., Yang, C.Y. 624, 634
Wittmann, D. see Praagh, J.P. et al.
Wolbarsht, M.L., Wagner, H.G., Bodenstein, D. 681, 697
Wolburg-Buchholz, K. see Hausen, K., Wolburg-Buchholz, K.
-- see Hausen, K. et al.
Wolf, R., Gebhardt, B., Gademann, R., Heisneberg, M. 64, 71, 73, 75, 76, 77, 90, 102, 114
-- Heisenberg, M. 620
-- see Heisenberg, M., Wolf, R.
-- see Heisenberg, M. et al.
Wolken, J.J. 777, 786
-- Florida, R.G. 409, 438
-- Shih, E. 119, 130
Wolsky, A. 343, 399, 681, 683, 697
-- Huxley, J.S. 681, 697
-- Wolsky, M. 681, 683, 697
Wolsky, M. see Wolsky, A., Wolsky, M.
Wonneberger, R. see Heisenberg, M. et al.
Wood, D.C. 116, 124, 130, 145, 178
Wood, W.B. see Edgar, R.S., Wood, W.B.
Woolever, R., Pipa, R.L. 681, 698
Woollacott, R.M., Eakin, R.M. 156, 178
-- Zimmer, R.L. 132, 134, 136, 178
-- see Brandenburger, J.L. et al.
-- see Hughes, R.L., Woollacott, R.M.

-- see Zimmer, R.L., Woollacott, R.M.
Woolsey, T.A. see Durham, D. et al.
Wright, D.E. see Barber, V.C., Wright, D.E.
Wright, D.G.S. 231, 232, 239
-- Lavigne, D.M., Ronald, K. 233, 239
Wright, K.A. 179, 193, 215
Wunderer, H., Smola, U. 65, 72, 73, 74, 78, 81, 114, 448, 449, 455, 485, 508, 522
-- see Smola, U., Wunderer, H.
Wurdak, E. see Clément, P., Wurdak, E.
-- see Clément, P. et al.
-- see Cornillac, A. et al.
-- see Coulon, J. et al.
Wyman, R.J. see Koto, M. et al.
-- see Tanouye, M., Wyman, R.J.

Yagi, N., Koyama, N. 686, 698
Yamada, E. see Usukura, J., Yamada, E.
Yamaguchi, T. 88, 114
-- Katagiri, Y. Ochi, K. 88, 114
-- see Wiersma, C.A.G., Yamaguchi, T.
Yamamoto, M., Yoshida, M. 161, 178, 756, 757, 758, 760, 761, 771
Yamanouti, T. 677, 698
Yamashita, S., Tateda, H. 376, 377, 399
Yamasu, T., Yoshida, M. 139, 140, 141, 144, 155, 178
Yanase, T., Fujimoto, K., Nishimura, T. 310, 334
-- Sakamoto, S. 707, 725
-- see Fujimoto, K. et al.
Yang, C.Y. see Witkovsky, P., Yang, C.Y.
Yasumasu, I. 744, 771
Yasuroaka, K. 233, 239
Yerkes, W.A. 324, 334
Yingst, D.R., Fernandez, H.R., Bishop, L.G. 313, 314, 315, 316, 334

Yochelson, E.I. 700, 725
Yoshida, M. 142, 169, 178, 744, 746, 747, 750, 751, 753, 754, 771
-- Millott, N. 747, 771
-- Ohtsuki, H. 751, 752, 754, 755, 771
-- Takasu, N., Tamotsu, S. 417, 438, 742-743, 774, 778, 788
-- see Millott, N., Yoshida, M.
-- see Takasu, N., Yoshida, M.
-- see Takasu, N. et al.
-- see Yamamoto, M., Yoshida, M.
-- see Yamasu, T., Yoshida, M.
Yoshikawa, S. see Kato, M. et al.
Yoschikura, M. 353, 399
Yoshizawa, T. see Hubbard, R. et al.
-- Shichida, Y. 20, 27, 28, 61
-- see Shichida, Y. et al.
-- see Tokunaga, F. et al.
Young, A. see Fletcher, A. et al.
Young, E.C. 675, 677, 698
Young, J.Z. 717, 721, 725, 777, 788
Young, M., Wanless, F. 384, 399
Young, R.E. 719, 725
Young, R.W. 87, 114
Young, S. 179, 215, 424, 438
-- Downing, A.C. 423, 438

Zaagman, W.H., Mastebroek, H.A.K., Boyse, I., Kuiper, J.W. 539, 547, 559
-- Mastebroek, H.A.K., Kuiper, J.W. 541, 542, 547, 559
-- see Mastebroek, H.A.K. et al.
Zahid, Z.R., Golding, D.W. 294, 295, 296, 299, 334
Zanker, J. see Wehrhahn, C. et al.

Zantema, A. see Stavenga, D.G. et al.
Zar, J.H. 201, 215
Zaretsky, M., Rowell, C.H.F. 605, 621
Zeil, J. 71, 114, 602, 609, 621
Zettler, F., Autrum, H. 465, 481
-- Järvilehto, M. 465, 466, 481, 489, 491, 522
-- see Järvilehto, M., Zettler, F.
Zimmer, R.L., Woollacott, R.M. 132, 178
-- see Woollacott, R.M., Zimmer, R.L.
Zonana, H.V. 720, 725
Zrenner, E. 66, 114
-- Gouras, P. 166, 178
-- see Gouras, P., Zrenner, E.
Zvarguilis, J.E. see DeVoe, R.D. et al.
Zwerger, H. see Wagner, H.J. et al.
Zwicky, K. 384, 399
Zyznar, E.J., Nicol, J.A.C. 432, 438

SPECIES INDEX

Acanthephyra purpurea 780
Acanthopachylus aculeatus 339, 346, 348, 349, 378-381
Acanthopleura 707
Acronycta rumicis 778
Acropsopilio chilensis 360
Aedes 674, 677
-- *aegypti* 12, 21, 667
-- *nigromaculus* 2
Aeshna 664, 680, 690
-- *cyanea* 668, 677, 684
Agelena 384
-- *gracilens* 383
-- *gracilis* 362
-- *labyrinthica* 382
Agrion virgo 675
Allograpta 779
Alopecosa 354
Amoeba proteus 126
Aname 351
Anaspides 404
Anax 675
Ancylostoma tubaeforme 192
Androctonus australis 358, 360, 367, 368, 370, 371, 380, 383, 384, 778
Anomalocera 416
Antedon bifida 751
Antheraea pernyi 778

Apatemon sp. 231
Apis 69, 71, 72, 75, 76, 78-82, 97, 99-101, 637, 667, 669, 689
-- *mellifera* 17, 20, 24, 663
Aplysia 701, 710, 778
Apodemia mormo 20
Araeolaimus elegans 136
Aranea 339
Araneus diadematus 358, 360, 362, 376
Araneus umbricatus 384
Arca 417, 699, 700, 702, 706, 707, 709
Architeuthis 703, 719
Arctosa 382
-- *variana* 358, 359, 382
Argiope bruennichi 376
Armadillidium 424
Armandia brevis 224, 291, 294, 301
-- *cirrosa* 301
Artemia 404, 422, 424, 643
-- *salina* 41
Ascalaphus 27, 29, 41, 47, 73, 779
-- *macaronius* 12, 20, 27
Ascaris 191
Asplanchna 243, 244, 253-255, 265, 266, 270, 274, 281, 282

Asplanchna brightwelli 242,
  244-247, 253, 254, 257,
  260-262, 267, 268, 270-275,
  277, 282
-- *girodi* 270
-- *priodonta* 270
Astacus 86, 432
-- *fluviatilis* 19
Asterias 417
-- *amurensis* 751, 752, 754,
  755, 758, 759, 762, 763,
  765-767
Asterina gibbosa 768
-- *pectinifera* 764
Asteropecten scoparius 764
-- *polyacanthus* 764
Atelophlebia 47
Atlanta 715
Atta 2
Aurelia aurita 139
Autolytus pictus 298, 300
-- *sp.* 290, 292

Baëtis 664, 669
Balanus amphitrite 19
-- *eburneus* 19
Barbatia 705, 706
-- *cancellaria* 706
Bibio 81, 82
Blepharisma 124, 125
Bombyx 669, 681, 686
-- *mori* 672, 778
Bougainvillia principis 138-140,
  157
Branchiomma 290
-- *vesiculosum* 301, 302, 317,
  320, 322-324
Brachionus 243, 253-255, 266,
  274
-- *calyciflorus* 249-251, 253,
  267-269, 271, 272, 275, 282
-- *plicatilis* 251-253
-- *rubens* 279, 280
Branchiomma 417, 709
Bugula neritina 134, 136
-- *pacifica* 132, 133
-- *simplex* 132, 133, 138, 156
-- *stolonifera* 133

-- *turrita* 134, 135
Bunodera mediovitellata 233
Buthus occitanus 377, 378

Caenorhabditis elegans 136, 170,
  191-193, 195-200, 206
Calanella 406
-- *mediterranea* 403
Callinectes 27, 84, 425, 426
Calliphora 73, 75-78, 80, 81,
  448, 489, 514, 650
-- *erythrocephala* 12, 21, 24,
  501, 529, 532
Carausius 664, 667, 680-682
-- *morosus* 343, 667
Cardium 707
Carinaria 714
Carcinus 84, 85, 87, 417, 426
Carinaria 715
Carpocapsa pomonella 778
Cataglyphis 71, 76, 78, 79, 82,
  86, 94, 97, 779
Centrostephanus longispinus 748
Centruroides sculpturatus 377
Ceratogyrus darlingii 358, 359,
  371, 372, 374
Cherax 435
Chlamydomonas 120, 121, 145, 149,
  159
-- *reinhardii* 146, 147
Chlorella 125, 126
Chlorophanus 582
Chlorosarcinopsis gelatinosa 147
Chone ecaudata 291, 303
Chorthippus curtipennis 778
Chromadorina sp. 136
-- *bioculata* 136, 196
Chromulina psammobia 148, 149,
  156, 266
Chroomonas 121
Ciniflo similis 384
Cirolana borealis 421, 422, 424,
  425
Cladonema radiatum 138, 140, 157
Clubiona terrestris 384
Copilia 404, 409, 415
Corixa 675
Corixidae 677

SPECIES INDEX

Corycaeus 409
Corythalia 385
-- *xanthopa* 385, 386
Cryptocatyle lingua 231
Cryptomonas 121, 122
Cucumaria sykion 751
Culex 664
Cyrtodiopsis whitei 567

Dalyella viridis 226
Daphnia 93-96, 179, 404, 417, 422-424, 435, 562, 637, 638, 643
Dasychone 290, 301, 302, 304
-- *bombyx* 292, 300, 302, 317
-- *lucullana* 301, 321
Deilephila 32, 41
-- *elpenor* 20, 30
Dendrocoelum dorotocephala 155, 194
Dendrocoelum lacteum 229
Dendryphantes morsitans 359, 371, 374
Deontostoma californicum 136, 137
Diadema antillarum 746, 747, 749, 751
-- *setosum* 746, 747, 749, 750
Dinophilus 305, 306
Dinopis 44, 364, 367, 375, 376, 378, 779
-- *subrufus* 359, 363, 365, 366, 375, 376
Diplostomum spathaceum 226
Diplozoon homoion gracile 234
-- *paradoxum* 224, 230
Discocyrtus cornutus 339
Dolomedes 367
-- *aquaticus* 360
-- *fimbriatus* 348
Drosophila 32, 33, 35, 71, 73-76, 81, 90, 102, 150, 343, 484, 492, 496, 498, 532, 563, 565, 566, 594-598, 602, 605-607, 609, 623, 627-629, 636-640, 668, 673, 674, 681, 683, 778
-- *melanogaster* 12, 21, 24, 440, 623, 625

Dugesia 219, 229
-- *dorotocephla* 229
-- *japonica* 768
-- *lugubris* 159

Echinostoma togoensis 223-225
Eisenia 307, 309
-- *foetida* 308, 321
Eledone moschata 12, 18, 24
Elydia 701
Enchytraeus 309
Enoplus anisospiculus 136, 137, 196
Entobdella hippoglossi 234
-- *soleae* 226, 231, 233
Ephestia 669, 677, 681
Epiphanes senta 243
Eristalis 31, 41, 484
-- *tenax* 21, 24
Erpobdella 311
-- *octoculata* 312
Erythropsis 142
Euglena 117, 119, 184. 188, 255
Eukrohnia 727, 729, 737
-- *fowleri* 736
-- *hamata* 734, 736, 737
-- *proboscidae* 736
Eulalia 289, 296
Eunice viridis 301, 780
Euphausia pacifica 780
-- *superba* 19
Euphrostenops 339
Euscorpius carpathicus 358, 367
Eutreptiella 119
Euzetrema knoepffleri 221, 227, 230
Evarcha falcota 386
Eyleis 338

Fannia 532
Fasciola hepatica 226, 230
Filinia 255, 271
-- *longiseta* 271
Formica 94
-- *pratensis* 663
Formicina 685

Galleria melonella 20
Gastropacha rubi 689

Gelastocoris 667
Gerris lacustris 20
Gigantocypris 142, 401, 404, 407, 408, 416, 713
-- *mülleri* 407
Glomeris 336
Glossiphonia 311
Glycera convoluta 299
Gonodactylus 84, 427
-- *chiragra* 427
Grapholitha molesta 778
Grapsus 84, 426
Gyrinus 664, 674

Haemadipsa zeylanica 310, 313
Haemopis 321, 322
-- *sanguisuga* 312
Haliotis 701, 703, 716
Halobacterium 144
Harmothoë 305
Helix 699, 701, 703, 706, 709
-- *aspersa* 150
-- *pomatia* 703
Helobdella 309, 310, 312, 326
-- *stagnalis* 311, 312
Hemicentrotus pulcherrimus 744, 745
Hemicordulia 68, 73, 668
Hemigrapsus edwardsii 19
Henricia leviuscula 762
Hermissenda 701, 707
Hersilia 338
Heterodera spp. 186, 198, 203
Heterometrus 384
-- *fulvipes* 377
-- *gravimanus* 377
Heteronereis 299
Heteropoda 339
Hirudo 289, 309, 310, 312, 313, 318, 319
-- *medicinalis* 290, 293, 310-313
Histioteuthis 719
Homarus americanus 19, 24
Hyalaphora cecrophia 778
Hydroides dianthus 324
-- *uncinata* 320
Hypericum 125

Ischyropsalis 336, 369, 681
-- *luteipes* 339, 342, 344-346, 348-350, 354, 355, 358, 360, 369, 378, 382
-- *müllneri* 358, 360
-- *pyreneae* 349
-- *strandi* 355, 356

Labidocera 409, 411-413, 416, 714, 718
-- *acutifrons* 410, 411
Lampyris splendicula 663
Leodice fucata 320
Lepisma 672
Leptasterias pusilla 762
Leptinotarsa decemlineata 778
Leptograpsus 420-423, 426, 427
Lethocerus 670
Lethrus 67
Leucophaea maderae 778
Leucopsis cylindrica 123
Leukartiara octona 138, 139, 152, 156, 157, 168
Libellula 664
Libinia emarginata 17, 19
Lima 707, 709
Limax flavus 768
Limnodrilus 309
Limulus 44, 169, 417-422, 426, 428, 429, 471, 643, 709
-- *polyphemus* 19, 348
Liphistius 339
Lithobius 387, 388, 390
-- *forficatus* 336, 388, 390
Littorina 699-701, 703, 713
-- *littorea* 705
Locusta 664, 668, 683
Loligo japonica 18
-- *pealii* 12, 17, 18
Lucilia 639, 648
Luidia quinaria 764
Lumbricus 307, 309
-- *rubellus* 321
-- *terrestris* 290, 308, 319
Lycosa 367, 369
-- *baltimoriana* 376
-- *bedelli* 358, 259, 371, 373, 374
-- *carolinensis* 376

SPECIES INDEX

Lycosa lenta 376
-- *miami* 376
Lymantria 681
Lymnea 701, 714
Lytechinus variegatus 751, 752

Macrocypridina 404
-- *castanea* 407
Macrovestibulum 224
Manduca sexta 20
Matta 338
Meganyctiphanes 422, 429-431
-- *norvegica* 19, 430
Megoura viciae 778, 779
Menemerus confusus 376
Menneus 367
Mercenaria 706
Mermis 137
-- *nigrescens* 136
Metaphidippus aeneolus 361-363, 370
-- *harfordi* 357, 358
Micrommata virescens 378
Musca 31, 35, 37, 73, 75-77, 81, 442, 443, 445, 448, 463, 484, 532, 572, 583, 585, 590, 592, 597-600, 602, 608, 623, 630, 637, 639-643, 647, 648, 650
Musca domestica 20, 441, 447, 486, 572, 590, 603, 623, 630
Myrmecia 82
Mytilus 701, 702

Nasonia vitripennis 778
Nautilus 700, 703, 706, 716-718, 779
Neanthes succinea 304
Nebalia 404
Nemastoma lugubre 357, 358
Neopilina 701
Nepanthia belcheri 758, 759
Nephtys 293-296, 299, 320
Nereis 290, 323
-- *diversicolor* 313, 314, 323, 324, 326
-- *mediator* 290, 313-316
-- *pelagica* 305, 323-325, 327

-- *virens* 295, 297, 313
-- *vexillosa* 296, 297
Nerilla 305, 306
Nops 338, 339
Notodromas 404
-- *monarchus* 416
Notommata codonella 276
-- *copeus* 242, 267, 276-279
Notonecta 69, 73

Octopus 701, 703, 706, 717-719, 721
-- *ocellatus* 18
-- *vulgaris* 12, 17, 18, 765
Ocypode 83, 427, 428
Odontodactylus 427
Odontosyllis ctenostoma 297, 298
Oligolophus tridens 358
Onchidium 702, 707
-- *floridanum* 707
-- *verruculatum* 707
Oncholaimus vesicarius 136, 137, 156, 166, 168, 192, 195, 196
Oncopeltus 636, 675, 684
Onithochiton 707
Opheodesoma spectabilis 754-757, 760, 761
Opilio parietinus 357, 358
Opisthacanthus 378
-- *validus* 377
Oplophorus 422, 432
Orconectes rusticus 17, 19, 24
Oxygyrus 411
-- *keraudreini* 705, 715

Pachylus quinamavidensis 339-341, 378, 379
Pacifastacus 86, 643
Palaemon serratus 434
Palaemonetes 422, 434, 435
-- *palludosus* 19
-- *varians* 434
Pandinus imperator 358
Panuliris argus 19
Panulirus 84
Papilio 86, 87, 90
-- *xanthus* 690
Paramecium 117, 124, 125, 145
-- *bursaria* 125, 126

Pardosa 378
Parechinus angulosus 751
Patella 701
Patiria miniata 762, 764
Pecten 142, 145, 220, 416, 699, 702, 706-709, 711-713
-- *irradians* 18, 708
-- *maximus* 221, 705, 706, 711
Pectunculus 706
Penaeus 84, 85
-- *duorarum* 19
Pentatoma 681
Perga 690
Peridinium 121
Peripatopsis 182
Peripatus 343
Periplaneta 675, 677, 681, 684, 689
-- *americana* 671
Petrolisthes 84, 86
Phacus 121
Phaenicia 532, 639
-- *sericata* 343
Phalangium opilio 348
Phascolosoma agassizii 160
Pheretima agrestis 319
Phiale 385
Phiddipus 376
-- *johnsoni* 361, 362, 375
-- *regius* 376
Philodina 255, 265
-- *roseola* 254-257, 262, 264
Philophthalmus 222
Pholcus podophthalmus 338, 340
Photinus pyralis 780
Phronima 401, 421, 422, 425
Phrasina semilunata 421
Pieris 671, 673
-- *brassicae* 778
-- *rapae* 665
Piscicola 311
-- *geometra* 292, 310, 312
Placobdella rugosa 312
Planaria 189
-- *maculata* 189, 190
Platycnemis 682
Platymonas 145

Platynereis dumerilii 294, 296, 297, 313, 314, 324
Plexippus 378
-- *validus* 376
Polybothrus 388
-- *fasciatus* 387, 388
Polydora ligni 304
Polygordius 307
-- *appendiculatus* 305, 306
Polyorchis penicillatus 138, 139, 152, 157
Polystoma integerrimum 225, 234
-- *pelobatis* 230
Pontella 409, 413-417
-- *securifer* 415
-- *spinipes* 410, 413
Portia fimbriata 359
Posthodiplostomum cuticola 231, 233
Potamilla 290
-- *reniformis* 302, 317
Procambarus 86-89, 91, 150, 426, 643
-- *clarkii* 17, 19, 24, 157, 768, 778
Proterotrema macrostoma 231
Protodrilus 305, 306
Pseudoceros canadensis 158, 189, 220, 224
Pterotrachea 701, 706, 715
-- *coronata* 704, 705
Ptilogyna 81

Rhinoglena 255
-- *frontalis* 159, 257-259
Ribeiroia marini 234

Sabella spallanzanii 320
Saccocirrus 305, 306
Sagitta 727-729, 731, 736, 737
-- *bipunctata* 738, 739
-- *crassa* 727, 729-732, 734-736, 738-741
-- *elegans* 738, 739
-- *enflata* 731
-- *minima* 731
-- *serratodentata* 731
-- *scrippsae* 732, 733, 735
-- *setosa* 731, 739

Sagitta tasmanica 734, 737
Salpa democratica 145
Salticus scenicus 384
Sapphirina 406, 409
Sarcophaga bullata 21, 343
Sarsia tubulosa 138, 139, 157
Schistocerca 671, 677, 690
Schistosoma haematobium 232
-- *mansoni* 226, 230-233
Scutigera 336, 389, 390, 417
Scylla 87, 89, 92
Scytodes 339
Segestria senoculata 353
Sepia 719, 721
-- *esculenta* 18
-- *officinalis* 12, 17, 18
Sepiella japonica 18
Serpula vermicularis 324
Solea solea 234
Spadella 737
Sparassus mygalinus 359, 371-374
Sphodromantis 677
Spisula 706, 709
Spodoptera 32
-- *exempta* 20
Squilla 83
Stentor 116, 124, 145
-- *coeruleus* 124
-- *niger* 124
-- *polymorphus* 124
Streetsia 425
Strombus 700, 701, 703, 706, 714, 715, 718
-- *luhuanus* 701, 704, 705
Stylaria 309
Stylocheiron maximum 431
-- *microphthalma* 431
-- *suhmii* 431
Stylodrilus 309
Syllis 290
-- *amica* 298, 300, 304
-- *krohnii* 299
-- *spongicola* 301, 303
Sympetrum 72, 637, 642-644, 648
Synchaeta 243
Syritta 605, 609

Tabanus 484
Tamoya bursaria 140, 141, 144, 155, 157
Tasmanopilio megalops 360
Tegenaria 378
-- *parietina* 376
Telema 338
Teleogryllus commodus 778
Telyphonus schimkewitschi 360
-- *sepiaris* 358, 360
Temnopleurus toreumaticus 754
Tetranychus urticae 377
Tetrablemma 338
Theromyzon 309, 320
Thysanopoda tricuspidata 431
Tiaropsis multicirrata 139
Timandra 669
Todarodes 25
-- *pacificus* 12, 17, 18, 24
Torrea 290, 313, 314, 316
-- *candida* 290, 314-316
Trachelomonas 119
Tricellaria occidentalis 134
Trichocerca 253-255, 266
-- *rattus* 247, 248, 253, 262, 263, 276
Trichonema sp. 192, 199
Tridacna 189
Triops 404
Trite planiceps 386
Tritonia 701
Trochopus pini 226

Urodacus novae-hollandiae 384
Uroproctus assamensis 360

Vanadis 289, 290, 295, 296, 313, 314
-- *formosa* 295
-- *tagensis* 295
Vanessa cardui 20
Vejovis 378
-- *spinigerus* 377
Viciria 339
Volvox 119, 121, 123, 145
-- *aureus* 123

Watesinia scintillans 12, 17, 18

SUBJECT INDEX

Absorbance spectrum 26, 48, 145
Absorption spectrum 12, 14, 16, 25, 26, 30, 34, 37, 39, 42, 45, 77, 124, 169, 271, 316, 442-446
Acanthocephala 4, 146
Acarid 335, 337, 338, 352, 357, 369, 377
Accumulation behavior 186, 194, 200, 209
Acoelomata 4
Actinic light 752, 768
Action potential 124, 125, 536, 537
-- spectrum 35, 44, 125, 126, 145, 751
Acuity, visual 71, 74, 87, 367, 417, 423, 597
Adaptation of behavior 199
-- to dark and light See also Chromatophore dark, light adapted, Eye dark, light adapted 87
-- to dark, animals 422, 424, 426, 755, 757, 758, 762, 765-768
-- -- -- cellular 150, 389, 390, 749, 750, 760-763, 765
-- pool 491
-- to light, animals 422, 424, 426, 754, 755, 758, 762, 765-768
-- -- -- cellular 150, 389, 390, 491, 750, 760-763, 765
Adaptive radiation 2, 4, 7, 403
Aesthetes (chiton) 707
Agelenid 362
Alciopid annelid 132, 142
Agelenidae 351, 369, 376, 377
Alciopidae 290, 293, 295, 296, 313, 315, 714
Algae 118-120, 125, 126, 147
Amacrine cell 489, 491, 492, 494, 526, 528, 635, 642, 784
Amaurobiidae 351
Amblypygid 335, 337, 338, 352, 370
Ambulacral factor 750
Ammoxenidae 338
Amoeba 115, 126
Amphipod See also hyperiid amphipods 401, 402, 421, 424
Amphipoda 404, 422
Amplification of signal 468, 470, 477
-- matched 470, 471
Aneural organism 6, 115

SUBJECT INDEX

Animalia 2, 4
Annelid 8, 132, 141, 142, 224, 226, 227, 262, 266, 289-328, 417, 709
Annelida 3, 4, 143, 146, 157, 161, 162, 281
Ant 2, 3, 64, 67, 72, 77-79, 82, 94, 97, 150, 663, 779
-- bulldog 82
-- desert 71, 94
-- forest 94
Ants, formicine 82
-- myrmecine 82
Antenna, light 119-123
-- photostable 76
Anthomedusa 138
Anthozoa 137
Aphid 690, 778
Aphragmophora 728, 737
Aplacophora 701, 702
Aphroditidae 297, 304
Apoidae 663
Apposition eye See also Eye, apposition
Apterygota 5
Arachnid 8, 132, 142, 335-390, 417
Arachnida 5, 335-390
Araneidae 351, 368
Archiannelid 221, 305, 306, 328
Arcoid 706
Argentea See also Tapetum 710
Argiopid 362
Argiopidae 358, 359, 376, 377
Ark shells 706
Arrow worm 727-741
Arthropod 3, 6, 70, 71, 74, 80, 87, 141, 143-145, 152, 154, 155, 157, 161, 169, 221, 227, 226, 278, 281, 317, 343, 348, 349, 388, 424, 439, 457, 635-639, 642, 669, 674, 679, 681, 709, 718, 720, 737
Arthropoda 5, 143, 146, 157, 161, 162, 169, 182
Aschelminthes 142, 146, 265

Ascidian 161, 169, 709
Asteroid 142, 743, 746, 753, 755, 756, 758
Astigmatism (arachnids) 371
Auto-fluorescence 33, 73, 441, 445, 447, 767, 779
Autolytinae 296-299
Autoradiography of Drosophila brain 623, 625, 626, 628-630
Axonal projections of neurons 448, 458

Bacteria 144, 179, 181, 182, 185
Barnacles 19, 38, 46, 707
Bathorhodopsin 28, 29, 32
Bathynellacea 404
Bdelloid rotifers 243, 254, 255, 262, 264
Bee 3, 41, 44, 47, 63, 66, 70, 72, 77-79, 81, 90, 94, 98-100, 462-465, 472, 475, 476, 496, 562, 635, 643, 650, 664, 671, 672, 688, 706, 720, 783
-- dances 97-102
Beetle See also Carabid bettle, scarabeid beetle 3, 428, 498, 582
Behavioral responses 6, 7, 11, 71, 73, 75, 76, 80, 87, 92-102, 116, 179-210, 267-281, 384-386, 402, 413, 439, 448, 449, 523, 552, 582, 595, 596, 598, 601, 699, 701, 702, 706, 707, 737, 764, 781, 785
Biantinae 338
Bilateria 4, 146, 155, 162
Binnenkörper See Phaosome
Binocular depth perception 566
-- overlap 495, 511
-- sensitivity to motion 545
Bioluminescence See also spectral emission, bioluminescence 46, 69, 780
Bipolar cell 369, 458, 784
Bivalve 417, 699-701, 706, 707, 709
Bivalvia 18, 700, 702, 705

SUBJECT INDEX

Blowfly 7, 12-17, 21, 24, 27 31-34, 150, 483-518, 529, 532, 597, 775
Blurring of neural image 459-461
Brachiopoda 5, 146
Branchiopod 41
-- crustacean 637
Branchiopoda 404, 422
Branchiura 404
Brightness, image 708, 713, 718
-- ratio, retinal 435
Bryozoa 5, 132, 133, 135, 136, 146, 156, 161
Bryozoan 8, 132-136, 156, 166, 168
Bug aquatic 670
Butterfly 3, 32, 43, 47, 86, 90, 420, 498, 666, 671
-- metalmark 20
-- Pierid 779

Calcium (Ca^{++}) channel 145, 150, 164
Calliphora 15, 31, 34, 41, 466, 469, 525, 530-532, 534, 537, 538, 541, 542-545, 549, 583, 598, 668, 681, 683, 690
Carabid beetles 566
Carabidae 681
Carotenoids 119, 125, 145, 243, 244, 246, 255, 271, 442-444
Catecholamines 752
Centre-surround 413, 491, 552
Centipeds 417
Cephalocarida 404
Cephalochorda 2
Cephalochordata 146
Cephalochordate 161
Cephalopod See also Coleoid cephalopods, mollusc, Cephalopoda 12, 25, 30, 44-47, 69, 123, 132, 142, 157, 293, 413, 435, 699, 700, 703, 706, 710, 714, 716, 718-721
Cephalopoda 6, 12, 17, 18, 700, 702, 705

Cercariae 218, 219, 221, 223, 224, 226, 227, 230-234
-- opistorchioidean 221
Cestoda 217, 218
Cestodaria 217
Chaetognatha 5, 7, 146, 727-741
Chaetognaths 5, 7, 727-741
Channel activation 468, 491
-- functional 491, 550, 551
-- ion 781
-- orthogonal dichroic input 89
-- visual input See also Polarisation channel and colour channel 69, 74, 86, 88
Chelicerata 5, 19, 337, 348
Chemoreception See Chemosensitivity
Chemosensitivity 151, 156, 161, 164, 166, 168-170
Chemotaxis 93, 170, 197, 198, 210
Chiasma, external 524, 541, 546, 552
-- internal 524, 528, 592
Chilopod 336, 387
Chilopoda 5, 336, 387
Chiton 701, 702, 706, 707
Chlorophyceae 120, 121
"Chocolate drop", oceanic 407
Chordata 2, 3, 5, 145, 151
Chordates 442, 735
Chromatophore 743, 746-748
-- dark adapted 746, 747
-- light adapted 747
Chromophore 12, 13, 15, 20, 28, 77, 442-444
Chromoprotein 12
Chrysophyceae 119
Cicacids 665
Cilia See also Receptors ciliary 124, 132-134, 136, 138, 139, 141, 143, 144, 148-153, 155, 156, 158, 160-162, 165, 166, 168
-- in annelid receptors 289, 296, 299, 301, 303, 305, 307, 309, 312, 326, 328
-- in arachnid and myriapod receptors 335, 344-346

Cilia, in chaetognath receptors 732, 735, 736
-- -- echinoderm receptors 743, 744, 756-759, 766
-- -- mollusca receptors 701, 707-709, 716
-- -- platyhelminthes receptors 224, 230
-- -- rotifer receptors 255, 257, 258, 261-264, 266, 274, 276, 281, 282
Ciliates 124-126, 145, 182
Cirripedia 19, 404
Citon 751
Clam 189, 699, 706
Clitellats 327
Closed-loop experiment 561, 568, 569, 572-575, 577, 590, 601, 607-609
Cnidaria 3, 4, 132, 137, 139-142, 144, 146, 148, 150, 155, 157, 159, 162
Cnidarian 5, 6, 156
Coccinellidae 281
Cockle 707
Cockroach 643, 664, 676, 677, 683, 684, 778
Coelenterata 4
Coelenterates 709
Coelomata 4
Coleoid cephalopods 703, 714, 716, 718
Coleoidea 702
Coleoptera 3, 498, 664, 665, 667
Coleopteran *See also* Coleoptera 69
Collembola 667
Collothecacea 243, 255
Colour adaptation experiments 597, 600
Colour channel 90
-- coding 476
-- discrimination 41, 42, 46, 65, 188
-- processing (fly) 449
-- vision 43, 47, 63, 66, 169, 187, 386, 387, 423, 426

Columnar cell or columnar neuron 492, 493, 496, 511, 526, 527, 535, 542, 551, 552
Component-selection model (evolution) 163-170
Compound eye *See* eye compound,
-- -- -- eye apposition compound,
-- -- -- eye superposition
Conductance in cell 36, 317, 491, 708
-- ionic...model 80
-- channels 468
Conical body 733-735
Contrast 43, 47, 69, 449, 459-462, 466-468, 470-472, 475, 477, 490, 491, 528, 537, 540, 561, 583, 584, 586, 592, 595, 597, 605, 713
-- channel 472, 476, 477
-- coding 465, 466, 470, 473
-- element 181, 185, 186
-- frequency 541, 543, 544, 547, 582, 584, 586, 605, 606
-- temporal modulation 550
-- transfer 594, 595
Copepod *See also* Pontellids 401-403, 405, 406, 408, 409, 416, 417, 779
Copepoda 404
Cornea 123, 141, 349, 350, 355-357, 375, 389, 402, 426, 431, 434, 441, 446, 484, 487, 594, 665, 688, 690
-- aquatic 713
-- negative response 290, 314, 316, 383
-- positive response 290, 314, 316
Corneal cells 758, 759, 762, 764
-- facets 402, 440, 445
-- filters 43
Corner reflector 432, 433
Covering reaction 751, 752
Crab 17, 32, 43, 44, 46, 84, 417, 420, 423, 426, 427, 432, 434, 778, 779
-- anomuran 84, 86
-- blue 27, 84
-- brachyuran 84, 401, 404, 418, 424-426

Crab, ghost 83, 427
-- green 85
-- horseshoe 19
-- mud 19, 89, 92
-- paguran 404
-- rock 19, 84
-- shore 422
-- spider 19
-- true 424, 426
Crayfish 17, 19, 24, 30, 35, 42, 43, 46, 48-51, 82, 86, 88-90, 92, 147, 150, 157, 390, 402, 426, 428, 432, 434, 435, 462, 463, 472, 643, 778
-- northern 19
-- swamp 19
Cribellata 350
Cricket 475, 535, 778
Crinoid 751, 753
Criterion response 36, 37
Critical fusion frequency CFF See also Flicker fusion frequency 378, 390
Crustacea 5, 17, 157, 158, 317, 390, 401-435, 458, 636, 718, 778
Crustacean 6, 31, 32, 46, 64, 65, 69, 77, 82, 83, 88, 90, 92, 132, 157, 336, 337, 364, 401-436, 440, 643, 713, 714, 758, 779, 780
Cryptophiceae 121, 122
Cryptostigmata 338
Crystalline cone, corneal 389, 390, 418, 419, 421, 424-426, 428-432, 434, 440
-- shreads See Tract-crystalline shreads
-- pyramids 432
Ctenophora 4, 146
Cubomedusan 140, 142
Cumacea 404
Cuttlefish 12, 17, 18, 703, 718, 721
Cyrtodiopsid 566
Cytochalasin B 746

Dalyelliid flatworm 779
Dalyellidae 220, 227, 228
Damsel fly 80
Daphnia 179
Dartnall nomogram 27, 316
Decapod 77, 82, 85, 88, 403, 424, 426, 432, 636
Decapoda 404, 422
Decapods long-bodied 403
-- macruran 428
-- macrurous 402
Dendritic density 533, 534, 540, 544, 656
Deoxyglucose radioactive technique 7, 498, 503, 601, 623-631
Dermal sensitivity 702, 706, 746, 753, 754
Dermatoptical sense 228, 229
Deuterostome 265, 756, 774
Deuterostomia 5, 7
Diadematid See Sea urchins, diadematid
Diadematidae 746
Dichroic ratio 66, 86
Dichroism of visual and accessory pigments 63, 70, 76
Diclidophoridae 221
Dictynidae 369
Difference spectrum 23, 26, 31
Dimorphism, sexual ocular 74, 77, 340, 401, 409-411, 416, 417, 463, 493, 511, 526, 551, 601, 774, 779, 780
Dinoflagellates 117, 121, 122, 142
Dinophyceae 121
Dinopid spiders 154, 336, 363
Dinopidae 340, 359, 361, 367, 376, 377, 387
Diopsidae 567
Diopsids 663
Diplopod 336
Diplopoda 5, 336, 387, 390
Diptera 3, 21, 72, 484, 487, 498, 567, 640, 641, 664, 665, 667, 672, 683, 779

Dipteran *See also* Fly 73, 74, 76, 78, 80, 100, 439, 440, 563, 623
-- nematoceran *See also* Nematocera 81
Direction-selectivity 503, 505
-- sensitivity (or directional sensitivity) 132, 138, 147, 151, 505, 538, 539
-- specific nervous activity 623, 626-628, 631
Dolichopodid fly 43
Dragonfly 47, 68, 72, 73, 462, 463, 465, 472, 487, 491, 637, 642-644, 647, 648, 689, 779, 783
Dronefly 21, 24, 31, 41
Drossidae 369
Dye injection 545, 624
Dynamogenic action 228
Dysderidae 338, 369

E-vector *See also* polarisation or polarised light 526, 540
-- -- orientation (or direction) (ϕ) 63, 64, 66-68, 71, 72, 75, 77, 80, 86, 88-94, 97-102
-- -- rotation 65, 88, 90, 91
-- -- U.V. 69, 72, 77
Early receptor potential (ERP) 38
Earthworm 2
Ecclisis 182, 185, 195, 196, 209
Ecclitic response 182, 183, 185, 186, 193-201, 203, 205, 206, 209
Ecdysone 677
Echinodera 4
Echinoderm 7, 9, 142, 777, 779
Echinodermata 3, 4, 5, 7, 146, 150, 161
Echinoid 743, 746, 751, 753, 755
Echinoplutei 743-745
Echiura 6, 146

Ectoparasites 218, 226
Ectoprocta 5
Electroretinogram (ERG) 36, 38, 69, 313-315, 321, 377, 378, 381, 383, 384, 390, 707
Elemental motion detection *See* Movement detector
Emission spectrum 34, 35
Enchytraeidae 307, 309
Endoparasites, 7, 218, 226
Endoprocta 4
Energy-transfer efficiency 442
Enterozoa 2, 4
Entoproct 156
Entrainment 778
Epeiridae 369
Ephemeroptera 664, 665, 673
Epifluorescence 440, 445, 446, 449
Eucarida 404
Euglena *See also* Euglena 119
Euglenophyceae 119, 120
Eukaryotes 115
Eumalacostraca 404
Euphausiacea 404, 422
Euphausiids 82, 402, 403, 418, 420, 428-432
Eurypterids 718
Eustigmatophyceae 120
Eusyllinae 296-299
Evolution *See* Eye evolution, photoreceptors evolution, lens evolution
Excitation spectrum 34, 35
Eye accessory 293, 335, 350-354, 357, 377
-- apposition (compound) 88, 401-403, 417, 418, 421-429, 431, 432, 434, 464, 487, 672-674, 699, 706
-- branchial 290
-- camera 5, 439, 777
-- cellular differentiation 343-350
-- cephalic 699, 701-704, 706, 709, 710, 716, 720
-- -- Polychaete 294

SUBJECT INDEX

Eye cerebral *See also* Eye cephalic 161-162, 219, 243-257, 262, 265, 266, 232, 233, 235, 241, 242, 289
-- compound *See also* eye: apposition, superposition, reflecting and refracting superposition 7, 31, 33, 41, 68-75, 77, 78, 82, 84, 88, 91, 97, 141, 142, 157, 317, 337, 401-403, 407, 417, 423, 428, 432, 435, 439-449, 457, 458, 524-526, 538, 547, 550, 562, 566, 594, 596, 600, 635, 639, 643, 661-663, 675, 678, 687, 689, 699, 700, 702, 705, 709, 718, 775, 778, 780, 783
-- corneal 405
-- curvature 493, 495
-- dark-adaptation 37, 226, 229, 314-316, 319, 325, 336, 364, 367, 368, 376, 378, 380, 381, 383, 387, 422, 424, 428, 432, 434, 435, 459, 461, 462, 467, 475, 598, 673, 731, 732
-- development 7, 335, 336, 338-343, 354, 492, 635-656, 661-663, 675-691
-- direct type 335, 337, 338, 352
-- double 71, 73, 431, 664
-- everted type 138, 727, 729, 736
-- evolution *See also* Eye phylogeny 131-143, 376, 699, 700, 710, 718, 774, 775, 777
-- green-adapted 314, 316
-- growth 661, 676-691
-- hemispherical 663, 719
-- hyperopic 371, 375
-- indirect type 335, 337, 338, 352
-- inverted type 138, 727, 729, 730-736
-- "lens" 131, 132, 137, 142, 143, 405, 435, 699, 700, 702, 703, 705, 710, 717, 783
-- -- aquatic 713

-- lensless 416, 703, 705, 716, 727, 739
-- light adaptation 42, 155, 319, 325, 336, 364-368, 376, 378, 380, 381, 383, 422, 424, 428, 434, 435, 461, 462, 466, 467, 469, 470, 666, 673, 731
-- "mirror" *See also* Mirror (in eye) 221, 225, 227, 403, 405, 700, 702, 709
-- nauplius 401-409, 416, 417, 435, 778
-- neural superposition 74, 88, 487
-- non-cephalic 705-707, 709
-- non-pigmented 244
-- nuchal 219
-- photopic 672
-- phylogeny *See also* Eye evolution 387, 417, 435
-- "pigment-cup" 131, 132, 134, 135, 138, 139, 141, 142, 148, 151, 158-160, 189, 190, 195
-- "pigment-spot" 186, 191-193
-- pigmentation 341, 342
-- "pinhole" 702, 703, 705, 716, 717, 777
-- -- camera Nautilus 777
-- pit 700-702
-- reflecting superposition (compound) 405, 422, 432-434, 705
-- refracting superposition (compound) 405, 422, 428, 432, 433
-- scanning 411, 715, 716
-- scotopic 672
-- simple 9, 156, 189, 335-337, 401, 403, 405, 417, 643, 662, 689, 775
-- single 243, 388
-- -- chambered 718
-- -- -- camera 417, 703, 713
-- supernumerary 243, 255
-- superposition (compound) 43, 82, 88, 401-403, 418, 424, 426, 429, 431, 432, 434, 672-674
-- telephoto 409, 425, 719
-- tentacular 219

Eye tubular 719
-- vertebrate *See also* vertebrate vision & vertebrates 131
-- cup 388, 406, 409, 411-414, 417
-- spot 146, 707, 772
Eyespot compound 417, 709
Eyespots quarterwave reflecting 120
-- simple screen 119

F-number 370, 375, 408, 419, 429, 435, 713, 718
Facet density 71
-- size gradient 70-71
Figure-ground discrimination 550, 551
File-shells 707
Filistatidae 338, 350
Firefly 46, 69, 401, 420, 428, 780
Fish *See* Vertebrates, fish
Flabelligeridae 291, 294
Flagellates 117, 145, 193, 255
Flatworms *See also* Dalyelliid flatworm 265, 768
Flavin 126
Fleshfly 21
Flicker fusion frequency FFF *See also* Critical fusion frequency 379, 380, 382, 390
Floscularicea 243, 255
Flowfield 561, 563-566
Fluorescence *See also* Auto-fluorescence, epi-fluorescence 32-35, 86, 384, 446, 765, 767
Fluoroscopy, ommatidial fundus 445
Fly *See also* blow-, damsel-, dipteran-, dolichopodid-, fire-, flesh-, fruit-, house-, hover-, may-, muscoid-, owl-, tabanid-, 3, 7, 12, 14, 15, 17, 31-34, 38, 39, 41-45, 48, 64, 73, 75-77, 424, 439-449, 460, 462-465, 467, 471-475, 484, 490, 491, 493, 496-498, 510, 523-576, 561, 563, 565, 567, 569-577, 580, 583-587, 591, 592, 595-597, 602-609, 623-631, 636, 638, 639, 641-648, 651, 655, 671, 672, 779, 781, 783, 784
Focal length 370, 375, 412, 413, 416, 418, 420, 424, 699, 705, 713, 714, 719
Forcipulata 762
Fourier spectroscopy 445
Frequency fusion *See also* Flicker frequency fusion 336
-- transfer function FTF 379, 380, 382
Frog 785
Fruitfly *See also* *Drosophila* 12, 21, 24, 32, 38, 74, 76, 562, 594, 608, 625, 664, 679, 683, 686, 687
Füllmasse 139, 140
Furcilia larva 422, 431

Gain 460, 468, 469, 543, 546, 583, 602-604
-- control *See* Gain
Galatheids, anomuran 432
Galatheoid 84
Gamasidae 335, 337
Ganglion cell 43, 458, 470, 476, 634, 637, 708, 784
-- optical or ganglion optic or ganglion visual 88, 89, 441, 448, 592, 626, 627, 670, 682, 717, 720
Gastropod *See also* Mollusc, prosobranch gastropod, pulmonates 132, 142, 699-701, 703, 706, 707, 709, 710, 713, 716, 718, 720
Garypidae 337
Gastropoda 700, 702, 705
Gastrotricha 4, 146
Gene mutation 167-170, 774
-- regulation 163-170
Geometric interference 586
Geophilidae 336, 387
Geotaxis 233

SUBJECT INDEX

Geotropism 738
Glossiphoniidae 309
Gnathostomulida 146
Gnesiotroque 255
Gonyleptidae 339, 341, 342
Graded potential 124, 536, 537
Grasshopper 43, 778

Habituation, response 124, 199, 200, 475, 610
Hadrotarsidae 351
Haemadipsidae 313
Harvestman 337, 339, 344, 348-350, 354, 357, 364, 378
Hemichorda 2
Hemichordata 146, 162
Hemichordate 152
Hemiptera 672
Heterometabolous insect 661-664, 675-679, 681-684
Heteropod 700-703, 709, 715, 716
Heteroptera 20, 665
Hirudidae 309, 313
Hirudinae 292, 309-312, 317, 319, 321, 322, 326, 327
Holometabolous insect 638, 661-663, 676-677, 681-685, 689, 690, 691
Holopeltid 335, 337, 338
Holothuria 7
Holothurian 755, 756, 758, 760, 761
Holothuroid 743, 746, 751, 753
Honeybee 64, 67, 68, 71, 76, 79, 90, 97, 99, 100, 101, 779
-- drone 17, 20, 24, 31, 41, 47
Hookworm 192, 195
Hoplocarida 404
Horizontal cells 484, 496-498, 503, 507, 508, 546, 548, 551, 784
Housefly 21, 31, 32, 35, 37, 42, 77, 446, 485, 562, 566, 572, 590, 598, 601, 603, 609, 630
Hoverfly 721
Hydromedusa 140, 145, 150, 152
Hydrozoa 137-139

Hymenoptera 3, 20, 665, 685
Hymenopteran *See also* Hymenoptera 67, 68, 77, 80, 449
Hyperiid amphipods 404, 425, 431

Imago, insect 492, 639, 661, 663, 676
Induced resonance 442
Infrared reflectometry 72
Inhibition, shunt- 550, 587
-- lateral 86, 464, 466, 474, 491, 541, 546
-- self 491
-- subtractive 466, 467, 471
-- temporal 466
Insecta 4, 5, 17, 20, 158, 773
Insects *See also* Heterometabolous-, holometabolous-, imago-, 2, 3, 6-8, 12, 31, 35, 41, 46-48, 64, 65, 67-69, 72, 74, 80, 82, 87, 88, 90, 99-101, 142, 157, 181, 182, 317, 336, 337, 343, 348, 349, 389, 390, 401, 402, 417, 423, 427, 428, 435, 439, 440, 444, 449, 458, 461, 484, 491, 498, 534, 561-601, 623-631, 635, 656, 661-691, 699, 779, 781, 783-785
-- diurnal 401, 403, 417, 563, 661, 672
-- nocturnal 661, 672
-- parasites 663
Intensity absolute light 460
-- distribution 460
-- light 23, 25, 36-38, 43, 67, 80, 88, 91, 123, 124, 186, 188, 189, 193, 195-197, 228, 229, 231, 232, 267, 268, 275, 277, 278, 318, 319, 321-327, 378, 381, 390, 431, 458-462, 464-467, 470, 471, 487, 490, 499, 579, 673, 674, 744-746
-- relative light 231, 268, 460
-- response 461, 674
Inter-ommatidal angle 421, 423, 424, 428, 495
-- receptor angle *See* Receptor angular separation

Intramembranous particles (IMP) 762, 764-766, 768
Inverted responses 598, 607
Iris diaphragm 140
-- variable 421, 702
Ischyropsalidae 339
Isopod 401, 402, 421, 424, 425, 780
Isopoda 404, 422

Jellyfish See also Medusa 137

Kamptozoa 446, 456, 461
Kinesis See also Klinokinesis, photokinesis, photoklinokinesis, orthokinesis 93, 116, 179-185, 193, 203, 209
Kinorhyncha 6, 146
Klinokinesis 116, 125, 180-187, 194, 195, 198, 200, 201, 203, 205-207, 209, 229, 230, 273, 276
Klinotaxis 185-188, 190, 192, 193, 197, 198, 205, 210
Krill 19, 422, 428, 430, 435

Lamina 7, 38, 74, 76-78, 448, 457, 458, 461-466, 471-474, 476, 477, 484, 485, 487-490, 492, 493, 495, 496, 498, 499, 511, 524, 526, 527, 540, 597, 601, 626, 628, 635-656, 671, 781, 783
Larva See also Furcilia larva, miracidium-, nauplius-, tornaria-, trocophore-, pluteus-
-- arachnids 340-350, 354
-- ascidian tadpole 161, 169
-- bryozoan 132-136, 156, 166, 168
-- crustacean 404, 422, 434
-- echinoderm 744, 745
-- insect 64, 142, 442, 444, 492, 635, 656, 661-663, 670, 675-685, 689, 690
-- nematode 192, 199

-- pelagic 186
-- platyhelminthes 8, 158, 189, 217-221, 224, 226, 230-234
-- (Spionidae) 304, 306
-- sponge 182
Late receptor potential (LRP) 36-38, 781
Leech 144, 179, 289, 290, 317-321
Lens 31, 123, 141, 142, 221, 341, 342, 349-355, 357, 387, 389, 401-403, 408, 409, 411-414, 416, 417, 423, 424, 432, 434, 484, 485, 487, 490, 495, 699, 705, 707, 710-713, 780, 783
-- cellulary 291, 295, 299, 327
-- corneal 42, 43, 388, 390, 418, 440, 663
-- cylinder 418, 420, 425, 426, 428, 429, 432, 667
-- -- Exner 401, 418, 432
-- design 141, 713
-- double 406
-- evolution 139, 140, 141
-- extracellulary 304, 667
-- hard 699, 701
-- insect 665-667
-- Matthiessen 714, 717, 718
-- secreted 291, 296-298, 301, 304, 327, 357, 667
-- secretion 335, 349, 350
-- simple 428
-- soft 699, 701
-- triplet 413, 414
Lenslet corneal 441, 445
Lepidoptera 3, 20, 68, 72, 498, 663, 665, 667, 674, 683
Lightguide (rhabdom) 31, 401, 421, 425, 426, 428, 432, 594
Limpets 699, 700
Limulus 672
Linyphiidae 351, 384
Liphistiomorphae 338
Lobster 24, 32, 46, 402, 420, 428, 432, 778
-- american 19
-- spiny 19, 84

SUBJECT INDEX

Lobula 457, 474, 475, 477, 483, 484, 492, 493, 495, 496, 498, 499, 501, 503, 511, 513, 523, 524, 526-528, 530, 551, 552, 601, 626-628, 630, 780
-- complex 7, 523, 526-528, 530, 783
-- plate 483-485, 492, 493, 495-499, 501, 502-509, 511, 512, 523-552, 597, 623, 627, 628, 630, 631
Locust 44, 475, 496, 510, 535, 565, 604, 649, 671, 675, 682, 683, 690
Locomotor responses 267, 268, 276
Lumbricidae 307
Lumirhodopsin 27-29, 32
Lycosid 362
Lycosidae 351, 358, 359, 366-368, 376, 377

Machilids 668
Macruran 432, 434
Malacostraca 401, 403, 404, 424, 426, 427
Mammalia 3
Mandibulata 5
Mantids 565, 665
Mantis 44
Matthiessen's ratio 140, 413, 699, 700, 713, 714, 718
Mayfly 47, 779
Mechanoreceptors 508, 510, 512, 753
Mechanosensitivity 151, 161, 164, 169, 170
Medulla 74, 77, 78, 448, 449, 457, 458, 465, 472-477, 484, 485, 487, 489, 491-493, 495, 496, 498, 499, 501, 503, 511, 524, 526-528, 530, 536, 546, 552, 623, 626-628, 630, 631, 638, 671, 783, 784
-- column 485, 491, 493, 498, 501, 513, 516

Medusa *See also* Anthomedusa, scyphomedusa 137-139, 144, 148, 156
Membrane chemosensory 151, 152, 162
-- mosaicism 647-649
-- photoreceptor 14, 15, 25, 30, 36, 43, 44, 65, 70, 87, 144, 147-149, 151, 152, 154, 155, 426
-- potential 36-38, 465
-- recycling *See also* membrane renewal, membrane turnover 762, 764
-- renewal 778
-- turnover 43, 70, 87, 152, 154, 155, 426, 758, 778, 779
-- vertebrate photoreceptive 152-155
Menotaxis 101, 701
Merostomata 5
Mesorhodopsin 27, 28
Mesostigmata 338
Mesozoa 146
Metal heavy, effect of 785
Metarhodopsin 12-14, 21, 22, 23-30, 32-36, 38-42, 47-49, 51, 86
Metazoa 6, 146, 162
Microspectrofluorometry MSF 32, 440
Microspectrophotometry MSP 30-32, 35, 37, 73, 77, 86, 271, 420, 440, 444, 445, 447, 783
Microvilli *See also* Receptors rhabdomeric 44, 70, 73, 75, 77, 78, 79, 80, 81, 82, 87, 133, 138-141, 143, 144, 148, 150-152, 154, 155, 157-160, 162, 166, 168, 169, 441, 485, 783
-- in annelid receptors 289, 295, 296, 299, 303-307, 312, 326, 328
-- -- arachnid and myriapod receptors 344-346, 349, 350, 355, 361, 364, 367, 368, 388-390
-- -- chaetognath receptors 727, 737

Microvilli in crustacean
receptors 84, 85, 86, 89,
403, 423, 426
-- -- echinoderm receptors 743,
756, 758-765, 768
-- -- insects 669, 681
-- -- molluscan receptors 702,
703, 707-709, 716, 717,
720
-- -- platyhelminthes receptors
222, 226, 227
-- -- rotifer receptors 262,
265
Migration (behavior) 185-187,
195, 200, 201, 203, 205-
207, 209
-- klinokinetic 180, 183, 195,
200, 207
-- pigment See Pigment migration
-- vertical 728, 739-741, 779
Miracidia 219, 221, 222, 226,
230-233
Miracidium larva 219, 227
Mirror (in eye) 244, 247, 367,
401-403, 405, 407, 416, 417,
432-434, 699, 706, 708, 711-
713, 720
Mollusc 7, 65, 141, 143, 145,
154, 157, 179, 219, 221,
227, 230, 233, 411, 417,
699-722
Mollusca 4, 5, 7, 18, 143, 145,
146, 161, 162, 699-722
Monera 4
Monogenea 217, 218, 226, 228
Monogenean 218, 220, 221, 224-
227, 233
Monogonont (Rotifers) 243, 253-
255, 265, 266, 274
Monopisthocotylea 221, 224, 226
Monoplacophora 701, 702
Monopolar cell 448, 459, 461,
463-465, 471, 473, 484, 485,
488-492, 498, 499, 501, 503,
597, 637, 638, 640, 642-646,
648, 655
Mosquito 2, 3, 12, 21, 779

Moth 3, 30, 32, 41, 401, 420,
428, 429, 435, 498, 778, 783
-- African army-worm 20
-- bee 20
-- Sphingid 20
-- tobacco hornworm 20
Motion *See also* Movement
-- parallax 565, 604
-- pitch 88, 483, 510-512
-- responses 193
-- roll 88, 483, 484, 490, 511,
512
-- sensitive neurons 483, 497,
498, 503-505, 545
-- vector 562, 563
-- yaw 88, 483, 484, 490-492
Motor inhibition 228
Movement angular acceleration or
deceleration 483, 508, 512
-- (behavior) 179-210
-- cell 116
-- ciliary 744
-- detection *See also* relative
movement detection 189, 241,
449, 476, 477, 491, 496, 497,
503, 528, 535, 536, 539, 540,
542, 561, 566, 567, 577-583,
586-588, 591-593, 596, 597,
600, 602, 605-607, 610, 611,
623, 631, 710, 784, 785
-- detectors 474-476, 484, 496,
498, 504, 505, 535, 539, 541,
542, 549, 550, 578-580, 582,
587, 592-598, 602, 604, 606,
784
-- detecting neurons 475
-- eye 385, 423, 424, 700, 718,
721
-- membrane 312
-- optokinetic *See* Optokinetic
movements
-- rotatory or rotational 181,
184, 497, 563, 565, 572, 605,
607
-- saccades 385
-- sensitivity 476, 483, 488-490,
495, 503-505, 510

SUBJECT INDEX

Movement spontaneous activity of retina 385
-- shielding pigments 426
-- translatory, translational 181, 184, 563, 565, 572, 604, 605
-- specific nervous activity 623, 626, 627, 630, 631
Muscoid fly 464, 465
Myriapods 335, 387-390
Mysid 82, 402, 403, 418, 428, 429, 431
Mysidacea 404
Mystacocarida 404

Naididae 307
Nauplius *See* Eye nauplius
-- larva 402
Nautiloidea 702, 718
Nematoceran 690
Nematoda 3, 4, 132, 146, 159, 161
Nematode 2, 135-137, 142, 155, 156, 179-183, 186, 191-193, 195, 196, 199, 200, 205, 207, 780
Nematomorpha 4, 146
Nemertea 4
Nemertinea 4
Nemertini 146
Neoptera 5
Nephthydidae 294
Nereidae 289-291, 294-297, 299, 304, 313, 314, 320
Nesticidae 351
Neural image 458-460, 462, 466, 471, 476
-- pooling 449
-- superposition 75, 76, 563, 601
Neurite 638, 641, 644, 649, 652-655
-- resorption 653, 655
Neurogenetics 440
Neurons brain 710
-- columnar *See also* columnar cell 483, 484, 492, 495, 498, 513

Neurons descending 483, 508, 511
-- depolarisation 458-461, 465, 475, 476, 709, 720
-- giant lobula plate 496, 513, 516
-- hyperpolarisation 465, 466, 475, 709, 720
-- lamina differentiation 635-656
-- small-field 489, 501, 503, 509, 511, 526, 536, 551, 552
-- spiking 475, 491
-- wide-field 497-499, 506
Neuropils *See* Lamina, medulla, lobula plate lobula
Neuroptera 20
Neuropteran 27, 73
Neurotransmitter 247, 611
Newt 151
Noctuids 676
Noise 116, 418, 460, 463-465, 468, 471, 472, 477, 572, 574, 576, 583, 586, 602, 779
See also Signal-to-noise ratio
-- quantal 443, 449
-- synaptic 490
Notocotylidae 221, 223
Notommatidae 243
Notonecta *See also* Water bug 675, 683
Nudibranch 701

Ocellar cup 758
-- photoreception 744
-- potential 145
Ocelli 19, 132, 189, 483, 485, 509-512, 549, 610, 643, 662, 688-690, 701, 707, 734
-- in annelids 290, 300, 306, 310, 311, 321, 328
-- in arachnids and myriapods 335-338, 387-390
-- in echinoderms 743, 751, 753, 754, 756-760, 763, 765, 766, 768

Ocelli in platyhelminthes 219-222, 224, 228, 231
-- in rotifers 241, 242, 255, 257-259, 262, 263, 271, 275, 276, 281, 282
Ocelloid 122, 123
Ocellus compound 758
-- extracerebral 300
Octopus 12, 17, 18, 24, 25, 28, 699, 703, 705, 713, 716, 718, 765
Odonata 490, 663, 665, 666, 675, 679, 682, 683
Odonate See also Odonata 72, 73
Oecobiidae 338
Off-response See also On-off 232, 233, 267, 321, 390, 473, 528, 702, 706-709, 750
Oil droplets coloured 443, 781
Oligochaete (worms) 289, 307, 308, 319, 321, 326, 327
Ommatidium 68, 71-74, 78, 81, 89, 157, 390, 403, 417, 423-426, 432, 440, 441, 445, 447, 449, 484-488, 492, 495, 499, 504, 710
-- acceptance angle 425
Ommine (pigment) 348
Ommochrome granules 335, 348, 349
On-off (response) See also off-response, on-response 390, 459, 465, 473, 474, 476, 528, 551, 592, 623, 627, 700
On-response See also On-off 321, 390, 473, 528, 707-709
On sustained cells 528, 541, 546, 552
Onchidiacea 706
Orcomiracidia 218, 219, 221, 224, 226, 230, 231, 234
Ontogenic regression (Arachnid eye) 354-356, 379
Onychophora 6, 146, 182, 343

Open-loop experiment 179, 186, 187, 191, 192, 208-210, 219, 220, 223, 226-228
Opheliidae 291, 294, 301
Ophiuroids 749
Opilionid 335, 338, 343-345, 357, 358, 360, 369, 387, 681
Opisthobranch 701, 706, 707
Opisthobranchia 702
Opsin 12, 13, 15, 21, 26, 153, 165, 167, 442, 444
Optic cartridge 485-489, 492, 493, 498, 510
-- chiasma 369, 465, 485, 491, 493, 495, 524, 527, 638
-- cushion 758, 765, 767
-- foci 527
-- nerve 92, 318, 341, 347, 354, 369, 390, 703, 704, 708, 711, 717, 730, 732
-- -- interneuron responses 88-91
-- vesicle 335, 340, 341, 343, 344, 347, 349
Optical guide 672, 674
Optics lens 408, 415, 416
-- -- cylinder 418
-- -- mirror 712-714
-- ommatidial 80
-- mirror 142, 403, 405, 407, 415
-- pinhole 700
-- under-focussed 423
-- refracting 402
-- -- superposition 429
-- superposition 432
Optokinetic equilibrium 604
-- movements 75, 721
-- reflex 621, 783
-- nystagmus 424
Optomotor responses 75, 523, 547-551, 604, 716
Optophysiology 35
Oribatid 338
Orientation (behavior) 63, 66, 70-72, 92, 97, 99, 101, 102, 186, 382, 426, 498
-- azimuth-finding 71, 72
-- landmark 97, 101
-- piloting 70

Orientation polarised light
   compass 67-72, 78, 79, 82,
   88, 93, 97-101
-- selectivity 505
-- sky compass *See* Orientation
   polarised light compass
-- sun compass 82, 97, 98, 99,
   101
Orthokinesis 116, 180, 181,
   184, 186, 187, 195, 199,
   203, 206, 209, 229, 231,
   273, 276
Orthoptera 72, 551
Orthotaxis 188
Ostracod 142, 401, 713
Ostracoda 404, 405, 407, 416
Otoplanid species 224
Owlfly 12, 20, 29, 47
Oxyopidae 351

Paleoptera 5
Palolo worm 780
Palpigrad 337
Palpigradida 335
Palpimanidae 351
Parallel channels 7, 457, 458,
   460, 461, 463, 465, 473,
   477, 499
-- input channels 474
-- spatial channels 474, 475
-- -- units 462
Paramecium 115, 125, 126
Parazoa 1, 2, 4
Pauropod 336
Pauropoda 388
Pentastomulida 146
Peracarida 404
Phanerozonia 764
Phaosome 156, 226, 227, 232,
   262, 264, 265, 281, 289, 303,
   304, 307-309, 312, 326, 327
Phase drifting 599, 600
Phasic response 528
Phasme 343, 680
Phasmoptera 665
Phobic response *See also*
   Photophobic response 116,
   181, 182, 209

Pholcidae 340, 350
Phoronida 5, 146
Photobacteria 186
Photoecclisis 195
Photoecclitic response 195
Photochromism 16
Photoconversion 23-27, 30, 33-36,
   43, 50
Photodetector 119
-- directional 118
Photoequilibrium 25, 34, 39, 40
-- spectrum 22, 23, 25, 49, 50
Photokinesis 179, 193, 228, 229,
   241, 268, 271, 272, 275
Photoklinokinesis 136, 217, 229,
   230
Photoklinotaxis 188, 191
Photolysis flash 28, 32
Photon flux curve sky 68
Photonegative response *See also*
   Phototaxis negative 116, 228,
   230-233, 267, 290, 319-321
Photoperiod 276-279, 777-779
Photoperiodic induction 779
Photophobic response *See also*
   phobic response 123-125
Photopigment *See also* Pigment
   3, 5, 6, 74, 121, 124, 144,
   145, 147, 149, 150, 166, 312,
   403, 413, 426, 669, 749, 779,
   781
-- absorption coefficient 420
-- bleaching 32, 709
-- offset 69
Photopositive response *See also*
   Phototaxis positive 116, 228,
   230, 233, 242, 267, 269, 270,
   272-274, 277, 290, 319
Photoproduct 16
Photoreception extra-ocular or
   Photoreceptor extra-ocular
   336, 384, 709, 755
Photoreceptor *See also* Receptor
-- acilious 159
-- cell differentiation 165-167,
   169
-- cone or rod *See* Vertebrate
   vision
-- dichroic 70, 187-189

Photoreceptor directional 118, 119
-- epigerous or epigenic 143, 146, 159, 161, 164, 168
-- extra-retinal organised 777, 778
-- non-directional 117, 118
-- slab waveguide 122
Photoreceptors evolution See also Photoreceptors, phylogenesis 143-170, 709, 774
-- phylogenesis See also Photoreceptor evolution 2, 3, 4, 115, 121, 265, 266, 281, 402, 756
Photoregeneration 27, 32, 43, 48
Photosensitivity See also Sensitivity and spectral sensitivity 22, 24, 43, 137, 144, 146, 148-151, 155, 162, 167, 170, 281, 675, 746, 753-755
-- dermal See also Sensitivity dermal 756
-- in arthropods See also in crustacean & insects 424, 457-461, 463, 476
-- in cephalopods 716-717
-- in crustacean 42, 43, 46, 408, 418, 424-426, 430, 435
-- in echinoderms 746, 751
-- in platyhelminthes 217, 218, 227, 228, 231-233
-- myoneural 751
-- neuronal 749
-- spectrum 25, 35, 37, 38, 50
Phototactic responses See Phototaxis
Phototaxis 78, 101, 116-119, 123-125, 147, 156, 191, 195, 196, 200, 231, 233, 241, 267, 268, 271, 275, 276, 319-321, 377, 402, 707, 718, 743

Phototaxis negative See also Photonegative response 132, 136, 190, 753, 754
-- positive See also Photopositive response 123, 125, 132, 137, 184, 186, 738, 753, 754
Phototransduction See also Transduction 38, 39, 150, 165, 168
Phototropism 228
Phragmophora 729, 737
Phyllocarida 404
Phyllodocidae 289, 293, 295-297, 299
Phytoflagellate 117, 118, 123, 124, 126, 145-149, 155, 156, 262, 265, 266
Pigment See also Rhodopsin, Bathorhodopsin, lumirhodopsin, mesorhodopsin, metarhodopsin, ommine, ommochrome granules, ptérine
-- accessory 76, 77
-- cell 134, 136, 138-140, 142, 148, 155, 157, 160, 162, 347-350, 352, 361, 362, 364, 366, 367, 388-390, 703, 727, 729-732, 736, 739, 758, 762, 763
-- cellular differentiation 348, 349
-- accessory 443
-- chloroplast 125
-- cup 121, 134, 136-138, 189, 219, 222, 226, 231, 243, 244, 247, 249, 251, 253, 254, 257, 258, 266, 271-276, 281, 289, 291, 293, 303-305, 307, 310, 311, 313, 403, 758, 777
-- migration 35, 72, 226, 364, 368, 389, 434, 673, 674, 732, 779
-- ocular See Eye pigmentation
-- photochromic 16
-- photosensitive 443, 445
-- photosensitizing See Pigment sensitizing
-- photostable 443, 445
-- plant 125, 126, 442
-- receptor 69

Pigment reflecting 415
-- respiratory 126
-- screening 23, 41, 42, 48, 72, 76, 426-428, 434, 443-445, 447, 484, 639, 668, 781
-- sensitizing See also Pigment U.V. sensitizing 41, 42, 76, 124, 125, 440-444, 447
-- shading 133, 134, 136, 138, 146, 147, 187, 188, 195
-- shielding 402, 426
-- spot 117, 132, 135, 136, 138, 147, 148, 156, 188-190, 192, 196, 243, 255, 780
-- U.V. sensitizing 76, 77, 442-444
-- visual See also Photopigment 11-51, 63, 69, 76, 77, 86, 122, 151, 387, 423, 441, 445, 449, 470, 768, 780, 781, 783
Pigmentary cell See Pigment cell
Pill-bug See Woodlouse
Pinocytotic vesicles 153, 155, 219
Pisauridae 360, 366, 367, 378
Piscicolidae 310
Planaria 190, 194
Planarian 145, 189, 190, 194, 195, 217, 228, 229
Platyhelminthes See also Larva platyhelminthes 142, 159
Platyhelminthes 3, 4, 6, 141, 145, 146, 150, 154, 156, 158, 159, 161, 162, 189, 190, 194, 217-234, 244, 262, 264, 266, 281, 729, 735
-- parasitic 217, 218, 224, 226-228, 230, 232, 233
Plexiform layer 526, 784
Pluteus larva 744, 745
Pogonophora 146, 157
Pogonophores 262, 266
Pointed lady (Lepidoptera) 20

Polarisation 5, 63-102, 187, 188, 463, 720
-- channel 89-91
-- degree of (p) 63, 64, 66, 67, 68, 69, 88, 90, 93, 94, 101
-- map or pattern of sky 67, 71, 97, 100, 101
-- sensitivity See also Polarised light detection 63-102, 119, 123, 158, 159, 188, 423, 426, 457, 458, 463, 510, 525, 526, 540
-- -- Vs spectral sensitivity 63, 65, 66
-- underwater 69-70
Polarised light analysis 73, 88, 90, 383, 384
-- -- detection or polarised light perception 78, 336, 382, 384, 449, 485, 508, 512, 783
Polarotaxis 66, 92, 94, 96, 101-102, 188
Pollutants effect on vision 785
Polycephalia 300
Polychaete worms 142, 156, 224, 226, 289-299, 302, 305, 307, 313, 316, 320, 321, 327, 328, 714, 780
-- errantes 293, 296, 321, 324
-- sedentary 293, 295, 320, 321
Polyclada 218, 220
Polyclads 219, 220, 224, 277
Polydesmidae 336, 387
Polygordiidae 305
Polyopisthocatylea 221
Polyplacophora 701, 702, 706
Polystomatidae 221, 225, 227
Pontellid copepods 401, 404, 409, 410, 413, 416, 716
Pooling (convergence) 781
Porcellanid galatheoid 84
Porifera 3, 4, 146
Position detection 561, 567, 574, 577, 587-589, 592, 600, 610, 611
-- detectors 562, 578, 592, 600, 601, 604
-- stabilizer 88
Potassium channels 145, 709
Prawn 19, 46, 779

Predictive coding 470, 471
Presynaptic ribbon 641, 650-652
Priapula 6
Priapulida 146
Processing neural 458, 471
-- parallel 462
-- visual 88-92, 458
Prokaryotes 115
Proseriata 218, 224
Prosobranch gastropod 142, 701, 714, 715
Prosobranchia 702
Protista 2, 142, 144-149, 151, 162
Protostigmata 338
Protostome 144, 265, 774
Protostoma 141
Protostomia 4, 6, 7
Protozoa 1, 3, 4, 6, 115-126
Protozoan unpigmented 126
Pseudocoelomata 6
Pseudocone 389, 484, 487
Pseudopupil 31-33, 71, 421, 423, 425, 427, 428, 434, 666
Pseudoscorpion 335, 337, 352, 357, 369
Pterine pigment 348
Pterygota 5
Pulmonata 702
Pulmonates (gastropods) 701, 714
Pupal development 639-641, 645, 649-652, 654
Pupil 42, 43, 371, 375, 389, 431, 435, 705, 716-720
-- intracellular 37
Pupillary responses 72, 79
Pupils mobile 700
Pycnogonida 5

Quantum bumps 540
-- catch 443
-- efficiency 22, 24, 25, 35, 37, 50, 51, 752
-- flux 22, 65
-- spectrum 68
Quarter-wave multilayer reflector *See* Reflection multilayer

Radiata 4, 146
Radiates 265
Raster 187, 189, 190, 195
Rayleigh scattering 63, 67, 68, 97, 98
Receptive mechanism non-ocular 744
Receptor *See also* Photoreceptor
-- acceptance angle 420, 580, 594
-- -- function 586, 594
-- angular separation 421, 597, 599, 705, 706
-- packing distance 703
-- potentials 66, 71, 88, 89, 125, 145, 203-205, 440, 443, 445, 781
-- projection 463
Receptors ciliary 143-146, 155, 156, 159, 161, 162, 164, 166, 169, 170, 217, 224, 227, 232, 247, 265, 327, 681, 699, 702, 707, 710, 756, 774, 783
-- depolarising 145, 490, 536, 699, 700, 707-709, 781
-- dermal 307, 701
-- extra-ocular 169, 710
-- hyperpolarising 145, 536, 700, 707-709, 781
-- long 485, 487, 488
-- microvillous 65, 699, 702, 707, 720
-- mixed 146, 152, 161, 162, 224, 227
-- ocellar 707
-- off 699, 707
-- on 699, 707
-- retinal 408, 777, 778
-- rhabdomeric 84, 143-146, 150-152, 154, 155, 157, 159, 161, 162, 164, 166, 168-170, 217, 219, 223, 224, 227, 254, 257, 265, 327, 681, 701, 703, 707, 716, 774, 783, 785
-- short 484, 485, 487-489, 495
Recruitment 601, 602, 606, 607
Reflectance of light 68
Reflecting cups 707
-- layer 780
-- -- quarterwave 121

SUBJECT INDEX

Reflection Fresnel 68
-- multilayer 711, 712
-- specular 68, 73
Reflector *See also* corner-reflector 142, 408, 411, 416, 711
-- interference 120, 121
Reflex movements 751
Refocussed image 415
Refractive index 42, 140, 141, 121, 217, 244, 346, 371, 402, 413-415, 418-420, 429, 430, 432, 701, 710, 712-714, 718
Relative movement detection 600, 602, 604
Relaxation spectrum 23, 25, 50
Resolution 140-143, 148, 156, 189, 375, 376, 405, 408, 417, 418, 422-427, 431, 434, 435, 442, 443, 445, 499, 582, 596, 597, 700, 701, 705, 711, 716, 718, 719
-- angular 405, 448, 527, 566
Resolvable angle, minimum 705, 706
Resolving power 189, 485
Response 184, 189
-- directed 184, 186
-- tonic 528
-- unitary 184
Retina cup-shaped 337, 707
-- indirect type 350, 351
Retinal cells 141, 341, 346, 347, 352
-- chromophore 12, 13, 17, 18, 21, 25, 26, 44-46, 77, 153, 165, 765
-- clubs 224
-- mosaic 8, 71, 439-449, 499, 783
-- motion 548
Retinoid crystalline 123
Retinol 13, 442, 443
Retinomotor response 226
Retinotopic mosaic 490, 492, 493, 495, 499, 501, 505, 507, 510

Retinula 71, 73, 74, 77-79, 81, 84, 86, 87, 91, 150, 426, 484, 487, 783
Retinular cell 68, 71, 72, 74-77, 79-82, 84-91, 335, 337, 344, 346-353, 355, 361, 362, 367, 368, 383, 389, 390
Rhabdocoela 218-220, 226, 227
Rhabdom acceptance angle 423-426, 435
-- "ball" 414
-- "-- and doughnut" 413, 414
-- "banded" 426
-- closed 77, 81, 82, 344, 389, 445, 670
-- fused *See* Rhabdom closed
-- light guide *See* Light guide
-- open 74, 81, 670, 672
-- rectangular 423
-- twisting 70, 80-84
Rhabdoms 31, 32, 35, 41, 43, 44, 48, 70, 71, 78-85, 87, 89, 150, 157, 707, 717, 737, 758, 768
-- in arachnid and myriapod receptors 335, 337, 341, 344, 347, 349-352, 361, 364, 366-368, 382, 387-390
-- in crustacean receptors 401, 403, 411-418, 420, 421, 423-432, 434
-- in insect receptors 440, 661, 669, 672-675, 678, 679, 681, 689, 690
-- in rotifer receptors 257, 281
Rhabdomere toothed 82, 85
-- twisting 81-82, 85
Rhabdomeres 65, 73-75, 77-89, 134, 150, 152, 154, 157, 158, 169, 190, 403, 405, 777, 779, 783
-- in arachnid & myriapod receptors 335, 344-346, 361, 364, 366-368, 382, 383, 388, 390
-- in insect receptors 440, 441, 443-449, 484, 485, 487, 489
Rhodopsin 12-15, 22, 25-41, 43-48, 50, 51, 65, 77, 145, 147-151, 153, 159, 165, 169, 233, 441-444, 447, 669, 762, 765, 780

Rhynchocoela 4
Ricinulei 335
Ricinuleid 337
Rotifera 4, 146, 156, 159, 161, 162
Rotifers 6, 142, 156, 159, 179, 221, 226-228, 241-283
Rythm control 710
-- circadian 335, 364, 380, 383, 387
Rhythms biological 233, 234, 774, 777-779

Sabellid tube worm 699, 709
Sabellidae 289-291, 300-304, 317
Salticid spiders 335, 357, 361, 363, 369, 370, 376, 377, 385, 387, 643
Salticidae 338, 350, 358, 359, 362, 366-369, 376, 377, 384-387
Saturation of response 91, 468, 540, 542-544, 548, 602
-- spectrum 23
Scallop 18, 38, 48, 142, 145, 403, 416, 699, 700, 706, 707, 709-711, 713, 720
Scanning *See* Vision scanning
Scaphopoda 702
Scarabeid beetle 67
Schaltzone 388, 389
Schiff's base 12, 13, 44, 45
Schizopeltids *See* Uropygids
Scolopendridae 387
Scorpion 335-337, 352, 357-361, 368, 370, 377, 378, 380-382, 384, 387, 780
-- north-African fat-tailed 383
-- silurian 337, 367
Scutigeromorpha 336, 387
Scyphomedusa 157
Scyphozoa 137, 140, 141
Scyphozoan 142
Sea anemone 137
-- urchin 748, 751-753

Sea urchins, diadematid 743, 746, 747, 749, 750
Seisonidae 243, 255
Self-screening 38, 39
Self-shading 117, 118, 126
Self-shunting 460
Semper cell 484, 485, 665, 667, 672
Senoculidae 351
Sensitivity *See also* Photosensitivity and spectral sensitivity 42, 43, 44
-- absolute 39, 426, 443, 525, 597, 779
-- and receptor potential 203
-- and resolution 408, 418, 423, 435, 718, 721
-- and summation 463
-- angular 44, 460, 461, 472, 525, 597, 779
-- binocular 527, 544-546
-- blue 73, 77-79, 781, 783
-- blue-green 442, 778
-- changes in 41-43, 77, 90, 203, 205, 316, 442-444
-- contrast 43, 457, 460, 491, 541
-- dermal *See* Dermatoptical sense, dermal sensitivity
-- distribution 586, 594
-- effect of absence of lens 716-718
-- -- -- coloured oil droplets 781
-- -- -- diet 442, 444
-- -- -- gain-control 602
-- flicker 528
-- function 40, 41
-- generalised 2
-- green 68, 72, 78-80, 347, 377, 426, 444, 778, 781
-- in ocean depths 417, 430
-- measurement of 40, 41, 419
-- orange 72
-- patterns 476
-- polarisation *See* Polarisation sensitivity
-- relative 442
-- selective depression of 41
-- spatial 542, 543, 546, 579, 580, 594

SUBJECT INDEX

Sensitivity U.V. *See* U.V. sensitivity
-- violet 86, 90, 91, 92
-- wavelength 90-92
-- yellow-green 90-92
Sensory complex 261, 262
Septate junction 151, 160
Sexual dimorphism *See* Dimorphism, sexual
Shading (device) 119, 120, 122, 123
Shadowing 405
Shape perception 385
Shrimp 46, 82, 84, 85, 402, 428
-- brine 424
-- carid 432
-- deep-sea 432, 780
-- mantis 427
-- pennaeid 432
-- pink 19
-- shallow-water 422
Sicariidae 338
Signal-to-noise-ratio 31, 47, 425, 448, 472, 487, 490, 711
Silkmoth 778
Simulidae 663
Sipuncula 4
Sipunculid worm 159
Sipunculida 146, 160
Sky polarisation compass *See* Orientation, polarised light compass
Slip speed (retinal) 573, 575, 576
Slug 768
Snail 154, 699
-- land 142, 150, 701
-- sea 700
Snell's window 424
Sodium channels 145, 709
-- conductance 720
Solpugids 337, 338, 352
Spatial integration 543, 546
Sparassidae 352, 359
Spectral emission, bioluminescence 780

Spectral sensitivity *See also* Photosensitivity and sensitivity 11, 39, 45, 46, 87, 188, 777
-- -- and bioluminescence 46, 780
-- -- -- entrainment 778
-- -- definition 36, 65
-- -- -- effect of methyl mercury 785
-- -- -- -- mutations 167
-- -- -- -- screening pigments on 42, 426, 427, 443, 444, 781
-- -- -- -- sensitizing pigments on 77, 443, 444
-- -- function of photoreceptor cell 36, 39, 41
-- -- in annelid photoreceptors 290, 314-316, 321
-- -- -- -- arachnid and myriapod photoreceptors 336, 363, 367, 368, 376-378, 380, 381, 384, 387
-- -- -- -- insect receptors 35, 42, 46, 76, 77, 78, 79, 80, 90, 440, 441, 443-445, 447-449, 460, 463, 465, 468, 470, 472, 476, 491, 496, 506, 525, 526, 540, 541, 545, 602, 779, 781
-- -- -- molluscan receptors 46
-- -- maxima 76, 441, 442, 444, 447, 525, 540, 778
-- -- measurement 445, 449
-- -- waveguide effect on 42
Spectrophotometry 25, 27, 30, 31
Spherical aberration 140, 141, 413, 431, 713, 714
Sphingids 676
Spiders 2, 44, 334-390, 402, 413, 416, 426, 643, 713, 714
-- diurnal 336, 361, 367, 376
-- hunting 351, 357
-- jumping 386, 415
-- nocturnal 336, 363, 376, 779
Spike response 71, 88-92, 317, 543, 592
-- discharge, sustained 473
Spinulosa 764
Spionidae 304, 306

Spirotrich (ciliates) 124, 125
Sponges 1
Squid 2, 12, 18, 21, 24, 25, 27-29, 32-35, 140, 144, 169, 439, 699, 703, 718-720
Starfish 417, 709, 743, 751, 752, 754, 755, 758, 762-768
Statocysts 704, 716, 721
Stemmata 662, 676, 683, 688, 690, 691
Stentor 124
Step down response or step off response 116, 183
-- up response or step on response 116, 183
Stigma 117-122, 149, 184
Stimulus definition 183, 184, 208
-- field definition 184, 185
-- gradient 179-184, 210
Stomatopod 82-84, 88, 401, 404, 424, 427
Strombid 699, 701, 702
Summation 460, 463-465, 468, 476, 537, 539, 546, 549
Superposition eye See Eye superposition
-- neural 464
Surface factor 750
Syllidae 296, 298, 299
Syllinae 297, 298, 301, 303, 304
Symphyla 336, 387
Synapse chemical 470
-- feed-back 466
-- frequency (A/N ratio) 651, 652
-- input-output function 469, 470
Synapses multiple-contact 642, 644, 647
Synaptic amplification 457
Synaptid holothurian 754, 756, 757
Synaptogenesis 7, 635, 656
Syncarida 404
Synchaetidae 243

Tabanid fly 43
Tanaiadacea 404
Tangential cell 483, 491, 493, 496, 511, 526, 528-547, 550
Tapetum 43, 350-354, 366, 367, 384, 403, 416, 424, 434, 710, 711, 713, 720, 728
Tardigrada 146
Taxis See also Geotaxis, klino-, photoklino-, meno-, ortho-, polaro-, telo-, thermo-, tropo- 93, 180, 181, 183, 185, 192, 197, 198, 200, 201, 203, 205-207, 209, 210
Telotactism or Telophototactism 727, 739
Telotaxis 186, 187, 191, 192, 210
Temnocephala 217
Temporal interference 587
Texture discrimination 547, 550
Theridiidae 351, 369
Thermotaxis 197, 199, 200, 210
Therophosidae 358, 359
Thomisidae 340, 352, 368, 384
Thrombidium 338
Thysanopoda 422
Tomopteridae 780
Tornaria (hemichordate larva) 152
Torque (response) 75, 547, 549, 550, 571-577, 588, 590-592, 599, 602-604, 607-610
Tract-crystalline shread 667
Transduction (mechanism) See also Phototransduction 115, 145, 146, 163, 164, 166, 169, 458, 460, 468, 709, 781
Transmission non-spiking 491
-- spiking 491
Transmitter production 11
-- release 490
Trematod 8, 241, 242, 244, 245, 251, 256, 257
Trematoda 217, 218, 223-225, 228, 234
Trichoceridae 243
Trichoptera 665
Triclada 218

SUBJECT INDEX

Tricladida 154
Triclads 155, 219, 220, 226, 229
Trilobites 417, 418
Trochelminthes 4
Trocophore larva 304, 306, 307, 328
Trogulidae 338
Tropotaxis 185-188, 192, 193, 197, 210, 701
Tunicata 145, 146, 169
Turbellaria 145, 158, 159, 189, 190, 194, 217, 219, 220, 222, 224, 226-230, 233
Turbellarian 152, 154, 155, 217, 221

Urochorda 2
Urocteidae 350
Uropygids 335, 337, 338, 352, 358, 360, 370
U.V. receptors 68, 69, 72, 73, 78, 79, 377, 444, 476, 783
-- reflectance patterns 68, 69
-- sensitivity *See also* Pigment U.V. sensitizing and e-vector U.V. 41, 42, 47, 67, 69, 73, 77, 374, 441-444, 449, 783

Vertebrata 145, 146, 161, 162, 169
Vertebrate vision 1, 2, 12, 14, 16, 25, 27, 31, 32, 43, 44, 64, 65, 87, 123, 131, 142, 143, 145, 147, 151-155, 157, 161, 165, 168, 169, 182, 218, 219, 266, 281, 293, 257, 402, 403, 416, 440, 443, 449, 471, 535, 552, 623, 625, 626, 643, 647, 669, 681, 699, 700, 709, 713, 716, 718, 720, 721, 734, 758, 762, 774, 778, 780-785
-- -- mammals 181, 443, 497, 624
-- -- man 43, 63, 90, 188, 435, 439, 592, 597, 605, 606, 706, 710, 716, 781, 783

Vertebrates, amphibian 781
-- bird 93, 781
-- bullfrog rod 783
-- cat 458
-- fish 46, 69, 70, 140-142, 413, 435, 470, 699, 700, 713, 714, 718-721, 781
-- -- blind cave 718
-- -- cyprinid goldfish 70
-- -- herring 720
-- -- trout pineal organ 145
-- -- -- rainbow 785
-- monkey 783, 784
-- rabbit retina 587, 588
-- reptile 781
Vertical cells 484, 497, 498, 503, 506-508
Vision defects 785
-- direct 291, 296, 310, 327
-- diurnal 781
-- fixation of object 75, 93, 448, 523, 547, 550, 561, 592, 608, 609, 623
-- flicker detection 561, 590-592, 599, 600
-- indirect 291, 294, 303, 310, 327
-- initiation (Heisenberg) 609
-- learning 609, 610
-- polarisation 63-102
-- scanning 118, 385, 386, 409, 411, 412, 423, 424, 700, 775
-- scotopic 781
-- stereoscopic 187
-- tracking 93, 385, 386, 523, 547, 550, 561, 575, 590, 592, 601, 609, 610, 775, 777
Visual cell 340, 484, 490
-- deprivation 534, 535
-- surround 483
-- tracts 527
Vitreous body 123, 341, 350, 351, 353, 357, 361, 387, 389
-- cell 335, 349, 351-353, 355

Warnowiaceae 122
Warnowiidae 123
Wasp 100, 720

Waterbug 69
Water-flea 417
Water strider 20, 31
Waveguide modes 42
Wavelength selectivity 781
Winkle 699
Wood louse (Pill-bug) 424
Worm *See also* Polychaete worm, sipunculid -, palolo -, arrow -, sabellid tube -, dalyelliid flatworm, earth-, hook-, 137, 160, 190, 195-200, 203

Xanthophyceae 119, 120
Xiphosura 422
Xiphosuran 417, 718
Xiphosurida 19

Yaw torque response 547-551, 567, 568, 570, 583, 590, 591, 600

Zeitgeber's cycle 382
Zodariidae 351
Zoochlorellae 124, 125
Zooplancton 122, 241, 727, 739